D0620886

INTELLIGENT SIGNAL PROCESSING

IEEE Press
445 Hoes Lane, P.O. Box 1331
Piscataway, NJ 08855-1331

10/2007

INTELLIGENT SIGNAL PROCESSING

Edited by

Simon Haykin
McMaster University
Hamilton, Ontario, Canada

Bart Kosko
University of Southern California
Los Angeles, CA, USA

A Selected Reprint Volume

The Institute of Electrical and Electronics Engineers, Inc., New York

This book and other books may be purchased at a discount
from the publisher when ordered in bulk quantities. Contact:

IEEE Press Marketing
Attn: Special Sales
445 Hoes Lane
P.O. Box 1331
Piscataway, NJ 08855-1331
Fax: +1 732 981 9334

For more information about IEEE Press products, visit the
IEEE Online Catalog and Store: http://www.ieee.org/store.

Printed in the United States of America.

10 9 8 7 6 5 4 3 2 1

ISBN 0-7803-6010-9

IEEE Order No. PC5860

Library of Congress Cataloging-in-Publication Data

Intelligent signal processing / edited by Simon Haykin, Bart Kosko.
p. cm.
"A selected reprint volume."
Includes index.
ISBN 0-7803-6010-9
1. Signal processing—Digital techniques. 2. Intelligent control systems. 3. Adaptive signal processing. I. Haykin, Simon S., 1931– II. Kosko, Bart.

TK5102.9.I5455 2000
621.382′2—dc21

00-061369

This book is dedicated to Bernard Widrow
for laying the foundations of adaptive filters

Contents

Preface

This expanded reprint volume is the first book devoted to the new field of intelligent signal processing (ISP). It grew out of the November 1998 ISP special issue of the *IEEE Proceedings* that the two of us coedited. This book contains new ISP material and a fuller treatment of the articles that appeared in the ISP special issue.

What Is ISP?

ISP uses learning and other "smart" techniques to extract as much information as possible from incoming signal and noise data. It makes few if any assumptions about the statistical structure of signals and their environment. ISP seeks to let the data set tell its own story rather than to impose a story on the data in the form of a simple mathematical model.

Classical signal processing has largely worked with mathematical models that are linear, local, stationary, and Gaussian. These assumptions stem from the precomputer age. They have always favored closed-form tractability over real-world accuracy, and they are no less extreme because they are so familiar.

But real systems are nonlinear except for a vanishingly small set of linear systems. Almost all bell-curve probability densities have infinite variance and infinite higher order moments. The set of bell-curve densities itself is a vanishingly small set in the space of all probability densities. Real-world systems are often highly nonlinear and can depend on many partially correlated variables. The systems can have an erratic or impulsive statistical structure that varies in time in equally erratic ways. Small changes in the signal or noise structure can lead to qualitative global changes in how the system filters noise or maintains stability.

ISP has emerged recently in signal processing in much the same way that intelligent control has emerged from standard linear control theory. Researchers have guessed less at equations to model a complex system's throughput and have instead let so-called "intelligent" or "model-free" techniques guess more for them.

Adaptive neural networks have been the most popular black box tools in ISP. Multilayer perceptrons and radial-basis function networks extend adaptive linear combiners to the nonlinear domain but require vastly more computation. Other ISP techniques include fuzzy rule-based systems, genetic algorithms, and the symbolic expert systems of artificial intelligence. Both neural and fuzzy systems can learn with supervised and unsupervised techniques. Both are (like polynomials) universal function approximators: They can uniformly approximate any continuous function on a compact domain, but this may not be practical in many real-world cases. The property of universal approximation justifies the term "model free" to describe neural and fuzzy systems even though equations describe their own throughput structure. They are one-size-fits-all approximators that can model any process if they have access to enough training data.

But ISP tools face new problems when we apply them to more real-world problems that are nonlinear, nonlocal, nonstationary, non-Gaussian, and of high dimension. Practical neural systems may require prohibitive computation to tune the values of their synaptic weights for large sets of high-dimensional data. New signal data may require total retraining or may force the neural network's vast and unfathomable set of synapses to forget some of the signal structure it has learned. Blind fuzzy approximators need a number of if-then rules that grows exponentially with the dimension of the training data. This volume explores how the ISP tools can address these problems.

Organization of the Volume

The 15 chapters in this book give a representative sample of current research in ISP and each has helped extend the ISP frontier. Each chapter passed through a full peer-review filter:

1. Steve Mann describes a novel technique that lets one include human intelligence in the operation of a wearable computer.
2. Sanya Mitaim and Bart Kosko present the noise processing technique of stochastic resonance in a signal processing framework and then show how neural or fuzzy or other model-free systems can adaptively add many types of noise to nonlinear dynamical systems to improve their signal-to-noise ratios.
3. Malik Magdon-Ismail, Alexander Nicholson, and Yaser S. Abu-Mostafa explore how additive noise affects information processing in problems of financial engineering.
4. Partha Niyogi, Fredrico Girosi, and Tomaso Poggio show how prior knowledge and virtual sampling can expand the size of a data set that trains a generalized supervised learning system.
5. Kenneth Rose reviews how the search technique of deterministic annealing can optimize the design of unsupervised and supervised learning systems.
6. Jose C. Principe, Ludong Wang, and Mark A. Motter use the neural self-organizing map as a tool for the local modeling of a nonlinear dynamical system.
7. Lee A. Feldkamp and Gintaras V. Puskorius describe how time-lagged recurrent neural networks can perform difficult tasks of nonlinear signal processing.
8. Davide Mattera, Francesco Palmieri, and Simon Haykin describe a semiparametric form of support vector machine for nonlinear model estimation that uses prior knowledge that comes from a rough parametric model of the system under study.
9. Yann LeCun, Léon Bottou, Yoshua Bengio, and Patrick Haffner review ways that gradient-descent learning can train a multilayer perceptron for handwritten character recognition.
10. Shigeru Katagiri, Biing-Hwang Juang, and Chin-Hui Lee show how to use the new technique of generalized probabilistic gradients to solve problems in pattern recognition.
11. Lee A. Feldkamp, Timothy M. Feldkamp, and Danil V. Prokhorov present an adaptive classification scheme that combines both supervised and unsupervised learning.
12. J. Scott Goldstein, J.R. Guescin and I.S. Reed describe an algebraic procedure based on reduced rank modeling as a basis for intelligent signal processing
13. Simon Haykin and David J. Thomson discuss an adaptive procedure for the difficult task of detecting a nonstationary target signal in a nonstationary background with unknown statistics.
14. Robert D. Dony and Simon Haykin describe an image segmentation system based on a mixture of principal components.
15. Aapo Hyvärinen, Patrik Hoyar and Erkki Oja discuss how sparse coding can denoise images.

These chapters show how adaptive systems can solve a wide range of difficult tasks in signal processing that arise in highly diverse fields. They are a humble but important first step on the road to truly intelligent signal processing.

Simon Haykin
McMaster University
Hamilton, Ontario, Canada

Bart Kosko
University of Southern California
Los Angeles, CA

List of Contributors

Chapter 1
Steve Mann
University of Toronto
Department of Electrical and Computer Engineering
10 King's College Road, S.F. 2001
Toronto, Ontario, M5S 3G4 CANADA

Chapter 2
Bart Kosko
Signal and Image Processing Institute
Department of Electrical Engineering-Systems
University of Southern California
Los Angeles, California 90089-2564

Sanya Mitaim
Signal and Image Processing Institute
Department of Electrical Engineering-Systems
University of Southern California
Los Angeles, California 90089-2564

Chapter 3
Malik Magdon-Ismail
Department of Electrical Engineering
California Institute of Technology
136-93 Pasadena, CA 91125

Alexander Nicholson
Department of Electrical Engineering
California Institute of Technology
136-93 Pasadena, CA 91125

Yaser S. Abu-Mostafa
Department of Electrical Engineering
California Institute of Technology
136-93 Pasadena, CA 91125

Chapter 4
Partha Niyogi
Massachusetts Institute of Technology
Center for Biological and Computational Learning
Cambridge, MA 02129

Fredrico Girosi
Massachusetts Institute of Technology
Center for Biological and Computational Learning
Cambridge, MA 02129

Tomaso Poggio
Massachusetts Institute of Technology
Center for Biological and Computational Learning
Cambridge, MA 02129

Chapter 5
Kenneth Rose
Department of Electrical and Computer Engineering
University of California
Santa Barbara, CA 93106

Chapter 6
Jose C. Principe
Computational NeuroEngineering Laboratory
University of Florida
Gainsville, FL 32611

Ludong Wang
Computational NeuroEngineering Laboratory
University of Florida
Gainsville, FL 32611

Mark A. Motter
Computational NeuroEngineering Laboratory
University of Florida
Gainsville, FL 32611

Chapter 7
Lee A. Feldkamp
Ford Research Laboratory
PO Box 2053
Dearborn, MI 48121-2053

Gintaras V. Puskorius
Ford Research Laboratory
PO Box 2053
Dearborn, MI 48121-2053

Chapter 8
Davide Mattera
Dipartimento di Ingegneria Elettronica e delle Telleco-municazioni
Università degli Studi di Napoli Federico II
Via Claudio, 21, 80125
Napoli, ITALY

Francesco Palmieri
Dipartimento di Ingegneria Elettronica e delle Telecomun-icazioni
Università degli Studi di Napoli Federico II
Via Claudio, 21, 80125
Napoli, ITALY

Simon Haykin
McMaster University
Electrical and Computer Engineering
1280 Main Street West
Hamilton, Ontario, L8S 4K1 CANADA

Chapter 9
Yann LeCun
Speech and Image Processing Services Research Laboratory
AT&T Labs–Research
100 Schulz Drive
Red Bank, NJ 07701

Léon Bottou
Speech and Image Processing Services Research Laboratory
AT&T Labs–Research
100 Schulz Drive
Red Bank, NJ 07701

Yoshua Bengio
Speech and Image Processing Services Research Laboratory
AT&T Labs–Research
100 Schulz Drive
Red Bank, NJ 07701
and
Département d'Informatique et de Recherche Opératonelle
Université de Montréal
C.P. 6128
Succ. Centre-Ville
2920 Chemin de la Tour
Montréal, Québec, H3C 3J7 CANADA

Patrick Haffner
Speech and Image Processing Services Research Laboratory
AT&T Labs–Research
100 Schulz Drive
Red Bank, NJ 07701

Chapter 10
Shigeru Katagiri
ATR Human Information Processing Research Laboratories
2-2 Hikaridai, Seika-cho, Soraku-gun
Kyoto 619-02 JAPAN

Biing-Hwang Juang
Bell Laboratories, Lucent Technologies
600–700 Mountain Avenue
Murray Hill, NJ 07974-0636

Chin-Hui Lee
Bell Laboratories, Lucent Technologies
600–700 Mountain Avenue
Murray Hill, NJ 07974-0636

Chapter 11
Lee A. Feldkamp
Powertrain Control Systems Department
Ford Research Laboratory
Ford Motor Company
Dearborn, MI 48121-2053

Timothy M. Feldkamp
Powertrain Control Systems Department
Ford Research Laboratory
Ford Motor Company
Dearborn MI 48121-2053

Danil V. Prokhorov
Powertrain Control Systems Department
Ford Research Laboratory
Ford Motor Company
Dearborn, MI 48121-2053

Chapter 12
J. Scott Goldstein
SAIC
Adaptive Signal Exploitation
4001 Fairfax Drive, Suite 675
Arlington, VA 22203

J. R. Guerci
SAIC
Adaptive Signal Exploitation
4001 Fairfax Drive, Suite 675
Arlington, VA 22203

I. S. Reed
Department of Electrical Engineering—Systems
University of Southern California
Los Angeles, CA 90089-2564

Chapter 13

Simon Haykin
McMaster University
Electrical and Computer Engineering
1280 Main Street West
Hamilton, Ontario, L8S 4K1 CANADA

David J. Thomson
Lucent Technologies
600 Mountain Avenue
Murray Hill, NJ 07974

Chapter 14

Robert D. Dony
School of Engineering
University of Guelph
Guelph, ON N1G 2W1 CANADA

Simon Haykin
McMaster University
Electrical and Computer Engineering
1280 Main Street West
Hamilton, Ontario, L8S 4K1 CANADA

Chapter 15

Aapo Hyvärinen
Helsinki University of Technology
Laboratory of Computer and Information Science
P.O.B. 5400
Espoo 02015 FINLAND

Patrik Hoyer
Neural Networks Research Centre
Helsinki University of Technology
PO Box 5400, FIN-02015 HUT
FINLAND

Erkki Oja
Helsinki University of Technology
Rakentajanaukio 2C
Espoo 02015 FINLAND

Chapter 1

Humanistic Intelligence: 'WearComp' as a new framework and application for intelligent signal processing

Steve Mann

Abstract

Humanistic Intelligence (HI) is proposed as a new signal processing framework in which the processing apparatus is inextricably intertwined with the natural capabilities of our human body and mind. Rather than trying to emulate human intelligence, HI recognizes that the human brain is perhaps the best neural network of its kind, and that there are many new signal processing applications, within the domain of personal cybernetics, that can make use of this excellent but often overlooked processor. The emphasis of this chapter is on personal imaging applications of HI, to take a first step toward an intelligent wearable camera system that can allow us to effortlessly capture our day-to-day experiences, help us remember and see better, provide us with personal safety through crime reduction, and facilitate new forms of communication through collective connected HI. The wearable signal processing hardware, which began as a cumbersome backpack-based photographic apparatus of the 1970s, and evolved into a clothing-based apparatus in the early 1980s, currently provides the computational power of a UNIX workstation concealed within ordinary-looking eyeglasses and clothing. Thus it may be worn continuously during all facets of ordinary day-to-day living, so that, through long-term adaptation, it begins to function as a true extension of the mind and body.

Keywords

Signals, Image processing, Human factors, Mobile communication, Machine vision, Photoquantigraphic imaging, Cybernetic sciences, Humanistic property protection, Consumer electronics

I. Introduction

WHAT is now proposed, is a new form of "intelligence" whose goal is to not only work in extremely close synergy with the human user, rather than as a separate entity, but more importantly, to arise, in part, because of the very **existence** of the human user. This close synergy is achieved through a *user-interface* to signal processing hardware that is both in *close physical proximity* to the user, and is *constant.*

The constancy of user-interface (interactional constancy) is what separates this signal processing architecture from other related devices such as pocket calculators and Personal Digital Assistants (PDAs).

Not only is the apparatus operationally constant, in the sense that although it may have power saving (sleep) modes, it is never completely shut down (dead as is typically a calculator worn in a shirt pocket but turned off most of the time). More important is the fact that it is also interactionally constant. By interactionally constant, what is meant is that the inputs and outputs of the device are always potentially active. Interactionally constant implies operationally constant, but operationally constant does not necessarily imply interactionally constant. Thus, for example, a pocket calculator, worn in a shirt pocket, and left on all the time is still not interactionally constant, because it cannot be used in this state (e.g. one still has to pull it out of the pocket to see the display or enter numbers). A wrist watch is a borderline case; although it operates constantly in order to continue to keep proper time, and it is conveniently worn on the body, one must make a conscious effort to orient it within one's field of vision in order to interact with it.

S. Mann is with the department of Electrical and Computer Engineering, University of Toronto, 10 King's College Road, S.F. 2001, Canada, M5S 3G4, E-mail: mann@eecg.toronto.edu http://wearcomp.org .

Special thanks to Kodak, Digital Equipment Corporation, Xybernaut, and ViA

A. Why Humanistic Intelligence

It is not, at first, obvious why one might want devices such as pocket calculators to be operationally constant. However, we will later see why it is desirable to have certain personal electronics devices, such as cameras and signal processing hardware, be on constantly, for example, to facilitate new forms of intelligence that assist the user in new ways.

Devices embodying HI are not merely intelligent signal processors that a user might wear or carry in close proximity to the body, but instead, are devices that turn the user into part of an intelligent control system where the user becomes an integral part of the feedback loop.

B. Humanistic Intelligence does not necessarily mean "user-friendly"

Devices embodying HI often require that the user learn a new skill set, and are therefore not necessarily easy to adapt to. Just as it takes a young child many years to become proficient at using his or her hands, some of the devices that implement HI have taken years of use before they began to truly behave as if they were natural extensions of the mind and body. Thus, in terms of Human-Computer Interaction [1], the goal is not just to construct a device that can model (and learn from) the user, but, more importantly, to construct a device in which the user also must learn from the device. Therefore, in order to facilitate the latter, devices embodying HI should provide a constant user-interface — one that is not so sophisticated and intelligent that it confuses the user. Although the device may implement very sophisticated signal processing algorithms, the cause and effect relationship of this processing to its input (typically from the environment or the user's actions) should be clearly and continuously visible to the user, even when the user is not directly and intentionally interacting with the apparatus. Accordingly, the most successful examples of HI afford the user a very tight feedback loop of system observability (ability to perceive how the signal processing hardware is responding to the environment and the user), even when the controllability of the device is not engaged (e.g. at times when the user is not issuing direct commands to the apparatus). A simple example is the viewfinder of a wearable camera system, which provides framing, a photographic point of view, and facilitates the provision to the user of a general awareness of the visual effects of the camera's own image processing algorithms, even when pictures are not being taken. Thus a camera embodying HI puts the human operator in the feedback loop of the imaging process, even when the operator only wishes to take pictures occasionally. A more sophisticated example of HI is a biofeedback-controlled wearable camera system, in which the biofeedback process happens continuously, whether or not a picture is actually being taken. In this sense, the user becomes one with the machine, over a long period of time, even if the machine is only directly used (e.g. to actually take a picture) occasionally.

Humanistic Intelligence attempts to both build upon, as well as re-contextualize, concepts in *intelligent signal processing* [2][3], and related concepts such as neural networks [2][4][5], fuzzy logic [6][7], and artificial intelligence [8]. Humanistic Intelligence also suggests a new goal for signal processing hardware, that is, in a truly personal way, to directly assist, rather than replace or emulate human intelligence. What is needed to facilitate this vision is a simple and truly personal computational signal processing framework that empowers the human intellect. It should be noted that this framework which arose in the 1970s and early 1980s is in many ways similar to Engelbart's vision that arose in the 1940s while he was a radar engineer, but that there are also some important differences. Engelbart, while seeing images on a radar screen, envisioned that the cathode ray screen could also display letters of the alphabet, as well as computer generated pictures and graphical content, and thus envisioned computing as an interactive experience for manipulating words and pictures. Engelbart envisioned the mainframe computer as a tool for augmented intelligence and augmented communication, in which a number of people in a large amphitheatre could interact with one another using a large mainframe computer[9] [10].

While Engelbart himself did not realize the significance of the personal computer, his ideas are certainly embodied in modern personal computing. What is now described is a means of realizing a similar

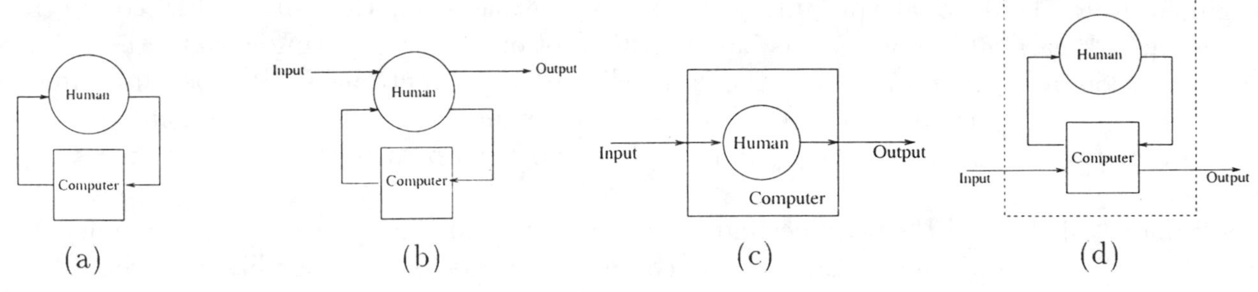

Fig. 1. The three basic operational modes of WearComp. (a) Signal flow paths for a computer system that runs continuously, constantly attentive to the user's input, and constantly providing information to the user. Over time, constancy leads to a symbiosis in which the user and computer become part of each other's feedback loops. (b) Signal flow path for augmented intelligence and augmented reality. Interaction with the computer is secondary to another primary activity, such as walking, attending a meeting, or perhaps doing something that requires full hand–to eye coordination, like running down stairs or playing volleyball. Because the other primary activity is often one that requires the human to be attentive to the environment as well as unencumbered, the computer must be able to operate in the background to augment the primary experience, for example, by providing a map of a building interior, or providing other information, through the use of computer graphics overlays superimposed on top of the real world. (c) WearComp can be used like clothing, to encapsulate the user and function as a protective shell, whether to protect us from cold, protect us from physical attack (as traditionally facilitated by armour), or to provide privacy (by concealing personal information and personal attributes from others). In terms of signal flow, this encapsulation facilitates the possible mediation of incoming information to permit solitude, and the possible mediation of outgoing information to permit privacy. It is not so much the absolute blocking of these information channels that is important; it is the fact that the wearer can control to what extent, and when, these channels are blocked, modified, attenuated, or amplified, in various degrees, that makes WearComp much more empowering to the user than other similar forms of portable computing. (d) An equivalent depiction of encapsulation (mediation) redrawn to give it a similar form to that of (a) and (b), where the encapsulation is understood to comprise a separate protective shell.

vision, but with the computing re–situated in a different context, namely the truly personal space of the user. The idea here is to move the tools of augmented intelligence and augmented communication directly onto the body, giving rise to not only a new genre of truly personal computing, but to some new capabilities and affordances arising from direct physical contact between the computational apparatus and the human body. Moreover, a new family of applications arises, such as "personal imaging", in which the body–worn apparatus facilitates an augmenting of the human sensory capabilities, namely vision. Thus the augmenting of human memory translates directly to a visual associative memory in which the apparatus might, for example, play previously recorded video back into the wearer's eyeglass mounted display, in the manner of a so–called *visual thesaurus*[11].

II. 'WearComp' as means of realizing Humanistic Intelligence

WearComp [12] is now proposed as an apparatus upon which a practical realization of HI can be built, as well as a research tool for new studies in intelligent signal processing.

A. *Basic principles of WearComp*

WearComp will now be defined in terms of its three basic modes of operation.

A.1 Operational modes of WearComp

The three operational modes in this new interaction between human and computer, as illustrated in Fig 1 are:

- Constancy: The computer runs continuously, and is "always ready" to interact with the user. Unlike a hand-held device, laptop computer, or PDA, it does not need to be opened up and turned on prior to use. The signal flow from human to computer, and computer to human, depicted in Fig 1(a) runs continuously to provide a constant user–interface.

• Augmentation: Traditional computing paradigms are based on the notion that computing is the primary task. WearComp, however, is based on the notion that computing is NOT the primary task. The assumption of WearComp is that the user will be doing something else at the same time as doing the computing. Thus the computer should serve to augment the intellect, or augment the senses. The signal flow between human and computer, in the augmentational mode of operation, is depicted in Fig 1(b).
• Mediation: Unlike hand held devices, laptop computers, and PDAs, WearComp can encapsulate the user (Fig 1(c)). It doesn't necessarily need to completely enclose us, but the basic concept of mediation allows for whatever degree of encapsulation might be desired, since it affords us the possibility of a greater degree of encapsulation than traditional portable computers. Moreover, there are two aspects to this encapsulation, one or both of which may be implemented in varying degrees, as desired:
– Solitude: The ability of WearComp to mediate our perception can allow it to function as an information filter, and allow us to block out material we might not wish to experience, whether it be offensive advertising, or simply a desire to replace existing media with different media. In less extreme manifestations, it may simply allow us to alter aspects of our perception of reality in a moderate way rather than completely blocking out certain material. Moreover, in addition to providing means for blocking or attenuation of undesired input, there is a facility to amplify or enhance desired inputs. This control over the input space is one of the important contributors to the most fundamental issue in this new framework, namely that of user empowerment.
– Privacy: Mediation allows us to block or modify information leaving our encapsulated space. In the same way that ordinary clothing prevents others from seeing our naked bodies, WearComp may, for example, serve as an intermediary for interacting with untrusted systems, such as third party implementations of digital anonymous cash, or other electronic transactions with untrusted parties. In the same way that martial artists, especially stick fighters, wear a long black robe that comes right down to the ground, in order to hide the placement of their feet from their opponent, WearComp can also be used to clothe our otherwise transparent movements in cyberspace. Although other technologies, like desktop computers, can, to a limited degree, help us protect our privacy with programs like Pretty Good Privacy (PGP), the primary weakness of these systems is the space between them and their user. It is generally far easier for an attacker to compromise the link between the human and the computer (perhaps through a so-called Trojan horse or other planted virus) when they are separate entities. Thus a personal information system owned, operated, and controlled by the wearer, can be used to create a new level of personal privacy because it can be made much more personal, e.g. so that it is always worn, except perhaps during showering, and therefore less likely to fall prey to attacks upon the hardware itself. Moreover, the close synergy between the human and computers makes it harder to attack directly, e.g. as one might look over a person's shoulder while they are typing, or hide a video camera in the ceiling above their keyboard[1].
Because of its ability to encapsulate us, e.g. in embodiments of WearComp that are actually articles of clothing in direct contact with our flesh, it may also be able to make measurements of various physiological quantities. Thus the signal flow depicted in Fig 1(a) is also enhanced by the encapsulation as depicted in Fig 1(c). To make this signal flow more explicit, Fig 1(c) has been redrawn, in Fig 1(d), where the computer and human are depicted as two separate entities within an optional protective shell, which may be opened or partially opened if a mixture of augmented and mediated interaction is desired.
Note that these three basic modes of operation are not mutually exclusive in the sense that the first is embodied in both of the other two. These other two are also not necessarily meant to be implemented in isolation. Actual embodiments of WearComp typically incorporate aspects of both augmented and

[1]For the purposes of this paper, privacy is not so much the absolute blocking or concealment of personal information, but it is the ability to control or modulate this outbound information channel. Thus, for example, one may wish certain people, such as members of one's immediate family, to have greater access to personal information than the general public. Such a family–area–network may be implemented with an appropriate access control list and a cryptographic communications protocol.

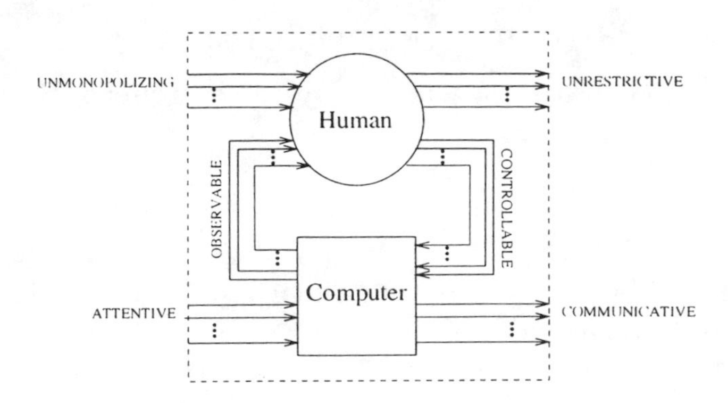

Fig. 2. The six signal flow paths for the new mode of human–computer interaction provided by WearComp. These six signal flow paths each define one of the six attributes of WearComp.

mediated modes of operation. Thus WearComp is a framework for enabling and combining various aspects of each of these three basic modes of operation. Collectively, the space of possible signal flows giving rise to this entire space of possibilities, is depicted in Fig 2. The signal paths typically comprise vector quantities. Thus multiple parallel signal paths are depicted in this figure to remind the reader of this vector nature of the signals.

B. The six basic signal flow paths of WearComp

There are six informational flow paths associated with this new human–machine symbiosis. These signal flow paths each define one of the basic underlying principles of WearComp, and are each described, in what follows, from the human's point of view. Implicit in these six properties is that the computer system is also operationally constant and personal (inextricably intertwined with the user). The six basic properties are:

1. **UNMONOPOLIZING of the user's attention**: it does not necessarily cut one off from the outside world like a virtual reality game or the like does. One can attend to other matters while using the apparatus. It is built with the assumption that computing will be a secondary activity, rather than a primary focus of attention. In fact, ideally, it will provide enhanced sensory capabilities. It may, however, facilitate mediation (augmenting, altering, or deliberately diminishing) these sensory capabilities.
2. **UNRESTRICTIVE to the user**: ambulatory, mobile, roving, one can do other things while using it, e.g. one can type while jogging, running down stairs, etc.
3. **OBSERVABLE by the user**: It can get the user's attention continuously if the user wants it to. The output medium is constantly perceptible by the wearer. It is sufficient that it be almost–always–observable, within reasonable limitations such as the fact that a camera viewfinder or computer screen is not visible during the blinking of the eyes.
4. **CONTROLLABLE by the user**: Responsive. The user can take control of it at any time the user wishes. Even in automated processes the user should be able to manually override the automation to break open the control loop and become part of the loop at any time the user wants to. Examples of this controllability might include a "Halt" button the user can invoke as an application mindlessly opens all 50 documents that were highlighted when the user accidently pressed "Enter"
5. **ATTENTIVE to the environment**: Environmentally aware, multimodal, multisensory. (As a result this ultimately gives the user increased situational awareness).
6. **COMMUNICATIVE to others**: WearComp can be used as a communications medium when the user wishes. Expressive: WearComp allows the wearer to be expressive through the medium, whether as a direct communications medium to others, or as means of assisting the user in the production of expressive or communicative media.

III. Philosophical issues

There are many open questions in this new area of research. Some of these include the following:
- Is WearComp good?
- Do we need it?
- Could it be harmful to the user?
- Could it be harmful to society?
- Are humans prepared for such a close synergy with machines, or will nature "strike back"?
- Will the apparatus modify the behaviour of the wearer in an undesirable way?
- Will it become a cause of irritation to others?

As with many new inventions, such as clothing, the bicycle, hot air balloons, etc., there has been an initial rejection, followed by scientific study and experimentation, followed by either acceptance or modification of the invention to address specific problems.

For example, the adverse effects of the bright screen constantly projected onto the eye were addressed by building the apparatus into very dark sunglasses so that a much lower brightness could be used.

More importantly, is perhaps the danger of becoming dependent on the technology in the same way that we become dependent on shoes, clothing, the automobile, etc.. For example, the fact that we cannot survive naked in the wilderness, or that we have become sedentary because of the automobile, must have its equivalent problems within the context of the proposed invention.

Many of these issues are open philosophical questions that will only be answered by further research. However, the specific framework and some of the various ideas surrounding it, will hopefully form the basis for a further investigation of some of these questions.

A. *Fundamental issues of WearComp*

The most fundamental paradigm shift that WearComp has to offer is that of personal empowerment. In order to fully appreciate the magnitude of this paradigm shift, some historical examples of tools of empowerment will now be described to place WearComp in this historical context.

A.1 Historical context

In early civilization, individuals were all roughly equal, militarily. Wealth was generally determined by how many head of cattle, or how many "mounts" (horses) a person owned. In hand-to-hand combat, fighting with swords, each individual was roughly an equal. Since it was impossible to stay on a horse while fighting, horses provided little in the way of military power, so that even those too poor to afford to keep a horse were not at a tremendous disadvantage to others from a fighting standpoint.

It was the invention of the stirrup, however, that radically changed this balance. With the stirrup, it became possible to stay on a horse while fighting. Horses and heavy armour could only be afforded by the wealthy, and even a large group of unruly peasants was no match for a much smaller group of mounted cavalry. However, toward the middle ages, more and more ordinary individuals mastered the art of fighting on horseback, and eventually the playing field leveled out.

Then, with the invention of gunpowder, the ordinary civilian was powerless against soldiers or bandits armed with guns. It was not until guns became cheap enough that everyone could own one — as in the "old west". The Colt 45, for example, was known as the "equalizer" because it made everyone roughly equal. Even if one person was much more skilled in its use, there would still be some risk involved in robbing other civilians or looting someone's home.

A.2 The shift from guns to cameras and computers

In today's world, the hand gun has a lesser role to play. Wars are fought with information, and we live in a world in which the appearance of thugs and bandits is not ubiquitous. While there is some crime, we spend most of our lives living in relative peace. However, surveillance and mass media have become the new instruments of social control. Department stores are protected with security cameras rather than by owners keeping a shotgun under the counter or hiring armed guards to provide a visible deterrent. While some department stores in rough neighbourhoods may have armed guards, there has been a paradigm shift where we see less guns and more surveillance cameras.

A.3 The shift from draconian punishment to micro management

There has also been a paradigm shift, throughout the ages, characterized by a move toward less severe punishments, inflicted with greater certainty. In the middle ages, the lack of sophisticated surveillance and communications networks meant that criminals often escaped detection or capture, but when they were captured, punishments were extremely severe. Gruesome corporeal punishments where criminals might be crucified, or whipped, branded, drawn and quartered, and then burned at the stake, were quite common in these times.

The evolution from punishment as a spectacle in which people were tortured to death in the village square, toward incarceration in which people were locked in a cell, and forced to attend church sermons, prison lectures, etc., marked the first step in a paradigm shift toward less severe punishments[13]. Combined with improved forensic technologies like fingerprinting, this reduction in the severity of punishment came together with a greater chance of getting caught.

More recently, with the advent of so–called "boot camp", where delinquent youths are sent off for mandatory military–style training, the trend continues by addressing social problems earlier before they become large problems. This requires greater surveillance and monitoring, but at the same time is characterized by less severe actions taken against those who are deemed to require these actions. Thus there is, again, still greater chance of being affected by smaller punishments.

If we extrapolate this trend, what we arrive at is a system of social control characterized by total surveillance and micro–punishments. At some point, the forces applied to the subjects of the social control are too weak to even justify the use of the word "punishment", and perhaps it might be better referred to as "micro management".

This "micro management" of society may be effected by subjecting the population to mass media, advertising, and calming music played in department stores, elevators, and subway stations.

Surveillance is also spreading into areas that were generally private in earlier times. The surveillance cameras that were placed in banks have moved to department stores. They first appeared above cash registers to deal with major crimes like holdups. But then they moved into the aisles and spread throughout the store to deal with petty theft. Again, more surveillance for dealing with lesser crimes.

In the U.K., cameras installed for controlling crime in rough areas of town spread to low crime areas as well, in order to deal with problems like youths stealing apples from street markets, or patrons of pubs urinating on the street. The cameras have even spread into restaurants and pubs — not just above the cash register, but throughout the pub, so that going out for pints, one may no longer have privacy.

Recently, electronic plumbing technology, originally developed for use in prisons, for example, to prevent all inmates from flushing the toilets simultaneously, has started to be used in public buildings. The arguments in favor of it go beyond human hygiene and water conservation, as proponents of the technology argue that it also reduces vandalism. Their definition of vandalism has been broadened to include deliberately flooding a plumbing fixture, and deliberately leaving faucets running. Thus, again, what we see is greater certainty of catching or preventing people from committing lesser transgressions of the social order.

One particularly subtle form of social control using this technology, is the new hands free electronic showers developed for use in prisons where inmates would otherwise break off knobs, levers, and pushbuttons. These showers are just beginning to appear in government buildings, stadiums, health clubs, and schools. The machine watches the user, from behind a tiled wall, through a small dark glass window. When the user steps toward the shower, the water comes on, but only for a certain time, and then it shuts off. Obviously the user can step away from the viewing window, and then return, to receive more water, and thus defeat the timeout feature of the system, but this need to step away and move back into view is enough of an irritant as to effect a slight behavioural modification of the user. Thus what we see is that surveillance has swept across all facets of society, but is being used to deal with smaller and smaller problems. From dealing with mass murderers and bank robbers, to people who threaten the environment by taking long showers, the long arm of surveillance has reached into even the most private of places, where we might have once been alone. The peace and solitude of the shower, where our greatest inspirations might come to us, has been intruded upon with not a major punishment, but a very minor form of social control, too small in fact to even be called a punishment.

These surveillance and social control systems are linked together, often to central computer systems. Everything from surveillance cameras in the bank, to electronic plumbing networks is being equipped with fiber optic communications networks. Together with the vast array of medical records, credit card purchases, buying preferences, etc., we are affected in more ways, but with lesser influence. We are no longer held at bay by mounted cavalry. More often than being influenced by weapons, we are influenced in very slight, almost imperceptible ways, for example, through a deluge of junk mail, marketing, advertising, or a shower that shuts off after it sees that we've been standing under it for too long.

While there are some (the most notable being Jeremy Bentham[13]) who have put forth an argument that a carefully managed society results in maximization of happiness, there are others who argue that the homogenization of society is unhealthy, and reduces humans to cogs in a larger piece of machinery, or at the very least, results in a certain loss of human dignity. Moreover, just as nature provides biodiversity, many believe that society should also be diverse, and people should try to resist ubiquitous centralized surveillance and control, particularly to the extent where it homogenizes society excessively. Some argue that micro-management and utilitarianism, in which a person's value may often be measured in terms of usefulness to society, is what led to eugenics, and eventually to the fascism of Nazi Germany. Many people also agree that, even without any sort of social control mechanism, surveillance, in and of itself, still violates their privacy, and is fundamentally wrong.

As with other technologies, like the stirrup and gunpowder, the electronic surveillance playing field is also being leveled. The advent of the low-cost personal computer has allowed individuals to communicate freely and easily among themselves. No longer are the major media conglomerates the sole voice heard in our homes. The World Wide Web has ushered in a new era of underground news and alternative content. Thus centralized computing facilities, the very technology that many perceived as a threat to human individuality and freedom, have given way to low cost personal computers that many people can afford. This is not to say that home computers will be as big or powerful as the larger computers used by large corporations or governments, but simply that if a large number of people have a moderate degree of computational resources, there is a sense of balance in which people are roughly equal in the same sense that two people, face to face, one with a 0.22 calibre handgun and the other with a Colt 0.45 are roughly equal. A large bullet hole or a small one, both provide a tangible and real risk of death or injury.

It is perhaps modern cryptography that makes this balance even more pronounced, for it is so many orders of magnitude easier to encrypt a message than it is to decrypt it. Accordingly, many governments have defined cryptography as a munition and attempted, with only limited success, to restrict its use.

A.4 Fundamental issues of WearComp

The most fundamental issue in WearComp is no doubt that of personal empowerment, through its ability to equip the individual with a personalized, customizable information space, owned, operated, and controlled by the wearer. While home computers have gone a long way to empowering the individual, they only do so when the user is at home. As the home is perhaps the last bastion of space not yet touched by the long arm of surveillance — space that one can call one's own, the home computer, while it does provide an increase in personal empowerment, is not nearly so profound in its effect as the WearComp which brings this personal space — space one can call one's own — out into the world.

Although WearComp, in the most common form we know it today (miniature video screen over one or both eyes, body worn processor, and input devices such as a collection of pushbutton switches or joystick held in one hand and a microphone) was invented in the 1970s for personal imaging applications, it has more recently been adopted by the military in the context of large government–funded projects.

However, as with the stirrup, gunpowder, and other similar inventions, it is already making its way out into the mainstream consumer electronics arena.

An important observation to make, with regards to the continued innovation, early adopters (military, government, large multinational corporations), and finally ubiquity, is the time scale. While it took a relatively longer time for the masses to adopt the use of horses for fighting, and hence level the playing field, later, the use of gunpowder became ubiquitous in a much shorter time period.

Then, sometime after guns had been adopted by the masses, the spread of computer technology, which in some situations even replaced guns, was so much faster still. As the technology diffuses into society more quickly, the military is losing its advantage over ordinary civilians. We are entering a pivotal era in which consumer electronics is surpassing the technological sophistication of some military electronics. Personal audio systems like the SONY Walkman are just one example of how the ubiquity and sophistication of technology feed upon each other to the extent that the technology begins to rival, and in some ways, exceed, the technical sophistication of the limited–production military counterparts such as two–way radios used in the battlefield.

Consumer technology has already brought about a certain degree of personal empowerment, from the portable cassette player that lets us replace the music piped into department stores with whatever we would rather hear, to small hand held cameras that capture police brutality and human rights violations. However, WearComp is just beginning to bring about a much greater paradigm shift, which may well be equivalent in its impact to the invention of the stirrup, or that of gunpowder. Moreover, this leveling of the playing field may, for the first time in history, happen almost instantaneously, should the major consumer electronics manufacturers beat the military to raising this invention to a level of perfection similar to that of the stirrup or modern handguns. If this were to happen, this decreasing of the time scale over which technology diffuses through society will have decreased to zero, resulting in a new kind of paradigm shift that society has not yet experienced. Evidence of this pivotal shift is already visible, in, for example, the joint effort of Xybernaut Corp. (a major manufacturer of wearable computers) and SONY Corp. (a manufacturer of personal electronics) to create a new personal electronics computational device.

B. Aspects of WearComp and personal empowerment

There are several aspects and affordances of WearComp. These are:

- Photographic/videographic memory: Perfect recall of previously collected information, especially visual information (*visual memory*[14]).
- Shared memory: In a collective sense, two or more individuals may share in their collective consciousness, so that one may have a recall of information that one need not have experienced personally.
- Connected collective humanistic intelligence: In a collective sense, two or more individuals may collaborate while one or more of them is doing another primary task.
- Personal safety: In contrast to a centralized surveillance network built into the architecture of the city, a personal safety system is built into the architecture (clothing) of the individual. This framework has the potential to lead to a distributed "intelligence" system of sorts, as opposed to the centralized "intelligence" gathering efforts of traditional video surveillance networks.

• Tetherless operation: WearComp affords and requires mobility, and the freedom from the need to be connected by wire to an electrical outlet, or communications line.
• Synergy: Rather than attempting to emulate human intelligence in the computer, as is a common goal of research in Artificial Intelligence (AI), the goal of WearComp is to produce a synergistic combination of human and machine, in which the human performs tasks that it is better at, while the computer performs tasks that it is better at. Over an extended period of time, WearComp begins to function as a true extension of the mind and body, and no longer feels as if it is a separate entity. In fact, the user will often adapt to the apparatus to such a degree, that when taking it off, its absence will feel uncomfortable, in the same way that we adapt to shoes and clothing to such a degree that being without them most of us would feel extremely uncomfortable (whether in a public setting, or in an environment in which we have come to be accustomed to the protection that shoes and clothing provide). This intimate and constant bonding is such that the combined capability resulting in a synergistic whole far exceeds the sum of its components.
• Quality of life: WearComp is capable of enhancing day–to–day experiences, not just in the workplace, but in all facets of daily life. It has the capability to enhance the overall quality of life for many people.

IV. Practical embodiments of WearComp

The WearComp apparatus consists of a battery-powered wearable Internet-connected [15] computer system with miniature eyeglass-mounted screen and appropriate optics to form the virtual image equivalent to an ordinary desktop multimedia computer. However, because the apparatus is tetherless, it travels with the user, presenting a computer screen that either appears superimposed on top of the real world, or represents the real world as a video image[16].

Due to advances in low power microelectronics [17], we are entering a pivotal era in which it will become possible for us to be inextricably intertwined with computational technology that will become part of our everyday lives in a much more immediate and intimate way than in the past.

Physical proximity and constancy were simultaneously realized by the 'WearComp' project[2] of the 1970s and early 1980s (Fig 3) which was a first attempt at building an intelligent "photographer's assistant" around the body, and comprised a computer system attached to the body, a display means constantly visible to one or both eyes, and means of signal input including a series of pushbutton switches and a pointing device (Fig 4) that the wearer could hold in one hand to function as a keyboard and mouse do, but still be able to operate the device while walking around. In this way, the apparatus re-situated the functionality of a desktop multimedia computer with mouse, keyboard, and video screen, as a physical extension of the user's body. While the size and weight reductions of WearComp over the last 20 years, have been quite dramatic, the basic qualitative elements and functionality have remained essentially the same, apart from the obvious increase in computational power.

However, what makes WearComp particularly useful in new and interesting ways, and what makes it particularly suitable as a basis for HI, is the collection of other input devices, not all of which are found on a desktop multimedia computer.

In typical embodiments of 'WearComp' these measurement (input) devices include the following:
• ultra-miniature cameras concealed inside eyeglasses and oriented to have the same field of view as the wearer, thus providing the computer with the wearer's "first-person" perspective.
• one or more additional cameras that afford alternate points of view (e.g. a rear-looking camera with a view of what is directly behind the wearer).
• sets of microphones, typically comprising one set to capture the sounds of someone talking to the wearer (typically a linear array across the top of the wearer's eyeglasses), and a second set to capture the wearer's own speech.

[2]For a detailed historical account of the WearComp project, and other related projects, see [18][19].

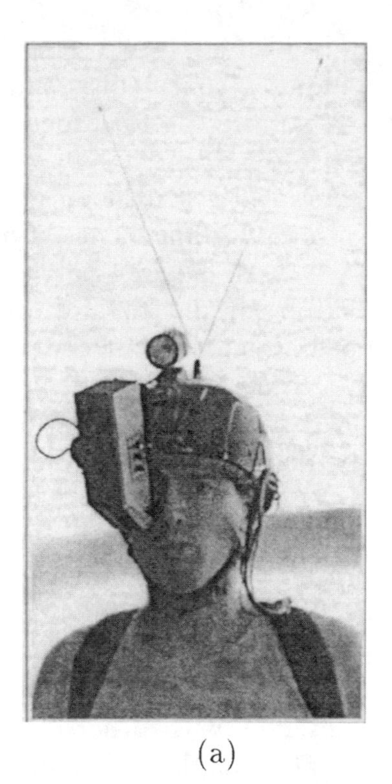

(a) (b)

Fig. 3. Early embodiments of the author's original "photographer's assistant" application of Personal Imaging. (a) Author wearing WearComp2, an early 1980s backpack-based signal processing and personal imaging system with right eye display. Two antennas operating at different frequencies facilitated wireless communications over a full-duplex radio link. (b) WearComp4, a late 1980s clothing-based signal processing and personal imaging system with left eye display and beam splitter. Separate antennas facilitated simultaneous voice, video, and data communication.

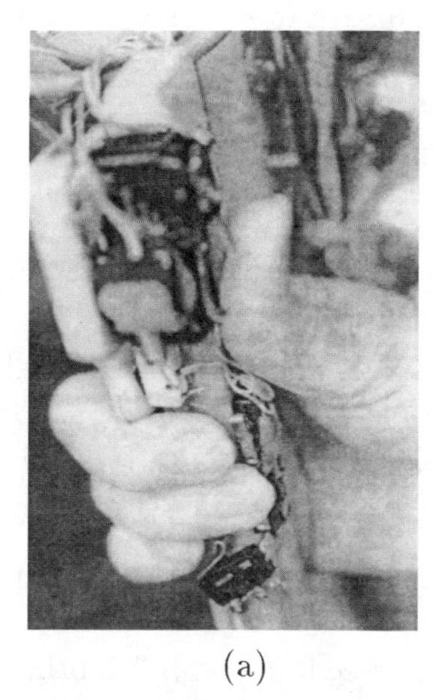

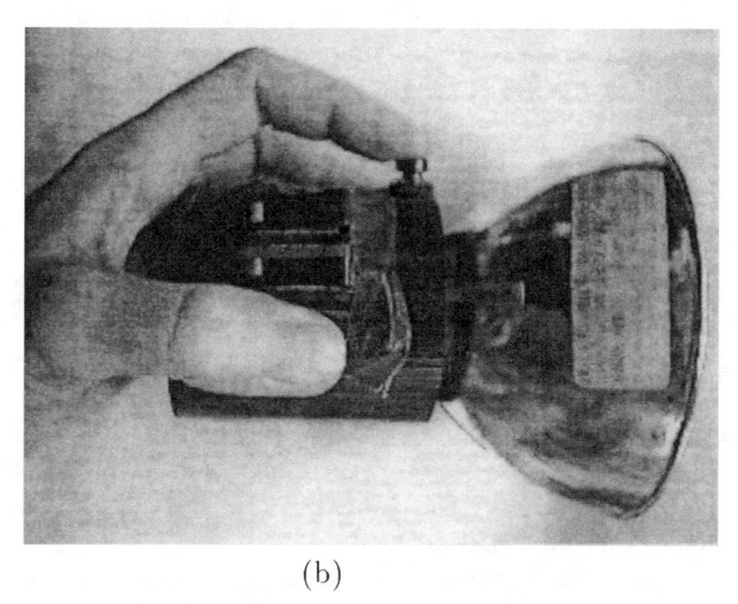

(a) (b)

Fig. 4. Author using some early input devices ("keyboards" and "mice") for WearComp: (a) 1970s: input device comprising pushbutton switches mounted to a wooden hand-grip (b) 1980s: input device comprising microswitches mounted to the handle of an electronic flash. These devices also incorporated a detachable joystick (controlling two potentiometers), designed as a pointing device for use in conjunction with the WearComp project.

• biosensors, comprising not just heart rate but full ECG waveform, as well as respiration, skin conductivity, sweat level, and other quantities [20] each available as a continuous (sufficiently sampled) time-varying voltage. Typically these are connected to the wearable central processing unit through an eight-channel analog to digital converter.
• footstep sensors typically comprising an array of transducers inside each shoe.
• wearable radar systems in the form of antenna arrays sewn into clothing. These typically operate in the 24.36GHz range.
The last three, in particular, are not found on standard desktop computers, and even the first three, which often are found on standard desktop computers, appear in a different context here than they do on a desktop computer. For example, in WearComp, the camera does not show an image of the user, as it does typically on a desktop computer, but, rather, it provides information about the user's environment. Furthermore, the general philosophy, as will be described in Sections V and VI, will be to regard all of the input devices as measurement devices. Even something as simple as a camera will be regarded as a measuring instrument, within the signal processing framework.

Certain applications use only a subset of these devices, but including all of them in the design facilitates rapid prototyping and experimentation with new applications. Most embodiments of WearComp are modular, so that devices can be removed when they are not being used.

A side-effect of this 'WearComp' apparatus is that it replaces much of the personal electronics that we carry in our day-to-day living. It enables us to interact with others through its wireless data communications link, and therefore replaces the pager and cellular telephone. It allows us to perform basic computations, and thus replaces the pocket calculator, laptop computer and personal data assistant (PDA). It can record data from its many inputs, and therefore it replaces and subsumes the portable dictating machine, camcorder, and the photographic camera. And it can reproduce ("play back") audiovisual data, so that it subsumes the portable audio cassette player. It keeps time, as any computer does, and this may be displayed when desired, rendering a wristwatch obsolete. (A calendar program which produces audible, vibrotactile, or other output also renders the alarm clock obsolete.)

However, it goes beyond replacing all of these items, because not only is it currently far smaller and far less obtrusive than the sum of what it replaces, but these functions are interwoven seamlessly, so that they work together in a mutually assistive fashion. Furthermore, entirely new functionalities, and new forms of interaction arise, such as enhanced sensory capabilities, as will be discussed in Sections V and VI.

A. Building signal-processing devices directly into fabric

The wearable signal processing apparatus of the 1970s and early 1980s was cumbersome at best, so an effort was directed toward not only reducing its size and weight, but, more importantly, reducing its undesirable and somewhat obtrusive appearance. Moreover, an effort was also directed at making an apparatus of a given size and weight more comfortable to wear and bearable to the user [12], through bringing components in closer proximity to the body, thereby reducing torques and moments of inertia. Starting in 1982, Eleveld and Mann[19] began an effort to build circuitry directly into clothing. The term 'smart clothing' refers to variations of WearComp that are built directly into clothing, and are characterized by (or at least an attempt at) making components distributed rather than lumped, whenever possible or practical.

It was found [19] that the same apparatus could be made much more comfortable by bringing the components closer to the body which had the effect of reducing both the torque felt bearing the load, as well as the moment of inertia felt in moving around. This effort resulted in a version of WearComp called the 'Underwearable Computer' [19] shown in Fig 5.

Typical embodiments of the underwearable resemble an athletic undershirt (tank top) made of durable mesh fabric, upon which a lattice of webbing is sewn. This facilitates quick reconfiguration in the layout of components, and re-routing of cabling. Note that wire ties are not needed to fix cabling,

(a)

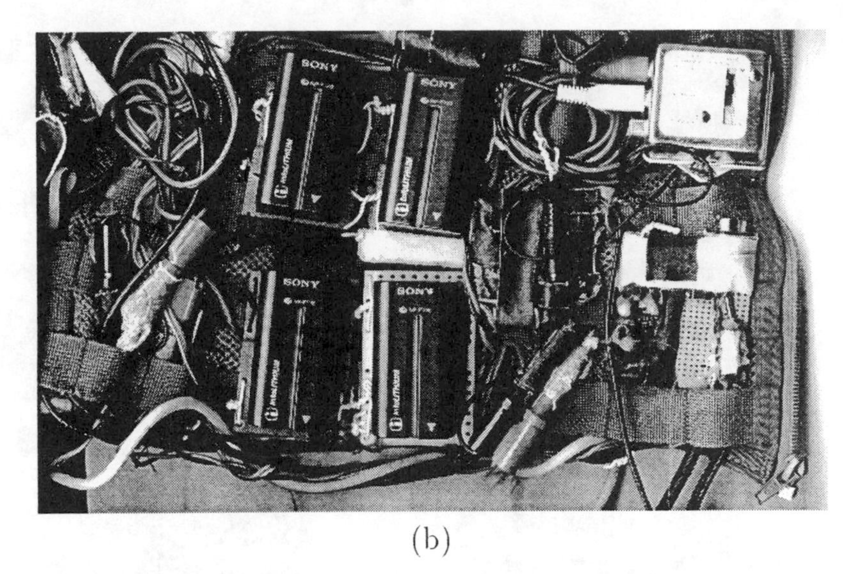

(b)

Fig. 5. The 'underwearable' signal processing hardware: (a) as worn by author (stripped to the undershirt, which is normally covered by a sweater or jacket); (b) close up of underwearable signal processor, showing webbing for routing of cabling.

as it is simply run through the webbing, which holds it in place. All power and signal connections are standardized, so that devices may be installed or removed without the use of any tools (such as soldering iron) by simply removing the garment and spreading it out on a flat surface.

Some more recent related work by others [21], also involves building circuits into clothing, in which a garment is constructed as a monitoring device to determine the location of a bullet entry. The underwearable differs from this monitoring apparatus in the sense that the underwearable is totally reconfigurable in the field, and also in the sense that it embodies HI (the apparatus reported in [21] performs a monitoring function but does not facilitate human interaction).

In summary, there were three reasons for the signal processing hardware being 'underwearable':

- By both distributing the components throughout the garment, and by keeping the components in close physical proximity to the body, it was found that the same total weight and bulk could be worn much more comfortably.
- Wearing the apparatus underneath ordinary clothing gave rise to a version of the WearComp apparatus which had a normal appearance. Although many of the prototypes were undetectable (covert) to visual inspection, early prototypes of course could be detected upon physical contact with the body by others. However, it was found that by virtue of social norms, the touching of the body by others (and therefore discovery of the apparatus) was seldom a problem. Thus making certain that the apparatus did not have an unsightly or unusual appearance was found to be sufficient to integrating into society in a normal way. Unobtrusiveness is essential so that the apparatus does not interfere with normal social situations, for one cannot truly benefit from the long-term adaptation process of HI unless it is worn nearly constantly for a period of many years. Two examples of underwearables, as they normally appear when worn under clothing, are depicted in Fig 6, where the normal appearance is quite evident.
- The close proximity of certain components to the body provided additional benefits, such as the ability to easily integrate measuring apparatus for quantities such as respiration, heart rate and full ECG waveform, galvanic skin resistance, etc., of the wearer. The fact that the apparatus is worn underneath clothing facilitated direct contact with the body, providing a much richer measurement space and facilitating new forms of intelligent signal processing.

A.1 Remaining issues of underwearable signal processing hardware

This work on "underwearable signal processing hardware" has taken an important first step toward solving issues of comfort, detectability (and detectability on physical contact, which remains as a

(a)

(b)

Fig. 6. Covert embodiments of WearComp suitable for use in ordinary day-to-day situations. Both incorporate fully functional UNIX-based computers concealed in the small of the back, with the rest of the peripherals, analog to digital converters, etc., also concealed under ordinary clothing. Both incorporate camera–based imaging systems concealed within the eyeglasses, the importance of which will become evident in Section V, in the context of Personal Imaging. While these prototype units are detectable by physical contact with the body, detection of the unusual apparatus was not found to be a problem, since normal social conventions are such that touching of the body is normally only the domain of those known well to the wearer. As with any prosthetic device, first impressions are important to normal integration into society, and discovery by those who already know the wearer well (e.g. to the extent that close physical contact may occur) typically happens after an initial acceptance is already established. Other prototypes have been integrated into the clothing in a manner that feels natural to the wearer and to others who might come into physical contact with the wearer. (a) lightweight black and white version completed in 1995. (b) full-color version completed in 1996 included special-purpose digital signal processing hardware based on an array of TMS 320 series processors connected to a UNIX-based host processor, concealed in the small of the back.

greater technological challenge). Issues such as shielding of electromagnetic radiation have already been largely solved through the use of conductive undergarments that protect the body from radiation (especially from the transmitting antennas that establish the connection to the Internet). In addition to protecting the wearer who would otherwise be in very close proximity to the transmitting antennas, such shielding also improves system operation, for example, by keeping RF from the transmitter as well as ambient environmental RF out of the biosensors and measurement instrumentation that is in direct contact with the body. Many of these practical issues will be dealt with in future design in moving from early prototypes of the "underwearable signal processing system" into production systems.

B. Multidimensional signal input for Humanistic Intelligence

The close physical proximity of WearComp to the body, as described earlier, facilitates a new form of signal processing[3]. Because the apparatus is in direct contact with the body, it may be equipped with various sensory devices. For example, a tension transducer (pictured leftmost, running the height of the picture from top to bottom, in Fig 7). is typically threaded through and around the underwearable, at stomach height, so that it measures respiration. Electrodes are also installed in such a manner that they are in contact with the wearer's heart. Various other sensors, such as an array of transducers

[3]The first wearable computers equipped with multichannel biosensors were built by the author during the 1980s inspired by a collaboration with Dr. Ghista of McMaster University. More recently, in 1995, the author put together an improved apparatus based on a Compaq Contura Aero 486/33 with a ProComp 8 channel analog to digital converter, worn in a Mountainsmith waist bag, and sensors from Thought Technologies Limited. The author subsequently assisted Healey in duplicating this system for use in trying to understand human emotions [22].

Fig. 7. Author's Personal Imaging system equipped with sensors for measuring biological signals. The sunglasses in the upper right are equipped with built in video cameras and display system. These look like ordinary sunglasses when worn (wires are concealed inside the eyeglass holder). At the left side of the picture is an 8 channel analog to digital converter together with a collection of biological sensors, both manufactured by Thought Technologies Limited, of Canada. At the lower right is an input device called the "twiddler", manufactured by HandyKey, and to the left of that is a Sony Lithium Ion camcorder battery with custom-made battery holder. In the lower central area of the image is the computer, equipped with special-purpose video processing/video capture hardware (visible as the top stack on this stack of PC104 boards). This computer, although somewhat bulky, may be concealed in the small of the back, underneath an ordinary sweater. To the left of the computer, is a serial to fiber-optic converter that provides communications to the 8 channel analog to digital converter over a fiber-optic link. Its purpose is primarily one of safety, to isolate high voltages used in the computer and peripherals (e.g. the 500 volts or so present in the sunglasses) from the biological sensors which are in close proximity, typically with very good connection, to the body of the wearer.

in each shoe [23] and a wearable radar system (described in Section VI) are also included as sensory inputs to the processor. The ProComp 8 channel analog to digital converter with some of the input devices that are sold with it is pictured in Fig 7 together with the CPU from WearComp6.

B.1 Safety first!

The importance of personal safety, in the context of this new paradigm in computing, must be emphasized. Since electrical connections are often in close proximity to the body, and are, in fact, often made directly to the body (such as when fitting the undergarment with electrodes that connect to the bare flesh of the user, in the chest area, for ECG waveform monitoring), the inputs to which these connections are made must be fully isolated. In the present prototypes, this isolation is typically accomplished by using a length of fiber optic cable between the measurement hardware and the host computer. This is particularly essential in view of the fact that the host computer may be connected to other devices such as head mounted displays containing voltages from 500 volts to as high as 9kV (e.g. in the case of head mounted displays containing cathode ray tubes).

Moreover, the presence of high voltages in the eyeglasses themselves, as well as in other parts of the system (as when it is interfaced to an electronic flash system which typically uses voltages ranging from 480V to 30kV), requires extra attention to insulation, since it is harder to free oneself from the apparatus when it is worn, than when it is merely carried. For example, a hand–held electronic flash can be dropped to the ground easier should it become unstable, while a wearable system embodies the danger of entrapment in a failing system.

Furthermore, since batteries may deliver high currents, there would be the risk of fire, and the prospect of being trapped in burning clothing, were it not for precautions taken in limiting current flow. Thus in addition to improved high voltage insulation, there is also the need to install current limiting fuses and the like, throughout the garment.

As an additional precaution, all garments are made from flame-proof material. In this regard, especially with the development of early prototypes over the last 20 years, it was felt that a healthy sense of paranoia was preferable to carelessness that might give rise to a dangerous situation.

B.2 More than just a health status monitor

It is important to realize that this apparatus is not merely a biological signal logging device, as is often used in the medical community, but, rather, enables new forms of real-time signal processing for HI. A simple example might include a biofeedback-driven video camera.

Picard also suggests its possible use to estimate human emotion [24].

The emphasis of this chapter will be on visual image processing with the WearComp apparatus. The author's dream of the 1970s, that of an intelligent wearable image processing apparatus, is just beginning to come to fruition.

V. The Personal Imaging application of Humanistic Intelligence

A. *Some simple illustrative examples*

A.1 **Always Ready:** From *point and click* to 'look and think'

Current day commercial personal electronics devices we often carry are just useful enough for us to tolerate, but not good enough to significantly simplify our lives. For example, when we are on vacation, our camcorder and photographic camera require enough attention that we often either miss the pictures we want, or we become so involved in the process of video or photography that we fail to really experience the immediate present environment [25].

One ultimate goal of the proposed apparatus and methodology is to "learn" what is visually important to the wearer, and function as a fully automatic camera that takes pictures without the need for conscious thought or effort from the wearer. In this way, it might summarize a day's activities, and then automatically generate a gallery exhibition by transmitting desired images to the World Wide Web, or to specific friends and relatives who might be interested in the highlights of one's travel. The proposed apparatus, a miniature eyeglass-based imaging system, does not encumber the wearer with equipment to carry, or with the need to remember to use it, yet because it is recording all the time into a circular buffer [19], merely overwriting that which is unimportant it is *always ready.* Although some have noted that the current embodiment of the invention, still in prototype stage, is somewhat cumbersome enough that one might not wear it constantly, it is easy to imagine how, with mass production, and miniaturization, smaller and lighter units could be built, perhaps with the computational hardware built directly into ordinary glasses. Making the apparatus small enough to comfortably wear at all times will lead to a truly constant user–interface.

In the context of the *always ready* framework, when the signal processing hardware detects something that might be of interest, recording can begin in a retroactive sense (e.g. a command may be issued to start recording from thirty seconds ago), and the decision can later be confirmed with human input. Of course this apparatus raises some important privacy questions discussed previously, and also addressed elsewhere in the literature [26][27].

The system might use the inputs from the biosensors on the body, as a multidimensional feature vector with which to classify content as important or unimportant. For example, it might automatically record a baby's first steps, as the parent's eyeglasses and clothing-based intelligent signal processor make an inference based on the thrill of the experience. It is often moments like these that we fail to capture on film: by the time we find the camera and load it with film, the moment has passed us by.

A.2 Personal safety device for reducing crime

A simple example of where it would be desirable that the device operate by itself, without conscious thought or effort, is in an extreme situation such as might happen if the wearer were attacked by a robber wielding a shotgun, and demanding cash.

In this kind of situation, it is desirable that the apparatus would function autonomously, without conscious effort from the wearer, even though the wearer might be aware of the signal processing activities of the measuring (sensory) apparatus he or she is wearing.

As a simplified example of how the processing might be done, we know that the wearer's heart rate, averaged over a sufficient time window, would likely increase dramatically[4] with no corresponding increase in footstep rate (in fact footsteps would probably slow at the request of the gunman). The computer would then make an inference from the data, and predict a high visual saliency. (If we simply take heart rate divided by footstep rate, we can get a first-order approximation of the visual saliency index.) A high visual saliency would trigger recording from the wearer's camera at maximal frame rate, and also send these images together with appropriate messages to friends and relatives who would look at the images to determine whether it was a false alarm or real danger[5].

Such a system is, in effect, using the wearer's brain as part of its processing pipeline, because it is the wearer who sees the shotgun, and not the WearComp apparatus (e.g. a much harder problem would have been to build an intelligent machine vision system to process the video from the camera and determine that a crime was being committed). Thus HI (intelligent signal processing arising, in part, because of the very existence of the human user) has solved a problem that would not be possible using machine-only intelligence.

Furthermore, this example introduces the concept of 'collective connected HI', because the signal processing systems also rely on those friends and relatives to look at the imagery that is wirelessly send from the eyeglass–mounted video camera and make a decision as to whether it is a false alarm or real attack. Thus the concept of HI has become blurred across geographical boundaries, and between more than one human and more than one computer.

A.3 The retro-autofocus example: Human in the signal processing loop

The above two examples dealt with systems which use the human brain, with its unique processing capability, as one of their components, in a manner in which the overall system operates without conscious thought or effort. The effect is to provide a feedback loop of which subconscious or involuntary processes becomes an integral part.

An important aspect of HI is that the conscious will of the user may be inserted into or removed from the feedback loop of the entire process at any time. A very simple example, taken from everyday

[4] Perhaps it may stop, or "skip a beat" at first, but over time, on average, in the time following the event, experience tells us that our hearts beat faster when frightened.

[5] It has been suggested that the robber might become aware that his or her victim is wearing a personal safety device and try to eliminate it or perhaps even target it for theft. In anticipation of these possible problems, personal safety devices operate by continuous transmission of images, so that the assailant cannot erase or destroy the images depicting the crime. Moreover, the device itself, owing to its customized nature, would be unattractive and of little value to others, much as are undergarments, a mouthguard, or prescription eyeglasses. Furthermore, devices may be protected by a password embedded into a CPLD that functions as a finite state machine, making them inoperable by anyone but the owner. To protect against passwords being extracted through torture, a personal distress password may be provided to the assailant by the wearer. The personal distress password unlocks the system but puts it into a special tracking and distress notification mode.

experience, rather than another new invention, is now presented.

One of the simplest examples of HI is that which happens with some of the early autofocus Single Lens Reflex (SLR) cameras in which autofocus was a retrofit feature. The autofocus motor would typically turn the lens barrel, but the operator could also grab onto the lens barrel while the autofocus mechanism was making it turn. Typically the operator could "fight" with the motor, and easily overpower it, since the motor was of sufficiently low torque. This kind of interaction is particularly useful, for example, when shooting through a glass window at a distant object, where there are two or three local minima of the autofocus error function (e.g. focus on particles of dust on the glass itself, focus on a reflection in the glass, and focus on the distant object). Thus when the operator wishes to focus on the distant object and the camera system is caught in one of the other local minima (for example, focused on the glass), the user merely grasps the lens barrel, swings it around to the approximate desired location (as though focusing crudely by hand, on the desired object of interest), and lets go, so that the camera will then take over and bring the desired object into sharp focus.

This very simple example illustrates a sort of humanistic intelligent signal processing in which the intelligent autofocus electronics of the camera work in close synergy with the intellectual capabilities of the camera operator.

It is this aspect of HI, that allows the human to step into and out of the loop at any time, that makes it a powerful paradigm for intelligent signal processing.

B. Mathematical framework for personal imaging

The theoretical framework for HI is based on processing a series of inputs from various wearable sensory apparatus, in a manner that regards each one of these as belonging to a measurement space; each of the inputs (except for the computer's keyboard–like input device comprised of binary pushbutton switches) is regarded as a measurement instrument to be linearized in some meaningful continuous underlying physical quantity.

Since the emphasis of this chapter is on personal imaging, the treatment here will focus on the wearable camera (discussed here in Section V) and the wearable radar (discussed later in Section VI). The other measurement instruments are important, but their role is primarily to facilitate harnessing and amplification of the human intellect for purposes of processing data from the imaging apparatus.

The theoretical framework for processing video is based on regarding the camera as an array of light measuring instruments capable of measuring how the scene or objects in view of the camera respond to light[6]. This framework has two important special cases, the first of which is based on a quantifiable self-calibration procedure and the second of which is based on algebraic projective geometry as a means of combining information from images related to one-another by a projective coordinate transformation.

These two special cases of the theory are now presented in Sections V.B.1 and V.B.2 respectively, followed by bringing both together in Section V.B.3. The theory is applicable to standard photographic or video cameras, as well as to the wearable camera and personal imaging system.

B.1 Quantigraphic Imaging and the Wyckoff principle

Quantigraphic imaging is now described.

It should be noted that the goal of quantigraphic imaging, to regard the camera as an array of light measuring instruments, is quite different from the goals of other related research[29] in which there is an attempt to separate the reflectance of objects from their illumination. Indeed, while Stockham's effort was focused on separately processing the effects due to reflectance and scene illumination[29], quantigraphic imaging takes a camera–centric viewpoint, and does not attempt to model the cause of the light entering the camera, but merely determines, to within a single unknown scalar constant, the quantity of light entering the camera from each direction in space. Quantigraphic imaging measures neither radiometric *irradiance* nor photometric *illuminance* (since the camera will not necessarily have

[6]This 'lightspace' theory was first written up in detail in 1992 [28].

the same spectral response as the human eye, or, in particular, that of the photopic spectral luminous efficiency function as determined by the CIE and standardized in 1924). Instead, quantigraphic imaging measures the quantity of light integrated over the particular spectral response of the camera system, in units that are quantifiable (e.g. linearized), in much the same way that a photographic light meter measures in quantifiable (linear or logarithmic) units. However, just as the photographic light meter imparts to the measurement its own spectral response (e.g. a light meter using a selenium cell will impart the spectral response of selenium cells to the measurement) quantigraphic imaging accepts that there will be a particular spectral response of the camera, which will define the quantigraphic unit of measure.

A field of research closely related to Stockham's approach is that of colorimetry, in the context of the so-called *color constancy problem*[30] [31] in which there is an attempt made to determine the true color of an object irrespective of the color of the illuminant. Thus solving the *color constancy problem*, for example, might amount to being able to recognize the color of an object and ignore the orange color cast of indoor tungsten lamps or ignore the relatively bluish cast arising from viewing the same object under the illumination of the sky outdoors. Quantigraphic imaging, on the other hand, makes no such attempt, and provides a true measure of the light arriving at the camera, without any attempt to determine whether color effects are owing to the natural color of an object or the color of the illumination.

Quantigraphic imaging may, however, be an important first step to solving the *color constancy problem* — once we know how much light is arriving from each direction in space, in each of the spectral bands of the sensor, and we have a measure of these quantities that is linearized, we can then apply to these quantigraphic images, any of the traditional mathematical image processing frameworks such as those of Stockham[29], those of Venetsanopoulos[32], or those from color theory. Quantigraphic imaging may also be a first step to other uses of the image data, whether they be for machine vision or simply for the production of a visually pleasing picture.

The special case of quantigraphic imaging presented here in Section V.B.1 pertains to a fixed camera (e.g. as one would encounter in mounting the camera on tripod). Clearly this is not directly applicable to the wearable camera system, except perhaps in the case of images acquired in very rapid succession. However, this theory, when combined with the Video Orbits theory of Section V.B.2, is found to be useful in the context of the personal imaging system, as will be described in Section V.B.3.

Fully automatic methods of seamlessly combining differently exposed pictures to extend dynamic range have been proposed[33][34], and are summarized here.

Most everyday scenes have a far greater dynamic range than can be recorded on a photographic film or electronic imaging apparatus (whether it be a digital still camera, consumer video camera, or eyeglass-based personal imaging apparatus as described in this paper). However, a set of pictures, that are identical except for their exposure, collectively show us much more dynamic range than any single picture from that set, and also allow the camera's response function to be estimated, to within a single constant scalar unknown.

A set of functions,

$$E_n(\mathbf{x}) = f(k_n q(\mathbf{x})), \tag{1}$$

where k_n are scalar constants, is known as a Wyckoff set [35][19], and describes a set of images, E_n, when $\mathbf{x} = (x, y)$ is the spatial coordinate of a piece of film or the continuous spatial coordinates of the focal plane of an electronic imaging array, q is the quantity of light falling on the sensor array, and f is the unknown nonlinearity of the camera's response function (assumed to be invariant to $\mathbf{x} \in \mathbb{R}^2$.

Because of the effects of noise (quantization noise, sensor noise, etc.), in practical imaging situations, the dark (often underexposed) pictures show us highlight details of the scene that might have been overcome by noise (e.g. washed out) had the picture been properly exposed. Similarly, the light pictures show us some shadow detail that might not have appeared above the noise threshold had the picture been properly exposed.

A means of simultaneously estimating f and k_n, given a Wyckoff set E_n, has been proposed [33][35][19]. A brief outline of this method follows. For simplicity of illustration (without loss of generality), suppose that the Wyckoff set contains two pictures, $E_1 = f(q)$ and $E_2 = f(kq)$, differing only in exposure (e.g. where the second image received k times as much light as the first). Photographic film is traditionally characterized by the so-called "D logE" (Density versus log Exposure) characteristic curve [36][37]. Similarly, in the case of electronic imaging, we may also use logarithmic exposure units, $Q = \log(q)$, so that one image will be $K = log(k)$ units darker than the other:

$$\log(f^{-1}(E_1)) = Q = \log(f^{-1}(E_2)) - K \tag{2}$$

The existence of an inverse for f follows from the semimonotonicity assumption [35][19]. (We expect any reasonable camera to provide a semimonotonic relation between quantity of light received, q, and the pixel value reported.) Since the logarithm function is also monotonic, the problem is reduced to that of estimating the semimonotonic function $F() = \log(f^{-1}())$ and the scalar constant K, given two pictures E_1 and E_2:

$$F(E_2) = F(E_1) + K \tag{3}$$

Thus:

$$E_2 = F^{-1}(F(E_1) + K) \tag{4}$$

provides a recipe for "registering" (appropriately lightening or darkening) the second image to match the first. This registration procedure differs from the image registration procedure commonly used in image resolution enhancement (to be described in Section V.B.2) because it operates on the *range* (tonal range) of the image $E(\mathbf{x})$ as opposed to its *domain* (spatial coordinates) $\mathbf{x} = (x, y)$. (In Section V.B.3, registration in both *domain* and *range* will be addressed.)

Once f is determined, each picture becomes a different estimate of the same true quantity of light falling on each pixel of the image sensor:

$$\hat{q}_n = \frac{1}{k_n} f^{-1}(E_n) \tag{5}$$

Thus one may regard each of these measurements (pixels) as a light meter (sensor element) that has some nonlinearity followed by a quantization to a measurement having typically 8-bit precision.

It should be emphasized that most image processing algorithms incorrectly assume that the camera response function is linear (e.g. almost all current image processing, such as blurring, sharpening, unsharp masking, etc., operates linearly on the image) while in fact it is seldom linear. Even Stockham's homomorphic filtering [29], which advocates taking the log, applying linear filtering, and then taking the antilog, fails to capture the correct nonlinearity [19][35], as it ignores the true nonlinearity of the sensor array. It has recently been shown [35][19] that, in the absence of any knowledge of the camera's nonlinearity, an approximate one-parameter parametric model of the camera's nonlinear response is far better than assuming it is linear or logarithmic. Of course, finding the true response function of the camera allows one to do even better, as one may then apply linear signal processing methodology to the original light falling on the image sensor.

B.2 Video Orbits

A useful assumption in the domain of 'personal imaging', is that of zero parallax, whether this be for obtaining a first-order estimate of the yaw, pitch, and roll of the wearer's head [20], or making an important first step in the more difficult problem of estimating depth and structure from a scene[7]. Thus, in this section, the assumption is that most of the image motion arises from that of generating an environment map, zero-parallax is assumed.

[7] first modeling the motion as a projective coordinate transformation, and then estimating the residual epipolar structure or the residual epipolar structure or the like[38][39][40][41][42][43]

The problem of assembling multiple pictures of the same scene into a single image commonly arises in mapmaking (with the use of aerial photography) and photogrammetry[44], where zero-parallax is also generally assumed. Many of these methods require human interaction (e.g. selection of features), and it is desired to have a fully automated system that can assemble images from the eyeglass-based camera. Fully automatic featureless methods of combining multiple pictures have been previously proposed[45][46], but with an emphasis on subpixel image shifts; the underlying assumptions and models (affine, and pure translation, respectively) were not capable of accurately describing more macroscopic image motion. A characteristic of video captured from a head-mounted camera is that it tends to have a great deal more macroscopic image motion, and a great deal more perspective 'cross-chirping' between adjacent frames of video, while the assumptions of static scene content and minimal parallax are still somewhat valid. This assumption arises for the following reasons:

- Unlike the heavy hand-held cameras of the past, the personal imaging apparatus is very lightweight.
- Unlike the hand-held camera which extends outward from the body, the personal imaging apparatus is mounted close to the face. This results in a much lower moment of inertia, so that the head can be rotated quickly. Although the center of projection of the wearable camera is not located at the center of rotation of the neck, it is much closer than with a hand-held camera.

It was found that the typical video generated from the personal imaging apparatus was characterized by rapid sweeps or pans (rapid turning of the head), which tended to happen over much shorter time intervals and therefore dominate over second-order effects such as parallax and scene motion [19]. The proposed method also provides an indication of its own failure, and this can be used as a feature, rather than a "bug" (e.g. so that the WearComp system is aware of scene motion, scene changes, etc., by virtue of its ability to note when the algorithm fails). Thus the projective group of coordinate transformations captures the essence of video from the WearComp apparatus[8].

Accordingly, two featureless methods of estimating the parameters of a projective group of coordinate transformations were first proposed in [33], and in more detail in [35], one direct and one based on optimization (minimization of an objective function). Although both of these methods are multiscale (e.g. use a coarse to fine pyramid scheme), and both repeat the parameter estimation at each level (to compute the residual errors), and thus one might be tempted to call both *iterative*, it is preferable to refer to the direct method as *repetitive* to emphasize that does not require a nonlinear optimization procedure such as Levenberg-Marquardt[48] [49], or the like. Instead, it uses repetition with the correct law of composition on the projective group, going from one pyramid level to the next by application of the group's law of composition. A method similar to the optimization-method was later proposed in [35][50]. The direct method has also been subsequently described in more detail [51].

The direct featureless method for estimating the 8 scalar parameters[9], of an exact projective (homographic) coordinate transformation is now described. In the context of personal imaging, this result is used to multiple images to seamlessly combine images of the same scene or object, resulting in a single image (or new image sequence) of greater resolution or spatial extent.

Many papers have been published on the problems of motion estimation and frame alignment. (For review and comparison, see [52].) In this Section the emphasis is on the importance of using the "exact" 8-parameter projective coordinate transformation [51], particularly in the context of the head-worn miniature camera.

The most common assumption (especially in motion estimation for coding, and optical flow for computer vision) is that the coordinate transformation between frames is translation. Tekalp, Ozkan, and Sezan [46] have applied this assumption to high-resolution image reconstruction. Although translation is less simpler to implement than other coordinate transformations, it is poor at handling large changes

[8]The additional one-time download of a lens distortion map, into WearComp's coordinate transformation hardware, eliminates its lens distortion which would otherwise be very large, owing to the covert, and therefore small, size of the lens, and in engineering compromises necessary to its design. The Campbell method [47] is used to estimate the lens distortion for this one-time coordinate transformation map.

[9]published in detail in [51]

due to camera zoom, rotation, pan and tilt.

Zheng and Chellappa [53] considered the image registration problem using a subset of the affine model — translation, rotation and scale. Other researchers [45][54] have assumed affine motion (six parameters) between frames.

The only model that properly captures the "keystoning" and "chirping" effects of projective geometry is the projective coordinate transformation. However, because the parameters of the projective coordinate transformation had traditionally been thought to be mathematically and computationally too difficult to solve, most researchers have used the simpler affine model or other approximations to the projective model.

The 8-parameter *pseudo-perspective* model [39] does, in fact, capture both the converging lines and the chirping of a projective coordinate transformation, but not the true essence of projective geometry.

Of course, the desired "exact" eight parameters come from the projective group of coordinate transformations, but they have been perceived as being notoriously difficult to estimate. The parameters for this model have been solved by Tsai and Huang [55], but their solution assumed that features had been identified in the two frames, along with their correspondences. The main contribution of the result summarized in this Section is a simple featureless means of automatically solving for these 8 parameters.

A group is a set upon which there is defined an associative law of composition (*closure*, *associativity*), which contains at least one element (*identity*) who's composition with another element leaves it unchanged, and for which every element of the set has an *inverse*.

A *group* of operators together with a *set* of operands form a so-called *group operation*[10]. In the context of a Lie group of spatial coordinate transformation operators acting on a set of visual images as operands, such a group is also known as a *Lie group of transformations*[57].

Note that Hoffman's use of the term "transformation" is not synonymous with "homomorphism"[56] as is often the case in group theory, such as when a transformation T acts on a law of composition between elements $g, h \in G$, such that $T(gh) = T(g)T(h)$. Instead, what is meant by "transformation", in the context of this paper, is a change in coordinates of a picture (image). Thus, in the context of this paper, transformations act on images not on elements of the group (which happens to be a group of transformation operators).

As in Hoffman' s work[57], coordinate transformations are operators selected from a group, and the set of images are the operands. This group of operators and set of images thus form the group action in the sense defined in[56]. When the coordinate transformations form a group, then two such coordinate transformations, $\mathbf{p}_1$ and $\mathbf{p}_2$, acting in succession, on an image (e.g. $\mathbf{p}_1$ acting on the image by doing a coordinate transformation, followed by a further coordinate transformation corresponding to $\mathbf{p}_2$, acting on that result) can be replaced by a single coordinate transformation. That single coordinate transformation is given by the *law of composition* in the group.

The *orbit* of a particular element of the set, under the group operation [56] is the new set formed by applying to it, all possible operators from the group.

Thus the orbit is a collection of pictures formed from one picture through applying all possible projective coordinate transformations to that picture. This set is referred to as the 'video orbit' of the picture in question [51]. Equivalently, we may imagine a static scene, in which the wearer of the personal imaging system is standing at a single fixed location. He or she generates a family of images in the same orbit of the projective group of transformations by looking around (rotation of the head)[11].

The coordinate transformations of interest, in the context of this paper,

$$\mathbf{x}' = \begin{bmatrix} x' \\ y' \end{bmatrix} = \frac{\mathbf{A}[x,y]^T + \mathbf{b}}{\mathbf{c}^T[x,y]^T + d} = \frac{\mathbf{A}\mathbf{x} + \mathbf{b}}{\mathbf{c}^T\mathbf{x} + d} \tag{6}$$

[10]also known as a *group action* or *G-set* [56].

[11]The resulting collection of images may be characterized by fewer parameters, through application of the Hartley constraint [58][35] to an estimate of a projective coordinate transformations.

define an operator, $\mathbf{P_1}$ that acts on the images as follows:

$$P_{\mathbf{A_1},\mathbf{b_1},\mathbf{c_1},d_1} \circ E(\mathbf{x}) = E(\mathbf{x}') = E\left(\frac{\mathbf{A_1}\mathbf{x} + \mathbf{b_1}}{\mathbf{c_1}^T\mathbf{x} + d_1}\right) \tag{7}$$

where the operator $\mathbf{P_1}$ is parameterized by $\mathbf{A}_1 \in \mathbb{R}^{2\times2}$, $\mathbf{b}_1 \in \mathbb{R}^{2\times1}$, $\mathbf{c}_1 \in \mathbb{R}^{2\times1}$, and $d_1 \in \mathbb{R}$.

These operators may be applied to an image in succession, this succession being defined, for example, with such operators, $\mathbf{P_1}$ and $\mathbf{P_2}$, as:

$$\mathbf{P_2} \circ \mathbf{P_1} \circ E(\mathbf{x}) = \mathbf{P}_{\mathbf{A}_2,\mathbf{b}_2,\mathbf{c}_2,\mathbf{d}_2} \circ \mathbf{P}_{\mathbf{A}_1,\mathbf{b}_1,\mathbf{c}_1,\mathbf{d}_1} \circ E(\mathbf{x}) \tag{8}$$

$$\mathbf{P_2} \circ \mathbf{P_1} \circ E(\mathbf{x}) = E\left(\frac{A_2(\frac{\mathbf{A_1}\mathbf{x}+\mathbf{b_1}}{\mathbf{c_1}^T\mathbf{x}+d_1}) + b_2}{c_2^T(\frac{\mathbf{A_1}\mathbf{x}+\mathbf{b_1}}{\mathbf{c_1}^T\mathbf{x}+d_1}) + d_2}\right) \tag{9}$$

$$= E\left(\begin{bmatrix}\mathbf{A_2}, \mathbf{b_2} \\ \mathbf{c_2}^T, d_2\end{bmatrix}\begin{bmatrix}\mathbf{A_1}, \mathbf{b_1} \\ \mathbf{c_1}^T, d_1\end{bmatrix}\begin{bmatrix}x \\ y\end{bmatrix}\right) \tag{10}$$

from which we can see that the operators can be represented as 2×2 matrixes, and that the law of composition defined on these operators can be represented by matrix multiplication. Associativity follows from matrix multiplication, and the matrix

$$\begin{bmatrix}\mathbf{I}_2, & \mathbf{0} \\ \mathbf{0}, & 1\end{bmatrix} \tag{11}$$

where I_2 is the identity matrix $[1, 0; 0, 1]$.

The 'video orbit' of a given 2-D frame is defined to be the set of all images that can be produced by applying operators from the 2-D projective group of coordinate transformations (6) to the given image. Hence, the problem may be restated: Given a set of images that lie in the same orbit of the group, find for each image pair, that operator in the group which takes one image to the other image.

If two frames of the video image sequence, say, f_1 and f_2, are in the same orbit, then there is an group operation $\mathbf{p}$ such that the mean-squared error (MSE) between f_1 and $f_2' = \mathbf{p} \circ f_2$ is zero. In practice, however, the element of the group that takes one image "nearest" the other is found (e.g. there will be a certain amount of error due to violations in the assumptions, due to noise such as parallax, interpolation error, edge effects, changes in lighting, depth of focus, etc).

The brightness constancy constraint equation [59] which gives the flow velocity components, is:

$$\mathbf{u_f}^T\mathbf{E_x} + E_t \approx 0 \tag{12}$$

As is well-known [59] the optical flow field in 2-D is under constrained The model of *pure translation* at every point has two parameters, but there is only one equation (12) to solve, thus it is common practice to compute the optical flow over some neighborhood, which must be at least two pixels, but is generally taken over a small block, 3×3, 5×5, or sometimes larger (e.g. the entire image, as in the Video Orbits algorithm described here).

However, rather than estimating the 2 parameter translational flow, the task here is to estimate the eight parameter projective flow (6) by minimizing:

$$\begin{aligned}\varepsilon_{flow} &= \sum\left(\mathbf{u_m}^T\mathbf{E_x} + E_t\right)^2 \\ &= \sum\left((\frac{\mathbf{A}\mathbf{x} + \mathbf{b}}{\mathbf{c}^T\mathbf{x} + d} - \mathbf{x})^T\mathbf{E_x} + E_t\right)^2\end{aligned} \tag{13}$$

Although a sophisticated nonlinear optimization procedure, such as Levenberg-Marquardt, may be applied to solve (13), it has been found that solving a slightly different but much easier problem, allows us to estimate the parameters more directly and accurately for a given amount of computation [51]:

$$\varepsilon_w = \sum \left((\mathbf{A}\mathbf{x} + \mathbf{b} - (\mathbf{c}^T\mathbf{x} + d)\mathbf{x})^T \mathbf{E}_\mathbf{x} + (\mathbf{c}^T\mathbf{x} + d)E_t \right)^2 \tag{14}$$

(This amounts to weighting the sum differently.)

Differentiating (13) with respect to the free parameters $\mathbf{A}, \mathbf{b},$ and $\mathbf{c}$, and setting the result to zero gives a linear solution:

$$\begin{aligned} \left(\sum \phi\phi^T\right) [a_{11}, a_{12}, b_1, a_{21}, a_{22}, b_2, c_1, c_2]^T \\ = \sum (\mathbf{x}^T \mathbf{E}_\mathbf{x} - E_t)\phi \end{aligned} \tag{15}$$

where $\phi^T = [E_x(x, y, 1), E_y(x, y, 1), xE_t - x^2E_x - xyE_y, yE_t - xyE_x - y^2E_y]$

In practice this process has been further improved by making an initial estimate using methods such as described in [60], [61], [62], as well as [63].

B.3 *Dynamic Range* and 'Dynamic Domain'

The contribution of this Section is a simple method of "scanning" out a scene, from a fixed point in space, by panning, tilting, or rotating a camera, whose gain (automatic exposure, electronic level control, automatic iris, AGC, or the like[12]) is also allowed to change of its own accord (e.g. arbitrarily).

Nyquist showed how a signal can be reconstructed from a sampling of finite resolution in the domain (e.g. space or time), but assumed infinite dynamic range (e.g. infinite precision or *word length* per sample). On the other hand, if we have infinite spatial resolution, but limited dynamic range (even if we have only 1 bit of image depth), Curtis and Oppenheim [64] showed that we can also obtain perfect reconstruction using an appropriate modulation function. In the case of the personal imaging system, we typically begin with images that have very low spatial resolution and very poor dynamic range (video cameras tend to have poor dynamic range, and this poor performance is especially true of the small CCDs that the author uses in constructing unobtrusive lightweight systems). Thus, since we lack both spatial and tonal resolution, we are not at liberty to trade some of one for more of the other. Thus the problem of 'spatiotonal' (simultaneous spatial and tonal) resolution enhancement is of particular interest in personal imaging.

In Section V.B.1, a new method of allowing a camera to self-calibrate was proposed. This methodology allowed the tonal range to be significantly improved. In Section V.B.2, a new method of resolution enhancement was described. This method allowed the spatial range to be significantly enhanced.

In this Section (V.B.3), a method of enhancing both the tonal range and the spatial domain resolution of images is proposed. It is particularly applicable to processing video from miniature covert eyeglass-mounted cameras, because it allows very noisy low quality video signals to provide not only high-quality images of great spatiotonal definition, but also to provide a rich and accurate photometric measurement space which may be of significant use to intelligent signal processing algorithms. That it provides not only high quality images, but also linearized measurements of the quantity of light arriving at the eyeglasses from each possible direction of gaze, follows from a generalization of the photometric measurement process outlined in Section V.B.1.

Most notably, this generalization of the method no longer assumes that the camera need be mounted on a tripod, but only that the images fall in the same orbit of a larger group, called the 'projectivity+gain' group of transformations.

[12]For simplicity, all these methods of automatic exposure control are referred to as AGC in this paper, whether or not they are actually implemented using an Automatic Gain Control (AGC) circuit or otherwise.

Thus the apparatus can be easily used without conscious thought or effort, which gives rise to new intelligent signal processing capabilities. The method works as follows: As the wearer of the apparatus looks around, the portion of the field of view that controls the gain (usually the central region of the camera's field of view) will be pointed toward different objects in the scene. Suppose for example, that the wearer is looking at someone so that their face is centered in the frame of the camera, f_1. Now suppose that the wearer tips his or her head upward so that the camera is pointed at a light bulb up on the ceiling, but that the person's face is still visible at the bottom of the frame, f_2. Because the light bulb has moved into the center of the frame, the camera's AGC causes the entire image to darken significantly. Thus these two images, which both contain the face of the person the wearer is talking to, will be very differently exposed. When *registered* in the spatial sense (e.g. through the appropriate projective coordinate transformation), they will be identical, over the region of overlap, except for exposure, if we assume that the wearer swings his or her head around quickly enough to make any movement in the person he is talking to negligible. While this assumption is not always true, there are certain times that it is true (e.g. when the wearer swings his or her head quickly from left to right and objects in the scene are moving relatively slowly). Because the algorithm can tell when the assumptions are true (by virtue of the error), during the times it is true, it use the multiple estimates of $\hat{q}_n$, the quantity of light received, to construct a high definition environment map.

An example of an image sequence captured with a covert eyeglass-based version of the author's WearComp7, and transmitted wirelessly to the Internet, appears in Fig 8.

Clearly, in this application, AGC, which has previously been regarded as a serious impediment to machine vision and intelligent image processing, becomes an advantage. By providing a collection of images with differently exposed but overlapping scene content, additional information about the scene, as well as the camera (information that can be used to determine the camera's response function, f) is obtained. The ability to have, and even benefit from AGC is especially important for WearCam contributing to the hands-free nature of the apparatus, so that one need not make any adjustments when, for example, entering a dimly lit room from a brightly lit exterior.

The group of spatiotonal image transformations of interest is defined in terms of of projective coordinate transformations, taken together with the one-parameter group of gain changes (image darkening/lightening) operations:

$$p_{\mathbf{A},\mathbf{b},\mathbf{c},d,k} \circ f(q(x)) = g_k(f(q(\frac{\mathbf{A}\mathbf{x}+\mathbf{b}}{\mathbf{c}^T\mathbf{x}+d}))) = f(kq(\frac{\mathbf{A}\mathbf{x}+\mathbf{b}}{\mathbf{c}^T\mathbf{x}+d})) \tag{16}$$

where g_k characterizes the gain operation, and admits a group representation:

$$\begin{bmatrix} \mathbf{A} & \mathbf{b} & \mathbf{0} \\ \mathbf{c} & d & 0 \\ \mathbf{0} & 0 & k \end{bmatrix}, \tag{17}$$

giving the law of composition defined by matrix multiplication.

Two successive frames of a video sequence are related through a group-action that is near the identity of the group, thus the Lie algebra of the group as provides the structure locally. As in previous work[34] an approximate model which matches the 'exact' model in the neighbourhood of the identity is used, together with the law of composition in the group (e.g. all of the manipulation and composition of multiple coordinate transformations is based on the algebra of the 'exact' model, even though an approximate model is used in the innermost loop of the computation.

To construct a single floating-point image of increased spatial extent and increased dynamic range, first the images are spatiotonally registered (brought not just into register in the traditional 'domain motion' sense, but also brought into the same tonal scale through quantigraphic gain adjustment). This form of spatiotonal transformation is illustrated in Fig 9 where all the images are transformed into the coordinates of the first image of the sequence, and in Fig 10 where all the images are transformed into

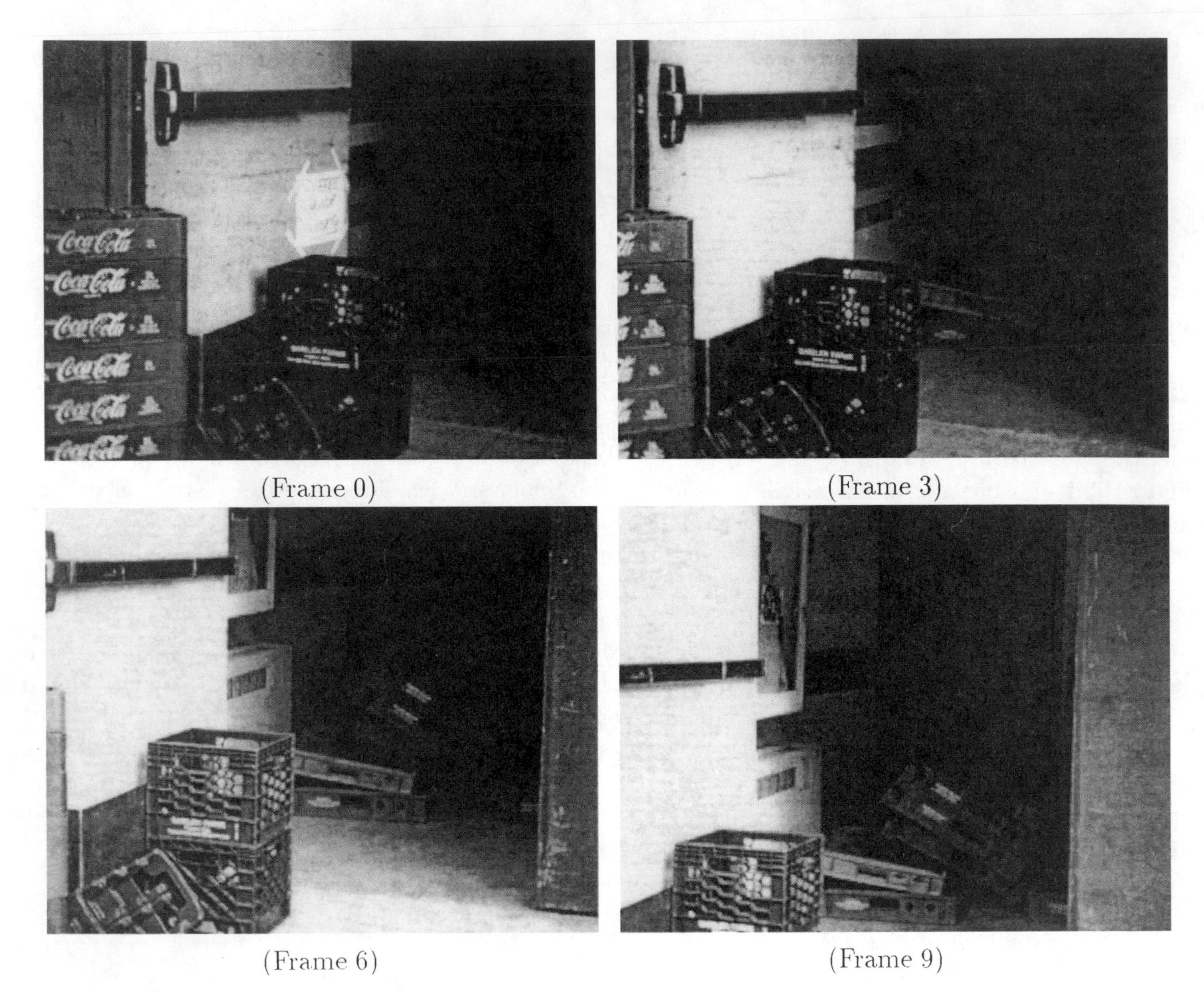

Fig. 8. The 'fire–exit' sequence, captured using a covert eyeglass-based personal imaging system with automatic gain control (AGC). Here, every third frame of the ten–frame image sequence is shown. As the camera pans across to take in more of the open doorway, the image brightens up showing more of the interior, while, at the same time, clipping highlight detail. Frame 0 shows the writing on the white paper taped to the door very clearly, but the interior is completely black. In Frame 3 the paper is obliterated — it is so "washed out" that we can no longer read what is written on it. Although the interior is getting brighter at this point, it is still not discernible in Frame 3 — we can see neither what is written on the paper, nor what is at the end of the dark corridor. However, as the author turns his head to the right, pointing the camera into the dark corridor, more and more detail of the interior becomes visible as we proceed through the sequence, revealing the inner depths of the long dark corridor, and showing that the fire exit is blocked by the clutter inside.

the coordinates of the last frame in the image sequence. It should be noted that the final quantigraphic composite can be made in the coordinates of any of the images. The choice of *reference frame* is arbitrary since the result is a floating point image array (not quantized)! Furthermore, the final composite need not even be expressed in the spatiotonal coordinates of any of the incoming images. For example quantigraphic coordinates (linear in the original light falling on the image array) may be used, to provide an array of measurements that linearly represent the quantity of light, to within a single unknown scalar constant for the entire array.

Once spatiotonally registered, each pixel of the output image is constructed from a weighted sum of the images whose coordinate-transformed bounding boxes fall within that pixel. The weights in the weighted sum are the so-called 'certainty functions', which are found by evaluating the derivative of

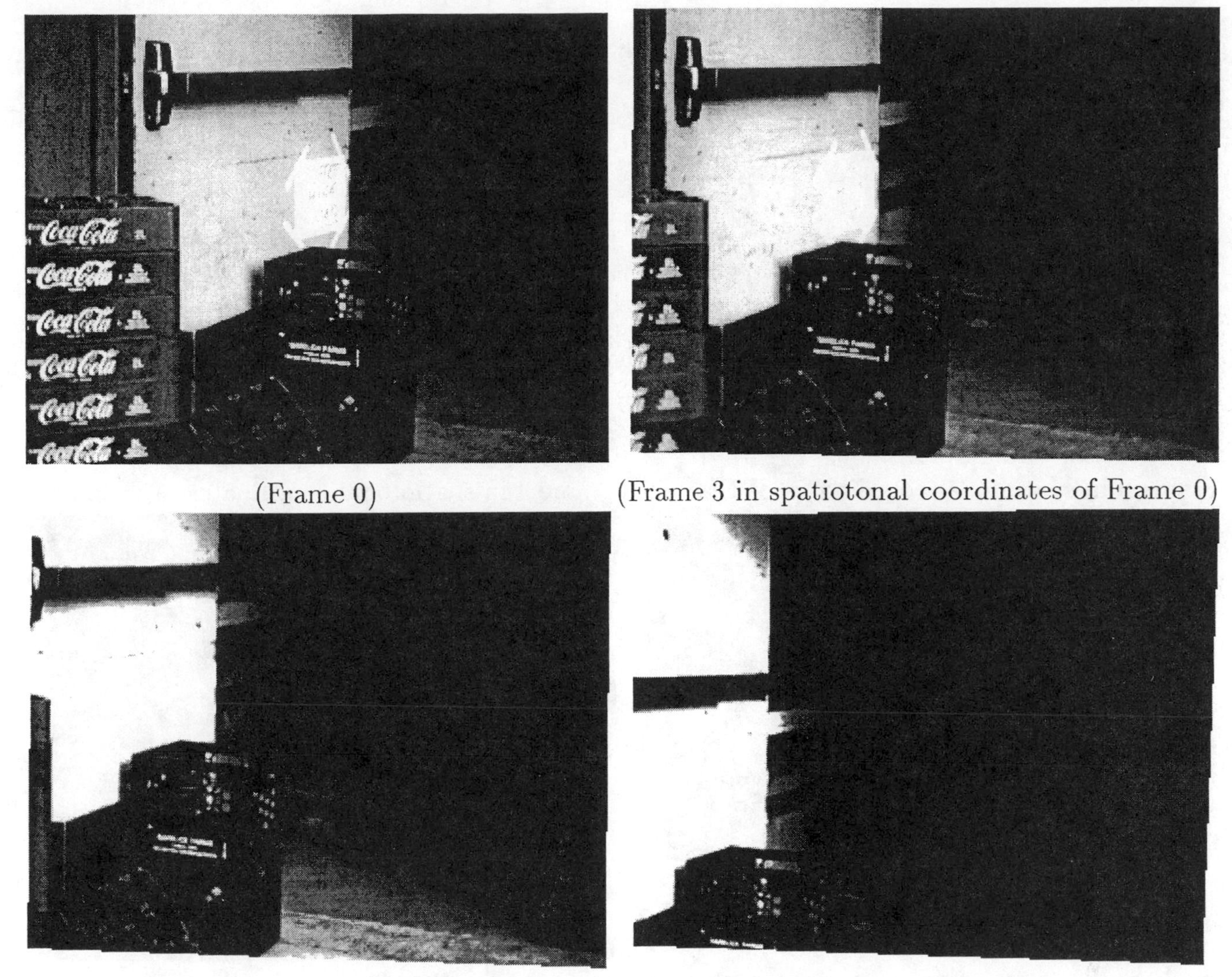

(Frame 0) (Frame 3 in spatiotonal coordinates of Frame 0)

(Frame 6 in spatiotonal coordinates of Frame 0) (Frame 9 in spatiotonal coordinates of Frame 0)

Fig. 9. Images of Fig 8 expressed in the spatiotonal coordinates of the first image in the sequence. Note both the keystoning, and chirping effect of the images toward the end of the sequence, indicating the spatial coordinate transformation, as well as the darkening, indicating the tone scale adjustment, both of which make the images match (Frame 0). Prior to quantization for printing in this figure, the images that were most severely darkened (e.g. (Frame 6) and (Frame 9)) to match frame 10 contained a tremendous deal of shadow detail, owing to the fact that the quantigraphic step sizes are much smaller when compressed into the range of (Frame 0).

the corresponding estimated effective characteristic function at the pixel value in question[65].

Although the response function, $f(q)$, is fixed for a given camera, the 'effective response function', $f(k_i(q))$ depends on the exposure, k_i, associated with frame, i, in the image sequence.

The composite image may be explored interactively on a computer system (Fig 11). This makes the personal imaging apparatus into a remote camera in which viewers on the World Wide Web experience something similar to a QuickTime VR environment map [66], except with some new additional controls allowing them to move around in the environment map both spatially and tonally.

It should be noted that the environment map was generated by a covert wearable apparatus, simply by looking around, and that no special tripod or the like was needed, nor was there significant conscious thought or effort required. In contrast to this proposed method of building environment maps, consider what must be done to build an environment map using QuickTime VR:

Despite more than twenty years photographic experience, Charbonneau needed to learn new approaches for this type of photography. First, a special tripod rig is required, as the camera must be completely

(Frame 0 in spatiotonal coordinates of Frame 9)

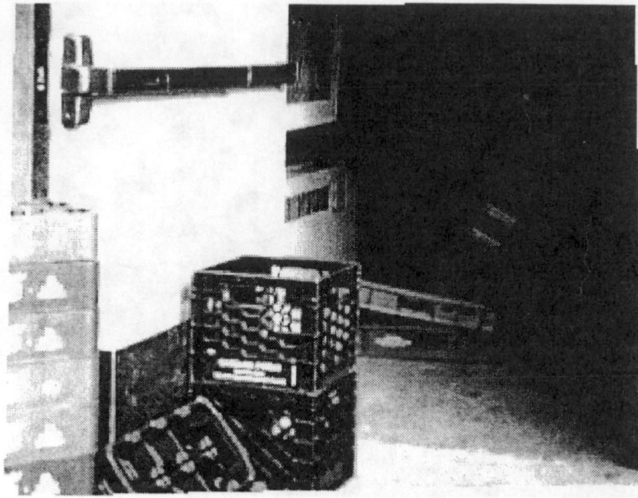

(Frame 3 in spatiotonal coordinates of Frame 9)

(Frame 6 in spatiotonal coordinates of Frame 9)

(Frame 9)

Fig. 10. Images of Fig 8 expressed in the spatiotonal coordinates of the last image in the sequence. Before re-quantization for printing in this figure, (Frame 10) had the highest level of highlight detail, owing to is very small quantigraphic quantization step size in the bright areas of the image.

level for all shots. A 35 mm camera ... with a lens wider than 28 mm is best, and the camera should be set vertically instead of horizontally on the tripod. ... Exposure is another key element. Blending together later will be difficult unless identical exposure is used for all views. [66]

The constraint of the QuickTime VR method and many other methods reported in the literature [50][41][43], that all pictures be taken with identical exposure, is undesirable for the following reasons:

- It requires a more expensive camera as well as a non-standard way of shooting (most low cost cameras have automatic exposure that cannot be disabled, and even on cameras where the AGC can be disabled, AGC is still used so the methods will seldom work with pre-existing video that was not shot in this special manner).
- Imposing that all pictures be taken with the same exposure means that those images shot in bright areas of the scene will be grossly overexposed, while those shot in dark areas will be grossly under-exposed. Normally the AGC would solve this problem and adjust the exposure as the camera pans around the scene, but since it must be shut off, shooting all the pictures at the same exposure will mean that most scenes will not record well. Thus special studio lighting is often required to carefully ensure that everything in the scene is equally illuminated.

In contrast to the prior art, the proposed method allows natural scenes of extremely high dynamic

Fig. 11. Virtual camera: Floating point projectivity+gain image composite constructed from the fire-exit sequence. The dynamic range of the image is far greater than that of a computer screen or printed page. The quantigraphic information may be interactively viewed on the computer screen, however, not only as an environment map (with pan, tilt, and zoom), but also with control of 'exposure' and contrast.

range to be captured from a covert eyeglass-mounted camera, by simply looking around. The natural AGC of the camera ensures that (1) the camera will adjust itself to correctly expose various areas of the scene, so that no matter how bright or dark (within a very large range) objects in the scene are, they will be properly represented without saturation or cutoff, and (2) the natural ebb and flow of the gain, as it tends to fluctuate, will ensure that there is a great deal of overlapping scene content that is differently exposed, and thus the same quantities of light from each direction in space will be measured with a large variety of different quantization steps. In this way, it will not be necessary to deliberately shoot at different apertures in order to obtain the Wyckoff effect.

Once the final image composite, which reports, up to a single unknown scalar, the quantity of light arriving from each direction in space, it may also be reduced back to an ordinary (e.g. non-quantigraphic) picture, by evaluating it with the function f. Furthermore, if desired, prior to evaluating it with f, a lateral inhibition similar to that of the human visual system, may be applied, to reduce its dynamic range, so that it may be presented on a medium of limited display resolution, such as a printed page (Fig 12). It should be noted that this quantigraphic filtering process (that of producing 12) would reduce to a variant of homomorphic filtering, in the case of a single image, $E(\mathbf{x})$, in the sense that E would be treated to a global nonlinearity f^{-1} (to obtain q) then linearly processed (e.g. with unsharp masking or the like), and then the nonlinearity, f^{-1} would be undone, by applying f:

$$E_c = f(L(f^{-1}(E))) \tag{18}$$

where E_c is the output (or composite) image and L is the linear filtering operation. Images sharpened in this way tend to have a much richer, more pleasing and natural appearance [19], than those that are sharpened according to either a linear filter, or the variant of homomorphic filtering suggested by Stockham [29].

Perhaps the greatest value of quantigraphic imaging, apart from its ability to capture high quality pictures that are visually appealing, is its ability to *measure* the quantity of light arriving from each direction in space. In this way, quantigraphic imaging turns the camera into an array of accurate light meters.

Furthermore, the process of making these measurements is *activity driven* in the sense that areas of interest in the scene will attract the attention of the human operator, so that he or she will spend more

Fig. 12. Fixed-point image made by tone-scale adjustments that are only locally monotonic, followed by quantization to 256 greylevels. Note that we can see clearly both the small piece of white paper on the door (and even read what it says — "COFFEE HOUSE CLOSED"), as well as the details of the dark interior. Note that we could not have captured such a nicely exposed image using an on-camera "fill-flash" to reduce scene contrast, because the fill-flash would mostly light up the areas near the camera (which happen to be the areas that are already too bright), while hardly affecting objects at the end of the dark corridor which are already too dark. Thus, one would need to set up additional photographic lighting equipment to obtain a picture of this quality. This image demonstrates the advantage of a small lightweight personal imaging system, built unobtrusively into a pair of eyeglasses, in that an image of very high quality was captured by simply looking around, without entering the corridor. This might be particularly useful if trying to report a violation of fire-safety laws, while at the same time, not appearing to be trying to capture an image. Note that this image was shot from some distance away from the premises (using a miniaturized tele lens I built into my eyeglass-based system) so that the effects of perspective, although still present, are not as immediately obvious as with some of the other extreme wide-angle image composites presented in this thesis. The success of the covert, high definition image capture device suggests possible applications in investigative journalism, or simply to allow ordinary citizens to report violations of fire safety without alerting the perpetrators.

time looking at those parts of the scene. In this way, those parts of the scene of greatest interest will be observed with the greatest variety of quantization steps (e.g. with the richest collection of differently quantized measurements), and will therefore, without conscious thought or effort on the part of the wearer, be automatically emphasized in the composite representation. This natural *foveation* process arises, not because the *Artificial Intelligence* (AI) problem has been solved and built into the camera, so that it knows what is important, but simply because the camera is using the operator's brain as its guide to visual saliency. Because the camera does not take any conscious thought or effort to operate, it resides on the human host without presenting the host with any burden, yet it benefits greatly from this form of HI.

B.4 Bi-foveated WearCam

The natural foveation, arising from the symbiotic relationship between human and machine (HI) described in Section V.B.3 may be further accentuated by building a camera system that is itself foveated.

Accordingly, the author designed and built a number of WearComp embodiments containing more than one electronic imaging array. One common variant, with a wide-angle camera in *landscape* orientation combined with a telephoto camera in *portrait* orientation was found to be particularly useful for HI: The wide camera provided the overall contextual information from the wearer's perspective, while the other (telephoto) provided close-up details, such as faces.

This 'bi-foveated' scheme was found to work well within the context of the spatiotonal model described in the previous Section (V.B.3).

One realization of the apparatus comprised two cameras concealed in a pair of ordinary eyeglasses, is depicted in Fig 13. It should be noted that there are precedents for display–only systems, such as

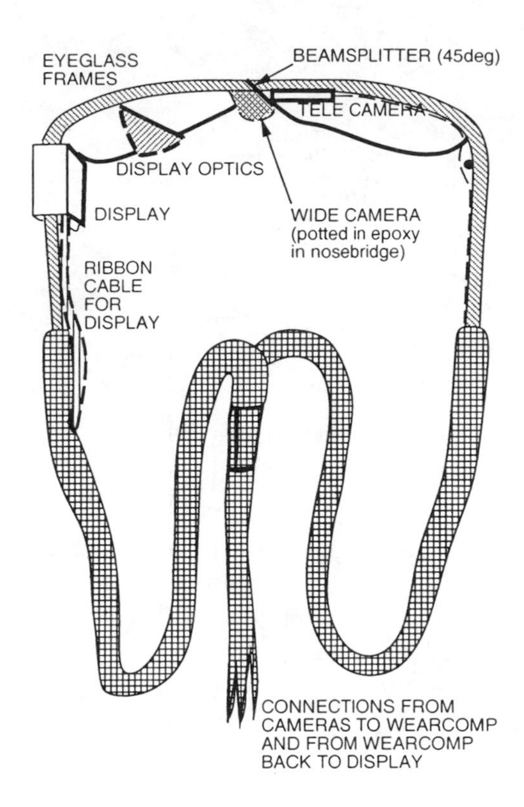

Fig. 13. A multicamera personal imaging system with two miniature cameras and display built into ordinary eyeglasses. This bi-foveated scheme was found to be useful in a host of applications ranging from crime-reduction (personal safety/personal documentary), to situational awareness and shared visual memory.

Kaiser ElectroOptical's head mounted display product, but that the use of multiple resolution levels within the current invention is new.

Signal processing with respect to bi-foveated cameras is a special consideration. In particular, since the geometry of one camera is fixed (in epoxy or the like) with respect to the other, there exists a fixed coordinate transformation that maps any image captured on the wide camera to one that was captured on the foveal camera at the same time. Thus when there is a large jump between images captured on the foveal camera — a jump too large to be considered in the neighbourhood of the identity — signal processing algorithms may look to the wide camera for a contextual reference, owing to the greater overlap between images captured on the wide camera, apply the estimation algorithm to the two wide images, and then relate these to the two foveal images. Furthermore, additional signal inputs may be taken from miniature wearable radar systems, inertial guidance, or electronic compass, built into the eyeglasses or clothing. These extra signals typically provide ground-truth, as well as cross-validation of the estimates reported by the proposed algorithm. The procedure (described in more detail in [20]) is illustrated in Fig 14.

B.5 Lightspace modeling for HI

The result of quantigraphic imaging is that, with the appropriate signal processing, WearComp can *measure* the quantity of light arriving from each angle in space. Furthermore, because it has display capability (usually the camera sensor array and display element are both mounted in the same eyeglass frame), it may also direct rays of light into the eye. Suppose that the display element has a response function h. The entire apparatus (camera, display, and signal processing circuits) may be used to create an 'illusion of transparency', through display of the quantity $h^{-1}(f^{-1}(E_c)) = h^{-1}(q)$ where E_c is the image from the camera. In this way, the wearer sees "through" (e.g. by virtue of) the camera[13],

[13]In some embodiments of WearComp, only a portion of the visual field is mediated in this way. Such an experience is referred to as 'partially mediated reality' [16].

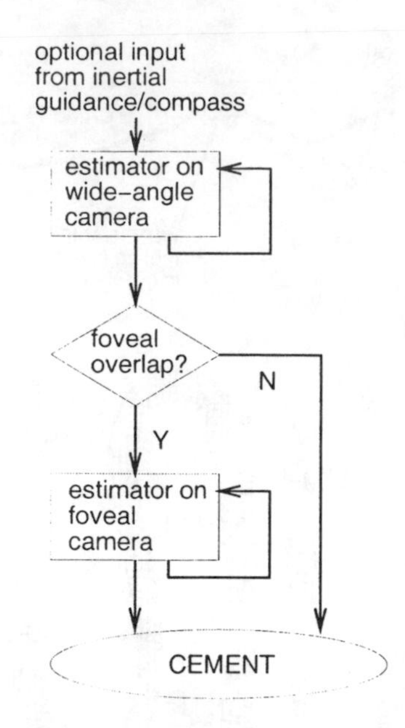

Fig. 14. Signal processing approach for bi-foveated 'WearCam'. Note also that the spatial coordinates are propagated according to the projective group's law of composition while the gain parameters between the wide-camera and foveal-camera are not directly coupled.

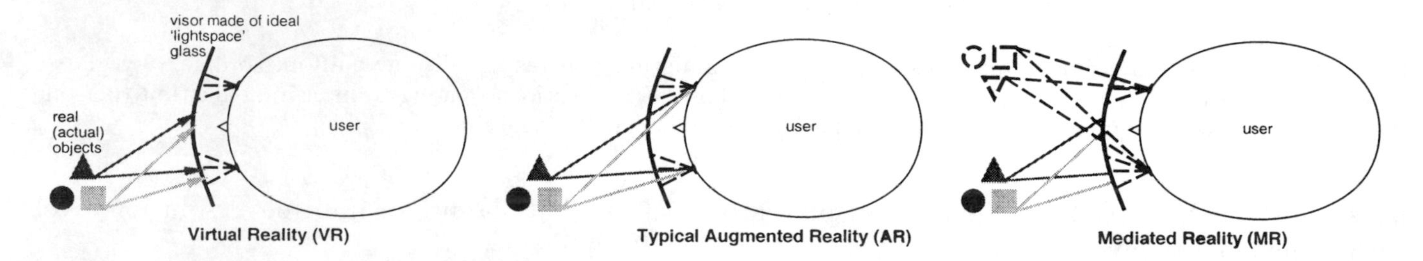

Fig. 15. **Lightspace modeling**: The WearComp apparatus, with the appropriate quantigraphic signal processing, may be thought of as a hypothetical glass that absorbs and quantifies every ray of light that hits it, and is also capable of generating any desired bundle of rays of light coming out the other side. Such a glass, made into a visor, could produce a virtual reality (VR) experience by ignoring all rays of light from the real world, and generating rays of light that simulate a virtual world. Rays of light from real (actual) objects indicated by solid shaded lines; rays of light from the display device itself indicated by dashed lines. The device could also produce a typical augmented reality (AR) [69][70] experience by creating the 'illusion of transparency' and also generating rays of light to make computer-generated "overlays". Furthermore, it could 'mediate' the visual experience, allowing the perception of reality itself to be altered. In this figure, a simple but illustrative example is shown: objects are *left-right reversed* before being presented to the viewer.

and would be blind to outside objects in the region over which the apparatus operates, but for the camera.

Now suppose that a filter, L, is inserted into the 'reality stream' by virtue of the appropriate signal processing on the incoming images E_c prior to display on h:

$$E_m = h^{-1}(L(f^{-1}(E_c))) \tag{19}$$

In this context, L is called the 'visual filter' [16], and may be more than just a linear spatial filtering operation. As a trivial but illustrative example, consider L such that it operates spatially to flip the image left-right. This would make the apparatus behave like the left-right reversing glasses that Kohler [67] and Dolezal [68] made from prisms for their psychophysical experiments. (See Fig 15 (MR).) In general, through the appropriate selection of L, the perception of visual reality may be *augmented*, deliberately diminished (e.g. to emphasize certain objects by diminishing the perception of all but those objects), or otherwise altered.

Fig. 16. Shared environment maps are one obvious application of WearComp. Multiple images transmitted from the author's 'Wearable Wireless Webcam' [71] may be seamlessly combined together onto a WWW page so that others can see a first-person-perspective point of view, as if looking out through the eyes of the person wearing the apparatus. However, because the communication is bidirectional, others can also communicate with the wearer by altering the wearer's visual perception of reality. This might, for example, allow one to recognize people one has never met before. Thus personal imaging allows the individual to go beyond vicarious experience, toward a more symbiotic relation to a networked collective HI environment within a mediated visual reality [16]. (C) Steve Mann, 1995. (Picture rendered at higher-than-normal screen resolution for use as cover for Presence journal.)

One feature of this wearable tetherless computer-mediated reality system is that the wearer can choose to allow others to alter his or her visual perception of reality over an Internet connected wireless communications channel. An example of such a shared environment map appears in Fig 16). This map not only allows others to vicariously experience our point of view (e.g. here a spouse can see that the wearer is at the bank, and send a reminder to check on the status of a loan, or pay a forgotten bill), but can also allow the wearer to allow the distant spouse to mediate the perception of reality. Such mediation may range from simple annotation of objects in the 'reality stream', to completely altering the perception of reality.

Other examples of computer-mediated reality include lightspace modeling, so that the response of everyday objects to light can be characterized, and thus the objects can be recognized as belonging to the same orbit of the group of transformations described in this paper. This approach facilitated such efforts as a way-finding apparatus that would prevent the wearer from getting lost, as well as an implementation of Feiner's Post-It-note metaphor using a wearable tetherless device, so that messages could be left on everyday objects.

VI. Beyond video: Synthetic synesthesia and Personal Imaging

The manner in which WearComp, with its rich multidimensional measurement and signal processing space, facilitates enhanced environmental awareness, is perhaps best illustrated by way of the author's effort of the 1980s at building an system to assist the visually challenged. This device, which used radar, rather than video, as the input modality, is now described.

A. Synthetic Synesthesia: Adding new sensory capabilities to the body

The addition of a new sensory capability for assisting the visually challenged is now described. It has also been found to be of great use to the sighted as well. For example, the author found that increased situational awareness using the system resulted in greater safety in many ordinary day–to–day activities such as riding a bicycle on a busy street.

Mediated reality may include, in addition to video, an audio reality mediator, or, more generally, a 'perceptual reality mediator'. This generalized mediated perception system may include deliberately induced synesthesia[14]. Perhaps the most interesting example of synthetic synesthesia was the addition of a new human sensory capability based on miniature wearable radar systems combined with intelligent signal processing. In particular, the author developed a number of vibrotactile wearable radar systems in the 1980s, of which there were three primary variations:

- 'CorporealEnvelope': baseband output from the radar system was envelope-detected to provide a vibrotactile sensation which was proportional to the overall energy of the return [15]. This provided the sensation of an extended 'envelope' around the body, in which one could feel objects at a distance. In later (late 1980s) embodiments of 'CorporealEnvelope', envelope detection was done after splitting the signal into three or four separate frequency bands, each driving a separate vibrotactile device, so that each would convey a portion of the Doppler spectrum (e.g. each corresponding to a range of velocities of approach). In another late 1980s embodiment, variously colored lamps were used, attached to the wearer's eyeglasses to provide a visual synesthesia of the radar sense. In one particular embodiment, red, green, and blue lamps were used, such that objects moving toward the wearer illuminated the blue lamp, while objects moving away illuminated the red lamp. Objects not moving relative to the wearer, but located near the wearer appeared green. This work was inspired by using the metaphor of the natural Doppler shift colors.
- 'VibroTach' (vibrotactile tachometer): the speed of objects moving toward or away from the wearer was conveyed, but not the magnitude of the Doppler return (e.g. it was not possible to distinguish between objects of small radar cross section and those of large radar cross section). This was done by having a Doppler return drive a motor, so that the faster an object moved toward or away from the wearer, the faster the motor would spin. The spinning motor could be felt as a vibration having frequency proportional to that of the dominant Doppler return.
- 'Electric Feel Sensing': the entire doppler signal (not just a single dominant speed or amplitude) was conveyed to the body. Thus if there were two objects approaching at different speeds, one could discern them separately from a single vibrotactile sensor. Various embodiments of 'electric feel sensing' included direct electrical stimulation of the body, as well as the use of a single broadband vibrotactile device.

One of the problems with this work was the processing, which, in the early 1980s embodiments of WearComp was very limited. However, Today's wearable computers, far more capable of computing the chirplet transform[16] in real time, suggest a renewed hope for the success of this effort to assist the visually impaired.

[14]Synesthesia [72] is manifest as the crossing of sensory modalities, as, for example, the ability (or as some might call a disability) to taste shapes, see sound, etc..

[15]Strictly speaking the actual quantity measured in early systems was that of a single homodyne channel, which only approximated energy. Later in some systems energy was measured properly with separate I and Q channels.

[16]The chirplet transform [73] characterizes the acceleration signature of Doppler returns, so that objects can be prioritized, e.g. those accelerating faster toward the wearer can be given higher priority, predicting eminent collision, etc..

A.1 Safety first!

Again, with direct connection to the body, there must of course be extra attention devoted to user safety. For example, direct electrical stimulation of the body has certain risks associated with it, such as nerve damage from excessively strong signals, as well as nerve damage from weaker signals that are not properly conditioned (such as may happen with excessive D.C. offset). Similarly, vibrotactile devices may also afflict the user with long-term damage, as one might experience with any sensory device in the extreme (just as loud music can cause deafness, and excessively bright head mounted displays can cause blindness, excessive vibrotactile stimulation can cause loss of feeling). While the radar signals themselves tend to be less of a concern, owing to their very low power levels (often below the level of background radiation), there should still be the obvious precautions taken with radar as with any other radio signals (such as the much more powerful transmitters used to establish an Internet connection). The potential problem of radio frequency noise pollution has been addressed through the use of very low power transmissions from the radar. Because of the homodyne nature of the receiver, a very weak signal may be sent out, since it is known exactly what signal will be expected back. Even in pulsed modes of operation, by virtue of pulse compression, signal levels can often remain below the ambient signal levels. Radar systems operate according to an inverse fourth power law (since what is transmitted falls off as the square of the distance, and what is received back falls off as the square of the distance also). Distances to objects of interest with a personal radar system are typically on the order of only a few meters, in contrast to traditional radar, where distances of interest are many kilometers. Thus, because power output needed is proportional to the fourth exponent of the distance from the objects of interest, power output is very low, and thus has not been a hazard.

One must also consider the safety issues of both the effects of synthetic synesthesia, as well as the development of a reliance upon it.

Synthetic synesthesia involves a remapping of the human perceptual system, the long-term effects of which should still be studied carefully.

Another concern, which might not at first occur to one practicing this art, is that of an acquired dependence. In many ways, the author discovered that, after many years, the device began to function as a true extension of the mind and body, as if it were an additional sense. Much like a person who is blind at birth, but has his or her sight restored later in life, due to medical advancements, there is a strange sense of confusion when a new sense is introduced to the body, without the support infrastructure within the brain. After many years of use, one begins to learn the new sense, and internalize the meanings of the new sensory modalities. Together with this remapping of the human perceptual system, and the obvious dangers it might pose (and the question as to whether learning is damage, since learning permanently alters the brain), is the deeper philosophical question as to whether or not acquiring a dependence on a new sensory modality is a good idea.

As a specific example, in the form of a personal anecdote, the author had one time found himself on a long bike trip (180 mile trip along major busy highways), relying greatly on this new sense, when at some point along the trip there was a major thunderstorm which required removal and shutting down of the extra sensory capabilities. In many ways the author felt at great risk, owing to the heavy traffic, and the acquired need to have enhanced situational awareness. Removal of a new sensory capability, after an acquired dependency, was much like removal of an existing sense (e.g. like suddenly having to ride a bicycle while blindfolded or wearing ear plugs).

Thus an important safety issue, as we enhance our intellectual and sensory capabilities, will be risks involved should this apparatus ever quit functioning, after we have become dependent upon it.

Acquired dependency is nothing new, of course. For example, we have acquired a dependency on shoes and clothing, and would doubtless have much greater difficulty surviving naked in the wilderness, than might those indigenous to the wilderness, especially those who had not invented shoes or clothing.

Thus as we build prostheses of the mind and body, we must consider carefully their implications, especially as they pertain to personal safety. This new computational framework will therefore give a whole new meaning to the importance of reliability.

A.2 A true extension of the mind and body

Such simple early prototypes as those discussed already suggest a future in which intelligent signal processing, through the embodiment of HI, may allow the wearer to experience increased situational awareness. It will then be misleading to think of the wearer and the computer with its associated input/output apparatus as separate entities. Instead it will be preferable to regard the computer as a second brain, and its sensory modalities as additional senses, which through synthetic synesthesia are inextricably intertwined with the wearer's own biological sensory apparatus.

VII. Conclusions

A new form of intelligent signal processing, called 'Humanistic Intelligence' was proposed. It is characterized by processing hardware that is inextricably intertwined with a human being to function as a true extension of the user's mind and body. This hardware is **constant** (always on, therefore its output is always observable), **controllable** (e.g. is not merely a monitoring device attached to the user, but rather, it takes its cues from the user), and is **corporeal** in nature (e.g. tetherless and with the point of control in close proximity to the user so as to be perceived as part of the user's body).

Furthermore, the apparatus forms a symbiotic relationship with its host (the human), in which the high-level intelligence arises on account of the existence of the host (human), and the lower-level computational workload comes from the signal processing hardware itself.

The emphasis of this chapter was on Personal Imaging, to which the application of HI gave rise to a new form of intelligent camera system. This camera system was found to be of great use in both photography and documentary video making. Its success arose from the fact that it (1) was simpler to use than even the simplest of the so-called "intelligent point and click" cameras of the consumer market (even though many of these embody sophisticated neural network architectures), and (2) afforded the user much greater control than even the most versatile and fully-featured of professional cameras.

This application of HI took an important first step in moving from the 'point and click' metaphor, toward the 'look and think' metaphor — toward making the camera function as a true visual memory prosthetic which operates without conscious thought or effort, while at the same time affording the visual artist a much richer and complete space of possibilities. Moreover, this work sets forth the basic principles of a photo–videographic memory system.

A focus of HI was to put the human intellect into the loop but still maintain facility for failsafe mechanisms operating in the background. Thus the personal safety device was proposed.

What differentiates HI from environmental intelligence (*ubiquitous computing* [74], reactive rooms [75], and the like), is that there is no guarantee environmental intelligence will be present when needed, or that it will be in control of the user. Instead, HI provides a facility for intelligent signal processing that travels with the user. Furthermore, because of the close physical proximity to the user, the apparatus is privy to a much richer multidimensional information space than that obtainable by environmental intelligence.

Furthermore, unlike an intelligent surveillance camera that people attempt to endow with an ability to recognize suspicious behaviour, WearComp takes its task from the user's current activity, e.g. if the user is moving, the apparatus is continually rendering new composite pictures, while if the user is not moving it is no longer taking in new orbits. This activity–based response is based on the premise that the viewpoint changes cause a change in orbit, etc.

Systems embodying HI are:

- ACTIVITY DRIVEN and ATTENTION DRIVEN: Salience is based on the computer's taking information in accordance with human activity. Video orbits is activity driven (starts when wearer stops at a fixed orbit). In other words the visual salience comes from the human; the computer is doing the processing but taking cue from the wearer's activity. For example, if the wearer is talking to a bank clerk, but takes brief glances at the periphery, the resulting image will reveal the wearer's clerk in high resolution, while the other clerks to the left and right will be quantified at much lesser certainty. Further processing on the image measurements thus reflect this saliency, so that the system adapts to the manner in which it is used.

- ENVIRONMENTALLY AWARE: Situated awareness arises in the context of both the wearer's environment and his/her own biological quantities which, through the wearer's own mind, body, and perceptual system, also depend on the environment.
- INEXTRICABLY INTERTWINED with the human; SITUATED: If the user is aroused the system will take more pictures. In this way, the computation makes a departure from that of traditional artificial intelligence. The processor automatically uses the wearer's sense of salience to help it, so that the machine and human are always working in parallel.

VIII. Acknowledgements

The author wishes to thank Simon Haykin, Rosalind Picard, Steve Feiner, Charles Wyckoff, Woodrow Barfield, Hiroshi Ishii, Thad Starner, Jeffrey Levine, Flavia Sparacino, Ken Russell, Richard Mann, Ruth Mann, Bill Mann, and Steve Roberts (N4RVE), for much in the way of useful feedback, constructive criticism, etc., as this work has evolved, and Zahir Parpia for making some important suggestions for the presentation of this material. Thanks is due also to individuals the author has hired to work on this project, including Nat Friedman, Chris Cgraczyk, Matt Reynolds (KB2ACE), etc., who each contributed to this effort.

Dr. Carter volunteered freely of his time to help in the design of the interface to WearComp2 (the author's 6502-based wearable computer system of the early 1980s), and Kent Nickerson similarly helped with some of the miniature personal radar units and photographic devices involved with this project throughout the mid 1980s.

Much of the early work on biosensors and wearable computing was done with, or at least inspired by work the author did with Dr. Nandegopal Ghista, and later refined with suggestions from Dr. Hubert DeBruin, both of McMaster University. Dr. Max Wong of McMaster university supervised a course project in which the author chose to design an RF link between two 8085-based WearComp systems which had formed part of the author's "photographer's assistant" project.

Much of the inspiration towards making truly wearable (also comfortable and even fashionable) signal processing systems was through collaboration with Jeff Eleveld during the early 1980s.

Bob Kinney of US Army Natick Research Labs assisted in the design of a tank top, based on a military vest, which the author used for a recent (1996) embodiment of the WearComp apparatus worn underneath ordinary clothing.

Additional Thanks to Xybernaut Corp., Digital Equipment Corp., ViA, VirtualVision, HP labs, Compaq, Kopin, Colorlink, Ed Gritz, Miyota, Chuck Carter, and Thought Technologies Limited for lending or donating additional equipment that made these experiments possible.

References

[1] William A.S. Buxton Ronald M. Baecker, *Readings in human-computer interaction a multidisciplinary approach*, chapter 1,2, 1987.
[2] Simon Haykin, *Neural Networks: A comprehensive foundation*, McMillan, New York, 1994.
[3] Simon Haykin, ""radar vision"," Second International Specialist Seminar on Parallel Digital Processors, Portugal, April 15–19 1992.
[4] D. E. Rumelhart and Eds. J. L. McMlelland, *Parallel distributed processing*, MIT Press, Cambridge, MA, 1986.
[5] Tomaso Poggio and Federico Girosi, "Networks for approximation and learning," *Proc. IEEE*, vol. 78, no. 9, 1990, Sept.
[6] Bart Kosko, *Fuzzy thinking: the new science of fuzzy logic*, Hyperion, New York, 1993.
[7] Bart Kosko and Satoru Isaka, *Fuzzy logic*, vol. 269, New York, July 1993.
[8] Phillip Laplante editor Marvin Minsky, "Steps toward artificial intelligence," in *Great papers on computer science*, West Publishing Company, Minneapolis/St. Paul, 1996 (paper in IRE 1960).
[9] D. C. Engelbart, "Augmenting human intellect. a conceptual framework," Research Report AFOSR-3223, Stanford Research Institute, Menlo Park, 1962, http://www.histech.rwth-aachen.de/www/quellen/engelbart/ahi62index.html.
[10] Douglas C. Engelbart, "A conceptual framework for the augmentation of man's intellect," in *Vistas in Information Handling*, P.D. Howerton and D.C. Weeks, Eds. Spartan Books, Washington, D.C., 1963.

[11] ," Method and apparatus for relating and combining multiple images of the same scene or object(s), Steve Mann and Rosalind W. Picard, U. S. Pat. 5706416.
[12] Steve Mann, "Wearable computing: A first step toward personal imaging," *IEEE Computer*, vol. 30, no. 2, Feb 1997, http://hi.eecg.toronto.edu/ieeecomputer/index.html.
[13] Michel Foucault, *Discipline and Punish*, 1977, Translated from "Surveiller et punir".
[14] Stephen M. Kosslyn, *Image and Brain The Resolution of the Imagery Debate*, M.I.T. Press, 1994.
[15] N1NLF S. Mann, "'wearstation': With today's technology, it is now possible to build a fully equipped ham radio station, complete with internet connection, into your clothing.," January, 1997.
[16] S. Mann, "'mediated reality'," TR 260, M.I.T. Media Lab Perceptual Computing Section, Cambridge, Massachusetts, http://wearcam.org/mediated-reality/index.html, 1994.
[17] James D. Meindl, "Low power microelectronics: Retrospect and prospect," *Proc. IEEE*, vol. 83, no. 4, Apr. 1995.
[18] Steve Mann, "An historical account of the 'WearComp' and 'WearCam' projects developed for 'personal imaging'," in *International Symposium on Wearable Computing*, Cambridge, Massachusetts, October 13-14 1997, IEEE.
[19] Steve Mann, *Personal Imaging*, Ph.D. thesis, Massachusetts Institute of Technology (MIT), 1997.
[20] Steve Mann, "Further developments on 'headcam': Joint estimation of camera rotation + gain group of transformations for wearable bi-foveated cameras," in *Proceedings of the International Conference on Acoustics, Speech and Signal Processing*, Munich, Germany, Apr 1997, IEEE, vol. 4.
[21] Eric J. Lind and Robert Eisler, "A sensate liner for personnel monitoring application," in *International Symposium on Wearable Computing*. October 13-14 1997, IEEE.
[22] R. W. Picard and J. Healey, "Affective wearables," in *Proceedings of the First International Symposium on Wearable Computers*, Cambridge, MA, Oct. 1997.
[23] Steve Mann, "Smart clothing: The wearable computer and wearcam," *Personal Technologies*, March 1997, Volume 1, Issue 1.
[24] R. W. Picard, *Affective Computing*, The MIT Press, Cambridge, MA, Sep. 1997.
[25] Don Norman, *Turn signals are the facial expressions of automobiles*, Addison Wesley, 1992.
[26] Steve Mann, "'smart clothing': Wearable multimedia and 'personal imaging' to restore the balance between people and their intelligent environments," Boston, MA, Nov. 18-22 1996, Proceedings, ACM Multimedia 96.
[27] Steve Mann, "Humanistic intelligence," *Proceedings of Ars Electronica*, Sep 8-13 1997, Invited plenary lecture, Sep. 10, http://wearcam.org/ars/ http//www.aec.at/fleshfactor.
[28] Steve Mann, "Lightspace," Unpublished report (Paper available from author). Submitted to SIGGRAPH 92. Also see example images in http://wearcam.org/lightspace, July 1992.
[29] T. G. Stockham, Jr., "Image processing in the context of a visual model," *Proc. IEEE*, vol. 60, no. 7, pp. 828–842, July 1972.
[30] E. H. Land, "The retinex theory of color vision," *Scientific American*, pp. 108–129, 1977.
[31] D. Forsyth, "A novel algorithm for color constancy," *Int. J. Comput. Vision*, vol. 5, pp. 5–36, 1990.
[32] Anastasios N. Venetsanopoulos, "Nonlinear mean filters in image processing," *IEEE Trans. ASSP*, vol. 34, no. 3, pp. 573–584, June 1986.
[33] S. Mann, "Compositing multiple pictures of the same scene," in *Proceedings of the 46th Annual IS&T Conference*, Cambridge, Massachusetts, May 9-14 1993, The Society of Imaging Science and Technology.
[34] Steve Mann, "'pencigraphy' with AGC: Joint parameter estimation in both domain and range of functions in same orbit of the projective-Wyckoff group," Tech. Rep. 384, MIT Media Lab, Cambridge, Massachusetts, December 1994; http://hi.eecg.toronto.edu/icip96/index.html, Also appears in: Proceedings of the IEEE International Conference on Image Processing (ICIP–96), Lausanne, Switzerland, September 16–19, 1996, pages 193–196.
[35] S. Mann and R. W. Picard, "Virtual bellows: constructing high-quality images from video," in *Proceedings of the IEEE first international conference on image processing*, Austin, Texas, Nov. 13-16 1994.
[36] Charles W. Wyckoff, "An experimental extended response film," *S.P.I.E. NEWSLETTER*, JUNE-JULY 1962.
[37] Charles W. Wyckoff, "An experimental extended response film," Tech. Rep. NO. B-321, Edgerton, Germeshausen & Grier, Inc., Boston, Massachusetts, MARCH 1961.
[38] O. D. Faugeras and F. Lustman, "Motion and structure from motion in a piecewise planar environment," *International Journal of Pattern Recognition and Artificial Intelligence*, vol. 2, no. 3, pp. 485–508, 1988.
[39] G. Adiv, "Determining 3D Motion and structure from optical flow generated by several moving objects," *IEEE Trans. Pattern Anal. Machine Intell.*, pp. 304–401, July 1985.
[40] Amnon Shashua and Nassir Navab, "Relative Affine: Theory and Application to 3D Reconstruction From Perspective Views.," *Proc. IEEE Conference on Computer Vision and Pattern Recognition*, June 1994, 1994.
[41] H.S. Sawhney, "Simplifying motion and structure analysis using planar parallax and image warping," *ICPR*, vol. 1, October 1994, 12th IAPR.
[42] Nassir Navab and Steve Mann, "Recovery of relative affine structure using the motion flow field of a rigid planar patch.," *Mustererkennung 1994, Tagungsband.*, 1994.
[43] R. Kumar, P. Anandan, and K. Hanna, "Shape recovery from multiple views: a parallax based approach," *ARPA image understanding workshop*, 10 Nov 1994.
[44] William A. Radlinski American Society of Photogrammetry. Editor-in-chief: Morris M. Thompson. Associate editors: Robert C. Eller and Julius L. Speert, *Manual of photogrammetry*, Schaum's Outline Series. McGraw–Hill Book Company, Falls Church, Va., 3d ed edition, 1966, 2 v. (xx, 1199 p.) illus. 27 cm.
[45] M. Irani and S. Peleg, "Improving Resolution by Image Registration," *CVGIP*, vol. 53, pp. 231–239, May 1991.
[46] A.M. Tekalp, M.K. Ozkan, and M.I. Sezan, "High-resolution image reconstruction from lower-resolution image sequences and space-varying image restoration," in *Proc. of the Int. Conf. on Acoust., Speech and Sig. Proc.*, San Francisco, CA, Mar. 23-26, 1992, IEEE, pp. III–169.
[47] Lee Campbell and Aaron Bobick, "Correcting for radial lens distortion: A simple implementation," TR 322, M.I.T. Media Lab Perceptual Computing Section, Cambridge, Massachusetts, Apr 1995.

[48] K. Levenberg, "A method for the solution of certain nonlinear problems in least squares," *Quart. J. of Appl. Math.*, pp. 164–168, 1944, v.2.
[49] D.W. Marquardt, "An algorithm for the estimation of non-linear parameters," *SIAM J.*, pp. 431–441, 1963, v.11.
[50] R. Szeliski and J. Coughlan, "Hierarchical spline-based image registration," *CVPR*, pp. 194–201, 1994.
[51] S. Mann and R. W. Picard, "Video orbits of the projective group; a simple approach to featureless estimation of parameters," TR 338, Massachusetts Institute of Technology, Cambridge, Massachusetts, See http://n1nlf-1.eecg.toronto.edu/tip.ps.gz 1995, Also appears IEEE Trans. Image Proc., Sept 1997, Vol. 6 No. 9.
[52] S.S. Beauchemin J.L. Barron, D.J. Fleet, "Systems and experiment performance of optical flow techniques," *International journal of computer vision*, pp. 43–77, 1994.
[53] Qinfen Zheng and Rama Chellappa, "A Computational Vision Approach to Image Registration," *IEEE Transactions Image Processing*, July 1993, pages 311-325.
[54] L. Teodosio and W. Bender, "Salient video stills: Content and context preserved," *Proc. ACM Multimedia Conf.*, August 1993.
[55] R. Y. Tsai and T. S. Huang, "Estimating Three-Dimensional Motion Parameters of a Rigid Planar Patch," *Trans. Accoust., Speech, and Sig. Proc.*, 1981.
[56] M. Artin, *Algebra*, Prentice Hall, 1991.
[57] W.C. Hoffman, "The Lie Algebra of Visual Perception," *Journal of Mathematical Psychology*, 1967, Vol. 4, pp.348-349.
[58] Richard I. Hartley, "Self-calibration of stationary cameras, g.e. crd, schenectady, ny, 12301,"
[59] B. Horn and B. Schunk, "Determining Optical Flow," *Artificial Intelligence*, 1981.
[60] Bernd Girod and David Kuo, "Direct estimation of displacement histograms," *OSA Meeting on IMAGE UNDERSTANDING AND MACHINE VISION*, June 1989.
[61] Yunlong Sheng, Claude Lejeune, and Henri H. Arsenault, "Frequency-domain Fourier-Mellin descriptors for invariant pattern recognition," *Optical Engineering*, May 1988.
[62] S. Mann, "Wavelets and chirplets: Time–frequency perspectives, with applications," in *Advances in Machine Vision, Strategies and Applications*, Petriu Archibald, Ed. World Scientific, Singapore . New Jersey . London . Hong Kong, world scientific series in computer science - vol. 32 edition, 1992.
[63] R. Wilson, A. D. Calway, and E. R. S. Pearson, "A generalized wavelet transform for fourier analysis: the multiresolution fourier transform and its application to image and audio signal analysis," *IEEE Transactions on Information Theory*, vol. 38, no. 2, pp. 674–690, March 1992.
[64] S. R. Curtis and A. V. Oppenheim, "Signal reconstruction from Fourier transform sign information," Technical Report No. 500, MIT Research Laboratory of Electronics, May 1984.
[65] S. Mann and R.W. Picard, "Being 'undigital' with digital cameras: Extending dynamic range by combining differently exposed pictures," Tech. Rep. 323, M.I.T. Media Lab Perceptual Computing Section, Boston, Massachusetts, 1994, Also appears, IS&T's 48th annual conference, pages 422-428, May 1995.
[66] June Campbell, "QuickTime VR," October 18 1996, http://www.webreference.com/content/qtvr/.
[67] Ivo Kohler, *The formation and transformation of the perceptual world*, vol. 3 of *Psychological issues*, International university press, 227 West 13 Street, 1964, monograph 12.
[68] Hubert Dolezal, *Living in a world transformed*, Academic press series in cognition and perception. Academic press, Chicago, Illinois, 1982.
[69] S. Feiner, B. MacIntyre, and D. Seligmann, "Knowledge-based augmented reality," Jul 1993, Communications of the ACM, 36(7).
[70] S. Feiner, Webster, Krueger, B. MacIntyre, and Keller, "Architectural anatomy," 1995, Presence, 4(3), 318-325.
[71] S. Mann, "Wearable Wireless Webcam," 1994, http://wearcam.org.
[72] R. E. Cytowic, *Synesthesia: A union of the senses*, Springer-Verlag, New York, 1989.
[73] Steve Mann and Simon Haykin, "The chirplet transform: Physical considerations," *IEEE Trans. Signal Processing*, vol. 43, no. 11, November 1995.
[74] Mark Weiser, "Ubiquitous computing," 1988-1994, http://sandbox.parc.xerox.com/ubicomp.
[75] Jeremy Cooperstock, "Reactive room," http://www.dgp.toronto.edu/ ~rroom/research/papers/.

Chapter 2

Adaptive Stochastic Resonance

Sanya Mitaim and Bart Kosko
Signal and Image Processing Institute
Department of Electrical Engineering—Systems
University of Southern California
Los Angeles, California 90089-2564

Abstract

This chapter shows how adaptive systems can learn to add an optimal amount of noise to some nonlinear feedback systems. Noise can improve the signal-to-noise ratio of many nonlinear dynamical systems. This "stochastic resonance" effect occurs in a wide range of physical and biological systems. The SR effect may also occur in engineering systems in signal processing, communications, and control. The noise energy can enhance the faint periodic signals or faint broadband signals that force the dynamical systems. Most SR studies assume full knowledge of a system's dynamics and its noise and signal structure. Fuzzy and other adaptive systems can learn to induce SR based only on samples from the process. These samples can tune a fuzzy system's if-then rules so that the fuzzy system approximates the dynamical system and its noise response. The chapter derives the SR optimality conditions that any stochastic learning system should try to achieve. The adaptive system learns the SR effect as the system performs a stochastic gradient ascent on the signal-to-noise ratio. The stochastic learning scheme does not depend on a fuzzy system or any other adaptive system. The learning process is slow and noisy and can require heavy computation. Robust noise suppressors can improve the learning process when we can estimate the impulsiveness of the learning terms. Simulations test this SR learning scheme on the popular quartic-bistable dynamical system and on other dynamical systems. The driving noise types range from Gaussian white noise to impulsive noise to chaotic noise. Simulations suggest that fuzzy techniques and perhaps other adaptive "black box" or "intelligent" techniques can induce SR in many cases when users cannot state the exact form of the dynamical systems. The appendix derives the basic additive fuzzy system and the neural-like learning laws that tune it.

1 Stochastic Resonance and Adaptive Function Approximation

Noise can sometimes enhance a signal as well as corrupt it. This fact may seem at odds with almost a century of effort in signal processing to filter noise or to mask or cancel it. But noise is itself a signal and a free source of energy. Noise can amplify a faint signal in some feedback nonlinear systems even though too much noise can swamp the signal. This implies that a system's optimal noise level need not be zero noise. It also suggests that nonlinear signal systems with nonzero-noise optima may be the rule rather than the exception. Figure 1 shows how uniform pixel noise can improve our subjective perception of an image. A small level of noise sharpens the image contours and helps fill in features. Too much noise swamps the image and degrades its contours.

Stochastic resonance (SR) [11, 12, 13, 23, 25, 28, 70, 85, 94, 136, 170, 171, 189, 193, 194, 201, 255] occurs when noise enhances an external forcing signal in a nonlinear dynamical system. SR occurs in a signal system if, and only if, the system has a nonzero noise optimum. The classic SR signature, is a signal-to-noise ratio (SNR) that is not monotone. Figure 2 shows the SR effect for the popular quartic bistable dynamical system [13, 28, 187]. The SNR rises to a maximum and then falls as the variance of the additive white noise grows. More complex systems may have multimodal SNRs and so show stochastic "multiresonance" [84, 252].

SR holds promise for the design of engineering systems in a wide range of applications. Engineers may want to shape the noise background of a fixed signal pattern to exploit the SR effect. Or they may want to adapt their signals to exploit a fixed noise background. Engineers now add noise to some systems to improve how humans perceive signals. These systems include audio compact discs [159], analog-to-digital devices [10], video images [231], schemes for visual perception [223, 224, 237], and cochlear implants [71, 186, 190]. Some control and quantization schemes add a noise-like dither to improve system performance [10, 156, 159, 206, 231]. Additive noise can sometimes stabilize chaotic attractors [18, 82, 177]. Noise can also improve human tactile response [50, 226], muscle contraction [44], and coordination [51]. This suggests that SR designs may improve how robots grasp objects [53] or balance themselves. SR designs might also improve how virtual or augmented

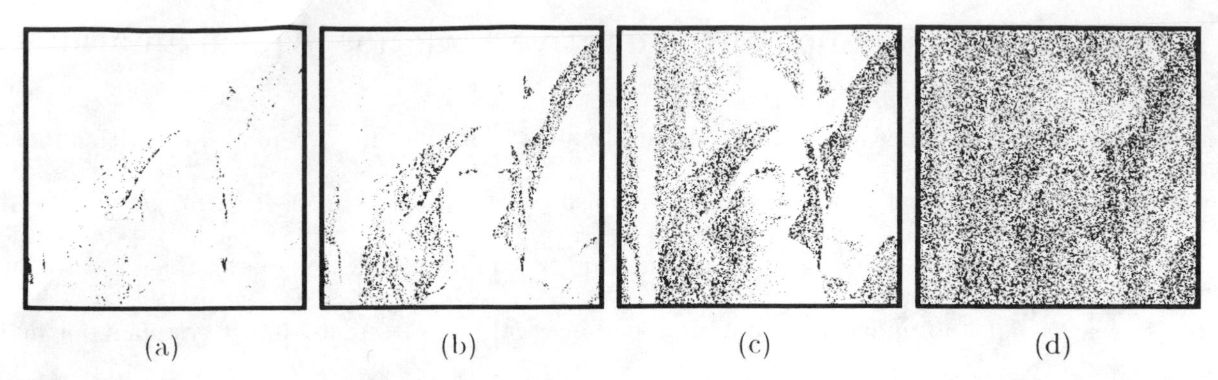

Figure 1: Uniform pixel noise can improve the subjective response of our nonlinear perceptual system. The noise gives a nonmonotonic response: A small level of noise sharpens the image features while too much noise degrades them. These noisy images result when we apply a pixel threshold to the popular "Lena" image of signal processing [195]: $y = g((x+n)-\Theta)$ where $g(x) = 1$ if $x \geq 0$ and $g(x) = 0$ if $x < 0$ for an input pixel value $x \in [0,1]$ and output pixel value $y \in \{0,1\}$. The input image's gray-scale pixels vary from 0 (black) to 1 (white). The threshold is $\Theta = 0.05$. We threshold the original "Lena" image to give the faint image in (a). The uniform noise n has mean $m_n = -0.02$ for images (b)-(d). The noise variance σ_n^2 grows from (b)-(d): $\sigma_n^2 = 1.67 \times 10^{-3}$ in (b), $\sigma_n^2 = 2.34 \times 10^{-2}$ in (c), and $\sigma_n^2 = 1.67 \times 10^{-1}$ in (d).

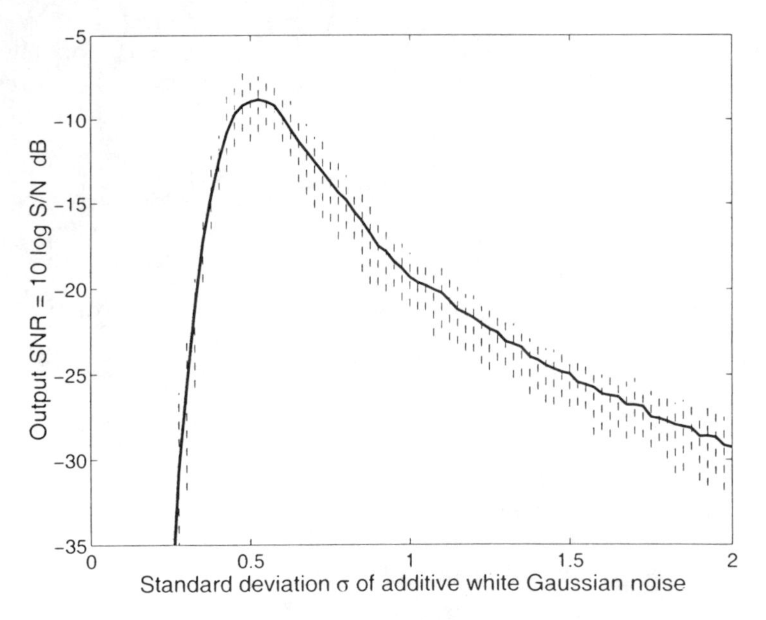

Figure 2: The non-monotonic signature of stochastic resonance. The graph shows the smoothed output signal-to-noise ratio of a quartic bistable system as a function of the standard deviation of additive white Gaussian noise n. The vertical dashed lines show the absolute deviation between the smallest and largest outliers in each sample average of 20 outcomes. The system has a nonzero noise optimum and thus shows the SR effect. The noisy signal-forced quartic bistable dynamical system has the form $\dot{x} = f(x) + s(t) + n(t) = x - x^3 + \varepsilon \sin \omega_0 t + n(t)$. The Gaussian noise $n(t)$ adds to the external forcing narrowband signal $s(t) = \varepsilon \sin \omega_0 t$. Other systems can use multiplicative noise [9, 29, 69, 79, 83, 88, 244] or use non-Gaussian noise [38, 40, 41, 84, 214].

reality systems [34, 113] can create or enhance the sensations of touch and balance.

SR designs might lead to better schemes to filter or multiplex the faint signals found in spread-spectrum communication systems [76, 236]. These systems transmit and detect faint signals in noisy backgrounds across wide bands of frequencies. SR designs might also exploit the signal-based crosstalk noise found in cellular systems [151, 238], Ethernet packet flows [152], or Internet congestion [120].

The study of SR has emerged largely from physics and biology. The awkward term "stochastic resonance" stems from a 1981 article in which physicists observed "the cooperative effect between internal mechanism and the external periodic forcing" in some nonlinear dynamical systems [13]. Scientists soon explored SR in climate models [203] to explain how noise could induce periodic ice ages [11, 12, 201, 202]. They conjectured that global or other noise sources could amplify small periodic variations in the Earth's orbit. This might explain the observed 100,000 year primary cycle of the Earth's ice ages. This SR conjecture remains the subject of debate [78, 202, 257]. Physicists have since found stronger evidence of SR in ring lasers [179, 248], threshold hysteretic Schmitt triggers [74, 180], Chua's electrical circuit [4, 5], bistable magnetic systems [103], electron paramagnetic resonance [86, 89, 225], magnetoelastic ribbons [239], superconducting quantum interference devices (SQUIDs) [110, 124, 229], Ising systems [22, 196, 235], coupled diode resonators [160], tunnel diodes [174, 175], Josephson junctions [24, 111], optical systems [9, 63, 128, 247], chemical systems [64, 77, 105, 112, 137, 154, 188], and quantum-mechanical systems [99, 100, 101, 102, 162, 173, 213, 222, 246].

Some biological systems may have evolved to exploit the SR effect. Most SR studies have searched for the SR effect in the sensory processing of prey and predators. Noisy or turbulent water can help the mechanoreceptor hair cells of the crayfish *Procambarus clarkii* detect faint periodic signals of predators such as a bass's fin motion [60, 61, 194, 210, 216, 218, 255]. Noise helps the mechanosensors of the cricket *Acheta domestica* detect small-amplitude low-frequency air signals from predators [155, 181, 245]. Dogfish sharks use noise in their mouth sensors when they detect periodic signals from prey [19]. The SR effect appears in the mechanoreceptors in a rat's skin [49] and in the neurons in a rat's hippocampus [95]. The SR effect occurs in a wide range of models of neurons [27, 29, 46, 47, 48, 108, 215, 240] and neural networks [26, 27, 29, 31, 32, 43, 46, 47, 48, 121, 122, 158, 163, 164, 165, 166, 167, 168, 191, 197, 214].

Research in SR has grown from the study of external periodic signals in simple dynamical systems to the study of external aperiodic and broadband signals in more complex dynamical systems [37, 38, 43, 46, 47, 48, 49, 72, 73, 108, 115, 155, 217, 240]. Below we review examples of these dynamical systems and the performance measures involved in the SR effect. There is no consensus on which signal-to-noise performance measure best measures the SR effect. The breadth of SR systems suggests that the SR effect may occur in still more complex dynamical systems for still more complex signals and noise types. These signal systems may prove too complex to model with simple closed-form techniques. This suggests, in turn, that we might use "intelligent" or adaptive model-free techniques to learn or approximate the SR effects.

Below we explore how to learn the SR effect with adaptive systems in general and with adaptive fuzzy function approximators [141, 142, 143, 144, 145] in particular. Adaptive fuzzy systems approximate functions with "if-then rules" that relate tunable fuzzy subsets of input and outputs. Each rule defines a fuzzy patch or subset of the input-output state space. The fuzzy system approximates a function as its rule patches cover the graph of the function. These systems resemble the radial-basis function networks found in neural networks [106, 184, 145]. Neural-like learning laws tune and move the fuzzy rule patches as they tune the shape of the fuzzy sets that make up the rule patches. The learning laws in the appendix use input-output data from the sampled noisy dynamical system. The rule patches move quickly to cover optimal or near-optimal regions of the function (such as its extrema). Experts can also state verbal if-then rules in some cases and add them to the fuzzy patch covering. These rules offer a simple way to endow a fuzzy approximator with prior knowledge or "hints" [1, 2] that can improve how well a fuzzy system approximates a function or how well it generalizes from training samples [205]. Fuzzy systems achieve their patch-covering approximation at the high cost of rule explosion [144, 145]. The number of rules grows exponentially with the state-space dimension of the fuzzy system. We stress that our SR learning laws can also tune non-fuzzy adaptive systems.

Adaptive fuzzy systems offer a balance between the structured and symbolic rulebased expert systems found in artificial intelligence [230] and the unstructured but numeric approximators found in modern neural networks [106, 107, 141]. These or other adaptive model-free approximators might better model the SR effect in some dynamical systems. Our first goal was to show that adaptive systems can learn to shape the input noise and perhaps shape other terms to achieve SR in the main closed-form dynamical systems that scientists have shown produce the SR effect. Our second goal was to suggest through these simulation experiments that adaptive fuzzy systems or other model-free approximators might achieve SR in more complex dynamical systems that defy easy math modeling or measurement.

This paper presents three main results. The first and central result is that a system can learn the SR effect if it performs a stochastic gradient ascent on the signal-to-noise ratio SNR $= S/N$. Then the random noise gradient $\frac{\partial \text{SNR}}{\partial \sigma}$ can tune the parameters in any adaptive system through a slow type of stochastic approximation [228]. We derive these learning laws in terms of discrete Fourier transforms. The idea behind the gradient-ascent learning is that such hill climbing is nontrivial if, and only if, the SNR surface shows some form of SR. The second result is that the SNR first-order condition for an extremum has the ratio form $\frac{S}{N} = \frac{S'}{N'}$ for $S' = \frac{\partial S}{\partial \sigma}$. The term $\frac{S'}{N'}$ can produce impulsive or even Cauchy noise that can destabilize the stochastic gradient ascent. Time lags in the training process can compound this impulsiveness. The third result is that a Cauchy-based noise suppressor from the theory of robust statistics can often reduce the impulsiveness of the noise gradient $\frac{\partial \text{SNR}}{\partial \sigma}$ and so improve the learning process.

The paper reviews the main math models involved in SR to date and reviews the adaptive fuzzy rule structure that can implicitly approximate these models and produce a like SR effect. The next two sections review these dynamical systems and the competing performance measures that scientists have used to detect SR in them. We used a standard signal-to-noise ratio SNR $= S/N$ based on discrete Fourier spectra. Most SR research has focused on the quartic bistable dynamical system. We worked with that signal system in detail and also applied the stochastic learning scheme to other dynamical systems. The learning scheme converged in most cases to the SR effect or the SNR mode in all of these systems. The SR learning scheme still converged for the quartic bistable system when we replaced the forcing additive Gaussian white noise with other additive random noise, with infinite-variance noise, and with chaotic noise from a chaotic logistic dynamical system. Sections 5 and 6 derive the SR optimality conditions and the stochastic learning law and then test the learning scheme in SR simulations of the quartic bistable dynamical system and other dynamical systems. The appendix derives the supervised learning laws for the fuzzy function approximator where the fuzzy sets have the shape of sinc, that is $\sin\pi x/x$, functions.

2 SR Dynamical Systems

This section reviews the main known dynamical systems that show SR. These models involve only simple nonlinearities. They also simply add a random noise term to a differential equation rather than use a formal Ito stochastic differential [45, 62, 91]. There are so far no theorems or formal taxonomies that tell which dynamical systems show SR and which do not. A dynamical system relates its input-output response through a differential equation of the form

$$\dot{x} = f(x) + u(x,t) \quad (1)$$

$$y(t) = g(x(t)) \quad (2)$$

The input u may depend on both time t and on the system's state x. The system is unforced or autonomous when $u(x,t) = 0$ for all x and t. The system output or measurement y depends on the state x through $y = g(x)$. The output of a simple model neuron may be a signum function: $y = \text{sgn}(x)$.

- **Quartic bistable system** [13, 58, 80, 87, 116, 129, 187, 261]. The quartic bistable system is the most studied model that shows SR. It has the form

$$\dot{x} = -\frac{\partial}{\partial x}U(x,t) + s(t) + n(t) \quad (3)$$

$$= ax - bx^3 + s(t) + n(t) \quad (4)$$

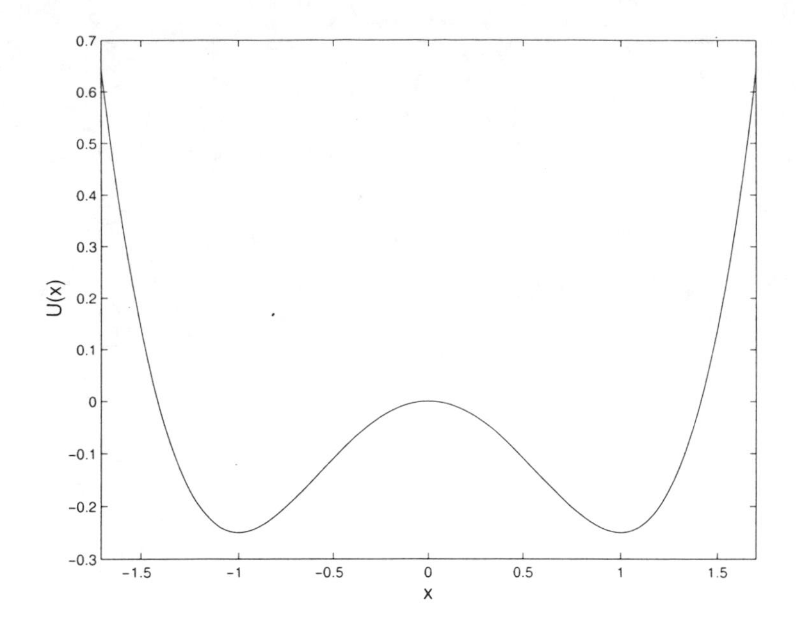

Figure 3: Unforced quartic potential: $U(x,t) = -\frac{1}{2}x^2 + \frac{1}{4}x^4$

for a quartic potential $U(x,t) = -\frac{a}{2}x^2 + \frac{b}{4}x^4$ with $a > 0$, $b > 0$, input signal s, and white Gaussian noise n with zero mean and variance σ^2: $E[n] = 0$ and $Var(n) = \sigma^2 < \infty$. Researchers sometimes include the forcing functions s and n in the potential function: $U(x,t) = -\frac{a}{2}x^2 + \frac{b}{4}x^4 + x(t)[s(t) + n(t)]$. The unforced version of (4) has the form $\dot{x} = ax - bx^3$. It has two stable fixed points at $x = \pm c = \pm\sqrt{a/b}$ and one metastable fixed point at $x = 0$. These fixed points are the minima and the local maximum of the potential $U(x,t) = -\frac{a}{2}x^2 + \frac{b}{4}x^4$. Figure 3 shows the quartic potential for $a = b = 1$. The two minima are at $x = \pm 1$. Figure 3 shows the potential at rest and hence with no input force. Figure 4 shows the potential $U(x,t)$ when the external sinusoidal input modulates it at each time instant t.

- **Threshold systems** [38, 84, 93, 96, 109, 130, 131, 209]. Threshold systems are among the simplest SR systems. They show the SR effect for many of the performance measures in the next section. A simple threshold system can take the form

$$y(t) = \operatorname{sgn}(x(t)) = \begin{cases} -1 & \text{if } x(t) < \Theta \\ 1 & \text{if } x(t) \geq \Theta \end{cases} \tag{5}$$

for the signal $x(t) = s(t) + n(t)$ and a threshold $\Theta \in R$. Thresholds quantize signals. So we state the general forms of uniform infinite quantizers with gain $G > 0$. A uniform mid-tread quantizer

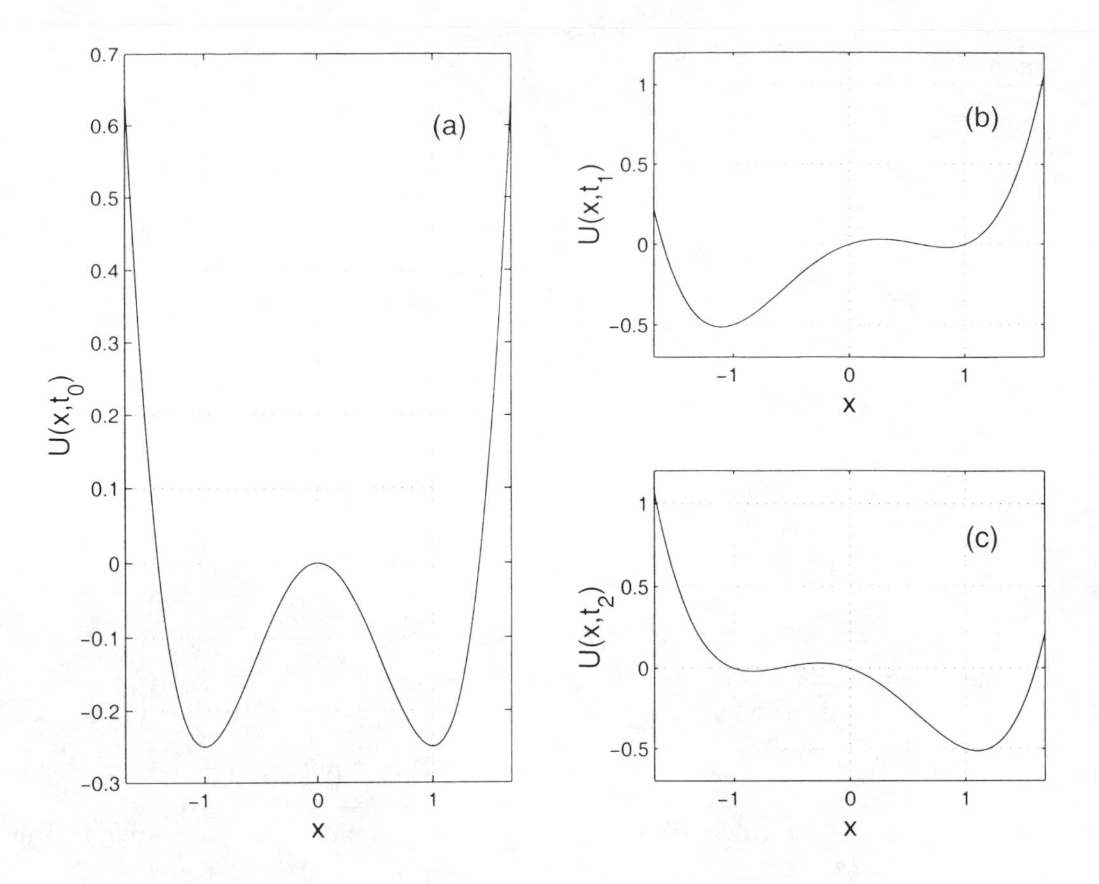

Figure 4: Forced evolution of the noise-free quartic potential system: $U(x,t) = -\frac{1}{2}x^2 + \frac{1}{4}x^4 + \frac{1}{4}x \sin 2\pi t$. (a) Unforced potential surface at $t = 0$ when the sinusoidal forcing term is zero. (b) Surface $U(x,t)$ at time $t = \frac{1}{4}$. (c) Surface $U(x,t)$ at time $t = \frac{3}{4}$.

with step size Δ has the form

$$y(t) = Q(x(t)) = G\Delta \left\lfloor \frac{x(t)}{\Delta} + \frac{1}{2} \right\rfloor. \tag{6}$$

A mid-riser quantizer has the form

$$y(t) = Q(x(t)) = G\Delta \left\lfloor \frac{x(t)}{\Delta} \right\rfloor + \frac{G\Delta}{2}. \tag{7}$$

The floor operator $\lfloor\rfloor$ gives the greatest integer less than or equal to its argument. Researchers have studied the SR effect in M-level quantizers that approximate some dynamical systems [211].

- **Bistable potential neuron model** [29]. This neuron model is a bistable system of the form

$$\dot{x} = -x + (\eta_0 + n_m(t)) \tanh x + n_a(t) + s(t). \tag{8}$$

The multiplicative and additive noises n_m and n_a are zero mean and uncorrelated. The term η_0 is a constant.

• **Monostable Systems** [65, 66, 68, 104, 241, 250]. These systems have no potential barriers as do bistable and multistable systems. They have only one stable fixed point. A special case is the single-well Duffing oscillator:

$$\ddot{x} + 2\Gamma\dot{x} + \omega_1^2 x + \gamma x^3 = s(t) + n(t) = \varepsilon \cos \omega_0 t + n(t) \tag{9}$$

where $\Gamma, |\delta\omega| << \omega_0$ and $\gamma\delta\omega > 0$ for $\delta\omega = \omega_0 - \omega_1$. These systems show the SR effect in the small signal limit with an approximate linear response.

• **Hodgkin-Huxley neuron model** [46, 165, 217]: The Hodgin-Huxley model is among the most studied models in the neural literature:

$$C\dot{x} = -g_{Na}m^3h(x - x_{Na}) - g_K p^4(x - x_K) - g_L(x - x_L) + I + s(t) + n(t) \tag{10}$$

$$\dot{m} = \alpha_m(x)(1 - m) - \beta_m(x)m \tag{11}$$

$$\dot{h} = \alpha_h(x)(1 - h) - \beta_h(x)h \tag{12}$$

$$\dot{p} = \alpha_p(x)(1 - p) - \beta_p(x)p \tag{13}$$

Here x is the membrane potential or activation and m is the sodium activation. The term h is the sodium inactivation, p is the potassium activation, C is the membrane capacitance. x_L is the leakage reversal potential, g_L is the leakage conductance, x_K is the potassium reversal potential. $\bar{g}_K$ is the maximal potassium conductance, ρ_K is the potassium ion-channel density. x_{Na} is the sodium reversal potential, $\bar{g}_{Na}$ is the maximal sodium conductance, ρ_{Na} is the sodium ion-channel density, I is an input current, and s is a subthreshold aperiodic input signal. These systems use a neural threshold signal function $S(x)$ that lets the neuron rest or retract after firing. SR occurs when a low level of noise n brings the input signal above the neuron's firing threshold.

• **FitzHugh-Nagumo (FHN) neuron model** [37, 46, 47, 48, 108, 163, 164, 191, 215, 256]. The FHN neuron model is a two-dimensional limit cycle oscillator that has the form

$$\epsilon\dot{x} = x(x - a)(1 - x) - w + A + s(t) + n(t) \tag{14}$$

$$\dot{w} = x - w - b \tag{15}$$

Here x is a fast (voltage) variable. w is a slow (recovery) variable, A is a constant (tonic) activation signal, s is an input signal, and n is noise. Sample constants for the SR effect are $\epsilon = 0.005$, $a = 0.5$, $A = -5/12\sqrt{3} = -0.24056$, and $b = 0.15$ [48].

• **Integrate-fire neuron model** [27, 33, 39, 41, 46, 79, 219, 240]. This neuron model has linear activation dynamics:

$$\dot{x} \;=\; \lambda(u_r - x) + \mu + s(t) + n(t) \tag{16}$$

where x is cell membrane voltage, μ is a positive drift, λ is a decay constant rate, and u_r is a resting level. A threshold function governs the neuron's output pulse firing and gives the nonlinear system that shows the SR effect.

• **Array and coupled systems** [26, 29, 31, 32, 43, 46, 47, 48, 98, 117, 121, 122, 125, 132, 139, 158, 163, 164, 165, 166, 167, 168, 185, 191, 196, 197, 198, 214, 220, 223, 224]. These systems combine many units of the above systems. They include neural networks and other coupled systems. A special case is the Cohen-Grossberg [141] feedback neural network:

$$C_i\dot{x}_i = -\frac{x_i}{R_i} + \sum_{j=1}^{N} m_{ij} \tanh x_j + s(t) + n(t) \qquad \text{for } i = 1, \ldots, N \tag{17}$$

for neural activation potential x_i, synaptic efficacy m_{ij}, and hyperbolic neural firing function $S_j(x_j) = \tanh x_j$. Simulations show that the SR profile grows more peaked as the number N of neurons grows [122]. One study [122] found that the SR effect goes away for $N \geq 10$.

• **Chaotic systems** [3, 5, 13, 35, 54, 169, 204, 253, 254, 263]. Some chaotic systems show the SR effect. These models include Chua's electric circuit, the Henon map, the Lorenz system, and the following forced Duffing oscillator:

$$\ddot{x} \;=\; -\,\delta\dot{x} + x + x^3 + \varepsilon \sin(\omega_0 t) + n(t). \tag{18}$$

At least one researcher [92] has argued that noise-induced chaos-order transitions need not be SR.

• **Random Systems** [15, 16, 17, 21, 30, 70, 153, 264]. These systems include many classical random processes such as random walks and Poisson processes. They also include the pulse system [15, 16, 17] whose response is a random train of pulses with a pulse probability r that depends on an input signal V through

$$r(V(t)) = r(0)\exp(V(t)). \tag{19}$$

The input V is the signal plus noise: $V(t) = s(t) + n(t)$. This model includes many kT-driven physio-chemical systems [15, 16, 17].

Other systems show SR in the literature [7, 11, 14, 22, 57, 114, 135, 176, 178, 196, 201, 221, 234, 251, 258, 260]. Special issues of physics journals [25, 189] also present other systems that show SR. Most use the SR measures in the next section.

3 SR Performance Measures

This section reviews the most popular measures of SR. These performance measures depend on the forcing signal and noise and can vary from system to system. There is no consensus in the SR literature on how to measure the SR effect.

Some researchers study a stochastic dynamical system in terms of the Fokker-Planck (or forward Kolmogorov) equation [59, 133, 192, 227]:

$$\frac{\partial p}{\partial t} = -\frac{\partial}{\partial x}(a(x,t)\,p) + \frac{1}{2}\frac{\partial^2}{\partial x^2}(b(x,t)\,p) \tag{20}$$

for drift term $a(x,t)$ and diffusion term $b(x,t)$. This partial differential equation stems from a Taylor series and shows how a probability density function p of a Markov system's states evolves in time. System nonlinearities often preclude closed-form solutions. Approximations and assumptions such as small noise and small signal effects can give closed-form solutions in some cases. These solutions motivate some of the performance measures below. SR dynamical systems in general need not be Markov processes [83, 200].

• **Signal-to-noise ratio**. The most common SR measure is some form of a signal-to-noise ratio (SNR) [74, 80, 90, 118, 178, 261]. This seems the most intuitive measure even though there are many ways to define a SNR.

Suppose the input signal is the sinewave $s(t) = \varepsilon \sin \omega_0 t$. Then the SNR measures how much

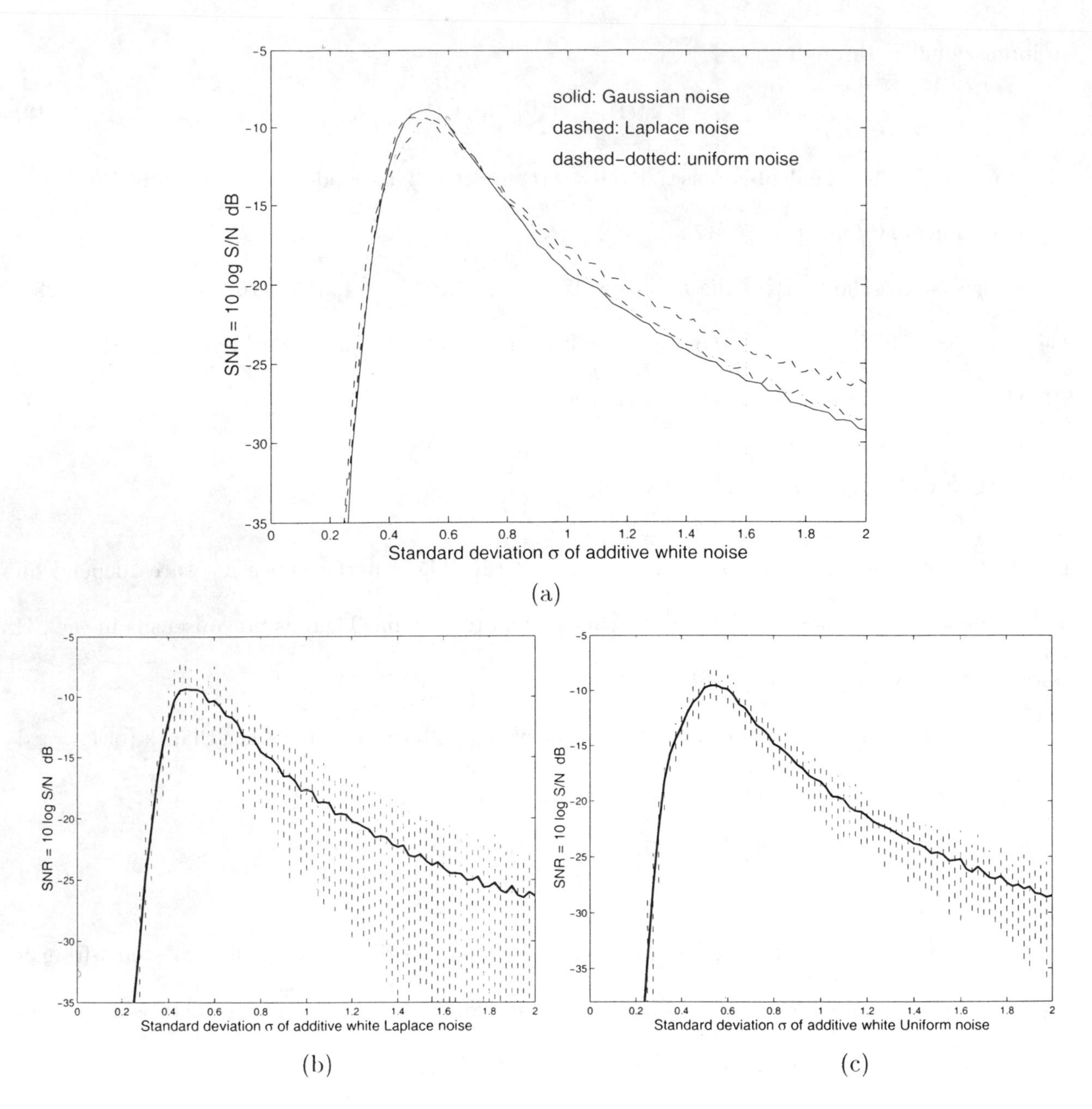

Figure 5: SNR measure of the quartic bistable system $\dot{x} = x - x^3 + s(t) + n(t)$ with output $y(t) = \text{sgn}(x(t))$. The signal s is the sinewave $s(t) = \varepsilon \sin 2\pi f_0 t$ where $\varepsilon = 0.1$ and $f_0 = 0.01$ Hz. (a) SNR-noise profiles of zero-mean white noise from Gaussian, Laplace, and uniform probability densities. The simulation ran over 20 distinct noise seeds over 10,000 seconds with time step $\Delta T = 10000/1000000 = 0.01$ seconds in the forward Euler formula of numerical analysis. (b) Average SNR-noise profile and its spread for Laplace noise. (c) Average SNR-noise profile and its spread for uniform noise. Figure 2 shows a like SR profile for Gaussian noise. Figure 18 shows the SR profile for the quartic bistable system when chaotic noise drives the system. The plots show distinct spreads of SNR for each kind of noise.

the system output $y = g(x)$ contains the input signal frequency ω_0:

$$\text{SNR} = 10 \log \frac{S}{N} \tag{21}$$

$$= 10 \log \frac{S(\omega_0)}{N(\omega_0)} \quad \text{dB}. \tag{22}$$

The signal power $S = |Y(\omega_0)|^2$ is the magnitude of the output power spectrum $Y(\omega)$ at the input frequency ω_0. The background noise spectrum $N(\omega_0)$ at input frequency ω_0 is some average of $|Y(\omega)|^2$ at nearby frequencies [123, 178, 261]. The discrete Fourier transform (DFT) $Y[k]$ for $k = 0, \ldots, L-1$ is an exponentially weighted sum of elements of a discrete-time sequence $\{y_0, y_1, \ldots, y_{L-1}\}$ of output signal samples

$$Y[k] = \sum_{t=0}^{L-1} y_t \, e^{-\frac{2\pi k t}{L}}. \tag{23}$$

The signal frequency ω_0 corresponds to bin k_0 in the DFT for integer $k_0 = L\Delta T f_0$ and for $\omega_0 = 2\pi f_0$. This gives the output signal in terms of a DFT as $S = |Y[k_0]|^2$. The noise power $N = N[k_0]$ is the average power in the adjacent bins $k_0 - M, \ldots, k_0 - 1, k_0 + 1, \ldots, k_0 + M$ for some integer M [6, 261]:

$$N = \frac{1}{2M} \sum_{j=1}^{M} \left(|Y[k_0 - j]|^2 + |Y[k_0 + j]|^2\right) \tag{24}$$

We expand this noise term in Section 5 to include all energy not due to the signal.

An adiabatic approximation [178] can give an explicit signal-to-noise ratio R for the quartic bistable system in (4) with sinewave input $s(t) = \varepsilon \sin \omega_0 t$:

$$R = \frac{S}{N} = \left[\frac{\sqrt{2} a \varepsilon^2 c^2}{(\sigma^2)^2} e^{-2U_0/\sigma^2}\right] \left[1 - \frac{\frac{4a^2\varepsilon^2c^2}{\pi^2(\sigma^2)^2} e^{-4U_0/\sigma^2}}{\frac{2a^2}{\pi^2} e^{-4U_0/\sigma^2} + \omega_0^2}\right] \tag{25}$$

$$\approx \frac{\sqrt{2} a \varepsilon^2 c^2}{(\sigma^2)^2} e^{-2U_0/\sigma^2}. \tag{26}$$

Here $U_0 = a^2/4b$ is the barrier height when $\varepsilon = 0$, $x = \pm c = \pm\sqrt{a/b}$ defines the potential minima, and σ^2 is the variance of the additive white Gaussian noise n. This result stems from Kramers rate [148] if the signal amplitude ε is small and if its frequency is smaller than the characteristic rate or curvature at the minimum $U''(\pm c)$ [178]. The SNR approximation (26) is zero for zero noise

$\sigma^2 = 0$. It grows from zero as σ^2 grows and reaches a maximum at $\sigma^2 = U_0$ before it decays. So the optimum noise intensity is $\sigma^2 = U_0 = a^2/4b$.

There is no standard definition of system-level signal and noise in nonlinear systems. We work with a SNR that is easy to compute and that depends on standard spectral power measures in signal processing. We start with a sinewave input and view the output state $y(t) = g(x(t))$ of the dynamical system as a mixture of signal and noise. We arrange the DFT computation so that the energy of the sine term lies in frequency bin k_0. The squared magnitude of this energy spectrum $Y[k_0]$ acts as the system-level signal: $S = 2|Y[k_0]|^2$. We view all else in the spectrum as noise: $N = P - S = P - 2|Y[k_0]|^2$ where the total energy is $P = \sum_{k=0}^{L-1} |Y[k]|^2$. We ignore the factor L that scales S and P since the ratio S/N cancels its effect. Figure 2 shows the SR profile with this SNR measure for the quartic bistable system with forcing sinewave input signal and Gaussian noise. Figure 5 shows the SR profiles of the quartic bistable system with forcing Gaussian, uniform, and Laplace noise. Figure 17 shows the SR profiles of the quartic bistable system for impulsive noise with infinite variance. Figure 18 shows the SR profile of the quartic bistable system for chaotic noise from a logistic dynamical system.

- **Cross-correlation measures**. These 'shape matchers"can measure SR when inputs are not periodic signals. Researchers coined the term "aperiodic stochastic resonance" (ASR) [43, 46, 47, 108] for such cases. They defined cross-correlation measures for the input signal s and the system response in terms of the mean transition rate r in the FHN model in (14)-(15):

$$C_0 = \max\{\overline{s(t)r(t+\tau)}\} \tag{27}$$

$$C_1 = \frac{C_0}{[\overline{s^2(t)}]^{1/2}\{\overline{[r(t) - \overline{r(t)}]^2}\}^{1/2}} \tag{28}$$

where $\overline{x}$ is the time average: $\overline{x} = \dfrac{1}{T}\displaystyle\int_0^T x(t)dt$.

- **Probability of residence time and escape rate**. This approach looks at the probability $P(T)$ of the time T that a dynamical system spends in a stable state between consecutive switches between the stable states [57, 70, 87, 129, 262]. So $P(T)$ depends on the input noise intensity. Data can give a histogram of this $P(T)$ to estimate the actual probability for each input noise intensity σ_n^2. The probability of residence time relates to the first passage time density function (FPTDF) or the interspike interval histogram (ISIH) found in the neurophysiological literature [21, 27, 30, 36, 81, 166, 167, 163, 164, 165, 172]. The symmetric bistable system (4) with input $s(t) = \varepsilon \sin \omega_0 t$ gives a system that tends to stay at or wander about one stable state for $T = T_0/2 = 2\pi/\omega_0$ seconds and then hops to a new stable state as it tracks the input.

• **Information and probability of detection**. Tools from information theory can also measure SR. The information rate of a threshold system shows the SR effect for subthreshold inputs [33, 38, 39, 240]. The FitzHugh-Nagumo (FHN) neuron model (14)-(15) shows SR for aperiodic input waveforms when we measure the cross-correlation between input and output or the information rate [46, 48, 108]. Noise can also sometimes maximize the mutual information [52]:

$$I(X;Y) \;=\; H(X) - H(X|Y) \;=\; \sum_{x,y} p(x,y) \log \frac{p(x,y)}{p(x)p(y)}. \tag{29}$$

The mutual (Kullback) information $I(X;Y)$ and Fisher information [52] can measure SR in some neuron models [33, 199, 240]. Probability of correct detection and other statistics can also measure SR [97, 115, 123, 240].

• **Complexity measures**. Researchers have suggested other ways to measure SR. These include Lyapunov exponents, Shannon entropy, fluctuation complexity that measures the net information gain, and ϵ-complexity for first-order Markov stochastic automata [169, 259].

Other forms of SR measures also occur in the SR literature. They include the other signal-to-noise ratios [66, 131, 140, 157, 161], the amplification characteristic of a system like those found in electronic devices [9, 42, 100, 101, 134], susceptibility [67, 68, 185, 242], "crisis" measure in chaos [35], and prediction error of spike rates [37]. The number of SR performance measures will likely grow as researchers explore how noise and signals drive other systems in the vast function space of nonlinear dynamical systems.

4 Additive Fuzzy Systems and Function Approximation

This section reviews the basic structure of additive fuzzy systems. The appendices review and extend the more formal math structure that underlies these adaptive function approximators.

A fuzzy system $F : R^n \to R^p$ stores m rules of the word form "If $X = A_j$, then $Y = B_j$" or the patch form $A_j \times B_j \subset X \times Y = R^n \times R^p$. The if-part fuzzy sets $A_j \subset R^n$ and then-part fuzzy sets $B_j \subset R^p$ have set functions $a_j : R^n \to [0, 1]$ and $b_j : R^p \to [0, 1]$. Generalized fuzzy sets map to intervals other than [0,1]. The scalar sinc set functions in Figure 23 map real inputs to "membership degrees" in the bipolar range [-0.217,1]. The system design must take care when these negative set values enter the SAM ratio in (31). The system can use the joint set function a_j or some factored form such as $a_j(x) = a_j^1(x_1) \cdots a_j^n(x_n)$ or $a_j(x) = \min(a_j^1(x_1), \ldots, a_j^n(x_n))$ or any other conjunctive form for input vector $x = (x_1, \ldots, x_n) \in R^n$ [141].

An additive fuzzy system [141, 142] sums the "fired" then-part sets B'_j :

$$B(x) \quad = \quad \sum_{j=1}^{m} w_j B'_j \quad = \quad \sum_{j=1}^{m} w_j a_j(x) B_j. \tag{30}$$

Figure 6a shows the parallel fire-and-sum structure of the standard additive model (SAM). These nonlinear systems can uniformly approximate any continuous (or bounded measurable) function f on a compact domain [142, 145]. Engineers often apply fuzzy systems to problems of control [127] but fuzzy systems can also apply to problems of communication [208] and signal processing [138] and other fields.

Figure 6b shows how three rule patches can cover part of the graph of a scalar function f : $R \to R$. The patch-cover structure implies that fuzzy systems $F : R^n \to R^p$ suffer from *rule explosion* in high dimensions. A fuzzy system F needs on the order of k^{n+p-1} rules to cover the graph and thus to approximate a vector function $f : R^n \to R^p$. Optimal rules can help deal with the exponential rule explosion. Lone or local mean-squared optimal rule patches cover the extrema of the approximand f [144, 145]. They "patch the bumps" as in Figure 7. Better learning schemes move rule patches to or near extrema and then fill in between extrema with extra rule patches if the rule budget allows.

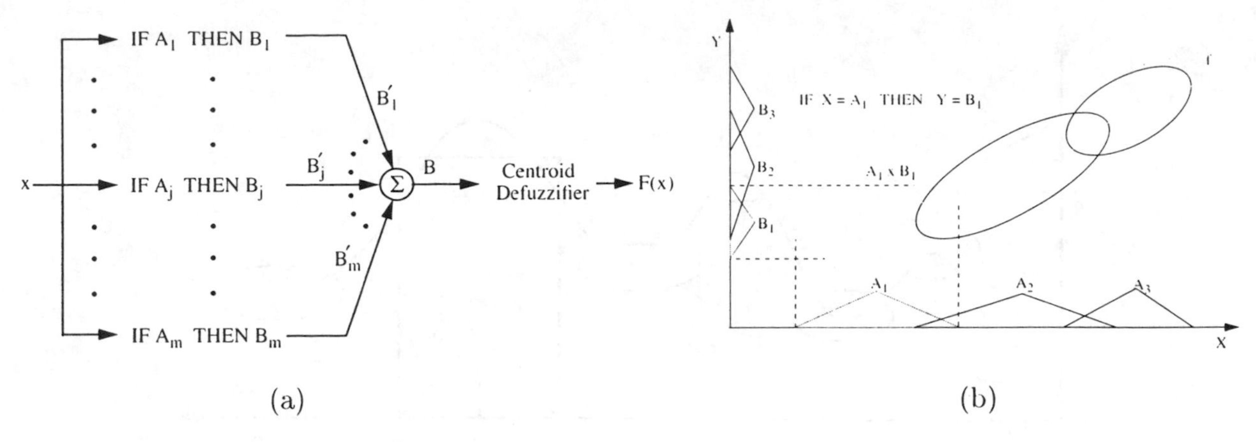

Figure 6: Feedforward fuzzy function approximator. (a) The parallel associative structure of the additive fuzzy system $F : R^n \to R^p$ with m rules. Each input $x_0 \in R^n$ enters the system F as a numerical vector. At the set level x_0 acts as a delta pulse $\delta(x - x_0)$ that combs the if-part fuzzy sets A_j and gives the m set values $a_j(x_0) = \int_{R^n} \delta(x - x_0) a_j(x) dx$. The set values "fire" or scale the then-part fuzzy sets B_j to give B'_j. A standard additive model (SAM) scales each B_j with $a_j(x)$. Then the system sums the B'_j sets to give the output "set" B. The system output $F(x_0)$ is the centroid of B. (b) Fuzzy rules define Cartesian rule patches $A_j \times B_j$ in the input-output space and cover the graph of the approximand f. This leads to exponential rule explosion in high dimensions. Optimal lone rules cover the extrema of the approximand as in Figure 7.

The scaling choice $B'_j = a_j(x)B_j$ gives a *standard additive model* or SAM. Appendix A shows that taking the centroid of $B(x)$ in (30) gives the following SAM ratio [141, 142, 143, 144]

$$F(x) \quad = \quad \frac{\sum_{j=1}^{m} w_j a_j(x) V_j c_j}{\sum_{j=1}^{m} w_j a_j(x) V_j} \quad = \quad \sum_{j=1}^{m} p_j(x) c_j. \tag{31}$$

Here V_j is the finite positive volume or area of then-part set B_j and c_j is the centroid of B_j or its center of mass. The convex weights $p_1(x), \ldots, p_m(x)$ have the form $p_j(x) = \dfrac{w_j a_j(x) V_j}{\sum_{i=1}^{m} w_i a_i(x) V_i}$. The convex coefficients $p_j(x)$ change with each input vector x.

Figure 8 shows how supervised learning moves and shapes the fuzzy rule patches to give a finer approximation as the system samples more input-output data. Appendix B derives the supervised SAM learning algorithms for the sinc set functions [145, 182, 183] in Figure 23 that we use in the SR simulations. Supervised gradient ascent changes the SAM parameters with performance data. The learning laws update each SAM parameter to maximize the performance measure P of the SR

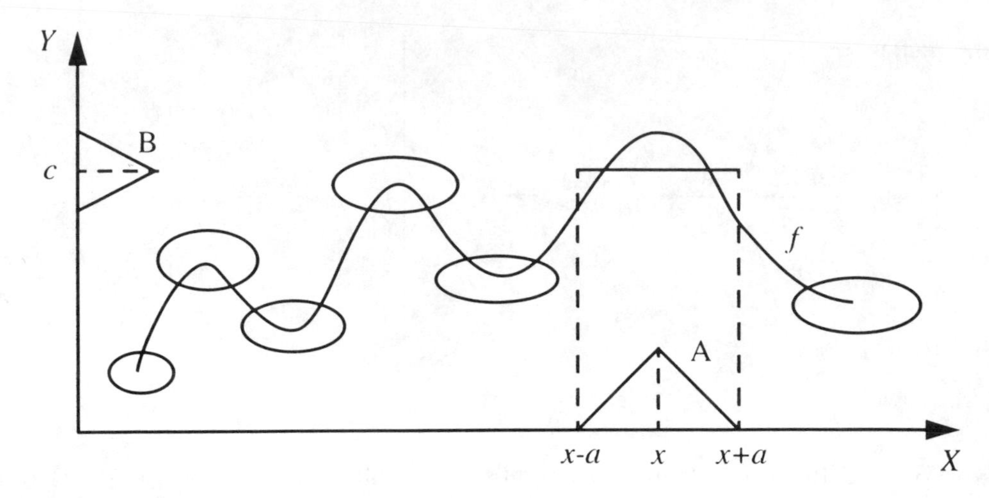

Figure 7: Lone optimal fuzzy rule patches cover the extrema of approximand f. A lone rule defines a flat line segment that cuts the graph of the local extremum in at least two places. The mean value theorem implies that the extremum lies between these points. This can reduce much of fuzzy function approximation to the search for zeroes $\hat{x}$ of the derivative map $f' : f'(\hat{x}) = 0$.

dynamical system. This process repeats as needed for a large number of sample data pairs (x_t, y_t). Figure 8(f) displays the absolute error of the sinc-based fuzzy function approximation.

5 SR Learning and Equilibrium

The scalar SAM fuzzy system $F : R^n \rightarrow R$ can learn the SR pattern of optimum noise of an unknown dynamical system if it uses enough rules and if it samples enough data from a dynamical system that stochastically resonates. Below we derive a gradient-based learning law that tunes the SAM parameters to achieve SR from samples of system dynamics. It can also tune the parameters in other adaptive systems. We first define a practical SNR measure in terms of discrete Fourier transforms. Other SR measures can give other learning laws.

5.1 The Signal-to-Noise Ratio in Nonlinear Systems

Suppose a nonlinear dynamical system has a sinewave forcing function $s(t)$ of known frequency f_0 Hertz. We search the sinusoidal part $r(t)$ of the output $y(t)$ for the known frequency f_0 but unknown amplitude and phase in the system output response $y(t)$. The "noisy signal" $y(t)$ has the

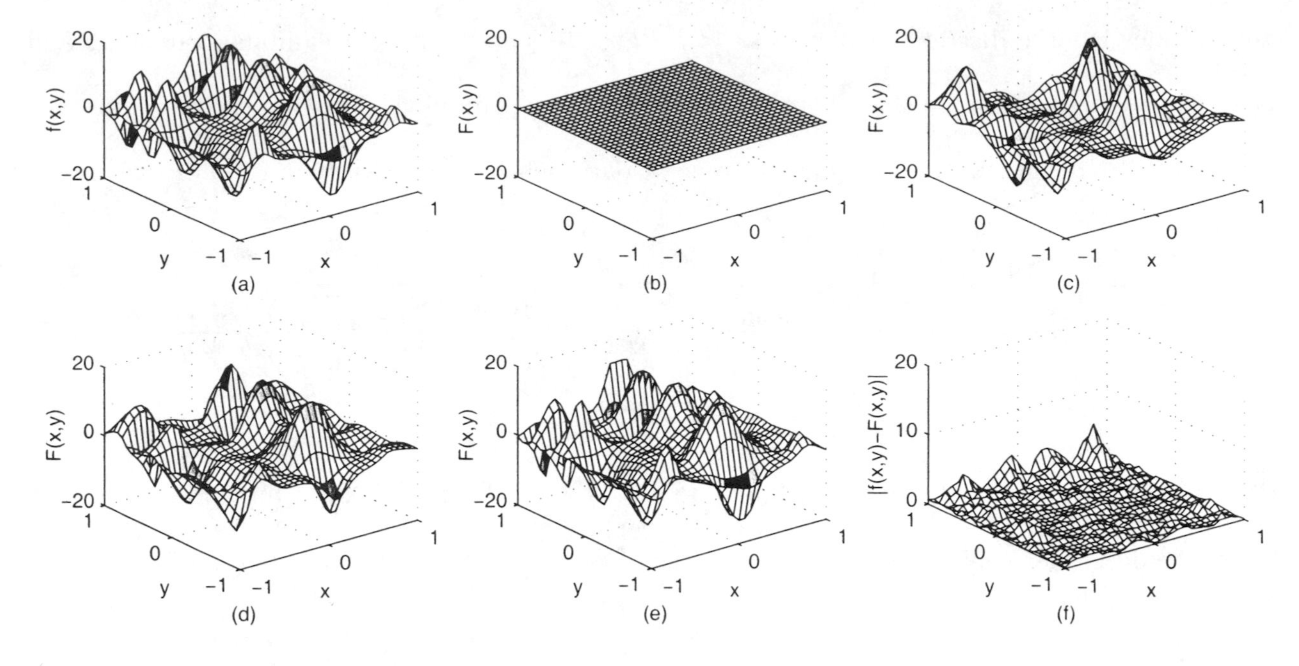

Figure 8: Fuzzy function approximation. 2-D Sinc standard additive model (SAM) function approximation with 100 fuzzy if-then rules and supervised gradient descent learning. (a) Desired function or approximand f. (b) SAM initial phase as a flat sheet or constant approximator F. (c) SAM approximator F after it initializes its centroids to the samples: $c_j = f(m_j)$. (d) SAM approximator F after 100 epochs of learning. (e) SAM approximator F after 6000 epochs of learning. (f) Absolute error of the fuzzy function approximation ($|f - F|$).

form of "signal" plus "noise":

$$y_t = r_t + n_t. \tag{32}$$

The signal-to-noise ratio (SNR) at the output is the spectral ratio of the energy of $\{r_t\}$ to the energy of $\{n_t\}$. We assume that the signal $s(t)$ is always present. This ignores the important problem of signal detection but lets us focus on learning the SR effect.

We define the SNR measure as

$$\text{SNR} = \frac{S}{N} = \frac{S}{P - S}. \tag{33}$$

Here $S = 2\,|Y[k_0]|^2$, $P = \sum_{k=0}^{L-1} |Y[k]|^2$, and $Y[k]$ is the L-point discrete Fourier transform (DFT) of y_n:

$$Y[k] = \sum_{t=0}^{L-1} y_t\, e^{-i\frac{2\pi k}{L}t}. \tag{34}$$

We assume that the discrete frequency $k_0 = f_0 L T_s > 0$ is an integer for sampling rate $1/T_s$ and $\omega_0 = 2\pi f_0$. We also assume that there is no aliasing due to sampling. Then we can show that for large L the SNR measure in (33) tends to the standard definition of SNR as a ratio of variances:

$$\textit{Theorem:} \qquad \text{SNR} = \frac{2\,|Y[k_0]|^2}{\sum_{k=0}^{L-1} |Y[k]|^2 - 2\,|Y[k_0]|^2} \longrightarrow \frac{\sigma_r^2}{\sigma_n^2} = \frac{A^2/2}{\sigma_n^2} = \frac{\frac{1}{L}\sum_{t=0}^{L-1} r_t^2}{\frac{1}{L}\sum_{t=0}^{L-1} n_t^2}. \tag{35}$$

Here $\sigma_r^2 = \frac{1}{T}\int_0^T (A\sin\omega_0 t)^2 dt = A^2/2$ and $\sigma_n^2 = Var(n) = E[n^2]$. We need further assumptions to derive (35). First consider the "energy" in each frequency bin k of the transform $Y[k]$:

$$|Y[k]|^2 = Y[k]Y^*[k] \tag{36}$$

$$= (R[k] + N[k])\,(R[k] + N[k])^* \tag{37}$$

$$= R[k]R^*[k] + R[k]N^*[k] + R^*[k]N[k] + N[k]N^*[k] \tag{38}$$

$$= |R[k]|^2 + |N[k]|^2 + 2Re\{R[k]N^*[k]\} \tag{39}$$

where $R[k]$ and $N[k]$ are the DFTs of r_t and n_t in (32). Suppose the sinusoidal term has the form

$$r_t = A\cos(2\pi f_0 T_s t + \phi) \tag{40}$$

for $t = 0, \ldots, L-1$. Its DFT has the form [207]

$$R[k] = \sum_{t=0}^{L-1} r_t\, e^{-i\frac{2\pi k}{L}t} \tag{41}$$

$$= \sum_{t=0}^{L-1} A\cos(2\pi f_0 T_s t + \phi)\, e^{-i\frac{2\pi k}{L}t} \tag{42}$$

$$= e^{ik\Omega(\phi - a)}\,\frac{A}{T_s}\left[\frac{\sin a\Omega(k-k_0)}{\Omega(k-k_0)} + \frac{\sin a\Omega(k+k_0)}{\Omega(k+k_0)}\right] \tag{43}$$

$$= e^{ik\Omega(\phi - a)}\,\frac{A\,a}{T_s}\Big(\delta[k-k_0] + \delta[k-(L-k_0)]\Big) \tag{44}$$

where $k_0 = f_0 L T_s > 0$ is an integer, $\Omega = \frac{2\pi}{LT_s}$ is a frequency band, $a = \frac{LT_s}{2}$, and δ is the Kronecker delta function. So $R[k]$ vanishes when both $k \neq k_0$ and $k \neq L - k_0$. This gives

$$\sum_{k=0}^{L-1} |R[k]|^2 = |R[k_0]|^2 + |R[L-k_0]|^2 = 2\,|R[k_0]|^2 = 2\left(\frac{a\,A}{T_s}\right)^2 = \frac{L^2 A^2}{2}. \tag{45}$$

So $R[k_0]$ and $R[L-k_0]$ contain all the energy of the sinusoidal signal r_t. We define the noise power as $\sigma_n^2 = E\{n_t^2\}$ and assume that n_t is stationary and ergodic with zero mean. Then Parseval's theorem gives

$$\sum_{k=0}^{L-1} |N[k]|^2 = L \sum_{t=0}^{L-1} |n_t|^2 \tag{46}$$

$$\approx L\,(L\sigma_n^2) \tag{47}$$

$$= L^2 \sigma_n^2. \tag{48}$$

The ergodicity of n_t gives (47). Now consider the total output spectrum P:

$$P = \sum_{k=0}^{L-1} |Y[k]|^2 \tag{49}$$

$$= |Y[k_0]|^2 + |Y[L-k_0]|^2 + \sum_{k=0, k\neq k_0, L-k_0}^{L-1} |Y[k]|^2 \tag{50}$$

$$= 2\,|Y[k_0]|^2 + \sum_{k=0, k\neq k_0, L-k_0}^{L-1} |N[k]|^2 \tag{51}$$

$$= 2\,|R[k_0]|^2 + 2\,|N[k_0]|^2 + 4\,Re\{R[k_0]N^*[k_0]\} + \sum_{k=0, k\neq k_0, L-k_0}^{L-1} |N[k]|^2 \tag{52}$$

$$= 2\,|R[k_0]|^2 + \left(\sum_{k=0}^{L-1} |N[k]|^2\right) + 4Re\{R[k_0]N^*[k_0]\}. \tag{53}$$

Then (53) and (39) give

$$P - 2\,|Y[k_0]|^2 = \sum_{k=0}^{L-1} |N[k]|^2 - 2\,|N[k_0]|^2. \tag{54}$$

Then the SNR structure in (33) follows:

$$\text{SNR} = \frac{S}{N} = \frac{S}{P-S} \tag{55}$$

$$= \frac{2|R[k_0]|^2 + 2|N[k_0]|^2 + 4Re\{R[k_0]N^*[k_0]\}}{\left(\sum_{k=0}^{L-1} |N[k]|^2\right) - 2|N[k_0]|^2} \tag{56}$$

$$\approx \frac{2|R[k_0]|^2}{\sum_{k=0}^{L-1} |N[k]|^2} = \frac{L^2\,A^2/2}{L^2\sigma_n^2} = \frac{A^2/2}{\sigma_n^2} \tag{57}$$

for large L and for small (or null) $|N[k_0]|$ and $|N[L-k_0]|$. Note that $|N[k_0]| = |N[L-k_0]|$ for $k_0 \neq 0$ due to the symmetry of the DFT.

The result (57) also holds if the zero-mean noise sequence n_t is not correlated in time and does not correlate with r_t. Then we can take expectations of $2|Y[k_0]|^2$ and $P - 2|Y[k_0]|^2$ to get

$$E\Big[\,|Y[k_0]|^2\,\Big] = E\Big[\,|R[k_0]|^2 + |N[k_0]|^2 + 2Re\{R[k_0]N^*[k_0]\}\,\Big] \tag{58}$$

$$= E[\,|R[k_0]|^2\,] + E[\,|N[k_0]|^2\,] + 2Re\{E[\,R[k_0]\,]\,E[\,N^*[k_0]\,]\}\,\Big] \tag{59}$$

$$= E[\,|R[k_0]|^2\,] + E\Big[\Big(\sum_{t=0}^{L-1} n_t\, e^{-i\frac{2\pi k}{L}t}\Big)\Big(\sum_{\tau=0}^{L-1} n_\tau\, e^{-i\frac{2\pi k}{L}\tau}\Big)^*\Big]$$

$$+\, 2Re\{R[k_0]\,E\Big[\sum_{\tau=0}^{L-1} n_\tau\, e^{-i\frac{2\pi k}{L}\tau}\Big]^*\} \tag{60}$$

$$= |R[k_0]|^2 + \sum_{t=0}^{L-1}\sum_{\tau=0}^{L-1} E[\,n_t\, n_\tau\,]\, e^{-i\frac{2\pi k}{L}(t-\tau)}$$

$$+\, 2Re\{E[\,R[k_0]\,]\Big(\sum_{\tau=0}^{L-1} E[\,n_\tau\,]\, e^{i\frac{2\pi k}{L}\tau}\Big)\} \tag{61}$$

$$= |R[k_0]|^2 + \sum_{t=0}^{L-1}\sum_{\tau=0}^{L-1} \sigma_n^2\, \delta[t-\tau]\, e^{-i\frac{2\pi k}{L}(t-\tau)} + 0 \tag{62}$$

$$= |R[k_0]|^2 + L\sigma_n^2 \tag{63}$$

$$= L^2\frac{A^2}{4} + L\sigma_n^2 \tag{64}$$

and

$$E[\,P - 2\,|Y[k_0]|^2\,] = E\Big[\,\sum_{k=0}^{L-1} |N[k]|^2 \; - 2|N[k_0]|^2\,\Big] \tag{65}$$

$$= \sum_{k=0}^{L-1} L\sigma_n^2 \; - 2L\sigma_n^2 \tag{66}$$

$$= L^2\,\sigma_n^2 - 2L\sigma_n^2. \tag{67}$$

Putting (64) and (67) into (33) gives

$$\text{SNR} = \frac{2\,E[\,|Y[k_0]|^2\,]}{E[\,P - 2|Y[k_0]|^2\,]} \tag{68}$$

$$= \frac{2(\,L^2\frac{A^2}{4} + L\sigma_n^2\,)}{L^2\sigma_n^2 - 2L\sigma_n^2}. \tag{69}$$

Then SNR $\rightarrow \dfrac{A^2/2}{\sigma_n^2}$ as $L \rightarrow \infty$.

5.2 Supervised Gradient Learning and SR Optimality

An adaptive system can learn a SR noise pattern that maximizes a dynamical system's SNR. The learning law updates a parameter m_j of a SAM fuzzy system (or of any other adaptive system) at time step n with the deterministic law

$$m_j(n+1) \;=\; m_j(n) + \mu_n \frac{\partial E[\mathrm{SNR}]}{\partial m_j}. \tag{70}$$

for learning coefficients $\{\mu_n\}$. This is gradient *ascent* learning. We assume that the first-order moment of the SNR exists. We seldom know the probability structure or the expectation of the SNR. So we estimate this expectation with its random realization at each time step: $E[\mathrm{SNR}] \approx \mathrm{SNR}$. This gives the *stochastic* gradient learning law

$$m_j(n+1) \;=\; m_j(n) + \mu_n \frac{\partial\, \mathrm{SNR}}{\partial m_j} \tag{71}$$

or simple random hill climbing. We assume the chain rule holds (at least approximately) to give

$$\frac{\partial\, \mathrm{SNR}}{\partial m_j} \;=\; \frac{\partial\, \mathrm{SNR}}{\partial \sigma}\frac{\partial \sigma}{\partial m_j}. \tag{72}$$

Here σ is the noise level or standard deviation of the forcing noise term $n(t)$. We want the SAM or other adaptive system F to approximate the optimum noise level $\hat{\sigma}$ for any input signal or initial condition of the dynamical system: $F \approx \hat{\sigma}$. We then use σ and F interchangeably:

$$\frac{\partial\, \mathrm{SNR}}{\partial m_j} \;=\; \frac{\partial\, \mathrm{SNR}}{\partial \sigma}\frac{\partial F}{\partial m_j}. \tag{73}$$

The term $\frac{\partial F}{\partial m_j}$ shows how any adaptive system F depends on its jth parameter m_j. We again assume that the chain rule holds to get

$$\frac{\partial\, \mathrm{SNR}}{\partial \sigma} \;=\; \frac{\partial\, \mathrm{SNR}}{\partial S}\frac{\partial S}{\partial \sigma} + \frac{\partial\, \mathrm{SNR}}{\partial N}\frac{\partial N}{\partial \sigma}. \tag{74}$$

Then $\mathrm{SNR} = S/N$ implies that

$$\frac{\partial\, \mathrm{SNR}}{\partial S} \;=\; \frac{\partial}{\partial S}\frac{S}{N} \;=\; \frac{1}{N} \tag{75}$$

$$\frac{\partial\, \mathrm{SNR}}{\partial N} \;=\; \frac{\partial}{\partial N}\frac{S}{N} \;=\; -\frac{S}{N^2} \;=\; -\frac{\mathrm{SNR}}{N}. \tag{76}$$

Like results hold for the decibel definition SNR $= 10 \log S/N$ dB for the base-10 logarithm:

$$\frac{\partial \, \text{SNR}}{\partial S} = \frac{\partial}{\partial S} 10 \log \frac{S}{N} = (10 \log e) \frac{1}{S} \tag{77}$$

$$\frac{\partial \, \text{SNR}}{\partial N} = \frac{\partial}{\partial N} 10 \log \frac{S}{N} = -(10 \log e) \frac{1}{N}. \tag{78}$$

We next put (75)-(78) into (74) to get the log term that drives SR learning:

$$\frac{\partial \, \text{SNR}}{\partial \sigma} = \begin{cases} \dfrac{1}{N}\dfrac{\partial S}{\partial \sigma} - \dfrac{\text{SNR}}{N}\dfrac{\partial N}{\partial \sigma} & \text{if SNR} = \dfrac{S}{N} \\ \\ (10 \log e)\left(\dfrac{1}{S}\dfrac{\partial S}{\partial \sigma} - \dfrac{1}{N}\dfrac{\partial N}{\partial \sigma}\right) & \text{if SNR} = 10 \log \dfrac{S}{N}. \end{cases} \tag{79}$$

The right side of (79) leads to the first-order condition for an SNR extremum:

$$\frac{1}{S}\frac{\partial S}{\partial \sigma} - \frac{1}{N}\frac{\partial N}{\partial \sigma} = 0 \tag{80}$$

or simply

$$\frac{S}{N} = \frac{S'}{N'}. \tag{81}$$

We can rewrite this optimality condition as

$$\left.\frac{S}{N}\right|_{\sigma_{opt}} = \left.\frac{\partial S/\partial \sigma}{\partial N/\partial \sigma}\right|_{\sigma_{opt}} \tag{82}$$

when the partial derivatives of S and N with respect to σ are not zero at $\sigma = \sigma_{opt}$. Equations (80) and (82) give a necessary condition for the SR maximum. The result (82) says that at SR the ratio of the rate of changes of S and N must equal the ratio of S and N. This has the same form as the result in microeconomics [149] that the marginal rates of substitution of two goods must at optimality equal the partial derivatives of the utility function with respect to each good. But (81) and (82) hold only in a stochastic sense for sufficiently well-behaved random processes.

We find the second-order condition for an SR maximum when SNR $= 10 \log S/N$ from

$$0 > \frac{\partial^2 \, \text{SNR}}{\partial \sigma^2} = \frac{\partial}{\partial \sigma}\frac{\partial \, \text{SNR}}{\partial \sigma} \tag{83}$$

$$= \frac{\partial}{\partial \sigma}(10 \log e)\left[\frac{1}{S}\frac{\partial S}{\partial \sigma} - \frac{1}{N}\frac{\partial N}{\partial \sigma}\right] \tag{84}$$

$$= (10 \log e)\left[\left(\frac{1}{S}\frac{\partial^2 S}{\partial \sigma^2} + \frac{\partial S}{\partial \sigma}\left(-\frac{1}{S^2}\frac{\partial S}{\partial \sigma}\right)\right) - \left(\frac{1}{N}\frac{\partial^2 N}{\partial \sigma^2} + \frac{\partial N}{\partial \sigma}\left(-\frac{1}{N^2}\frac{\partial N}{\partial \sigma}\right)\right)\right] \tag{85}$$

$$= \quad (10 \log e) \left[\frac{1}{S} \frac{\partial^2 S}{\partial \sigma^2} - \frac{1}{S^2} \left(\frac{\partial S}{\partial \sigma} \right)^2 - \frac{1}{N} \frac{\partial^2 N}{\partial \sigma^2} + \frac{1}{N^2} \left(\frac{\partial N}{\partial \sigma} \right)^2 \right] \tag{86}$$

$$= \quad (10 \log e) \left[\frac{1}{S} \frac{\partial^2 S}{\partial \sigma^2} - \frac{1}{N} \frac{\partial^2 N}{\partial \sigma^2} \right] \tag{87}$$

or $\frac{S''}{S} < \frac{N''}{N}$. The last equality follows from the first-order condition $\frac{1}{S}\frac{\partial S}{\partial \sigma} - \frac{1}{N}\frac{\partial N}{\partial \sigma} = 0$ or $\frac{S'}{S} = \frac{N'}{N}$ since then $\frac{(S')^2}{S^2} = \frac{(N')^2}{N^2}$. A like result holds for SNR $= S/N$. We still get the second-order condition

$$\frac{1}{S} \frac{\partial^2 S}{\partial \sigma^2} - \frac{1}{N} \frac{\partial^2 N}{\partial \sigma^2} \quad < \quad 0. \tag{88}$$

These first- and second-order conditions show how the signal power S and noise power N relate to each other and to their derivatives at the SR maximum.

Much of the noisiness and complexity of the random learning law (71) stems from the probability structure that underlies the random optimality "error" process $\mathcal{E}$:

$$\mathcal{E} \quad = \quad \frac{S}{N} - \frac{\partial S/\partial \sigma}{\partial N/\partial \sigma} \tag{89}$$

near the optimum noise $\sigma = \sigma_{opt}$. The probability density of $\mathcal{E}$ depends on the statistics of the input noise, the differential equation that defines the dynamical system, and how we define the signal and noise terms S and N.

Below we test statistics of the random process $\mathcal{E}$ for the quartic bistable system in Figure 9. The results suggest that in some cases the density of $\mathcal{E}$ is Cauchy or otherwise belongs to the "impulsive" or thick-tailed family of symmetric alpha-stable bell curves with parameter α in the characteristic function $e^{-|\omega|^{\alpha}}$ [20, 75, 232, 233]. The parameter α lies in $0 < \alpha \leq 2$ and gives the Gaussian random variable when $\alpha = 2$ or $\phi(\omega) = e^{-\omega^2}$. It gives the thicker-tailed Cauchy bell curve when $\alpha = 1$ or $\phi(\omega) = e^{-|\omega|}$. The moments of stable distributions with $\alpha < 2$ are finite only up to the order k for $k < \alpha$. The Gaussian density alone has finite variance and higher moments. Alpha-stable random variables characterize the class of normalized sums that converge in distribution to a random variable [20] as in the famous Gaussian version of the central limit theorem. The noisiness or impulsiveness of the $\mathcal{E}$-based learning grows as α falls. Note also that the ratio X/Y is Cauchy if X and Y are jointly Gaussian [75, 146, 150, 212]. Our simulations found that the impulsiveness of $\mathcal{E}$ stemmed at least in part from the step size of the successive DFTs in (92).

We now derive the SR learning laws in terms of DFTs. We can approximate $\frac{\partial S}{\partial \sigma}$ and $\frac{\partial N}{\partial \sigma}$ with a ratio of time differences at each iteration n:

$$\frac{\partial S_n}{\partial \sigma_n} \approx \frac{\Delta S_n}{\Delta \sigma_n} = \frac{S_n - S_{n-1}}{\sigma_n - \sigma_{n-1}} \quad (90)$$

$$\frac{\partial N_n}{\partial \sigma_n} \approx \frac{\Delta N_n}{\Delta \sigma_n} = \frac{N_n - N_{n-1}}{\sigma_n - \sigma_{n-1}}. \quad (91)$$

The math model in (1)-(2) gives the exact learning laws. Recall that the L-point DFT [207] for a sequence of states $\{y_t\}$ has the form

$$Y_n[k] = \sum_{l=0}^{L-1} y_{l+(n+1-L)}\, e^{-i\frac{2\pi k}{L}l} \qquad k = 0, \ldots, L-1. \quad (92)$$

The time index n denotes the current time $t = nT_s$ for the sampling period T_s. Let $\frac{\partial S_n}{\partial y_j}$ denote the partial derivative of the signal energy S at iteration n with respect to the output y evaluated at time step j: $\frac{\partial S_n}{\partial y_j} = \frac{\partial S_n}{\partial y}[j]$. We likewise put $\frac{\partial N_n}{\partial y_j} = \frac{\partial N_n}{\partial y}[j]$ and $\frac{\partial y_j}{\partial \sigma} = \frac{\partial y}{\partial \sigma}[j]$. We assume some form of the chain rule holds to give

$$\frac{\partial S_n}{\partial \sigma} = \sum_{j=n+1-L}^{n} \frac{\partial S_n}{\partial y_j}\frac{\partial y_j}{\partial \sigma} \quad \text{and} \quad \frac{\partial N_n}{\partial \sigma} = \sum_{j=n+1-L}^{n} \frac{\partial N_n}{\partial y_j}\frac{\partial y_j}{\partial \sigma}. \quad (93)$$

We first derive $\frac{\partial S_n}{\partial y_j}$ and $\frac{\partial N_n}{\partial y_j}$ in (93). Consider the partial derivative of $|Y_n[k]|^2$ with respect to y at time step j:

$$\frac{\partial}{\partial y_j}\left|Y_n[k]\right|^2 = \frac{\partial}{\partial y_j} Y_n[k] Y_n^*[k] \quad (94)$$

$$= Y_n[k]\frac{\partial}{\partial y_j} Y_n^*[k] + Y_n^*[k]\frac{\partial}{\partial y_j} Y_n[k] \quad (95)$$

$$= Y_n[k]\, e^{i\frac{2\pi k}{L}(j-(n+1-L))} + Y_n^*[k]\, e^{-i\frac{2\pi k}{L}(j-(n+1-L))} \quad (96)$$

$$= 2\mathrm{Re}\{Y_n[k]\, e^{i\frac{2\pi k}{L}(j-(n+1-L))}\} \quad (97)$$

$$= 2\,\mathrm{Re}\{Y_n[k]\}\cos\left(\frac{2\pi k}{L}(j-(n+1-L))\right) - 2\,\mathrm{Im}\{Y_n[k]\}\sin\left(\frac{2\pi k}{L}(j-(n+1-L))\right) \quad (98)$$

So the partial derivative of the signal spectrum $S_n = 2\,|Y_n[k_0]|^2$ is

$$\frac{\partial S_n}{\partial y_j} = 4\,\mathrm{Re}\{Y_n[k_0]\}\cos\left(\frac{2\pi k_0}{L}(j-(n+1-L))\right) - 4\,\mathrm{Im}\{Y_n[k_0]\}\sin\left(\frac{2\pi k_0}{L}(j-(n+1-L))\right). \quad (99)$$

The partial derivative $\frac{\partial N_n}{\partial y_j}$ follows in like manner:

$$\frac{\partial N_n}{\partial y_j} = \frac{\partial}{\partial y_j}(P_n - S_n) \tag{100}$$

$$= \frac{\partial}{\partial y_j}\sum_{k=0}^{L-1}\left|Y_n[k]\right|^2 - \frac{\partial S_n}{\partial y_j} \tag{101}$$

$$= \frac{\partial}{\partial y_j} L \sum_{t=0}^{L-1} y_t^2 - \frac{\partial S_n}{\partial y_j} \quad \text{from Parseval's relation} \tag{102}$$

$$= 2\,L\,y_j - \frac{\partial S_n}{\partial y_j} \tag{103}$$

We can consider the term $\frac{\partial y_j}{\partial \sigma}$ in (93) as a sample of $\frac{\partial y}{\partial \sigma}$ at the time step j.

Recall the math model of the dynamical system (1)-(2) and let $G(x, u, t) = f(x) + u(x, t)$. Assume that $u(x, t) = s(t) + n(t) = s(t) + \sigma w(t)$ for the zero-mean white noise process $w(t)$ with unit variance $E[w^2] = 1$. So the model becomes

$$\dot{x} = G(x, s, \sigma, w) = f(x) + s(t) + \sigma w(t) \tag{104}$$

$$y(t) = g(x(t)). \tag{105}$$

The chain rule gives

$$\frac{\partial y}{\partial \sigma} = \frac{\partial g}{\partial x}\frac{\partial x}{\partial \sigma}. \tag{106}$$

Let $\eta(t)$ denote $\frac{\partial x}{\partial \sigma}$. Assume that G is sufficiently differentiable. Then differentiate η with respect to time [8] to get

$$\frac{d\eta}{dt} = \frac{d}{dt}\left(\frac{\partial x}{\partial \sigma}\right) = \frac{\partial \frac{dx}{dt}}{\partial \sigma} = \frac{\partial G(x, s, \sigma, w)}{\partial \sigma} \tag{107}$$

$$= \frac{\partial G}{\partial x}\frac{\partial x}{\partial \sigma} + \frac{\partial G}{\partial \sigma} = \frac{\partial G}{\partial x}\eta(t) + \frac{\partial G}{\partial \sigma}. \tag{108}$$

The last derivative $\frac{\partial G}{\partial \sigma}$ results from G's explicit dependence on σ. So the additive case $G(x, s, \sigma, w) = f(x) + s(t) + \sigma w(t)$ gives

$$\frac{\partial G}{\partial x} = \frac{\partial f}{\partial x} \tag{109}$$

$$\frac{\partial G}{\partial \sigma} = \frac{\partial}{\partial \sigma}[f(x) + s(t) + \sigma w(t)] = w(t). \tag{110}$$

We need to simulate the evolution (108) for $\frac{\partial x}{\partial \sigma}$ and obtain $\frac{\partial y}{\partial \sigma}$ from (106). Then we put (99), (103), and $\frac{\partial y}{\partial \sigma}$ into (93) to get the stochastic gradient learning law:

$$\begin{aligned}
\sigma(n+1) &= \sigma(n) + \mu_n \frac{\partial \,\mathrm{SNR}_n}{\partial \sigma} && (111)\\
&= \sigma(n) + \mu_n \Big(\frac{\partial \mathrm{SNR}_n}{\partial S_n}\frac{\partial S_n}{\partial \sigma} + \frac{\partial \mathrm{SNR}_n}{\partial N_n}\frac{\partial N_n}{\partial \sigma}\Big) && (112)\\
&= \sigma(n) + \mu_n \Big(\frac{1}{S_n}\sum_{l=n+1-L}^{n} \frac{\partial S_n}{\partial y_l}\frac{\partial y_l}{\partial \sigma} - \frac{1}{N_n}\sum_{l=n+1-L}^{n} \frac{\partial N_n}{\partial y_l}\frac{\partial y_l}{\partial \sigma}\Big). && (113)
\end{aligned}$$

Here we omit the constant factor $10 \log e$ from (75)-(78) or view it as part of the learning rate μ_n in (113). The learning law for the parameters m_j of a function approximator F that approximates the surface of optimal noise levels follows in like manner. Here F replaces the parameter σ so the learning law becomes

$$\begin{aligned}
m_j(n+1) &= m_j(n) + \mu_n \frac{\partial \,\mathrm{SNR}_n}{\partial m_j} && (114)\\
&= m_j(n) + \mu_n \Big(\frac{\partial \mathrm{SNR}_n}{\partial S_n}\frac{\partial S_n}{\partial m_j} + \frac{\partial \mathrm{SNR}_n}{\partial N_n}\frac{\partial N_n}{\partial m_j}\Big) && (115)\\
&= m_j(n) + \mu_n \Big(\frac{1}{S_n}\sum_{l=n+1-L}^{n} \frac{\partial S_n}{\partial y_l}\frac{\partial y_l}{\partial F}\frac{\partial F}{\partial m_j} - \frac{1}{N_n}\sum_{l=n+1-L}^{n} \frac{\partial N_n}{\partial y_l}\frac{\partial y_l}{\partial F}\frac{\partial F}{\partial m_j}\Big). && (116)
\end{aligned}$$

We get (113) if σ replaces F and m_j. Appendix B derives the last partial derivative $\frac{\partial F}{\partial m_j}$ in the chain-rule expansion (73) for all SAM fuzzy parameters m_j. This is again the step where users can insert other adaptive function approximators F and derive learning laws for their parameters m_j by expanding $\frac{\partial F}{\partial m_j}$. Formal stochastic approximation [228] further requires that the learning rate μ_n must decrease slowly but not too slowly:

$$\sum_{n=1}^{\infty} \mu_n^2 < \infty \quad \text{and} \quad \sum_{n=1}^{\infty} \mu_n = \infty. \qquad (117)$$

Linear decay terms $\mu_n = 1/n$ obey (117). We used small but constant learning rates in most simulations.

6 SR Learning: Simulation Results

This section shows how the stochastic SR learning laws in Section 5 tend to find the optimal noise levels in many dynamical systems. The learning process updates the noise parameter σ_n at each iteration n. The learning process is noisy and may not be stable due to the impulsiveness of the random gradient $\frac{\partial \text{SNR}_n}{\partial \sigma_n}$. We used a Cauchy noise suppressor from the theory of robust statistics [119] to stabilize the learning process. Then sample paths of σ_n converged and wander about the optimal values if the initial values were close to the optimum.

The response of a system depends on its dynamics and on the nature of its input signals. We applied the SNR measure to the quartic bistable and other dynamical systems with sinusoidal inputs. Future research may extend SR learning to wideband input signals. Figure 24a shows how the optimum noise level varies for each input sinewave in the quartic bistable system. The learning process samples the system's input-output response as it learns the optimum noise. It does not make direct use of the equation that underlies the system.

An adaptive fuzzy system can encode this pattern of optimum noise in its "if-then" rules when gradient learning tunes its parameters. The fuzzy system learns this optimum noise level as it varies the output of a random noise generator. More complex fuzzy systems can themselves act as adaptive random number generators [145, 208].

Consider the forced dynamical system in (1)-(2) with initial condition $x(0)$. We set up a discrete computer simulation with the stochastic version of Euler's method (the Euler-Maruyama scheme) [55, 91, 122]:

$$x_{t+1} = x_t + \Delta T \left(f(x_t) + s_t \right) + \sigma \sqrt{\Delta T} w_t \tag{118}$$

$$y_t = g(x_t) \tag{119}$$

with initial condition $x_0 = x(0)$. Here the zero-mean white noise sequence $\{w_t\}$ has unit variance $\sigma_w^2 = 1$. The term $\sqrt{\Delta T}$ scales w_t so that $\sqrt{\Delta T} w_t$ conforms with the Wiener increment [91, 122, 192]. The learning process itself does not use the system model in any calculation. It needs access only to the system's input-output responses. The learning process's sampling period T_s differs from the time step ΔT of the dynamical system's simulator in (118)-(119). The subsampling rate for the quartic bistable system is 1:50. We ignored all aliasing effects.

6.1 SR Test Case: The Quartic Bistable System

We tested the quartic bistable system (4) in detail because of its wide use in the SR literature as a benchmark SR dynamical system. The quartic bistable system for $a = b = 1$ with binary output has the form [193]

$$\dot{x} = x - x^3 + s(t) + n(t) \tag{120}$$

$$y(t) = \text{sgn}(x(t)) \tag{121}$$

or $y(t) = x(t)$ in the linear-output case. The sinewave input forcing term is $s(t) = \varepsilon \sin \omega_0 t$. The term $n(t) = \sigma w(t)$ is a zero-mean additive white Gaussian noise with variance σ_n^2 and where $E[w] = 0$ and $E[w^2] = 1$. The discrete version has the form (118)-(119):

$$x_{t+1} = x_t + \Delta T \left(x_t - x_t^3 + \varepsilon \sin 2\pi f_0 \Delta T t \right) + \sigma \sqrt{\Delta T} w_t \tag{122}$$

$$y_t = \text{sgn}(x_t) \quad \text{or} \quad y_t = x_t \tag{123}$$

with initial condition x_0. The time step is $\Delta T = 0.0195$. The sampling period is $T_s = 0.976$ with 1:50 subsampling.

We can freely choose the time length between the iteration step n and the step $n+1$. Longer time lengths can better show how the noise intensity σ at iteration n affects S_n, N_n, and SNR_n. We chose the time length $T_{n+1} - T_n = 2000$ seconds for the simulations of the quartic bistable system. The sampling period was $T_s = 0.976$ seconds. This yields 2048 samples per iteration. This long period of time allows for low-frequency signals such as $f_0 = 0.001$ Hz.

The simulations use Gaussian noise, Laplace noise, uniform noise, and impulsive alpha-stable noise. We also tested the quartic bistable system with the chaotic noise from the logistic map. Figures 2, 5, and 18 show the output SNR for input signal $s(t) = 0.1 \sin 2\pi(0.01)t$ for Gaussian noise, Laplace noise, uniform noise, and chaotic noise from the logistic map.

The Jacobian of the quartic bistable system has the form

$$\frac{\partial G}{\partial x} = \frac{\partial}{\partial x}[x - x^3 + s(t) + \sigma w(t)] \tag{124}$$

$$= 1 - 3x^2. \tag{125}$$

Then the partial derivative $\frac{\partial G}{\partial \sigma} = w(t)$ from (110) gives the evolution of $\eta(t) = \frac{\partial x}{\partial \sigma}$ for the quartic bistable system

$$\dot{\eta} \;=\; (1 - 3x^2)\,\eta(t) + w(t). \tag{126}$$

Its discrete version has the form

$$\eta(t+1) \;=\; \eta(t) + \Delta T\,(1 - 3x_t^2)\,\eta(t) + \sqrt{\Delta T} w_t. \tag{127}$$

We used the initial condition $\frac{\partial x_0}{\partial \sigma} = 0$ in simulations. Then we get $\frac{\partial y}{\partial \sigma}$ from (106) for use in the learning law (113). The linear output $y = g(x) = x$ has $\frac{\partial g}{\partial x} = 1$. We can approximate a binary output as $g(x) = \text{sgn}(x) \approx \tanh(cx)$ for a large positive $c >> 0$. Then $\frac{\partial g}{\partial x} = c(1 - \tanh^2(cx))$.

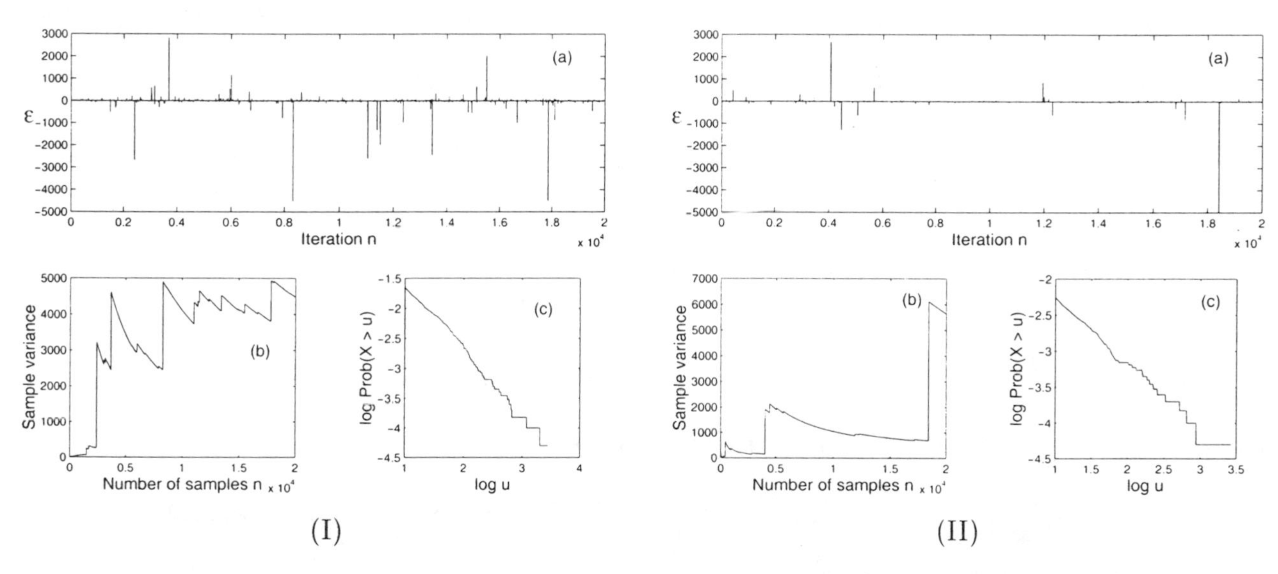

Figure 9: Visual display of samples from the equilibrium term $\mathcal{E}_n = \frac{S_n}{N_n} - \left(\frac{\partial S_n}{\partial \sigma} \Big/ \frac{\partial N_n}{\partial \sigma}\right)$. (a) Cauchy-like impulsive samples of $\mathcal{E}_n$ at each iteration n for the discretized version of the quartic bistable system $\dot{x} = x - x^3 + \varepsilon \sin 2\pi f_0 t + n(t)$ where $\varepsilon = 0.1$ and $f_0 = 0.01$ Hz. The system outputs are (I) $y_t = x_t$ and (II) $y_t = \text{sgn}(x_t)$. The noise intensity is the constant $\sigma_n^2 = 0.25$ that lies near the optimal level. (b) Converging variance test as a test for infinite variance. The sequence of sample variances will converge to a finite value if the underlying probability density has finite variance and diverges if it has infinite variance. (c) Log-tail test of the parameter α in an alpha-stable probability density. The test plots $\log \text{Prob}(X > u)$ versus $\log u$ for large u. If the density is alpha-stable with $\alpha < 2$ then the slope of this plot is approximately $-\alpha$. The test found $\alpha \approx 1$. So the probability density of $\mathcal{E}_n$ was approximately Cauchy.

The equilibrium term $\mathcal{E}_n$ in (89) helps gauge the noisiness of the learning process. We compute

$\mathcal{E}_n$ at each iteration n from

$$\mathcal{E}_n = \frac{S_n}{N_n} - \frac{\partial S_n}{\partial \sigma} \Big/ \frac{\partial N_n}{\partial \sigma}. \tag{128}$$

The statistics of $\mathcal{E}_n$ change with the noise level σ^2 and with the sinewave values ε and f_0. The empirical histogram of $\mathcal{E}_n$ is a bell curve. A key question is how thick are its tails. Figure 9 shows $\mathcal{E}_n$ samples from the quartic bistable system (122)-(123) with Gaussian noise $n(t) = \sigma w(t)$. The convergence of variance test [232] confirms that $\mathcal{E}_n$ had infinite variance in our simulations. The log-tail test [232] of parameter α in the family of alpha-stable probability densities leads to the estimate $\alpha \approx 1.0$. So the $\mathcal{E}_n$ density is approximately Cauchy. Recall also that $Z = X/Y$ is a Cauchy random variable if X and Y are Gaussian [75, 212] or if they obey certain more general statistical conditions [146, 150]. This suggests that much of the impulsive nature of $\mathcal{E}_n$ and hence of the learning process may stem from the ratio of derivatives in (128).

We also simulate the random gradient $\frac{\partial \text{SNR}_n}{\partial \sigma}$ with the partial derivatives from (99), (103), and $\frac{\partial y}{\partial \sigma}$ from (108):

$$\frac{\partial \text{SNR}_n}{\partial \sigma} = \frac{1}{S_n} \sum_{l=n+1-L}^{n} \frac{\partial S_n}{\partial y_l} \frac{\partial y_l}{\partial \sigma} - \frac{1}{N_n} \sum_{l=n+1-L}^{n} \frac{\partial N_n}{\partial y_l} \frac{\partial y_l}{\partial \sigma}. \tag{129}$$

The simulations confirm that the random gradient $\frac{\partial \text{SNR}_n}{\partial \sigma}$ is often impulsive and can destabilize the learning process (113) at or near the optimal noise level. The impulsiveness of $\frac{\partial \text{SNR}_n}{\partial \sigma}$ in Figure 10 suggests that $\frac{\partial \text{SNR}_n}{\partial \sigma}$ may have an alpha-stable probability density function with parameter $\alpha < 2$. A log-tail test found that $\alpha \approx 1$. So $\frac{\partial \text{SNR}_n}{\partial \sigma}$ again has an approximate Cauchy distribution.

We tested the learning law (113)

$$\sigma(n+1) = \sigma(n) + \mu_n \left(\frac{1}{S_n} \sum_{l=n+1-L}^{n} \frac{\partial S_n}{\partial y_l} \frac{\partial y_l}{\partial \sigma} - \frac{1}{N_n} \sum_{l=n+1-L}^{n} \frac{\partial N_n}{\partial y_l} \frac{\partial y_l}{\partial \sigma} \right). \tag{130}$$

Figure 11 shows the simulation results. It displays instability in the learning due to the impulsiveness of the random gradient $\frac{\partial \text{SNR}}{\partial \sigma}$.

The theory of robust statistics [119] suggests one way to reduce the impulsiveness of $\frac{\partial \text{SNR}_n}{\partial \sigma}$. We can replace the noisy random sample z_n with a Cauchy-like noise suppressor $\phi(z_n)$ [119]:

$$\phi(z_n) = \frac{2z_n}{1 + z_n^2}. \tag{131}$$

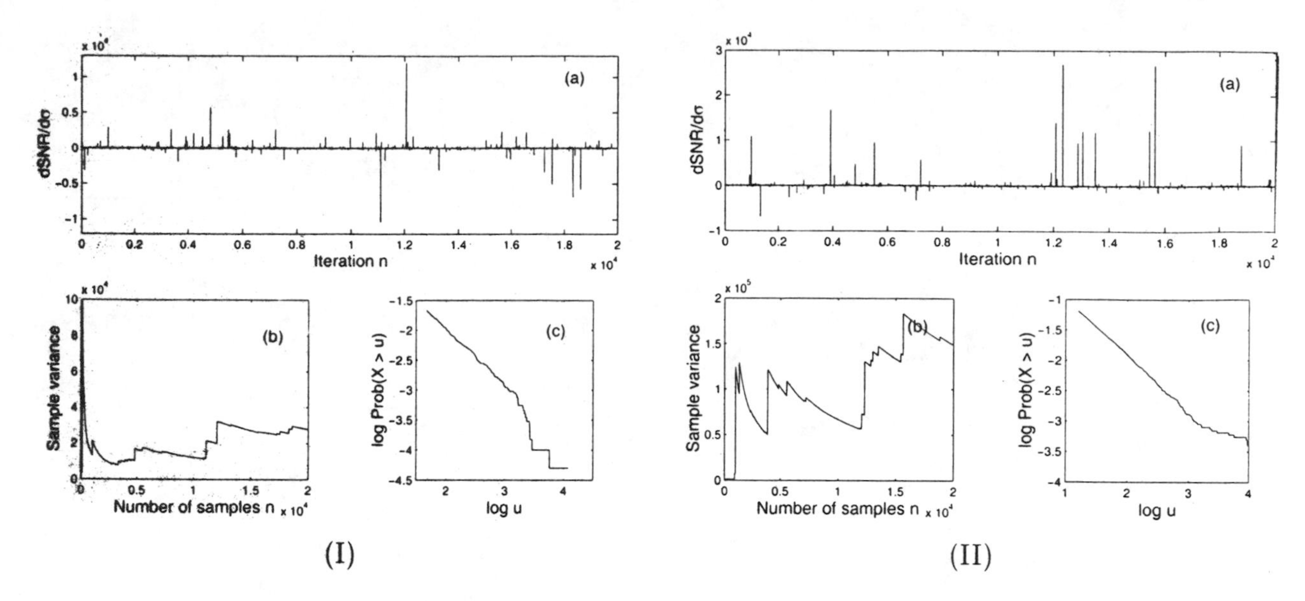

Figure 10: Visual display of $\frac{\partial \mathrm{SNR}_n}{\partial \sigma} = \frac{1}{S_n}\frac{\partial S_n}{\partial \sigma} - \frac{1}{N_n}\frac{\partial N_n}{\partial \sigma}$ for the quartic bistable system $\dot{x} = x - x^3 + s(t) + n(t)$ where $s(t) = \varepsilon \sin 2\pi f$ with $\varepsilon = 0.1$ and $f = 0.01$ Hz. The system has linear output $y(t) = x(t)$ in (I) and binary output $y(t) = \mathrm{sgn}(x(t))$ in (II). The noise variances are the constants $\sigma_n^2 = 0.25$. (a) Cauchy-like samples of $\frac{\partial \mathrm{SNR}_n}{\partial \sigma}$ at each iteration n. (b) Converging variance test as test of infinite variance. The sequence of sample variances converges to a finite value if the underlying probability density has finite variance; else it has infinite variance. (c) Log-tail test of the parameter α for an alpha-stable bell curve. The test looks at the plot of log Prob$(X > u)$ versus $\log u$ for large u. If the underlying density is alpha-stable with $\alpha < 2$ then the slope of this plot is approximately $-\alpha$. This test found that $\alpha \approx 1$ and so the density was approximately Cauchy.

So $\phi(\frac{\partial \mathrm{SNR}_n}{\partial \sigma})$ replaces the noise gradient $\frac{\partial \mathrm{SNR}_n}{\partial \sigma}$ in (129). This gives the robust SR learning law

$$\sigma(n+1) = \sigma(n) + \mu_n \, \phi\Big(\frac{\partial \mathrm{SNR}_n}{\partial \sigma}\Big). \tag{132}$$

Figure 12 shows the results of the SR learning law (132) with the gradient in (129). The σ_n learning paths in (113) converge near the optimal noise level.

The above learning law requires a complete knowledge of the math model that describes the dynamical system. It also needs accurate estimation of the evolution of (108). This may not be practical in many cases. So we instead sample S_n and N_n and use the approximation formulas (90) and (91). This gives the learning law

$$\sigma_{n+1} = \sigma_n + \mu_n \frac{\partial \mathrm{SNR}_n}{\partial \sigma} \tag{133}$$

$$= \sigma_n + \mu_n \Big(\frac{1}{S_n}\frac{\partial S_n}{\partial \sigma} - \frac{1}{N_n}\frac{\partial N_n}{\partial \sigma}\Big) \tag{134}$$

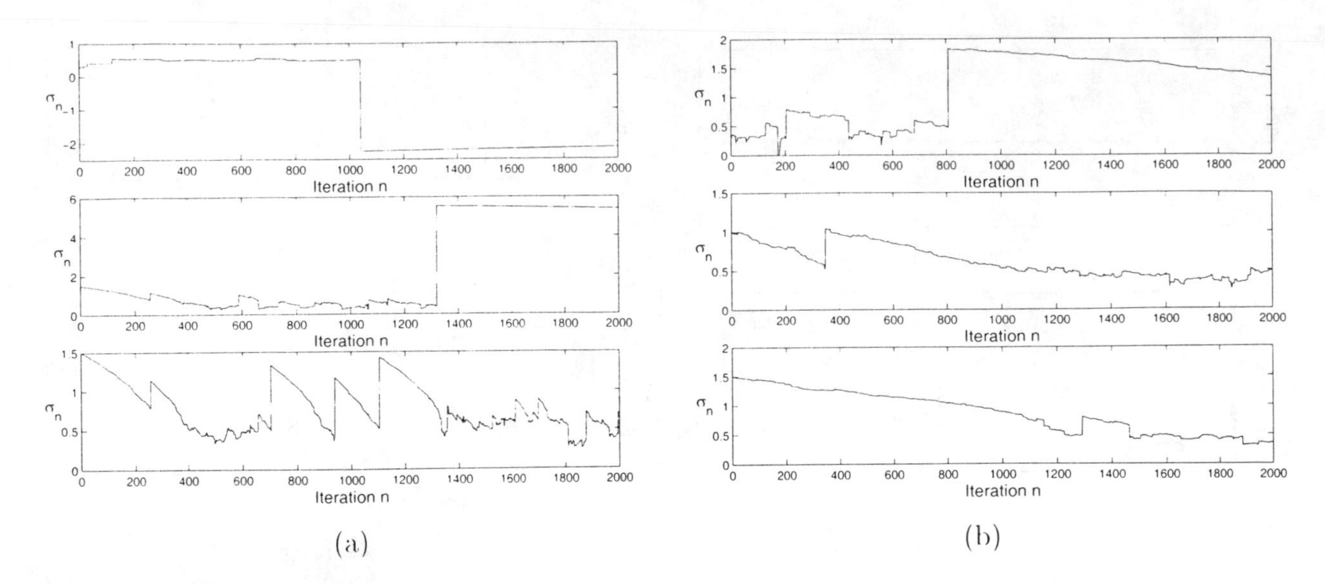

Figure 11: Learning paths for the quartic bistable system with output $y(t) = x(t)$ in (a) and $y(t) = \text{sgn}(x(t))$ in (b). The learning law takes the form (113). The optimal noise level is $\sigma \approx 0.5$ for both cases. The impulsiveness of the learning term $\frac{\partial \text{SNR}}{\partial \sigma}$ destabilizes the learning process near the optimal noise level.

$$\approx \quad \sigma_n + \mu_n \Big(\frac{1}{S_n} \frac{S_n - S_{n-1}}{\sigma_n - \sigma_{n-1}} - \frac{1}{N_n} \frac{N_n - N_{n-1}}{\sigma_n - \sigma_{n-1}} \Big). \tag{135}$$

We also replace the difference $\sigma_n - \sigma_{n-1}$ with its sign $\text{sgn}(\sigma_n - \sigma_{n-1})$ to avoid numerical instability. The gradient becomes

$$\frac{\partial \text{SNR}_n}{\partial \sigma} \quad \approx \quad \Big(\frac{S_n - S_{n-1}}{S_n} - \frac{N_n - N_{n-1}}{N_n} \Big) \text{sgn}(\sigma_n - \sigma_{n-1}). \tag{136}$$

This approximation gives the SR learning law

$$\sigma_{n+1} \quad = \quad \sigma_n + \mu_n \Big(\frac{S_n - S_{n-1}}{S_n} - \frac{N_n - N_{n-1}}{N_n} \Big) \text{sgn}(\sigma_n - \sigma_{n-1}). \tag{137}$$

This learning law does not require that we know the dynamical model. It depends only on samples from the system dynamics and from the input signal $s(t)$.

Figure 13a shows sample learning paths of σ_n for the quartic bistable system and approximation (136). Figure 13b shows the noise-SNR profile of the dynamical system. The σ_n learning paths converge to the optimum noise values only in some cases. The chance of path convergence is higher for larger sinewave amplitudes. The paths do not converge as often for small amplitudes. The simulations confirm that the random gradient $\frac{\partial \text{SNR}_n}{\partial \sigma_n}$ in (136) is often impulsive and can destabilize

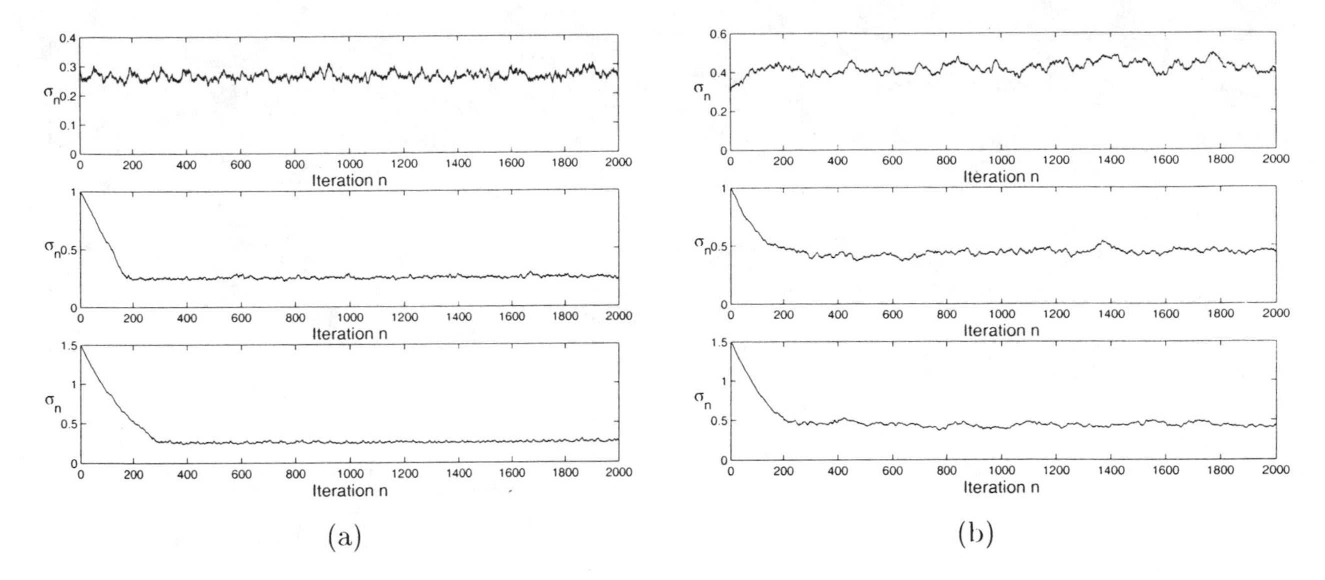

Figure 12: Learning paths for the quartic bistable system with output $y(t) = x(t)$. The learning law has the form (132). Optimal noise levels are (a) $\sigma \approx 0.35$ and (b) $\sigma \approx 0.5$. The learning paths converge close to the optimal levels.

the learning process (137) as in Figure 13. The impulsiveness of $\frac{\partial \mathrm{SNR}_n}{\partial \sigma}$ in Figure 14 suggests that $\frac{\partial \mathrm{SNR}_n}{\partial \sigma}$ may have an alpha-stable probability density function with parameter $\alpha < 2$. A log-tail test found that $\alpha \approx 1$. So $\frac{\partial \mathrm{SNR}_n}{\partial \sigma}$ in (136) also has an approximate Cauchy distribution.

We again apply the Cauchy-like noise suppressor from robust statistics [119] to reduce the impulsiveness of the approximated term $\frac{\partial \mathrm{SNR}_n}{\partial \sigma}$ in (136). So $\phi(\frac{\partial \mathrm{SNR}_n}{\partial \sigma})$ replaces the approximation of the noise gradient $\frac{\partial \mathrm{SNR}_n}{\partial \sigma}$ in (136) to give the robust SR learning law

$$\sigma_{n+1} = \sigma_n + \mu_n \phi\Bigg(\Big(\frac{S_n - S_{n-1}}{S_n} - \frac{N_n - N_{n-1}}{N_n}\Big) \operatorname{sgn}(\sigma_n - \sigma_{n-1})\Bigg). \tag{138}$$

Figure 15 shows the results of the SR learning law (138). The σ_n learning paths converge to the optimum noise level if the initial value lies close enough to it. Then σ_n wanders in a small Brownian-like motion about the optimum noise level.

Like results hold for other noise densities with finite variance such as Laplace and uniform noise. Figure 16 shows σ_n learning paths for the quartic bistable system (122)-(123) with Laplace noise and uniform noise. We also tested the quartic bistable system with alpha-stable noise. The very large noise impulses can force the state x of the quartic bistable system to a large value that can cause numerical instability in a discrete computer simulation. So we let the state x of the quartic

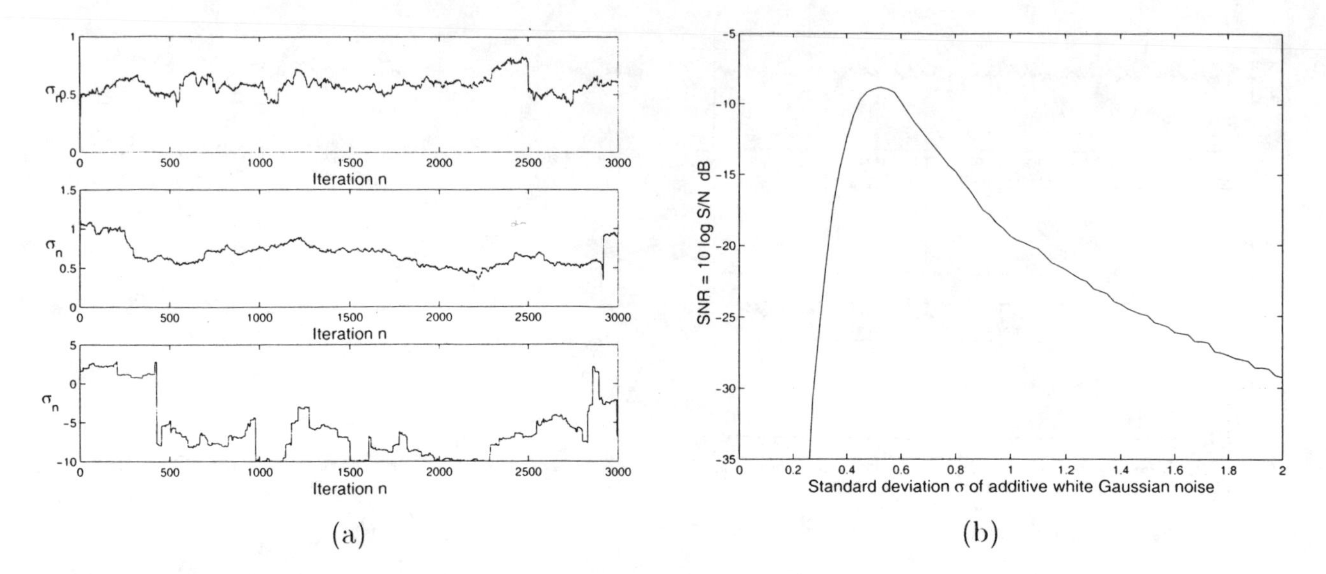

Figure 13: Impulsive effects on learning paths of noise intensity σ_n. The quartic bistable system has the form $\dot{x} = x - x^3 + s(t) + n(t)$ with binary output $y(t) = \text{sgn}(x(t))$ and initial condition $x(0) = -1$. The input sinusoid signal function is $s(t) = 0.1 \sin 2\pi(0.01)t$. (a) The sequence σ_n with different initial values that differ from the optimum noise intensity. (b) Noise-SNR profile of the quartic bistable system. The graph shows that the optimum noise intensity lies near $\sigma = 0.5$. The paths of σ_n do not converge to the optimum noise. This stems from the impulsiveness of the derivative term $\frac{\partial \text{SNR}_n}{\partial \sigma}$ in the approximate SR learning law (137).

bistable saturate at $|x| = 10$: $x = \text{sgn}(x) \min(|x|, 10)$. Figure 17 shows the paths of the optimal noise scale κ_n for $\alpha = 1.9$, 1.8, and 1. The learning degrades as α falls and the alpha-stable bell curves have thicker tails.

We also used a chaotic time series as the forcing noise n_t in the quartic bistable dynamical system [126]. The simple and popular logistic map created the noise sequence $\{z_t\}$:

$$z_{t+1} \;=\; 4\, z_t\, (1 - z_t) \tag{139}$$

from the initial value $z_0 = .123456789$ [126]. The positive sequence $\{z_t\}$ stays bounded within the unit interval: $z_t \in (0, 1)$. The chaotic noise n_t comes from

$$n_t \;=\; A\, (z_t - \frac{1}{2}). \tag{140}$$

The factor $A > 0$ acts as the scaled power or standard deviation if the term $(z_t - \frac{1}{2})$ is a zero-mean random variable with unit variance. Learning tunes A so that the dynamical system shows the SR

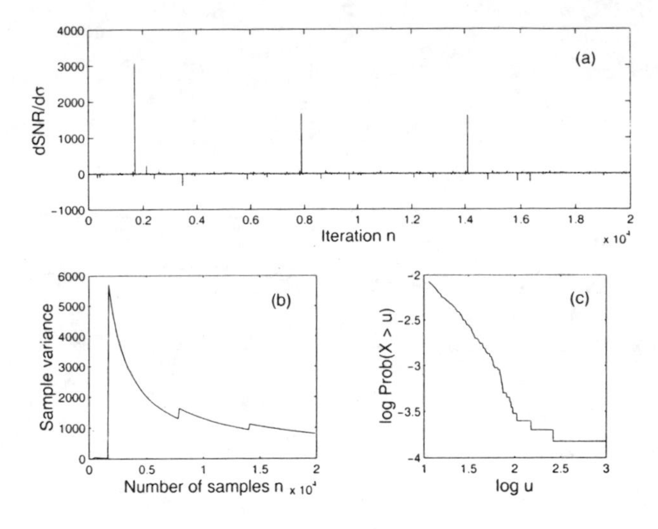

Figure 14: Visual display of sample statistics of approximated $\frac{\partial \mathrm{SNR}_n}{\partial \sigma}$. (a) Cauchy-like samples of $\frac{\partial \mathrm{SNR}_n}{\partial \sigma}$ at each iteration n for quartic bistable system with sinusoidal input of amplitude $\varepsilon = 0.1$ and frequency $f_0 = 0.01$ Hz. We compute $\frac{\partial \mathrm{SNR}_n}{\partial \sigma}$ at each iteration from $\frac{\partial \mathrm{SNR}_n}{\partial \sigma} \approx (\frac{S_n - S_{n-1}}{S_n} - \frac{N_n - N_{n-1}}{N_n})\mathrm{sgn}(\sigma_n - \sigma_{n-1})$ in (136). We vary the noise level σ_n between $\sigma_n = 0.50$ and $\sigma_n = 0.51$ so that $\mathrm{sgn}(\sigma_n - \sigma_{n-1})$ changes values between 1 and -1. The plot shows impulsiveness of the random variable $\frac{\partial \mathrm{SNR}_n}{\partial \sigma}$. (b) Converging variance test as test of infinite variance. The sequence of sample variances converges to a finite value if the underlying probability density has finite variance. Else it has infinite variance. (c) Log-tail test of the parameter α in for an alpha-stable bell curve. The test looks at the plot of log Prob$(X > u)$ versus $\log u$ for large u. If the underlying density is alpha-stable with $\alpha < 2$ then the slope of this plot is approximately $-\alpha$. This test found that $\alpha \approx 1$ and so the density was approximately Cauchy. The result is that we need to apply the Cauchy noise suppressor (131) to the approximate SR gradient $\frac{\partial \mathrm{SNR}_n}{\partial \sigma}$ in (136) as well as to the exact SR gradient in (129).

effect. Figure 18 shows a sample chaotic noise sequence and shows two A learning paths on their way to stochastic convergence.

6.2 Other SR Test Cases

The SR learning schemes also work for other SR models. We here show only the results for zero-mean white Gaussian noise. We first tested the discrete-time threshold neuron model

$$y_t = \begin{cases} 1 & \text{if } s_t + n_t \geq \Theta \\ -1 & \text{if } s_t + n_t < \Theta \end{cases} \tag{141}$$

for $t = 0, 1, 2, \ldots$ The threshold Θ sets the output of the neuron. The input sinewave has the form $s_t = \varepsilon \sin 2\pi f_0 \Delta T t$. The Gaussian noise n_t has variance σ_n^2. The threshold system is not a

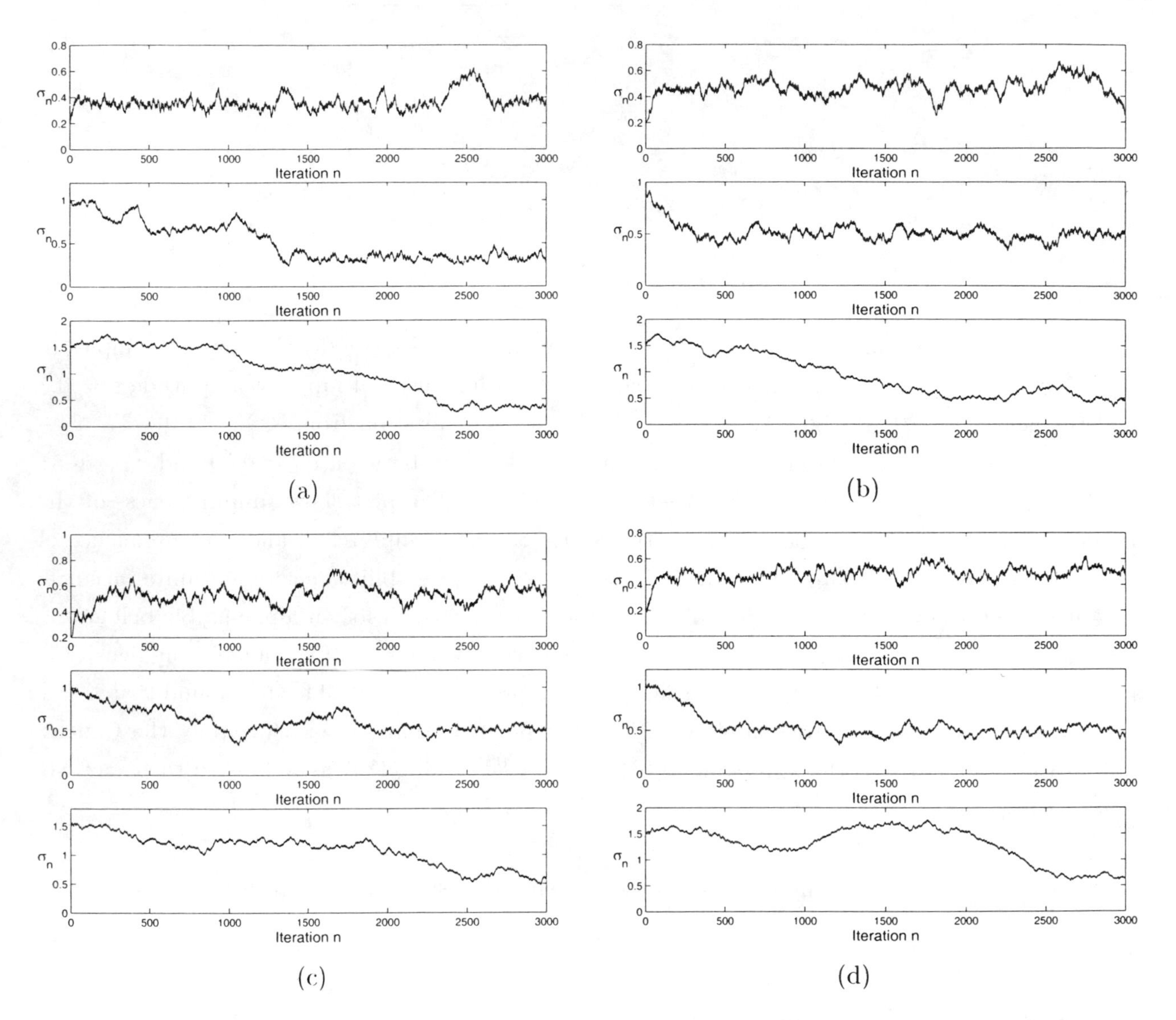

Figure 15: Learning paths of σ_n with the Cauchy noise suppressor $\phi(z) = 2z/(1+z^2)$ for the quartic bistable system with binary threshold output $y_t = \text{sgn}(x_t)$. The term $\phi(\frac{\partial \text{SNR}_n}{\partial \sigma})$ replaces $\frac{\partial \text{SNR}_n}{\partial \sigma}$ in the SR learning law (133). The paths of σ_n wander in a Brownian-like motion around the optimum noise. The suppressor function ϕ makes the learning algorithm more robust against impulsive shocks. The input signals are (a) $s(t) = 0.1 \sin 2\pi(0.001)t$, (b) $s(t) = 0.1 \sin 2\pi(0.005)t$, (c) $s(t) = 0.1 \sin 2\pi(0.01)t$, and (d) $s(t) = 0.2 \sin 2\pi(0.01)t$.

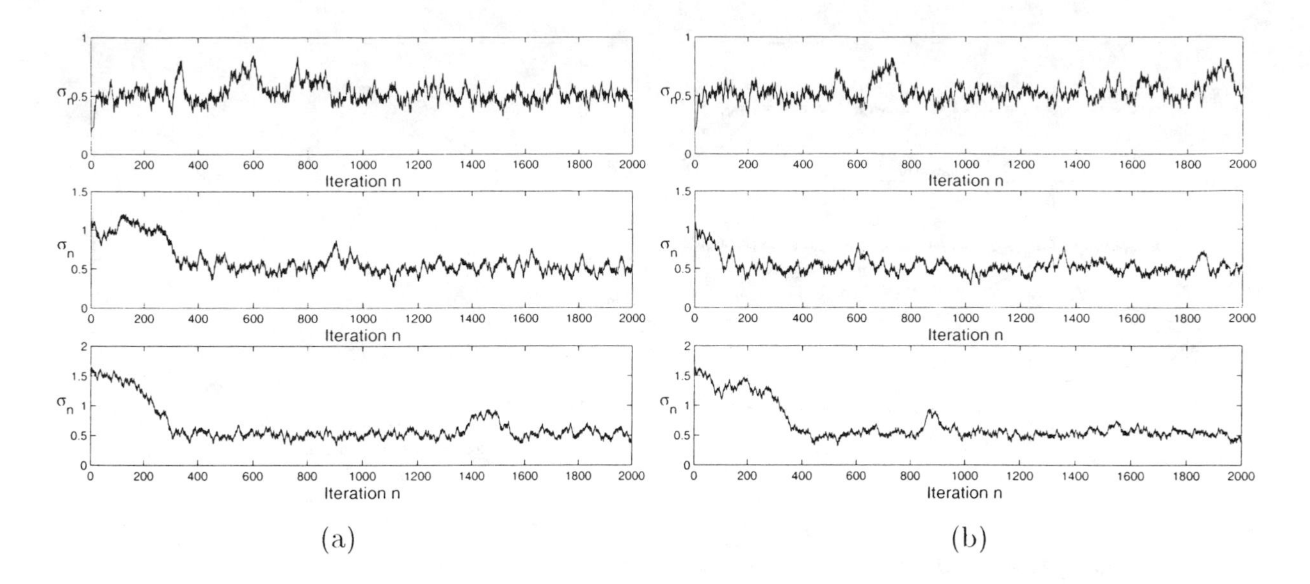

Figure 16: Learning paths of σ_n for other noise densities in the quartic bistable system with binary output $y_t = \text{sgn}(x_t)$. The input signal is $s(t) = 0.1 \sin 2\pi(0.01)t$. The optimal noise lies near $\sigma = 0.5$ for both cases of (a) Laplace noise and (b) uniform noise.

dynamical system but it does show SR. Figure 19 shows the result of learning when $f_0 = 0.001$, $\varepsilon = 0.1$, and $\Theta = 0.5$ and when $f_0 = 0.001$, $\varepsilon = 0.5$, and $\Theta = 1$. The sampling period is $T_s = \Delta T = 1$.

We next tested the bistable potential neuron model with Gaussian white noise [29]

$$\dot{x} = -x + 2\tanh x + s(t) + n(t) \tag{142}$$

$$y(t) = \text{sgn}(x(t)). \tag{143}$$

We ignored the multiplicative noise in (8). Figure 20 shows the SR learning paths of σ_n. The sinewave input is $s(t) = \varepsilon \sin 2\pi f_0 t$ where $f_0 = 0.01$ Hz and $\varepsilon = 0.1$ and the $\varepsilon = 0.3$. The time step in the discrete simulation is $\Delta T = 0.0195$. The sampling period is $T_s = 0.975$ or 50 times the time step ΔT.

We next tested the forced FitzHugh-Nagumo neuron model [191]. We rewrote (14)-(15) with $a = 0.5$ and with the changes of variables $x \to x + 0.5$, $w \to w - b + 0.5$, and $A \to A - b + 0.5$ [48]:

$$\epsilon\dot{x} = -x(x^2 - \frac{1}{4}) - w + A + s(t) + n(t) \tag{144}$$

$$\dot{w} = x - w \tag{145}$$

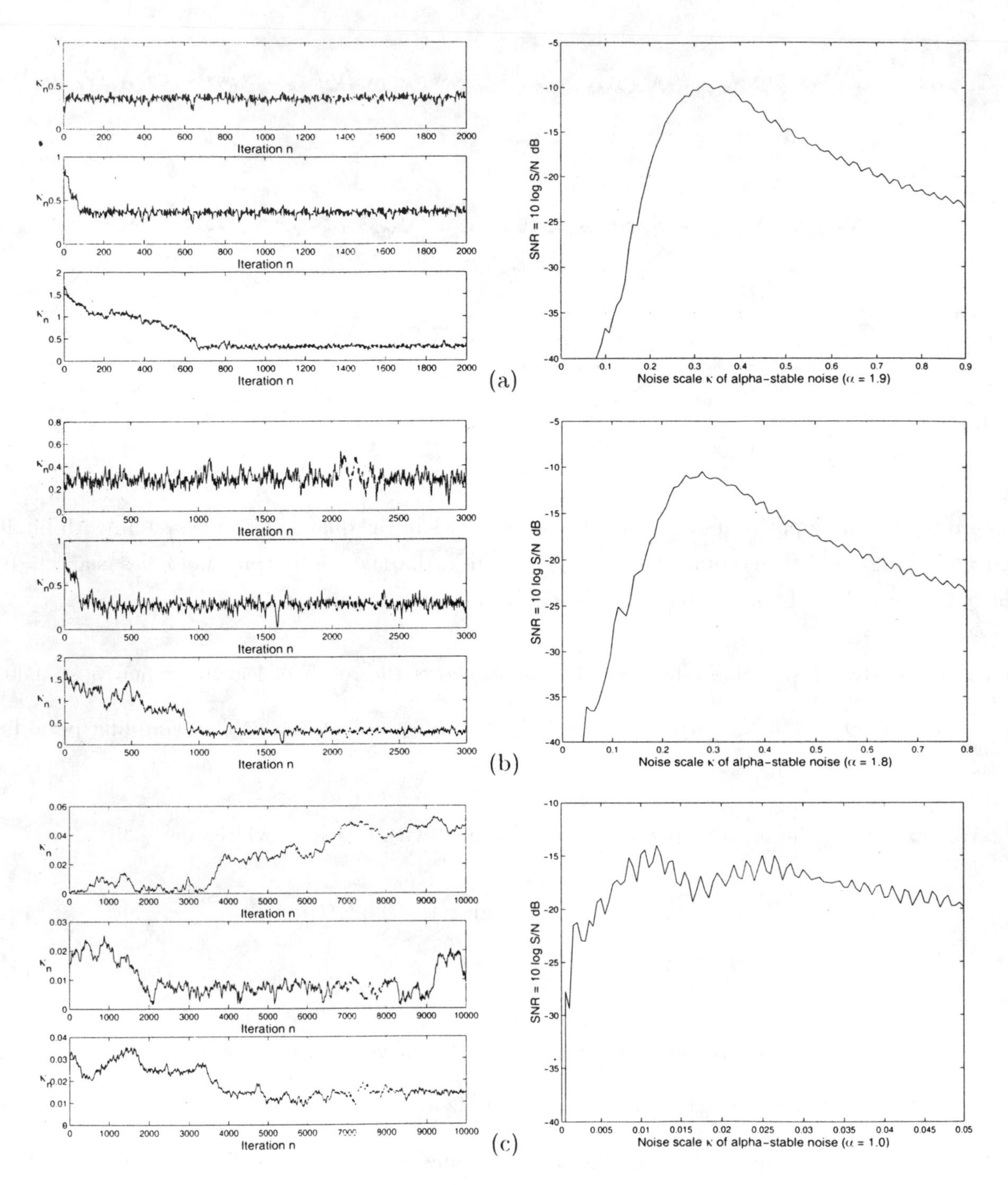

Figure 17: Learning paths of κ_n for alpha-stable noise in the quartic bistable system with binary output $y_t = \text{sgn}(x_t)$. The input signal is $s(t) = 0.1 \sin 2\pi(0.01)t$. (a) $\alpha = 1.9$. (b) $\alpha = 1.8$. (c) $\alpha = 1$. The noise scale κ acts like a standard deviation and controls the width of the alpha-stable bell curve through the dispersion $\gamma = \kappa^\alpha$. Learning becomes more difficult as α falls and the bell curves have thicker tails. The impulsiveness is so severe in the Cauchy case (c) that κ_n often fails to converge. Note the noisy multimodal nature of the SNR profiles.

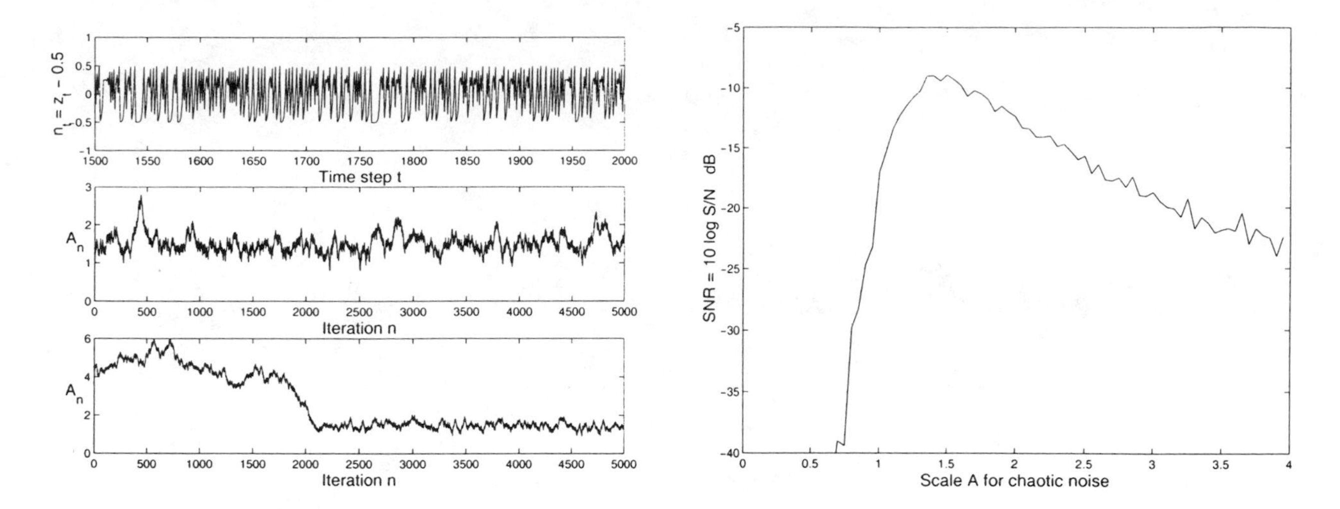

Figure 18: Learning paths of the scaling factor A_n in chaotic noise $n_t = A_n(z_t - \frac{1}{2})$ from the logistic dynamical system $z_{t+1} = 4z_t(1 - z_t)$. The dynamical system is the quartic bistable system with binary output $y_t = \text{sgn}(x_t)$. The input signal is $s(t) = \varepsilon \sin 2\pi f_0 t$ where $f_0 = 0.01$ Hz and $\varepsilon = 0.1$. The top figure shows a sample noise path n_t from the chaotic logistic map when $A_n = 1$.

$$y(t) \quad = \quad x(t). \tag{146}$$

The constants are $\epsilon = 0.005$, $a = 0.5$, and $A = -(5/12\sqrt{3} + 0.07) = -.31056$ as in [108]. The sinewave input is $s(t) = \varepsilon \sin 2\pi f_0 t$ with $\varepsilon = 0.01$, $f_0 = 0.1$ and 0.5 Hertz. The sampling period is $T_s = 0.01$ with $\Delta T = 0.001$. Figure 21 shows the learning paths of the standard deviation σ_n of the Gaussian white noise n.

We also showed SR learning in the forced Duffing oscillator with Gaussian white noise n [204]:

$$\ddot{x} \quad = \quad -0.15\dot{x} + x + x^3 + \varepsilon \sin(\omega_0 t) + n(t) \tag{147}$$

$$y(t) \quad = \quad x(t). \tag{148}$$

Figure 22 shows the learning paths of σ_n for input sinewave s with frequency $f_0 = 0.01$ Hz and with amplitudes $\varepsilon = 0.1$ and $\varepsilon = 0.3$. The sampling period is $T_s = 0.02$ with $\Delta T = 0.005$.

6.3 Fuzzy SR Learning: The Quartic Bistable System

We used a fuzzy function approximator $F : R^n \to R$ to learn and store the entire surface of optimal noise values for the quartic bistable system with input sinewaves. The fuzzy system had as its input the 2-D vector of sinewave amplitude ε and frequency f_0. We tested the system with the

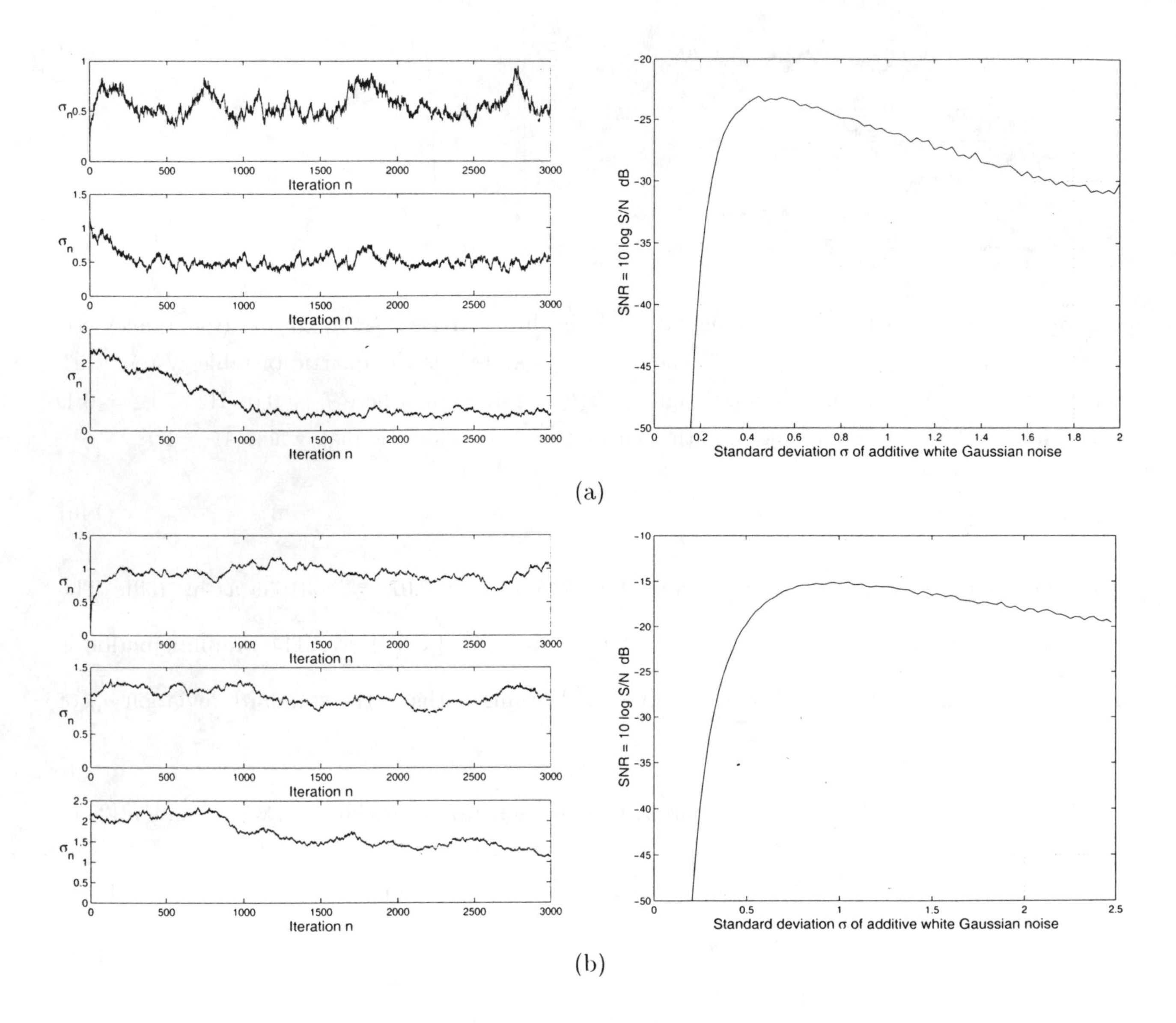

Figure 19: SR learning paths of σ_n for the threshold system $y_t = \text{sgn}(s_t + n_t - \Theta)$ where $\text{sgn}(x) = 1$ if $x \geq 0$ and $\text{sgn}(x) = -1$ if $x < 0$. The input sinewave is $s_t = \varepsilon \sin 2\pi f_0 t$ with additive white Gaussian noise sequence n_t. The parameters are (a) $f_0 = 0.001$, $\varepsilon = 0.1$, and $\Theta = 0.5$ and (b) $f_0 = 0.001$, $\varepsilon = 0.5$, and $\Theta = 1$.

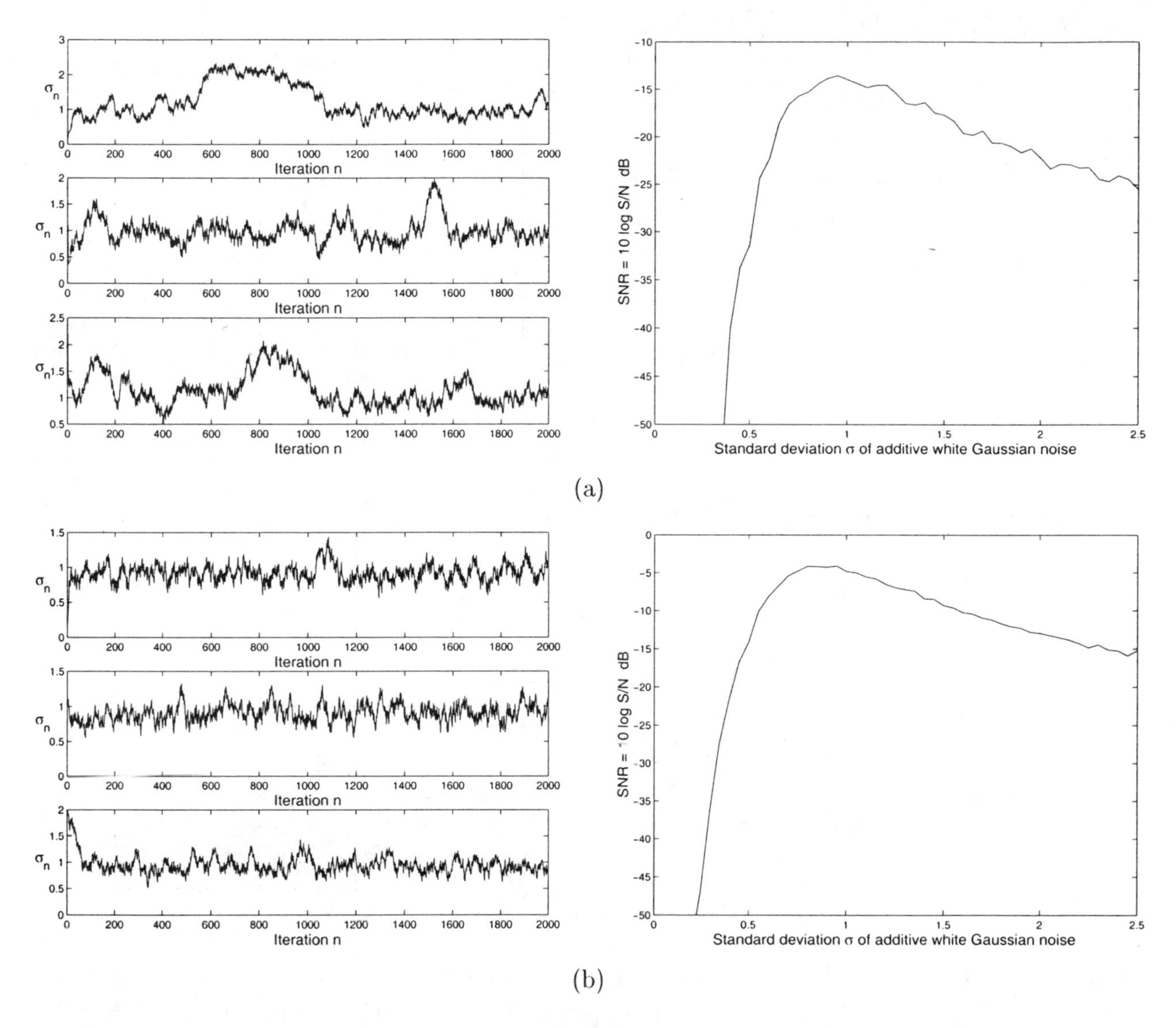

Figure 20: SR learning paths of σ_n for the forced bistable neuron model $\dot{x} = -x + 2\tanh x + \varepsilon \sin 2\pi f_0 t + n(t)$ with binary output $y(t) = \text{sgn}(x(t))$. The parameters of the input sinewaves are $f_0 = 0.01$ Hz and (a) $\varepsilon = 0.1$ and (b) $\varepsilon = 0.3$.

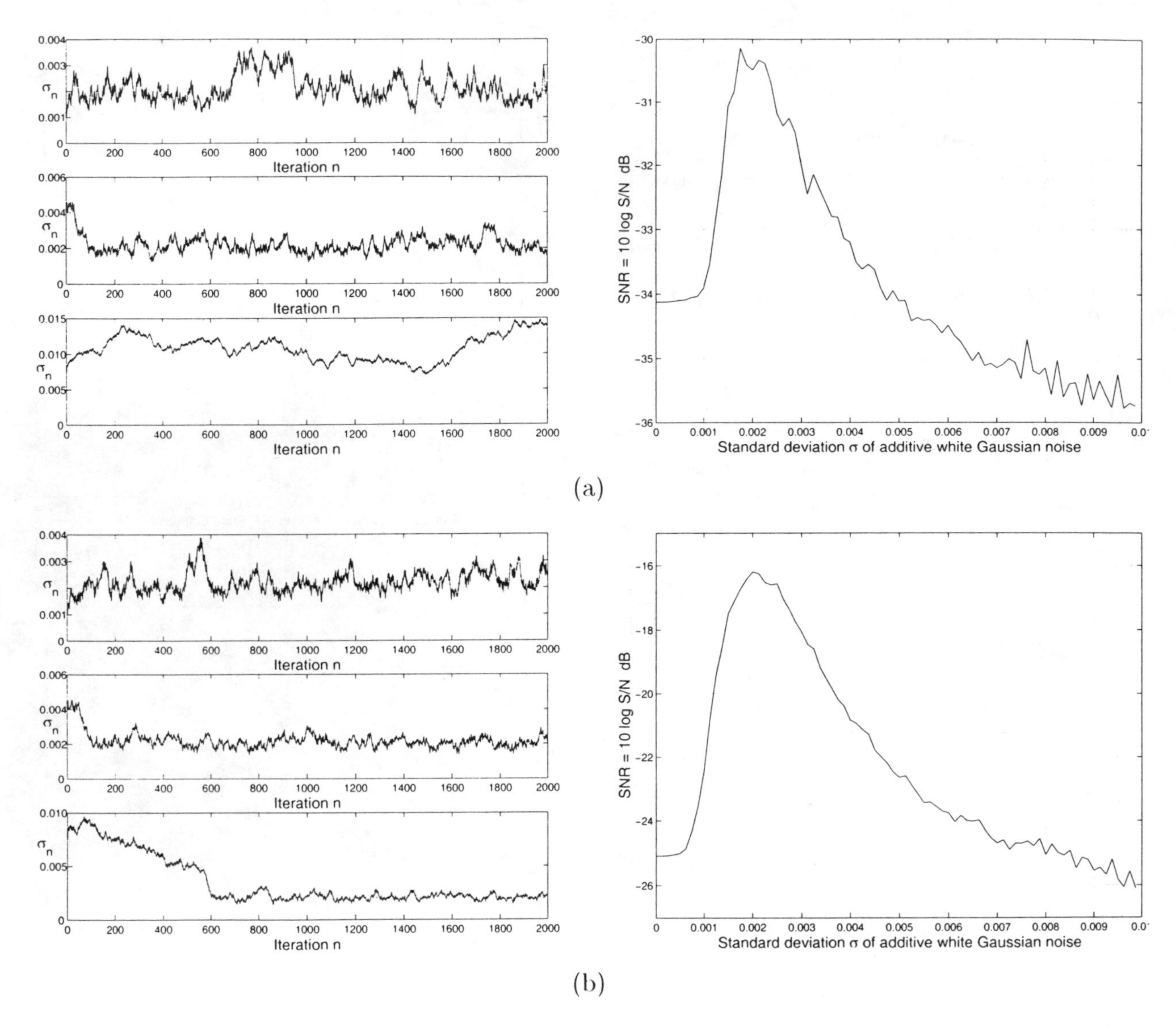

Figure 21: SR learning paths of σ_n for the FitzHugh-Nagumo neuron model $\epsilon\dot{x} = -x(x^2 - \frac{1}{4}) - w + A + s(t) + n(t)$ and $\dot{w} = x - w$ with output $y(t) = x(t)$. The parameters are $\epsilon = 0.005$ and $A = -(5/12\sqrt{3} + 0.07) = -.31056$. The sinewave input signal is $s(t) = \varepsilon \sin 2\pi f_0 t$ where (a) $\varepsilon = 0.01$ and $f_0 = 0.1$ Hz and (b) $\varepsilon = 0.01$ and $f_0 = 0.5$ Hz. Figures (a) and (b) show how SR learning convergence can depend on initial conditions. The distant starting point $\sigma_0 > 7.5 \times 10^{-3}$ leads to divergence in the third learning sample in (a) but leads to convergence in the third learning sample in (b).

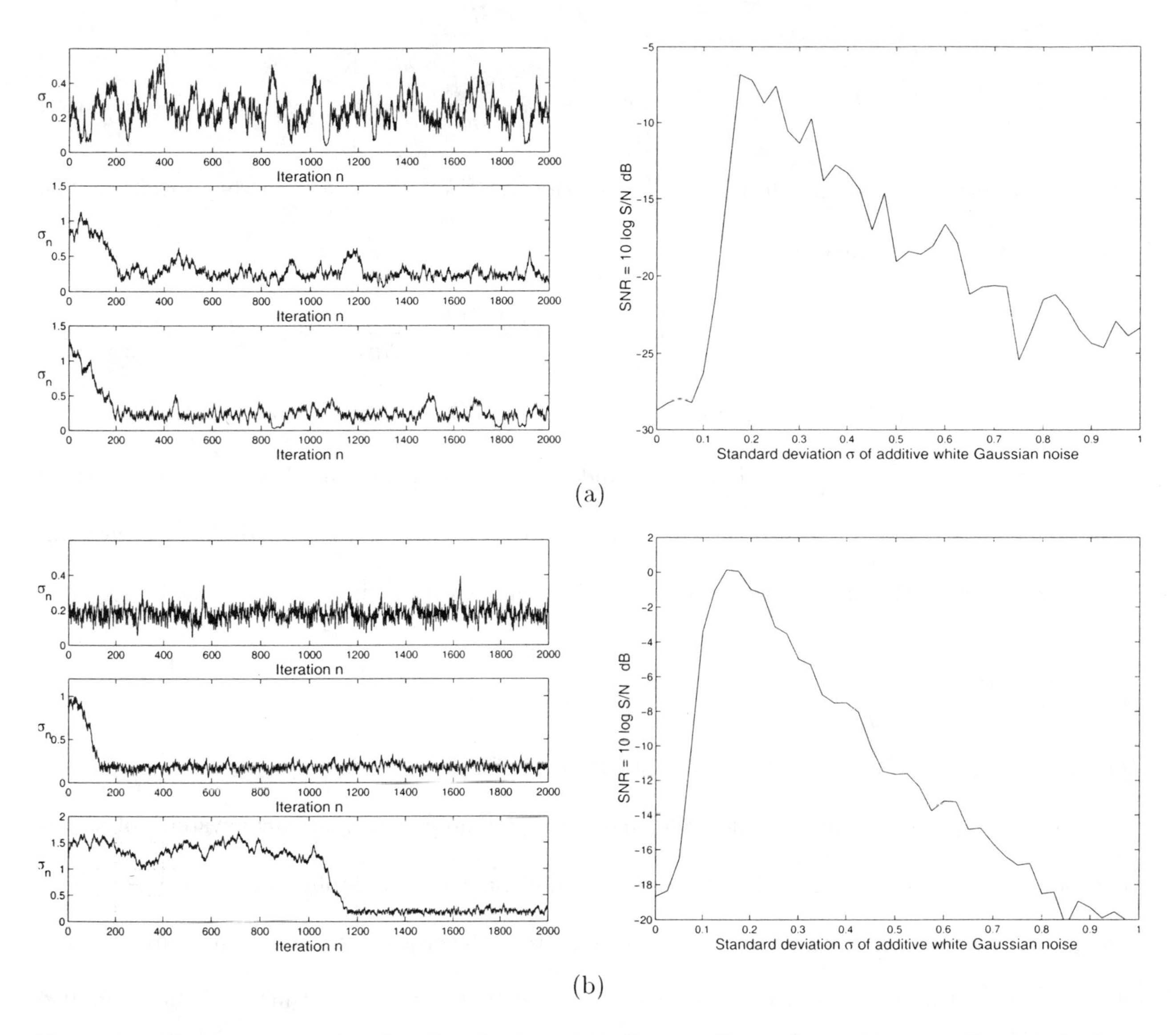

Figure 22: SR learning paths of σ_n for the forced Duffing oscillator $\ddot{x} = -\delta\dot{x} + x + x^3 + \varepsilon \sin 2\pi f_0 t + n(t)$ with output $y(t) = x(t)$ and $\delta = 0.15$. The parameters of the input sinewaves are $f_0 = 0.01$ Hz and (a) $\varepsilon = 0.1$ and (b) $\varepsilon = 0.3$.

fixed input initial value $x(0) = -1$. The fuzzy system itself defined a vector function $F : R^2 \to R$ and used 200 rules. The chain rule extended the learning laws in the previous sections to tune the fuzzy system's parameters m_j as in (71):

$$m_j(n+1) = m_j(n) + \mu_n \frac{\partial \, \mathrm{SNR}_n}{\partial m_j} \tag{149}$$

$$= m_j(n) + \mu_n \frac{\partial \, \mathrm{SNR}_n}{\partial \sigma} \frac{\partial F}{\partial m_j}. \tag{150}$$

Appendix B derives the partial derivative $\frac{\partial F}{\partial m_j}$ for the sinc SAM fuzzy system that we used. The Cauchy noise suppressor gives the learning law as

$$m_j(n+1) = m_j(n) + \mu_n \, \phi\Big(\frac{\partial \, \mathrm{SNR}_n}{\partial \sigma}\Big) \frac{\partial F}{\partial m_j}. \tag{151}$$

Figure 23 shows how we formed a first set of rules on the product space of the two variables ε and f_0. It also shows how the learning laws move and shape the width of the if-part sinc set. Figure 24 shows the results of SAM learning of the optimal noise pattern for the quartic bistable system. The sinc SAM used 200 rules. Fewer rules gave a coarser approximation.

7 Conclusions

Stochastic gradient ascent can learn to find the SR mode of at least some simple dynamical systems. This learning scheme may fail to scale up for more complex nonlinear dynamical systems of higher dimension or may get stuck in the local maxima of multimodal SNR profiles. Simulations showed that impulsive noise can destabilize the SR learning process even though the learning process does not minimize a mean-squared error. Simulations showed that the key learning term itself can give rise to strong impulsive shocks in the learning process. These shocks often approached Cauchy noise in intensity. A Cauchy noise suppressor gave a working SR learning scheme for the DFT-based SNR measure. Other SNR measures or other process statistics may favor other types of robust noise suppressors or may favor still other techniques to lessen the impulsiveness.

Fourier techniques may not extend well to the general case or broadband or nonperiodic forcing signals found in many nonlinear and nonstationary environments. Wavelet transforms [56, 147,

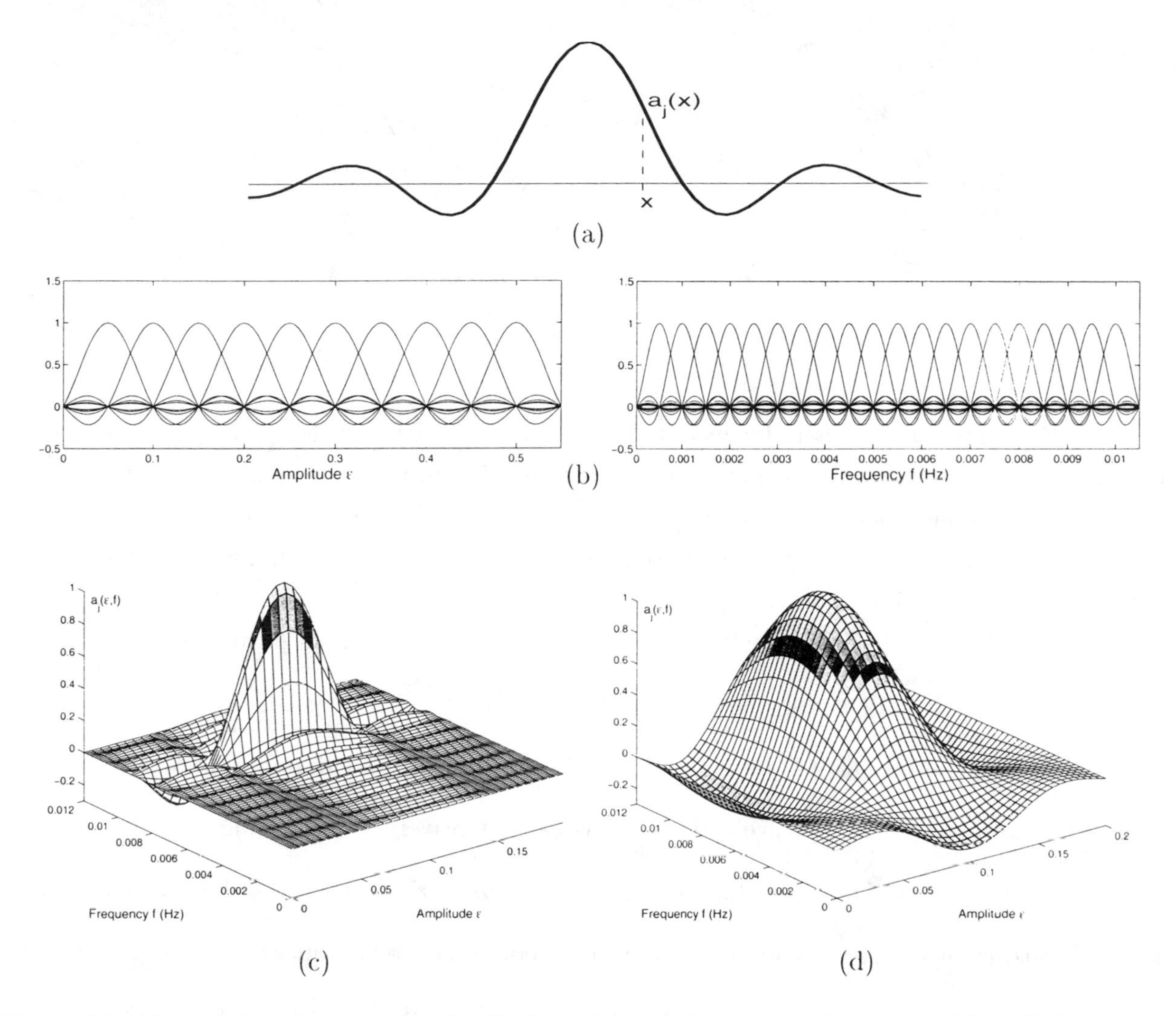

Figure 23: If-part sinc fuzzy sets. (a) Scalar sinc set function $a_j(x) = \sin x/x$. Sinc sets are generalized fuzzy sets with "membership values" in [-.217,1]. Element x belongs to set A_j to degree $a_j(x)$: Degree($x \in A_j$) = $a_j(x)$. (b) Initial subsets for sinewave amplitudes and frequencies. There are 10 fuzzy sets for amplitude ε and 20 fuzzy sets for frequency f_0. The product of two 1-D sets gives the 2-D joint sets: $a_j(x) = a_j(\varepsilon, f_0) = a_j^1(\varepsilon)\, a_j^2(f_0)$. So the product space gives $10 \times 20 = 200$ if-part sets in the if-then rules. (c) One of the 2-D if-part sinc sets in the 200 rules at the initial location. (d) Learning laws tune the location and width of the same set in (c) after 30 epochs of learning.

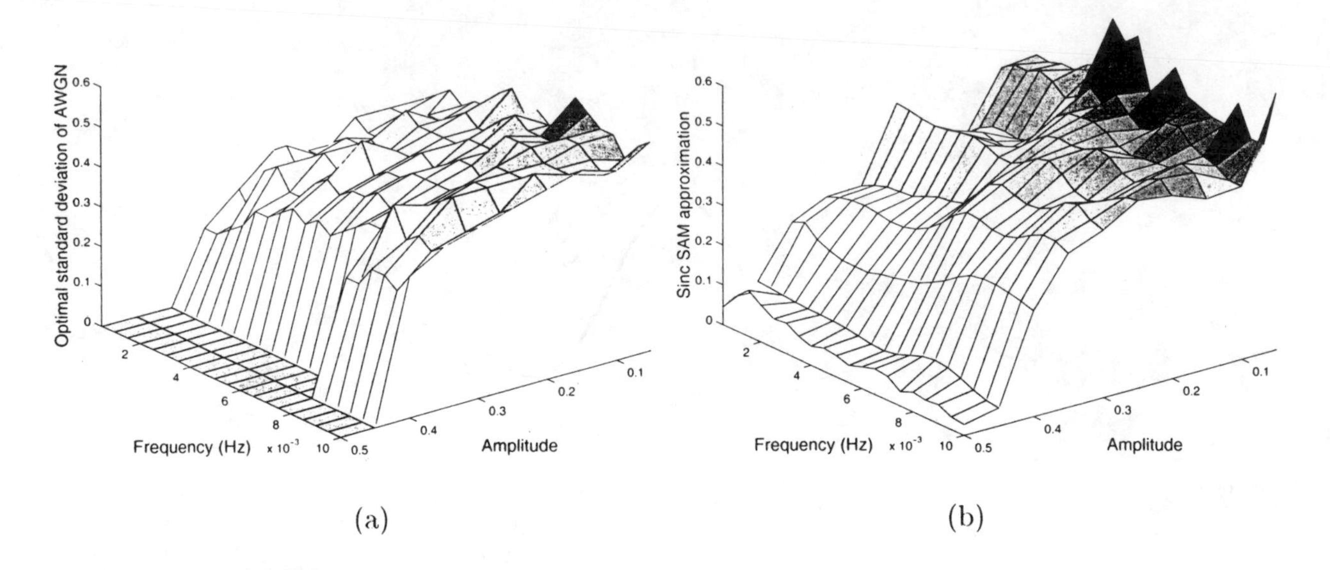

Figure 24: Optimal noise levels in terms of the signal-to-noise ratio for the quartic bistable system with binary output. (a) The optimum noise pattern when inputs are sinewaves with distinct amplitudes and frequencies. (b) SAM fuzzy approximation of the optimum noise after 30 epochs. The sinc SAM used 200 rules. One epoch used 20 iterations that trained on 200 input amplitudes and frequencies. The quartic bistable system has the form $\dot{x} = x - x^3 + s(t) + n(t)$ with initial condition $x(0) = -1$. The initialized SAM gave the output value 0.2 as its first estimate of the optimal noise level.

243, 249] may offer better ways to measure SR effects in these cases when nonperiodic signals drive nonlinear dynamical systems. Wavelet transforms can adaptively localize nonperiodic signals in both time and frequency. Fourier techniques tend to localize periodic signals either in frequency or in time. Arbitrary or random broadband signals may require new techniques to detect these signals and extract their key statistical features from their noisy dynamical backgrounds.

Gradient-ascent learning can find the SR mode of the main known dynamical models that show the SR effect and can do so in the presence of a wide range of noise types. This suggests that SR may occur in many multivariable dynamical systems in science and engineering and that simple learning schemes can sometimes measure or approximate this behavior. We lack formal results that describe when and how such SR learning algorithms will converge for which types of SR systems. This reflects the general lack of a formal taxonomy in this promising new field: Which noisy dynamical systems show what SR effects for which forcing signals?

References

[1] Y. S. Abu-Mostafa, "Learning from Hints in Neural Networks," *Journal of Complexity*, vol. 6, pp. 192–198, 1990.

[2] Y. S. Abu-Mostafa, "Hints," *Neural Computation*, vol. 7, pp. 639–671, 1995.

[3] V. S. Anishchenko, A. B. Neiman, and M. A. Safonova, "Stochastic Resonance in Chaotic Systems," *Journal of Statistical Physics*, vol. 70, no. 1/2, pp. 183–196, 1993.

[4] V. S. Anishchenko, M. A. Safonova, and L. O. Chua, "Stochastic Resonance in Chua's Circuit," *International Journal of Bifurcation and Chaos*, vol. 2, no. 2, pp. 397–401, 1992.

[5] V. S. Anishchenko, M. A. Safonova, and L. O. Chua, "Stochastic Resonance in Chua's Circuit Driven by Amplitude or Frequency Modulated Signals," *International Journal of Bifurcation and Chaos*, vol. 4, no. 2, pp. 441–446, 1994.

[6] A. S. Asdi and A. H. Tewfik, "Detection of Weak Signals Using Adaptive Stochastic Resonance," in *Proceedings of the 1995 IEEE International Conference on Acoustics, Speech, and Signal Processing (ICASSP-95)*, May 1995, vol. 2, pp. 1332–1335.

[7] P. Babinec, "Stochastic Resonance in the Weidlich Model of Public Opinion Formation," *Physics Letters A*, vol. 225, pp. 179–181, January 1997.

[8] P. Baldi, "Gradient Descent Learning Algorithm Overview: A General Dynamical Systems Perspective," *IEEE Transactions on Neural Networks*, vol. 6, pp. 182–195, January 1995.

[9] R. Bartussek, P. Hänggi, and P. Jung, "Stochastic Resonance in Optical Bistable Systems," *Physical Review E*, vol. 49, no. 5, pp. 3930–3939, May 1994.

[10] M. Bartz, "Large-Scale Dithering Enhances ADC Dynamic Range," *Microwave & RF*, vol. 32, pp. 192–194+, May 1993.

[11] R. Benzi, G. Parisi, A. Sutera, and A. Vulpiani, "Stochastic Resonance in Climatic Change," *Tellus*, vol. 34, pp. 10–16, 1982.

[12] R. Benzi, G. Parisi, A. Sutera, and A. Vulpiani, "A Theory of Stochastic Resonance in Climatic Change," *SIAM Journal on Applied Mathematics*, vol. 43, no. 3, pp. 565–578, June 1983.

[13] R. Benzi, A. Sutera, and A. Vulpiani, "The Mechanism of Stochastic Resonance," *Journal of Physics A: Mathematical and General*, vol. 14, pp. L453–L457, 1981.

[14] R. Benzi, A. Sutera, and A. Vulpiani, "Stochastic Resonance in the Landau-Ginzberg Equation," *Journal of Physics A: Mathematical and General*, vol. 18, pp. 2239–2245, 1985.

[15] S. M. Bezrukov, "Stochastic Resonance as an Inherent Property of Rate-Modulated Random Series of Events," *Physics Letters A*, vol. 248, pp. 29–36, November 1998.

[16] S. M. Bezrukov and I. Vodyanoy, "Stochastic Resonance in Non-dynamical Systems without Response Thresholds," *Nature*, vol. 385, pp. 319–321, January 1997.

[17] S. M. Bezrukov and I. Vodyanoy, "Stochastic Resonance in Thermally Activated Reactions: Application to Biological Ion Channels," *Chaos: Focus Issue on the Constructive Role of Noise in Fluctuation Driven Transport and Stochastic Resonance*, vol. 8, no. 3, pp. 557–566, 1998.

[18] Y. Braiman, J. F. Lindner, and W. L. Ditto, "Taming Spatiotemporal Chaos with Disorder," *Nature*, vol. 378, pp. 465–467, November 1995.

[19] H. A. Braun, H. Wissing, K. Schäfer, and M. C. Hirsch, "Oscillation and Noise Determine Signal Transduction in Shark Multimodal Sensory Cells," *Nature*, vol. 367, pp. 270–273, January 1994.

[20] L. Breiman, *Probability*, Addison-Wesley, 1968.

[21] J. J. Brey, J. Casado-Pascul, and B. Sánchez, "Resonant Behavior of a Poisson Process Driven by a Periodic Signal," *Physical Review E*, vol. 52, no. 6, pp. 6071–6081, December 1995.

[22] J. J. Brey and A. Prados, "Stochastic Resonance in a One-Dimension Ising Model," *Physics Letters A*, vol. 216, pp. 240–246, June 1996.

[23] K. S. Brown, "Noises On," *New Scientist*, vol. 150, pp. 28–31, June 1996.

[24] P. Bryant, K. Wiesenfeld, and B. McNamara, "The Nonlinear Effects of Noise on Parametric Amplification: An Analysis of Noise Rise in Josephson Junctions and Other Systems," *Journal of Applied Physics*, vol. 62, no. 7, pp. 2898–2913, October 1987.

[25] A. Bulsara, S. Chillemi, L. Kiss, P. V. E. McClintock, R. Mannella, F. Marchesoni, K. Nicolis, and K. Wiesenfeld, Eds., *Il Nuovo Cimento. Special Issue on Fluctuation in Physics and Biology: Stochastic Resonance, Signal Processing, and Related Phenomena*, vol. 17 D, no. 7-8, Luglio-Agosto 1995.

[26] A. R. Bulsara, R. D. Boss, and E. W. Jacobs, "Noise Effects in an Electronic Model of a Single Neuron," *Biological Cybernetics*, vol. 61, pp. 211–222, 1989.

[27] A. R. Bulsara, T. C. Elston, C. R. Doering, S. B. Lowen, and K. Lindenberg, "Cooperative Behavior in Periodically Driven Noisy Integrate-Fire Models of Neuronal Dynamics," *Physical Review E*, vol. 53, no. 4, pp. 3958–3969, April 1996.

[28] A. R. Bulsara and L. Gammaitoni, "Tuning in to Noise," *Physics Today*, pp. 39–45, March 1996.

[29] A. R. Bulsara, E. W. Jacobs, T. Zhou, F. Moss, and L. Kiss, "Stochastic Resonance in a Single Neuron Model: Theory and Analog Simulation ," *Journal of Theoretical Biology*, vol. 152, pp. 531–555, 1991.

[30] A. R. Bulsara, S. B. Lowen, and C. D. Rees, "Cooperative Behavior in the Periodically Modulated Wiener Process: Noise-Induced Complexity in a Model Neuron," *Physical Review E*, vol. 49, no. 6, pp. 4989–5000, June 1994.

[31] A. R. Bulsara, A. J. Maren, and G. Schmera, "Single Effective Neuron: Dendritic Coupling Effects and Stochastic Resonance," *Biological Cybernetics*, vol. 70, pp. 145–156, 1993.

[32] A. R. Bulsara and G. Schmera, "Stochastic Resonance in Globally Coupled Nonlinear Oscillators," *Physical Review E*, vol. 47, no. 5, pp. 3734–3737, May 1993.

[33] A. R. Bulsara and A. Zador, "Threshold Detection of Wideband Signals: A Noise-Induced Maximum in the Mutual Information," *Physical Review E*, vol. 54, no. 3, pp. R2185–R2188, September 1996.

[34] G. C. Burdea, *Force and Touch Feedback for Virtual Reality*, John Wiley & Sons, 1996.

[35] T. L. Carroll and L. M. Pecora, "Stochastic Resonance and Crises," *Physical Review Letters*, vol. 70, no. 5, pp. 576–579, February 1993.

[36] J. M. Casado and M. Morillo, "Distribution of Escape Times in a Driven Stochastic Model," *Physical Review E*, vol. 49, no. 2, pp. 1136–1139, February 1994.

[37] R. Castro and T. Sauer, "Chaotic Stochastic Resonance: Noise-Enhanced Reconstruction of Attractors," *Physical Review Letters*, vol. 79, no. 6, pp. 1030–1033, August 1997.

[38] F. Chapeau-Blondeau, "Noise-Enhanced Capacity via Stochastic Resonance in an Asymmetric Binary Channel," *Physical Review E*, vol. 55, no. 2, pp. 2016–2019, February 1997.

[39] F. Chapeau-Blondeau and X. Godivier, "Stochastic Resonance in Nonlinear Transmission of Spike Signals: An Exact Model and an Application to the Neuron," *International Journal of Bifurcation and Chaos*, vol. 6, no. 11, pp. 2069–2076, 1996.

[40] F. Chapeau-Blondeau and X. Godivier, "Theory of Stochastic Resonance in Signal Transmission by Static Nonlinear System ," *Physical Review E*, vol. 55, no. 2, pp. 1478–1495, February 1997.

[41] F. Chapeau-Blondeau, X. Godivier, and N. Chambet, "Stochastic Resonance in a Neuron Model that Transmits Spike Trains," *Physical Review E*, vol. 53, no. 1, pp. 1273–1275, January 1996.

[42] A. K. Chattah, C. B. Briozzo, O. Osenda, and M. O. Cáceres, "Signal-to-Noise Ratio in Stochastic Resonance," *Modern Physics Letters B*, vol. 10, no. 22, pp. 1085–1094, 1996.

[43] D. R. Chialvo, A. Longtin, and J. Müller-Gerking, "Stochastic Resonance in Models of Neuronal Ensembles," *Physical Review E*, vol. 55, no. 2, pp. 1798–1808, February 1997.

[44] F. Y. Chiou-Tan, K. N. Magee, S. M. Tuel, L. R. Robinson, T. A. Krouskop, M. R. Netson, and F. Moss, "Augmented Sensory Nerve Action Potentials During Distant Muscle Contraction," *American Journal of Physical Medicine and Rehabilitation*, vol. 76, no. 1, pp. 14–18, January/February 1997.

[45] K. L. Chung and R. J. Williams, *Introduction to Stochastic Integration*, Birkhäuser, second edition, 1990.

[46] J. J. Collins, C. C. Chow, A. C. Capela, and T. T. Imhoff, "Aperiodic Stochastic Resonance," *Physical Review E*, vol. 54, no. 5, pp. 5575–5584, November 1996.

[47] J. J. Collins, C. C. Chow, and T. T. Imhoff, "Aperiodic Stochastic Resonance in Excitable Systems," *Physical Review E*, vol. 52, no. 4, pp. R3321–R3324, October 1995.

[48] J. J. Collins, C. C. Chow, and T. T. Imhoff, "Stochastic Resonance without Tuning," *Nature*, vol. 376, pp. 236–238, July 1995.

[49] J. J. Collins, T. T. Imhoff, and P. Grigg, "Noise-Enhanced Information Transmission in Rat SA1 Cutaneous Mechanoreceptors via Aperiodic Stochastic Resonance," *Journal of Neurophysiology*, vol. 76, no. 1, pp. 642–645, July 1996.

[50] J. J. Collins, T. T. Imhoff, and P. Grigg, "Noise-Enhanced Tactile Sensation," *Nature*, vol. 383, pp. 770, October 1996.

[51] P. Cordo, J. T. Inglis, S. Vershueren, J. J. Collins, D. M. Merfeld, S. Rosenblum, S. Buckley, and F. Moss, "Noise in Human Muscle Spindles," *Nature*, vol. 383, pp. 769–770, October 1996.

[52] T. M. Cover and J. A. Thomas, *Elements of Information Theory*, John Wiley & Sons, 1991.

[53] J. J. Craig, *Introduction to Robotics: Mechanics and Control*, Addison-Wesley, second edition, 1989.

[54] A. Crisanti, M. Falcioni, G. Paladin, and A. Vulpiani, "Stochastic Resonance in Deterministic Chaotic Systems," *Journal of Physics A: Mathematical and General*, vol. 27, pp. L597–L603, 1994.

[55] G. Dahlquist and Å. Björck, *Numerical Methods*, Prentice Hall, 1974.

[56] I. Daubechies, *Ten Lectures on Wavelets*, Philadelphia: SIAM, 1992.

[57] G. Debnath, F. Moss, Th. Leiber, H. Risken, and F. Marchesoni, "Holes in the Two-Dimensional Probability Density of Bistable Systems Driven by Strongly Colored Noise," *Physical Review A*, vol. 42, no. 2, pp. 703–710, July 1990.

[58] G. Debnath, T. Zhou, and F. Moss, "Remarks on Stochastic Resonance," *Physical Review A*, vol. 39, no. 8, pp. 4323–4326, April 1989.

[59] J. L. Doob, *Stochastic Processes*, John Wiley & Sons, 1953.

[60] J. K. Douglass, L. Wilkens, E. Pantazelou, and F. Moss, "Noise Enhancement of Information Transfer in Crayfish Mechanoreceptors by Stochastic Resonance," *Nature*, vol. 365, pp. 337–340, September 1993.

[61] J. K. Douglass, L. Wilkens, E. Pantazelou, and F. Moss, "Stochastic Resonance in Crayfish Hydrodynamic Receptors Stimulated with External Noise," in *AIP Conference Proceedings 285: Noise in Physical Systems and 1/f Fluctuations*, P. H. Hanel and A. L. Chung, Eds., 1993, pp. 712–715.

[62] R. Durrett, *Stochastic Calculus: A Practical Introduction*, CRC Press, 1996.

[63] M. I. Dykman, G. P. Golubev, I. Kh. Kaufman, D. G. Luchinsky, P. V. E. McClintock, and E. A. Zhukov, "Noise-Enhanced Optical Heterodyning in an All-Optical Bistable System," *Applied Physics Letters*, vol. 67, no. 3, pp. 308–310, July 1995.

[64] M. I. Dykman, T. Horita, and J. Ross, "Statistical Distribution and Stochastic Resonance in a Periodically Driven Chemical System," *Journal of Chemical Physics*, vol. 103, no. 3, pp. 966–972, July 1995.

[65] M. I. Dykman, D. G. Luchinsky, R. Mannella, P. V. E. McClintock, S. M. Soskin, and N. D. Stein, "Resonant Subharmonic Absorption and Second-Harmonic Generation by a Fluctuating Nonlinear Oscillator," *Physical Review E*, vol. 54, no. 3, pp. 2366–2377, September 1996.

[66] M. I. Dykman, D. G. Luchinsky, R. Mannella, P. V. E. McClintock, N. D. Stein, and N. G. Stocks, "Nonconventional Stochastic Resonance," *Journal of Statistical Physics*, vol. 70, no. 1/2, pp. 479–499, January 1993.

[67] M. I. Dykman, D. G. Luchinsky, R. Mannella, P. V. E. McClintock, N. D. Stein, and N. G. Stocks, "Stochastic Resonance: Linear Response and Giant Nonlinearity," *Journal of Statistical Physics*, vol. 70, no. 1/2, pp. 463–478, January 1993.

[68] M. I. Dykman, D. G. Luchinsky, R. Mannella, P. V. E. McClintock, N. D. Stein, and N. G. Stocks, "Supernarrow Spectral Peaks and High-Frequency Stochastic Resonance in Systems with Coexisting Periodic Attractors," *Physical Review E*, vol. 49, no. 2, pp. 1198–1215, February 1994.

[69] M. I. Dykman, D. G. Luchinsky, P. V. E. McClintock, N. D. Stein, and N. G. Stocks, "Stochastic Resonance for Periodically Modulated Noise Intensity," *Physical Review A*, vol. 46, no. 4, pp. R1713–R1716, August 1992.

[70] J.-P. Eckmann and L. E. Thomas, "Remarks on Stochastic Resonances," *Journal of Physics A: Mathematical and General*, vol. 15, pp. L261–L266, 1982.

[71] K. Ehrenberger, D. Felix, and K. Svozil, "Stochastic Resonance in Cochlear Signal Transduction," *Acta Otolaryngol (Stockh)*, vol. 119, pp. 166–170, 1999.

[72] R. Fakir, "Nonstationary Stochastic Resonance," *Physical Review E*, vol. 57, no. 6, pp. 6996–7001, June 1998.

[73] R. Fakir, "Nonstationary Stochastic Resonance in a Single Neuronlike System," *Physical Review E*, vol. 58, no. 4, pp. 5175–5178, October 1998.

[74] S. Fauve and F. Heslot, "Stochastic Resonance in a Bistable System," *Physics Letters A*, vol. 97, no. 1,2, pp. 5–7, August 1983.

[75] W. Feller, *An Introduction to Probability Theory and Its Applications*, vol. II, John Wiley & Sons, 1966.

[76] P. G. Flikkema, "Spread-Spectrum Techniques for Wireless Communication," *Signal Processing Magazine*, vol. 14, no. 3, pp. 26–36, 1997.

[77] A. Förster, M. Merget, and F. W. Schneider, "Stochastic Resonance in Chemistry. 2. The Peroxidase-Oxidase Reaction," *Journal of Physical Chemistry*, vol. 100, pp. 4442–4447, 1996.

[78] A. M. Forte and J. X. Mitrovica, "A Resonance in the Earth's Obliquity and Precession over the Past 20 Myr Driven by Mantle Convection," *Nature*, vol. 390, pp. 676–680, December 1997.

[79] J. Foss, F. Moss, and J. Milton, "Noise, Multistability, and Delayed Recurrent Loops," *Physical Review E*, vol. 55, no. 4, pp. 4536–4543, April 1997.

[80] R. F. Fox, "Stochastic Resonance in a Double Well," *Physical Review A*, vol. 39, no. 8, pp. 4148–4153, April 1989.

[81] R. F. Fox and Y. N. Lu, "Analytic and Numerical Study of Stochastic Resonance," *Physical Review E*, vol. 48, no. 5, pp. 3390–3398, November 1993.

[82] W. J. Freeman, H.-J. Chang, B. C. Burke, P. A. Rose, and J. Badler, "Taming chaos: Stabilization of aperiodic attractors by noise," *IEEE Transactions on Circuits and Systems–I: Fundamental Theory and Applications*, vol. 44, no. 10, pp. 989–996, October 1997.

[83] A. Fuliński, "Relaxation, Noise-Induced Transitions, and Stochastic Resonance Driven by Non-Markovian Dichotomic Noise," *Physical Review E*, vol. 52, no. 4, pp. 4523–4526, October 1995.

[84] L. Gammaitoni, "Stochastic Resonance in Multi-Threshold Systems," *Physics Letters A*, vol. 208, pp. 315–322, December 1995.

[85] L. Gammaitoni, P. Hänggi, P. Jung, and F. Marchesoni, "Stochastic Resonance," *Reviews of Modern Physics*, vol. 70, no. 1, pp. 223–287, January 1998.

[86] L. Gammaitoni, F. Marchesoni, M. Martinelli, L. Pardi, and S. Santucci, "Phase Shifts in Bistable EPR Systems at Stochastic Resonance," *Physics Letters A*, vol. 158, no. 9, pp. 449–452, September 1991.

[87] L. Gammaitoni, F. Marchesoni, E. Menichella-Saetta, and S. Santucci, "Stochastic Resonance in Bistable Systems," *Physical Review Letters*, vol. 62, no. 4, pp. 349–352, January 1989.

[88] L. Gammaitoni, F. Marchesoni, E. Menichella-Saetta, and S. Santucci, "Multiplicative Stochastic Resonance," *Physical Review E*, vol. 49, no. 6, pp. 4878–4881, June 1994.

[89] L. Gammaitoni, M. Martinelli, L. Pardi, and S. Santucci, "Observation of Stochastic Resonance in Bistable Electron-Paramagnetic-Resonance Systems," *Physical Review Letters*, vol. 67, no. 13, pp. 1799–1802, September 1991.

[90] L. Gammaitoni, E. Menichella-Saetta, S. Santucci, F. Marchesoni, and C. Pressilla, "Periodically Time-Modulated Bistable Systems: Stochastic Resonance," *Physical Review A*, vol. 40, no. 4, pp. 2114–2119, August 1989.

[91] T. C. Gard, *Introduction to Stochastic Differential Equations*, Marcel Dekker, Inc., 1988.

[92] F. Gassmann, "Noise-Induced Chaos-Order Transitions," *Physical Review E*, vol. 55, no. 3, pp. 2215–2221, March 1997.

[93] Z. Gingl, L. B. Kiss, and F. Moss, "Non-Dynamical Stochastic Resonance: Theory and Experiments with White and Arbitrarily Coloured Noise," *Europhysics Letters*, vol. 29, no. 3, pp. 191–196, January 1995.

[94] J. Glanz, "Mastering the Nonlinear Brain," *Science*, vol. 277, pp. 1758–1760, September 1997.

[95] B. J. Gluckman, T. I. Netoff, E. J. Neel, W. L. Ditto, M. L. Spano, and S. J. Schiff, "Stochastic Resonance in a Neuronal Network from Mammalian Brain," *Physical Review Letters*, vol. 77, no. 19, pp. 4098–4101, November 1996.

[96] X. Godivier and F. Chapeau-Blondeau, "Noise-Assisted Signal Transmission in a Nonlinear Electronic Comparator: Experiment and Theory," *Signal Processing*, vol. 56, pp. 293–303, 1997.

[97] X. Godivier and F. Chapeau-Blondeau, "Stochastic Resonance in the Information Capacity of a Nonlinear Dynamic System," *International Journal of Bifurcation and Chaos*, vol. 8, no. 3, pp. 581–589, 1998.

[98] D. C. Gong, G. Hu, X. D. Wen, C. Y. Yang, G. R. Qin, R. Li, and D. F. Ding, "Experimental Study of the Signal-to-Noise Ratio of Stochastic Resonance Systems," *Physical Review A*, vol. 46, no. 6, pp. 3243–3249, September 1992.

[99] M. Grifoni, "Dynamics of the Dissipative Two-State System under AC Modulation of Bias and Coupling Energy," *Physical Review E*, vol. 54, no. 4, pp. R3086–R3089, October 1996.

[100] M. Grifoni and P. Hänggi, "Nonlinear Quantum Stochastic Resonance," *Physical Review E*, vol. 54, no. 2, pp. 1390–1401, August 1996.

[101] M. Grifoni, L. Hartmann, S. Berchtold, and P. Hänggi, "Quantum Tunneling and Stochastic Resonance," *Physical Review E*, vol. 53, no. 6, pp. 5890–5898, June 1996.

[102] M. Grifoni, M. Sassetti, P. Hänggi, and U. Weiss, "Cooperative Effects in the Nonlinearly Driven Spin-Boson System," *Physical Review E*, vol. 52, no. 4, pp. 3596–3607, October 1995.

[103] A. N. Grigorenko and P. I. Nikitin, "Stochastic Resonance in a Bistable Magnetic System," *IEEE Transactions on Magnetics*, vol. 31, no. 5, pp. 2491–2493, September 1995.

[104] A. N. Grigorenko, S. I. Nikitin, and G. V. Roschepkin, "Stochastic Resonance at Higher Harmonics in Monostable Systems," *Physical Review E*, vol. 56, no. 5, pp. R4907–4910, November 1997.

[105] A. Guderian, G. Dechert, K.-P. Zeyer, and F. W. Schneider, "Stochastic Resonance in Chemistry. 1. The Belousov-Zhabitinsku Reaction," *Journal of Physical Chemistry*, vol. 100, pp. 4437–4441, 1996.

[106] S. Haykin, *Neural Networks: A Comprehensive Foundation*, McMillan, 1994.

[107] R. Hecht-Nielsen, *Neurocomputing*, Addison-Wesley, Reading, MA, 1990.

[108] C. Heneghan, C. C. Chow, J. J. Collins, T. T. Imhoff, S. B. Lowen, and M. C. Teich, "Information Measures Quantifying Aperiodic Stochastic Resonance," *Physical Review E*, vol. 54, no. 3, pp. R2228–R2231, 1996.

[109] S. M. Hess and A. M. Albano, "Minimum Requirements for Stochastic Resonance in Threshold Systems," *International Journal of Bifurcation and Chaos*, vol. 8, no. 2, pp. 395–400, 1998.

[110] A. Hibbs, E. W. Jacobs, J. Bekkedahl, A. R. Bulsara, and F. Moss, "Signal Enhancement in a r.f. SQUID Using Stochastic Resonance," *Il Nuovo Cimento*, vol. 17 D, no. 7-8, pp. 811–817, Luglio-Agosto 1995.

[111] A. D. Hibbs, A. L. Singsaas, E. W. Jacobs, A. R. Bulsara, J. J. Bekkedahl, and F. Moss, "Stochastic Resonance in a Superconducting Loop with Josephson Junction," *Journal of Applied Physics*, vol. 77, no. 6, pp. 2582–2590, March 1995.

[112] W. Hohmann, J. Müller, and F. W. Schneider, "Stochastic Resonance in Chemistry. 3. The Minimal-Bromate Reaction," *Journal of Physical Chemistry*, vol. 100, pp. 5388–4492, 1996.

[113] R. Hollands, *The Virtual Reality Home Brewer's Handbook*, John Wiley & Sons, 1996.

[114] G. Hu, T. Ditzinger, C. Z. Ning, and H. Haken, "Stochastic Resonance without External Periodic Force," *Physical Review Letters*, vol. 71, no. 6, pp. 807–810, August 1993.

[115] G. Hu, D. C. Gong, X. D. Wen, C. Y. Yang, G. R. Qing, and R. Li, "Stochastic Resonance in a Nonlinear System Driven by an Aperiodic Force," *Physical Review A*, vol. 46, no. 6, pp. 3250–3254, September 1992.

[116] G. Hu, H. Haken, and C. Z. Ning, "A Study of Stochastic Resonance without Adiabatic Approximation," *Physics Letters A*, vol. 172, no. 1,2, pp. 21–28, December 1992.

[117] G. Hu, H. Haken, and F. Xie, "Stochastic Resonance with Sensitive Frequency Dependence in Globally Coupled Continuous Systems," *Physical Review Letters*, vol. 77, no. 10, pp. 1925–1928, September 1996.

[118] G. Hu, G. Nicolis, and N. Nicolis, "Periodically Forced Fokker-Planck Equation and Stochastic Resonance," *Physical Review A*, vol. 42, no. 4, pp. 2030–2041, August 1990.

[119] P. J. Huber, *Robust Statistics*, John Wiley & Sons, 1981.

[120] B. A. Huberman and R. M. Lukose, "Social Dilemmas and Internet Congestion," *Science*, vol. 277, pp. 535–537, July 1997.

[121] M. E. Inchiosa and A. R. Bulsara, "Coupling Enhances Stochastic Resonance in Nonlinear Dynamic Elements Driven by a Sinusoidal Plus Noise," *Physics Letters A*, vol. 200, pp. 283–288, April 1995.

[122] M. E. Inchiosa and A. R. Bulsara, "Nonlinear Dynamic Elements with Noisy Sinusoidal Forcing: Enhancing Response via Nonlinear Coupling," *Physical Review E*, vol. 52, no. 1, pp. 327–339, July 1995.

[123] M. E. Inchiosa and A. R. Bulsara, "Signal Detection Statistics of Stochastic Resonators," *Physical Review E*, vol. 53, no. 3, pp. R2021–R2024, March 1996.

[124] M. E. Inchiosa, A. R. Bulsara, and L. Gammaitoni, "Higher-Order Resonant Behavior in Asymmetric Nonlinear Stochastic Systems," *Physical Review E*, vol. 55, no. 4, pp. 4049–4056, April 1997.

[125] M. E. Inchiosa, A. R. Bulsara, K. A. Wiesenfeld, and L. Gammaitoni, "Nonlinear Signal Amplification in a 2D System Operating in Static and Oscillatory Regimes," *Physics Letters A*, vol. 252, pp. 20–26, February 1999.

[126] E. Ippen, J. Lindner, and W. L. Ditto, "Chaotic Resonance: A Simulation," *Journal of Statistical Physics*, vol. 70, no. 1/2, pp. 437–450, January 1993.

[127] J.-S. R. Jang, C.-T. Sun, and E. Mizutani, *Neurofuzzy and Soft Computing: A Computational Approach to Learning and Machine Intelligence*, Prentice Hall, 1996.

[128] B. M. Jost and B. E. A. Saleh, "Signal-to-Noise Ratio Improvement by Stochastic Resonance in a Unidirectional Photorefractive Ring Resonator," *Optics Letters*, vol. 21, no. 4, pp. 287–289, February 1996.

[129] P. Jung, "Thermal Activation in Bistable Systems under External Periodic Forces," *Zeitschrift für Physik B*, vol. 76, pp. 521–535, 1989.

[130] P. Jung, "Threshold Devices: Fractal Noise and Neural Talk," *Physical Review E*, vol. 50, no. 4, pp. 2513–2522, October 1994.

[131] P. Jung, "Stochastic Resonance and Optimal Design of Threshold Detectors," *Physics Letters A*, vol. 207, pp. 93–104, October 1995.

[132] P. Jung, U. Behn, E. Pantazelou, and F. Moss, "Collective Response in Globally Coupled Bistable Systems," *Physical Review A*, vol. 46, no. 4, pp. R1709–R1712, August 1992.

[133] P. Jung and P. Hänggi, "Resonantly Driven Brownian Motion: Basic Concepts and Exact Results," *Physical Review A*, vol. 41, no. 6, pp. 2977–2988, March 1990.

[134] P. Jung and P. Hänggi, "Amplification of Small Signals via Stochastic Resonance," *Physical Review A*, vol. 44, no. 12, pp. 8032–8042, December 1991.

[135] P. Jung and P. Talkner, "Suppression of Higher Harmonics at Noise Induced Resonances," *Physical Review E*, vol. 51, no. 3, pp. 2640–2643, March 1995.

[136] P. Jung and K. Wiesenfeld, "Too Quiet to Hear a Whisper," *Nature*, vol. 385, pp. 291, January 1997.

[137] S. Kádár, J. Wang, and K. Showalter, "Noise-supported travelling waves in sub-excitable media," *Nature*, vol. 391, pp. 770–772, February 1998.

[138] H. M. Kim and B. Kosko, "Neural Fuzzy Motion Estimation and Compensation," *IEEE Transactions on Signal Processing*, vol. 45, no. 10, pp. 2515–2532, October 1997.

[139] S. Kim, S. H. Park, and H.-B. Pyo, "Stochastic Resonance in Coupled Oscillator Systems with Time Delay," *Physical Review Letters*, vol. 82, no. 8, pp. 1620–1623, February 1999.

[140] L. B. Kiss, "Possible Breakthrough: Significant Improvement of Signal to Noise Ratio by Stochastic Resonance," in *AIP Conference Proceedings 375: Chaotic, Fractal, and Nonlinear Signal Processing, 1995*, R. A. Katz, Ed., 1996, pp. 382–396.

[141] B. Kosko, *Neural Networks and Fuzzy Systems: A Dynamical Systems Approach to Machine Intelligence*, Prentice Hall, 1991.

[142] B. Kosko, "Fuzzy Systems as Universal Approximators," *IEEE Transactions on Computers*, vol. 43, no. 11, pp. 1329–1333, November 1994.

[143] B. Kosko, "Combining Fuzzy Systems," in *Proceedings of the IEEE International Conference on Fuzzy Systems (IEEE FUZZ-95)*, March 1995, pp. 1855–1863.

[144] B. Kosko, "Optimal Fuzzy Rules Cover Extrema," *International Journal of Intelligent Systems*, vol. 10, no. 2, pp. 249–255, February 1995.

[145] B. Kosko, *Fuzzy Engineering*, Prentice Hall, 1996.

[146] I. Kotlasi, "On Random Variables Whose Quotient Follows the Cauchy Law," *Colloquium Mathematicum*, vol. VII, pp. 277–284, 1960.

[147] J. Kovačević and I. Daubechies, Eds., *Proceedings of the IEEE, Special Issue on Wavelets*, vol. 84, no. 4, April 1996.

[148] H. A. Kramers, "Brownian Motion in a Field of Force and the Diffusion Model of Chemical Reactions," *Physica*, vol. VII, no. 4, pp. 284–304, April 1940.

[149] D. M. Kreps, *A Course in Microeconomic Theory*. Princeton University Press, 1990.

[150] R. G. Laha, "On a Class of Distribution Functions Where the Quotients Follows the Cauchy Law," *Transactions of the American Mathematical Society*, vol. 93, pp. 205–215, November 1959.

[151] W. C. Y. Lee, *Mobile Cellular Telecommunications: Analog and Digital Systems*, McGraw-Hill, 1995.

[152] W. E. Leland, M. S. Taqqu, W. Willinger, and D. V. Wilson, "On the Self-Similar Nature of Ethernet Traffic," *Computer Communication Review: Proceedings of the SIGCOMM-93*, vol. 23, pp. 183–193, September 1993.

[153] D. S. Leonard, "Stochastic Resonance in a Random Walk," *Physical Review A*, vol. 46, no. 10, pp. 6742–6744, November 1992.

[154] D. S. Leonard and L. E. Reichl, "Stochastic Resonance in a Chemical Reaction," *Physical Review E*, vol. 49, no. 2, pp. 1734–1737, February 1994.

[155] J. E. Levin and J. P. Miller, "Broadband Neural Encoding in the Cricket Cercal Sensory System Enhanced by Stochastic Resonance," *Nature*, vol. 380, pp. 165–168, March 1996.

[156] P. H. Lewis and C. Yang, *Basic Control Systems Engineering*, Prentice Hall, 1997.

[157] R. Li, G. Hu, C. Y. Yang, X. D. Wen, G. R. Qing, and H. J. Zhu, "Stochastic Resonance in Bistable Systems Subject to Signal and Quasimonochromatic Noise," *Physical Review E*, vol. 51, no. 5, pp. 3964–3967, May 1995.

[158] J. F. Lindner, B. K. Meadows, W. L. Ditto, M. E. Inchiosa, and A. R. Bulsara, "Scaling Laws for Spatiotemporal Synchronization and Array Enhanced Stochastic Resonance," *Physical Review E*, vol. 53, no. 3, pp. 2081–2086, March 1996.

[159] S. P. Lipshitz, R. A. Wannamaker, and J. Vanderkooy, "Quantization and Dither: A Theoretical Survey," *Journal of Audio Engineering Society*, vol. 40, no. 5, pp. 355–374, May 1992.

[160] M. Löcher, G. A. Johnson, and E. R. Hunt, "Spatiotemporal Stochastic Resonance in a System of Coupled Diode Resonators," *Physical Review Letters*, vol. 77, no. 23, pp. 4698–4701, December 1996.

[161] K. Loerincz, Z. Gingl, and L. B. Kiss, "A Stochastic Resonator is Able to Greatly Improve Signal-to-Noise Ratio," *Physics Letters A*, vol. 224, pp. 63–67, December 1996.

[162] R. Löfstedt and S. N. Coppersmith, "Quantum Stochastic Resonance," *Physical Review Letters*, vol. 72, no. 13, pp. 1947–1950, March 1994.

[163] A. Longtin, "Stochastic Resonance in Neuron Models," *Journal of Statistical Physics*, vol. 70, no. 1/2, pp. 309–327, January 1993.

[164] A. Longtin, "Synchronization of the Stochastic Fitzhugh-Nagumo Equations to Periodic Forcing," *Il Nuovo Cimento*, vol. 17 D, no. 7-8, pp. 835–846, Luglio-Agosto 1995.

[165] A. Longtin, "Autonomous Stochastic Resonance in Bursting Neurons," *Physical Review E*, vol. 55, no. 1, pp. 868–876, January 1997.

[166] A. Longtin, A. R. Bulsara, and F. Moss, "Time-Interval Sequences in Bistable Systems and the Noise-Induced Transmission of Information by Sensory Neurons," *Physical Review Letters*, vol. 67, no. 5, pp. 656–659, July 1991.

[167] A. Longtin, A. R. Bulsara, D. Pierson, and F. Moss, "Bistability and the Dynamics of Periodically Forced Sensory Neurons," *Biological Cybernetics*, vol. 70, pp. 569–578, 1994.

[168] A. Longtin and K. Hinzer, "Encoding with Bursting, Subthreshold Oscillations, and Noise in Mammalian Cold Receptors," *Neural Computation*, vol. 8, pp. 215–255, 1996.

[169] V. Loreto, G. Paladin, and A. Vulpiani, "Concept of Complexity in Random Dynamical Systems," *Physical Review E*, vol. 53, no. 3, pp. 2087–2098, March 1996.

[170] J. Maddox, "Towards the Brain-Computer's Code?," *Nature*, vol. 352, pp. 469, August 1991.

[171] J. Maddox, "Bringing More Order out of Noisiness," *Nature*, vol. 369, pp. 271, May 1994.

[172] M. C. Mahato and S. R. Shenoy, "Hysteresis Loss and Stochastic Resonance: A Numerical Study of a Double-Well Potential," *Physical Review E*, vol. 50, no. 4, pp. 2503–2512, October 1994.

[173] D. E. Makarov and N. Makri, "Stochastic Resonance and Nonlinear Response in Double-Quantum-Well Structures," *Physical Review B*, vol. 52, no. 4, pp. R2257–R2260, July 1995.

[174] R. N. Mantegna and B. Spagnolo, "Stochastic Resonance in a Tunnel Diode," *Physical Review E*, vol. 49, no. 3, pp. R1792–R1795, March 1994.

[175] R. N. Mantegna and B. Spagnolo, "Stochastic Resonance in a Tunnel Diode in the Presence of White or Coloured Noise," *Il Nuovo Cimento*, vol. 17 D, no. 7-8, pp. 873–881, Luglio-Agosto 1995.

[176] F. Marchesoni, L. Gammaitoni, and A. R. Bulsara, "Spatiotemporal Stochastic Resonance in a ϕ^4 Model of Kink-Antikink Nucleation," *Physical Review Letters*, vol. 76, no. 15, pp. 2609–2612, April 1996.

[177] K. Matsumoto and I. Tsuda, "Noise-induced order," *Journal of Statistical Physics*, vol. 31, no. 1, pp. 87–106, 1983.

[178] B. McNamara and K. Wiesenfeld, "Theory of Stochastic Resonance," *Physical Review A*, vol. 39, no. 9, pp. 4854–4869, May 1989.

[179] B. McNamara, K. Wiesenfeld, and R. Roy, "Observation of Stochastic Resonance in a Ring Laser," *Physical Review Letters*, vol. 60, no. 25, pp. 2626–2629, June 1988.

[180] V. I. Melnikov, "Schmitt Trigger: A Solvable Model of Stochastic Resonance," *Physical Review E*, vol. 48, no. 4, pp. 2481–2489, October 1993.

[181] J. P. Miller, G. A. Jacobs, and F. E. Theunissen, "Representation of Sensory Information in the Cricket Cercal Sensory System. I. Response Properties of the Primary Interneurons," *Journal of Neurophysiology*, vol. 66, no. 5, pp. 1680–1689, November 1991.

[182] S. Mitaim and B. Kosko, "What is the Best Shape for a Fuzzy Set in Function Approximation?," in *Proceedings of the 5th IEEE International Conference on Fuzzy Systems (FUZZ-96)*, September 1996, vol. 2, pp. 1237–1243.

[183] S. Mitaim and B. Kosko, "Adaptive Joint Fuzzy Sets for Function Approximation," in *Proceedings of the 1997 IEEE International Conference on Neural Networks (ICNN-97)*, June 1997, vol. 1, pp. 537–542.

[184] J. Moody and C. Darken, "Fast Learning in Networks of Locally-Tuned Processing Unit," *Neural Computation*, vol. 1, no. 2, pp. 281–294, 1989.

[185] M. Morillo, J. Gómez-Ordo nez, and J. M. Casado, "Stochastic Resonance in a Mean-Field Model of Cooperative Behavior," *Physical Review E*, vol. 52, no. 1, pp. 316–320, July 1995.

[186] R. P. Morse and E. F. Evans, "Enhancement of Vowel Coding for Cochlear Implants by Addition of Noise," *Nature Medicine*, vol. 2, no. 8, pp. 928–932, August 1996.

[187] F. Moss, "Stochastic Resonance: From the Ice Ages to the Monkey's Ear," in *Contemporary Problems in Statistical Physics*, G. H. Weiss, Ed., chapter 5, pp. 205–253. SIAM, 1994.

[188] F. Moss, "Noisy waves," *Nature*, vol. 391, pp. 743–744, February 1998.

[189] F. Moss, A. Bulsara, and M. Shlesinger, Eds., *Journal of Statistical Physics, Special Issue on Stochastic Resonance in Physics and Biology (Proceedings of the NATO Advanced Research Workshop)*, vol. 70, no. 1/2, January 1993.

[190] F. Moss, F. Chiou-Tan, and R. Klinke, "Will There be Noise in Their Ears?," *Nature Medicine*, vol. 2, no. 8, pp. 860–862, August 1996.

[191] F. Moss, J. K. Douglass, L. Wilkens, D. Pierson, and E. Pantazelou, "Stochastic Resonance in an Electronic FitzHugh-Nagumo Model," in *Annals of the New York Academy of Sciences Volume 706: Stochastic Processes in Astrophysics*, J. R. Buchler and H. E. Kandrup, Eds., 1993, pp. 26–41.

[192] F. Moss and P. V. E. McClintock, Eds., *Noise in Nonlinear Dynamical Systems*, vol. I-III, Cambridge University Press, 1989.

[193] F. Moss, D. Pierson, and D. O'Gorman, "Stochastic Resonance: Tutorial and Update," *International Journal of Bifurcation and Chaos*, vol. 4, no. 6, pp. 1383–1397, 1994.

[194] F. Moss and K. Wiesenfeld, "The Benefits of Background Noise," *Scientific American*, vol. 273, no. 2, pp. 66–69, August 1995.

[195] D. C. Munson, "A Note on Lena," *IEEE Transactions on Image Processing*, vol. 5, no. 1, pp. 3, January 1996.

[196] Z. Néda, "Stochastic Resonance in Ising Systems," *Physical Review E*, vol. 51, no. 6, pp. 5315–5317, June 1995.

[197] A. Neiman and L. Schimansky-Geier, "Stochastic Resonance in Two Coupled Bistable Systems," *Physics Letters A*, vol. 197, pp. 379–386, February 1995.

[198] A. Neiman, L. Schimansky-Geier, and F. Moss, "Linear Response Theory Applied to Stochastic Resonance in Models of Ensembles of Oscillators," *Physical Review E*, vol. 56, no. 1, pp. R9–R12, July 1997.

[199] A. Neiman, B. Shulgin, V. Anishchenko, W. Ebeling, L. Schimansky-Geier, and J. Freund, "Dynamic Entropies Applied to Stochastic Resonance," *Physical Review Letters*, vol. 76, no. 23, pp. 4299–4302, June 1996.

[200] A. Neiman and W. Sung, "Memory Effects on Stochastic Resonance," *Physics Letters A*, vol. 223, pp. 341–347, December 1996.

[201] C. Nicolis, "Stochastic Aspects of Climatic Transitions–Response to a Periodic Forcing," *Tellus*, vol. 34, pp. 1–9, 1982.

[202] C. Nicolis, "Long-Term Climatic Transitions and Stochastic Resonance," *Journal of Statistical Physics*, vol. 70, no. 1/2, pp. 4–14, January 1993.

[203] C. Nicolis and G. Nicolis, "Stochastic Aspects of Climate Transitions–Additive Fluctuations," *Tellus*, vol. 33, pp. 225–234, 1981.

[204] G. Nicolis, C. Nicolis, and D. McKernan, "Stochastic Resonance in Chaotic Dynamics," *Journal of Statistical Physics*, vol. 70, no. 1/2, pp. 125–139, January 1993.

[205] P. Niyogi and F. Girosi, "On the Relationship between Generalization Errors Hypothesis Complexity and Sample Complexity for Radial Basis Functions," *Neural Computation*, vol. 8, pp. 819–842, 1996.

[206] K. Ogata, *Modern Control Engineering*, Prentice Hall, third edition, 1997.

[207] A. V. Oppenheim and R. W. Schafer, *Discrete-Time Signal Processing*, Prentice Hall, 1989.

[208] P. J. Pacini and B. Kosko, "Adaptive Fuzzy Frequency Hopper," *IEEE Transactions on Communications*, vol. 43, no. 6, pp. 2111–2117, June 1995.

[209] A. Restrepo (Palacios), L. F. Zuluaga, and L. E. Pino, "Optimal Noise Levels for Stochastic Resonance," in *Proceedings of the 1997 IEEE International Conference on Acoustics, Speech, and Signal Processing (ICASSP-97)*, April 1997, vol. III, pp. 2365–2368.

[210] E. Pantazelou, C. Dames, F. Moss, J. Douglass, and L. Wilkens, "Temperature Dependence and the Role of Internal Noise in Signal Transduction Efficiency of Crayfish Mechanoreceptors," *International Journal of Bifurcation and Chaos*, vol. 5, no. 1, pp. 101–108, 1995.

[211] H. C. Papadopoulos and G. W. Wornell, "A Class of Stochastic Resonance Systems for Signal Processing Applications," in *Proceedings of the 1996 IEEE International Conference on Acoustics, Speech, and Signal Processing (ICASSP-96)*, May 1996, pp. 1617–1620.

[212] A. Papoulis, *Probability and Statistics*, Prentice Hall, 1990.

[213] T. P. Pareek, M. C. Mahato, and A. M. Jayannavar, "Stochastic Resonance and Nonlinear Response in a Dissipative Quantum Two-State System," *Physical Review B*, vol. 55, no. 15, pp. 9318–9321, April 1997.

[214] B. R. Parnas, "Noise and Neuronal Populations Conspire to Encode Simple Waveforms Reliably," *IEEE Transactions on Biomedical Engineering*, vol. 43, no. 3, pp. 313–318, March 1996.

[215] X. Pei, K. Bachmann, and F. Moss, "The Detection Threshold, Noise and Stochastic Resonance in the Fitzhugh-Nagumo Neuron Model," *Physics Letters A*, vol. 206, pp. 61–65, October 1995.

[216] X. Pei, L. Wilkens, and F. Moss, "Light Enhances Hydrodynamic Signaling in the Multimodal Caudal Photoreceptor Interneurons of the Crayfish," *Journal of Neurophysiology*, vol. 76, no. 5, pp. 3002–3011, November 1996.

[217] X. Pei, L. Wilkens, and F. Moss, "Noise-Mediated Spike Timing Precision from Aperiodic Stimuli in an Array of Hodgkin-Huxley-Type Neurons," *Physical Review Letters*, vol. 77, no. 22, pp. 4679–4682, November 1996.

[218] D. Pierson, J. K. Douglass, E. Pantazelou, and F. Moss, "Using an Electronic FitzHugh-Nagumo Simulator to Mimic Noisy Electrophysiological Data from Stimulated Crayfish Mechanoreceptor Cells," in *AIP Conference Proceedings 285: Noise in Physical Systems and 1/f Fluctuations*, P. H. Hanel and A. L. Chung, Eds., 1993, pp. 731–734.

[219] H. E. Plesser and S. Tanaka, "Stochastic Resonance in a Model Neuron with Reset," *Physics Letters A*, vol. 225, pp. 228–234, February 1997.

[220] W.-J. Rappel and A. Karma, "Noise-Induced Coherence in Neural Networks," *Physical Review Letters*, vol. 77, no. 15, pp. 3256–3259, October 1996.

[221] W.-J. Rappel and S. H. Strogatz, "Stochastic Resonance in an Autonomous System with a Nonuniform Limit Cycle," *Physical Review E*, vol. 50, no. 4, pp. 3249–3250, October 1994.

[222] D. O. Reale, A. K. Pattanayak, and W. C. Schieve, "Semiquantal Corrections to Stochastic Resonance," *Physical Review E*, vol. 51, no. 4, pp. 2925–2932, April 1995.

[223] M. Riani and E. Simonotto, "Stochastic Resonance in the Perceptual Interpretation of Ambiguous Figures: A Neural Network Model," *Physical Review Letters*, vol. 72, no. 19, pp. 3120–3123, May 1994.

[224] M. Riani and E. Simonotto, "Periodic Perturbation of Ambiguous Figure: A Neural Network Model and a Non-Simulated Experiment," *Il Nuovo Cimento*, vol. 17 D, no. 7-8, pp. 903–913, Luglio-Agosto 1995.

[225] T. F. Ricci and C. Scherer, "Linear Response and Stochastic Resonance of Superparamagnets," *Journal of Statistical Physics*, vol. 86, no. 3/4, pp. 803–819, February 1997.

[226] K. A. Richardson, T. T. Imhoff, R. Grigg, and J. J. Collins, "Using Electrical Noise to Enhance the Ability of Humans to Detect Subthreshold Mechanical Cutaneous Stimuli," *Chaos: Focus Issue on the Constructive Role of Noise in Fluctuation Driven Transport and Stochastic Resonance*, vol. 8, no. 3, pp. 599–603, September 1998.

[227] H. Risken, *The Fokker-Planck Equation: Methods of Solution and Application*, Springer-Verlag, 1984.

[228] H. Robbins and S. Monro, "A Stochastic Approximation Method," *Annals of Mathematical Statistics*, vol. 22, pp. 400–407, 1951.

[229] R. Rouse, S. Han, and J. E. Lukens, "Flux Amplification Using Stochastic Superconducting Quantum Interference Devices," *Applied Physics Letters*, vol. 66, no. 1, pp. 108–110, January 1995.

[230] S. Russell and P. Norvig, *Artificial Intelligence: A Modern Approach*, Prentice Hall, 1995.

[231] D. W. E. Schobben, R. A. Beuker, and W. Oomen, "Dither and Data Compression," *IEEE Transactions on Signal Processing*, vol. 45, no. 8, pp. 2097–2101, August 1997.

[232] M. Shao and C. L. Nikias, "Signal Processing with Fractional Lower Order Moments: Stable Processes and Their Applications," *Proceedings of the IEEE*, vol. 81, pp. 984–1010, July 1993.

[233] M. Shao and C. L. Nikias, *Signal Processing with Alpha-Stable Distributions and Applications*, Wiley, 1995.

[234] V. A. Shneidman, P. Jung, and P. Hänggi, "Power Spectrum of a Driven Bistable System," *Europhysics Letters*, vol. 26, no. 8, pp. 571–576, June 1994.

[235] S. W. Sides, R. A. Ramos, P. A. Rikvold, and M. A. Novotny, "Kinetic Ising System in an Oscillating External Field: Stochastic Resonance and Residence-Time Distributions," *Journal of Applied Physics*, vol. 81, no. 8, pp. 5597–5599, April 1997.

[236] M. K. Simon, J. K. Omura, R. A. Scholtz, and B. K. Levitt, *Spread Spectrum Communications Handbook*, McGraw Hill, 1994.

[237] E. Simonotto, M. Riani, C. Seife, M. Roberts, J. Twitty, and F. Moss, "Visual Perception of Stochastic Resonance," *Physical Review Letters*, vol. 78, no. 6, pp. 1186–1189, February 1997.

[238] C. Smith and C. Gervelis, *Cellular System Design and Optimization*, McGraw-Hill 1996.

[239] M. L. Spano, M. Wun-Fogle, and W. L. Ditto, "Experimental Observation of Stochastic Resonance in a Magnetoelastic Ribbon," *Physical Review A*, vol. 46, no. 8, pp. 5253–5256, October 1992.

[240] M. Stemmler, "A Single Spike Suffices: The Simplest Form of Stochastic Resonance in Model Neurons," *Network: Computation in Neural Systems*, vol. 7, pp. 687–716, 1996.

[241] N. G. Stocks, N. D. Stein, and P. V. E. McClintock, "Stochastic Resonance in Monostable Systems," *Journal of Physics A: Mathematical and General*, vol. 26, pp. L385–L390, 1993.

[242] N. G. Stocks, N. D. Stein, S. M. Soskin, and P. V. E. McClintock, "Zero-Dispersion Stochastic Resonance," *Journal of Physics A: Mathematical and General*, vol. 25, pp. L1119–L1125, 1992.

[243] G. Strang and T. Q. Nguyen, *Wavelets and Filter Banks*, Wellesley-Cambridge Press, 1996.

[244] C. J. Tessone and H. S. Wio, "Stochastic Resonance in Bistable Systems: The Effect of Simultaneous Additive and Multiplicative Correlated Noises," *Modern Physics Letters B*, vol. 12, no. 28, pp. 1195–1202. 1998.

[245] F. E. Theunissen and J. P. Miller, "Representation of Sensory Information in the Cricket Cercal Sensory System. II. Information Theoretic Calculation of System Accuracy and Optimal Tuning-Curve Widths of Four Primary Interneurons," *Journal of Neurophysiology*, vol. 66, no. 5, pp. 1690–1703, November 1991.

[246] M. Thorwart and P. Jung, "Dynamical Hysteresis in Bistable Quantum Systems," *Physical Review Letters*, vol. 78, no. 13, pp. 2503–2506, March 1997.

[247] F. Vaudelle, J. Gazengel, G. Rivoire, X. Godivier, and F. Chapeau-Blondeau, "Stochastic Resonance and Noise-Enhanced Transmission of Spatial Signals in Optics: The Case of Scattering," *Journal of the Optical Society of America*, vol. 15, no. 11, pp. 2674–2680, November 1998.

[248] G. Vemuri and R. Roy, "Stochastic Resonance in a Bistable Ring Laser," *Physical Review A*, vol. 39, no. 9, pp. 4668–4674, May 1989.

[249] M. Vetterli and J. Kovačević, *Wavelets and Subband Coding*, Prentice Hall, 1995.

[250] J. M. G. Vilar and J. M. Rubí, "Divergent Signal-to-Noise Ratio and Stochastic Resonance in Monostable Systems," *Physical Review Letters*, vol. 77, no. 14, pp. 2863–2866, September 1996.

[251] J. M. G. Vilar and J. M. Rubí, "Spatiotemporal Stochastic Resonance in the Swift-Hohenberg Equation," *Physical Review Letters*, vol. 78, no. 15, pp. 2886–2889, April 1997.

[252] J. M. G. Vilar and J. M. Rubí, "Stochastic Multiresonance," *Physical Review Letters*, vol. 78, no. 15, pp. 2882–2885, April 1997.

[253] S. T. Vohra and F. Bucholtz, "Observation of Stochastic Resonance near a Subcritical Bifurcation," *Journal of Statistical Physics*, vol. 70, no. 1/2, pp. 413–421, January 1993.

[254] S. T. Vohra and L. Fabiny, "Induced Stochastic Resonance near a Subcritical Bifurcation," *Physical Review E*, vol. 50, no. 4, pp. R2391–2394, October 1994.

[255] K. Wiesenfeld and F. Moss, "Stochastic Resonance and the Benefits of Noise: From Ice Ages to Crayfish and SQUIDs," *Nature*, vol. 373, pp. 33–36, January 1995.

[256] K. Wiesenfeld, D. Pierson, E. Pantazelou, C. Dames, and F. Moss, "Stochastic Resonance on a Circle," *Physical Review Letters*, vol. 72, pp. 2125–2129, April 1994.

[257] I. J. Winograd, T. B. Coplen, J. M. Landwehr, A. C. Riggs, K. R. Ludwig, B. J. Szabo, P. T. Kolesar, and K. M. Revesz, "Continuous 500,000-Year Climate Record from Vein Calcite in Devils Hole, Nevada," *Science*, vol. 258, pp. 255–260. October 1992.

[258] H. S. Wio, "Stochastic Resonance in a Spatially Extended System," *Physical Review E*, vol. 54, no. 4, pp. R3075–3078, October 1996.

[259] A. Witt, A. Neiman, and J. Kurths, "Characterizing the Dynamics of Stochastic Bistable Systems by Measures of Complexity," *Physical Review E*, vol. 55, no. 5, pp. 5050–5059, May 1997.

[260] W. Yang, M. Ding, and G. Hu, "Trajectory (Phase) Selection in Multistable Systems: Stochastic Resonance, Signal Bias, and the Effect of Signal Phase," *Physical Review Letters*, vol. 74, no. 20, pp. 3955–3958, May 1995.

[261] T. Zhou and F. Moss, "Analog Simulations of Stochastic Resonance," *Physical Review A*, vol. 41, no. 8, pp. 4255–4264, April 1990.

[262] T. Zhou, F. Moss, and P. Jung, "Escape-Time Distributions of a Periodically Modulated Bistable System with Noise," *Physical Review A*, vol. 42, no. 6, pp. 3161–3169, September 1990.

[263] S. Zozor and P.-O. Amblard, "Stochastic Resonance in a Discrete Time Nonlinear SETAR(1,2,0,0) Model," in *Proceedings of the 1997 IEEE Workshop on Higher-Order-Statistics*, July 1997.

[264] U. Zürcher and C. R. Doering, "Thermally Activated Escape over Fluctuating Barriers," *Physical Review E*, vol. 47, no. 6, pp. 3862–3869, June 1993.

Appendix A. The Standard Additive Model (SAM) Theorem

This appendix derives the basic ratio structure (31) of a standard additive model (SAM) fuzzy system and review the local structure of optimal fuzzy rules.

SAM Theorem. Suppose the fuzzy system $F : R^n \rightarrow R^p$ is a standard additive model: $F(x) = \text{Centroid}(B(x)) = \text{Centroid}(\sum_{j=1}^{m} w_j a_j(x) B_j)$ for if-part joint set function $a_j : R^n \rightarrow [0,1]$, rule weights $w_j \geq 0$, and then-part fuzzy set $B_j \subset R^p$. Then $F(x)$ is a convex sum of the m then-part set centroids:

$$F(x) \quad = \quad \frac{\sum_{j=1}^{m} w_j a_j(x) V_j c_j}{\sum_{j=1}^{m} w_j a_j(x) V_j} \quad = \quad \sum_{j=1}^{m} p_j(x) c_j. \tag{152}$$

The convex coefficients or discrete probability weights $p_1(x), \ldots, p_m(x)$ depend on the input x through

$$p_j(x) \quad = \quad \frac{w_j a_j(x) V_j}{\sum_{i=1}^{m} w_i a_i(x) V_i}. \tag{153}$$

V_j is the finite positive volume (or area if $p = 1$) and c_j is the centroid of then-part set B_j:

$$V_j \quad = \quad \int_{R^p} b_j(y_1, \ldots, y_p)\, dy_1 \cdots dy_p \quad > \ 0 \tag{154}$$

$$c_j \quad = \quad \frac{\int_{R^p} y\, b_j(y_1, \ldots, y_p)\, dy_1 \cdots dy_p}{\int_{R^p} b_j(y_1, \ldots, y_p)\, dy_1 \cdots dy_p}. \tag{155}$$

Proof. There is no loss of generality to prove the theorem for the scalar-output case $p = 1$ when $F : R^n \rightarrow R^p$. This simplifies the notation. We need but replace the scalar integrals over R with the p-multiple or volume integrals over R^p in the proof to prove the general case. The scalar case $p = 1$ gives (154) and (155) as

$$V_j \quad = \quad \int_{-\infty}^{\infty} b_j(y)\, dy \tag{156}$$

$$c_j \quad = \quad \frac{\int_{-\infty}^{\infty} y\, b_j(y)\, dy}{\int_{-\infty}^{\infty} b_j(y)\, dy}. \tag{157}$$

Then the theorem follows if we expand the centroid of B and invoke the SAM assumption $F(x) = \text{Centroid}(B(x)) = \text{Centroid}(\sum_{j=1}^{m} w_j\, a_j(x)\, B_j)$ to rearrange terms:

$$
\begin{aligned}
F(x) &= \text{Centroid}(B(x)) && (158)\\
&= \frac{\int_{-\infty}^{\infty} y\, b(y)\, dy}{\int_{-\infty}^{\infty} b(y)\, dy} && (159)\\
&= \frac{\int_{-\infty}^{\infty} y \sum_{j=1}^{m} w_j\, b_j'(y)\, dy}{\int_{-\infty}^{\infty} \sum_{j=1}^{m} w_j\, b_j'(y)\, dy} && (160)\\
&= \frac{\int_{-\infty}^{\infty} y \sum_{j=1}^{m} w_j\, a_j(x)\, b_j(y)\, dy}{\int_{-\infty}^{\infty} \sum_{j=1}^{m} w_j\, a_j(x)\, b_j(y)\, dy} && (161)\\
&= \frac{\sum_{j=1}^{m} w_j\, a_j(x) \int_{-\infty}^{\infty} y\, b_j(y)\, dy}{\sum_{j=1}^{m} w_j\, a_j(x) \int_{-\infty}^{\infty} b_j(y)\, dy} && (162)\\
&= \frac{\sum_{j=1}^{m} w_j\, a_j(x)\, V_j \dfrac{\int_{-\infty}^{\infty} y\, b_j(y)\, dy}{V_j}}{\sum_{j=1}^{m} w_j\, a_j(x)\, V_j} && (163)\\
&= \frac{\sum_{j=1}^{m} w_j\, a_j(x)\, V_j\, c_j}{\sum_{j=1}^{m} w_j\, a_j(x)\, V_j}. && (164)
\end{aligned}
$$

Now we give a simple *local* description of optimal lone fuzzy rules [144, 145]. We move a fuzzy rule patch so that it most reduces an error. We look (locally) at a minimal fuzzy system $F : R \rightarrow R$ of just one rule. So the fuzzy system is constant in that region: $F = c$. Suppose that $f(x) \neq c$ for $x \in [a, b]$ and define the error

$$e(x) \;=\; (f(x) - F(x))^2 \;=\; (f(x) - c)^2. \qquad (165)$$

We want to find the best place $\hat{x}$. So the first-order condition gives $\nabla e = \mathbf{0}$ or

$$0 \;=\; \frac{\partial e(x)}{\partial x} \;=\; 2(f(x) - c)\frac{\partial f(x)}{\partial x}. \tag{166}$$

Then $f(x) \neq c$ implies that

$$\frac{\partial e(x)}{\partial x} = 0 \qquad \Longleftrightarrow \qquad \frac{\partial f(x)}{\partial x} = 0 \tag{167}$$

at $x = \hat{x}$. So the extrema of e and f coincide in this case. Figure 7 shows how fuzzy rule patches can "patch the bumps" and so help minimize the error of approximation.

Appendix B. SAM Gradient Learning

Supervised gradient ascent can tune all the parameters in the SAM model (31) [143, 145]. A gradient ascent learning law for a SAM parameter ξ has the form

$$\xi(t+1) \;=\; \xi(t) + \mu_t \frac{\partial P}{\partial \xi} \tag{168}$$

where μ_t is a learning rate at iteration t. We seek to maximize the performance measure P of the dynamical system $\dot{q} = h(q, u)$. Here the signal-to-noise ratio (SNR) defines the performance P.

Let ξ_j^k denote the kth parameter in the set function a_j. Then the chain rule gives the gradient of the SNR with respect to ξ_j^k, with respect to the then-part set centroid c_j, and with respect to the then-part set volume V_j:

$$\frac{\partial\,\mathrm{SNR}}{\partial \xi_j^k} \;=\; \frac{\partial\,\mathrm{SNR}}{\partial F}\frac{\partial F}{\partial a_j}\frac{\partial a_j}{\partial \xi_j^k}, \qquad \frac{\partial\,\mathrm{SNR}}{\partial c_j} \;=\; \frac{\partial\,\mathrm{SNR}}{\partial F}\frac{\partial F}{\partial c_j}, \qquad \text{and} \qquad \frac{\partial\,\mathrm{SNR}}{\partial V_j} \;=\; \frac{\partial\,\mathrm{SNR}}{\partial F}\frac{\partial F}{\partial V_j}. \tag{169}$$

We have derived the partial derivative $\frac{\partial\,\mathrm{SNR}}{\partial F} = \frac{\partial\,\mathrm{SNR}}{\partial \sigma}$ in Section 5.2. We next derive the partial derivatives for the SAM parameters:

$$\frac{\partial F}{\partial a_j} \;=\; \frac{\left(\sum_{i=1}^{m} w_i\, a_i(x)\, V_i\right)(w_j\, V_j\, c_j) - w_j\, V_j \left(\sum_{i=1}^{m} w_i\, a_i(x)\, V_i\, c_i\right)}{\left(\sum_{i=1}^{m} w_i\, a_i(x)\, V_i\right)^2} \tag{170}$$

$$\;=\; \frac{[c_j - F(x)]\, w_j\, V_j}{\sum_{i=1}^{m} w_i\, a_i(x)\, V_i} \;=\; [c_j - F(x)]\,\frac{p_j(x)}{a_j(x)}. \tag{171}$$

The SAM ratio (31) gives [143]

$$\frac{\partial F}{\partial c_j} = \frac{w_j\, a_j(x)\, V_j}{\sum_{i=1}^{m} w_i\, a_i(x)\, V_i} = p_j(x) \tag{172}$$

and

$$\frac{\partial F}{\partial V_j} = \frac{w_j\, a_j(x)\, [c_j - F(x)]}{\sum_{i=1}^{m} w_i\, a_i(x)\, V_i} = \frac{p_j(x)}{V_j}\, [c_j - F(x)]. \tag{173}$$

Then the learning laws for the centroid and volume have the final form

$$c_j(t+1) = c_j(t) + \mu_t \frac{\partial\, \mathrm{SNR}}{\partial \sigma}\, p_j(x) \tag{174}$$

and

$$V_j(t+1) = V_j(t) + \mu_t \frac{\partial\, \mathrm{SNR}}{\partial \sigma} \frac{p_j(x)}{V_j}\, [c_j - F(x)]. \tag{175}$$

Learning laws for set parameters depend on how we define the set functions. The partial derivatives for the scalar sinc set function $a_j(x) = \sin\left(\frac{x - m_j}{d_j}\right) / \left(\frac{x - m_j}{d_j}\right)$ have the form

$$\frac{\partial a_j}{\partial m_j} = \begin{cases} \left(a_j(x) - \cos\left(\frac{x - m_j}{d_j}\right)\right) \frac{1}{x - m_j} & \text{for } x \neq m_j \\ 0 & \text{for } x = m_j \end{cases} \tag{176}$$

$$\frac{\partial a_j}{\partial d_j} = \left(a_j(x) - \cos\left(\frac{x - m_j}{d_j}\right)\right) \frac{1}{d_j}. \tag{177}$$

So this scalar set function leads to the learning laws

$$m_j(t+1) = m_j(t) + \mu_t \frac{\partial\, \mathrm{SNR}}{\partial \sigma} [c_j - F(x)] \frac{p_j(x)}{a_j(x)} \left(a_j(x) - \cos\left(\frac{x - m_j}{d_j}\right)\right) \frac{1}{x - m_j} \tag{178}$$

$$d_j(t+1) = d_j(t) + \mu_t \frac{\partial\, \mathrm{SNR}}{\partial \sigma} [c_j - F(x)] \frac{p_j(x)}{a_j(x)} \left(a_j(x) - \cos\left(\frac{x - m_j}{d_j}\right)\right) \frac{1}{d_j}. \tag{179}$$

Like results hold for the learning laws of product n-D set functions. A factored set function $a_j(x) = a_j^1(x_1) \cdots a_j^n(x_n)$ leads to a new form for the performance gradient. The gradient with respect to the parameter m_j^k of the jth set function a_j has the form

$$\frac{\partial P}{\partial m_j^k} = \frac{\partial P}{\partial F} \frac{\partial F}{\partial a_j} \frac{\partial a_j}{\partial a_j^k} \frac{\partial a_j^k}{\partial m_j^k} \quad \text{where} \quad \frac{\partial a_j}{\partial a_j^k} = \prod_{i \neq k}^{n} a_j^i(x_i) = \frac{a_j(x)}{a_j^k(x_k)}. \tag{180}$$

Products of the scalar sinc set functions defined the if-part fuzzy sets $A_j \subset R^n$ in the SAM approximator. Simulations have shown [182, 183] that sinc set functions tend to perform at least as well as other popular set functions in supervised fuzzy function approximation.

Chapter 3

Learning in the Presence of Noise

Malik Magdon-Ismail
magdon@work.caltech.edu
Department of Electrical Engineering
California Institute of Technology
136-93 Pasadena, CA, 91125

Alexander Nicholson
zander@foot.caltech.edu
Department of Computer Science
California Institute of Technology
136-93 Pasadena, CA, 91125

Yaser S. Abu-Mostafa
yaser@caltech.edu
Department of Electrical Engineering
and Department of Computer Science
California Institute of Technology
136-93 Pasadena, CA, 91125

Abstract

We report results about the impact of noise on information processing, with application to financial markets. These results quantify the tradeoff between the amount of data and the noise level in the data. They also provide estimates for the performance of a learning system in terms of the noise level. We use these results to derive a method for detecting the change in market volatility from period to period. We successfully apply these results to the four major foreign exchange markets. The results hold for linear as well as non-linear learning models and algorithms, and for different noise models.

Keywords: Learning, Noise, Convergence, Bounds, Test Error, Generalization Error, Model Limitation, Volatility.

1 Introduction

Information processing of financial data entails the extraction of relevant information from overwhelming noise. The levels of noise in financial markets are such that the most one can hope for is 'getting it right' slightly better than 50% of the time [17]. To complicate matters further, one also needs to be reasonably sure that one is not being fooled by a finite set of examples from historical data into believing that the performance is

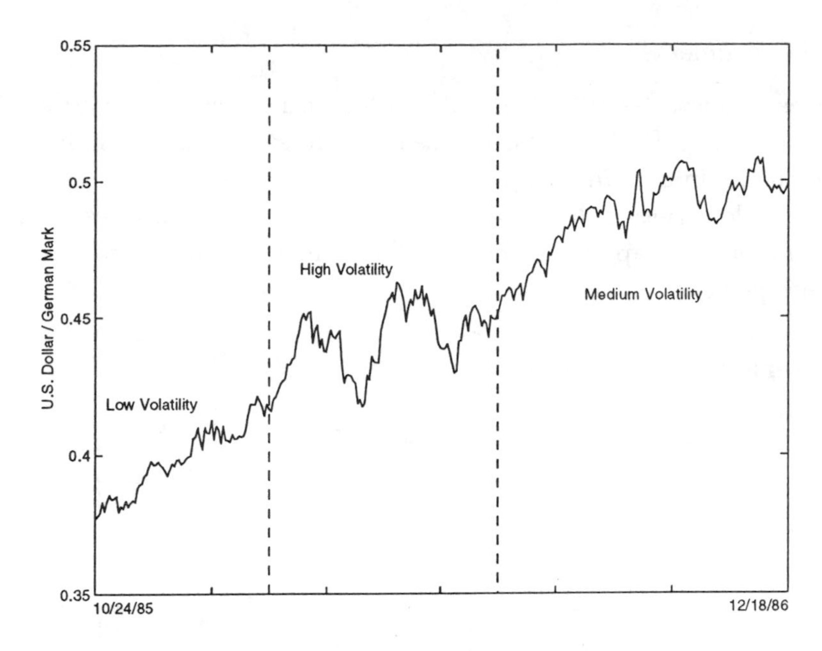

Figure 1: *The price curve for the U.S. Dollar vs. German Mark. The market volatility can change noticeably over the course of time.*

acceptable when it is actually (and disastrously) slightly worse than acceptable.

In addition to being a nuisance that complicates the processing of financial data, noise plays a role as a tradable commodity in its own right. Indeed, market volatility is the basis for a number of financial instruments, such as *options* [6], whose price explicitly depends on the level of volatility in the underlying market. For this reason, it is of economic value to be able to predict the changes in the noise level in financial time series as these changes are reflected in the price changes in tradable instruments. These changes can be significant as one can observe in figure 1 where the U.S. Dollar/German Mark market has undergone extreme changes in volatility.

In spite of the high levels of noise, financial data are among the best application domains for intelligent processing and advanced learning techniques. These data have been recorded very accurately for very long periods of time. They are available on different time scales, and simultaneously available in many different markets. This provides a very rich environment for analysis and experimentation using advanced processing techniques. Moreover, the payoff for even minute, but consistent, improvements in performance is huge.

In this paper, we tackle the question of how information processing is affected by the presence and variability of noise in the data. In doing so, we do not restrict the distribution or the time-varying nature of the noise, nor do we restrict the learning model or learning algorithm that we use. We report results that provide quantitative estimates of the optimal performance that can be achieved in the presence of noise. In financial markets, this provides a benchmark for the target performance given a set of data. We also quantify the tradeoff between the amount of data needed and the level of noise in the data. Our experiments with real foreign exchange data demonstrate that the results are applicable to the case of finite data, the only case of practical interest. They also provide a means of assessing the change in the level of noise in financial data that can be applied to volatility-based financial instruments.

The paper is organized as follows. Section 2 introduces financial time series and section 3 covers the main results about the impact of noise. It should be stressed that in section 3 we analyze a very general learning paradigm, and although the main domain of application here is to the processing of financial time series, the results are applicable to a wide variety of learning problems. These results are tested in the four major foreign exchange markets in section 4. The appendix includes the formal definitions, theorems, and complete proofs of all the results that we report.

2 Financial Time Series Prediction

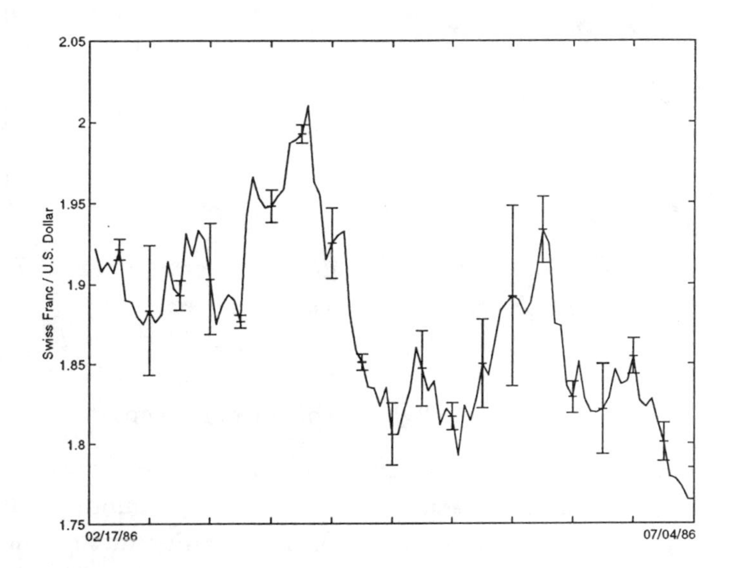

Figure 2: *Financial time series tend to be very volatile. Above is the realization of one such time series from the FX markets. Error bars are used merely to illustrate that each point is the outcome of a noisy event.*

Financial markets present us with data in the form of a time series. We might have the daily, hourly or tick-by-tick stock prices or foreign exchange rates. For financial time series, it is of economic interest to predict the value at some time in the future. Thus, we would like to extract as much information as possible from historical data with the hope of learning the underlying behavior. In general, we can consider the value of a time series $y(t)$ at any time t as a noisy data point $y = f(\mathbf{x}) + \epsilon$. Here f is a deterministic function of a vector $\mathbf{x}(t)$ of market indicators and $\epsilon(t)$ is noise. The task at hand is one of learning $f(\cdot)$ from a finite data set (the history of the series).

In modeling the time variation of a stock price S, the standard Black-Scholes model for pricing options based on volatility [6] assumes the variation to be of the form

$$\mathrm{d}S = \tilde{\mu} S \mathrm{d}t + \tilde{\sigma} S \eta \sqrt{\mathrm{d}t}$$

where $\tilde{\sigma}$ is the market volatility and η has a zero-mean normal distribution with variance 1. Thus, the Black-Scholes model uses only the previous price as the indicator vector $\mathbf{x}$. The price at different times has a deterministic dependence on the past (the $\tilde{\mu}$ term) and a noisy component (the $\tilde{\sigma}$ term). The variance of the noisy component is related to the volatility and need not be constant. The precise relation is given by

$$\sigma_i^2 = \tilde{\sigma}^2 \Delta S_i^2 \tag{1}$$

where i is a time index, and ΔS_i is increment in the stock price from time i to time $i+1$.

Extensive literature already exists on methods for extracting information from noisy time series ([7], [8], [10], [11], [14]). The details of such methods are not our present concern. We are interested in determining how our prediction performance depends on the amount of available data and the variability of the data (which is related to market volatility (1)) – what change in performance are we to expect if this year's market is more volatile than last years market? What change in performance relative to some benchmark are we to expect if the market changed recently and hence we only have few data points to learn from?

Pricing information is available on a variety of time scales, which presents us with a data set size vs. variability tradeoff. We could choose to use the tick-by-tick data because we will then have many data points, but the price we have to pay is that these data points are much noisier. The trade off will depend on how much noisier the tick-by-tick data is, and the details of the learning scheme. Market analysts would like to quantify this tradeoff by how it affects performance.

An estimate of the best performance that we can achieve with a given information extraction scheme might also be economically useful. As well as providing a criterion for selecting between different models, knowing the model limitation could be useful for determining whether even an unlimited amount of data will give a system that is financially worth the risk. This would allow analysts to compare trading strategies based on their model limitation.

It is to be expected that when markets are volatile, the performance of a learning system drops. However, the effects of the noise should become less pronounced with increasing data availability. In the next section, we quantify this intuition.

3 Impact of Noise on Learning

In this section we address the issues raised in section 2 in the context of learning theory. We begin by setting up the learning problem, restate the questions in the learning theory framework, and present theoretical and experimental results.

3.1 The Learning Problem

We assume the standard learning paradigm. The goal is to learn a target function $f : \mathbf{R}^d \to \mathbf{R}$. A training data set $\mathcal{D}_N$ is given, which consists of N input output pairs $\{\mathbf{x}_i, y_i\}_{i=1}^N$. Each $\mathbf{x}_i \in \mathbf{R}^d$ is drawn from some input probability measure $dF(\mathbf{x})$ which we assume to have compact support. Learning entails choosing a hypothesis function g from a collection of candidate functions $\mathcal{H}$. We will assume that the target function f and the candidate functions $g \in \mathcal{H}$ are continuous. The set $\mathcal{H}$ is called the *learning model* because it reflects how we choose to model the target function. The hypothesis function is chosen by a *learning algorithm* $\mathcal{A}$ based on some performance criterion on the data. We assume that $\mathcal{A}$ is a mapping $\mathcal{A} : \mathcal{D}_N \to \mathcal{H}$. A typical learning algorithm might be one that uses gradient descent to select the hypothesis which minimizes the mean squared error on the training set. Given a learning task, we select a particular *learning system*, which takes as input a data set and produces a hypothesis function (see figure 3).

Definition 3.1 *A Learning System* $\mathcal{L}$ *is a pair* $\{\mathcal{A}, \mathcal{H}\}$.

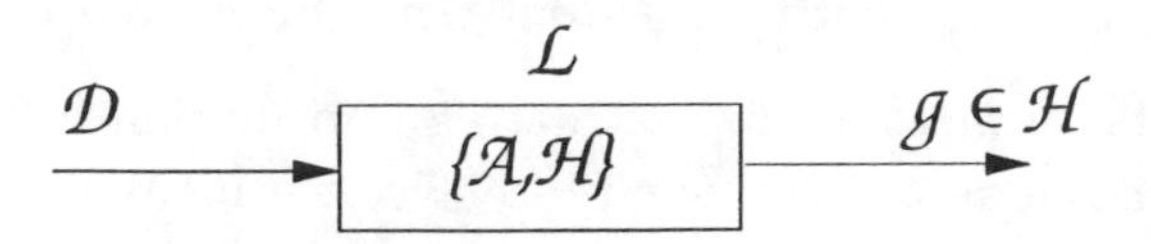

Figure 3: *The learning setup*

Additive noise is present in the training data,

$$y_i = f(\mathbf{x}_i) + \epsilon_i$$

We further assume that the noise realizations are independent and zero mean, so

$$\langle \epsilon \mid \mathbf{x} \rangle_\epsilon = \mathbf{0}, \qquad \langle \epsilon \epsilon^T \mid \mathbf{x} \rangle_\epsilon = diag[\sigma_1^2, \sigma_2^2, \ldots, \sigma_N^2]$$

(we use $\langle \cdot \rangle$ to denote expectations, $\sigma = [\sigma_1 \sigma_2 \ldots \sigma_N]$, and $diag[\cdot]$ denotes a diagonal matrix). It should be noted that we allow the noise variance to change from one data point to another, which is always the case in financial markets.

Define $g_{\mathcal{D}_N}(\mathbf{x}) \in \mathcal{H}$ as $\mathcal{A}(\mathcal{D}_N)$, the function that was chosen by the algorithm. We define the test error for $g_{\mathcal{D}_N}$ as the expectation of the squared deviation between $g_{\mathcal{D}_N}$ and $f(\mathbf{x})$ taken over the input space. Thus the test error measures the ultimate performance of our system after it has learned from the data. We denote the test error by $E[g_{\mathcal{D}_N}]$.

$$E[g_{\mathcal{D}_N}] = \left\langle (g_{\mathcal{D}_N}(\mathbf{x}) - f(\mathbf{x}))^2 \right\rangle_{\mathbf{x}} \tag{2}$$

We can further define the expected test error, $\mathcal{E}_N(\sigma)$ as the expectation of the test error taken over possible realizations of the noise and the data set.

$$\mathcal{E}_N(\sigma) = \langle E[g_{\mathcal{D}_N}] \rangle_{\epsilon, \mathcal{D}_N} \tag{3}$$

The goal is to minimize $\mathcal{E}_N(\sigma)$. $\mathcal{E}_N(\sigma)$ represents the expected test performance averaged over the choice of training examples. It is related to the "future profit" you expect to make having trained your learning system on the available data. $\mathcal{E}_N(\sigma)$ will depend on the detailed properties of the learning system and target function. It would be a daunting task to tackle the behavior of $\mathcal{E}_N(\sigma)$ in general, but as we shall see, under quite unrestrictive conditions, the changes in $\mathcal{E}_N(\sigma)$ as the noise or data set size change can be quantified. This will be related to the tradeoff in profit when attempting to learn and predict during more volatile stages of the market compared to less volatile stages.

A related quantity of interest is $\mathcal{N}$, the number of data points (with noise added) that are needed to attain a test error comparable to that attainable when N noiseless examples are available.

$$\mathcal{N}(\Delta, \sigma, N) \triangleq \min_{N_1}\{N_1 : \mathcal{E}_{N_1}(\sigma) - \mathcal{E}_N(0) \leq \Delta\} \tag{4}$$

$\mathcal{N}(\Delta, \sigma, N)$ is the number of noisy examples that are equivalent to N noiseless examples, and it describes the trade off between numerous, more volatile data, versus fewer and less volatile data. The answers to the questions posed in section 2 lie in the behavior of $\mathcal{E}_N(\sigma)$ and $\mathcal{N}(\Delta, \sigma, N)$. We address these questions analytically next.

3.2 Performance of a Learning System

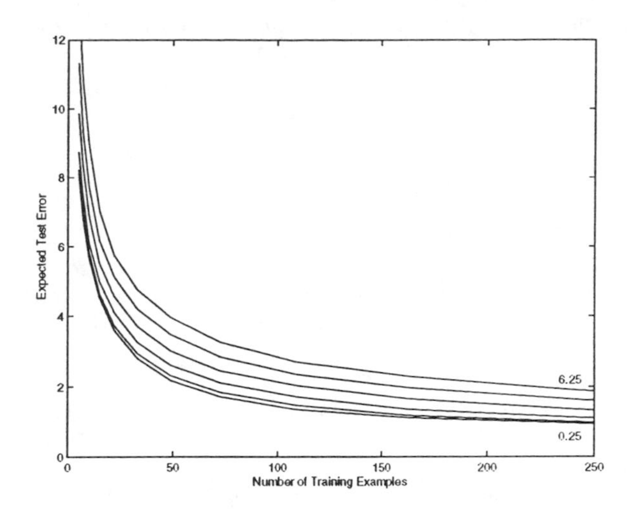

Figure 4: *Experiments illustrating the behavior of the test error as a function of N for various noise levels with variances ranging from 0.25–6.25. A non-linear neural network learning model was used with gradient descent on the squared error. Data was created using a non-linear target function.*

Intuition tells us that noisier data leads to worse test performance. This is because the learning system attempts to fit the noise (i.e. learn a random effect) at the expense of fitting the true underlying dependence. However, the more data we have, the less pronounced the impact of the noise will be. This intuition is illustrated in figure 4. We observe that the higher the noise, the higher the test error, but the curves appear to be getting closer to each other as we use more and more examples for the learning process. We would like to quantify this idea.

In order to be able to do so, we need to restrict ourselves to *stable learning systems*. Stable learning systems possess the two properties "continuity" and "unbiasedness". Continuity ensures that "close" data sets are mapped to "close" functions. For two data sets differing only by the addition of zero-mean noise, unbiasedness requires that at every point, the average value (with respect to the noise) of the functions resulting from the noisy data set is equal to the value of the function resulting from the noiseless data set. (Refer to the appendix for formal definitions.)

These properties are somewhat intuitive, and we note that, for any learning system $\mathcal{L}$, they can be checked directly. We would like our learning procedure to be robust towards small noise fluctuations in the data so we do not consider learning models that may yield discontinuous behavior. The unbiasedness property may seem fragile, especially given the extremely nonlinear nature of a learning algorithm. Nevertheless, we consider it an important and not overly restrictive condition on a learning system. If the noise is small, then the first order change in $\mathcal{A}(D_N)$ should be proportional to the noise parameter, so that the average change is zero with zero-mean noise. Indeed, experiments with neural networks show that learning with gradient descent and conjugate gradient descent on the mean squared error are unbiased with a reasonable noise level. Thus, linear and neural network learning models give learning systems that are stable.

We then have the following theorem.

Theorem 3.2 *Let $\mathcal{L}$ be stable. Then $\forall \epsilon > 0$, $\exists C_1$, C_2 such that using $\mathcal{L}$, it is at least possible to attain a test error bounded by*

$$\mathcal{E}_N(\sigma) < \mathcal{E}_N(0) + \frac{\overline{\sigma^2} C_1}{N} + \epsilon + O\left(\frac{1}{N^2}\right) \tag{5}$$

$$\mathcal{E}_N(0) < E_0 + \frac{C_2}{N} + \epsilon + o\left(\frac{1}{N}\right) \tag{6}$$

where $\lim_{N\to\infty} \mathcal{E}_N(0) = E_0$ *and* $\overline{\sigma^2} = \frac{1}{N}\sum_{i=1}^{N} \sigma_i^2$. C_1, C_2 *are constants that depend on the input distribution, target function and learning system.*

The proof can be found in the appendix (Theorem B.5). Furthermore, in certain cases we can combine (5) and (6) to get

$$\mathcal{E}_N(0) < E_0 + \frac{C_1\overline{\sigma^2} + C_2}{N} + o\left(\frac{1}{N}\right) \tag{7}$$

The essential content of the theorem is that the expected test error increases in proportion to $\overline{\sigma^2}$ holding everything else constant, and decreases in proportion to $1/N$ holding everything else constant. The conditions of Theorem 3.2 are quite general and are satisfied by a wide variety of learning models and algorithms. For learning models that are linear $C_1 = d + 1$. E_0 is the model limitation modulo the learning algorithm when tested on noiseless data. The limiting performance on noisy future data is $E_0 + \overline{\sigma^2}$. One expects that for more complex models, the model limitation (E_0) is lower than for less complex learning models. However, the convergence parameters (C_1, C_2) are expected to be larger for more complex models. Thus, for a given number of data points, there will be an optimal model complexity (eg. number of hidden units for a neural network) minimizing the bound of theorem 3.2. One can compare this tradeoff to the bias-variance tradeoff [19].

Experimentally we observe that the bounds of theorem 3.2 are quite tight even for small N (see figure 5) so combining (5) and (6) we expect the following dependence for $\mathcal{N}(\Delta, \sigma, N)$, the number of noisy examples that are equivalent to N noiseless examples.

$$\mathcal{N}(\Delta, \sigma, N) \sim \frac{\overline{\sigma^2} C_1 + C_2}{\frac{C_2}{N} + \Delta} \tag{8}$$

The results are illustrated in figure 5. Artificial data sets were created from a known target function. Figure 5(a) illustrates the results of fitting a linear model to nonlinear data. Shown is the residual error $\hat{\mathcal{E}}_N(\sigma) = \mathcal{E}_N(\sigma) - \mathcal{E}_N(0)$. The inputs are chosen from $\mathbf{R}^2$, and the dashed lines illustrate that $\hat{\mathcal{E}}_N(\sigma)$ quickly converges to $3\overline{\sigma^2}/N$ as expected from (5). Figure 5(b) shows similar results for a nonlinear learning model. Gradient descent was used to train the three hidden unit neural network model. Ideally we expect this algorithm/model pair to be continuously compatible, and it was empirically shown to be mean preserving. The residual errors very closely follow 20 $\overline{\sigma^2}/N$, showing that we have approximate equality in (5) for $C_1 \sim 20$.[1] Figures 5 (c) and (d) show that $\mathcal{E}_N(0)$ also behaves linearly in $1/N$ for both cases (i.e. it quickly approaches the bound in 6).

[1]This suggests that the condition in Corollary B.6 holds.

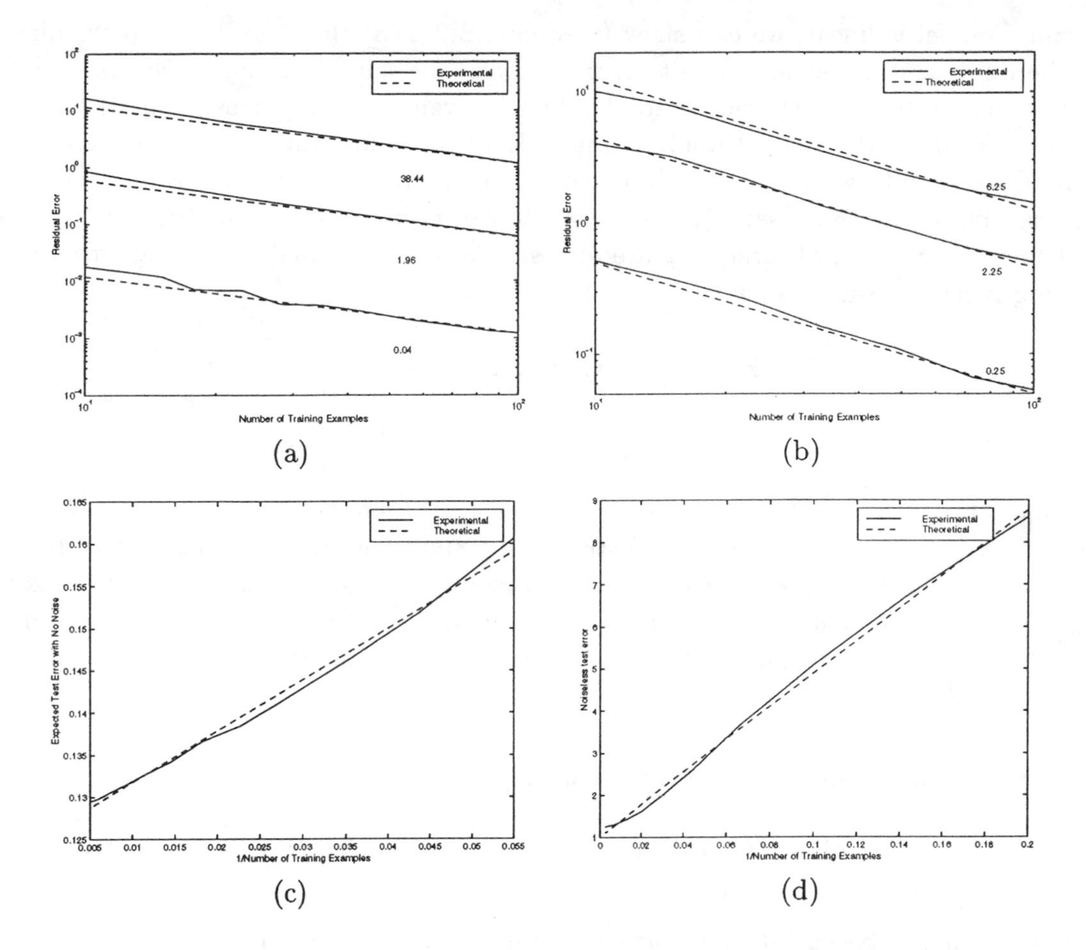

Figure 5: (a)*Nonlinear target function and linear learning model – The residual error is shown for a nonlinear target function trained using a linear model. Gaussian noise with* $\overline{\sigma^2}$ *ranging from 0.04 to 38.44 was added to training sets. The dashed lines show the error level predicted by Theorem B.2.* (b)*Nonlinear target function and nonlinear learning model – The residual error is shown for a nonlinear target function trained using a 2-3-1 network. Gaussian noise with* $\overline{\sigma^2}$ *ranging from 0.25 to 6.25 was added to training sets. The results correspond very closely to those predicted by Theorem 3.2 when* $C_1 = 20$ *(shown with dashed lines).* (c)*The behavior of the expected test error with no noise for the learning scenario in (a). We observe that for even small* N *there is close agreement between the theoretical* $1/N$ *decrease.* (d)*The behaviour of the expected test error with no noise for the non-linear learning scenario in (b). Once again for small* N *we observe the expected behavior.*

3.3 Estimating the Model Limitation

When the learning model is linear, we can show (theorems B.2, B.3) that the expected training error $\mathcal{E}_{tr}(\sigma)$ (the error on the data set) and expected test error approach the same limiting value from opposite sides as $N \to \infty$. Further the rates of convergence to this limiting value are the same. Amari [1] has obtained a similar asymptotic result in the case of nonlinear models when performing gradient descent on the training error. Using the Amari result, we can use our bound on the test error to bound the training error performance. The expected error on a noisy data set, $\mathcal{E}_{test}$ is related to $\mathcal{E}_N(\sigma)$ by $\mathcal{E}_{test}(\sigma) = \mathcal{E}_N(\sigma) + \overline{\sigma^2}$. The experiments demonstrate that the bounds of theorem 3.2 are almost saturated for small N, so, ignoring terms that are $o(1/N)$, and using Amari's result we have

$$E_0 + \overline{\sigma^2} \leq \ \mathcal{E}_{test}(\sigma) \approx \ E_0 + \overline{\sigma^2} + \frac{C_1\overline{\sigma^2} + C_2}{N} \tag{9}$$

$$E_0 + \overline{\sigma^2} \geq \ \mathcal{E}_{tr}(\sigma) \approx \ E_0 + \overline{\sigma^2} - \frac{C_1\overline{\sigma^2} + C_2}{N} \tag{10}$$

(in the case of linear learning models we can replace C_1 by $d+1$). From the data set of size N, for $N_1 < N$, we can randomly pick N_1 data points (perform Bootstrapping [18] on the training data). Thus, by varying N_1 in the training phase and observing the error on the training set, we obtain an estimate of the model limitation $E_0 + \overline{\sigma^2}$. This method also immediately furnishes an estimate of $C_1\overline{\sigma^2} + C_2$, so we can estimate the parameters that are needed for the bound (7). This is illustrated in the next section where we apply the results presented here to the case of financial time series.

4 Application to Financial Market Forecasting

We can apply the results of section 3 to real financial market data. Figure 4 illustrates the $1/N$ behavior of the residual error $\hat{\mathcal{E}}_N(\sigma)$ for foreign exchange rates.

Daily close exchange rates between 1984 and 1995 were used for the Swiss Franc (CHF), German Mark (DEM), British Pound Sterling (STG) and Japanese Yen (JPY). A linear model was used to learn the future price as a function of the close price of the previous five days.

We performed the following experiments. The last 1000 data points of each time series were held out as a test set. The remaining points were used to create a data set

$$\{\mathbf{x}_k = (S_{k-4}, \ldots, S_k), y_k = S_{k+1}\}$$

N_1 points were sampled from this set and used to learn. This was repeated to obtain an estimate of the expected test and training error. We show the dependence of the expected test error on the number of training examples in figure 4. Though it is not obvious that the assumptions made to derive the results hold, as with the results on artificial data, the test error seems to not only obey the bound of equation (5), but quickly assumes $1/N$ behavior. Assuming the bounds to be tight for both the test error and training error, we are able to estimate the best possible performance of the linear model by finding the line best fitting $\mathcal{E}_N(0)$ as a function of $1/N$. Table 4 summarizes these estimates.

We compare the model limitation to that of simply predicting the present value as the next value. We find that this simple strategy virtually attains the model limitation suggesting that today's price completely reflects

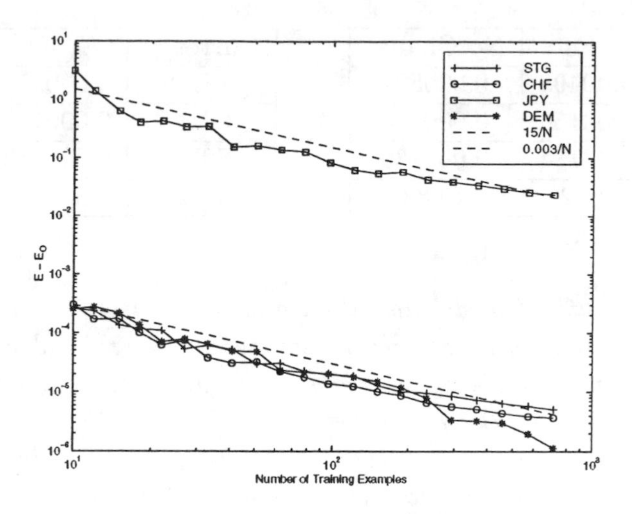

Figure 6: *The dependence of the test error-E_0 on N is depicted for the British Pound (STG), the Swiss Franc (CHF), the Japanese Yen (JPY) and the German Mark (DEM). Also shown are two lines that show $1/N$ behavior. We see that the test error curves follow the theory well.*

tomorrow's price – that's the best we can expect to achieve systematically. The results in table 4 are appealing on two accounts. Firstly, assuming that today's price is the best predictor of tomorrow's price, the technique we use to predict the model limitation is performing well. Secondly, because the model limitation estimates are slightly below the error of the simple strategy, we deduce that there is some information that can be extracted from previous prices.

By training on different time periods, we find that the model limitation may change. If we assume the underlying dependence to have remained constant so that E_0 has not changed, then the resulting change can only be due to a change in $\overline{\sigma^2}$ thus providing an estimate of the change in the volatility (since the volatility is related to the change in $\overline{\sigma^2}$ (1)). It appears from table 4 that of the four currencies, the British Pound's volatility seems to have increased while the remaining three markets display decreasing volatility.

We see that the results of section 3 apply to the problem of financial forecasting. Experiment bears out the fact that the answers to the questions posed in section 2 lie in the expressions for $\mathcal{E}_N(\sigma)$ and $\mathcal{N}(\Delta, \sigma, N)$ in equations (5) and (8).

5 Conclusion

The new results in section 2 are represented in theorem 3.2. The experimental results on artificial data amply support the theory. We have shown that the number of noisy examples required for comparable generalization with N noiseless examples increases as $\overline{\sigma^2}$. Explicitly, the main result bounds the test error for a noisy data set by $\mathcal{E}_N(\sigma) \leq E_0 + \frac{\overline{\sigma^2}C_1' + C_2'}{N} + o\left(\frac{1}{N}\right)$. We also obtained a result that bounds the expected test error relative to a benchmark test error (5). Experimentally we showed that this result applies to the non-asymptotic regime – the empirical results show that the bounds hold with almost equality for N as small as 20. Intuitively this is because the non-asymptotic effects that affect $\mathcal{E}_N(\sigma)$ also have a similar effect on $\mathcal{E}_N(0)$.

Currency	E_0 Est.	No Change
DEM	0.000499	0.000502
CHF	0.000158	0.000160
STG	0.000134	0.000136
JPY	1.082	1.083

(a)

Currency	E_0 Est.	No Change
DEM	0.000156	0.000152
CHF	0.000148	0.000151
STG	0.000153	0.000157
JPY	0.851	0.867

(b)

Table 1: *Estimate of model limitation and comparison to simple predictor. In (a) we use the training error to estimate* $E_0 + \overline{\sigma^2}$ *and compare to the performance on the training set when we use the simple system: predict no change in price. In (b) we use the test error curve to estimate* E_0. *Only (a) is possible in practice, but both yield very good estimates(if we assume that this simple strategy is close to the best you can do), thus verifying that the results of section 3 can be applied to this learning problem. The change in the estimate from (a) to (b) is due to the fact that the test and training sets are taken from different time intervals, and hence the estimates reflect a change in the market volatility over time.*

We began with the goal of answering two questions (initially posed in the context of financial time series): Relative to a benchmark scenario (that of learning with no noise), how does the performance change as the noise and number of examples changes? This dependence is represented by the expression for $\mathcal{E}_N(\sigma)$ above. This expression is a similar result to those derived by Amari [1] and Moody [4]. However the differences are significant. Amari compares the training error when descending on a given error function to the expectation of *that* error when you have finished learning. The learning algorithm is specific but the form of the error function may vary. Moody considers minimization of an error term plus a complexity term and assumes that the input distribution is a sum of delta functions at the training data points. In this paper, we derive a convergence result for the expected squared error without severely restricting the learning algorithm or the input distribution. The results were presented in the context of financial time series analysis, but we note that they are applicable to the general learning problem, independent of most of the details of the learning model and learning algorithm. In particular, we do not require the learning algorithm to minimize a simple training error measure – optimizing a generalized regularized training error (as in [5]) should produce an algorithm that still satisfies the conditions of theorem 3.2.

We provided an estimate of the model limitation which we used to estimate the best possible performance when learning in the FX markets. The results were consistent with the assumption that today's price reflects all the information about tomorrow's price. Using this method for predicting the model limitation, we could detect changes in the market volatility, which is of economic use.

It would be useful to explore the relationship between the constants (E_0, C_1, C_2) that parametrize the expected test error dependence.

Acknowledgements

We would like to thank Amir Atiya, Joseph Sill, Zehra Cataltepe and Xubo Song for their helpful comments.

6 Appendix

A Definitions

One expects that if one has "close" data sets $\mathcal{D}_N = \{\mathbf{x}_i, f(\mathbf{x}_i)\}$ and $\mathcal{D}'_N = \{\mathbf{x}_i, f(\mathbf{x}_i) + e(\mathbf{x}_i)\}$ where $e(\mathbf{x}_i)$ is small, then $\mathcal{A}(\mathcal{D}_N)$ should be "close" to $\mathcal{A}(\mathcal{D}'_N)$. For $\mathcal{A}$ to have this property, $\mathcal{H}$ should be able to implement the two "close" functions.[2] We formalize this notion by defining the class of learning systems that are n^{th} order–*continuously compatible* ($\mathcal{CC}_n$) with respect to the probability measure $dF(\mathbf{x})$. We will use the following notation. Let $\mathcal{S}$ be the compact support for $dF(\mathbf{x})$. Let $\mathcal{D} = \{\mathbf{x}_i, y_i\}$, $\mathcal{D}' = \{\mathbf{x}'_i, y'_i\}$ be any two data sets on $\mathcal{S}$ such that for every $\{x_i, y_i\}$ in $\mathcal{D}$ there is a $\{x'_j, y'_j\}$ in $\mathcal{D}'$ such that $\|\mathbf{x}_i - \mathbf{x}'_j\| \le \epsilon_{max}$ and $|y_i - y'_j| \le \epsilon_{max}$ and the same is true for every $\{x'_j, y'_j\}$ in $\mathcal{D}'$, i.e. there is a $\{x_i, y_i\}$ in $\mathcal{D}$ such that $\|\mathbf{x}_i - \mathbf{x}'_j\| \le \epsilon_{max}$ and $|y_i - y'_j| \le \epsilon_{max}$. Let $\mathcal{A}(\mathcal{D}) = g(\mathbf{x})$, $\mathcal{A}(\mathcal{D}') = g(\mathbf{x}) + \eta(\mathbf{x})$.

Definition A.1 *$\mathcal{L}$ is <u>n^{th} order–continuously compatible</u> if $\exists C$ such that*

$$\langle |\eta(\mathbf{x})|^n \rangle_{\mathbf{x}} \le (C\epsilon_{max})^n$$

with probability 1 (i.e. for almost every $\mathcal{D}$). We will write $\mathcal{L} \in \mathcal{CC}_n$.

We would like $\mathcal{A}$ to be "unbiased" in the following sense. If we have a data set $\mathcal{D}$ with $\mathcal{A}(D) = g_0$ and we add independent, zero mean noise to the targets to get a new data set $\mathcal{D}'$ then we would like $\langle \mathcal{A}(\mathcal{D}') \rangle_{noise} = g_0$, where this average of functions is taken pointwise. This motivates the following definition.

Definition A.2 *Let $\mathcal{D}$ and $\mathcal{D}'$ be two data sets related by $\mathbf{y}' = \mathbf{y} + \epsilon$ where the ϵ_i's are independent and zero mean. Then $\mathcal{L}$ is <u>mean preserving</u> or <u>unbiased</u> if $\langle \mathcal{A}(\mathcal{D}') \rangle_{\epsilon} = \mathcal{A}(\mathcal{D})$ with probability 1 (i.e. for almost every $\mathcal{D}$).*

Definition A.3 *A learning system $\mathcal{L}$ is <u>stable</u> if it is in $\mathcal{CC}_2$ and it is mean preserving.*

B Proofs of Results

Proposition B.1 *If $\mathcal{L} \in \mathcal{CC}_n$ then $\mathcal{L} \in \mathcal{CC}_m$ for $m = 1 \ldots n$.*

PROOF: By Jensen's inequality, $\langle |f(\mathbf{x})|^a \rangle_{\mathbf{x}} \le \langle |f(\mathbf{x})| \rangle_{\mathbf{x}}^a$ for $0 \le a \le 1$. Letting $f(\mathbf{x}) = \eta(\mathbf{x})^n$ as in Definition A.1 and $a = m/n$ for $m \le n$, the proposition now follows because $\mathcal{L} \in \mathcal{CC}_n$. ∎

We use the notation

$$\mathbf{X} = [\mathbf{x}_1 \;\; \mathbf{x}_2 \;\; \ldots \;\; \mathbf{x}_N], \qquad \mathbf{y} = \mathbf{f} + \epsilon$$
$$\Sigma \equiv \langle \mathbf{x}\mathbf{x}^T \rangle_{\mathbf{x}}, \qquad \mathbf{q} = \langle \mathbf{x} f(\mathbf{x}) \rangle_{\mathbf{x}}$$

The law of large numbers gives us that $\mathbf{X}\mathbf{X}^T \underset{N\to\infty}{\longrightarrow} N\Sigma$ and $\mathbf{X}\mathbf{f} \underset{N\to\infty}{\longrightarrow} N \langle \mathbf{x} f(\mathbf{x}) \rangle_{\mathbf{x}}$. where we assume that the conditions for this to happen are satisfied.

Theorem B.2 *Let $\mathcal{H}$, the learning model, be the set of linear functions $\mathbf{w} \cdot \mathbf{x} + w_0$ and let the learning algorithm be minimization of the squared error. Then*

$$\mathcal{E}_N(\sigma) = \mathcal{E}_N(0) + \frac{\overline{\sigma^2}(d+1)}{N} + O\left(\frac{1}{N^2}\right) \tag{11}$$

$$\mathcal{E}_N(0) = E_0 + \frac{B}{N} + O(\frac{1}{N^{\frac{3}{2}}}) \tag{12}$$

where $E_0 = \lim_{N\to\infty} \{\mathcal{E}_N(0)\}$ and B is a constant dependent on the input distribution. It follows that $\mathcal{N}(\Delta, \sigma, N) = \frac{\overline{\sigma^2}(d+1)+B}{\Delta + \frac{B}{N}} + O(\frac{1}{N^{\frac{1}{2}}})$.

[2]These conditions will often be satisfied in practice.

PROOF:. $g \in \mathcal{L} \Leftrightarrow g(\mathbf{x}) = \mathbf{x}^T\mathbf{w}$. The Least Squares estimate of $\mathbf{w}$ is given by

$$\hat{\mathbf{w}} = (\mathbf{X}\mathbf{X}^T)^{-1}\mathbf{X}\mathbf{y} \tag{13}$$

from which we calculate

$$\begin{aligned}
\mathcal{E}_N(\sigma) &= \langle \hat{\mathbf{w}}^T\mathbf{x}\mathbf{x}^T\hat{\mathbf{w}} - 2\hat{\mathbf{w}}^T\mathbf{x}f(\mathbf{x}) + f(\mathbf{x})^2\rangle_{\mathbf{x},\mathbf{X},\epsilon} \\
&= \langle f^2\rangle - 2\left\langle \mathbf{f}^T\mathbf{X}^T(\mathbf{X}\mathbf{X}^T)^{-1}\right\rangle_{\mathbf{X}}\mathbf{q} + \left\langle \mathbf{f}^T\mathbf{X}^T(\mathbf{X}\mathbf{X}^T)^{-1}\Sigma(\mathbf{X}\mathbf{X}^T)^{-1}\mathbf{X}\mathbf{f}\right\rangle_{\mathbf{X}} \\
&\quad + \left\langle \epsilon^T\mathbf{X}^T(\mathbf{X}\mathbf{X}^T)^{-1}\Sigma(\mathbf{X}\mathbf{X}^T)^{-1}\mathbf{X}\epsilon\right\rangle_{\mathbf{X},\epsilon} \\
&= \mathcal{E}_N(0) + \sum_i \sigma_i^2\left\langle \left[\mathbf{X}^T(\mathbf{X}\mathbf{X}^T)^{-1}\Sigma(\mathbf{X}\mathbf{X}^T)^{-1}\mathbf{X}\right]_{ii}\right\rangle_{\mathbf{X}}
\end{aligned}$$

where we have used (1) and $\mathbf{q} = \langle \mathbf{x}f(\mathbf{x})\rangle_{\mathbf{x}}$. By the law of large numbers, we note that $(\mathbf{X}\mathbf{X}^T)^{-1} \underset{N\to\infty}{\longrightarrow} N\Sigma$ and $\mathbf{X}\mathbf{f} \underset{N\to\infty}{\longrightarrow} N\mathbf{q}$, so we write

$$\mathbf{X}\mathbf{X}^T = N\Sigma + \sqrt{N}\mathbf{V}(\mathbf{X}), \qquad \mathbf{X}\mathbf{f} = N\mathbf{q} + \sqrt{N}\mathbf{a}(\mathbf{X}) \tag{14}$$

where $\langle \mathbf{V}\rangle_{\mathbf{X}} = \langle \mathbf{a}\rangle_{\mathbf{X}} = 0$ and Var($\mathbf{V}$) and Var($\mathbf{a}$) are $O(1)$. Using (14) and the identity $[1+\lambda\mathbf{A}]^{-1} = 1-\lambda\mathbf{A}+\lambda^2\mathbf{A}^2+O(\lambda^3)$

$$\mathbf{X}^T(\mathbf{X}\mathbf{X}^T)^{-1}\Sigma(\mathbf{X}\mathbf{X}^T)^{-1}\mathbf{X} = \frac{\mathbf{X}^T\Sigma^{-1}\mathbf{X}}{N^2} - 2\frac{\mathbf{X}^T\Sigma^{-1}\mathbf{V}\Sigma^{-1}\mathbf{X}}{N^{\frac{5}{2}}} + 3\frac{\mathbf{X}^T\Sigma^{-1}\mathbf{V}\Sigma^{-1}\mathbf{V}\Sigma^{-1}\mathbf{X}}{N^3} + O\left(\frac{1}{N^{\frac{7}{2}}}\right)$$

From the definition of $\mathbf{X}$ we find from the first term

$$[\langle \mathbf{X}^T\Sigma^{-1}\mathbf{X}\rangle]_{ii} = \left\langle \sum_{k,l}\Sigma_{kl}^{-1}(\mathbf{x}_i)_k(\mathbf{x}_i)_l\right\rangle = [\sum_{k,l}\Sigma_{kl}^{-1}\Sigma_{kl}]_{ii} = 1$$

by taking the expectation of the trace of both sides of the equation, the second term can be shown to be of the same order as the third term. So

$$\mathcal{E}_N(\sigma) = \mathcal{E}_N(0) + \frac{\sum_i \sigma_i^2}{N^2} + O(\frac{1}{N^2}) \tag{15}$$

The first part of the theorem now follows. Using similar techniques for $\mathcal{E}_N(0)$, we find

$$\begin{aligned}
\mathcal{E}_N(0) &= \langle f^2\rangle - \mathbf{q}^T\Sigma^{-1}\mathbf{q} + \frac{\overbrace{\langle \mathbf{q}^T\Sigma^{-1}\mathbf{V}\Sigma^{-1}\mathbf{q} - \mathbf{a}^T\Sigma^{-1}\mathbf{q}\rangle}^{0}}{N^{\frac{1}{2}}} \\
&\quad + \frac{\langle \mathbf{q}^T\Sigma^{-1}\mathbf{V}\Sigma^{-1}\mathbf{V}\Sigma^{-1}\mathbf{q} + \mathbf{a}^T\Sigma^{-1}\mathbf{a} - 2\mathbf{a}^T\Sigma^{-1}\mathbf{V}\Sigma^{-1}\mathbf{q}\rangle}{N} + O\left(\frac{1}{N^{\frac{3}{2}}}\right) \\
&= E_0 + \frac{B}{N} + O\left(\frac{1}{N^{\frac{3}{2}}}\right)
\end{aligned}$$

with $E_0 = \langle f^2\rangle - \mathbf{q}^T\Sigma^{-1}\mathbf{q}$ and B depending on the input distribution. This gives the N dependence of $\mathcal{E}_N(0)$. Finally we have

$$\mathcal{E}_N(\sigma) - \mathcal{E}_N(0) = \frac{\overline{\sigma^2}(d+1) + B}{\mathcal{N}} - \frac{B}{N} = \Delta$$

yielding the functional dependence $\mathcal{N}(\Delta, \sigma, N)$. ■

This result can immediately be generalized to the case where the learning model is linear in its parameter space. A similar technique can be used to derive a result on the expected mean squared residual itself which we will call $\mathcal{E}_r(\sigma)$.

Theorem B.3 *Let $\mathcal{H}$, the learning model, be the set of linear functions $\mathbf{w}\cdot\mathbf{x}+w_0$ and let the learning algorithm be minimization of the squared error. Then*

$$\mathcal{E}_r(\sigma) = \mathcal{E}_r(0) + \overline{\sigma^2} - \frac{\overline{\sigma^2}(d+1)}{N} \tag{16}$$

$$\mathcal{E}_r(0) = E_0 - \frac{B}{N} + O(\frac{1}{N^{\frac{3}{2}}}) \tag{17}$$

where E_0 and B are the same constants appearing in Theorem B.2. Thus we find

$$\mathcal{E}_N(\sigma) - \mathcal{E}_r(\sigma) = \frac{2(\overline{\sigma^2}(d+1)+B)}{N} + o(\frac{1}{N^2}) \tag{18}$$

PROOF: The residual error is given by

$$\begin{aligned}
\mathcal{E}_r(\sigma) &= \left\langle \frac{1}{N}(\hat{\mathbf{X}}^T\mathbf{w}-\mathbf{y})^2 \right\rangle = \left\langle \frac{1}{N}((\mathbf{X}^T(\mathbf{X}\mathbf{X}^T)^{-1}\mathbf{X}-\mathbf{1})y)^2 \right\rangle \\
&= \frac{\langle \mathbf{f}^T\mathbf{f}\rangle - \left\langle f^T\mathbf{X}(\mathbf{X}\mathbf{X}^T)^{-1}\mathbf{X}\mathbf{f}\right\rangle + \langle \epsilon^T\epsilon\rangle - \left\langle \epsilon^T\mathbf{X}^T(\mathbf{X}\mathbf{X}^T)^{-1}\mathbf{X}\epsilon\right\rangle}{N} \\
&= \underbrace{\langle f^2\rangle - \frac{\left\langle \mathbf{f}^T X^T(\mathbf{X}\mathbf{X}^T)^{-1}\mathbf{X}\mathbf{f}\right\rangle}{N}}_{\mathcal{E}_r(0)} + \overline{\sigma^2} - \frac{\overline{\sigma^2}(d+1)}{N}
\end{aligned}$$

from which the first part of the theorem follows. Using the techniques of Theorem B.2 we find that

$$\begin{aligned}
\mathcal{E}_r(0) &= \langle f^2\rangle - \mathbf{q}^T\Sigma^{-1}\mathbf{q} \\
&\quad - \frac{\langle \mathbf{q}^T\Sigma^{-1}\mathbf{V}\Sigma^{-1}\mathbf{V}\Sigma^{-1}\mathbf{q} + \mathbf{a}^T\Sigma^{-1}\mathbf{a} - 2\mathbf{a}^T\Sigma^{-1}\mathbf{V}\Sigma^{-1}\mathbf{q}\rangle}{N} + O(\frac{1}{N^{\frac{3}{2}}})
\end{aligned}$$

Comparing with Theorem B.2 we have the second part of the theorem. ∎

This result is similar to the results obtained by [1],[4].

We now consider the case of a non-linear learning model. The following proposition shows that $\lim_{N\to\infty}\mathcal{A}(\mathcal{D}_N) = g_\infty$ is well defined pointwise – i.e., $\forall\epsilon>0$, $\exists M$ such that if $N>M$ then $\max_{\mathbf{x}}|g_\infty - \mathcal{A}(\mathcal{D}_N)| < \epsilon$. This can be skipped if this fact is self evident or if one wishes to assume convergence and one is merely interested in the rate. It is included here purely for technical completeness.

Proposition B.4 *Let $\mathcal{L}\in\mathcal{CC}_2$. Then, the limit $\lim_{N\to\infty}\mathcal{A}(\mathcal{D}_N) = g_\infty$ for noiseless data sets is well defined point wise on sets of non-zero probability – i.e., $\forall\epsilon>0$, $\exists M$ such that if $N>M$ then $\max_{\mathbf{x}}|g_\infty - \mathcal{A}(\mathcal{D}_N)| < \epsilon$.*

PROOF: We will sketch the idea of the proof, the details can be filled in using exactly the same techniques as for the proof to theorem B.9. First we show that for any two infinite data sets, the learned functions are essentially identical. For any infinite data set, as the input support is compact (closed and bounded), any infinitesimal volume of non zero probability has an infinite number of data points. Consider two such data sets. The means of the targets in this small volume will be equal (by the law of large numbers). Because the target function is continuous on this compact support, the means for the two data sets are arbitrarily close to the true values for each data set (this can be attained by letting the the size of the volume be arbitrarily small). By continuous compatibility, these two data sets must both be mapped arbitrarily close to the data set with the means as targets. Therefore they must be mapped arbitrarily close to each other. Thus, we see that $\langle (g_1-g_2)^2\rangle$ is less than ϵ for arbitrary small ϵ, where the two different data sets drawn from the input distribution are mapped to g_i. So we conclude that $\langle (g_1-g_2)^2\rangle = 0$, therefore, $g_1 = g_2$ with probability 1. Thus, any two infinite noiseless data sets are mapped to the same function (as the functions are continuous), which we call g_∞.

Finally, consider a data set $\mathcal{D}_N$. For N large enough, this data set can be made arbitrarily close to an infinite data set using the argument above. Let $g_N = \mathcal{A}(\mathcal{D}_N)$. Therefore $\langle (g_N - g_\infty)^2 \rangle$ can be made arbitrarily small by choosing N large enough. In other words, $\lim_{N\to\infty} \langle (g_N - g_\infty)^2 \rangle = 0$, therefore g_N converges to g_∞ with probability 1. Further, because the functions are continuous and the support is compact, this convergence is uniform. ■

We have just shown that the limit $\mathcal{A}(\mathcal{D}_N)$ exists as $N \to \infty$. Thus, with noiseless data sets, we have convergence for stable learning systems. We now consider both the rate of convergence and what happens when noise is added.

Theorem B.5 *Let $\mathcal{L}$ be stable. Let the target function f be continuous. Let the probability measure on the input space have compact support $\mathcal{X}$. Then $\forall \epsilon > 0$, $\exists C_1 > 0$ such that using $\mathcal{L}$, it is at least possible to attain a test error bounded by*

$$\mathcal{E}_N(\sigma) < \mathcal{E}_N(0) + \frac{\overline{\sigma^2} C_1}{N} + \epsilon + O\left(\frac{1}{N^2}\right) \tag{19}$$

Corollary B.6 *If $C_1 \leq C_1' \; \forall \epsilon$ then*

$$\mathcal{E}_N(\sigma) \leq \mathcal{E}_N(0) + \frac{\overline{\sigma^2} C_1'}{N} + \left(\frac{1}{N^2}\right) \tag{20}$$

Corollary B.7 $\lim_{N\to\infty} \mathcal{E}_N(\sigma) = \lim_{N\to\infty} \mathcal{E}_N(0)$, *independent of σ.*

PROOF: By rescaling, we can assume that the input space $\mathcal{X} \subseteq \mathcal{S} = [0,1]^d$. f is continuous, so it is uniformly continuous on the compact set $\mathcal{S}$. Therefore, $\exists \delta_1$ such that

$$\mid \mathbf{x} - \mathbf{x}' \mid < \delta_1 \Rightarrow \mid f(\mathbf{x}) - f(\mathbf{x}') \mid < \delta_2$$

Divide $[0,1]$ into intervals of size $\delta_1/\sqrt{d}$. Thus we divide $\mathcal{S}$ into $\left(\sqrt{d}/\delta_1\right)^d$ cubes. Let $C_{\mathbf{i}} \equiv C_{i_1, i_2 \ldots i_d}$ define the cube with lowest coordinates $\mathbf{i}$. Let $N_{\mathbf{i}}$ be the number of data points in $C_{\mathbf{i}}$, and let $\mu_{\mathbf{i}} = \frac{1}{N_{\mathbf{i}}} \sum_{x_j \in C_{\mathbf{i}}} y_j$. Let $P_{\mathbf{i}} = Pr\{x \in C_{\mathbf{i}}\}$. We only need consider regions where $P_{\mathbf{i}} > 0$, as regions with $P_{\mathbf{i}} = 0$ are don't care regions. The following Lemma is easily obtained by noting that for $\mathbf{x}, \mathbf{x}' \in C_{\mathbf{i}}$, $|f(\mathbf{x}) - f(\mathbf{x}')| \leq \delta_2$.

Lemma B.8 *Let $\mathbf{x} \in C_{\mathbf{i}}$*

$$\left| \frac{1}{N_{\mathbf{i}}} \sum_{\mathbf{x}_j \in C_{\mathbf{i}}} y_j - f(\mathbf{x}) \right| = \mid \mu_{\mathbf{i}} - f(\mathbf{x}) \mid \leq \delta_2 + \frac{\mid \sum_{x_k \in C_i} \epsilon_k \mid}{N_{\mathbf{i}}}$$

Construct a new data set by replacing all the y's in $C_{\mathbf{i}}$ by $\mu_{\mathbf{i}}$. i.e., with no noise, the targets would be $f(\mathbf{x}_j)$ and with noise they are $\mu_{\mathbf{i}}$. $\forall \mathbf{x}_j \in C_{\mathbf{i}}$,

$$\mu_{\mathbf{i}} = f(\mathbf{x}_j) + \underbrace{\sum_{\mathbf{x}_k \neq \mathbf{x}_j} \frac{f(\mathbf{x}_k) - f(\mathbf{x}_j)}{N_{\mathbf{i}}}}_{|\cdot| < \delta_2} + \frac{\sum_{x_k \in C_i} \epsilon_k}{N_{\mathbf{i}}} = f(\mathbf{x}_j) + \eta_j + \xi_j$$

where $\eta_j = \sum_{\mathbf{x}_k \neq \mathbf{x}_j} \frac{f(\mathbf{x}_k) - f(\mathbf{x}_j)}{N_{\mathbf{i}}}$ and $\xi_j = \sum_{x_k \in C_i} \epsilon_k / N_{\mathbf{i}}$. We have that $\langle \eta_j \rangle_{\mathcal{D}_N} = 0$ and $\langle \xi_j \rangle_\epsilon = 0$. Let $\mathcal{A}$ map the noiseless data set to $g_0 \in \mathcal{H}$ and this noisy version of the data set to $g = g_0 + \eta$. So for the test error we have

$$\mathcal{E}_N(\sigma) = \langle (f-g)^2 \rangle_{\mathcal{D}_N, \mathbf{x}, \epsilon} = \underbrace{\langle (f-g_0)^2 \rangle_{\mathcal{D}_N, \mathbf{x}, \epsilon}}_{\mathcal{E}_N(0)} + \underbrace{2 \langle (f-g_0)(g_0-g) \rangle_{\mathcal{D}_N, \mathbf{x}, \epsilon}}_{T_1} + \underbrace{\langle (g_0-g)^2 \rangle_{\mathcal{D}_N, \mathbf{x}, \epsilon}}_{T_2}$$

We now examine T_1 and T_2.

$$|T_1| = \left| \left\langle (f-g_0) \langle g_0 - g \rangle_\epsilon \right\rangle_{\mathcal{D}_N, \mathbf{x}} \right|$$

$$
\begin{aligned}
&\overset{(a)}{=} \left|\left\langle (f-g_0)(\mathcal{A}(\{\mathbf{x}_k, f(\mathbf{x}_k)\}) - \mathcal{A}(\{\mathbf{x}_k, f(\mathbf{x}_k)+\eta_k\}))\right\rangle_{\mathcal{D}_N,\mathbf{x}}\right| \\
&\leq \left|\left\langle f\left\langle (\mathcal{A}(\{\mathbf{x}_k, f(\mathbf{x}_k)\}) - \mathcal{A}(\{\mathbf{x}_k, f(\mathbf{x}_k)+\eta_k\}))\right\rangle_{\mathcal{D}_N}\right\rangle_{\mathbf{x}}\right| \\
&\quad + \left|\langle g_0(\mathcal{A}(\{\mathbf{x}_k, f(\mathbf{x}_k)\}) - \mathcal{A}(\{\mathbf{x}_k, f(\mathbf{x}_k)+\eta_k\}))\rangle_{\mathcal{D}_N\mathbf{x}}\right| \\
&\overset{(b)}{\leq} \left\langle \max_{\mathbf{x}} |g_0| \langle |\mathcal{A}(\{\mathbf{x}_k, f(\mathbf{x}_k)\}) - \mathcal{A}(\{\mathbf{x}_k, f(\mathbf{x}_k)+\eta_k\})|\rangle_{\mathbf{x}} \right\rangle_{\mathcal{D}_N} \\
&\overset{(c)}{\leq} C\delta_2 \left\langle \max_{\mathbf{x}} |g_0| \right\rangle_{\mathcal{D}_N} \\
&\overset{(d)}{\leq} c_1\delta_2
\end{aligned}
$$

where (a) and (b) follow from the mean preserving assumption. (c) from continuous compatibility and (d) because we assume the limit g_∞ to exist pointwise. Similarly, for T_2 we get

$$
\begin{aligned}
|T_2| &= \left|\left\langle (\mathcal{A}(\{\mathbf{x}_k, f(\mathbf{x}_k)\}) - \mathcal{A}(\{\mathbf{x}_k, f(\mathbf{x}_k)+\eta_k+\xi_k\}))^2 \right\rangle_{\mathcal{D}_N,\mathbf{x},\epsilon}\right| \\
&\overset{(a)}{\leq} 2C^2\left(\delta_2^2 + \left\langle \sum_{\mathrm{i}} \frac{\sum_{\mathbf{x}_j,\mathbf{x}_k\in C_{\mathrm{i}}} \epsilon_j\epsilon_k}{N_{\mathrm{i}}^2} \right\rangle_{\mathcal{D}_N,\epsilon}\right) \\
&= 2C^2\left(\delta_2^2 + \left\langle \sum_{\mathrm{i}} \frac{\overline{\sigma_{\mathrm{i}}^2}}{N_{\mathrm{i}}} \right\rangle_{\mathcal{D}_N,\epsilon}\right) \\
&\overset{(b)}{=} 2C^2\left(\delta_2^2 + \overline{\sigma^2}\left\langle \sum_{\mathrm{i}} \frac{1}{N_{\mathrm{i}}} \right\rangle_{\mathcal{D}_N}\right) \\
&= 2C^2\delta_2^2 + 2C^2\overline{\sigma^2}\sum_{\mathrm{i}} \underbrace{\sum_{n=1}^{N} \frac{1}{n}\binom{N}{n} P_{\mathrm{i}}^n(1-P_{\mathrm{i}})^{N-n}}_{\frac{1}{NP_{\mathrm{i}}}+O(\frac{1}{N^2})} \\
&= 2C^2\delta_2^2 + \frac{\overline{\sigma^2}}{N}\underbrace{2C^2\sum_{\mathrm{i}}\frac{1}{P_{\mathrm{i}}}}_{C_1} + O(\frac{1}{N^2})
\end{aligned}
$$

(a) follows from the continuous compatibility assumption. (b) follows because the noise is chosen independently of the inputs. Choosing δ_1 such that $c_1\delta_2 + 2C^2\delta_2^2 < \epsilon$ we have

$$
\mathcal{E}_N(\sigma) \leq \mathcal{E}_N(0) + \frac{C_1(\epsilon)\overline{\sigma^2}}{N} + \epsilon + O\left(\frac{1}{N^2}\right)
$$

■

We note that it is easy to extend these theorems to the case where the noise variances are drawn from some distribution. By taking the expectation over that distribution, the same result with $\overline{\sigma^2}$ being the expected value of the variance parameter is obtained. Note also that the preceding proof is by no means suggesting a method to calculate C_1. It is simply a means to show its existence. Often, especially when the input distribution is bounded, Corollary B.6 will hold, and it might be possible to estimate these constants experimentally.

One might wonder what would happen if the mean preserving assumption is violated. We note that the only place where this is used is in the evaluation of T_1. Continuity could still be used however with the difference being that a term of order $\epsilon/\sqrt{N}$ would remain. in other words, one would have $\mathcal{E}_N(\sigma) \leq \mathcal{E}_N(0) + C''\sigma/\sqrt{N}$ + higher order. So if we do not have the mean preserving property then these methods do not guarantee $1/N$ convergence of the test error. Using identical methods, one can, however, get the following result using the continuity property alone: $\langle |f-g| \rangle \leq \langle |f-g_0| \rangle + \frac{C'''\sigma}{\sqrt{N}}$. This is very similar to Theorem B.5 where one measures test error by the expectation of the magnitude difference as opposed to the squared difference.

We now derive a theorem on the dependence of $\mathcal{E}_N(0)$.

Theorem B.9 *Let $\mathcal{L}$ be stable. Let the target function f be continuous. Let the probability measure on the input space have compact support. Then $\forall \epsilon > 0$, $\exists C_2 > 0$ such that using $\mathcal{L}$ it is at least possible to attain a noiseless test error bounded by*

$$\mathcal{E}_N(0) < E_0 + \frac{C_2}{N} + \epsilon + o\left(\frac{1}{N}\right) \tag{21}$$

Corollary B.10 *If $C_2 \leq C_2' \ \forall \epsilon$ then $\mathcal{E}_N(0) \leq E_0 + \frac{C_2'}{N} + o\left(\frac{1}{N}\right)$ where $E_0 = \lim_{N\to\infty} \mathcal{E}_N(0)$*

Before we proceed to the proof of the theorem, the following lemma is needed.

Lemma B.11 *Let N balls be independently be distributed into r cells according to the probabilities $p_1 \ldots p_r$. Then for every $m > 0, \exists A_m$ such that the probability, q, that at least one cell is empty is bounded by*

$$q \leq \frac{A_m}{N^m}$$

PROOF:

$$\begin{aligned} q &= Pr[\cup \, cell_i \text{ is empty}] \leq \sum_i Pr[cell_i \text{ is empty}] = \sum_i (1-p_i)^N \\ &\leq r(1-\min_i p_i)^N \leq \frac{A_m}{N^m} \end{aligned}$$

choosing $A_m \geq r(-m/ln(a))^m$, where $a = 1 - \min_i p_i$. ■

PROOF OF THEOREM B.9

Let $\mathcal{X}$, $\mathcal{S}$, δ_1, δ_2, C_i, P_i, N_i be as in the proof of Theorem B.2. We only consider those cubes with $P_i > 0$.

Suppose that we have an infinite noiseless data set, $\mathcal{D}_\infty$. For all i, let $\bar{y}_i = \langle f(\mathbf{x}) \rangle_{\mathbf{x}|\mathbf{x}\in C_i}$ and let $\mu_i = \frac{1}{N_i}\sum_{\mathbf{x}_j \in C_i} y_j$ if C_i is non-empty else, $\mu_i = 0$. Construct two data sets from the infinite one, $\mathcal{D}_1$ and $\mathcal{D}_2$, by replacing all the y's in C_i by $\bar{y}_i$, and μ_i respectively. $\mathcal{D}_1$ does not depend on $\mathcal{D}_N$ and $\mathcal{D}_2$ can be obtained from $\mathcal{D}_N$. $\mathcal{D}_\infty$ and $\mathcal{D}_1$ are close data sets because for $\mathbf{x} \in C_i$,

$$\left| f(\mathbf{x}) - \langle f(\mathbf{y}) \rangle_{\mathbf{y}|\mathbf{y}\in C_i} \right| = \left| \langle f(\mathbf{x}) - f(\mathbf{y}) \rangle_{\mathbf{y}|\mathbf{y}\in C_i} \right| \leq \langle |f(\mathbf{x}) - f(\mathbf{y})| \rangle_{\mathbf{y}|\mathbf{y}\in C_i} \leq \delta_2$$

Therefore by continuous compatibility, $\langle (g_\infty - g_1)^2 \rangle \leq C^2\delta_2^2$. Define ϵ_i by $\mu_i = \bar{y}_i + \epsilon_i$. Then $\langle \epsilon_i \rangle_{\mathcal{D}_N} = 0$ for all non-empty C_i. Let g_∞, g_1 and g_2 be $\mathcal{A}(\mathcal{D}_\infty)$, $\mathcal{A}(\mathcal{D}_1)$ and $\mathcal{A}(\mathcal{D}_2)$ respectively. Since we can construct $\mathcal{D}_2$, using $\mathcal{L}$ we can at least obtain a test error given by

$$\begin{aligned} \mathcal{E}_N(0) \leq\ & \overbrace{\langle (f-g_\infty)^2 \rangle}^{E_0} + \overbrace{\langle (g_\infty - g_1)^2 \rangle}^{\leq C^2\delta_2^2} + \langle (g_1-g_2)^2 \rangle \\ & +2\underbrace{|\langle (f-g_\infty)(g_\infty-g_1) \rangle|}_{\substack{\leq |f-g_\infty|_{max} C\delta_2 \\ \text{(by cont. comp.)}}} + 2|\langle (f-g_\infty)(g_1-g_2) \rangle| + 2|\langle (g_\infty - g_1)(g_1-g_2) \rangle| \end{aligned}$$

By the mean preserving property, $\langle (g_1 - g_2) \rangle_{\mathcal{D}_N} = 0$. Therefore,

$$|\langle (f-g_\infty)(g_1-g_2) \rangle| = |\langle (f-g_\infty)\langle (g_1-g_2) \rangle_{\mathcal{D}_N} \rangle_{\mathbf{x}}| = 0$$

Similar reasoning shows that $2|\langle(g_\infty - g_1)(g_1 - g_2)\rangle| = 0$. Let Q be the probability that at least one cell is empty. For the final term we have

$$\begin{aligned}
\langle(g_1 - g_2)^2\rangle &\overset{(a)}{\leq} (1-Q)C^2 \left\langle \sum_i \epsilon_i^2 \right\rangle_{\mathcal{D}_N|\forall i, N_i>0} + 4QC^2|f|^2_{max} \\
&\leq C^2 \left\langle \left\langle \sum_i \epsilon_i^2 \right\rangle_{y_i|N_i>0} \right\rangle_{N_i>0} + c_1 Q \\
&\leq C^2 \left\langle \sum_i \frac{\sigma_i^2}{N_i} \right\rangle_{N_i>0} + c_1 Q \\
&\overset{(b)}{\leq} \frac{1}{N} \underbrace{C^2 \sum_i \frac{\sigma_i^2}{P_i} x}_{C_2} + o(\frac{1}{N})
\end{aligned}$$

where $\sigma_i^2 = Var(y_i|y \in C_i)$. (a) follows by continuous compatibility because with probability $1 - Q$ the data sets are at most ϵ_i apart and $\sum_i \epsilon_i^2 \geq \max_i \epsilon_i^2$, and with probability Q they are at most $2|f|_{max}$ apart. (b) follows because $\langle 1/N_i \rangle = 1/(NP_i) + o(1/N)$ and Lemma B.11 can be used to yield $Q = o(1/N)$. Finally we have

$$\mathcal{E}_N(0) \leq E_0 + C^2\delta_2^2 + 2|f - g_\infty|_{max}\delta_2 + \frac{C_2}{N} + o(\frac{1}{N})$$

Choosing δ_2 small enough, we have the theorem because $|f - g_\infty|$ is bounded on the compact support $\mathcal{X}$. ■

References

[1] Murata, N., Yoshizawa, S. and Amari, S., "Learning Curves, Model Selection and Complexity of Neural Networks", *Advances in Neural Information Processing Systems* 5, 1993, pp 607–614.

[2] Amari, S., Murata, N.,Müller, K. R. and Yang, H., "Asymptotic Statistical Theory of Overtraining and Cross Validation", *IEEE Transactions on Neural Networks*, Vol 8, No. 5, pp 985–996 1997.

[3] Amari, S., Murata, N., "Statistical Theory of Learning Curves under Entropic Loss Criterion.", *Neural Computation*, 5, pp 140–153, 1992.

[4] Moody, J. "The Effective Number of Parameters: An Analysis of Generalization and Regularization in Nonlinear Learning Systems", *Advances in Neural Information Processing Systems* 4, 1992, pp. 847–854.

[5] Plaut, D., Nowlan, S. and Hinton, G. (1986), "Experiments on Learning by Backpropagation", *Technical Report CMU-CS-86-126*, Carnegie Mellon University.

[6] Black, F. and Scholes, M. S., "The Pricing of Options and Corporate Liabilities", *Journal of Political Economy* 3, 1973, pp. 637-654.

[7] Montgomery, D., Johnson, L. and J. Gardiner, *Forecasting and Time Series Analysis*, New York, McGraw-Hill, Inc., 1990.

[8] Abu-Mostafa, Y. S. and Atiya, A. F., "Introduction to Financial Forecasting", *Applied Intelligence*, 6, 1996, pp 205–213.

[9] Shiller, R. J., *Market Volatility*, Cambridge, MA, MIT Press, 1993.

[10] Trippi, R. R. and Turban, E., *Neural Networks in Finance and Investing*, Chicago, Probos Publishing Company, 1993.

[11] White, H., "Economic Prediction Using Neural Networks: The case of IBM Daily Returns", *Proceedings of the IEEE International Conference on Neural Networks*, 2, 1988, pp 451-458.

[12] Krogh, A. and Hertz, J. A., "Generalization in a Linear Perceptron in the Presence of Noise", *Journal of Physics A* 25, 1992, pp 1135–1147.

[13] Krogh, A., "Learning with Noise in a Linear Perceptron", *Journal of Physics A* 25, 1992, pp 1119–1133.

[14] Abu-Mostafa, Y. S., "Learning from Hints", Journal of Complexity 10, 1994, pp 165–178.

[15] Vapnik, V. N. and Chervonenkis, A., "On the Uniform Convergence of Relative Frequencies of Events to their Probabilities", *Theory Prob. Appl.* 16, 1971, pp 264–280.

[16] Leich, G. and Tanner, J. E., "Economic Forecast Evaluation: Profit versus the Conventional Error Measures", *American Economic Review*, 81, 1991, pp 580–590.

[17] Malkiel, B., *A Random Walk Down Wall Street*, New York, W. W. Norton & Co., 1985.

[18] Shao, J. and Tu, D., *The Jackknife and the Bootstrap*, New York, Springer-Verlag, 1996.

[19] Geman, S. and Bienenstock, E., "Neural Networks and the Bias Variance Dilemma", *Neural Computation*, 4, 1992, pp 1–58

Chapter 4

Incorporating Prior Information in Machine Learning by Creating Virtual Examples

P. Niyogi, F. Girosi, and T. Poggio
MIT Center for Biological and Computational Learning
Cambridge, MA 02139.

September 2, 1998

Abstract

One of the key problems in supervised learning is the insufficient size of the training set. The natural way for an intelligent learner to counter this problem and successfully generalize is to exploit prior information that may be available about the domain or that can be learned from prototypical examples. We discuss the notion of using prior knowledge by creating *virtual examples* and thereby expanding the effective training set size. We show that in some contexts, this idea is mathematically equivalent to incorporating the prior knowledge as a regularizer, suggesting that the strategy is well-motivated. The process of creating virtual examples in real world pattern recognition tasks is highly non-trivial. We provide demonstrative examples from object recognition and speech recognition to illustrate the idea.

1 Learning from Examples

Recently, machine learning techniques have become increasingly popular as an alternative to knowledge-based approaches to artificial intelligence problems in a variety of fields. The hope is that automatic learning from examples will eliminate the need for laborious handcrafting of domain-specific knowledge about the task at hand. However, analyses of the complexity of learning problems suggest that this hope might be overly optimistic — often the number of examples needed to solve the problem might be prohibitive. Clearly, a middle ground is needed and a useful direction of research is the study of how to incorporate prior world knowledge of the task within a learning from examples framework.

The current paper deals with this subject. We first begin by providing some background about how the problem of learning from examples is usually formulated. In the next section, we discuss briefly the complexity of the learning problem and why, in the absence of any prior knowledge, one might require a large number of examples to learn well. In section 3, we introduce the idea of virtual examples, i.e., creating additional examples from the current set of examples by utitilizing specific knowledge about the task at hand. While the overall framework is similar to learning from hints (Abu-Mostafa,1995), our emphasis in this paper is to describe some specific non-trivial transformations that allow us to create virtual examples for real world pattern recognition problems. We first show in section 4 that in certain function learning contexts, the framework of virtual examples is equivalent to imposing prior knowledge as a regularizer. Thus, the idea of virtual examples can be more than an ad hoc strategy. We then discuss in section 5, some specific examples from computer vision and speech recognition. Finally, we conclude by reiterating some of our main points in section 6.

1.1 Background: Learning as Function Approximation

The problem of learning from examples can be usefully modeled as trying to approximate some unknown target function f from (x, y) pairs that are consistent with this function (modulo noise). The target function f belongs to some *target class* of functions denoted by $\mathcal{F}$. The learner has access to a data set consisting of (say) n (x, y) pairs $((x_i, y_i) : i = 1, \ldots, n)$ and picks a function h chosen from some *hypothesis class* $\mathcal{H}$ on the basis of this data set. The hope is that if "enough" examples are drawn, the learner's hypothesis will be sufficiently close to the target resulting in successful generalization to novel unlabelled examples that the learner might encounter.

Numerous problems in pattern recognition, speech, vision, handwriting, finance, robotics, etc. can be cast within this framework and research typically focuses on different kinds of hypothesis classes ($\mathcal{H}$) and different ways of choosing an optimal function in this class (training algorithms). Thus, multilayer perceptrons (Rumelhart, Mclelland and Hinton, 1986), radial basis function networks (Poggio and Girosi, 1990; Moody and Darken, 1988), decision trees (Breiman etal, 1984), all correspond to different choices of hypothesis classes on which popular learning machines have been based. Similarly, different kinds of gradient descent schemes from backpropagation to the EM algorithm correspond to ways of choosing an optimal function from such a class given a finite data set. By varying the choices of hypothesis classes and training algorithms, a profusion of learning paradigms have emerged. The most significant issue of interest in each of these learning paradigms is how they generalize to new unseen data. In the next section, we discuss the factors upon which the generalization performance of a learning machine depends.

2 Prior Information and the Problem of Sample Complexity

In any learning from examples system, the number of examples (l) that needs to be collected for successful generalization is a key issue. This *sample complexity* is typically characterized by the theory of Vapnik and Chervonenkis (1982,1995) that describes the general laws that all probabilistically based learning machines have to obey. It turns out that if the learner picks a hypothesis ($\hat{h} \in \mathcal{H}$) on the basis of the example set, then the number of examples it needs in order to generalize well is of the order of $\sqrt{\frac{VC(\mathcal{H})}{l}}$. Here $VC(\mathcal{H})$ is the VC-dimension of the class $\mathcal{H}$ — a combinatorial measure of the complexity of the hypothesis class. Roughly speaking, the VC dimension (see Vapnik, 1982 for further details) is a measure of how many different kinds of functions there are in $\mathcal{H}$. For example, if $\mathcal{H}$ were the parametric class of univariate polynomials of degree n, its VC dimension[1] is $n+1$. In

[1]While in this case, the VC dimension is related in a simple way to the number of parameters, this need not be true in general. One can think of classes with many parameters having a small VC dimension and vice versa. Thus the VC dimension is a better and more direct measure of learning complexity than simply the number of parameters.

general, large or complex hypothesis classes that can accomodate many different data sets would have a higher VC dimension than smaller, restricted hypothesis classes. Thus we see that the number of examples needed is proportional to the VC-dimension, and in this sense, to the effective size of the hypothesis class. Consequently, it is in our interest to use small hypothesis classes in learning machines.

However, using a small hypothesis class is not enough. Recall that the target function f belongs to $\mathcal{F}$ and if our hypothesis class $\mathcal{H}$ is too small, then, even if we choose the best function in it, the distance from the target (generalization error) might be too high. To appreciate this point better, let us consider a situation of learning using neural networks in a least-squares setting. Recall that ideally, we would like to "learn" the target function that is given by the following (the expectation is with respect to the true probability distribution generating the data):

$$f_0 \equiv \arg\min_{f \in \mathcal{F}} E[(y - f(x))^2]$$

However, in practice, we don't know the true distribution and so cannot compute the true expectation; nor do we typically minimize over the class $\mathcal{F}$. For example, consider the typical situation if we were using neural networks to learn the function f_0. We draw a finite data set (x, y pairs), construct an empirical approximation to the objective function and then minimize this over a class of neural networks with a finite number of parameters. If we collected l data points and minimized over a neural network with n nodes in its hidden layer (say), we are in effect computing the following function $\hat{f}_{n,l}$:

$$\hat{f}_{n,l} = \arg\min_{h \in \mathcal{H}_n} E_{emp}[(y - h(x))^2] \equiv \min_{h \in \mathcal{H}_n} \frac{1}{l} \sum_{i=1}^{l} (y_i - h(x_i))^2$$

Thus, when we attempt to learn the function f_0 using a finite amount of data (l points) and a hypothesis class with a finite number of parameters ($\mathcal{H}_n$) then, the function we obtain in practice is given by $\hat{f}_{n,l}$. This is the function we use to predict future, unknown values and naturally, we would like to know how good this function is, i.e., how far this function is from the true target. In general, one can show that the generalization error ($\| f_0 - \hat{f}_{n,l} \|$) can be decomposed into an approximation component and an estimation component, i.e.,

$$\| f_0 - \hat{f}_{n,l} \| \leq e_{app}(n) + e_{est}(\frac{VC(\mathcal{H}_n)}{l})$$

The approximation error $e_{app}(n)$ is due to the finite size of the hypothesis class. As the number of hidden nodes, n, increases, the representational power of the hypothesis class increases and the approximation error goes to zero. The estimation error $e_{est}(\frac{VC(\mathcal{H}_n)}{l})$ is due to the finite amount of data that is available to the learner. It is a monotonically decreasing function and depends upon the VC dimension of the hypothesis class $\mathcal{H}_n$ and the amount of data (l). As the number of hidden nodes increases, the VC dimension of $\mathcal{H}_n$ increases and consequently the estimation error increases as well (keeping the data fixed). Thus to make the approximation error small, we need large sized networks (n large); to make the estimation error small, we need small sized networks (n small). This trade-off between the approximation error and estimation error arises in all learning paradigms and has been investigated for a number of different hypothesis classes ranging from multilayer perceptrons (Barron, 1994) to radial basis functions (Niyogi and Girosi, 1996). The following theorem states a canonical result for radial basis functions and fig. 1 below describes the generalization error surface as a function of the number of parameters and the number of data.

Theorem 1 (Niyogi and Girosi,1996) *Let H_n be the class of Gaussian Radial Basis Function networks with k input nodes and n hidden nodes, i.e.,*

$$H_n = \sum_{i=1}^{n} c_i G_i(\frac{\mathbf{x}_i - \mathbf{t}_i}{\sigma_i})$$

Let f_0 be an element of the Bessel potential space[2] *$\mathcal{L}_1^m(R^k)$ of order m, with $m > k/2$ (the class $\mathcal{F}$). Assume that a data set $\{(\mathbf{x}_i, y_i)\}_{i=1}^{l}$ has been obtained by randomly sampling the function f_0 in presence of noise, and that the noise distribution has compact support. Then, for any $0 < \delta < 1$, with probability greater than $1-\delta$, the following bound for the generalization error*

[2] This is defined as the set of functions f that can be written as $f = \lambda * G_m$, where $*$ stands for the convolution operation, $\lambda \in L_p$ and G_m is the Bessel-Macdonald kernel, i.e., the function whose Fourier transform is:

$$\tilde{G}_m(\mathbf{s}) = \frac{1}{(1 + 4\pi^2 \|\mathbf{s}\|^2)^{m/2}}$$

holds:

$$\|f_0 - \hat{f}_{n,l}\|^2_{L^2(P)} \leq O\left(\frac{1}{n}\right) + O\left(\left[\frac{nk\ln(nl) - \ln\delta}{l}\right]^{1/2}\right) \tag{1}$$

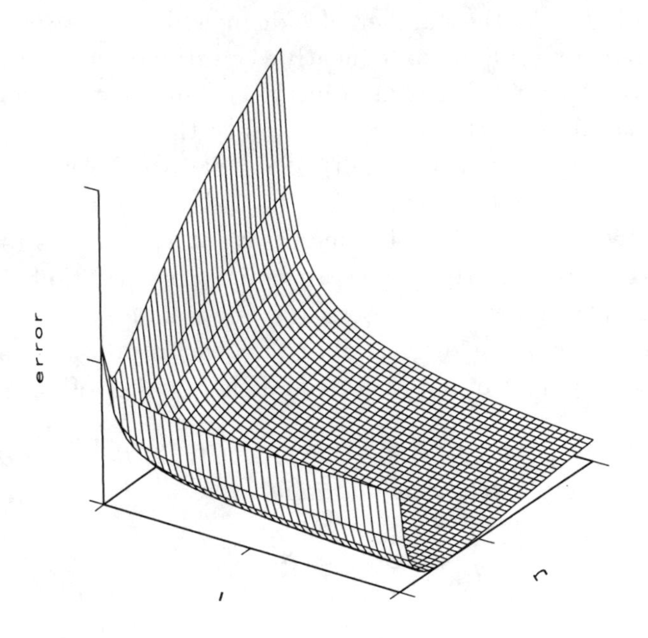

Figure 1: The generalization error, the number of examples (l) and the number of basis functions (n) as a function of each other.

What do we conclude from all this? First, that if we work with unconstrained hypothesis classes, the number of examples needed for low estimation error will almost certainly be prohibitive. On the other hand, if we work with highly constrained hypothesis classes, the approximation error will be too high for successful generalization. The only way around this dilemma is if the target class itself ($\mathcal{F}$) can be made small — but this is precisely the prior information we have about the problem being solved. Thus, if we have more prior information about the target function, it would correspond to a smaller target class and the problems with poor generalization would be ameliorated.

In essence, prior information is more than a good idea. The mathematics of generalization (from the no-free-lunch theorems of Wolpert (1995) to the statistical theory of Vapnik (1995)) all point to one thing — incorporation of prior knowledge might be the only way for learning machines to tractably generalize from finite data. One way of incorporating prior knowledge is the idea of virtual examples: utilizing prior information about the target function to generate novel examples from old data thereby enlarging our effective data set. We discuss this in the next section.

3 Virtual Examples: A Framework for Prior Information

As we have discussed above, a significant problem in learning from examples is the large amount of data (examples) needed for adequate learning. Consequently, it becomes crucial to exploit any form of prior knowledge we might have about the task at hand. A well known technique for incorporating prior information is to restrict the class of hypotheses — this would also reduce the data requirements by the Vapnik Chervonenkis theory as discussed in the previous section.

Another alternative might be to expand the set of available examples in some fashion so that the learner has access to an effectively larger set of examples resulting in more accurate learning. These additional examples, created from the existing ones by the application of prior knowledge, will be referred to as *virtual* examples (first introduced by Poggio and Vetter, 1992 and different from the notion of virtual examples introduced by Abu Mostafa (1990,1992,1993) as discussed later). We first lay the general framework for virtual examples in section 3.1. Of course, virtual examples is only one way of incorporating prior information and we will briefly review some alternate methods and their relationship to our approach in section 3.2.

At first, the idea of virtual examples might seem like an ad-hoc one but we show in section 4 the connection between the virtual examples approach and using regularization as a technique for incorporating prior information. For the example discussed, it is possible to prove that both techniques yield the same solution. The heart of the virtual examples idea involves the actual creation of the virtual examples —- this is the main focus of our paper and we discuss several substantive real world, practical demonstrations of this approach in later sections.

3.1 The General Framework

As discussed earlier, the primary goal of the learner is to approximate some unknown target function (f) from examples ((x, y) pairs) of this function. The unknown target function might be a real-valued, multivariate function (as in sec. 4), or even a characteristic funtion defined over some manifold (as in sec. 5).

In the absence of any prior information, the learner would attempt to fit a function from $\mathcal{H}$ to the data set and use it to predict future values. Suppose, however, that we have prior knowledge of a set of transformations that allow us to obtain new examples from old. For example, the target function might be invariant with respect to a particular group of transformations. A simple case is when the target function is known to be even or odd. A more complex case might be if the target function is a characteristic function defined over a manifold of 3D objects. Correspondingly, in some cases, obtaining the new examples might be easy, in other cases — as in the case for object recognition — it is quite difficult.

Thus, suppose we know some transformation T such that if $(x, f(x))$ is a valid example, then $(Tx, y_T(f(x)))$ is also a valid example. For an invariant transformation, y_T is the identity mapping. In general, of course, the relation of y_T to T depends upon the prior knowledge of the problem and might be quite complex. Then, given a set of n examples: $D = \{(x_1, y_1), \ldots, (x_n, y_n)\}$ and knowledge of this transformation T, we generate the set of virtual examples: $D' = \{(x'_1, y'_1), \ldots, (x'_n, y'_n)\}$ such that $x'_i = Tx$ and $y'_i = y_T(y_i)$.

$$(x, f(x)) \stackrel{T, y_T}{\longrightarrow} (Tx, y_T(f(x)))$$

For many interesting cases, prior knowledge of the problem might allow us to define a group of transformations $\mathcal{T}$ such that for every $T \in \mathcal{T}$, we can create new, *virtual* examples from the old data set. For example, rotations (in the image plane or in 3D) might define such a group for object recognition problems. Thus, the creation of virtual examples allows us to expand the example set and consequently move towards better generalization.

3.2 Techniques for Prior Information and Related Research

Needless to say, the idea of virtual examples is only one possible way of incorporating prior information. We discuss in this section various ways in which researchers have tried to utilize prior knowledge.

1. **Prior Knowledge in the Choice of Variables or Features:** Prior knowledge could be used in the choice of the variables or features that are used to represent the problem. Let us consider the case of object recognition. A simple form of prior knowledge is that the rotated version (in 2D) of an object still represents that object. Therefore one could think of using, as input to the network, features that are invariant under rotations in the image plane. In this case, rotation invariant recognition would be achieved with just one example. This approach is somehow limited in vision applications, because it is very difficult to find features that are invariant for "interesting" transformations. For example it does not seem likely that one can find features of face images that are invariant with respect to rotation in 3D space, apart from "trivial" ones such as the colour of the person's hair etc. (for more details on the possibilities of such an approach, see Mundy et al, 1992).

   Another kind of prior knowledge could be that certain features always appear in conjunction (or disjunction), or certain variables are always linked together in a certain form. In this case one could explicitly add these new variables to the set of original variables, making the learning task much easier. For example, in robotics it is known that for certain mechanical systems, the relation between torques and state variables is represented by certain combinations of trigonometric functions. Therefore, explicitly adding sine and cosine transformation of the state space variables usually makes the problem much easier to solve. This technique is also not uncommon in statistics, where often new variables are created by means of nonlinear transformation of the original ones.

2. **Prior Knowledge in the Learning Technique:** Another way to incorporate prior knowledge is to embed it in the learning technique. Examples of this are the recent **transformation distance** technique introduced by Simard, Le Cun and Denker (1993). The idea underlying this technique is the following: suppose a pattern classification

problem has to be solved, and we know that the outcome of the classification scheme should be invariant with respect to a certain transformation $R(\mathbf{w})$, where $\mathbf{w}$ is a set of parameters (for example the rotation angle in the image plane for object recognition). This means that for every input pattern $\mathbf{x}$ there is a manifold $S\mathbf{x}$ on which the output should be constant. Therefore, if we desire to use a classification technique such as Nearest Neighbors, that is based on a notion of distance, we should use as distance between two patterns $\mathbf{x}$ and $\mathbf{z}$ not the Euclidean distance between them, but the Euclidean distance between the manifolds $S(\mathbf{x})$ and $S(\mathbf{z})$. This quantity cannot be computed analytically, in general, but Simard, Le Cun and Denker (1993) show how to estimate it using a local approximation of the manifold $S(\mathbf{x})$ by its tangent plane that can be experimentally computed. In this case the prior knowledge has been embedded in the definition of distance, and therefore in the learning technique, rather than in the choice of the variables, as described above.

Another case in which prior knowledge is embedded in the learning technique is regularization theory, a set of mathematical tools introduced by Tikhonov in order to deal with ill-posed problems (Tikhonov, 1963; Tikhonov and Arsenin, 1977; Morozov, 1984; Bertero, 1986; Wahba, 1990; Poggio and Girosi, 1990; Girosi, Jones and Poggio, 1995). In regularization theory an ill-posed problem is transformed into a well-posed one using some prior knowledge. The most common form of prior knowledge is smoothness, whose role in the theory of learning from examples has been investigated at length by Poggio and Girosi (1990). However, other forms of prior knowledge can be used in the framework of regularization theory. This topic has been investigated by Verri and Poggio (1986), who gave sufficient conditions for a constraint to be embedded in the regularization framework. Examples of the prior knowledge they considered include monotonicity, convexity and positivity.

3. **Generating New Examples with Prior Knowledge:** Another form of utilising prior knowledge for learning is the idea of generating new examples from the existing data set. This is the idea of virtual examples (from Poggio and Vetter, 1992) that we consider in this paper. An example of a similar technique can be found in the work of Pomerleau (1989, 1991) on ALVINN, an autonomous navigation vehicle that learns to drive on a highway. The system consists of a camera mounted

on a vehicle, and a neural network that takes that the image of the road as an input and produces as output a set of steering commands. The examples are acquired by recording the actions of a human driver. Since humans are very good at keeping the vehicle in the right lane, the images of the road look all alike, and there are no examples of what action to take if the vehicle is in an "unusual attitude", that is, too far to the right or to the left. Therefore the network is not able to give correct answers if it finds itself in these kinds of situations, of which it has no examples. Pomerleau used prior knowledge on the geometry of the problem in order to create examples of what to do in the case of unusual attitudes. Knowing the location of the camera, with respect to the vehicle, and based on examples belonging to the data set created by the human driver, he was able to create images of what the road would look like if the vehicle were in an unusual attitude, say too close to the centerline. Given these new images and the corresponding locations of the vehicle he computed what the steering command should be for each one of them, creating a whole new set of images, now containing many examples of unusual attitudes, and allowing the system to achieve excellent performance.

4. **Incorporating Prior Knowledge as Hints:** Another technique is the one proposed by Abu-Mostafa (1990, 1992, 1993). Here we list very briefly the main points of his concept of *hints.* The approach overlaps to a good extent but not completely with our own ideas of virtual examples.

   Consider

   (a) a function f to be learned with domain, X, and range, Y;

   (b) the hypothesis g provided by the learning process, say by a Regularization Network approximation of f;

   (c) the functional $E(g, f)$ measuring the error.

   Then a *hint* H_m is a test that f must satisfy. One generates one example of the hint and measures e_m, the amount of error of g on that example (if the hint is that f is odd then one chooses an x and uses $e_m = (g(\mathbf{x}) + g(-\mathbf{x}))^2$). The total disagreement between g and H_m is then $E_m = E(e_m)$

   Here are some examples of hints

- Invariance hint: $f(\mathbf{x}) = f(\mathbf{x}')$ for certain examples $(\mathbf{x}, \mathbf{x}')$. The associated error can be $e_m = (g(\mathbf{x}) - g(\mathbf{x}'))^2$
- Monotonicity hint: $f(z) < f(z')$ for certain examples (z, z') for which $z \leq z'$. Then the associated error can be $e_m = (g(z) - g(z'))^2$ if $g(z) > g(z')$ and $e_m = 0$ otherwise.
- Example hint: the set of examples of f can be treated as a hint, H_0

Abu-Mostafa (1992, 1993) describes how to represent hints by virtual examples. It is important for us to distinguish our notion of virtual examples from that of Abu-Mostafa. For Abu-Mostafa, a virtual example is typically a pair, (x, x') that are related in some way by the hint. Minimization is then done over all virtual examples. On the other hand, Abu-Mostafa also introduces the notion of *duplicate* examples. These are (x, y) pairs in the traditional sense that are somehow created by knowledge of the hint. They are often associated with invariant sets and are essentially the same as our virtual examples. While Abu-Mostafa focuses on the learning mechanism (a kind of adaptive minimization scheme) to use the hint once it has been represented by the creation of virtual examples (or duplicate examples), our focus here is on the actual creation of the virtual examples for some non-trivial learning problems.

4 Virtual Examples and Regularization

We begin by showing that the idea of virtual examples can lead to a solution that is identical to that obtained by incorporating the prior knowledge as a regularizer. Related results have also been obtained by Leen (1995) and Bishop (1995).

4.1 Regularization Theory and RBF

Suppose that the set $D = \{(\mathbf{x}_i, y_i) \in R^d \times R\}_{i=1}^N$ is a random, noisy sample of some multivariate function h. The problem of recovering the function h from the data D is ill-posed, and can be formulated in the framework of regularization theory (Tikhonov, 1963; Wahba, 1990; Poggio and Girosi, 1990). In this framework the solution is found by minimizing a functional of the form:

$$H[f] = \sum_{i=1}^{N}(f(\mathbf{x}_i) - y_i)^2 + \lambda\phi[f] \ . \tag{2}$$

where λ is a positive number that is usually called the *regularization parameter* and $\phi[f]$ is a cost functional that constrains the space of possible solutions according to some form of prior knowledge. The most common form of prior knowledge is *smoothness*, that, in words, ensures that if two inputs are close the two corresponding outputs are also close. We consider here a very general class of rotation invariant smoothness functionals (Girosi, Jones and Poggio, 1995), defined as

$$\phi[f] = \int_{R^d} d\mathbf{s} \, \frac{|\tilde{f}(\mathbf{s})|^2}{\tilde{G}(\mathbf{s})}$$

where $\tilde{}$ indicates the Fourier transform, $\tilde{G}$ is some positive radial function that tends to zero as $||\mathbf{s}|| \to \infty$ (so that $\frac{1}{\tilde{G}}$ is a high-pass filter). We consider here for simplicity of subsequent notations the case in which G (the Fourier transform of $\tilde{G}$) is positive definite, rather than conditionally positive definite (Micchelli, 1986), and therefore is a bell-shaped function. It is possible to show (see the paper by Girosi, Jones and Poggio, 1995, for a sketch of the proof) that the function that minimizes the functional (2) is a classical Radial Basis Functions approximation scheme (Micchelli, 1986; Moody and Darken, 1989):

$$f(\mathbf{x}) = \sum_{i=1}^{N} c_i G(\mathbf{x} - \mathbf{x}_i) \tag{3}$$

where the vector of coefficients $(\mathbf{c})_i = c_i$ satisfies the following linear system:

$$(G + \lambda I)\mathbf{c} = \mathbf{y} \tag{4}$$

where I is the identity matrix, and we have defined the vector of output values $(\mathbf{y})_i = y_i$ and the matrix $(G)_{ij} = G(\mathbf{x}_i - \mathbf{x}_j)$. Classical examples of basis functions G include the Gaussian ($G(\mathbf{x}) = \exp(-||\mathbf{x}||^2)$) and the inverse multiquadric ($G(\mathbf{x}) = (1 + ||\mathbf{x}||^2)^{-\frac{1}{2}}$). In the next section we will show how to embed the prior knowledge about radial symmetry in this framework and we will derive the corresponding solution.

4.2 Regularization Theory in Presence of Radial Symmetry

In the standard regularization theory approach, the minimization of the functional $H[f]$ is usually done on the space of functions Φ for which $\phi[f]$ is finite. If additional knowledge of the solution is known, that can be used to further constrain the space of solutions. If we know that the solution is a function with radial symmetry, then we can restrict ourselves to minimize $H[f]$ over $\Phi \bigcap \mathcal{R}$, where $\mathcal{R}$ is the set of radial functions. The problem we have to solve now is therefore the following:

$$\min_{f \in \Phi \bigcap \mathcal{R}} H[f] = \min_{f \in \Phi \bigcap \mathcal{R}} \sum_{i=1}^{N} (f(\mathbf{x}_i) - y_i)^2 + \lambda \phi[f] \, . \tag{5}$$

We now notice that any function in $\mathcal{R}$ uniquely defines a one dimensional function f^* as follows

$$f(\mathbf{x}) \equiv f^*(\|\mathbf{x}\|) \, . \tag{6}$$

Using this notation and standard results from Fourier theory, we can represent elements of $\mathcal{R}$ by their Hankel transform (Dautray and Lions, 1988)

$$f(\mathbf{x}) = C \|\mathbf{x}\|^{-\frac{d}{2}+1} \int ds \; \tilde{f}^*(s) s^{\frac{d}{2}} J_{\frac{d}{2}-1}(s\|\mathbf{x}\|) \tag{7}$$

where C is a known number, $J_{\frac{d}{2}-1}$ is a Bessel function of the first kind (Gradshtein and Ryzhik, 1981) and $\tilde{f}^*(s)$ is defined by $\tilde{f}(\mathbf{s}) \equiv \tilde{f}^*(\|\mathbf{s}\|)$. The functional of eq. (5) can now be thought as a functional of $\tilde{f}^*(s)$, and the solution of the minimization problem can be found by imposing the stationarity condition $\frac{\delta H[f]}{\delta \tilde{f}^*(s)} = 0$. After some lengthy calculations it is found that the solution of the approximation problem can be written in the following form:

$$f(\mathbf{x}) = \sum_{i=1}^{N} c_i H(\|\mathbf{x}\|, \|\mathbf{x}_i\|) \tag{8}$$

where we have defined

$$H(\|\mathbf{x}\|, \|\mathbf{x}_i\|) = (\|\mathbf{x}\| \|\mathbf{x}_i\|)^{-\frac{d}{2}+1} \int ds \tilde{G}^*(s) s J_{\frac{d}{2}-1}(s\|\mathbf{x}\|) J_{\frac{d}{2}-1}(s\|\mathbf{x}_i\|) \tag{9}$$

Although the basis function H does not have a friendly look, notice the similarity of the solution (8) with the standard solution (3). In both cases the final approximating function is a linear superposition of basis functions, and there is one basis function for each data point. From a computational point of view, in both cases the coefficients c_i are found by solving a linear system, with the only difference that in the case (8) the matrix $(G)_{ij}$ of eq. (4) is replaced by the matrix $(H)_{ij} = H(||\mathbf{x}_i||, ||\mathbf{x}_j||)$. However, while it is clear that the standard solution is obtained by placing a "bump" function at each data point, this interpretation is not evident from the solution (8). As the following example shows, a very similar thing happens indeed, and this will become clearer in the next section, when we will discuss the creation of "virtual" examples.

Example Let us consider the very common case in which the basis function $G(\mathbf{x})$ is Gaussian. In this case its Fourier transform is also a Gaussian, and therefore $G^*(s) = \exp(-s^2)$. The integral of eq. (9) can be performed (Gradshtein and Ryzhik, 1981), to obtain the following form for H:

$$H(||\mathbf{x}||, ||\mathbf{x}_i||) = e^{-(||\mathbf{x}||^2+||\mathbf{x}_i||^2)} I_{\frac{d}{2}-1}(2||\mathbf{x}||||\mathbf{x}_i||) \tag{10}$$

where $I_{\frac{d}{2}-1}$ is the Bessel function of first kind of imaginary argument (Gradshtein and Ryzhik, 1981, par. 8.406). A plot of this function in 2 dimensions is presented in figure (2), where we have set $||\mathbf{x}_i|| = 2$. It is clear that this function is a radial "bump" function, whose bump is concentrated on a circle of radius $||\mathbf{x}_i||$. Any radial section of this function looks like a Gaussian function centered at $||\mathbf{x}_i||$, providing a local, radially symmetric, form of approximation.

4.3 Radial Symmetry and "Virtual" Examples

In this section, we use the prior knowledge to generate new, "virtual" examples, from the existing data set.

Let $D = \{(\mathbf{x}_i, y_i) \in R^d \times R\}_{i=1}^N$ be our data set, and let us assume that we know that the function h underlying the data has radial symmetry. This means that $f(\mathbf{x}) = f(R_\theta \mathbf{x})$ for all the possible rotation matrices R_θ in d dimensions. Here θ is a $d-1$ dimensional vector of parameters that represents a point of Σ_{d-1}, the surface of the d-dimensional unit sphere. This property implies that if $(\mathbf{x}_i, y_i)$ is an example of h, the points $(R_\theta \mathbf{x}_i, y_i)$, for all $\theta \in \Sigma_{d-1}$, are also examples of h, and we call these additional points the "virtual" examples.

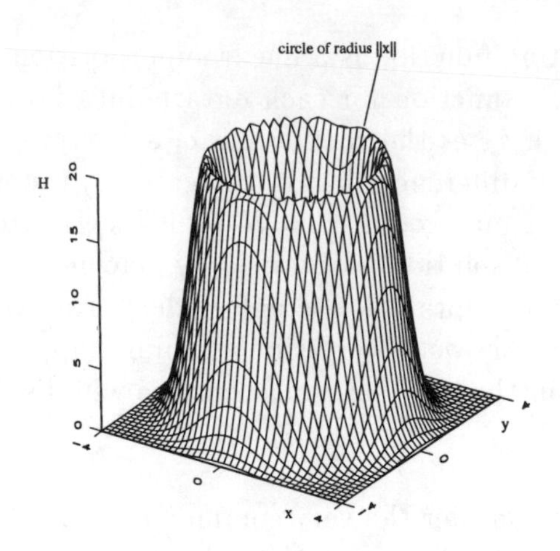

Figure 2: The basis function $H(\|\mathbf{x}\|, \|\mathbf{x}_i\|)$, for $\mathbf{x}_i = (2, 0)$.

Let us now consider a standard Radial Basis Functions approximation technique, of the form (3). Suppose for the moment that the function is invariant with respect to a *finite* number of rotations $R_{\theta_1}, \ldots, R_{\theta_n}$. Each example $\mathbf{x}_i$ will therefore generate n virtual examples $R_{\theta_1}\mathbf{x}_i, \ldots R_{\theta_n}\mathbf{x}_i$, that can now be included in the expansion (3) together with the regular examples. It is trivial to see that, because of the invariance property of h, the coefficients of the basis functions corresponding to the virtual examples will be equal to the coefficients of the corresponding, original example. As a result we have that eq. (3) has to be replaced by

$$f(\mathbf{x}) = \sum_{i=1}^{N} c_i \sum_{\alpha=0}^{n} G(\mathbf{x} - R_{\theta_\alpha}\mathbf{x}_i)$$

where we have defined $\theta_0 = 0$, so that $R_{\theta_0}\mathbf{x}_i = \mathbf{x}_i$. We now relax the assumption that the function is invariant with respect to only a finite number of rotations, and allow θ to span the entire surface Σ_{d-1}. The equation above suggests to replace eq. (3) with the following:

$$f(\mathbf{x}) = \sum_{i=1}^{N} c_i \int_{\Sigma_{d-1}} d\Omega_{d-1}(\theta)\; G(\mathbf{x} - R_\theta \mathbf{x}_i) \tag{11}$$

where $d\Omega_{d-1}(\theta)$ is the uniform measure over Σ_{d-1}. Using the Hankel representation (7) for the radial function G in eq. (11), the integral over Σ_{d-1} can be performed, and provides the result:

$$f(\mathbf{x}) = \sum_{i=1}^{N} c_i H(\|\mathbf{x}\|, \|\mathbf{x}_i\|)$$

where $H(\|\mathbf{x}\|, \|\mathbf{x}_i\|)$ is given precisely be expression (9)! From this derivation it is clear that the basis function $H(\|\mathbf{x}\|, \|\mathbf{x}_i\|)$ is an infinite superposition of Gaussian functions, whose centers uniformly cover the surface of the sphere of radius $\|\mathbf{x}_i\|$.

Therefore creating virtual examples seems to be, in a sense, the "right thing" to do, leading to the same result that one gets from the more "principled" and sophisticated approach of regularization theory. The appealing feature of the virtual examples technique is the fact that it can be applied in very general cases, in which it might be impossible to derive analytical results as the one derived in section 3.

5 Virtual Examples in Vision and Speech

The goal of this paper is to suggest the creation of virtual examples as a technique to incorporate prior information in machine learning problems. The previous section shows how creating virtual examples can be equivalent to incorporation of the prior information as a regularizer within a framework for function learning. Thus the virtual example strategy can often be more than a good heuristic. We now turn our attention to some real world problems that arise in computer vision and speech recognition and give examples of how one might generate virtual examples under certain conditions.

In the examples we are about to consider, the non-trivial part of the virtual example strategy is identifying the set of legal transformations that allow a new, valid, example to be created. In previous treatments of the machine learning problem that concentrated on function learning, the legal transformations were typically very simple and could easily be used to create examples. For example, whether the function is even or odd, or whether it has radial symmetry is easy to deal with. Imagine, instead, that one were interested in object recognition. How does one generate a new example? There are certain obvious cases. For example, by translating the image in the image plane or dilating the image (scale transformation) one could generate some trivial cases of new examples. However, there are some other non obvious ones like rotation in depth, or changing the expression of a face that are significantly harder to realize. In the next section, we discuss the problem of object recognition, how to view it within a function learning paradigm, and how to generate non-trivial virtual examples for it.

5.1 Virtual Views for Object Recognition

Consider the problem of recognizing 3D objects from their 2D images. A particular class of 3D objects (like cars, or cubes, or faces) can be defined in terms of pointwise features that have been put in correspondence with each other. If, for example, there are n features, and each feature is represented by its location in a 3D coordinate system (say by its x, y, z coordinates) then a particular view of a particular 3D object can be represented as a point in R^{3n}. However, note that not all points in R^{3n} correspond to valid views of 3D objects. Trying to learn this object class could be regarded as trying to learn a characteristic function in R^{3n}, i.e., a function of the form:

$$1_E(x) : R^{3n} \longrightarrow \{0,1\} = \begin{matrix} 1 & x \in E \subset R^{3n} \\ 0 & \text{otherwise} \end{matrix}$$

When the 3D view is projected to 2D, then each 2D view can now be represented as a characteristic function over R^{2n}. For problems such as these, one can rarely specify simple mathematical constraints (like radial symmetry etc.) on the characteristic functions. This makes the recognition problem particularly challenging. Consider, for example, the face recognition problem studied by Beymer (1994). The goal is to recognize faces of different people under a variety of views. One approach to this is to collect a large number of views from each person and train a classifier to recognize them. Shown in fig. 3 are fifteen views of one particular face that have been collected as training examples for that face. This relatively straightforward approach works but usually requires a large number of training examples.

In contrast, an alternative strategy is to use some kind of prior knowledge about the class of faces in order to generate virtual examples or virtual "views". One could then train a view independent system on the basis of these virtual examples. This raises the question: if we are given examples of images belonging to some class, then can we generate new examples of images belonging to the same class? In order to do this we need to uncover the set of *legal* transformations that allow us to take elements of E and come up with other elements of E. Prior knowledge about the class of objects allow us to uncover such a set of valid transformation.

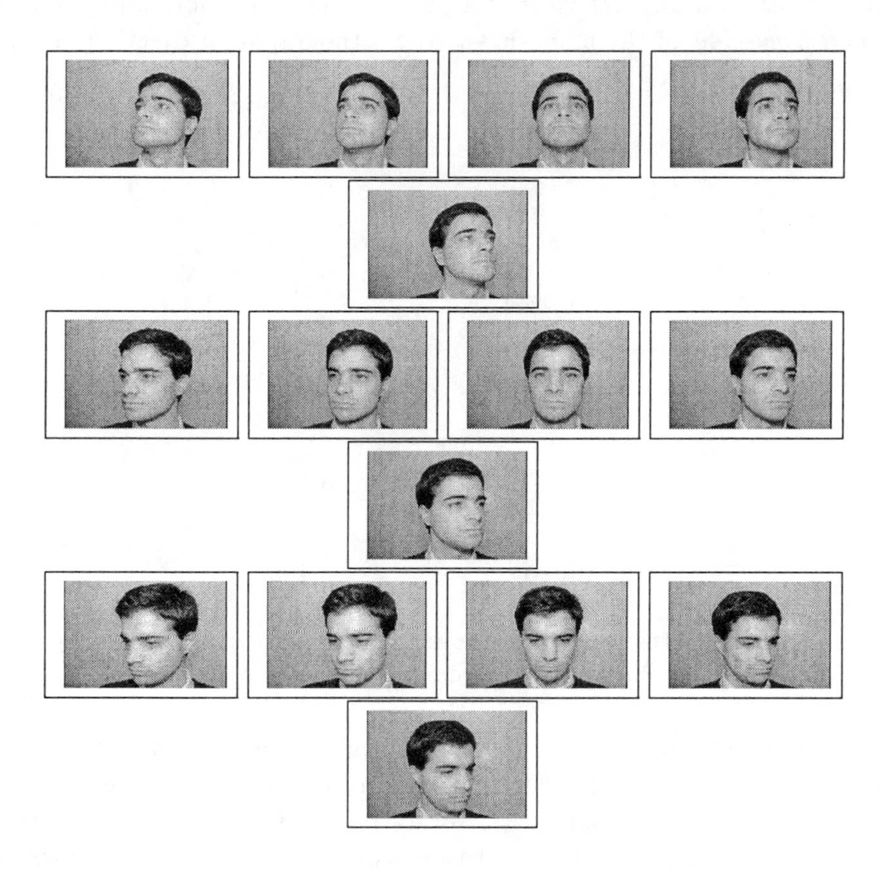

Figure 3: The pose-invariant, view-based face recognizer uses 15 views to model a person's face. From Beymer, 1994.

5.2 Symmetry as Prior Information

Poggio and Vetter (1992) examined in particular the case of bilateral symmetry of certain 3D objects, such as faces. Suppose that we have a model 2D view of an object and a pair of symmetric points in this 2D view. For our purposes, we can define an object to be *bilaterally symmetric* if the following transformation of any 2D view of a pair of symmetric points of the object yields a *legal view* of the pair, that is the orthographic projection of a rigid rotation of the object

$$D\mathbf{x}_{pair} = \mathbf{x}^*_{pair} \tag{12}$$

with

$$\mathbf{x}_{pair} = \begin{pmatrix} x_1 \\ x_2 \\ y_1 \\ y_2 \end{pmatrix} \qquad \mathbf{x}^*_{pair} = \begin{pmatrix} -x_2 \\ -x_1 \\ y_2 \\ y_1 \end{pmatrix}$$

and

$$D = \begin{pmatrix} 0 & -1 & 0 & 0 \\ -1 & 0 & 0 & 0 \\ 0 & 0 & 0 & 1 \\ 0 & 0 & 1 & 0 \end{pmatrix}.$$

Notice that symmetric pairs are the elementary features in this situations and points lying on the symmetry plane are degenerate cases of symmetric pairs.

Geometrically, this simply means that for bilaterally symmetric objects simple transformations of a 2D view yield other views that are *legal.* The transformations are similar to mirroring one view around an axis in the image plane, as shown in Figure 4 top (where the left image is "mirrored" into the right one) and correspond – but only for a bilaterally symmetric object – to proper rotations of a rigid 3D object and their orthographic projection on the image plane. Using the transformation of equation 12 an additional view is generated from the one model view. If the two views are linearly independent, then one can resort to the 1.5 views theorem[3]

[3] Using the notation introduced earlier, the set E defines the space of valid image views of a particular object. The 1.5 views theorem states essentially that E can be regarded as a 6-dimensional vector space. Furthermore this basis can be computed from two linearly independent views. For further details on this, see Poggio and Vetter (1992).

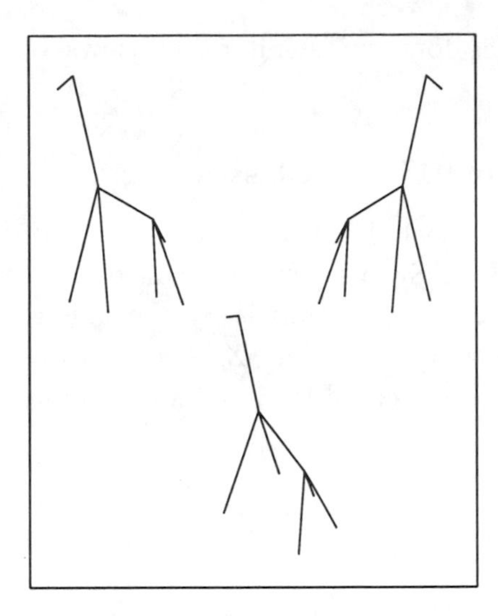

Figure 4: Given a single 2D view (upper left), a new view (upper right) is generated under the assumption of bilateral symmetry. The two views are sufficient to verify that a novel view (second row) corresponds to the same object as the first.

to compute a 3D basis that spans the space of the object. This allows us to compute a recognition function with just one true view. Bilateral symmetry has been used in face recognition systems (Beymer and Poggio, 1995) and psychophysical evidence supports its use by the human visual system (Schyns and Bulthoff, 1993; Troje and Bulthoff, 1995; Vetter, Poggio, and Bulthoff, 1994).

5.3 More General Transformations: Linear Object Classes

A more flexible way to acquire information about how images of objects of a certain class change under pose, illumination and other transformations, is to learn the possible pattern of variabilities and class-specific deformations from a representative training set of views of generic or prototypical objects of the same class – such as other faces. In particular, if the objects belong to a well behaved class known as a linear object class, the transformations can be easily learned. In this manner prior knowledge that the object class is linear can be utilized effectively to generate novel views that can be incorporated in the training process.

Although this approach of linear classes originates from the proposal of Poggio and Vetter (1992) for countering the curse-of-dimensionality in applications of supervised learning techniques, more powerful versions have been developed recently. Techniques based on non-linear learning networks have been developed by Beymer, Shashua and Poggio (1993) as well as Beymer and Poggio (1995). For our purposes here, we now provide a brief overview of the technique of linear classes for generating novel views of objects.

5.3.1 3D Objects, 2D Projections, and Linear Classes

Consider a 3D view of a three-dimensional object that is defined in terms of pointwise features (Poggio and Vetter, 1992). Such a 3D view can be represented by a vector $\mathbf{X} = (x_1, y_1, z_1, x_2, ..., y_n, z_n)^T$, that is by the x, y, z-coordinates of its n feature points. Further, assume that $\mathbf{X} \in \Re^{3n}$ is the linear combination of q 3D views $\mathbf{X}_i$ of *other* objects of the same dimensionality, such that:

$$\mathbf{X} = \sum_{i=1}^{q} \alpha_i \mathbf{X}_i. \quad (13)$$

Consider now some linear operator L associated with a desired uniform transformation such as for instance a specific rotation in 3D. Let us define $\mathbf{X}^r = L\mathbf{X}$ to be the rotated 3D view of object $\mathbf{X}$. Because of the linearity of the group of uniform linear transformations $\mathcal{L}$, it follows that

$$\mathbf{X}^r = \sum_{i=1}^{q} \alpha_i \mathbf{X}_i^r \quad (14)$$

Thus, *if a 3D view of an object can be represented as the weighted sum of views of other objects, its rotated view is a linear combination of the rotated views of the other objects with the same weights.* Of course for an arbitrary 2D view that is a projection of a 3D view, a decomposition like (13) does not in general imply a decomposition of the rotated 2D views (it is a necessary but not a sufficient condition).

Linear Classes

A natural question to ask, therefore, is: "Under what conditions do the 2D projections of 3D objects satisfy equation (13) to (14)?" The answer will clearly depend on the types of objects we use and also on the projections we allow. In a series of articles (Poggio and Vetter, 1992; Vetter and Poggio, 1995) the notion of linear classes has been introduced and developed — we provide a definition below:

A set of 3D views (of objects) $\{\mathbf{X}_i\}$ *is a* **linear object class** *under a linear projection* P *if* $dim\{\mathbf{X}_i\} = dim\{\mathbf{P}\mathbf{X}_i\}$ *with* $\mathbf{X}_i \in \Re^{3n}$ *and* $\mathbf{P}\mathbf{X}_i \in \Re^p$

and $p < 3n$

This is equivalent to saying that the minimal number of basis objects necessary to represent an object is not allowed to change under the projection. Note that the linear projection P is not restricted to projections from $3D$ to $2D$, but may also "drop" occluded points. Now assume $\mathbf{x} = P\mathbf{X}$ and $\mathbf{x}_i = P\mathbf{X}_i$ are the projections of elements of a linear object class with

$$\mathbf{x} = \sum_{i=1}^{q} \alpha_i \mathbf{x}_i \tag{15}$$

then $\mathbf{x}^r = P\mathbf{X}^r$ can be constructed without knowing $\mathbf{X}^r$ using α_i of equation (15) and the given $\mathbf{x}_i^r = P\mathbf{X}_i^r$ of the other objects.

$$\mathbf{x}^r = \sum_{i=1}^{q} \alpha_i \mathbf{x}_i^r. \tag{16}$$

5.3.2 Implications

The relations described earlier suggest that we can use "prototypical" 2D views (the projections of a basis of a linear object class) and their known transformations to synthesize an operator that will transform a 2D view into a new 2D view when the object is a linear combination of the prototypes. In other words we can compute a new 2D view of such an object without knowing explicitly its three-dimensional structure. Notice also, that knowledge of the correspondence between equation (15) and equation (16) is not necessary (rows in a linear equation system can be exchanged freely). Therefore, the technique does not require one to compute the correspondence between views from different viewpoints. In fact some points may be occluded. Fig. 5 shows a very simple example of a linear object class and the construction of a new view of an object. Since the dimension of the class of all cuboids is 3, any cuboid can be represented as a linear combination of 3 prototypical cuboids. Thus the class is linear under all orthographic projections that preserve the three dimensions.
Remark: Three-dimensional objects differ in shape as well as in texture. To truly apply the linear class idea to gray level images, we need to derive object representations that incorporate texture. This can be done by developing shape and texture vector representations that are in correspondence and using the linear class idea over both.

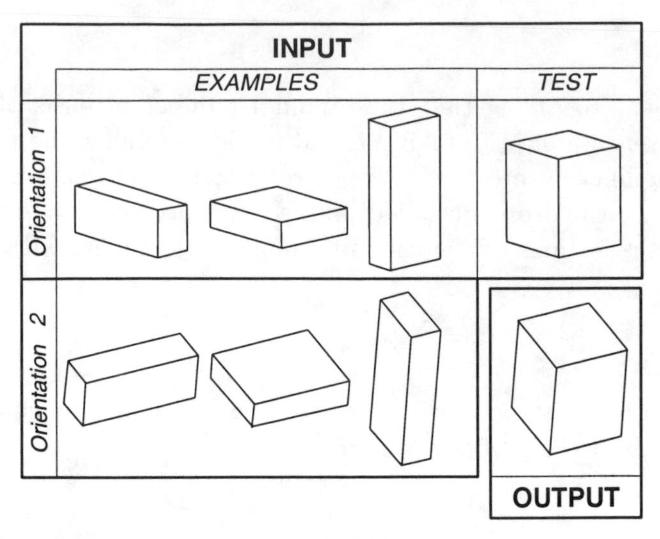

Figure 5: *Learning an image transformation according to a rotation of three-dimensional cuboids from one orientation (upper row) to a new orientation (lower row). The 'test' cuboid (upper row right) can be represented as a linear combination of the two-dimensional coordinates of the three example cuboids in the upper row. The linear combination of the three example views in the lower row, using the coefficients evaluated in the upper row, results in the correct transformed view of the test cuboid as output (lower row right). Notice that correspondence between views in the two different orientations is not needed and different points of the object may be occluded in the different orientations.*

5.3.3 Learning the Transformation

We finally complete the story by discussing how the transformation from a reference view to a novel view can be learned. Before we proceed, let us introduce a helpful change of coordinate systems in equations (15) and (16). Instead of using an absolute coordinate system, we represent the views as the difference to the view of a reference object of the same class. Subtracting the projection of a reference object from both sides of equations (15) and (16), we have

$$\Delta \mathbf{x} = \sum_{i=1}^{q} \alpha_i \Delta \mathbf{x}_i \tag{17}$$

and

$$\Delta \mathbf{x}^r = \sum_{i=1}^{q} \alpha_i \Delta \mathbf{x}_i^r. \tag{18}$$

After this change in the coordinate system, equation (18) now evaluates to the new difference vector to the rotated reference view. The new view of the object can be constructed by adding this difference to the reference view.

Steps in Constructing a Novel View:

Step 1: First, we compute the coefficients α_i for the optimal decomposition (in the sense of least square). We decompose an "initial" field $\Delta\mathbf{x}$ to a new object X into the "initial" fields $\Delta\mathbf{x}_i$ to the q given prototypes by minimizing

$$||\Delta \mathbf{x} - \sum_{i=1}^{q} \alpha_i \Delta \mathbf{x}_i||^2. \tag{19}$$

Rewriting this as $\Delta\mathbf{x} = \mathbf{\Phi}\boldsymbol{\alpha}$ (where $\mathbf{\Phi}$ is the matrix formed by the q vectors $\Delta\mathbf{x}_i$ arranged column-wise and $\boldsymbol{\alpha}$ is the column vector of the α_i coefficients) and minimizing equation (19) gives

$$\boldsymbol{\alpha} = (\mathbf{\Phi})^+ \Delta \mathbf{x}. \tag{20}$$

Step 2: The observation of the previous section implies that the operator L that transforms $\Delta\mathbf{x}$ into $\Delta\mathbf{x}^r$ through $\Delta\mathbf{x}^r = L\Delta\mathbf{x}$, is given by

$$\Delta \mathbf{x}^r = \mathbf{\Phi}^r \boldsymbol{\alpha} = \mathbf{\Phi}^r \mathbf{\Phi}^+ \Delta \mathbf{x} \qquad as \qquad L = \mathbf{\Phi}^r \mathbf{\Phi}^+ \tag{21}$$

and thus can be learned from the 2D example pairs $(\Delta\mathbf{x}_i, \Delta\mathbf{x}_i^r)$. In this case, a one-layer, linear network (compare Hurlbert and Poggio, 1988) can be used to learn the transformation L. L can then transform a view of a novel object of the same class. If the q examples are linearly independent $\mathbf{\Phi}^+$ is given by $\mathbf{\Phi}^+ = (\mathbf{\Phi}^T \mathbf{\Phi})^{-1} \mathbf{\Phi}^T$; in the other cases equation (19) can be solved by an SVD algorithm.
Step 3: Analogous steps have to be taken to deal with textures. Before decomposing the new texture into example textures, all textures are mapped onto a common basis — typically, the reference image using correspondences. In this representation the decomposition of the textures can be performed as described above.

Step 4: The final step is image rendering. The α coefficients that are computed for both texture and shape vectors are then applied to the prototype examples in the second orientation. The correspondence fields to the new image are combined with the reference image (often using forward warping; Wolberg, 1990) to generate the novel image.

Using this procedure, we can now generate novel views of images with prior knowledge the image belongs to a linear class. Fig. 5 shows a case where a new view has been generated for the class of cuboids for which the linear class assumption is correct. More interestingly, however, fig. 6 shows how novel views of a face can be generated from prototypical views using the linear class technique. While faces are not theoretically guaranteed to constitute a linear class, in practice, the assumption turns out to be quite good as the example shows. Thus, instead of collecting fifteen example patterns and training on them, one could generate virtual examples using the techniques described here and train on a combination of the real and virtual examples. A system based on this has been successfully implemented and described in Beymer and Poggio, 1995. A face recognition system using a single real view plus 14 virtual views as in figure 6 (per person) achieved a recognition rate of 85% correct while a system using 15 real views (per person) achieved 99% on the same database. Both systems were significantly better than a system using one real view per person (32%). Notice that in the approach outlined above, correspondence plays a key role. Interestingly, correspondence for the prototypes is only required between views from the same viewpoint. Correspondence is also required between the real image and one of the prototypes and can be computed automatically by optical flow techniques (Beymer and Poggio, 1996). A different approach (see Jones and Poggio, 1995) does not require an explicit correspondence between the real image and the prototypes. Vetter et al (1997) propose a technique that may also allow for an automatic correspondence between the prototype images.

5.4 Virtual Examples in Speech Recognition

Another example of the potential utility of the "virtual example" technique for incorporating prior information can be provided in the context of speech recognition. In the production of speech by humans, an important source of prior information lies in the physical constraints that the vocal tract and speech articulators (speech producing apparatus) necessarily have to obey. For example, different sounds (phonemes) produced by the same speaker might share some common characteristics related to the properties of the

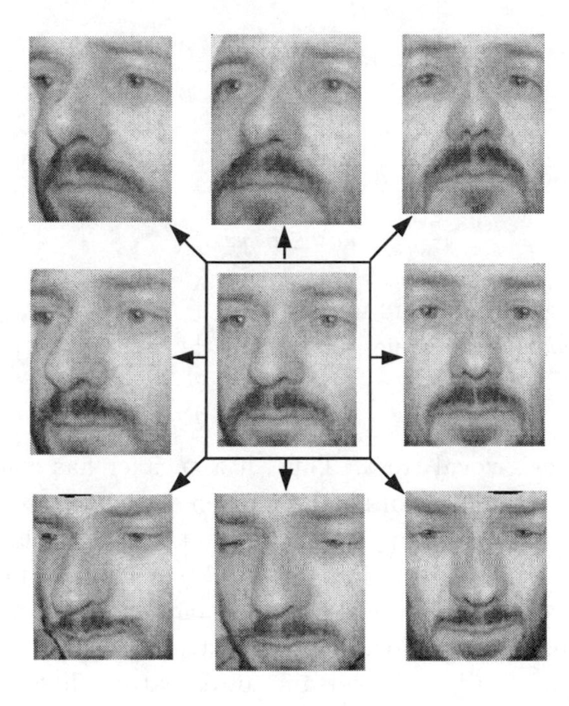

Figure 6: A real view (center) surrounded by virtual views derived from it a technique related to the linear class technique but even simpler. The correspondence between the real view and the prototypes is computed analytically. For details on the required process, see Beymer and Poggio, 1995.

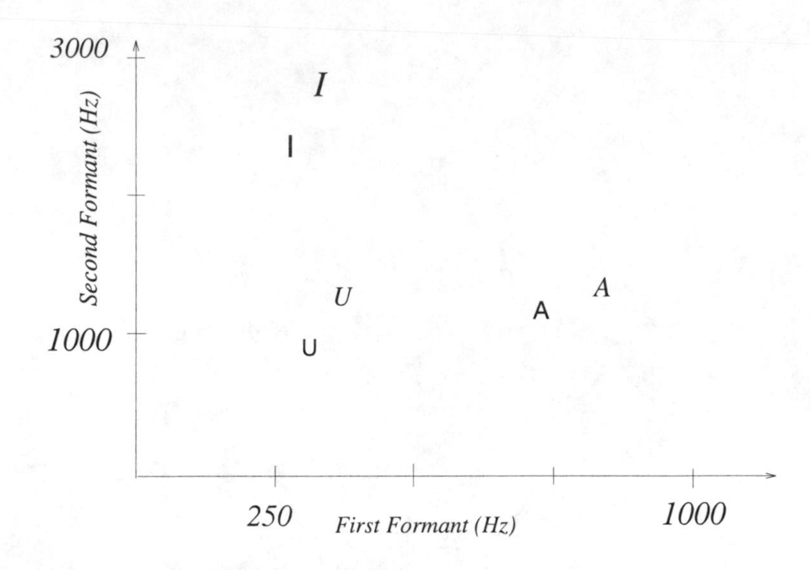

Figure 7: The first and second formants of "i" (as in *beat*), "a" (as in *palm*) and "u" (as in *boot*). The values for the female speaker are plotted in italics.

individual speaker's vocal tract. Thus, if a speaker has a high pitch, then this is likely to be the case for all the phonemes the speaker produces.

Consider fig. 7 which compares the formant values for three vowel sounds — "i" (as in *beat*, "a" (as in *palm*) and "u" (as in *boot*) for one male and one female speaker. The data has been obtained from Peterson and Barney (1952). In speech recognition, one will have to distinguish between the different vowels on the basis of certain vowel features like formants. Notice that while the broad pattern of formants for the two speakers are the same, the actual formant values differ — the female speaker has consistently higher formants in general. It turns out that the formant values are related to the length of the vocal tract — people with longer tracts have lower formants and vice-versa. This is the sort of prior information that one would like to capture while attempting to solve the speech recognition problem in a way that is invariant to the systematic speaker differences.

Roughly speaking, speech is produced when air is pumped into the vocal tract thereby exciting it and the corresponding acoustic waves are transmitted through the air to the hearer. Thus, the vocal tract (including the nasal tract) modulates the excitations produced from below resulting in speech. From an acoustic standpoint, the vocal tract has been modeled as a non-uniform tube and its resonances correspond to the formants described earlier. The length of the tube is inversely related to the frequency of the resonances and thus humans with longer vocal tracts have formants at lower frequencies. Each different sound corresponds to a different articulatory configuration, a correspondingly different vocal tract shape, a non-uniform tube of different shape, and therefore a correspondingly different set of formant values. All speakers producing an "a" would have roughly the same configuration — as a first order approximation, speaker differences would come about due to variations in scale, i.e., all the vocal tracts would have the same overall shape, the vocal tract of a large man would be longer than that of a child. This overall length, for example, is a quantity that is preserved across all sounds of the speaker. If it could be extracted automatically from a speaker's "a", then it could be used to scale a canonical speaker's "i" to produce a novel "i".

The acoustic problem can be modeled using electrical circuits and the most common formulation has taken a source-filter point of view. The vocal tract is viewed as a filter shaping the input provided by electrical sources. Sources are roughly of two types: (1) *periodic* sources that correspond to the glottal vibrations during voiced speech, and (2) *aperiodic* sources that correspond to various kinds of turbulent sources produced during unvoiced (or partially voiced) speech, e.g., during fricatives like "s" etc. Shown in fig. 8 is a schematic view of the speech production apparatus. The vocal tract filter $H(z)$ is shown to be parameterized by two kinds of parameters, p, that models the shape of the tract and depends upon the phonetic identity of the sound, and s, that models things like overall scale and depends upon the specific characteristics of the speaker. Similarly, the voiced periodic source is also parameterized by a set of parameters (denoted by w) that are speaker-specific and don't change from phoneme to phoneme. Typical examples of such speaker-specific parameters are pitch, voice quality, etc.

One approach to the virtual examples idea is to obtain a number of examples of a novel speaker's "a" sounds. For such sounds, we already know the phoneme-dependent parameter values p (since this depends only on the phoneme and is common to all speakers, such parameters can be estimated directly from the data collected from a reference speaker). From the novel speaker's sounds, we estimate the speaker-specific parameters, s. Now, to generate new instances of a speaker's "i", we can drive the speech production model using the phoneme-specific parameters (p that are derived from the reference speaker) and the speaker-specific parameters (s) that are

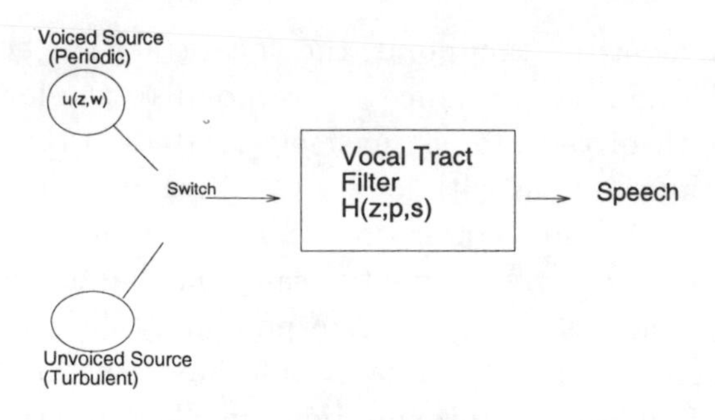

Figure 8: A schematic view of the speech production apparatus.

derived from the test speaker. Note that such a virtual example strategy would depend in large part upon the fidelity of the speech production model that one has in the first place.

A simpler and more direct strategy is to learn the mapping from "a" to "i" by collecting examples of each from a number of different speakers. By doing this, we can convert a speaker's "a" into a novel speaker's "i" using the functional mapping learned. Examine fig. 9 that shows the relationship between "ae" (as in *bat*) and "i" (as in *beat*) from the same speaker. Data from 360 speakers were collected and speakers were grouped into 9 speaker classes. A weighted spectral measure was computed for each of the speaker's "ae" and "i" sounds. The mean values of this measure for each of the speaker groups is plotted. Notice the strong correlation between the mean value of a group's "ae" and the mean value of that group's "i" suggesting strong predictability. Such an idea has been used successfully in incorporating speaker information into a speech recognition system (Niyogi and Zue, 1991; Niyogi, 1992).

The problem described here is analogous to the vision example we described. Different poses of the same face are like different sounds of the same speaker (equivalent to different "poses" of the vocal tract). In object recognition, we used knowledge of the relationship between different poses of prototypes to create a novel pose for a new speaker. In speech recognition, we used knowledge of the relationship between different sounds of prototypical speakers to create a novel sound for a new speaker. This particular strategy of creating virtual examples has rarely been used in speech recognition — and would be particularly useful if one wishes to adapt a speech

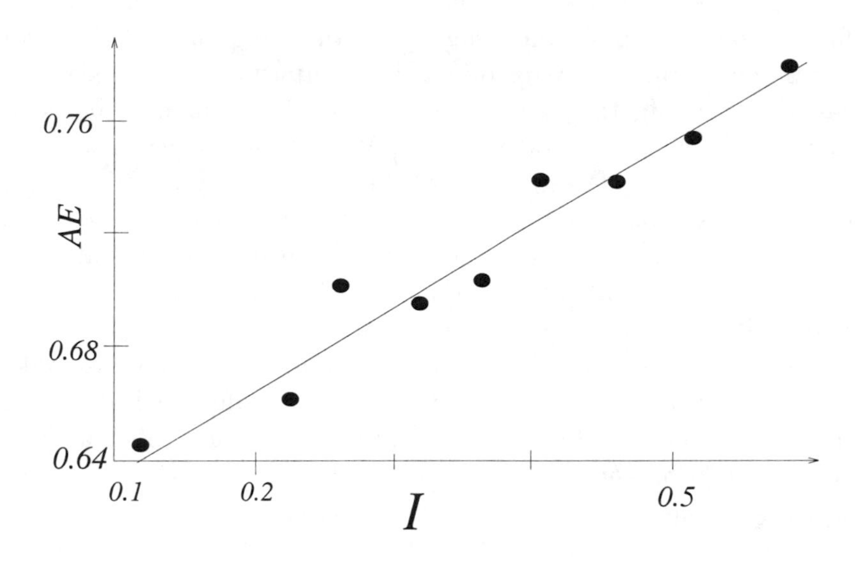

Figure 9: A weighted spectral average measure for a speaker's "i" and the same speaker's "ae" plotted against each other. Data from 360 speakers were collected and grouped into 9 classes and the class average for each of the two sounds has been plotted. Notice the strong correlation suggesting that it is possible to predict the "ae" sound from the "i" sound given the group to which a speaker belongs.

recognition system to a new speaker using extremely limited amounts of adaptation data.

6 Conclusions

In this paper, we introduced the idea of virtual examples as a possible strategy for incorporating prior knowledge about the target function in a learning from examples paradigm. We motivated the importance of using virtual examples by discussing the issue of sample complexity in machine learning. Specifically, using the general results of Vapnik and Chervonenkis, we argued that the number of examples needed for successful generalization would be prohibitive without adequate constraints on the hypothesis class. However, successful learning would result only if the such constraints properly reflected our prior knowledge of the problem being solved.

The creation of virtual examples is one way around the dilemma. We showed how the idea of virtual examples was mathematically equivalent to incorporating prior knowledge as a regularizer in function learning in certain restricted domains. In most substantive real world problems, however, it is rare that prior knowledge can be directly implemented as an elegant regularization constraint. For such cases, the creation of virtual examples might be more straightforward. However, this is not to say that virtual examples are easily created. In particular, in this paper, we spent a significant portion of time discussing the details of the creation of virtual examples for object recognition and speech recognition. The results obtained so far suggest that this might be a promising way to incorporate prior knowledge into an example-based learning framework leading to systems that generalize well from the limited amounts of data that is typically available in real world problems.

References

[1] Y. Abu-Mostafa. A method for learning from hints. In S. J. Hanson, Jack D. Cowan, and C. Lee Giles, editors, *Advances in Neural information processings systems 5*, San Mateo, CA, 1992. Morgan Kaufmann Publishers.

[2] Y.S. Abu-Mostafa. Learning from hints in neural networks. *Journal of Complexity*, 6:192–198, 1990.

[3] Y.S. Abu-Mostafa. Hints and the VC-dimension. *Neural Computation*, 5:278–288, 1993.

[4] Y.S. Abu-Mostafa. Hints. *Neural Computation*, 7:639–671, 1995.

[5] A.R. Barron. Approximation and estimation bounds for artificial neural networks. *Machine Learning*, 14:115–133, 1994.

[6] M. Bertero. Regularization methods for linear inverse problems. In C. G. Talenti, editor, *Inverse Problems*. Springer-Verlag, Berlin, 1986.

[7] D. Beymer and T. Poggio. Face recognition from one example view. In *Proceedings of the International Conference of Computer Vision*, Cambridge, MA, June 1995.

[8] D. Beymer and T. Poggio. Image representations for visual learning. *Science*, 272(5270):1905–1909, June 1996.

[9] D. Beymer, A. Shashua, and T. Poggio. Example based image analysis and synthesis. A.I. Memo No. 1431, Artificial Intelligence Laboratory, Massachusetts Institute of Technology, 1993.

[10] David J. Beymer. Face recognition under varying pose. In *Proceedings of IEEE Conf. on Computer Vision and Pattern Recognition*, Seattle, WA, 1994.

[11] C. M. Bishop. Training with noise is equivalent to tikhonov regularization. *Neural Computation*, 7:108–116, 1995.

[12] L. Breiman, J. H. Friedman, R.A. Olshen, and C. J. Stone. *Classification and Regression Trees*. Wadsworth, Belmont, CA, 1984.

[13] R. Dautray and J.L. Lions. *Mathematical Analysis and Numerical Methods for Science and Technology*, volume 2. Springer-Verlag, Berlin, 1988.

[14] D. Wolpert (ed.). *The Mathematics of Generalization*. Addison-Wesley, Reading, MA, 1995.

[15] F. Girosi, M. Jones, and T. Poggio. Regularization theory and neural networks architectures. *Neural Computation*, 7:219–269, 1995.

[16] I.S. Gradshteyn and I.M. Ryzhik. *Table of integrals, series, and products*. Academic Press, New York, 1981.

[17] A. Hurlbert and T. Poggio. Synthesizing a color algorithm from examples. *Science*, 239:482–485, 1988.

[18] M. Jones and T. Poggio. Model-based matching of line drawings by linear combination of prototypes. In *Proceedings of the International Conference on Computer Vision*, pages 531–536. IEEE, June 1995.

[19] T. K. Leen. From data distributions to regularization in invariant learning. *Neural Computation*, 7:974, 1995.

[20] C. A. Micchelli. Interpolation of scattered data: distance matrices and conditionally positive definite functions. *Constructive Approximation*, 2:11–22, 1986.

[21] J. Moody and C. Darken. Learning with localized receptive fields. In G. Hinton, T. Sejnowski, and D. Touretzsky, editors, *Proceedings of the 1988 Connectionist Models Summer School*, pages 133–143, Palo Alto, 1988.

[22] J. Moody and C. Darken. Fast learning in networks of locally-tuned processing units. *Neural Computation*, 1(2):281–294, 1989.

[23] V.A. Morozov. *Methods for solving incorrectly posed problems.* Springer-Verlag, Berlin, 1984.

[24] Joseph L. Mundy and Andrew Zisserman. *Geometric Invariance in Computer Vision.* MIT Press, Cambridge, MA, 1992.

[25] P. Niyogi. Modelling speaker variability and imposing speaker constraints in phonetic classification. Technical Report TR-533, MIT Laboratory for Computer Science, 1992.

[26] P. Niyogi and F. Girosi. On the relationship between generalization error, hypothesis complexity, and sample complexity for radial basis functions. *Neural Computation*, 8:819–842, 1996.

[27] P. Niyogi and V.W. Zue. Correlation analysis of vowels and its application to speech recognition. In *Proceedings of Eurospeech*, Genoa, Italy, 1991.

[28] G. E. Peterson and H. L. Barney. Control methods used in a study of the vowels. *Journal of the Acoustical Society of America*, 24(2):175–184, 1952.

[29] T. Poggio and F. Girosi. Networks for approximation and learning. *Proceedings of the IEEE*, 78(9), September 1990.

[30] T. Poggio and T. Vetter. Recognition and structure from one 2D model view: observations on prototypes, object classes and symmetries. A.I. Memo No. 1347, Artificial Intelligence Laboratory, Massachusetts Institute of Technology, 1992.

[31] D. Pomerleau. Alvinn: an autonomous land vehicle in a neural network. In *Advances in Neural Information Processing Systems, I*, San Mateo, CA, 1989. Morgan Kaufman.

[32] D. Pomerleau. Efficient training of artificial neural networks for autonomous navigation. *Neural Computation*, 3(1):88–97, 1991.

[33] D. E. Rumelhart, G. E. Hinton, and R. J. Williams. *Parallel Distributed Processing.* MIT Press, Cambridge, MA, 1986.

[34] P.G. Schyns and H. Bülthoff. Conditions for viewpoint dependent face recognition. A.I. Memo 1432, MIT Artificial Intelligence Lab., Cambridge, MA, August 1993.

[35] P. Simard, Y. LeCun, and J. Denker. Efficient pattern recognition using a new transformation distance. In *Advances in Neural Information Processing Systems V*, pages 50–58, San Mateo, CA, 1993. Morgan Kaufmann Publishers.

[36] P. Simard, B. Victorri, Y. LeCun, and J. Denker. Tangent-Prop - A formalism for specifying selected invariances in an adaptive network. In D. S. Touretzky, editor, *Advances in Neural Information Processing Systems IV*, pages 895–903, San Mateo, CA, 1992. Morgan Kaufmann Publishers.

[37] A. N. Tikhonov. Solution of incorrectly formulated problems and the regularization method. *Soviet Math. Dokl.*, 4:1035–1038, 1963.

[38] N. Troje and H.H. Bülthoff. Face recognition under varying pose: The role of texture and shape. *submitted*, 1995.

[39] V. Vapnik. *The Nature of Statistical Learning Theory.* Springer, New York, 1995.

[40] V. N. Vapnik. *Estimation of Dependences Based on Empirical Data.* Springer-Verlag, Berlin, 1982.

[41] A. Verri and T. Poggio. Regularization theory and shape constraints. A.I. Memo No. 916, Artificial Intelligence Laboratory, Massachusetts Institute of Technology, 1986.

[42] T. Vetter, M. Jones, and T. Poggio. A bootstrapping algorithm for learning linear models of object classes. In *Proceedings of the 1997 Computer Society Conference on Computer Vision and Pattern Recognition (CVPR)*, pages 40–46, June 1997.

[43] T. Vetter, T. Poggio, and H.H. Bülthoff. The importance of symmetry and virtual views in three-dimensional object recognition. *Current Biology*, 4:18–23, 1994.

[44] G. Wahba. *Splines Models for Observational Data.* Series in Applied Mathematics, Vol. 59, SIAM, Philadelphia, 1990.

[45] Georg Wolberg. *Image Warping.* IEEE Computer Society Press, Los Alamitos CA, 1990.

Chapter 5

Deterministic Annealing for Clustering, Compression, Classification, Regression, and Speech Recognition

Kenneth Rose*
Department of Electrical and Computer Engineering
University of California, Santa Barbara, CA 93106

November 7, 1999

Abstract

The deterministic annealing approach to clustering and its extensions have demonstrated substantial performance improvement over standard supervised and unsupervised learning methods in a variety of important applications including compression, estimation, pattern recognition and classification, and statistical regression. The method offers three important features: ability to avoid many poor local optima; applicability to many different structures/architectures; and ability to minimize the right cost function even when its gradients vanish almost everywhere as in the case of the empirical classification error. It is derived within a probabilistic framework from basic information theoretic principles (e.g., maximum entropy and random coding). The application-specific cost is minimized subject to a constraint on the randomness (Shannon entropy) of the solution, which is gradually lowered. We emphasize intuition gained from analogy to statistical physics, where this is an annealing process that avoids many shallow local minima of the specified cost and, at the limit of zero "temperature", produces a non-random (hard) solution. Alternatively, the method is derived within rate-distortion theory, where the annealing process is equivalent to computation of Shannon's rate-distortion function, and the annealing temperature is inversely proportional to the slope of the curve. This provides new insights into the method and its performance, as well as new insights into rate-distortion theory itself. The basic algorithm is extended by incorporating structural constraints to allow optimization of numerous popular structures including vector quantizers, decision trees, multilayer perceptrons, radial basis functions, mixtures of experts and hidden Markov models. Experimental results show considerable performance gains over standard structure-specific and application-specific training methods.

1 Introduction

There are several ways to motivate and introduce the material described in this paper. Let us place it within the neural network perspective, and particularly that of learning. The area of neural networks has greatly benefited from its unique position at the crossroads of several diverse scientific and engineering disciplines including statistics and probability theory, physics, biology, control and signal

*This work was supported in part by the National Science Foundation under grants no. NCR-9314335 and no. IIS-9978001, the University of California MICRO Program, Cisco Systems, Inc., Conexant Systems, Inc., Dialogic Corp., Fujitsu Laboratories of America, Inc., General Electric Co., Hughes Network Systems, Intel Corp., Lernout & Hauspie Speech Products, Lockheed Martin, Lucent Technologies, Inc., Panasonic Technologies, Inc., Qualcomm, Inc., and Texas Instruments, Inc.

processing, information theory, complexity theory, and psychology (see [59]). Neural networks have provided a fertile soil for the infusion (and occasionally confusion) of ideas, as well as a meeting ground for comparing viewpoints, sharing tools and renovating approaches. It is within the ill-defined boundaries of the field of neural networks that researchers in traditionally distant fields have come to the realization that they have been attacking fundamentally similar optimization problems.

This paper is concerned with such a basic optimization problem, and its important variants or derivative problems. The starting point is the problem of *clustering* which consists of optimal grouping of observed signal samples (i.e., a *training set*) for the purpose of designing a signal processing system. To solve the clustering problem one seeks the partition of the training set, or of the space in which it is defined, which minimizes a prescribed cost function (e.g., the average cluster variance). The main applications of clustering are in pattern recognition and signal compression. Given training samples from an unknown source, in the former application the objective is to characterize the underlying statistical structure (identify components of the mixture), while in the latter case a quantizer is designed for the unknown source. This paper describes the deterministic annealing approach to clustering, and its extension, via introduction of appropriate constraints on the clustering solution, to attack a large and important set of optimization problems.

Clustering belongs to the category of *unsupervised learning* problems, where during training we are only given access to input samples for the system under design. The desired system output is not available. The complementary category of *supervised learning* involves a "teacher" who provides, during the training phase, the desired output for each input sample. After training, the system is expected to emulate the teacher. Many important supervised learning problems can also be viewed as problems of grouping or partitioning, and fall within the broad class that we cover here. These include, in particular, problems of classification and regression. We shall further see that the methods described herein are also applicable to certain problems that do not, strictly speaking, involve partitioning.

The design of a practical system must take into account its complexity. Here we must, in general, restrict the complexity of the allowed partitions. This is typically done by imposing a particular structure for implementing the partition. Rather than allowing any arbitrary partition of the training set (or of the input space) we require that the partition be determined by a prescribed parametric function whose complexity is determined by the number of its parameters. For example, a vector quantizer structure implements a Voronoi (nearest neighbor) partition of space and its complexity may be measured by the number of codevectors, or prototypes. Another example is the partition obtained by a multilayer perceptron, whose complexity is determined by the number of neurons and synaptic weights. It is evident, therefore, that the design method will normally be specific to the structure, and this is indeed the case for most known techniques. However, the approach we describe here is applicable to a large and diverse set of structures and problems.

It is always instructive to begin with the simplest non-trivial problem instance in order to obtain an unobstructed insight into the essentials. We therefore start with the problem of clustering for quantizer design, where we seek the optimal partition into a prescribed number of subsets, which minimizes the average cluster variance, or the mean squared error. In this case, we need not even impose a structural constraint. The Voronoi partition is optimal and naturally emerges in the solution. (Structurally constrained clustering is still of interest whenever one wants to impose a different structure on the solution. One such example is the tree structured vector quantizer which is used when lower quantizer complexity is required.) Not having to explicitly impose the structure is a significant simplification, yet, even this problem is not easy. It is well documented (e.g., [55]) that basic clustering suffers from poor local minima that riddle the cost surface. A variety of heuristic approaches have been proposed to tackle this difficulty, and they range from repeated optimization with different initialization, and heuristics to obtain good initialization, to heuristic rules for cluster splits and merges, etc. Another approach was to use stochastic gradient techniques [23], particularly in conjunction with self-organizing feature maps, e.g., [139] and [32]. Nevertheless, there is a substantial margin of gains to be recouped by a methodical, principled attack on the problem as will be demonstrated in this paper for clustering, classification, regression, and other related problems.

The observation of annealing processes in physical chemistry motivated the use of similar concepts to avoid local minima of the optimization cost. Certain chemical systems can be driven to their low energy states by annealing, which is a gradual reduction of temperature, spending a long time at the vicinity of the phase transition points. In the corresponding probabilistic framework, a Gibbs distribution is defined over the set of all possible configurations, which assigns higher probability to configurations of lower energy. This distribution is parameterized by the temperature, and as the temperature is lowered it becomes more discriminating (concentrates most of the probability in a smaller subset of low energy configurations). At the limit of low temperature it assigns nonzero probability only to global minimum configurations. A known technique for nonconvex optimization that capitalizes on this physical analogy is stochastic relaxation or simulated annealing [72] based on the Metropolis algorithm [89] for atomic simulations. A sequence of random moves is generated and the random decision to accept a move depends on the cost of the resulting configuration, relative to that of the current state. However, one must be very careful with the annealing schedule, the rate at which the temperature is lowered. In their work on image restoration, Geman and Geman [48] have shown that, in theory, the global minimum can be achieved if the schedule obeys $T \propto 1/\log n$, where n is the number of the current iteration (see also the derivation of necessary and sufficient conditions for asymptotic convergence of simulated annealing in [56]). Such schedules are not realistic in many applications. In [129] it was shown that perturbations of infinite variance (e.g., the Cauchy distribution) provide better ability to escape from minima and allow, in principle, the use of faster schedules.

As its name suggests, deterministic annealing (DA) tries to enjoy the best of both worlds. On the one hand it is deterministic, meaning that we do not want to be wandering randomly on the energy surface, while making incremental progress on the average, as is the case for stochastic relaxation. On the other hand, it is still an annealing method and aims at the global minimum, instead of getting greedily attracted to a nearby local minimum. One can view DA as replacing stochastic simulations by the use of expectation. An effective energy function, which is parameterized by a (pseudo) temperature, is derived through expectation, and is deterministically optimized at successively reduced temperatures. This approach was adopted by various researchers in the fields of graph-theoretic optimization and computer vision [36, 127, 140, 46, 16, 128, 51]. Our starting point here is the early work on clustering by deterministic annealing which appeared in [114, 117, 116, 118]. Although strongly motivated by the physical analogy, the approach is formally based on principles of information theory and probability theory, and consists of minimizing the clustering cost at prescribed levels of randomness (Shannon entropy).

The DA method provides clustering solutions at different scales, where the scale is directly related to the temperature parameter. There are "phase transitions" in the design process, where phases correspond to the number of effective clusters in the solution, which grows via splits as the temperature is lowered. If a limitation on the number of clusters is imposed, then at zero temperature a hard clustering solution, or a quantizer, is obtained. The basic DA approach to clustering has since inspired modifications, extensions, and related work by numerous researchers including [119, 84, 21, 138, 11, 93, 94, 61, 91, 110, 132].

This paper begins with a tutorial review of the basic DA approach to clustering, and then goes into some of its most significant extensions to handle various partition structures [90], as well as hard supervised learning problems including classifier design [91], piecewise regression [106], and mixture of experts [110]. An important extension is to the use of DA for the design of speech recognizers that are based on hidden Markov models [108, 107]. Another important theoretical aspect is the connection with Shannon's rate distortion theory, which leads to better understanding of the method's contribution to quantization, and yields additional contributions to information theory itself [115]. Finally, some emerging new extensions, most notably in the context of multiple description quantizer design, are briefly discussed.

2 Deterministic Annealing for Unsupervised Learning

2.1 Clustering

Clustering can be informally stated as partitioning a given set of data points into subgroups, each of which should be as homogeneous as possible. The problem of clustering is an important optimization problem in a large variety of fields, such as pattern recognition, learning, source coding, image and signal processing. The exact definition of the clustering problem differs slightly from field to field, but in all of them it is a major tool for the analysis or processing of data without a priori knowledge of the distribution. The clustering problem statement is usually made mathematically precise by defining a cost criterion to be minimized. In signal compression it is commonly referred to as the distortion. Let x denote a source vector, and let $y(x)$ denote its best reproduction codevector from codebook Y. Denoting the distortion measure (typically, but not necessarily, the squared Euclidean distance) by $d(\cdot,\cdot)$, the expected distortion is

$$D = \sum_x p(x) d(x, y(x)) \approx \frac{1}{N} \sum_x d(x, y(x)) \tag{1}$$

where the right-hand side assumes that the source distribution may be approximated by a training set of N independent vectors[1]. In this case, the clustering solution is specified in terms of the codebook Y and an encoding rule for selecting the codevector which best matches an input vector. Virtually all useful distortion functions are *not* convex, and are instead riddled with poor local minima [55]. Thus, clustering is a nonconvex optimization problem. While exhaustive search will find the global minimum, it is hopelessly impractical for all nontrivial distributions and reasonably large data sets.

As the clustering problem appears in very diverse applications, solution methods have been developed in different disciplines. In the communications or information-theory literature, an early clustering method was suggested for scalar quantization, which is known as the Lloyd algorithm [79] or the Max quantizer [85]. This method was later generalized to vector quantization, and to a large family of distortion measures [78], and the resulting algorithm is commonly referred to as the generalized Lloyd algorithm (GLA). For a comprehensive treatment of the subject within the areas of compression and communications see [50]. In the pattern-recognition literature, similar algorithms have been introduced including the ISODATA [8] and the K-means [83] algorithms. Later, fuzzy relatives to these algorithms were derived [35][15]. All these iterative methods alternate between two complementary steps: optimization of the encoding rule for the current codebook, and optimization of the codebook for the encoding rule. When operating in "batch" mode (i.e., where the cost due to the entire training set is considered before adjusting parameters), it is easy to show that this iterative procedure is monotone non-increasing in the distortion. Hence, convergence to a local minimum of the distortion (or of its fuzzy variant, respectively) is ensured.

[1]The approximation of expected distortion by empirical distortion is practically unavoidable. In the sequel, whenever such approximation is obvious from the context, it will be used without repeating this explicit statement.

2.1.1 Principled Derivation of Deterministic Annealing

Various earlier versions of the principled derivation of DA appeared in [114], [117] and [118]. The derivation was revised here to include more recent insights and to provide the most natural foundation for the following sections. A probabilistic framework for clustering is defined here by *randomization of the partition*, or equivalently, *randomization of the encoding rule.* Input vectors are assigned to clusters in probability, which we call the *association probability.* This viewpoint bears similarity to fuzzy clustering, where each data point has partial membership in clusters. However, our formulation is purely probabilistic. While we consider clusters as regular (nonfuzzy) sets whose exact membership is the outcome of a random experiment, one may also consider the fuzzy sets obtained by equating degree of membership with the association probability in the former (probabilistic) model. It is, thus, possible to utilize DA for both fuzzy and "regular" clustering design. We will not, however, make any use of tools or methods from fuzzy sets theory in this paper. On the other hand, the traditional framework for clustering is the marginal special case where all association probabilities are either zero or one. In the pattern recognition literature this is called "hard" clustering in contradistinction with the more recent ("soft") fuzzy clustering.

For the randomized partition we can rewrite the expected distortion (1) as

$$D = \sum_x \sum_y p(x,y) d(x,y) = \sum_x p(x) \sum_y p(y|x) d(x,y) \tag{2}$$

where $p(x,y)$ is the joint probability distribution, and the conditional probability $p(y|x)$ is the association probability relating input vector x with codevector y. At the limit where the association probabilities are hard and each input vector is assigned to a unique codevector with probability one, (2) becomes identical with the traditional hard clustering distortion (1).

Minimization of D of (2) with respect to the free parameters $\{y, p(y|x)\}$ would immediately produce a hard clustering solution, as it is always advantageous to fully assign an input vector to the nearest[2] codevector. We, however, recast this optimization problem as that of seeking the distribution which minimizes D *subject to a specified level of randomness.* The level of randomness is, naturally, measured by the Shannon entropy:

$$H(X,Y) = -\sum_x \sum_y p(x,y) \log p(x,y). \tag{3}$$

This optimization is conveniently reformulated as minimization of the Lagrangian

$$F = D - TH \tag{4}$$

where T is the Lagrange multiplier, D is given by (2) and H is given by (3). Clearly, for large values of T we mainly attempt to maximize the entropy. As T is lowered we trade entropy for reduction in distortion, and as T approaches zero, we minimize D directly to obtain a hard (non-random) solution.

At this point, it is instructive to pause and consider an equivalent derivation based on the principle of maximum entropy. Suppose we fix the level of expected distortion D and seek to estimate the underlying probability distribution. The objective is to characterize the random solution at gradually diminishing levels of distortion until minimal distortion is reached. To estimate the distribution we appeal to Jaynes's maximum entropy principle [66] which states: Of all the probability distributions that satisfy a given set of constraints, choose the one that maximizes the entropy. The informal justification is that while this choice agrees with what is known (the given constraints), it maintains

[2]The term "nearest" is used in the sense of the distortion measure $d(\cdot,\cdot)$, which is not necessarily the Euclidean distance.

maximum uncertainty with respect to everything else. Had we chosen another distribution satisfying the constraints, we would have reduced the uncertainty and would have therefore implicitly made some extra restrictive assumption. For the problem at hand, we seek the distribution which maximizes the Shannon entropy while satisfying the expected distortion constraint. The corresponding Lagrangian to maximize is $H - \beta D$, with β the Lagrange multiplier. The equivalence of the two derivation is obvious – both Lagrangian functions are simultaneously optimized by the same solution configuration for $\beta = 1/T$.

To analyze further the Lagrangian F of (4) we note that the joint entropy can be decomposed into two terms: $H(X,Y) = H(X) + H(Y|X)$, where $H(X) = -\sum p(x) \log p(x)$ is the source entropy, which is independent of clustering. We may therefore drop the constant $H(X)$ from the Lagrangian definition, and focus on the conditional entropy:

$$H(Y|X) = -\sum_x p(x) \sum_y p(y|x) \log p(y|x). \tag{5}$$

Minimizing F with respect to the association probabilities $p(y|x)$ is straightforward and gives the Gibbs distribution:

$$p(y|x) = \frac{\exp(-\frac{d(x,y)}{T})}{Z_x}, \tag{6}$$

where the normalization is

$$Z_x = \sum_y \exp(-\frac{d(x,y)}{T}) \tag{7}$$

(which is the partition function of statistical physics.) The corresponding minimum of F is obtained by plugging (6) back into (4):

$$F^* = \min_{\{p(y|x)\}} F = -T \sum_x p(x) \log Z_x = -T \sum_x p(x) \log \sum_y \exp(-\frac{d(x,y)}{T}). \tag{8}$$

To minimize the Lagrangian with respect to the codevector locations $\{y\}$, its gradients are set to zero yielding the condition:

$$\sum_x p(x,y) \frac{d}{dy} d(x,y) = 0 \qquad \forall y \in Y. \tag{9}$$

Note that the derivative notation here stands, in general, for gradients. After normalization by $p(y) = \sum_x p(x,y)$ the condition can be rewritten as a centroid condition:

$$\sum_x p(x|y) \frac{d}{dy} d(x,y) = 0 \tag{10}$$

(where $p(x|y)$ denotes the posterior probability calculated using Bayes's rule), which for the squared error distortion case takes the familiar form:

$$y = \sum_x p(x|y) x. \tag{11}$$

While the above expressions convey most clearly the "centroid" aspect of the result, the practical approximation of the general condition (9) is

$$\frac{1}{N} \sum_x p(y|x) \frac{d}{dy} d(x,y) = 0 \qquad \forall y \in Y, \tag{12}$$

where $p(y|x)$ is the Gibbs distribution of (6).

The practical algorithm consists, therefore, of minimizing F^* with respect to the codevectors, starting at high value of T and tracking the minimum while lowering T. The central iteration consists of two steps:

- fix the codevectors and use (6) to compute the association probabilities;
- fix the associations and optimize the codevectors according to (12).

Clearly, the procedure is monotone non-increasing in F^*, and converges to a minimum. At high levels of T, the cost is very smooth and under mild assumptions[3] can be shown to be convex, which implies that the global minimum of F^* is found. As T tends to zero the association probabilities become hard and a hard clustering solution is obtained. In particular it is easy to see that the algorithm itself becomes the known GLA method [78] at this limit.

Some intuitive notion of the workings of the system can be obtained from observing the evolution of the association probabilities (6). At infinite T, these are uniform distributions, i.e., each input vector is equally associated with all clusters. These are extremely fuzzy associations. As T is lowered, the distributions become more discriminating, and the associations less fuzzy. At zero temperature the classification is hard with each input sample assigned to the nearest codevector with probability one[4]. This is the condition in which traditional techniques, such as GLA, work. From the DA viewpoint, standard methods are "zero temperature" methods. It is easy to visualize how the zero temperature system cannot "sense" a better optimum farther away, as each data point exercises its influence only on the nearest codevector. On the other hand, by starting at high T, and slowly "cooling" the system, we start with each data point equally influencing all codevectors, and gradually localize the influence. This gives us some intuition as to how the system senses, and settles into, a better optimum.

Another important aspect of the algorithm is seen if we view the association probability $p(y|x)$ as the expected value of the random binary variable $V_{xy} \in \{0, 1\}$ which takes the value 1 if input x is assigned to codevector y, and 0 if not. From this perspective one may recognize the known Expectation Maximization (EM) algorithm [31] in the above two step iteration. The first step, which computes the association probabilities, is the "expectation" step, and the second step which minimizes F^* is the "maximization" (of $-F^*$) step. Note further that the EM algorithm is applied here at each given level of T. The emergence of EM is not surprising given that for many choices of distortion measure, F^* can be given an interpretation as a negative likelihood function. For example, in the case of squared error distortion, the optimization of F^* is equivalent to maximum likelihood estimation of means in a normal mixture, where the assumed variance is determined by T. It is important, however, to note that in general we do not necessarily assume an underlying probabilistic model for the data. Our distributions are derived from the distortion measure. In compression applications, in particular, the distortion measure attempts to quantify the perceptual significance of reconstruction error, independent of the source statistics.

2.1.2 Statistical Physics Analogy

The above probabilistic derivation is largely motivated by analogies to statistical physics. In this section we develop this analogy and indicate more precisely how the method produces an annealing process.

[3] If $d(x, y)$ is a differentiable, convex function of y for all x, then F^* has a unique minimum, asymptotically, at high temperature T.

[4] More precisely, each input sample is uniformly associated with the set of equidistant nearest representatives. We will ignore the pathologies of encoding "ties" as they are of no significance in DA.

Moreover, we will demonstrate that the system undergoes a sequence of "phase transitions", and will thereby obtain further insights into the process.

Consider a physical system whose energy is our distortion D and whose Shannon entropy is H. The Lagrangian, $F = D - TH$, which is central to the DA derivation, is exactly the *Helmholtz free energy* of this system (strictly speaking it is the Helmholtz thermodynamic potential). The Lagrange multiplier T is accordingly the temperature of the system which governs its level of randomness. Note that our choice of notation (with the exception of D which stands for distortion) was made to agree with the traditional notation of statistical mechanics, and emphasizes this direct analogy. A fundamental principle of statistical mechanics (often called the principle of minimal free energy) states that the minimum of the free energy determines the distribution at thermal equilibrium. Thus, F^* is achieved by the system when it reaches equilibrium, at which point the system is governed by the Gibbs (or canonical) distribution. The chemical procedure of annealing consists of maintaining the system at thermal equilibrium while carefully lowering the temperature. Compare this with the computational procedure of DA: track the minimum of the free energy while gradually lowering the temperature! In chemistry, annealing is used to ensure that the ground state of the system, that is, the state of minimum energy, is achieved at the limit of low temperature. The method of simulated annealing [72] directly simulates the stochastic evolution of such a physical system. We, instead, derive its free-energy as the corresponding expectation, and deterministically (and quickly) optimize it to characterize the equilibrium at the given temperature.

In summary, the DA method performs annealing as it maintains the free energy at its minimum (thermal equilibrium) while gradually lowering the temperature; and it is deterministic because it minimizes the free energy directly rather than via stochastic simulation of the system dynamics.

But there is much more to the physical analogy. We shall next demonstrate that, as the temperature is lowered, the system undergoes a sequence of "phase transitions" which consists of natural cluster splits where the clustering model grows in size (number of clusters). This phenomenon is highly significant for a number of reasons. First, it provides a useful tool for controlling the size of the clustering model and relating it to the scale of the solution, as will be explained below. Second, these phase transitions are the critical points of the process where one needs to be careful with the annealing (as is the case in physical annealing). The "critical temperatures" are computable as will be shown next. This information allows us to accelerate the procedure in between phase transitions without compromising performance. Finally, the sequence of solutions at various phases, which are solutions of increasing model size, can be coupled with validation procedures to identify the optimal model size for performance outside the training set.

Let us begin by considering the case of very high temperature ($T \to \infty$). The association probabilities (6) are uniform, and the optimality condition (12) is satisfied by placing all codevectors at the same point – the centroid of the training set determined by

$$\frac{1}{N}\sum_{x}\frac{d}{dy}d(x,y) = 0. \tag{13}$$

(In the case of squared error distortion this optimal y is the sample mean of the training set.) Hence, at high temperature, the codebook Y collapses on a single point. We say, therefore, that there is effectively one codevector and one cluster – the entire training set. As we lower the temperature the cardinality of the codebook changes. We consider the effective codebook cardinality, or model size, as characterizing the *phases* of the physical system. The system undergoes phase transitions as the model size grows. An analysis of the phase transitions is fundamental to obtain an understanding of the evolution of the system.

In order to explicitly derive the "critical temperatures" for the phase transitions, we will assume the squared error distortion $d(x, y) = |x - y|^2$. The bifurcation occurs when a set of coincident codevectors splits into separate subsets. Mathematically, the existing solution above the critical temperature is no longer the minimum of the free energy as the temperature crosses the critical value. Although it is natural to define this as the point at which the Hessian of F^* loses its positive definite property, the notational complexity of working with this large and complex matrix motivates the equivalent approach of variational calculus. Let us denote by $Y + \epsilon\Psi = \{y + \epsilon\psi_y\}$ a perturbed codebook, where ψ_y is the perturbation vector applied to codevector y, and where the non-negative scalar ϵ is used to scale the magnitude of the perturbation. We can rewrite the necessary condition for optimality of Y:

$$\frac{d}{d\epsilon}F^*(Y + \epsilon\Psi)|_{\epsilon=0} = 0, \tag{14}$$

for all choices of finite perturbation Ψ. This variational statement of the optimality condition leads directly to the earlier condition of (9). But we must also require a condition on the second order derivative

$$\frac{d^2}{d\epsilon^2}F^*(Y + \epsilon\Psi)|_{\epsilon=0} \geq 0, \tag{15}$$

for all choices of finite perturbation Ψ. Bifurcation occurs when equality is achieved in (15) and hence the minimum is no longer stable.[5] Applying straightforward differentiation we obtain the following condition for equality in (15):

$$\sum_x \sum_y p(x, y)\psi_y^t[I - \frac{2}{T}(x - y)(x - y)^t]\psi_y + \sum_x p(x)[\sum_y p(y|x)\psi_y^t(x - y)]^2 = 0, \tag{16}$$

where I denotes the identity matrix. The first term can be rewritten more compactly and the equation becomes

$$\sum_y p(y)\psi_y^t[I - \frac{2}{T}C_{x|y}]\psi_y + \sum_x p(x)[\sum_y p(y|x)(x - y)^t\psi_y]^2 = 0, \tag{17}$$

where

$$C_{x|y} = \sum_x p(x|y)(x - y)(x - y)^t \tag{18}$$

is the covariance matrix of the posterior distribution $p(x|y)$ of the cluster corresponding to codevector y. We claim that the left-hand side of (17) is positive for all perturbations Ψ *if and only if* the first term is. The "if" part is trivial since the second term is obviously non-negative. To show the "only if" part, we first observe that the first term is non-positive only if there exists some $y_0 \in Y$ with positive probability, such that the matrix $I - \frac{2}{T}C_{x|y_0}$ is *not* positive definite. In fact, we assume that there are several coincident codevectors at this point to allow bifurcation. We next show that in this case there always exists a perturbation that makes the second term vanish. Select a perturbation Ψ satisfying:

$$\psi_y = 0, \qquad \forall y \neq y_0, \tag{19}$$

and

$$\sum_{y \in Y: y = y_0} \psi_y = 0. \tag{20}$$

With this perturbation the second term becomes

$$\sum_x p(x)[p(y_0|x)(x - y_0)^t \sum_{y \in Y: y = y_0} \psi_y]^2,$$

[5]For simplicity we ignore higher order derivatives, which should be checked for mathematical completeness, but which are of minimal practical importance. The result is a necessary condition for bifurcation.

which equals zero by (20). Thus, whenever the first term is non-positive, we can construct a perturbation such that the second term vanishes. Hence, *there is strict inequality in (15) if and only if the first term of (17) is positive* for all choices of finite perturbation Ψ.

In conclusion to the above derivation, the condition for phase transition requires that the coincident codevectors at some y_0 have a (posterior) data distribution $p(x|y)$ satisfying

$$\det[I - \frac{2}{T} C_{x|y_0}] = 0. \tag{21}$$

The critical temperature T_c is therefore determined as

$$T_c = 2\lambda_{max} \tag{22}$$

where λ_{max} is the largest eigenvalue of $C_{x|y_0}$. In other words, phase transitions occur as the temperature is lowered to twice the variance along the principal axis of the cluster. It can be further shown [118] that the split (separation of codevectors) is along this principal axis. We summarize this result in the form of a theorem.

Theorem 1 *For the squared error distortion measure, a cluster centered at codevector y undergoes a splitting phase transition when the temperature reaches the critical value $T_c = 2\lambda_{max}$, where λ_{max} is the cluster's principal component.*

Figure 1 illustrates the annealing process with its phase transitions on a simple example. The training set is generated from a mixture of six randomly displaced, equal variance Gaussians, whose centers are marked by X. At high temperature, there is only one effective cluster represented by one codevector, marked by O, at the center of mass of the training set. As the temperature is lowered, the system undergoes phase transitions which increase the number of effective clusters as shown in the figure. Note that as the partition is random, there are no precise boundaries. Instead, we give "isoprobability curves", or contours of equal probability (typically 1/3) of belonging to the cluster. In Figure 2 we give the corresponding "phase diagram" which describes the variation of average distortion (energy) with $\beta = 1/T$. Note, in particular, that when the temperature reaches the value which corresponds to the variance of all the isotropic Gaussians we get an "explosion" as there is no preferred direction (principal component) for the split. More on the analysis of phase transitions is given in a later section on rate distortion theory, but for a deeper treatment of the condition for explosion, or continuum of codevectors, and the special role of Gaussian distribution, see [115].

2.1.3 Mass-constrained clustering

The mass constrained clustering approach [119] is the preferred implementation of the DA clustering algorithm, and a detailed sketch of the algorithm will be given at the end of this subsection. We shall show that the annealing process, as described so far, has a certain dependence on the number of coincident codevectors in each effective cluster. This weakness is not desirable and can be eliminated, leading to a method that is totally independent of initialization.

We start by recalling a central characteristic of the DA approach: no matter how many codevectors are "thrown in", the effective number emerges at each temperature. This number is the model size, and defines the phase of the system. For example, even if we have thousands of codevectors, there is only one single effective codevector at very high temperature. However, after a split occurs, the result may differ somewhat depending on the number of codevectors in each of the resulting subgroups. Clearly,

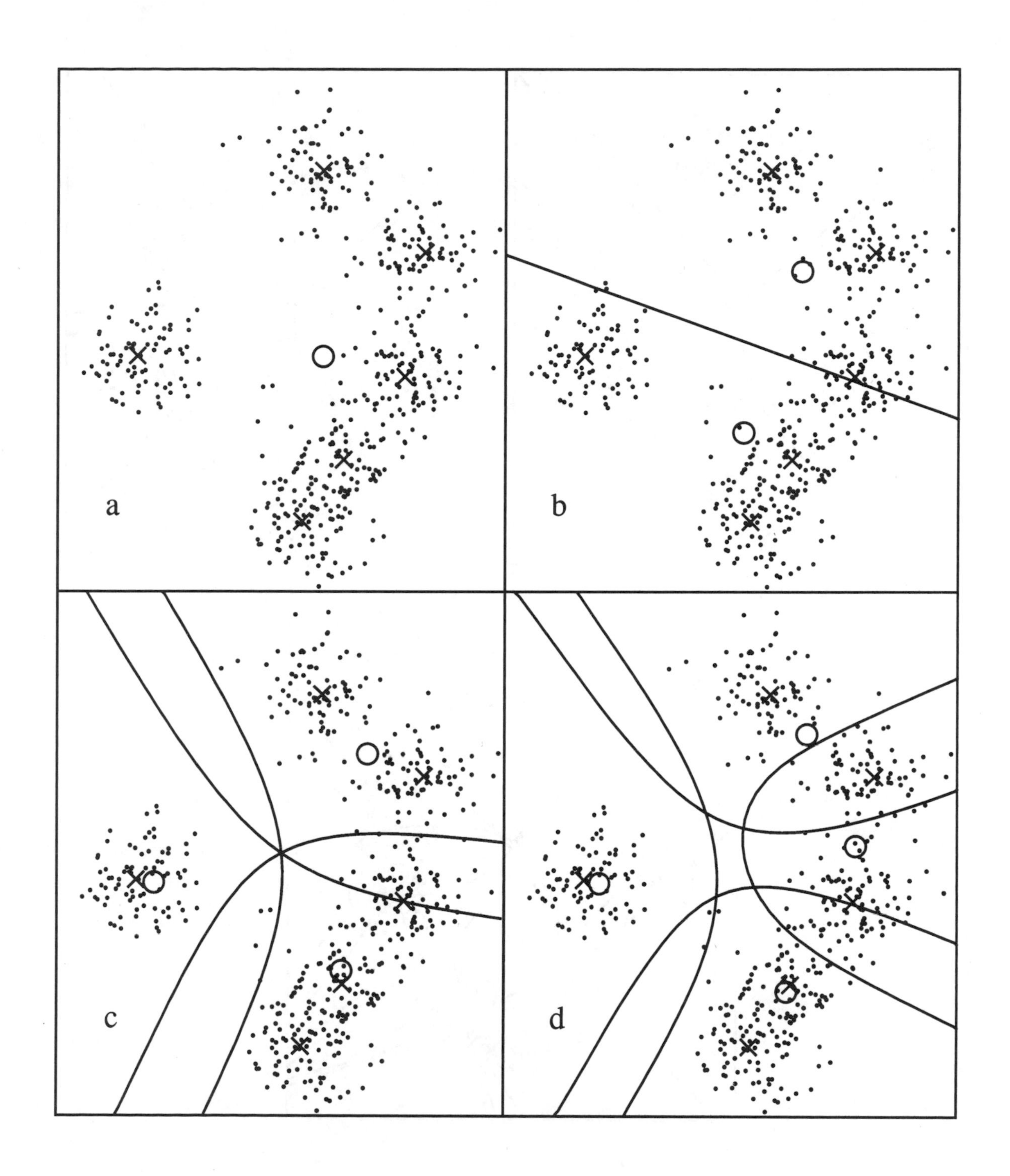

Figure 1: Clustering at various phases. The lines are equiprobable contours, $p = 1/2$ in (b), and $p = 1/3$ elsewhere. (a) 1 cluster ($\beta = 0$), (b) 2 clusters ($\beta = 0.0049$), (c) 3 clusters ($\beta = 0.0056$), (d) 4 clusters ($\beta = 0.0100$), (e) 5 clusters ($\beta = 0.0156$), (f) 6 clusters ($\beta = 0.0347$), and (g) 19 clusters ($\beta = 0.0605$). From [118].

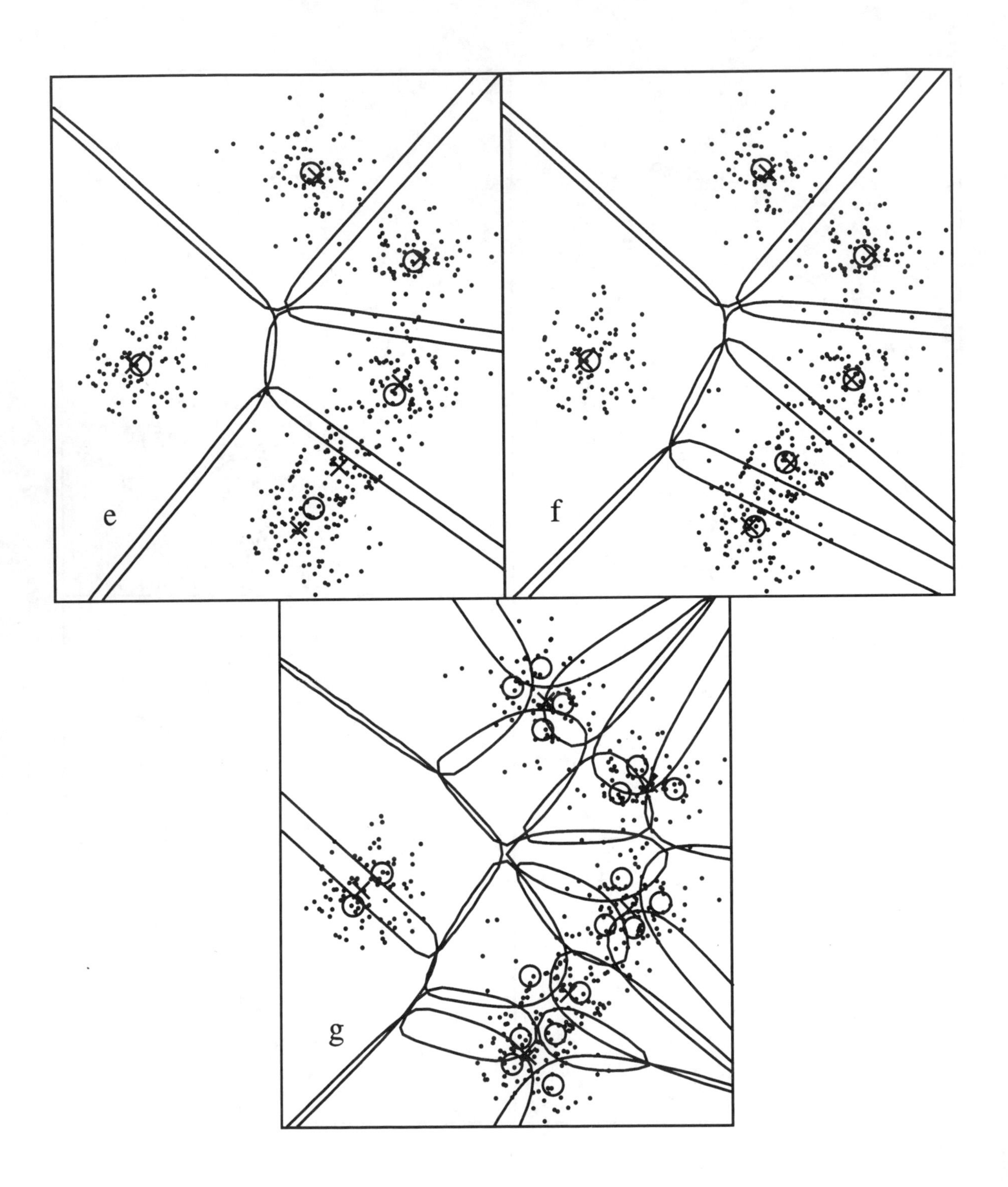
e
f
g

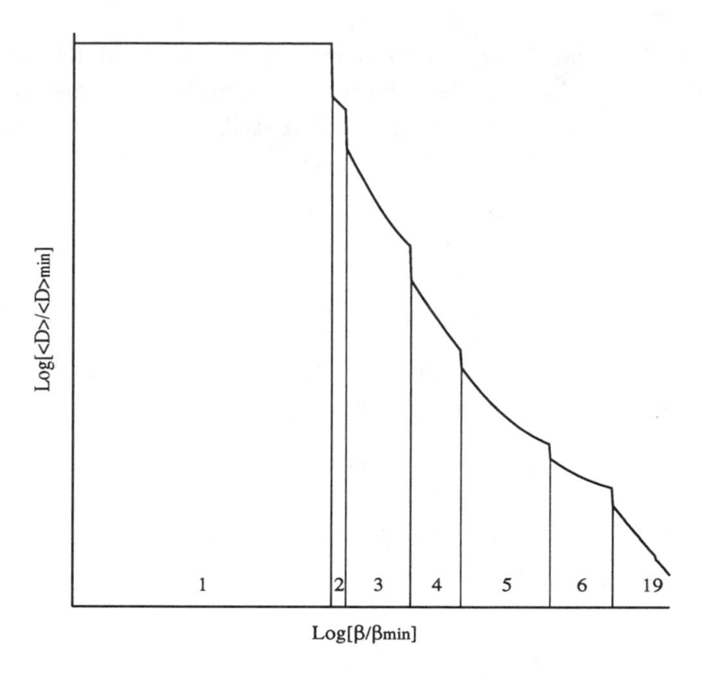

Figure 2: Phase diagram for the distribution shown in Figure 1. The number of effective clusters is shown for each phase. From [118].

the initial partition into subgroups depends on the perturbation. In order to fix this shortcoming, let us reformulate the method in terms of effective clusters (or distinct codevectors). Let us assume that there is an unlimited supply of codevectors, and let p_i denote the fraction of codevectors which represent effective cluster i, and are therefore coincident at position y_i. Using this notation, the partition function of (7) is rewritten equivalently as

$$Z_x = \sum_i p_i e^{-\frac{d(x,y_i)}{T}}, \tag{23}$$

where the summation is over distinct codevectors. The probability of association (6) with distinct codevector y_j is the so-called tilted distribution

$$p(y_j|x) = \frac{p_j e^{-\frac{d(x,y_j)}{T}}}{Z_x}, \tag{24}$$

and the free energy (8) is

$$F^* = -T \sum_x p(x) \log Z_x = -T \sum_x p(x) \log \sum_i p_i e^{-\frac{d(x,y_i)}{T}}. \tag{25}$$

The free energy is to be minimized under the obvious constraint that $\sum_i p_i = 1$. The optimization is performed as unconstrained minimization of the Lagrangian

$$F' = F^* + \lambda \left(\sum_i p_i - 1 \right), \tag{26}$$

with respect to the cluster parameters y_i and p_i. Note that although we started with a countable number of codevectors to be distributed among the clusters, we effectively view them now as possibly uncountable, and p_i is not required to be rational. One may therefore visualize this as a "mass of codevectors" which is divided among the effective clusters, or simply as a distribution over the codevector space. (The notion of inducing a possibly continuous distribution over the codevector space provides a direct link to rate-distortion theory which is pursued in a later section).

The optimal set of codevectors $\{y_j\}$ must satisfy,

$$\frac{\partial}{\partial y_j}F' = \frac{\partial}{\partial y_j}F^* = 0, \tag{27}$$

where the left equality is because the constraint is independent of the positions $\{y_j\}$. We thus get again the condition of (9); i.e.,

$$\sum_x p(x)p(y_j|x)\frac{\partial}{\partial y_j}d(x, y_j) = 0, \tag{28}$$

with the important distinction that now the association probabilities are tilted according to (24).

On the other hand, the set $\{p_i\}$ which minimizes F' satisfies

$$\frac{\partial}{\partial p_i}F' = -T\sum_x p(x)\frac{e^{-\frac{d(x,y_i)}{T}}}{Z_x} + \lambda = 0, \tag{29}$$

which yields

$$\frac{\lambda}{T} = \sum_x p(x)\frac{e^{-\frac{d(x,y_i)}{T}}}{Z_x}. \tag{30}$$

Taking the expectation of (30) with respect to the distribution p_i we obtain

$$\frac{\lambda}{T} = \sum_i p_i \sum_x p(x)\frac{e^{-\frac{d(x,y_i)}{T}}}{Z_x} = 1, \tag{31}$$

where the last equality uses the definition of Z_x in (23). Thus,

$$\lambda = T. \tag{32}$$

Substituting (32) in (30) we see that the optimal distribution p_i must satisfy

$$\sum_x p(x)\frac{e^{-\frac{d(x,y_i)}{T}}}{Z_x} = 1, \tag{33}$$

where $\{p_i\}$ are implicit in Z_x (23). Equation (33) is thus the equation we solve while optimizing over $\{p_i\}$. This equation also arises from the Kuhn-Tucker conditions of rate-distortion theory [12, 18, 54].

It is instructive to point out that (33) and (24) imply that

$$p_i = \sum_x p(x)\frac{p_i e^{-\frac{d(x,y_i)}{T}}}{Z_x} = \sum_x p(x)p(y_i|x) = p(y_i). \tag{34}$$

In other words, the optimal codevector distribution mimics the training data set partition into the clusters. The distribution p_i is identical to the probability distribution induced on the codevectors via the encoding rule which we have denoted by $p(y) = \sum_x p(x)p(y|x)$. As an aside note that the results

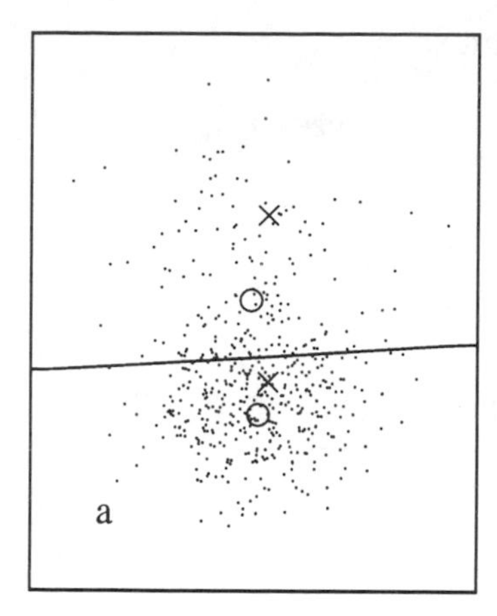

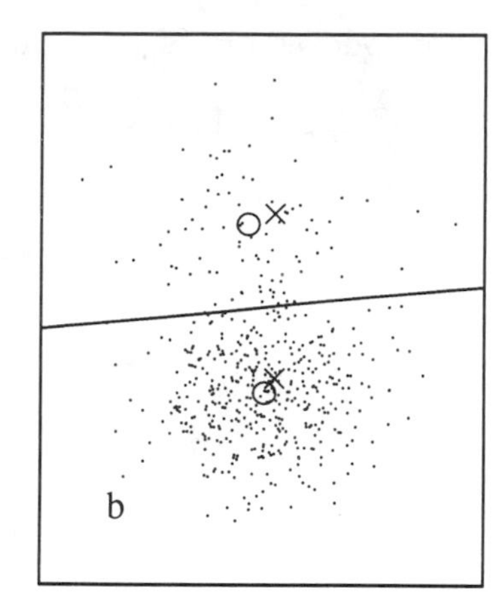

Figure 3: The effect of cluster mass (population) at intermediate β. The data is sampled from two normal distributions whose centers are marked by X. The computed representatives are marked by O. (a) Unconstrained clustering. (b) Mass-constrained clustering. From [119].

are also in perfect agreement with estimation of priors (mixing coefficients) in parametric estimation of mixture of densities. However, it must be kept in mind that the mass-constrained algorithm is applicable to solving the simple vector quantizer or clustering design problem where minimum distortion is the ultimate objective.

The mass-constrained formulation only needs as many codevectors as there are effective clusters at a given temperature. The process is computationally efficient, and increases the model size only when it is needed, i.e., when a critical temperature has been reached. The mechanism can be implemented by maintaining and perturbing pairs of codevectors at each effective cluster, so that they separate only when a phase transition occurs. Another possibility is to compute the critical temperature and supply an additional codevector only when the condition is satisfied.

It should also be noted that at the limit of low temperature ($T = 0$), both the unconstrained DA method and the mass-constrained DA method, converge to the same descent process, namely, GLA [78] (or basic ISODATA [8] for the sum of squared distances). This is so because the association probabilities of the two DA methods are identical at the limit, and they assign each data point to the nearest codevector with probability one. The difference between the two is in their behavior at intermediate T, where the mass-constrained clustering method takes the cluster populations into account. This is illustrated by a simple example in Figure 3.

2.1.4 Preferred Implementation of the DA Clustering Algorithm

We conclude the treatment of the basic clustering problem with a sketch of a preferred implementation of the clustering algorithm. This version incorporates the mass-constrained approach. The squared error distortion is assumed for simplicity, but the description is extendible to other distortion measures. It is also assumed that the objective is to find the hard (non-fuzzy) clustering solution for a given number of clusters.

1. Set limits: number of codevectors K_{max}, minimum temperature T_{min}

2. Initialize: $T > 2\lambda_{max}(C_x)$, $K = 1$, $y_1 = \sum_x xp(x)$ and $p(y_1) = 1$

3. Update for $i = 1, \ldots, K$:

$$y_i = \frac{\sum_x xp(x)p(y_i|x)}{p(y_i)}$$

   where,

$$p(y_i|x) = \frac{p(y_i)e^{-\frac{(x-y_i)^2}{T}}}{\sum_{j=1}^{K} p(y_j)e^{-\frac{(x-y_j)^2}{T}}}$$

$$p(y_i) = \sum_x p(x)p(y_i|x)$$

4. Convergence test. If not satisfied go to 3).

5. If $T \leq T_{min}$, perform last iteration for $T = 0$ and STOP.

6. Cooling step: $T \leftarrow \alpha T$, $(\alpha < 1)$.

7. If $K < K_{max}$, check condition for phase transition for $i = 1, \ldots, K$. If critical T is reached for cluster j, add a new codevector $y_{K+1} = y_j + \delta$, $p(y_{K+1}) = p(y_j)/2$, $p(y_j) \leftarrow p(y_j)/2$, and increment K.

8. Go to 3).

Note that the test for critical T in step 7, if considered expensive for high dimensions, may be replaced by a simple perturbation. In this case we always keep two codevectors at each location, and perturb them when we update T. Until the critical T is reached they will be merged together by the iterations. At phase transition they will move further apart.

The relation to the Lloyd algorithm for quantizer design is easy to see. At a given T, the iteration is a generalization of the nearest neighbor and the centroid conditions. The relation to maximum likelihood estimation of parameters in normal mixtures is also obvious. For a treatment of the problem of covariance matrix estimation in DA see [73].

While the algorithm sketch was given for the typical case of vector quantizer design, it is easy to modify it to produce cluster analysis solutions. In particular, fuzzy clustering solutions are produced naturally at a sequence of scales (as determined by the temperature). One simple approach is to combine the algorithm with cluster validation techniques to select a scale (and solution) from this sequence. It is also easy to produce hard clustering solutions at the different scales by adding quick "quenching" at each phase to produce the required hard solutions which can then be processed for validation.

2.1.5 Illustrative Examples

To further illustrate the performance of mass-constrained clustering, we consider the example shown in Figure 4. This is a mixture of six Gaussian densities of different masses and variances. We compare the result of DA with the well known generalized Lloyd algorithm (GLA) method. Since GLA yields results that depend on the initialization, we have run it twenty-five times, each time with a different initial set

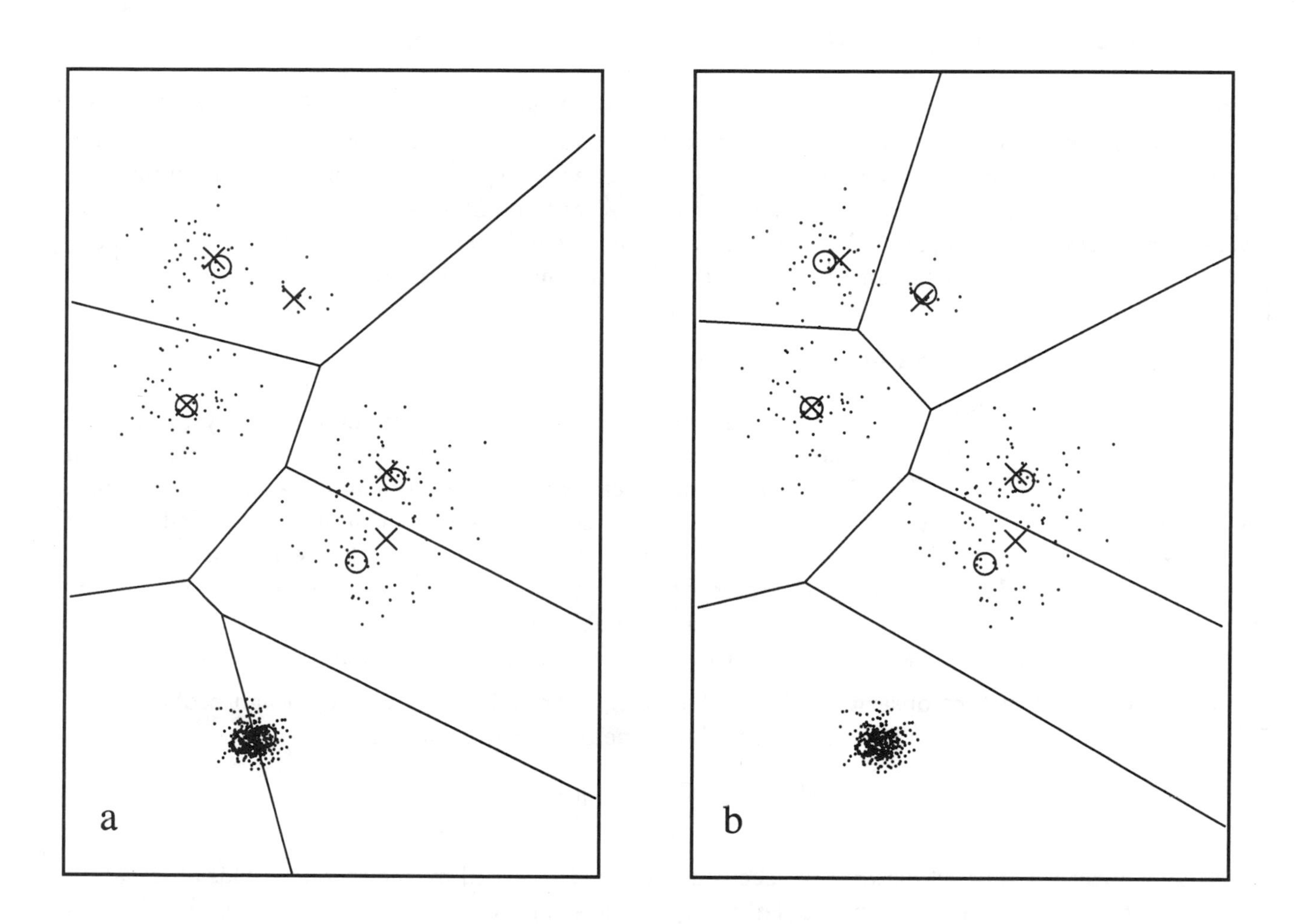

Figure 4: GLA vs. DA. (a) Best result of GLA out of twenty five runs with random initialization: $D = 6.4$. (b) Mass-constrained clustering: $D = 5.7$. From [119].

of representatives (randomly extracted from the training set). In Figure 4 (a) we show the best result obtained by GLA, where the mean squared error is 6.4. This result was obtained only once, while for $\approx 80\%$ of the runs it got trapped in local optima of ≈ 12.5 MSE. In Figure 4 (b) we show the result obtained by DA. The MSE is 5.7, and this solution is, of course, independent of the initialization. The process of "annealing" is illustrated in Figure 5. Here we have a mixture of nine overlapping Gaussian densities. The process undergoes a sequence of phase transitions as $\beta = 1/T$ increases. We show the results at some of the phases. Equiprobable contours are used to emphasize the fuzzy nature of the results at intermediate β. At the limit of high β, the MSE is 32.854. Repeated runs of GLA on this example yielded a variety of local optima with MSE from 33.5 to 40.3.

2.2 Extensions and Applications

In this section we consider several direct extensions of the DA clustering method. First, we consider extensions motivated by compression and communications applications including vector quantizer design for transmission through noisy channels, entropy-constrained vector quantization, and structurally constrained clustering which addresses the encoding and storage complexity problem. Finally, we briefly discuss straightforward extensions via constraints on the codevectors and identify as special cases approaches to the traveling salesman problem and self-organizing feature maps.

2.2.1 Vector Quantization for Noisy Channels

The area of source-channel coding is concerned with the joint design of communication systems while taking into account the distortion due to both compression and transmission over a noisy channel. In the particular case of vector quantizer-based communications systems, it is advantageous to optimize the quantizer while taking into account the effects of the channel. A noisy channel is specified here by its transition probabilities, $p(z|y)$, which denote the probability that the decoder decides on codevector z given that the encoder transmitted the index of codevector y.[6] (As an aside it may be noted that there exist applications [81] [82], where such noise models are used to model the distortion due to hierarchical or topological constraints rather than a real communication channel.)

A simple but important observation is the following: The noisy-channel VQ design problem is in fact identical to the regular VQ design, but with the modified distortion measure

$$d'(x,y) = \sum_z p(z|y)d(x,z), \tag{35}$$

which measures the expected distortion when codevector y is selected by the encoder. This observation allows direct extension of the known VQ design algorithms to this case. There is a long history of noisy-channel quantizer design. In the 60's, a basic method was proposed for scalar quantizers [77], and was extended in many papers since [34, 76, 4, 141, 39, 41]. These papers basically describe GLA-type methods which alternate between enforcing the encoder and centroid (decoder) optimality conditions. One can similarly extend the DA approach to the noisy channel case [21] [93].

We can write the expected overall source-channel distortion as

$$D = \sum_x \sum_y p(x,y)d'(x,y) = \sum_x \sum_y \sum_z p(x)p(y|x)p(z|y)d(x,z), \tag{36}$$

[6] The implicit simplifying assumption is that the channel is memoryless, or at least does not have significant temporal dependencies reaching beyond the transmission of a codevector index.

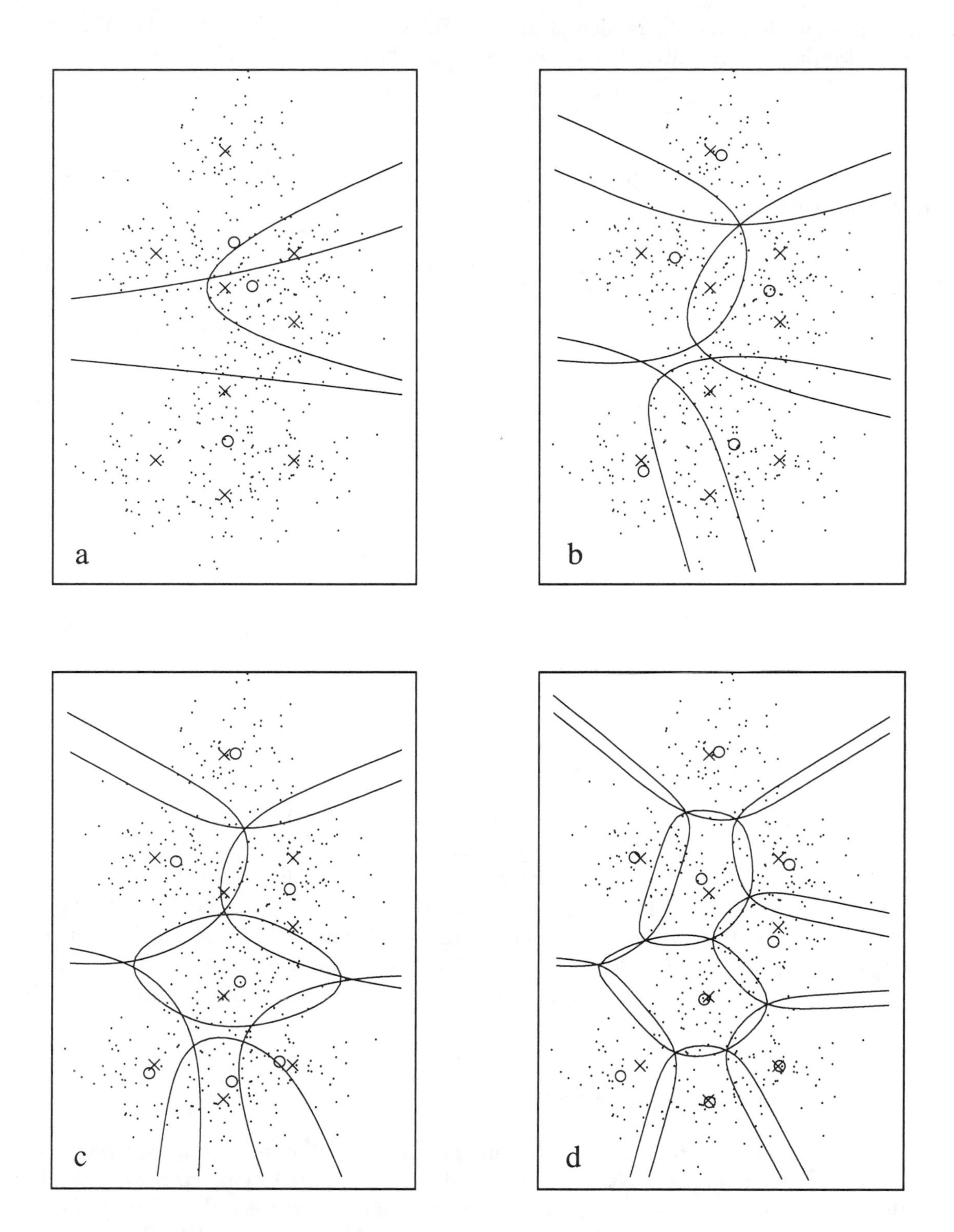

Figure 5: Various phases in the annealing process. Random partition represented by equiprobable contours (p) to emphasize the fuzzy nature of the results at intermediate β. (a) 3 clusters at $\beta = 0.005$ contours at $p = 0.45$, (b) 5 clusters at $\beta = 0.009$, $p = 0.33$, (c) 7 clusters at $\beta = 0.013$, $p = 0.33$, and (d) 9 clusters at $\beta = 0.03$, $p = 0.33$. From [119].

where $p(y|x)$ defines the encoder, which is random during the DA design. Note that we exploit the fact that (X, Y, Z) form a Markov chain. Alternatively, we may write that

$$D = \sum_x \sum_z p(x)p(z|x)d(x,z), \tag{37}$$

where $p(z|x) = \sum_y p(y|x)p(z|y)$. The entropy is defined over the encoding probabilities (the probabilities which are under our control):

$$H = -\sum_x \sum_y p(x,y) \log p(y|x). \tag{38}$$

Following the standard DA derivation we obtain the optimal encoding probability at a given temperature T

$$p(y|x) = \frac{\exp(-\frac{d'(x,y)}{T})}{Z_x} \tag{39}$$

and the free-energy

$$F^* = -T \sum_x p(x) \log \sum_y \exp(-\frac{d'(x,y)}{T}). \tag{40}$$

Optimizing F^* with respect to the parameters yields the centroid rule

$$\sum_x p(x)p(z|x)\frac{d}{dz}d(x,z) = 0, \tag{41}$$

which simplifies for the squared error distortion to

$$z = \sum_x p(x|z)x. \tag{42}$$

For a phase transition analysis of the DA process in the noisy-channel VQ case see [52]. In [93] it is shown that the DA approach avoids many local optima of the design and outperforms standard design. Of particular interest is how the design process converges to explicit error control coding (ECC) at the limit of very noisy channels. At this limit, certain encoding regions become empty, so that less codevectors are used by the encoder. Some of the available bit rate is thus reserved for channel protection. Note that the system itself finds the optimal error correcting code which need not (and does not, in general) satisfy any algebraic constraints such as linearity, as typically required in ECC design.

2.2.2 Entropy-Constrained Vector Quantizer Design

Variable-length coding (also called entropy coding) is commonly used to further reduce the rate required to transmit quantized signals. It results in rates that are close to the quantized signal entropy, which is the fundamental limit. The basic VQ design approach, however, assumes fixed-length coding, thereby simplifying the optimization problem to that of minimizing distortion for the given codebook size. It is, however, of much interest to derive a method for optimizing the VQ for use in conjunction with variable-length coding. Here the optimization must take into account both the distortion and the rate costs of selecting a codevector. This fundamental extension of the VQ problem was obtained by incorporation of an entropy constraint within the design to produce quantizers optimized for subsequent entropy coding. The earlier work was concerned with scalar quantizers [13] [40]. The VQ design method was proposed by Chou *et al.* [27]. We refer to this paradigm as the entropy-constrained vector quantizer (ECVQ).

The cost function is the weighted cost:

$$D + \lambda H = \sum_x \sum_y p(x, y)[d(x, y) - \lambda \log p(y)], \tag{43}$$

where λ determines the penalty for increase in rate relative to increase in distortion. It can also be used as a Lagrange multiplier for imposing a prescribed rate constraint while minimizing the distortion (or, alternatively, imposing a prescribed distortion while minimizing the rate). It follows that this problem reverts to the regular VQ problem with a modified cost function

$$d'(x, y) = d(x, y) - \lambda \log p(y), \tag{44}$$

subject to the additional constraint $\sum_y p(y) = 1$. This observation leads directly to the ECVQ algorithm of Chou *et al.* [27].

It can also be incorporated in a DA method for ECVQ design, as was first pointed out by Buhmann and Kuhnel [21]. The free-energy is

$$F^* = -T \sum_x p(x) \log \sum_y \exp(-\frac{d'(x, y)}{T}). \tag{45}$$

Note that in ECVQ we do not use a mass-constrained version of DA since the masses are already implicit in the modified distortion measure (44). Moreover, the above becomes equivalent to mass-constrained operation in the special case $\lambda = T$.

The resulting update rules are derived from the free-energy. The following assumes, for simplicity, the squared error distortion measure. The encoding rule is

$$p(y|x) = \frac{\exp(-\frac{d(x,y) - \lambda \log p(y)}{T})}{\sum_y \exp(-\frac{d(x,y) - \lambda \log p(y)}{T})}. \tag{46}$$

The centroid rule is

$$y = \sum_x p(x|y) x \tag{47}$$

where $p(x|y) = p(x)p(y|x)/p(y)$, and the mass rule is

$$p(y) = \sum_x p(x) p(y|x). \tag{48}$$

At $T = 0$ the DA iteration becomes identical to the standard ECVQ algorithm of [27].

2.2.3 Structurally Constrained Vector Quantizer Design

A major stumbling block in the way of VQ applications is the problem of encoding complexity. The size of the codebook grows exponentially with the vector dimension and rate (in bits per sample). As the encoder has to find the best codevector in the codebook, its complexity grows linearly with the codebook size, and hence exponentially with dimension and rate. In many practical applications, such encoding complexity is not acceptable, and low-complexity alternatives are needed. The most common approach for reducing encoding complexity involves the imposition of a structural constraint on the

VQ partition. A tree-structured partition is a typical such structure consisting of nested decision boundaries which can be represented by a decision tree. Sibling nodes in the tree define a vector quantizer that partitions the region associated with their parent node. The reproduction codebook of the tree-structured VQ (TSVQ) is, in fact, the set of leaves. The role of the internal nodes is to provide a mechanism for fast encoding search. The encoding operation is not an exhaustive search through the leaves. Instead, one starts at the root of the tree and determines a path to a leaf by a sequence of local decisions. At each layer the decision is restricted to selecting one of the descendants of the winning node in the previous layer. Thus, the encoding search grows linearly, rather than exponentially, with the dimension and the rate.

The design of TSVQ is, in general, a harder optimization problem than the design of regular VQ. Typical approaches [58] [112] [22] employ a greedy sequential design, optimizing a local cost to grow the tree one node (or layer) at a time. The reason for the greedy nature of standard approaches is that whereas in the unstructured case an optimal partition design step is readily specified by the nearest neighbor rule, in the tree structured case an optimal partition is determined only by solving a formidable multiclass risk discrimination problem [33]. Thus, the heuristically determined high level boundaries may severely constrain the final partition at the leaf layer, yet they are not re-adjusted when lower layers are being designed.

The DA approach to clustering offers a way to optimize the partition (i.e., all the parameters which define the final partition) directly and, moreover, escape many shallow local minima traps. It should be noted that the new difficulty here is the need to impose the structure on the partition. Earlier work on this problem [94] appealed to the principle of minimum cross-entropy (or minimum divergence) which is a known generalization of the principle of maximum entropy [126]. Minimum cross-entropy provides a probabilistic tool to gradually enforce the desired consistency between the leaf layer, where the quantization cost is calculated, and the rest of the tree – thereby imposing the structural constraint on the partition at the limit of zero temperature. This approach provided consistent substantial gains over the standard approaches. Although this method worked very well in all tests, it has two theoretical disadvantages. First, alternate minimization of the cross-entropy and of the cost at the leaf layer is not ensured to converge, though in practice this has not been a problem. The second undesired aspect is that it lacks the direct simplicity of basic DA. More recent developments of the DA approach in the context of supervised learning [90, 91] have since opened the way for a simpler, more general way to impose a structure on the partition, and which is also a more direct extension of the basic DA approach to clustering. A detailed description of this extension will be given in the section on supervised learning. Here we shall only cover the minimum required to develop the DA approach for TSVQ design. It is appropriate to focus on the latter derivation since it has none of the theoretical flaws of the earlier approach. However, no simulation results for it exist as yet at the time of writing this paper. To illustrate the annealing, and the type of gains achievable we will include some simulation results of the earlier approach [94] which, in spite of its theoretical shortcomings, approximates closely the optimal annealing process, and achieves (apparently) globally optimal solutions for these non-trivial examples.

We replace the hard encoding rule with a randomized decision. The probability of encoding input x with a leaf (codevector) y is in fact the probability of choosing the entire path starting at the root of the tree and leading to this leaf. This is a sequence of decisions where each node on the path competes with its siblings. The probability of choosing node s given that its parent was chosen is Gibbs[7]:

$$p(s|x, parent(s)) = \frac{\exp(-\gamma d(x,s))}{\sum_{s' \in sibling(s)} \exp(-\gamma d(x,s'))}, \tag{49}$$

where γ is a scale parameter. Thus, the selection of nodes at sequential layers is viewed as a Markov chain, where the transition probabilities obey the Gibbs distribution. Note in particular, that as $\gamma \to \infty$ this Markov chain becomes a hard decision tree, and the resulting partition corresponds to a standard TSVQ. We have thus defined a randomized tree partition which, at the limit enforces the desired structure on the solution.

We next wish to minimize the objective which is the overall distortion at a specified level of randomness. We again define the Lagrangian (the Helmholtz free energy):

$$F = D - TH \tag{50}$$

where D is the distortion at the leaf layer, and H is the entropy of the Markov chain, both computed while employing the explicit Gibbs form of (49). Then, by minimizing the free energy over the tree parameters $\{s, y, \gamma\}$ we obtain the optimal random tree at this temperature. As the temperature is lowered, reduction in entropy is traded for reduction in distortion and at $T = 0$ the tree becomes hard. This process also involves a sequence of phase transitions as the tree grows, similarly to the case of unconstrained VQ design. In Figure 6 we show the performance of the DA method from [94] on a mixture example, as well as the sequence of phase transitions and the manner in which the tree grows. In Figure 7 we show how TSVQ designed by DA outperforms the unstructured VQ designed by standard methods. This demonstrates the significant impact of poor local optima which cause worse degradation than the structural constraint itself.

Beside the tree-structure, on which we focused here, there are other important structures that are used in signal compression and for which DA design methods have been developed. One commonly used structure, particularly in speech coding, is the multi-stage vector quantizer (see [92] for an early DA approach). Another very important structure is the trellis quantizer (as well as the trellis vector quantizer) for which a DA approach has been proposed in [95].

2.2.4 Graph-Theoretic and Other Optimization Problems

In the deterministic annealing clustering algorithm, if we throw in enough codevectors and let $T \to 0$, then each data point will become a natural cluster. This can be viewed as a process of data association, where each data point is exclusively associated with a "codevector". As it stands, there is no preference as to which codevector is associated with which data point. However, by adding appropriate constraints, which are easy to incorporate in the Lagrangian derivation, we can encourage the process to obtain associations that satisfy additional requirements which embody the actual data assignment problem we wish to solve. This allows exploiting DA as a framework for solving a variety of hard graph-theoretic problems. As an example, when applied to the famous "traveling salesman" problem, such DA derivation becomes identical to the "elastic net" method [37][36]. The approach has been applied to various data assignment problems such as the module placement problem in CAD and graph partitioning. Another variant with a different constraint yields a DA approach for batch optimization of self-organizing feature maps. For more details see [119] [120].

[7] The choice of the Gibbs distribution is not arbitrary and will be explained in a fundamental and general setting in the supervised learning section. At this point let it simply be noted that it is directly obtainable from the maximum entropy principle.

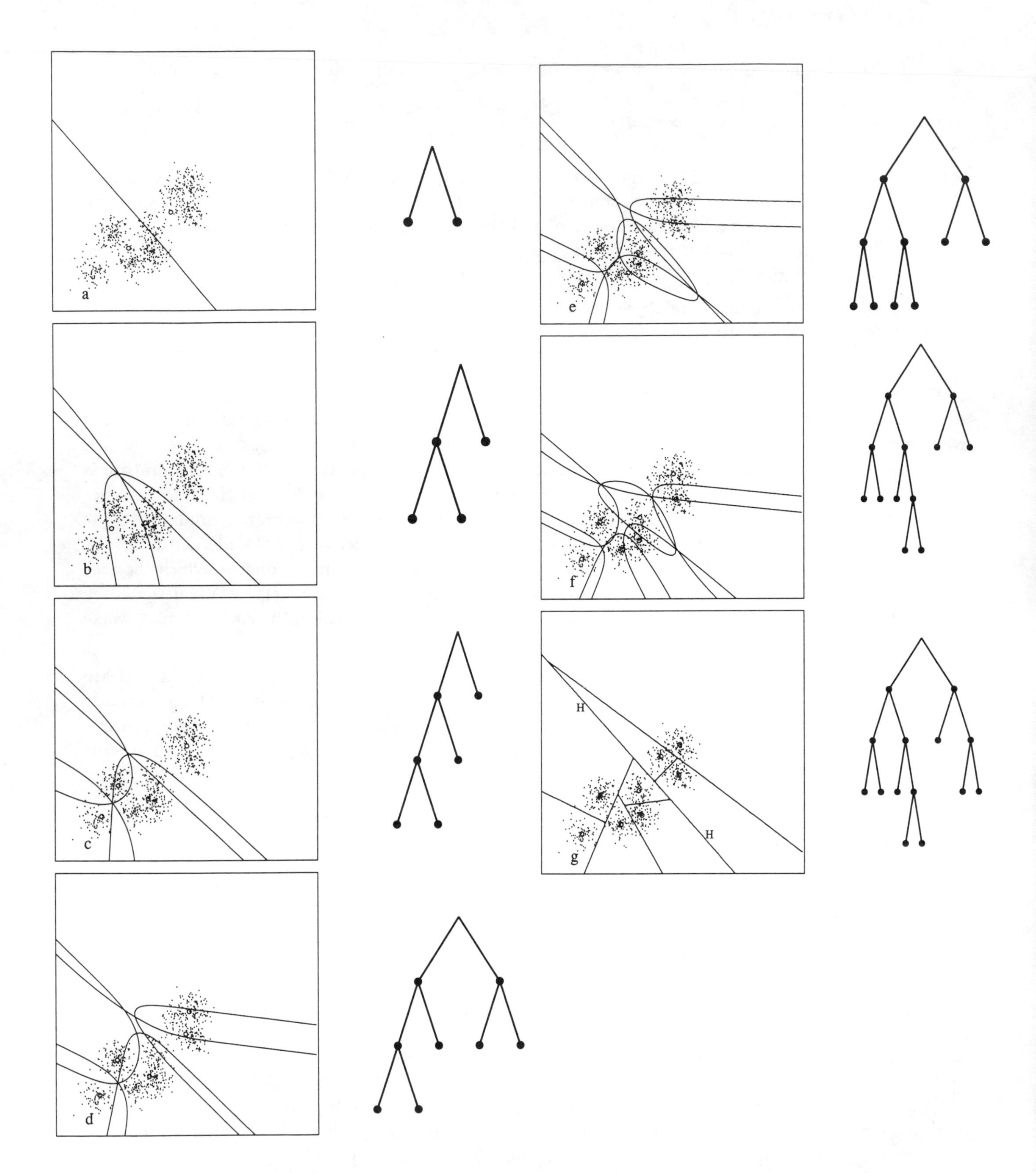

Figure 6: A hierarchy of tree-structured solutions generated by the annealing method for increasing β. The source is a Gaussian mixture with eight components. To the right of each figure is the associated tree structure. The lines in the figure are equiprobable contours with membership probability of $p = 0.33$ in a given partition region, except for Figure a and g, for which $p = 0.5$. "H" denotes the highest level decision boundary in Figure g. From [94].

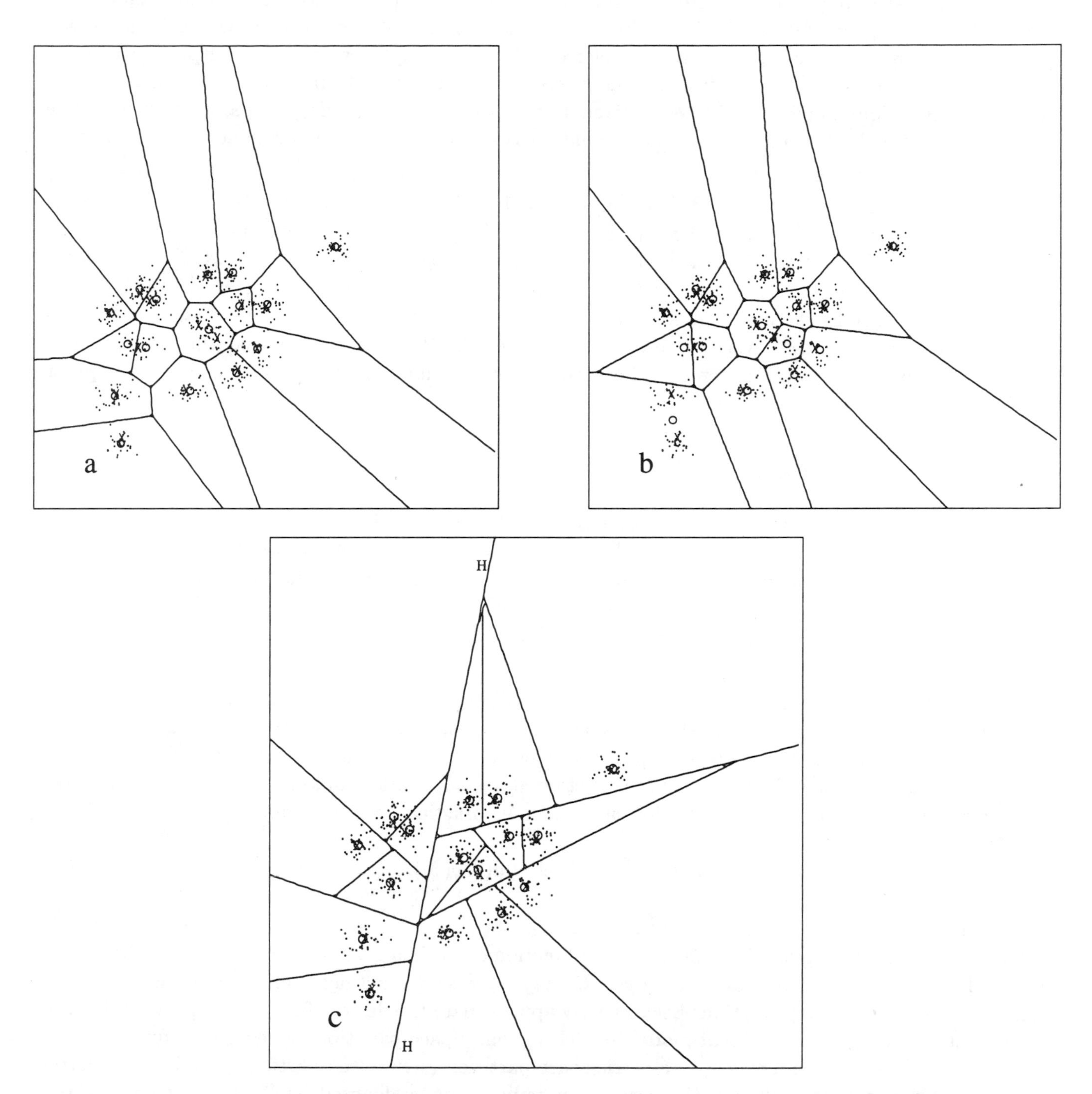

Figure 7: A Gaussian mixture example with sixteen components. a) The best *unconstrained* GLA solution after thirty random initializations within the data set, with $D = 0.51$. b) A typical unconstrained GLA solution, with $D = 0.72$. c) The unbalanced tree-structured DA solution with maximal depth of five and $D = 0.49$. From [94].

3 Deterministic Annealing for Supervised Learning

In this section we develop the deterministic annealing approach for problems of supervised learning. We consider first the general supervised learning, or function approximation, problem where from a given parametric class of functions, a function is selected which best captures the input-output statistics of a training set. In the learning literature, the given parametric class of functions is viewed as the set of transfer functions implementable on a given system structure by varying its parameters (e.g., the weights of a multi-layer perceptron). The ultimate performance evaluation is commonly performed on an independent test set of samples.

The fundamental extension of DA to supervised learning by inclusion of structural constraints [90, 91] led first to the DA method for classification [91], then to the DA method for piecewise regression [106], and was later extended to address regression by mixture of experts [110]. Here, after formulating the problem in its general setting, and deriving a DA method for its solution, we show how it specializes to various classical regression and classification problems by defining the structure (parametric form of the function/system), and the cost criterion. We demonstrate that the DA approach results in a powerful technique for classifier and regression function design for a variety of popular system structures.

3.1 Problem formulation

We define the objective of supervised learning as that of approximating an unknown function from the observation of a limited sequence of (typically noise-corrupted) input-output data pairs. Such function approximation is typically referred to as regression if the output is continuous, and as classification if the output is discrete or categoric. Regression and classification are important tools in diverse areas of signal processing, statistics, applied mathematics, business administration, computer science and the social sciences.

For concreteness, the problem formulation here will employ the terminology and notation of the regression problem. We will later show how the solution is, in fact, applicable to classification and other related problems. The regression problem is usually stated as the optimization of an expected cost that measures how well the regression function $g(X)$, applied to random vector X, approximates the output Y, over the joint distribution $p(x, y)$, or in practice, over a sample set $\{(x, y)\}$. Let us reformulate the cost as

$$D = \sum_x \sum_y p(x, y) d(y, g(x)), \tag{51}$$

where the distortion measure $d(\cdot, \cdot)$ is general, though the squared error is most often in use. The cost is optimized by searching for the best regression function g within a given parametric class of functions. We shall first restrict ourselves to ***space-partitioning*** regression functions. These functions are often called piecewise regression functions because they approximate the desired function by partitioning the space and matching a simple local model to each region. Space-partitioning regression functions are constructed of two components: a parametric space partition (structured partition); and a parametric local model per partition cell. Let the partition parameter set be denoted by Ω. It may consist of the nodes of a decision tree, the codevectors or prototypes of a vector quantizer, the weights of a multilayer perceptron, etc. Let $\Lambda = \{\Lambda_j\}$ denote the set of local model parameters, where Λ_j is the subset of parameters specifying the model for region R_j.

We can now write the regression function as

$$g(x) = f(x, \Lambda_k), \qquad \forall x \in R_k, \tag{52}$$

where $f(x, \Lambda_k)$ denotes the local model. Note that Ω is implicit in the partition into cells $\{R_k\}$. Typically, the local parametric model, $f(x, \Lambda_k)$ has a simple form such as constant, linear or Gaussian. The average regression error measured over a sample set is then

$$D = \frac{1}{N} \sum_j \sum_{(x,y):x \in R_j} d(y, f(x, \Lambda_j)). \tag{53}$$

The regression function $g(\cdot)$ is *learned* by minimizing the design cost, D, measured over a *training set*, $\mathcal{T} = \{x, y\}$, but with the ultimate performance evaluation based on the generalization cost, which is the error D measured over a *test set*. The mismatch between the design cost and the generalization cost is a fundamental difficulty which is the subject of much current research in statistics in general, and in neural networks in particular. It is well-known that for most choices of D, the cost measured during design decreases as the complexity (size) of the learned regression model is allowed to increase, while the generalization cost will start to increase when the model size grows beyond a certain point. In general, the optimal model size, or even a favorable regime of model sizes, is unknown prior to training the model. Thus, the search for the correct model size must naturally be undertaken as an integral part of the training. Most techniques for improving generalization in learning are inspired by the well-known principle of Occam's razor[8], which essentially states that the simplest model that accurately represents the data is most desirable. From the perspective of the learning problem, this principle suggests that the design should take into account some measure of the simplicity, or parsimony, of the solution, in addition to performance on the training set. In one basic approach, penalty terms are added to the training cost, either to directly favor the formation of a small model [2],[113], or to do so indirectly via regularization/smoothness constraints or other costs which measure overspecialization. A second common approach is to build a large model, overspecialized to the training set, and then attempt to "undo" some of the training by retaining only the vital model structure, removing extraneous parameters that have only learned the nuances of a particular noisy training set. This latter approach is adopted in the pruning methods for classification and regression trees [20] and in methods such as optimal brain surgeon [143] in the context of neural networks.

While these techniques provide a way of generating parsimonious models, there is an additional serious difficulty that most methods do not address directly, which can also severely limit the generalization achieved by learning. This difficulty is the problem of nonglobal optima of the cost surface, which can easily trap descent-based learning methods. If the designed regression function performs poorly as a result of a shallow, local minimum trap, the typical recourse is to optimize a larger model, under the assumption that the model was not sufficiently powerful to characterize the data well. The larger model will likely improve the design cost but may result in overspecialization to the training set and suboptimal performance outside the training set. Clearly, a superior optimization method that finds better models of smaller size will enhance the generalization performance of the regression function. While conventional techniques for parsimonious modeling control the model size, they do not address this optimization difficulty. In particular, standard methods such as CART [20] for tree-structured classification and regression employ greedy heuristics in the "growing" phase of the model design which might lead to poorly designed trees. The subsequent pruning phase is then restricted in its search for parsimonious models to choosing pruned subtrees of this initial, potentially suboptimal tree. Techniques which add penalty terms to the cost can also suffer from problems of local minima. In fact, in many cases the addition of a penalty term can actually increase the complexity of the cost surface and exacerbate the local minimum problem (e.g., [143]).

As an alternative approach, let us consider the DA optimization technique for regression modeling

[8]William of Occam (1285-1349): "Causes should not be multiplied beyond necessity."

which, through its formulation of the problem, simultaneously embeds the search for a parsimonious solution and for one that is optimal in the design cost.

3.2 Basic derivation

The design objective is minimization of the expected regression cost D of (53) which is repeated here for convenience

$$D = \frac{1}{N} \sum_j \sum_{(x,y):x \in R_j} d(y, f(x, \Lambda_j)), \tag{54}$$

over the partition parameters Ω (which are implicit in the partition $\{R_j\}$), and the local model parameters Λ. To begin the derivation let us assume that the local model parameters Λ are known and fixed, and focus on the more difficult problem of optimizing the partition. The partition is structurally constrained, that is, it is selected from a family of partitions by assigning values to the set of parameters Ω. Note that a partition is, in fact, a classifier as it assigns to each input a label indicating to which partition cell it belongs. There are many popular structures for the partition such as the vector quantizer, the decision tree, the multilayer perceptron, and the radial basis functions partitions. The operation of each of the above distinct structures (or classifiers) is consistent with that of the general (canonical) maximum discriminant model [33]: given input x the system produces competing outputs (one per partition cell) via the *discriminant functions* $\{F_j(x)\}$, and the input is assigned to the the largest, "winning" output. It thus uses the "winner-take-all" partition rule:

$$R_j \equiv \{x : F_j(x) \geq F_k(x), \quad \forall k\}. \tag{55}$$

Any partition can be represented by this model, albeit possibly with complicated discriminant functions. Note that the discriminant functions $F_j(x)$ in our case are specified by the set of parameters Ω, although we have suppressed this dependence in the notation of (55). We thus employ the maximum discriminant model to develop a general optimization approach for regression. We will later specialize the results to specific popular structures and learning costs and give experimental results to demonstrate the performance.

Let us write an objective function whose maximization determines the hard partition for given Ω:

$$S_h = \frac{1}{N} \sum_{j \in \mathcal{I}} \sum_{x \in R_j} F_j(x). \tag{56}$$

Note, in particular, that the winner-take-all rule (55) is optimal in the sense of S_h. Specifically, maximizing (56) over all possible partitions captures the decision rule of (55).

To derive a DA approach we wish to randomize the partition similar to the earlier derivation for the problem of clustering. The probabilistic generalization of (56) is

$$S = \frac{1}{N} \sum_x \sum_j p(j|x) F_j(x), \tag{57}$$

where the partition is now represented by association probabilities, $\{p(j|x)\}$, and the corresponding entropy is

$$H = -\frac{1}{N} \sum_x \sum_j p(j|x) \log p(j|x). \tag{58}$$

It is emphasized that H measures the average level of uncertainty *in the partition decisions.* We determine our assignment distribution at a given level of randomness as the one which maximizes S while maintaining H at a prescribed level $\hat{H}$:

$$\max_{\{p(j|x)\}} S \text{ subject to } H = \hat{H}. \tag{59}$$

The result is the *best* probabilistic partition, in the sense of the structural objective S, *at the specified level of randomness.* For $\hat{H} = 0$ we naturally revert to the hard partition which maximizes (56) and thus employs the winner-take-all rule. At any positive $\hat{H}$, the solution of (59) is the Gibbs distribution

$$p(j|x) = \frac{e^{\gamma F_j(x)}}{\sum_k e^{\gamma F_k(x)}}, \tag{60}$$

where γ is the Lagrange multiplier controlling the level of entropy. For $\gamma \to 0$, the associations become increasingly uniform, while for $\gamma \to \infty$, they revert to the hard partition, equivalent to application of the rule in (55). Thus, (60) is a probabilistic generalization of the winner-take-all rule which satisfies its structural constraint, specified by (55), for the choice $\gamma \to \infty$. Note that beside the obvious dependence on the parameter γ, the discriminant functions $\{F_j(x)\}$ are determined by $\Omega = \{\Omega_j\}$.

So far, we have formulated a controlled way of introducing randomness into the partition while enforcing its structural constraint. However, the derivation assumed that the model parameters were given, and thus produced only the *form* of the distribution $p(j|x)$, without actually prescribing how to choose the values of its parameter set. Moreover the derivation did not consider the ultimate goal of minimizing the expected regression cost D. We next remedy both shortcomings.

To apply the basic principles of DA design similar to our treatment of clustering, we need to introduce randomness into the partition while enforcing the required structure, only now we must also explicitly minimize the expected regression cost. A priori, satisfying these multiple objectives may appear to be a formidable task, but the problem is greatly simplified by restricting the choice of random partitions to the set of distributions $\{p(j|x)\}$ as given in (60) – these random partitions naturally enforce the structural constraint of (55) through γ, as explained earlier. Thus, from the parameterized set $\{p(j|x)\}$ (determined by the implicit Ω), we seek that distribution which minimizes the expected regression cost while constraining the entropy:

$$\min_{\Omega,\Lambda} D \equiv \min_{\Omega,\Lambda} \frac{1}{N} \sum_{(x,y)} \sum_j p(j|x) d(y, f(x, \Lambda_j)), \tag{61}$$

subject to

$$H = \hat{H}. \tag{62}$$

The solution yields the best random partition and model parameters in the sense of minimum D for a given entropy level $\hat{H}$. At the limit of zero entropy, we should get a *hard* partition which minimizes D, yet has the desired structure, as specified by (55).

We naturally reformulate (61) and (62) as minimization of the unconstrained Lagrangian, or free energy

$$F = D - TH, \tag{63}$$

where the Lagrange parameter, T, is the "temperature" and emphasizes the intuitively compelling analogy to statistical physics, in parallel to the DA derivation in the earlier sections. Virtually all the discussion on the analogy to statistical physics which appeared in the context of clustering holds here too, and provides strong motivation for use of the DA method. For conciseness we shall not elaborate on the analogy here.

We initialize the algorithm at $T \to \infty$ (in practice, T is simply chosen large enough to be above the first critical temperature). It is clear from (63) that the goal at this temperature is to maximize the entropy of the partition. The distributions, $\{p(j|x)\}$, are consequently uniform. The same parameters, Λ_j, are used for the local regression models in all the regions- effectively, we have a single, global regression model. As the temperature is gradually lowered, optimization is carried out at each temperature to find the partition parameters $\{\Omega_j\}$, local model parameters, $\{\Lambda_j\}$ that minimize the Lagrangian, F. As $T \to 0$, the Lagrangian reduces to the regression cost, D. Further, since we have forced the entropy to go to zero, the randomized space partition that we obtain becomes a hard partition satisfying the imposed structure. In practice, we anneal the system to a low temperature, where the entropy of the random partition is sufficiently small. Further annealing will not change the partition parameters significantly. Hence we fix the partition parameters at this point and jump (quench) to $T = 0$ to perform a "zero entropy iteration", where we partition the training set according to the "hard" partition rule and optimize the parameters of the local models $\{\Lambda_j\}$ to minimize the regression cost, D. This approach is consistent with our ultimate goal of optimizing the cost constrained on using a (hard) structured space partition.

A brief sketch of the DA algorithm is as follows:

1. Initialize : $T = T_i$.
2. $\min_{\Omega,\Lambda}\{F = D - TH\}$
3. lower temperature: $T \leftarrow q(T)$.
4. If $H \geq H_f$ goto step 2.
5. Zero entropy iteration :
   Partition using Hard partition rule,
   $\min_{\{\Lambda_j\}} D$

In our simulations we used an exponential schedule for reducing T, i.e., $q(T) = \alpha T$, where $\alpha < 1$, but other annealing schedules are possible. The parameter optimization of step 2) may be performed by any local optimization method.

3.3 Generality and Wide Applicability of the DA Solution

3.3.1 Regression, Classification, and Clustering

In the previous subsection we derived a DA method to design a regression function subject to structural constraints on the partition. In this subsection we pause to appreciate the general applicability of the DA solution. We show that special cases of the problem we defined include the problems of clustering and vector quantization, as well as statistical classifier design. These special cases are obtained by specifying appropriate cost functions and local models. We also review a number of popular structures from data compression and neural networks and show how they are special cases of the general maximum discriminant structure, and hence directly handled by the DA approach we have derived.

Let us first restate the learning problem. Given a training set of pairs (x, y) we wish to design a function which takes in x and estimates y. The estimator function is constructed by partitioning the input space and fitting a local model within each region. The learning cost is defined as

$$D = \frac{1}{N} \sum_{j} \sum_{(x,y):x \in R_j} d(y, f(x, \Lambda_j)), \tag{64}$$

where $\{R_j\}$ is the set of partition regions, and $\Lambda = \{\Lambda_j\}$ is the set of parameters which determine the local models. Beside the obvious and direct interpretation of the above as a regression problem with applications in function approximation and curve fitting, it is easy to see that the important problem of classifier design is another special case of this learning problem: If the local model $f(x, \Lambda_j)$ is simply the class label assigned to the region, and if we define the distortion measure as the error indicator function $d(u, v) = 1 - \delta(u, v)$, where the δ function takes the value 1 when its arguments are equal and vanishes otherwise, then the learning cost is exactly the *rate of misclassification*, or *error rate* of the classifier. Thus, statistical classifier design is a special case of the general learning problem we considered, albeit with a particularly difficult cost to optimize due to its discrete nature. This is a very important problem with numerous "hot" applications in conjunction with various structures, and we will devote more space to it in the sequel.

A somewhat contrived, yet important special case of the above regression problem is that of unsupervised clustering. Here we consider the degenerate case of $y = x$ and where the local regression models are constant. In other words, we approximate the training set with a piecewise-constant function. We partition the space into regions, and each region is represented by a constant vector (the codevector) so as to minimize the cost (e.g., mean squared error). This is clearly the vector quantization problem. If we apply our DA regression method to this problem, and assume a vector quantizer structure, we will get exactly the clustering approach we had derived directly, and more simply, earlier on. The simpler derivation of a DA clustering method was only possible because the VQ structure emerges by itself from minimization of the clustering distortion, and need not be externally imposed as in the case of the general DA regression method. Although the latter derivation seems unnecessarily cumbersome for clustering problems, it does in fact open the door to important clustering applications. We are often interested in solving the clustering problem while imposing a different structure. The typical motivation is that of reducing encoding or storage complexity, but may also be that of discerning hierarchical information, or because a certain structure better fits prior information on the underlying distribution. We have already considered in detail tree-structured clustering within the unsupervised learning section, however we had to postpone the complete description of the mechanism for enforcing the structure until the supervised learning section.

3.3.2 Structures

We next consider the applicability of the approach to a variety of structures. Recall that the approach was generally derived for the maximum discriminant partition structure which is defined by

$$R_j \equiv \{x : F_j(x) \geq F_k(x) \ \ \forall k\}. \tag{65}$$

This general structure can be specialized to specific popular structures such as the vector quantizer (or nearest prototype classifier), the multilayer perceptron, and the radial basis functions classifier. It is important to note that known design methods are structure specific, while the DA approach is directly applicable to virtually all structures. In the remainder of this section we describe these three structures. The choice of presentation is such that the applicability of the general design procedure is evident. For detailed structure-specific derivation see [91].

The VQ Classifier

The VQ structure is shown in Figure 8. The partition is specified by the parameter set $\Omega = \{\mathbf{x}_{jk}\}$ where $\mathbf{x}_{jk} \in \mathcal{R}^n$ is the kth prototype associated with class j. The VQ classifier maps a vector in $\mathcal{R}^n$ to the class associated with the nearest prototype, specifying a partition of $\mathcal{R}^n$ into the regions:

$$R_j \equiv \bigcup_k S_{jk} \;\text{ with }\; S_{jk} \equiv \{\mathbf{x} \in \mathcal{R}^n : d(\mathbf{x}, \mathbf{x}_{jk}) \leq d(\mathbf{x}, \mathbf{x}_{lm}) \;\; \forall l, m\}, \tag{66}$$

i.e., each region R_j is the union of Voronoi cells S_{jk}. Here, $d(\cdot,\cdot)$ is the "distance measure" used for classification. For consistency with the maximum discriminant classifier ("winner takes all") we note trivially that the classification rule can also be written as

$$R_j \equiv \bigcup_k S_{jk} \;\text{ with }\; S_{jk} \equiv \{\mathbf{x} \in \mathcal{R}^n : F_{jk}(\mathbf{x}) \geq F_{lm}(\mathbf{x}) \;\; \forall l, m\}, \tag{67}$$

by choosing $F_{jk}(\mathbf{x}) \equiv -d(\mathbf{x}, \mathbf{x}_{jk})$.

The Radial Basis Functions (RBF) Classifier

The RBF classifier structure is shown in Figure 9. The classifier is specified by a set of Gaussian receptive field functions, $\{e^{-\frac{|x-\mu_k|^2}{\sigma_k^2}}\}$, and by a set of scalar weights $\{\omega_{kj}\}$ which connect each of the receptive fields to the class outputs of the network. Thus, $\Omega = \{\{\mu_k\}, \{\sigma_k{}^2\}, \{\omega_{kj}\}\}$. The parameter μ_k is the "center" vector for the receptive field and $\sigma_k{}^2$ is its "width". In the "normalized" representation for RBFs [96] which we will adopt here, the network output for each class is written in the form

$$F_j(x) = \sum_k \omega_{kj}\phi_k(x), \tag{68}$$

where

$$\phi_k(x) = \frac{e^{-\frac{|x-\mu_k|^2}{\sigma_k^2}}}{\sum_l e^{-\frac{|x-\mu_l|^2}{\sigma_l^2}}}. \tag{69}$$

Since $\phi_k(\cdot)$ can be viewed as a probability mass function, each network output is effectively an average of weights emanating from each of the receptive fields. The classifier maps the vector x to the class with the largest output:

$$R_j \equiv \{x \in \mathcal{R}^n : F_j(x) \geq F_k(x) \;\; \forall k \in \mathcal{I}\}. \tag{70}$$

The Multilayer Perceptron (MLP) Classifier

The MLP classifier structure is shown in Figure 10. We restrict ourselves to the MLP structure with a binary output unit per class. The classification rule for MLPs is the same as that for RBFs (70), but the output functions $\{F_j(\cdot)\}$ are parameterized differently.

The input x passes through K layers with M_k neurons in layer k. We define u_{kj} to be the output of hidden unit j in layer k, with the convention that layer 0 is the input layer $u_{0j} = x_j$, and layer K is the output layer $u_{Kj} = F_j(\mathbf{x})$. To avoid special notation for thresholds, we define the augmented output vector of layer k as $\tilde{\mathbf{u}}_k = [u_{k1} \;\; u_{k2} \;\; \ldots \;\; u_{kM_k} \;\; 1]^T$. This is a standard notation allowing us to replace thresholds by synaptic weights which multiply a fixed input value of unity.

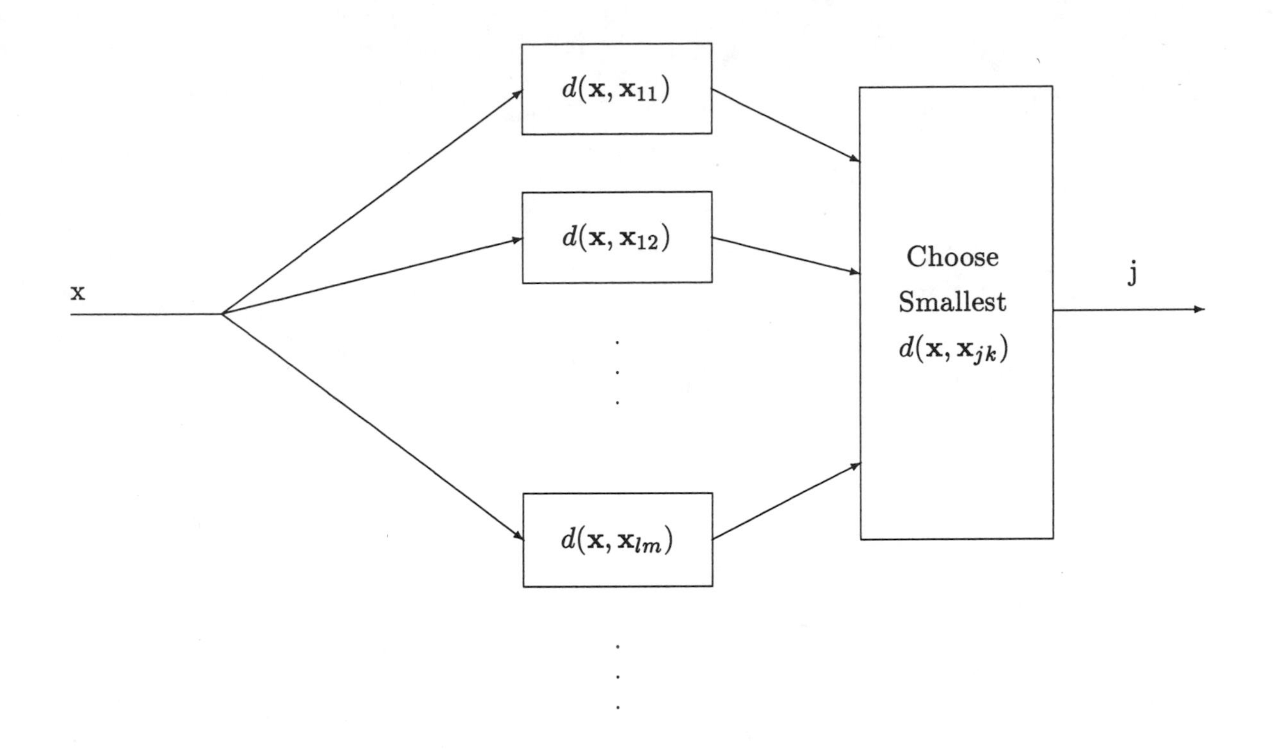

Figure 8: The VQ classifier architecture. From [91].

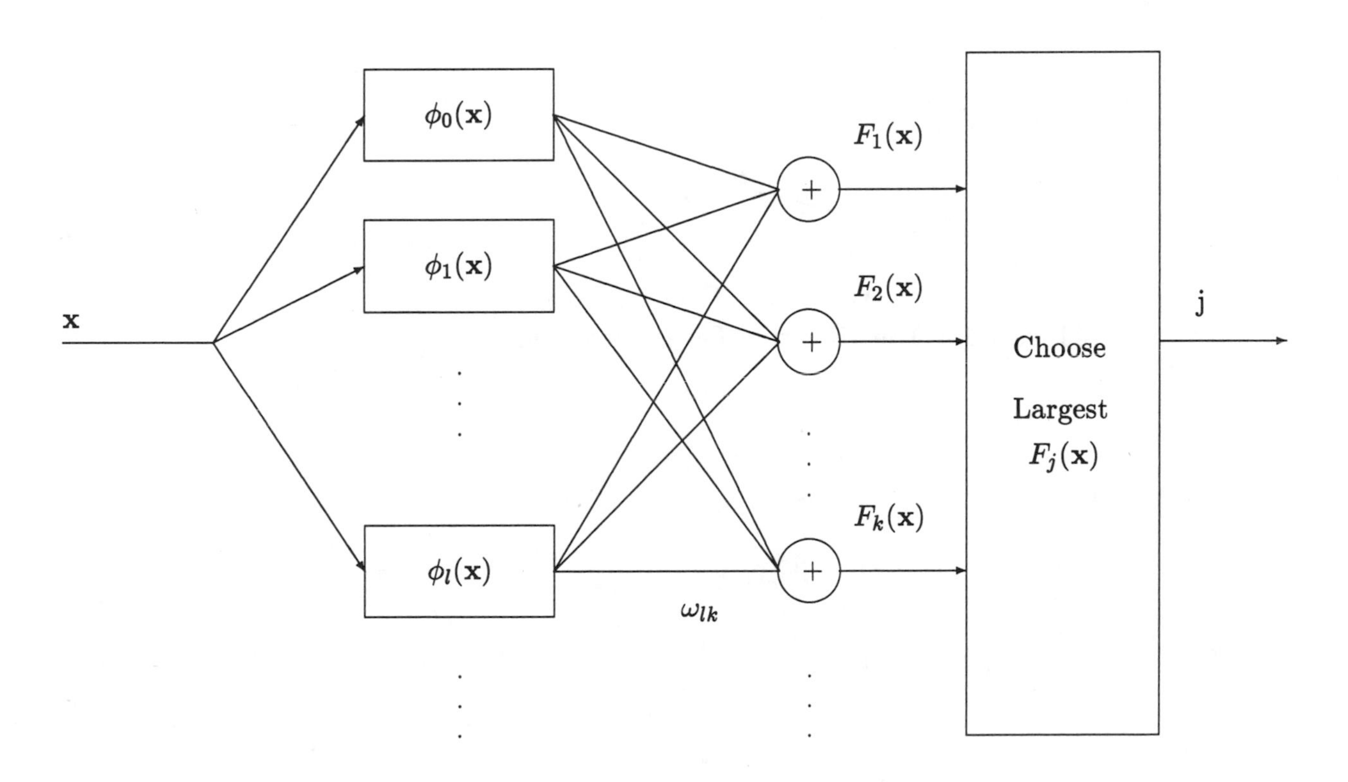

Figure 9: The RBF classifier architecture. From [91].

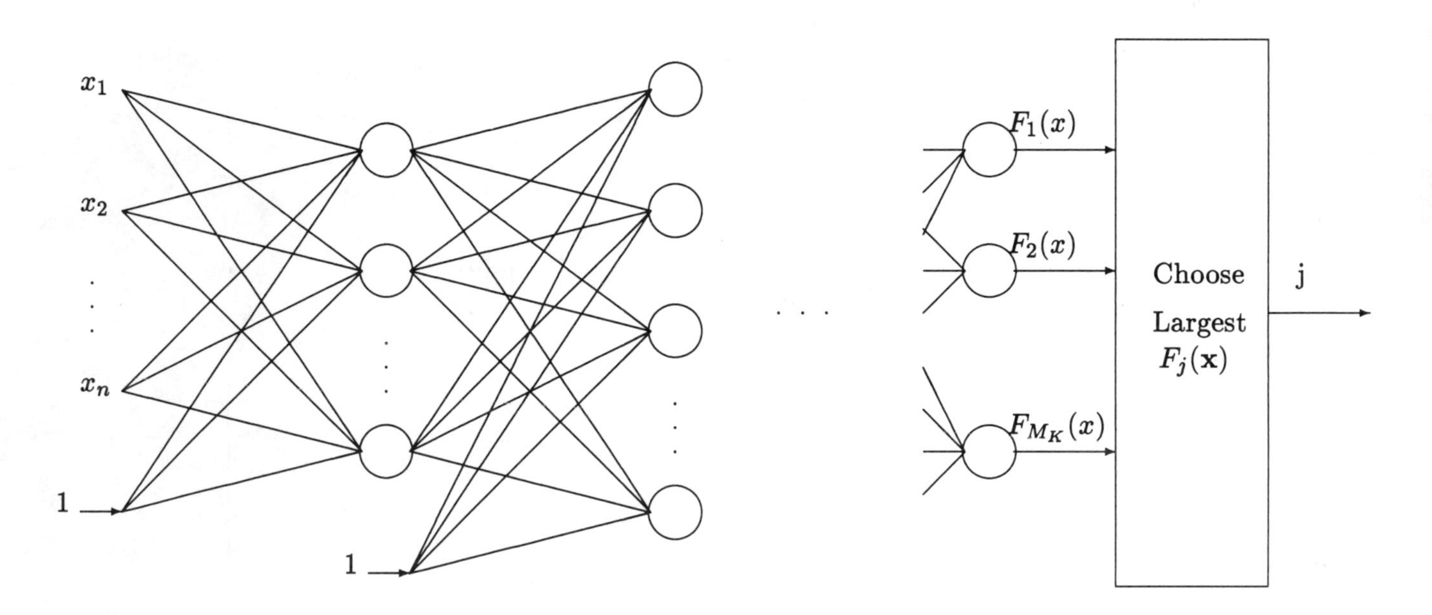

Figure 10: The MLP classifier architecture. From [91].

The weight matrix $\mathbf{W}_k$ connects the augmented outputs of layer $k-1$ and the neurons of layer k. The activation function of the kth layer is the vector valued function $\mathbf{f}_k : \mathcal{R}^{M_k} \rightarrow \mathcal{R}^{M_k+1}$defined as $\mathbf{f}_k(\mathbf{v}) = [f_k(v_1) \;\; f_k(v_2) \;\; \dots \;\; f_k(v_{M_k}) \;\; 1]^T$ where $f_k(\cdot)$ is the scalar activation function used by all neurons in the kth layer. In our experiments, we used the logistic activation function ($f_k(v) = \frac{1}{1+e^{-v}}$) for the hidden layers $k = 1, \dots, K-1$, and the linear activation function ($f_K(v) = v$) for the output layer. The *activity level* at the input of the kth layer is given by

$$\mathbf{v}_k = \mathbf{W}_k \tilde{\mathbf{u}}_{k-1}. \tag{71}$$

Thus, the network's operation can be described by the following recursion formula:

$$\tilde{\mathbf{u}}_k = \mathbf{f}_k(\mathbf{v}_k) = \mathbf{f}_k(\mathbf{W}_k \tilde{\mathbf{u}}_{k-1}) \qquad k = 1, 2, \dots, K. \tag{72}$$

3.4 Experimental Results

The general DA method for supervised learning has been specialized to specific design problems and tested. In particular, results are given for classifier design for the VQ, RBF, and MLP structures, piecewise regression, and mixture of experts regression. The exact equations used in the iteration depend on the structure, and can be derived in a straightforward manner from the general design approach. For more details on specific DA design refer to [91] for classifier design, [106] for piecewise regression, and [110] for mixture of experts.

3.4.1 VQ Classifier Design

The DA approach to VQ classifier design [91] is compared with the learning vector quantizer (LVQ) method [74]. Note that here LVQ will refer narrowly to that design method, not to the structure itself which we call VQ. The first simulation result is on the "synthetic" example from [111], where DA design achieved $P_e = 8.9\%$ on the test set using eight prototypes and $P_e = 8.6\%$ using twelve prototypes, in comparison to LVQ's $P_e = 9.5\%$ based on twelve prototypes. [For general reference, an MLP with six hidden units achieved $P_e = 9.4\%$.] For complicated mixture examples, with possibly twenty or more overlapping mixture components and multiple classes, the DA method was found to consistently achieve substantial performance gains over LVQ. As an example, consider the training data for a four-class problem involving twenty-four overlapping, non-isotropic mixture components in two dimensions, shown in Figure 11 and Figure 12. VQ-classifiers with 16 prototypes (four per class) were designed using both LVQ and DA. Figures 11a and 12a display the data and partitions formed by the two methods. Figures 11b and 12b display the prototype locations along with the partitions. The best LVQ solution based on ten random initializations, shown in Figure 11, achieved $P_e = 31\%$. Note that the method has failed to distinguish a component of class 0 in the upper left of Figure 11a, as well as a component of class 1 near the lower right of the figure. By contrast the DA solution shown in Figure 12 succeeds in discriminating these components, and achieves $P_e = 23\%$.

Another benchmark test data is the Finnish phoneme data set that accompanies the standard LVQ package. The training set consists of 1962 vectors of 20 dimensions each. Each vector represents speech attributes extracted from a short segment of continuous Finnish speech. These vectors are labeled according to the phoneme uttered by the speaker. There are 20 classes of phonemes in the training set. In both LVQ and DA approaches, The number of prototypes associated with a particular class was set to be proportional to the relative population of that class in the training set. This is referred to as the *propinit* initialization in the standard LVQ package. The experimental results are shown in Table 1. Note that the DA method consistently outperformed LVQ over the entire range.

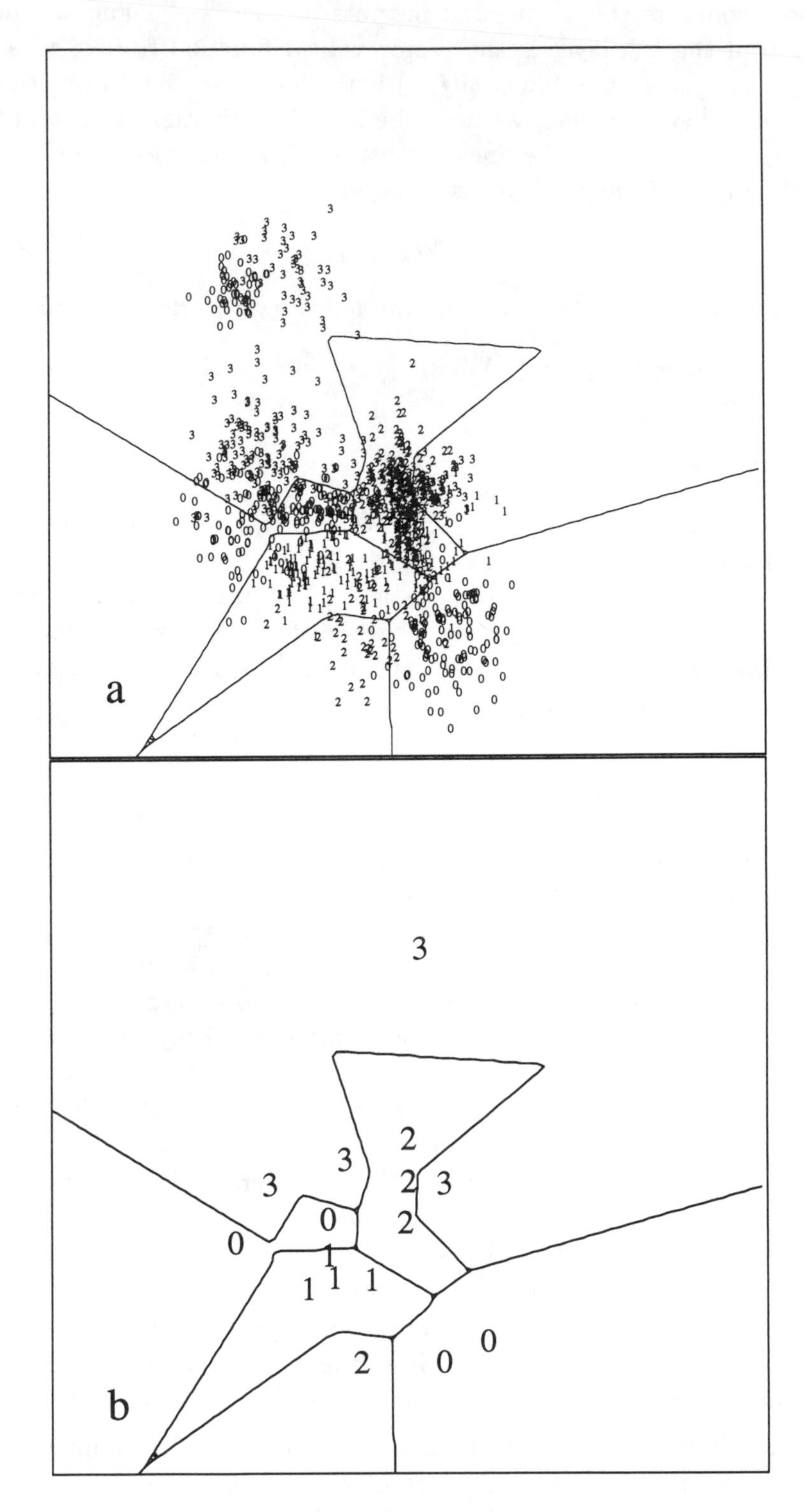

Figure 11: a) A four-class Gaussian mixture training set for a VQ classifier design and the partition produced by LVQ. b) The LVQ partition, with the sixteen class prototype. Locations of prototypes shown. The error rate is $P_e = 31\%$. From [91].

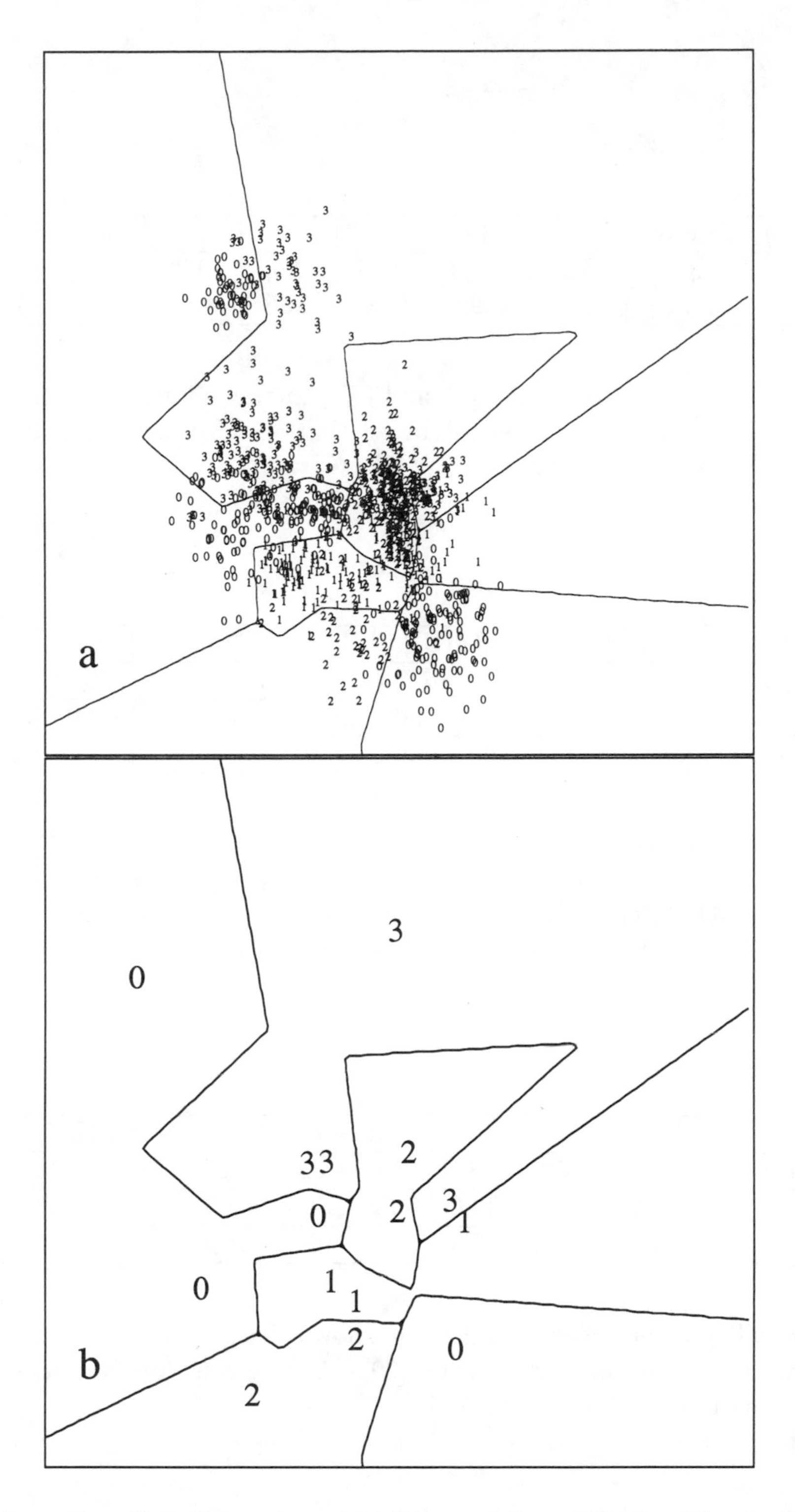

Figure 12: a) The four-class Gaussian mixture training set for a VQ classifier design and the partition produced by DA. b) The DA partition with the sixteen class prototypes shown. The error rate is $P_e = 23\%$. From [91].

M (# of units)	20	30	40	50	80	100	200
P_e (LVQ)	13.25	12.44	11.47	10.96	10.09	8.17	6.78
P_e (DA)	11.67	9.99	8.36	5.55	4.83	4.23	3.26

Table 1: Error probability comparison of the DA and LVQ methods for the design of VQ classifiers on the 20-dimensional, 20 class Finnish phoneme data set that accompanies the standard LVQ package. M represents the total number of prototypes.

Method	DA	TR-RBF				MD-RBF		G-RBF		
M (# of units)	3	3	5	15	25	10	30	5	10	15
P_e (training set)	14.0	38.0	15.0	14.0	10.0	25.0	18.0	48.0	21.0	14.0
P_e (test set)	16.0	38.0	22.0	18.0	17.0	26.0	19.0	47.0	19.0	16.0

Table 2: Error probability comparison of DA and other design techniques for RBF classification, on the 21-dimensional waveform data from [20]. M is the number of receptive fields. DA is compared with TR-RBF [130], MD-RBF [96], and with G-RBF which is gradient descent on $< P_e >$. The test set performance results have 95 % confidence intervals of half-length less than 2%.

Method	DA		TR-RBF				MD-RBF		G-RBF
M (# of units)	4	30	4	10	30	50	10	50	10
P_e (training set)	11.0	2.8	33.0	16.2	14.5	12.9	30.0	19.0	18.0
P_e (test set)	13.0	16.7	35.0	16.5	16.8	17.9	37.0	18.0	20.0

Table 3: Error probability comparison of DA with known design techniques for RBF classification on the 40-dimensional noisy waveform data from [20]. The test set performance results have 95 % confidence intervals of half-length less than 3.0 %.

3.4.2 RBF Classifier Design

The DA approach to RBF design [91] is compared here with the method of Moody and Darken [96] (MD-RBF), with a method described by Tarassenko and Roberts [130] (TR-RBF), and with the gradient method of steepest descent on $< P_e >$ (G-RBF). MD-RBF combines unsupervised learning of receptive field parameters with supervised learning of $\{\omega_{kj}\}$ to minimize the squared distance to target class outputs. The primary advantage of this approach is its modest design complexity. However, the receptive fields are not optimized in a supervised fashion, which can cause performance degradation. TR-RBF, one of the methods described in [130], optimizes all of the RBF parameters to approximate target class outputs in a squared error sense. This design is more complex than MD-RBF and achieves better performance for a given model size (the number of receptive fields the classifier uses). However, the TR-RBF design objective is not equivalent to minimizing P_e, but as in the case of back propagation, effectively aims to approximate the Bayes-optimal discriminant. While direct descent on $< P_e >$ may minimize the "right" objective, problems of local optima may be quite severe. In fact, we have found that the performance of all of these methods can be quite poor without a judicious initialization. For all of these methods, we have employed the unsupervised learning phase described in [96] (based on ISODATA clustering and variance estimation) as model initialization. Then, steepest descent was performed on the respective cost surface. We have found that the complexity of our design is typically 1-5 times that of TR-RBF or G-RBF (though occasionally our design is actually faster than G-RBF). Accordingly, we have chosen the best results based on five random initializations for these techniques, and compared with the single DA design run.

To illustrate that increasing M may not help to improve performance on the *test* set, we compared DA with the results reported in [97], for two and eight dimensional mixture examples. For the 2D example, DA achieved $P_{e_{\text{train}}} = 6.0\%$ for a 400 point training set and $P_{e_{\text{test}}} = 6.1\%$ on a 20,000 point test set, using $M = 3$ units (These results are near-optimal, based on the Bayes rate.). By contrast, the method in [97] used 86 receptive fields and achieved $P_{e_{\text{test}}} = 9.26\%$. For the 8D example and $M = 5$, DA achieved $P_{e_{\text{train}}} = 8\%$ and $P_{e_{\text{test}}} = 9.4\%$ (again near-optimal), while the method in [97] achieved $P_{e_{\text{test}}} = 12.0\%$ using $M = 128$.

More comprehensive tests on higher-dimensional data have also been performed. Two examples reported here are the 21D waveform data and the 40D "noisy" waveform data used in [20] (obtained from the UC-Irvine machine learning database repository.) Rather than duplicate the experiments conducted in [20], we split the 5000 vectors into equal size training and test sets. Our results in Table 2 and Table 3 demonstrate quite substantial performance gains over all the other methods, and performance quite close to the estimated Bayes rate of 14% [20]. Note in particular that the other methods perform quite poorly for small M, and need to increase M to achieve training set performance comparable to our approach. However, performance on the test set does not necessarily improve, and may degrade for large M.

3.4.3 MLP Classifier Design

The DA approach for designing MLPs [91] is compared with two other approaches – the standard back propagation (BP) algorithm of [122] and gradient descent on the $< P_e >$ cost surface (G-MLP). The BP weights were initialized to random numbers uniformly distributed between ± 0.01. A total of 50,000 epochs of a batch gradient descent algorithm were run to minimize the mean-squared error between the desired and actual outputs of the MLP. BP, however, descends on a cost surface mismatched to the minimum P_e objective. Further, its performance is dependent on the choice of initial weights. In G-MLP, the performance of BP is improved by taking the BP solution as initialization and then descending on $< P_e >$. However, in practice, the gains achieved by G-MLP over BP are only marginal, as the optimization performance sensitively depends on the choice of initialization.

The performance is tested on the 19-dimensional, 7-class image segmentation data from the UC-Irvine machine learning database. The training set contains 210 vectors and the test set contains 2100 vectors, each 19-dimensional. The features represent various attributes of a 3x3 block of pixels. The classes correspond to textures (brickface, sky, foliage, cement, window, path, grass). A sequence of single hidden layer neural networks was designed based on this data set. Table 4 summarizes the results for various hidden layer sizes (M). Networks designed by DA significantly outperformed the other approaches over the entire range of network sizes.

An important concern is the issue of design complexity. In the above experiments the DA learning complexity was roughly 4-8 times higher than that of back propagation and roughly the same as that of G-MLP. This suggests that the potential for performance improvement would, in typical applications, greatly outweigh the somewhat higher design complexity of the DA approach.

3.4.4 Piecewise Regression

Here we summarize experiments comparing the performance of the DA approach for VQ-based piecewise regression [106] with the conventional piecewise regression approach of CART [20]. Note that regular CART is severely restricted in that its partition is constrained to be tree-structured with partition boundaries that are parallel to the co-ordinate axes. The latter restriction which prevents regular CART from exploiting dependencies between the features of $\mathbf{X}$ can be overcome by adopting an extension of CART that allows the boundaries between regions to be arbitrary linear hyperplanes. While this extension allows better partitioning of the input space and hence smaller approximation error, the complexity of the design method for the extended structure [60] grows as N^2, where N is the size of the training set. Consequently, the extended form of CART is impractical unless the training set is short. In this section, we will refer to regular CART as CART1, and its extended form as CART2. Our implementation of CART consists of growing a large "full tree" and then pruning it down to the root node using the Breiman-Friedman-Olshen-Stone algorithm (see e.g., [26].) The sequence of CART regression trees is obtained during the pruning process. It is known that the pruning phase is optimal given the fully grown tree. Unlike CART2, the complexity of the DA method is linear in the size of the training set. Further, the DA algorithm optimizes all the parameters of the regression function simultaneously, while avoiding many shallow local minima that trap greedy methods.

In all the following comparisons, the simpler piecewise constant models are employed. In our implementation of the DA method we used an annealing schedule given by $q(T) = 0.95T$.

The first experiment involves synthetic datasets, where the regression input $x = (x_0, x_1)$ is two-dimensional and the output y is one-dimensional. The input components x_0, x_1 are uniformly distributed in the interval, $(0, 1)$. The output y is a weighted sum of six normalized Gaussian-shaped functions of x each with an individual center, variance and weight. By choosing different centers, variances and weights for the Gaussians, we created a number of data sets, each consisting of training and validation sets of size 1000 each, and a test set of size 3000. The output samples were corrupted by a zero-mean Gaussian noise with variance 10.0.

To compare the design approaches, DA and CART were applied to design regression functions for each dataset using the training and validation sets (validation was used to select the best model size for generalization) and the performance was evaluated on the independent test sets. The experiments were conducted over more than 40 different data sets. Table 5 provides a randomly selected subset of the results. Note that in this case DA is only compared with standard CART1, since CART2 is

	DA			BP				G-MLP			
M	4	6	8	4	6	8	10	4	6	8	10
P_e (training set)	19.1	8.1	6.2	45.7	31.9	20.0	6.1	45.7	31.4	19.6	6.0
P_e (test set)	20.1	11.2	10.5	48.1	31.7	25.3	13.3	47.2	34.4	23.2	13.0

Table 4: Error probability comparison of the DA, BP, and G-MLP design approaches on the 19-dimensional 7-class segmentation data example. The test set performance results have 95 % confidence intervals of half-length less than 2.1%.

Dataset	CART (Model Order)	DA (Model Order)
1	12.0 (21)	11.1 (8)
2	12.7 (30)	11.7 (10)
3	11.5 (22)	10.7 (13)
4	12.0 (33)	11.6 (14)
5	15.1 (59)	14.4 (9)
6	13.6 (47)	12.9 (11)
7	13.5 (46)	11.1 (20)
8	11.9 (27)	11.1 (14)

Table 5: Mean squared approximation error measured over the test set, and model order, for the best solutions produced by CART1 and by DA for multi-modal Gaussian data sets.

too complex for training sets of this size. Clearly, for all the examples, DA demonstrates consistent improvements over CART1.

We next compare CART with DA over a few data sets from real-world regression applications. This data is taken from the StatLib database[9] and has been extensively used by researchers in benchmarking the relative performance of competing regression methods. In some of the following experiments, due to the unavailability of sufficient data for proper validation, we simply compare the performance of the two regression models versus model size. Where enough data is available we show both training and test results. Note, however, that we mainly wish to demonstrate the optimization power of the approach.

One benchmark problem is concerned with predicting the value of homes in the Boston area from a variety of parameters [57]. The training set consists of data from 506 homes. The output in this case is the median price of a home, with the input consisting of a vector of 13 scalar features believed to influence the price. The objective is to minimize the average squared error in price prediction. Since the features have different dynamic ranges, they were normalized to unit variance prior to application of DA and CART. Piecewise constant regression models of different model sizes were generated by the design methods. Each sequence of CART functions was obtained by growing a full tree and optimally pruning this tree using data from the training set. The sequence of DA solutions was obtained via the phenomenon of phase transitions. Table 6 compares the squared-error in predicting house prices using the standard CART1, its extended form CART2, and DA. The training errors are tabulated for each method for a sample of model orders between 1 and 10. Clearly, the DA method substantially outperforms both CART1 and CART2 over the entire range of model sizes. This example illustrates that DA find substantially better solutions for the design objective. Also note that CART1 outperforms CART2 in several cases, despite the fact that CART2 is a potentially more powerful regression structure. These results are indicative of the difficulties due to local minima.

The data set for the next example was taken from the environmental sciences. This problem is concerned with predicting the age-adjusted mortality rate per 100,000 inhabitants of a locality, from 15 factors that have presumably influenced it. Some of these factors are related to the levels of environmental pollution in the locality, while others are measurements of non-environmental, mainly social, parameters. This data set has been used by numerous researchers since its introduction in the early 1970s [87]. As data are only available for 60 localities, they were not divided into separate training and test sets. We only show performance on the training set. Table 7 shows that the VQ-based regression function designed by DA offers a consistent and substantial advantage over CART for the entire range of model sizes.

The third regression data set is drawn from an application in the food sciences. The problem is that of efficient estimation of the fat content of a sample of meat. (Techniques of analytical chemistry can be used to measure this quantity directly, but it is a slow and time-consuming process.) We used a data set of quick measurements by the Tecator Infratec Food and Feed Analyzer which measures the absorption of electro-magnetic waves in 100 different frequency bands, and the corresponding fat content as determined by analytical chemistry. As suggested by the data providers, we divided the data into a training set of size 172 and a test set of size 43. We then applied CART1, CART2, and DA to the training set for different model sizes. Table 8 compares the mean-squared approximation error obtained over the training and test sets for all the methods. DA significantly outperformed the CART regression functions for the same number of regions in the input space. The performance gains of DA outside the training set confirm its expected good generalization capabilities. Note also that the CART2 method exhibits overfitting with deterioration of the test set performance from $K = 5$ to

[9]Available on the World-Wide Web at http://lib.stat.cmu.edu/data sets/ .

K	CART1	CART2	DA
1	84.4	84.4	84.4
2	46.2	43.2	34.4
3	31.8	33.0	25.0
4	25.7	26.1	16.9
5	20.7	23.2	14.4
6	17.9	21.9	11.0
7	15.6	20.8	10.8
8	13.6	19.7	10.7
9	12.5	18.8	8.6
10	11.8	18.1	8.5

Table 6: Average squared prediction error for housing prices in the Boston area. Comparison of the standard CART1, its extension CART2 and DA. K is the number of partition regions for each model.

Method	K=1	K=2	K=3	K=4	K=5	K=6	K=7
CART1	3805.13	2427.40	1786.90	1381.08	1122.68	938.91	792.91
CART2	3805.13	2087.0	1532.19	1323.50	1174.17	1050.55	917.2
DA	3805.13	2003.4	976.18	775.36	694.27	603.46	551.85

Table 7: Mean-squared prediction error for the age-adjusted mortality rate per 100,000 inhabitants from various environmental factors. Comparison of CART1, CART2 and DA. K is the number of regions allowed for each model.

Method	TR/TE	K=1	K=2	K=3	K=4	K=5	K=10
CART1	TR	159.89	113.86	106.93	101.66	85.27	48.20
	TE	168.25	141.34	142.74	140.54	109.54	107.28
CART2	TR	159.89	67.35	49.84	37.15	28.08	16.22
	TE	168.25	84.98	69.92	57.99	39.44	50.63
DA	TR	159.89	38.05	38.05	26.67	15.55	8.11
	TE	168.25	39.85	37.03	26.47	14.27	14.10

Table 8: Mean-squared approximation error for the fat content of a meat sample from 100 spectroscopic measurements. The performance of CART1, CART2 and DA is compared both inside (TR) and outside (TE) the training set. K is the number of regions used to represent the data.

$K = 10$.

3.4.5 Mixture of Experts

Mixture of experts is an important class of structures whose introduction was inspired by mixture models from statistics [88],[131]. This class includes specific structures commonly known as "mixture of experts" [65] and "hierarchical mixture of experts" (HME) [68], as well as "normalized radial basis functions" (NRBF) [96]. We refer to this class generally as *mixture of experts* (ME) models. ME has been suggested for a variety of problems, including classification [62],[65], control [64][68], and regression tasks [68],[134],[135].

We define the "local expert" regression function, $f(x, \Lambda_j)$, where Λ_j is the set of model parameters for local model j. The ME regression function is defined as :

$$g(x) = \sum_j p[j|x] f(x, \Lambda_j), \tag{73}$$

where, $p[j|x]$ is a non-negative weight of association between input x and expert j that effectively determines the degree to which expert j contributes to the overall model output. In the literature, these weights are often called *gating units* [65], and obey some prespecified parametric form. We further impose $\sum_j p[j|x] = 1$, which leads to the natural interpretation of the weight of association or gating unit as a probability of association.

ME is an effective compromise between purely piecewise regression models discussed earlier, such as CART [20], and "global" models such as the MLP [122]. By "purely piecewise" it is meant that the input space is hard-partitioned to regions, each with its own exclusive expert model. Effectively, the piecewise regression function is composed of a patchwork of local regression functions that collectively cover the input space. In addition to partitioning the input space, the model parameter set is partitioned into sub-models which are only "active" for a particular local input region. By contrast, in global models such as MLPs, there is a single regression function that must fit the data well everywhere, with no explicit partitioning of the input space nor sub-division of the parameter set. ME exploits a partition of the space but produces a function which combines the contributions of the various experts with some appropriate weighting.

The DA design approach [110] is based on controlling the entropy of the association probabilities $p[j|x]$. In this case, these probabilities are part of the problem definition rather than an artificial addition to avoid nonglobal minima. It is important to note that as we approach zero temperature, the entropic constraint simply disappears and we find the solution which minimizes the cost *regardless* of its entropy. Annealing consists of starting at high temperature (high entropy) and gradually allowing the entropy to drop to its optimal level (where the cost is minimized).

The following results compare the DA approach with conventional design methods for NRBF and HME regression functions. The experiments were performed over some popular benchmark data sets from the regression literature. In each experiment, we compare the average squared-error obtained over the training set using the DA design method and the alternative design methods. The comparisons are repeated for different network sizes. The network size, K, refers to the number of local experts used in the mixture model. For the case of binary HME trees with l levels, $K = 2^l$ and for the case of NRBF regression functions, K is the number of Gaussian basis functions used. Following the most common implementation, the local models are constant functions in the NRBF case and linear functions in the HME case. The alternative design approaches used for comparing our HME design algorithm are :

- "GD": a gradient descent algorithm to simultaneously optimize all HME parameters for the squared-error cost.
- "ML": Jordan and Jacobs's Maximum Likelihood approach [68].

For the NRBF regression function, we have compared the DA design approach [110] with the gradient descent ("GD") algorithm which is an enhanced version of the method suggested in [96] (see [110] for details). For fair comparison, we make the conservative (worst case) assumption that the complexity of DA is ten times greater than that of the competing methods (in fact the complexity of DA was higher by a factor that varied from 2 to 10 in the experiments). To compensate for the complexity, we allow each competing method to generate results based on ten different random initializations, with the best result obtained among those runs selected for comparison with the DA result. Since the regression function obtained by DA is generally independent of initialization, a single DA run sufficed.

Let us first consider the results for the "real-world" examples that had been used in the piecewise regression subsection. Results for the Boston home-value prediction problem are given in Table 9 and Table 10, and demonstrate that for both mixture models, the DA approach achieves a significantly smaller regression error compared with the other approaches over a variety of network sizes. Results for the mortality rate example are given in Table 11 and Table 12; and the meat fat content results (measurements was obtained by the Tecator Infratec Food and Feed Analyzer) are given in Table 13 and Table 14.

Finally, results are given for synthetic data. Here, $x = (x_0, x_1)$ is two-dimensional and the training set is generated according to a uniform distribution in the unit square. The output y is scalar. We created five different data sets based on the functions $(f_1(), f_2(), ... f_5())$ specified in [25],[63]. Each function was used to generate both a training set and test set of size 225. We designed NRBF and HME regression estimates for each data set using both DA and the competitive design approaches. The results shown in Table 15 and Table 16 show improved performance of the DA method that is consistent with the results obtained for the other benchmark sets.

4 Speech Recognition

In the late 1960s, Baum and his colleagues presented a series of papers (including [9, 10]) investigating the mathematical structure and practical usefulness of hidden Markov models (HMMs). In the years that followed, it became generally known that HMMs can be usefully employed for speech recognition. This important realization is due to the pioneering work of several researchers, notably Jelinek [67], Baker [7], Ferguson [42] and Rabiner [100, 101]. Since then, the HMM-based classifier has steadily replaced template matching as the main paradigm in use for speech recognition.

In HMM-based speech recognition, the user's speech is divided into segments, each of which is classified to a member of a finite length *dictionary* of *speech units*. Speech units may be words (in isolated word recognizers) or sub-word phones (in continuous speech recognizers). Classification is performed on each segment by selecting the winner in a competition between HMMs that represent speech units in the dictionary. In isolated word recognizers, the segments correspond to words, and the task of dividing the input speech into segments is usually performed independent of the classification task. In continuous speech recognizers, segments correspond to sub-word units and the tasks of segmentation and classification are performed jointly.

For superior performance, the HMM speech recognizer must be trained on a large speech database. Since recognition is performed by competition between HMMs, ideally, the training program should

K	DA	GD
1	87.7	87.7
2	19.7	23.8
4	12.9	19.3
6	12.6	15.7
10	6.5	13.7

Table 9: Comparison of regression error obtained for NRBF desigr by GD and DA on the Boston home value problem. K is the number of Gaussian basis functions used.

K	DA	GD	ML
4	5.7	5.9	7.5
8	3.4	3.6	5.6

Table 10: Comparison of regression error obtained by DA, GD and ML for HME function design for the Boston home value problem. K is the number of leaves in the binary tree.

K	DA	GD
1	3805.1	3805.1
2	1148.8	2154.0
4	720.8	1256.8
6	439.1	566.5
8	299.6	564.5
10	261.4	438.2

Table 11: Comparison of regression error obtained by DA and GD for NRBF design for the mortality rate prediction problem. K is the number of Gaussian basis functions used.

K	DA	GD	ML
4	18.2	121.8	70.4
8	2.1	12.3	41.8

Table 12: Comparison of regression error obtained using DA, GD and ML for HME design for the mortality rate prediction problem. K is the number of leaves in the binary tree.

	DA		GD	
K	TR	TE	TR	TE
1	159.9	168.2	159.9	168.2
2	52.9	58.8	131.4	159.7
4	28.6	32.9	119.8	138.0
6	27.3	40.1	74.9	83.7

Table 13: Comparison of regression error obtained using DA and GD for NRBF design for the fat content prediction problem. K is the number of Gaussian basis functions used. 'TR' and 'TE' refer to training and test sets respectively.

	DA		GD		ML	
K	TR	TE	TR	TE	TR	TE
4	8.3	11.5	14.1	18.1	15.1	23.9
8	6.9	9.8	12.8	17.2	12.5	39.7

Table 14: Comparison of regression error obtained using DA, GD and ML for HME function design for the fat content prediction problem. K is the number of leaves in the binary tree. 'TR' and 'TE' refer to training and test sets respectively.

Method	K	$f_1()$	$f_2()$	$f_3()$	$f_4()$	$f_5()$
DA(TR)	8	0.001	0.008	0.01	0.08	0.13
DA(TE)	8	0.001	0.009	0.01	0.09	0.13
GD(TR)	8	0.02	0.044	0.16	0.19	0.24
GD(TE)	8	0.02	0.049	0.17	0.17	0.23
DA(TR)	16	0.001	0.003	0.01	0.05	0.02
DA(TE)	16	0.001	0.005	0.01	0.05	0.03
GD(TR)	16	0.02	0.012	0.14	0.06	0.24
GD(TE)	16	0.02	0.017	0.12	0.07	0.23

Table 15: Comparison of regression error obtained by DA and GD on the training (TR) and test (TE) sets, for NRBF design to approximate functions, $f_1()..f_5()$. (K denotes the number of basis functions.)

Method	K	$f_1()$	$f_2()$	$f_3()$	$f_4()$	$f_5()$
DA(TR)	4	0.0006	0.02	0.18	0.20	0.19
DA(TE)	4	0.0006	0.02	0.18	0.25	0.21
GD(TR)	4	0.0079	0.06	0.39	0.36	0.35
GD(TE)	4	0.0082	0.06	0.47	0.43	0.38
ML(TR)	4	0.026	0.08	0.86	0.36	0.43
ML(TE)	4	0.039	0.12	0.79	0.46	0.51
DA(TR)	8	0.0003	0.01	0.09	0.08	0.17
DA(TE)	8	0.0003	0.02	0.09	0.01	0.16
GD(TR)	8	0.0063	0.05	0.12	0.35	0.28
GD(TE)	8	0.0079	0.05	0.12	0.40	0.30
ML(TR)	8	0.011	0.03	0.12	0.09	0.32
ML(TE)	8	0.016	0.04	0.14	0.14	0.44

Table 16: Regression error obtained by DA, GD and ML on the training (TR) and test (TE) sets, for HME design to approximate functions, $f_1()..f_5()$. (K denotes the number of leaves in the binary tree.)

jointly optimize all competing HMMs to minimize the misclassification error rate calculated over the training set. In isolated word recognition, this error rate is the fraction of training words that are misclassified. In continuous speech recognition, since the words are not clearly separated, the error rate can be defined as the fraction of training sentences that contain any recognition errors.

One important design difficulty is the complex nature of the error rate cost surface. This surface which represents the classifier error as a function of the HMM parameters is piecewise constant and is riddled with shallow local minima. Although we seek the system parameters that result in the global minimum on this surface, standard optimization methods such as gradient descent will normally fail to produce the optimal solution. The prevailing approach to speech recognizer design circumvents this difficulty by discarding the minimum error rate objective, and adopting instead the potentially mismatched maximum likelihood (ML) objective. The choice of ML leads to a smoother cost surface and also facilitates independent optimization of each HMM via the efficient Baum-Welch algorithm [9, 10]

The common justification for using ML is the asymptotic equivalence of the ML objective and the minimum error rate objective, which is valid if infinite training data is available and if the HMM is precisely the correct model for speech. Unfortunately, in practical speech recognition problems, neither of the above assumptions hold. Thus, there is an (at least theoretical) advantage to direct optimization of the HMM parameters to achieve the minimum error rate objective rather than the ML objective.

The sub-optimality of the ML objective has been recognized by several researchers (*e.g.* [5, 6, 19, 69, 24]) who have provided strong reasons for discarding it in favor of *discriminative design methods* [102, Sec 5.6] that jointly optimize all the HMMs in a classifier. One promising discriminative design method is Generalized Probabilistic Descent (GPD) that was proposed and extended by Katagiri, Juang, and others in a series of papers that are reviewed in [71]. Not surprisingly, GPD and other discriminative methods are generally applicable to all structured pattern classifiers, and have been accordingly discussed in a previous subsection. The central idea in GPD is to approximate the piecewise constant cost surface by a smooth and differentiable function. Following smoothing, gradient methods that seek a local minimum on the smoothed surface can be used to optimize the classifier's parameter set. GPD and other discriminative design methods have been applied to design speech recognition systems based on both template matching [24, 86] and HMMs [28, 69].

In the context of HMMs, it was shown in [28] and [69] that GPD provides a significant improvement in recognition accuracy over ML design. However, it has been our observation that while cost surface smoothing allows the use of gradient descent optimization, the smoothed cost surface, is nevertheless extremely complex with numerous shallow local minima traps. Consequently, GPD may often converge to a poor local minimum and yield suboptimal recognition performance.

From the preceding arguments, it is clear that direct minimization of the classifier error rate requires the use of a non-convex optimization method that can avoid shallow local minima traps on the cost surface. It is therefore a natural application for DA. The DA speech recognizer design method to be described next is general and can be applied to design both isolated word and continuous speech recognizers that use continuous, semi-continuous or discrete observation HMMs. However, for completeness and simplicity, we will focus the formulation and experiments on the simplest case of isolated word recognizers based on discrete HMM systems. The basic approach we describe here was proposed in [109, 107].

4.1 Problem Formulation

An HMM classifier is to be designed given a *labeled training set*,

$$\mathcal{T} \equiv \{(\mathbf{x}_1, c_1), (\mathbf{x}_2, c_2), ..(\mathbf{x}_N, c_N)\}, \tag{74}$$

where *training pattern* $\mathbf{x}_i$ is known to be an utterance of speech unit c_i, which is an entry in the given dictionary of speech unit labels $\mathcal{C} \equiv \{1, 2, ..J\}$. The pattern $\mathbf{x}_i$ is in fact a sequence of feature vectors of observations extracted from a segment of l_i speech frames, $\mathbf{x}_i = (\mathbf{x}_i(1), \mathbf{x}_i(2), \cdots \mathbf{x}_i(l_i))$.

The exact nature of the observation feature vector $\mathbf{x}_i(t)$ is application dependent. In many practical implementations, $\mathbf{x}_i(t)$ consists of cepstral coefficients or linear prediction coefficients and their derivatives extracted from the speech frame. Since these features are continuous quantities ($\mathbf{x}_i(t) \in \mathcal{R}^n$), such classification is performed by continuous observation HMMs. In other applications however, the high computational complexity involved in modeling continuous observations is not acceptable and the classification is implemented with discrete observation HMMs. Here, the feature vector is extracted from the speech frame and then vector-quantized to an entry in a pre-designed codebook of K *prototype vectors*. The sequence of l_i quantization indexes obtained by this process is the discrete observation vector or training pattern, $\mathbf{x}_i$. For simplicity we assume here the discrete observation case, $\mathbf{x}_i(t) \in \mathcal{K} \equiv \{1, 2, ..K\}$. However, the method is general and extendible to the case of continuous observations [47].

The HMM recognition system for discrete observations consists of a set of HMMs, $\{H_j,\ j = 1, 2..J\}$, which correspond to the J words in the dictionary. The model H_j has S_j states and is fully specified by the parameter set $\Lambda_j \equiv (A_j, B_j, \Pi_j)$ where, following the standard notation, A_j is the $(S_j \times S_j)$ state transition probability matrix, B_j is the $(S_j \times K)$ emission probability matrix and Π_j is the $(S_j \times 1)$ initial state probability vector.

We consider HMM classifier systems that use the common "best path" discriminant approach. Note, however, that this assumption is not required, and the design method can be modified to the case where the discriminant is obtained by appropriate averaging of the likelihood over all paths in the HMM.

The best path classifier operates as follows: Given a training pattern, $\mathbf{x}_i$, for each HMM, H_j, and for each sequence (length l_i) of states, $\mathbf{s} \equiv (s(1), s(2), \cdots, s(l_i))$ in the trellis of H_j, we determine the quantity ("path score")

$$l(\mathbf{x}_i, \mathbf{s}, H_j) = \frac{1}{l_i}\{\log \Pi_j[s(1)] + \sum_{t=1}^{l_i-1} \log A_j[s(t), s(t+1)] + \sum_{t=1}^{l_i} \log B_j[s(t), \mathbf{x}_i(t)]\}, \tag{75}$$

which is the normalized log of the joint probability of observation $\mathbf{x}_i$ and the state sequence, $\mathbf{s}$ given the parameters of H_j. We use the conventional notation: $Q[\cdot,\cdot]$ denotes an element of matrix Q; and $q[\cdot]$ denotes an element of the vector q. We note that although normalization of the likelihood (by the length of the observation) does not change the problem definition and is not commonly used, we find it useful in the DA algorithm.

Next, we maximize the path score over all paths in the trellis of H_j and determine the score of model H_j:

$$d_j(\mathbf{x}_i) = \max_{\mathbf{s} \in \mathcal{S}_{l_i}(H_j)} l(\mathbf{x}_i, \mathbf{s}, H_j). \tag{76}$$

where $\mathcal{S}_l(H_j)$ is the set of all state sequences of length l in the trellis of H_j. The quantity $d_j(\mathbf{x}_i)$ approximates the likelihood of model H_j given the observation, $\mathbf{x}_i$. Interpreting $d_j(\cdot)$ as the discriminant for class j, we adopt the traditional discriminant-based classification rule:

$$C(\mathbf{x}_i) = arg \max_j d_j(\mathbf{x}_i). \tag{77}$$

The classification procedure can be viewed as a competition between paths. The HMM containing the path with the highest score is declared as the "winner" and the classifier assigns pattern $\mathbf{x}_i$ to the corresponding class, which corresponds to a word in the dictionary. A known advantage of the "best path" discriminant classifier is that the search for the winning path can be reduced to a sequential optimization problem that can be solved via an efficient dynamic programming algorithm.

The HMM-based classifier should, in principle, be optimized by adjusting the HMM parameters $\{\Lambda_j\}$ to minimize the empirical misclassification rate measured over the training set:

$$\min_{\{\Lambda_j\}} P_e = \min_{\{\Lambda_j\}}\{1 - \frac{1}{N}\sum_{i=1}^{N}\delta(C(\mathbf{x}_i), c_i)\}. \tag{78}$$

Here δ is the Kronecker delta function:

$$\delta(u, v) = \begin{cases} 1 & \text{if } u = v \\ 0 & \text{otherwise} \end{cases} \tag{79}$$

We have already discussed the difficulties in solving the above optimization problem, which are due to the piecewise constant nature of the cost function, P_e, and the abundance of shallow local minima.

4.2 Design by Deterministic Annealing

As in the general DA approach to classification we apply the fundamental principles: (a) Introduce randomness in the classification rule during the design process; (b) Minimize the expected misclassification rate of the random classifier while controlling the level of randomness via a constraint on the Shannon entropy; and (c) Gradually relax the entropy constraint so that the effective cost converges to the misclassification cost at the limit of zero entropy (non-random classification).

Thus, we replace the original (non-random) best path classification rule with a randomized classification rule. While the non-random rule associates a pattern $\mathbf{x}_i$ to a unique winning state sequence, the randomized rule associates each pattern, $\mathbf{x}_i$, with every state sequence, $\mathbf{s}$, in the trellis of every model, H_j, with probability $P[\mathbf{s}, j|\mathbf{x}_i]$. Naturally, these conditional probabilities are normalized functions such that $\sum_j \sum_{\mathbf{s}\in S_{l_i}(H_j)} P[\mathbf{s}, j|\mathbf{x}_i] = 1$.

The probabilities, $P[\mathbf{s}, j|\mathbf{x}_i]$, are in fact, the representation of the *randomized classification rule* and should not be confused with the probabilities characterizing the hidden Markov model itself. $P[\mathbf{s}, j|\mathbf{x}_i]$ is the probability that the classifier will select $\mathbf{s}$ as the winning path and consequentially, H_j as the winning HMM. Following the prior treatment of DA for classification, we propose to derive the classification probabilities from basic principles. We first note that the non-random classifier takes in pattern $\mathbf{x}_i$ and finds the state sequence, $\mathbf{s}_i$ with the highest score among all state sequences in all the HMMs in order to determine the class. We may trivially formulate this operation via the criterion function

$$D_e = \frac{1}{N}\sum_i l(\mathbf{x}_i, \mathbf{s}_i, H_j). \tag{80}$$

where $\mathbf{s}_i \in \bigcup_j \mathcal{S}_{l_i}(H_j)$. Clearly, this function is maximized by applying to each $\mathbf{x}_i$ the best-path classification rule to determine $\mathbf{s}_i$:

$$\mathbf{s}_i = \arg \max_{\mathbf{s} \in \bigcup_j \mathcal{S}_{l_i}(H_j)} l(\mathbf{x}_i, \mathbf{s}, H_j) \tag{81}$$

We define the optimal random classifier as the choice of the distribution that maximizes :

$$< D_e >= \frac{1}{N} \sum_i \sum_j \sum_{\mathbf{s} \in \mathcal{S}_{l_i}(H_j)} P[\mathbf{s}, j | \mathbf{x}_i] l(\mathbf{x}_i, \mathbf{s}, H_j), \tag{82}$$

which is the immediate probabilistic generalization of D_e in (80). To maintain randomness in the classifier decision, we maximize $< D_e >$ subject to a constraint on the level of randomness in the classification rule, which we measure by the Shannon entropy:

$$H = -\frac{1}{N} \sum_i \sum_j \sum_{\mathbf{s} \in \mathcal{S}_{l_i}(H_j)} P[\mathbf{s}, j | \mathbf{x}_i] \log P[\mathbf{s}, j | \mathbf{x}_i]. \tag{83}$$

The probability distribution obtained via this constrained optimization problem is the Gibbs distribution,

$$P[\mathbf{s}, j | \mathbf{x}_i] = \frac{e^{\gamma l(\mathbf{x}_i, \mathbf{s}, H_j)}}{\sum_{j'} \sum_{\mathbf{s}' \in \mathcal{S}_{l_i}(H_{j'})} e^{\gamma l(\mathbf{x}_i, \mathbf{s}', H_{j'})}}. \tag{84}$$

The level of Shannon entropy, H_0, corresponding to this Gibbs distribution is determined by the positive *scale parameter*, γ. For $\gamma = 0$, the distribution over paths is uniform. For finite, positive values of γ, the Gibbs distribution indicates that we assign higher probabilities of winning to state sequences with higher likelihood scores. In the limiting case of $\gamma \to \infty$, the random classification rule reverts to the non-random "best path" classifier, which assigns a non-zero probability of winning only to the path with the highest likelihood score as in (81).

So far we have derived a framework for randomizing the classifier, which captures the "best path" classification rule in the limiting (zero entropy) case. The framework is next employed for minimization of the classifier error rate. The average misclassification rate of the random classifier is given by:

$$< P_e >= 1 - \frac{1}{N} \sum_{i=1}^{N} P[c_i | \mathbf{x}_i], \tag{85}$$

which is a straightforward randomization of (78). The quantity $P[c_i|\mathbf{x}_i]$ is the probability that the correct class c_i will be selected as the winner and is computed by the summation:

$$P[c_i | \mathbf{x}_i] = \sum_{\mathbf{s} \in \mathcal{S}_{l_i}(H_{c_i})} P[\mathbf{s}, c_i | \mathbf{x}_i] \tag{86}$$

Direct minimization of (85) would lead to a non-random ($\gamma \to \infty$) distribution. We, therefore, pose the problem of minimizing $< P_e >$ while maintaining a level of randomness in the classifier through a constraint on the entropy: $H = H_0$. This constrained optimization problem is, equivalently, the minimization of the familiar Lagrangian cost function,

$$\min_{\{\Lambda_j\}, \gamma} \{F \equiv < P_e > -TH\}, \tag{87}$$

which we recognize as the free energy of statistical physics. From the viewpoint of our optimization problem, we are ultimately interested in thermal equilibrium at $T = 0$ which corresponds to direct minimization of $< P_e >$, our ultimate objective.

The annealing process yields a sequence of solutions at decreasing levels of entropy and $< P_e >$ leading to a "best-path" classifier in the limit. The optimization of the Lagrangian, F, at each temperature is achieved by a series of gradient descent steps with the following expressions for the gradients:

$$\frac{\partial F}{\partial \Lambda_j} = \frac{\gamma}{N} \sum_i \sum_{\mathbf{s} \in \mathcal{S}_{l_i}(H_j)} \frac{\partial l(\mathbf{x}_i, \mathbf{s}, H_j)}{\partial \Lambda_j} P[\mathbf{s}, j|\mathbf{x}_i]\{f(\mathbf{x}_i, \mathbf{s}, H_j) - < f(\mathbf{x}_i, \mathbf{s}, H_j) >\} \tag{88}$$

and

$$\frac{\partial F}{\partial \gamma} = \frac{1}{N} \sum_i \sum_j \sum_{\mathbf{s} \in \mathcal{S}_{l_i}(H_j)} l(\mathbf{x}_i, \mathbf{s}, H_j) P[\mathbf{s}, j|\mathbf{x}_i]\{f(\mathbf{x}_i, \mathbf{s}, H_j) - < f(\mathbf{x}_i, \mathbf{s}, H_j) >\} \tag{89}$$

Here, $f(\mathbf{x}_i, \mathbf{s}, H_j) = T\gamma l(\mathbf{x}_i, \mathbf{s}, H_j) - \delta(j, c_i)$ where $\delta(\cdot, \cdot)$ is the Kronecker delta function. The operation $< h(\cdot) >$ represents the expectation of function, $h(\cdot)$, over all state sequences in the trellises of all the HMMs. Hence,

$$< f(\mathbf{x}_i, \mathbf{s}, H_j) > = \sum_j \sum_{\mathbf{s} \in \mathcal{S}_{l_i}(H_j)} P[\mathbf{s}, j|\mathbf{x}_i] f(\mathbf{x}_i, \mathbf{s}, H_j). \tag{90}$$

In our experiments, we used the following simple exponential annealing and quenching schedules: $\alpha(T) = 0.9T$, and $q(\gamma) = 1.2\gamma$. An analytical treatment of the question of annealing schedules has not been attempted as yet.

The DA formulation is supplemented with the derivation of a low complexity forward-backward implementation, which may be viewed as a generalization of the Baum-Welch re-estimation algorithm. For details see [107].

4.3 Experimental Results

This section summarizes experiments comparing DA with two conventional design approaches - standard maximum likelihood (ML) and Generalized Probabilistic Descent (GPD).

The ML method was implemented using the Baum-Welch algorithm. The observations were uniformly segmented into states, and the state dependent emission probabilities were initialized from the segmental histogram of observations in each state. The transition probabilities were initialized randomly, using a uniform distribution over the allowed interval (0 to 1). Fifty iterations of the maximum likelihood algorithm were run. During optimization, the values of all probabilities were constrained to lie between 10^{-6} and $1 - 10^{-6}$.

DA was implemented with the following parameters: initial value of the scale parameter, $\gamma_{init} = 0.1$, initial temperature, $T_i = 1.0$, final temperature, $T_f = 10^{-6}$, annealing schedule, $\alpha(T) = 0.9T$, quenching schedule, $q(\gamma) = 1.2\gamma$, and minimum entropy, $H_{min} = 10^{-6}$.

GPD method was implemented as a special case of DA[10] - the temperature was set to zero, the value of γ was fixed by the level of smoothing desired, and free energy minimization steps were performed via gradient descent.

[10]It is easy to show that when $T = 0$ and γ is fixed, the free energy of DA is equivalent to a particular implementation of the GPD cost function.

To improve the GPD results, we allowed 20 different values for the parameter γ and two different initializations for the HMM parameters. The allowed values of γ were chosen according to the geometric series, 1.0, 2.0, 4.0, 8.0, .. 524288.0 which allows for a large variation in its value, and which effectively represents a "post-optimization" of γ. Two initial sets of ML-designed models were each obtained from different random initializations of the transition probabilities. For each initialization, the same emission probabilities were chosen (state-specific histograms following uniform segmentation). ML was implemented as fifty steps of the Baum-Welch algorithm. In the literature, ML-designed models have often been used as to initialize GPD. Each GPD model was thus obtained by running 40 different GPD experiments - using 20 values of γ and 2 initial sets of HMMs - and choosing the best of the 40 solutions in terms of performance on the training set.

The experiment focuses on isolated word recognition. The recognition of spoken English letters appears as an important subproblem within several applications [70][80] including automatic car navigation, automated directory assistance and voice activated call forwarding. Speech recognizers based on HMMs [104][70], dynamic time warping (DTW) [103], neural networks [38] and knowledge-based classifiers [30][80] have been proposed to tackle this important problem. A comparison of different approaches on benchmark data sets is given in [104] .

The task of recognizing spoken English letters is known to be challenging due to the high confusability of the alphabet. Four subsets of English letters are highly confusable: the E-set $\{b, c, d, e, g, p, t, v, z\}$, the A-set, $\{a, k, j\}$, the I-set $\{i, r, y\}$, and the M-set $\{m, n\}$. Most notorious for its confusability is the E-set. In real-world situations, the difficulty is further aggravated by the presence of background noise and possibly by the presence of channel distortion.

The speech used in our experiments is drawn from the ISOLET database [29]. ISOLET consists of utterances by 150 native English speakers (75 male and 75 female, ages 14 to 70 years), where each speaker utters twice, the 26 English letters in isolation. The data is divided into five sets (ISOLET1 to ISOLET5), each containing utterances by 30 speakers. The ISOLET speech was recorded with a Sennheiser HMD 224 microphone that was low-pass filtered at 7.6 kHz and sampled at 16 kHz. The database was used to design and test speech recognizers in the "multi-speaker" mode - one set of utterances of all the letters by the 60 speakers in ISOLET1 and ISOLET2 was used as the training set and the other set of utterances by the same speakers were used as the test set.

The sampled speech signal was divided into 20ms frames, where consecutive frames overlap by 10ms. A set of features were extracted from each speech file by using a public domain software for end-point detection and feature extraction [11]. The end-point detector was employed to eliminate silent frames at the beginning and the end of the speech file. A 28-dimensional feature vector consisting of 14 Mel-scaled FFT cepstral coefficients (MFCC) [99] and their first-order time derivatives (ΔMFCC coefficients) was extracted in each frame. The MFCC coefficients are believed to be relatively robust to noise and can be easily computed via an FFT. The feature vector in each frame was quantized using a codebook of "prototypes", thus resulting in a sequence of discrete features for each isolated word. In our experiments, we used 16 prototypes. The codebook of prototypes was designed from features extracted from all speech frames in the training set using a vector quantizer design method based on successive splitting of prototypes [50, Chapter 11]. Each word model consisted of a 5-state left-to-right HMM.

Performance comparison of the three design methods is presented in Table 17. We designed and tested classifiers for each confusable set using only tokens from that confusable set (the separation between the sets is not a difficult task for which ML techniques largely suffice). The design was performed independently for each confusable set using ML, GPD and DA. The objective of this comparison is to demonstrate the level of improvement achievable in each confusable set.

[11] We used Jialong He's speech recognition research tool which can be obtained via anonymous ftp from ftp.informatik.uni-ulm.de/pub/NI/jialong/spchtool.zip

	ML		GPD		DA	
Set	TR	TE	TR	TE	TR	TE
E	49.4	60.2	47.6	58.9	25.2	42.8
A	12.2	18.3	6.7	20.0	3.9	9.4
I	10.6	12.8	4.4	8.9	0.6	7.2
M	15.0	26.7	12.5	21.7	7.5	15.0

Table 17: Error rates obtained by ML, GPD, and DA for each confusable set (E,A,I,M).

5 The Rate-Distortion Connection

Rate-distortion (RD) theory is the branch of information theory which is concerned with source coding. Its fundamental results are due to Shannon [124], [125]. These are the coding theorems which provide an (asymptotically) achievable bound on the performance of source coding methods. This bound is often expressed as a rate-distortion function $R(D)$ for a given source, whose curve separates the region of feasible operating points (R, D), from the region that cannot be attained by any coding system. Important extensions of the theory to more general classes of sources than those originally considered by Shannon, have been developed since (see, e.g., [44] and [12]).

Explicit analytical evaluation of the function $R(D)$ has been generally elusive, except for very few examples of sources and distortion measures. Two main approaches were taken to address this problem. The first was to develop bounds on $R(D)$. An important example is the Shannon lower bound [125] which is useful for difference distortion measures. The second main approach was to develop a numerical algorithm, the Blahut-Arimoto (BA) algorithm [17], [3], to evaluate RD functions. The power of the second approach is in that the function can be approximated arbitrarily closely at the cost of complexity. The disadvantage is that the complexity may become overwhelming, particularly in the case of continuous alphabets, and even more so for continuous vector alphabets where the complexity could grow exponentially with the dimension. Another disadvantage is, of course, that no closed-form expression is obtained for the function, even if a simple one happens to exist.

We shall restrict our attention here to continuous alphabet sources. The RD curve is obtained by minimizing the mutual information subject to an average distortion constraint. Formally stated, given a continuous source alphabet $\mathcal{X}$, random variable X with a probability measure given by the density $p(x)$, and a reproduction alphabet $\mathcal{Y}$, the problem is of that optimizing the mutual information

$$I(X;Y) = H(Y) - H(Y|X) = \int dx\, p(x) \int dy\, p(y|x) \log[\frac{p(y|x)}{\int dx\, p(x) p(y|x)}], \tag{91}$$

over the random encoders $p(y|x)$, subject to

$$\int dx\, p(x) \int dy\, p(y|x) d(x,y) \leq D, \tag{92}$$

where $d(\cdot,\cdot)$ is the distortion measure. By replacing the above minimization with parametric variational equations (see [44], [12], [18], or [54]), this problem, can be reformulated as a problem of optimization over the reproduction density $p(y)$. The functional to be minimized is

$$F[p(y)] = -\frac{1}{\beta} \int dx\, p(x) \log \int dy\, p(y) e^{-\beta d(x,y)}, \tag{93}$$

where β is a positive parameter that is varied to compute different points on the RD curve. *But this criterion is easily recognizable as a continuous version of the free-energy (25) we have developed in our (mass constrained) DA derivation!* Much intuition can be obtained from this realization. In particular:

- The computation of the RD function is equivalent to a process of annealing.
- The effective reproduction alphabet is almost always discrete and grows via a sequence of phase transitions.
- An efficient DA method can be used to compute the RD curve.

Thus, the result is of importance to both rate-distortion theory and the basic DA approach itself. A detailed treatment of the relations between RD theory and DA is given in [115]. Here we only give a superficial outline.

To see more clearly the connection with the DA derivation we note that the objective of the optimization in (93) is to determine a probability measure on the reproduction space $\mathcal{Y}$. We may consider an alternative "mapping" approach which, instead of searching for the optimal $p(y)$ directly, searches for the optimal mapping $y : [0,1] \rightarrow \mathcal{Y}$, where to the unit interval we assign the Lebesgue measure denoted by μ. The equivalence of the approaches is ensured by the theory of general measures in topological spaces (see for example [121, ch. 15] or [53, ch. 2–3]). We thus have to minimize the functional

$$F(y) = -\frac{1}{\beta} \int dx\, p(x) \log \int_{[0,1]} d\mu(u)\, e^{-\beta d[x,y(u)]} \tag{94}$$

over the mapping $y(u)$. We replace direct optimization of a density defined over the reproduction space with mapping of "codevectors" with their probabilities onto this space. This is exactly what the mass-constrained DA method does.

Recall that in the basic DA derivation, at high temperature (small β) no matter how many codevectors are "thrown in" they all converge to a single point and are viewed as one effective codevector. In the RD case we have a "continuum of codevectors", yet it is easy to see that they all collapse on the centroid of the source distribution. The reproduction support (or effective alphabet) is therefore of cardinality 1. Moreover, as we lower the temperature, the output remains discrete and its cardinality grows via a sequence of phase transitions exactly as we have seen in our treatment of DA for clustering. Using this approach, it was shown in [115] for the RD problem that the reproduction random variable, X, is continuous only if the Shannon lower bound (see e.g., [12]) is tight, which for the case of squared error distortion happens only when the source is Gaussian (or a sum of Gaussians). This is a surprising result in rate-distortion theory because the only analytically solved cases were exactly those where the Shannon lower bound is tight, which led to the implicit assumption that the optimal reproduction random variable is always continuous. (It should, however, be noted that the result was anticipated by Fix in an early paper [43] that, unfortunately, went relatively unnoticed.) From the DA viewpoint this is an obvious direct observation. It is summarized in a theorem [115]:

Theorem 2 *If the Shannon lower bound does not hold with equality, then the support of the optimal reproduction random variable consists of isolated singularities. Further, if this support is bounded (as is always the case in practice) then Y is discrete and finite.*

For the practical problem of RD computation, we see two approaches, namely BA and DA, whose equivalence follows from the Borel isomorphism theorem. However, these approaches are substantially different in their computational complexity and performance if we need to discretize (as we always do). When using BA, discretization means defining a grid $\{y_i\}$ on the output space $\mathcal{Y}$. In DA we "discretize the unit interval" (i.e., replace it by a set of indices) and induce an adaptive grid on $\mathcal{Y}$ by our mapping. Instead of a fixed grid in the output space, DA effectively optimizes a codebook of codevectors with their respective masses. This difference between the approaches is crucial because the output distributions are almost always discrete and finite. This gives DA the theoretical capability of producing *exact solutions* at finite model complexity, while BA can only approach exact solutions at the limit of infinite resolution. The mass-constrained DA algorithm (given in Section 2) can be used to compute the RD curve [115].

It is a known result from rate-distortion theory that the parameter $\beta = 1/T$ as defined above is simply related to the slope of the (convex) rate distortion curve:

$$\beta = -\frac{dR}{dD}. \tag{95}$$

This gives a new interpretation of the DA approach to clustering, and to the temperature parameter. The process of annealing is simply the process of RD computation which is started at maximum distortion and consists of "climbing" up the RD curve by optimally trading decrease in distortion for increase in rate. The position on the curve is determined by the temperature level which specifies the slope at this point. The process follows the RD curve as long as there are as many available codevectors as needed for the output cardinality. If the number of codevectors is a priori limited (as is the case in standard VQ design) then DA separates from, but attempts to stay as close as possible to, the RD curve after reaching the phase corresponding to the maximum allowed codebook size. Another important aspect of the annealing process, which is raised and demonstrated through the rate-distortion analysis, is the existence of two types of continuous phase transitions. One type is the cluster splitting transition which we have analyzed and computed its critical temperature. The other kind is that of "mass growing" where a cluster is born first with zero mass, and gradually gains in mass. The latter type of phase transition is more difficult to handle, and only preliminary results exist at this point. If, or when, we will be able to ensure that such phase transitions are always detected as well, we will have ensured that DA finds the global optimum. Note that, in practice, if a mass-growing phase transition is "missed" by the algorithm, this is often compensated by a corresponding splitting phase transition which occurs shortly afterwards, and optimality is regained.

6 Recent DA Extensions

In this section a couple of recent extensions of the DA approach are briefly mentioned.

One important extension is to a method for the design of multiple-description (MD) quantizers. The notion of multiple-description, or multiple-terminal communications, has its early roots in information theory and compression. If data is transmitted over several channels that may fail independently, then it is of interest to encode it in a way that allows "good" reconstruction when some of the channels fail. The fundamental problems are the derivation of performance bounds, and optimal system design. The former has been of substantial interest in information theory for the last two decades [98, 136, 137, 45, 14, 1, 142], and the latter has recently become "hot" due to its impact on communication via packet-switched networks where individual packets are viewed as "channels" that may fail independently. Vaishampayan [133] has recently derived an approach to scalar MD quantizer design, and has shown its asymptotic (high-resolution) optimality. (The approach was later extended to lattice quantizer design [123].) This work triggered much research effort in the area, particularly in application to image and video communications. From our perspective here we note that while [133] provides an asymptotically optimal quantizer, the optimality is only ensured for scalar quantizers and under various symmetry assumptions on the channels and the source statistics. The problem is a very hard optimization problem due to complicated index assignment considerations. Vaishampayan discovered a heuristic index assignment scheme that is indeed optimal in certain restricted cases, but is not extendible to the general vector quantizer case. It is therefore a very promising target for a DA approach. Preliminary work shows substantial gains of a recently developed DA-based design over GLA-type approaches [75]. Experiments also demonstrate that the problem of poor local optima is severe in this application due to the intricacies of index assignment.

Another extension was the application of DA to the problem of generalized vector quantization (GVQ). GVQ extends the VQ problem to handle joint quantization and estimation [49]. The GVQ observes random vector X and provides a quantized value for a statistically related, but unobservable, random vector Y. Of course, the special case $X = Y$ is the regular VQ problem. One typical application is in noisy source coding (often referred to as remote source coding in the information theory literature). Another application is concerned with the need to combine VQ with interpolation (e.g., when the vectors were down-sampled for complexity or other reasons). Preliminary results showing substantial gains due to the use of DA are given in [105].

7 Summary

Deterministic annealing is a useful approach to clustering and related optimization problems. The approach is strongly motivated by analogies to statistical physics, but is formally derived within information theory and probability theory. It enables escaping many poor local optima that plague traditional techniques, without the slow schedules typically required by stochastic methods. The solutions obtained by DA are totally independent of the choice of initial configuration. The main objectives of this paper were: to derive DA from basic principles; to emphasize its generality; to illustrate its wide applicability to problems of supervised and unsupervised learning; and to demonstrate its ability to provide substantial gains over existing methods that were specifically tailored to the particular problem.

Most problems addressed were concerned with data assignment, via supervised or unsupervised learning, and the most basic of all is the problem of clustering. A probabilistic framework was constructed by randomization of the partition, which is based on the principle of maximum entropy at a given level of distortion, or equivalently, minimum expected distortion at a given level of entropy. The Lagrangian was shown to be the Helmholtz free energy in the physical analogy, and the Lagrange multiplier T is the temperature. The minimization of the free energy determines isothermal equilibrium and yields the solution for the given temperature. The resulting association probabilities are Gibbs distributions parameterized by T. Within this probabilistic framework, annealing was introduced by controlling the Lagrange multiplier T. This annealing is interpreted as gradually trading entropy of the associations for reduction in distortion. Phase transitions were identified in the process, which are, in fact, cluster splits. A sequence of phase transitions produces a hierarchy of fuzzy-clustering solutions. Critical temperatures T_c for the onset of phase transitions were derived. At the limit of zero temperature, DA converges to a known descent method, the generalized Lloyd algorithm or K-means, which in standard implementations is arbitrarily or heuristically initialized. Consistent substantial performance gains were obtained.

The method was first extended to a variety of related unsupervised learning problems by incorporating constraints on the clustering solutions. In particular, DA methods were derived for noisy-channel VQ, entropy constrained VQ, and structurally-constrained VQ design. Additional constraints may be applied to address graph-theoretic problems.

A highly significant extension is to supervised learning problems. The DA approach was re-derived while allowing the imposition of structures on the partition, and while optimizing the ultimate optimization cost. This extension enables the DA approach to optimize complicated discrete costs for a large variety of popular structures. The method's performance was demonstrated on the problem of classification with the vector quantizer, radial basis functions, and multilayer perceptron structures, and on the problem of regression with the VQ, hierarchical mixture of experts, and normalized radial basis functions. For each one of the examples, the DA approach significantly outperformed standard design methods that were developed for the specific structure.

A major application of the DA approach is in speech recognition. Here the structure to optimize consists of hidden Markov model units which compete to determine the class. The complexity of the recognizer poses a considerable challenge for optimization techniques. A DA approach was derived for this problem and its potential gains over existing techniques were demonstrated in the simpler context of discrete HMM systems. Extension to continuous-density HMM has been derived and preliminarily tested, and further extensions to tied-mixture HMM recognizers are under investigation.

The relations to information theory, and in particular to rate-distortion theory were discussed. It was shown that the DA method for clustering is equivalent to the computation of the RD function. This observation led to contributions to rate-distortion theory itself, and to further insights into the workings of DA.

A couple of extensions, which are currently under investigation, were briefly introduced. One extension is to the design of multiple description vector quantizers, a problem strongly motivated by packet-switched networks such as the Internet. Another extension is concerned with the problem of generalized vector quantizer design.

References

[1] R. Ahlswede. The rate-distortion region for multiple descriptions without excess rate. *IEEE Trans. Inform. Theory*, IT-31:721–726, Nov. 1985.

[2] H. Akaike. A new look at statistical model identification. *IEEE Trans. Automatic Control*, 19:713–723, Dec. 1974.

[3] S. Arimoto. An algorithm for calculating the capacity of an arbitrary discrete memoryless channel. *IEEE Trans. Inform. Theory*, IT-18:14–20, Jan. 1972.

[4] E. Ayanoglu and R. M. Gray. The design of joint source and channel trellis waveform coders. *IEEE Trans. Inform. Theory*, IT-33:855–865, Nov. 1987.

[5] L. R. Bahl, P. F. Brown, P. V. DeSouza, and R. L. Mercer. Maximum mutual information estimation of hidden Markov model parameters. In *Proc. IEEE Conf. Acoustics Speech Signal Proc.*, pages 49–52, 1986.

[6] L. R. Bahl, P. F. Brown, P. V. DeSouza, and R. L. Mercer. A new algorithm for the estimation of hidden Markov model parameters. In *Proc. IEEE Conf. Acoustics Speech Signal Proc.*, pages 493–496, 1988.

[7] J. K. Baker. Stochastic modeling for automatic speech recognition. In D. R. Reddy, editor, *Speech Recognition*, pages 521–542. Academic Press, New York, 1975.

[8] G. Ball and D. Hall. A clustering technique for summarizing multivariate data. *Behavioral Science*, 12:153–155, Mar. 1967.

[9] L. E. Baum and T. Petrie. Statistical inference for probabilistic functions of finite state Markov chains. *Annals of Mathematics and Statistics*, 37:1554–1563, 1966.

[10] L. E. Baum, T. Petrie, G. Soules, and N. Weiss. A maximization technique occurring in the statistical analysis of probabilistic functions of Markov chains. *Annals of Mathematics and Statistics*, 41:164–171, 1970.

[11] G. Beni and X. Liu. A least biased fuzzy clustering method. *IEEE Trans. Pattern Analysis and Machine Intelligence*, 16:954–960, Sept. 1994.

[12] T. Berger. *Rate Distortion Theory.* Prentice-Hall, Englewood Cliffs, NJ, 1971.

[13] T. Berger. Minimum entropy quantizers and permutation codes. *IEEE Trans. Inform. Theory*, IT-28:149–157, Mar. 1982.

[14] T. Berger and Z. Zhang. Minimum breakdown degradation in binary source encoding. *IEEE Trans. Inform. Theory*, IT-29:807–814, Nov. 1983.

[15] J. C. Bezdek. A convergence theorem for the fuzzy ISODATA clustering algorithms. *IEEE Trans. Patt. Anal. Machine Intell.*, PAMI-2:1–8, Jan. 1980.

[16] G. L. Bilbro, W. E. Snyder, S. J. Garnier, and J. W. Gault. Mean field annealing: A formalism for constructing GNC-like algorithms. *IEEE Trans. on Neural Networks*, 3:131–138, Jan. 1992.

[17] R. E. Blahut. Computation of channel capacity and rate-distortion functions. *IEEE Trans. Inform. Theory*, IT-18:460–473, July 1972.

[18] R. E. Blahut. *Principles and Practice of Information Theory.* Addison-Wesley, Reading, MA, 1987.

[19] H. Bourlard, Y. Konig, N. Morgan, and C. Ris. A new training algorithm for hybrid HMM/ANN speech recognition systems. In *Proc. EUSIPCO*, pages 101–104, 1996.

[20] L. Breiman, J.H. Friedman, R.A. Olshen, and C.J. Stone. *Classification and Regression Trees.* The Wadsworth Statistics/Probability Series, Belmont,CA., 1980.

[21] J. Buhmann and H. Kuhnel. Vector quantization with complexity costs. *IEEE Trans. Inform. Theory*, 39:1133–1145, July 1993.

[22] A. Buzo, A. H. Gray Jr., R. M. Gray, and J. D. Markel. Speech coding based on vector quantization. *IEEE Trans. Acoust., Speech, Signal Processing*, 28:562–574, Oct. 1980.

[23] P.-C. Chang and R. M. Gray. Gradient algorithms for designing predictive vector quantizers. *IEEE Trans. Acoust., Speech, Signal Processing*, ASSP-34:679–690, Aug. 1986.

[24] P. C. Chang and B.-H. Juang. Discriminative training of dynamic programming based speech recognizers. *IEEE Trans. Speech Audio Proc.*, 1:135–143, Apr. 1993.

[25] V. Cherkassky, Y. Lee, and H. Lari-Najafi. Self-organizing network for regression: efficient implementation and comparative evaluation. In *Proc. International Joint Conference on Neural Networks*, volume 1, pages 79–84, 1991.

[26] P. A. Chou. Optimal partitioning for classification and regression trees. *IEEE Trans. Patt. Anal. Machine Intell.*, 13:340–354, Apr. 1991.

[27] P. A. Chou, T. Lookabaugh, and R. M. Gray. Entropy-constrained vector quantization. *IEEE Trans. Acoust., Speech, Signal Processing*, ASSP-37:31–42, Jan. 1989.

[28] W. Chou, B.-H. Juang, and C.-H. Lee. Segmental GPD training of HMM based speech recognizer. In *Proc. IEEE Conf. Acoustics Speech Signal Proc.*, pages 473–476, 1992.

[29] R. Cole, Y. Muthuswamy, and M. Fanty. The ISOLET spoken letter database. Technical Report 90-004, Oregon Graduate Institute, 1990.

[30] R. Cole, R. Stern, and M. Lasry. Performing fine phonetic distinctions: templates vs. frames. In *Proc. IEEE Conf. Acoustics Speech Signal Proc.*, pages 558–561, 1982.

[31] A.P. Dempster, N.M. Laird, and D.B. Rubin. Maximum-likelihood from incomplete data via the EM algorithm. *Journal of the Royal Stat. Society*, 39(1):1–38, 1977.

[32] R. D. Dony and S. Haykin. Neural network approaches to image compression. *Proceedings of the IEEE*, 83:288–303, Feb. 1995.

[33] R. O. Duda and P. E. Hart. *Pattern Classification and Scene Analysis.* Wiley-Interscience, New York, NY, 1974.

[34] J. G. Dunham and R. M. Gray. Joint source and channel trellis encoding. *IEEE Trans. Inform. Theory*, IT-27:516–519, July 1981.

[35] J. C. Dunn. A fuzzy relative of the ISODATA process and its use in detecting compact well-separated clusters. *J. Cybern.*, 3(3):32–57, 1974.

[36] R. Durbin, R. Szeliski, and A. Yuille. An analysis of the elastic net approach to the travelling salesman problem. *Neural Computation*, 1(3):348–358, 1989.

[37] R. Durbin and D. Willshaw. An analogue approach to the travelling salesman problem using an elastic net method. *Nature*, 326:689–691, 1987.

[38] M. Fanty and R. Cole. Spoken letter recognition. In *Proc. Neural Inform. Proc. Syst. Conf.*, pages 45–51, 1990.

[39] N. Farvardin. A study of vector quantization for noisy channels. *IEEE Trans. Inform. Theory*, 36:799–809, July 1990.

[40] N. Farvardin and J.W. Modestino. Optimum quantizer performance for a class of non-gaussian memoryless sources. *IEEE Trans. Inform. Theory*, IT-30:485–497, May 1984.

[41] N. Farvardin and V. Vaishampayan. On the performance and complexity of channel-optimized vector quantizers. *IEEE Trans. Inform. Theory*, 37:155–160, Jan. 1991.

[42] J. D. Ferguson. Hidden Markov analysis: an introduction. In *Hidden Markov Models for Speech.* Institute for Defense Analysis, Princeton, NJ, 1980.

[43] S. L. Fix. Rate distortion functions for squared error distortion measures. In *Proc. 16'th Annual Allerton Conf. on Commun. Contr. and Comput.*, pages 704–711, Oct. 1978.

[44] R. G. Gallager. *Information Theory and Reliable Communication.* John Wiley & Sons, New York, NY, 1968.

[45] A. A. El Gamal and T. M. Cover. Achievable rates for multiple descriptions. *IEEE Trans. Inform. Theory*, IT-28:851–857, Nov. 1982.

[46] D. Geiger and F. Girosi. Parallel and deterministic algorithms from MRFs: Surface reconstruction. *IEEE Trans. Patt. Anal. Machine Intell.*, 13:401–412, May 1991.

[47] C. Gelin-Huet, K. Rose, and A. Rao. The deterministic annealing approach for discriminative continuous HMM design. In *Proc. Eurospeech*, volume 7, pages 2717–2720, 1999.

[48] S. Geman and D. Geman. Stochastic relaxation, Gibbs distribution, and the Bayesian restoration of images. *IEEE Trans. Patt. Anal. Machine Intell.*, 6:721–741, Nov. 1984.

[49] A. Gersho. Optimal nonlinear interpolative vector quantizers. *IEEE Trans. Commun.*, COM-38:1285–1287, Sept. 1990.

[50] A. Gersho and R. M. Gray. *Vector Quantization and Signal Compression.* Kluwer Academic Publishers, Boston, MA, 1992.

[51] S. Gold and A. Rangarajan. A graduated assignment algorithm for graph matching. *IEEE Trans. Pattern Anal. Machine Intell.*, 18:377–388, Apr. 1996.

[52] T. Graepel, M. Burger, and K. Obermayer. Phase transitions in stochastic self-organizing maps. *Phys. Rev. E*, 56(4):3876–3890, 1997.

[53] R. M. Gray. *Probability, Random Processes, and Ergodic Properties.* Springer-Verlag, New York, NY, 1988.

[54] R. M. Gray. *Source Coding Theory.* Kluwer Academic Press, Boston, MA, 1990.

[55] R. M. Gray and E. D. Karnin. Multiple local minima in vector quantizers. *IEEE Trans. Inform. Theory*, IT-28:256–261, Mar. 1982.

[56] B. Hajek. A tutorial survey of theory and applications of simulated annealing. In *24th IEEE Conf. Decision and Control*, pages 755–760, 1985.

[57] D. Harrison and D. L. Rubinfeld. Hedonic prices and the demand for clean air. *J. Environ. Economics and Management*, 5:81–102, 1978.

[58] J. A. Hartigan. *Clustering Algorithms.* John Wiley, New York, 1975.

[59] S. Haykin. *Neural Networks: A Comprehensive Foundation.* Prentice-Hall, New Jersey, 2nd edition, 1998.

[60] G. E. Hinton and M. Revow. Using pairs of data points to define splits for decision trees. In *Proc. Neural Inform. Proc. Syst. Conf.*, volume 8, pages 507–513, 1995.

[61] T. Hofmann and J. M. Buhmann. Pairwise data clustering by deterministic annealing. *IEEE Transactions on Patt. Analysis and Machine Intell.*, 19:1–14, Jan. 1997.

[62] Y. H. Hu, S. Palreddy, and W. J. Tompkins. Customized ECG beat classifier using mixture of experts. In *Proc. IEEE Workshop on Neural Networks for Signal Processing*, pages 459–464, 1995.

[63] J.-N. Hwang, S.-R. Lay, M. Maechler, R. D. Martin, and J. Schimert. Regression modeling in back-propagation and projection pursuit learning. *IEEE Transactions on Neural Networks*, 5:342–353, May 1994.

[64] R. A. Jacobs and M. I. Jordan. Learning piecewise control strategies in a modular neural network architecture. *IEEE Transactions on Systems, Man and Cybernetics*, 23:337–345, Mar.-Apr. 1993.

[65] R. A. Jacobs, M. I. Jordan, S. J. Nowlan, and G. E. Hinton. Adaptive mixtures of local experts. *Neural Computation*, 3(1):79–87, 1991.

[66] E. T. Jaynes. Information theory and statistical mechanics. In R. D. Rosenkrantz, editor, *Papers on Probability, Statistics and Statistical Physics.* Kluwer Academic Publishers, Dordrecht, The Netherlands, 1989. (Reprints of the original 1957 papers in Physical Review).

[67] F. Jelinek. Continuous speech recognition by statistical methods. *Proc. IEEE*, 64:532–556, Apr. 1972.

[68] M. I. Jordan and R. A. Jacobs. Hierarchical mixtures of experts and the EM algorithm. *Neural Computation*, 6(2):181–214, 1994.

[69] B.-H. Juang, W. Chou, and C.-H. Lee. Minimum classification error rate methods for speech recognition. *IEEE Trans. Speech Audio Proc.*, 5:257–265, Mar. 1997.

[70] J.-C. Junqua. SmarTspelL: a multipass recognition system for name retrieval over the telephone. *IEEE Trans. Speech Audio Proc.*, 5:173–182, Mar. 1997.

[71] S. Katagiri, B.-H. Juang, and C.-H. Lee. Pattern recognition using a family of design algorithms based upon the generalized probabilistic descent method. *Proc. IEEE*, 86:2345–2373, Nov. 1998.

[72] S. Kirkpatrick, C. D. Gelatt, and M. P. Vecchi. Optimization by simulated annealing. *Science*, 220:671–680, 1983.

[73] M. Kloppenburg and P. Tavan. Deterministic annealing for density estimation by multivariate normal mixtures. *Phys. Rev. E*, 55(3):2089–2092, 1997.

[74] T. Kohonen, G. Barna, and R. Chrisley. Statistical pattern recognition with neural networks: benchmarking studies. In *Proc. IEEE International Conference on Neural Networks*, volume I, pages 61–68, 1988.

[75] P. Koulgi, S. L. Regunathan, and K. Rose. Multiple description quantization by deterministic annealing. In *Proc. IEEE Intl. Symposium on Information Theory*, p. 177, 2000.

[76] H. Kumazawa, M. Kasahara, and T. Namekawa. A construction of vector quantizers for noisy channels. *Electron. Commun. Japan*, 67(4):39–47, 1984.

[77] A. Kurtenbach and P. Wintz. Quantizing for noisy channels. *IEEE Trans. Communications*, COM-17:291–302, Apr. 1969.

[78] Y. Linde, A. Buzo, and R. M. Gray. An algorithm for vector quantizer design. *IEEE Trans. Commun.*, COM-28:84–95, Jan. 1980.

[79] S. P. Lloyd. Least squares quantization in PCM. *IEEE Trans. Inform. Theory*, IT-28:129–137, Mar. 1982. (Reprint of the 1957 paper).

[80] P. C. Loisou and A. S. Spanias. High performance alphabet recognition. *IEEE Trans. Speech Audio Proc.*, 4:430–445, Nov. 1996.

[81] S. P. Luttrell. Hierarchical vector quantization (image compression). *Proc. Inst. Elect. Eng. I (Commun. Speech and Vision)*, 136(6):405–413, 1989.

[82] S. P. Luttrell. Derivation of a class of training algorithms. *IEEE Trans. Neural Networks*, 1:229–232, June 1990.

[83] J. MacQueen. Some methods for classification and analysis of multivariate observations. In *Proc. Fifth Berkeley Symposium on Math. Stat. and Prob.*, pages 281–297, 1967.

[84] T. M. Martinetz, S.G. Berkovich, and K. J. Schulten. 'Neural-gas' network for vector quantization and its application to time-series prediction. *IEEE Trans. on Neural Networks*, 4:558–569, July 1993.

[85] J. Max. Quantizing for minimum distortion. *IRE Trans. Inform. Theory*, IT-6:7–12, Mar. 1960.

[86] E. McDermott and S. Katagiri. Prototype-based minimum classification error / generalized probabilistic descent training for various speech units. *Computer Speech and Language*, 8(4):351–368, 1994.

[87] G.C. McDonald and R.C. Schwing. Instabilities of regression estimates relating air pollution to mortality. *Technometrics*, 15:463–482, 1973.

[88] G. J. McLachlan and K. E. Basford. *Mixture models: Inference and Application to Clustering.* Marcel Dekker, 1988.

[89] N. Metropolis, A. W. Rosenbluth, M. N. Rosenbluth, A. H. Teller, and E. Teller. Equations of state calculations by fast computing machines. *J. Chem. Phys.*, 21(6):1087–1091, 1953.

[90] D. Miller. *An Information-theoretic Framework for Optimization with Applications in Source Coding and Pattern Recognition.* PhD thesis, University of California, Santa Barbara, 1995.

[91] D. Miller, A. V. Rao, K. Rose, and A. Gersho. A global optimization technique for statistical classifier design. *IEEE Trans. Signal Processing*, 44:3108–3122, Dec. 1996.

[92] D. Miller and K. Rose. An improved sequential search multistage vector quantizer. In *Proc. IEEE Data Compression Conf.*, pages 12–21, Mar. 1993.

[93] D. Miller and K. Rose. Combined source-channel vector quantization using deterministic annealing. *IEEE Trans. Commun.*, 42:347–356, Feb. 1994.

[94] D. Miller and K. Rose. Hierarchical, unsupervised learning with growing via phase transitions. *Neural Computation*, 8(2):425–450, 1996.

[95] D. Miller, K. Rose, and P. A. Chou. Deterministic annealing for trellis quantizer and HMM design using Baum-Welch re-estimation. In *Proc. IEEE Intl. Conf. Acoustics Speech Signal Proc.*, volume V, pages 261–264, 1994.

[96] J. Moody and C. J. Darken. Fast learning in locally-tuned processing units. *Neural Computation*, 1(2):281–294, 1989.

[97] M. T. Musavi, W. Ahmed, K. H. Chan, K. B. Faris, and D. M. Hummels. On the training of radial basis function classifiers. *Neural Networks*, 5(4):595–604, 1992.

[98] L. Ozarow. On a source coding problem with two channels and three receivers. *Bell Syst. Tech. Journal*, 59:1909–1921, Dec. 1980.

[99] J. W. Picone. Signal modeling techniques in speech recognition. *Proc. IEEE*, 81:1215–1247, 1993.

[100] L. R. Rabiner. A tutorial on hidden Markov models and selected applications in speech recognition. *Proc. IEEE*, 77:257–285, Feb. 1989.

[101] L. R. Rabiner and B.-H. Juang. An introduction to hidden Markov models. *IEEE ASSP Magazine*, pages 4–16, Jan. 1986.

[102] L. R. Rabiner and B.-H. Juang. *Fundamentals of Speech Recognition.* Prentice Hall, New Jersey, 1993.

[103] L. R. Rabiner, S. Levinson, A. Rosenberg, and J. Wilpon. Speaker independent recognition of isolated words using clustering techniques. *IEEE Trans. Acoust. Speech Signal Proc.*, ASSP-27:336–349, Aug. 1979.

[104] L. R. Rabiner and J. Wilpon. Some performance benchmarks for isolated word speech recognition systems. *Computer Speech Language*, 2:343–357, 1987.

[105] A. Rao, D. Miller, K. Rose, and A. Gersho. A generalized VQ method for combined compression and estimation. In *Proc. IEEE Intl. Conf. Acoustics Speech Signal Proc.*, volume IV, pages 2032–2035, 1996.

[106] A. Rao, D. Miller, K. Rose, and A. Gersho. A deterministic annealing approach for parsimonious design of piecewise regression models. *IEEE Trans. Pattern Analysis and Machine Intelligence*, 21:159–173, Feb. 1999.

[107] A. Rao and K. Rose. Deterministically annealed design of hidden Markov model speech recognizers. To appear in *IEEE Trans. Speech Audio Proc.*

[108] A. Rao, K. Rose, and A. Gersho. Design of robust HMM speech recognizer using deterministic annealing. In *Proc. IEEE Workshop on Automatic Speech Recognition and Understanding*, pages 466–473, Dec. 1997.

[109] A. V. Rao. *Design of Pattern Recognition Systems Using Deterministic Annealing: Applications in Speech Recognition, Regression, and Data Compression.* PhD thesis, University of California, Santa Barbara, 1998.

[110] A. V. Rao, D. Miller, K. Rose, and A. Gersho. Mixture of experts regression modeling by deterministic annealing. *IEEE Trans. Signal Processing*, 45:2811–2820, Nov. 1997.

[111] B. D. Ripley. Neural networks and related methods for classification. *Journal of the Royal Stat. Soc., Ser. B*, 56(3):409–456, 1994.

[112] E. A. Riskin and R. M. Gray. A greedy tree growing algorithm for the design of variable rate vector quantizers. *IEEE Trans. Signal Processing*, 39:2500–2507, Nov. 1991.

[113] J. Rissanen. Stochastic complexity and modeling. *Annals of Statistics*, 14:1080–1100, 1986.

[114] K. Rose. *Deterministic Annealing, Clustering, and Optimization.* PhD thesis, California Institute of Technology, 1991.

[115] K. Rose. A mapping approach to rate-distortion computation and analysis. *IEEE Trans. Inform. Theory*, 40:1939–1952, Nov. 1994.

[116] K. Rose, E. Gurewitz, and G. C. Fox. A deterministic annealing approach to clustering. *Pattern Recognition Letters*, 11(9):589–594, 1990.

[117] K. Rose, E. Gurewitz, and G. C. Fox. Statistical mechanics and phase transitions in clustering. *Physical Review Letters*, 65(8):945–948, 1990.

[118] K. Rose, E. Gurewitz, and G. C. Fox. Vector quantization by deterministic annealing. *IEEE Trans. Inform. Theory*, 38:1249–1257, July 1992.

[119] K. Rose, E. Gurewitz, and G. C. Fox. Constrained clustering as an optimization method. *IEEE Trans. Patt. Anal. Machine Intell.*, 15:785–794, Aug. 1993.

[120] K. Rose and D. Miller. Constrained clustering for data assignment problems with examples of module placement. In *Proc. IEEE Intl. Symp. Circuits and Syst.*, pages 1937–1940, May 1992.

[121] H. L. Royden. *Real Analysis.* Macmillan Publishing Co., New York, NY, 3rd edition, 1988.

[122] D. E. Rumelhart, G. E. Hinton, and R. J. Williams. *Parallel Distributed Processing.* MIT Press, Cambridge, MA, 1986.

[123] S. D. Servetto, V. A. Vaishampayan, and N. J. A. Sloane. Multiple description lattice vector quantization. In *Proc. IEEE Data Compression Conf.*, pages 13–22, 1999.

[124] C. E. Shannon. A mathematical theory of communication. *Bell Systems Technical Journal*, 27:379–423, 623–656, 1948.

[125] C. E. Shannon. Coding theorems for a discrete source with a fidelity criterion. *IRE National Convention Record*, Part 4:142–163, 1959.

[126] J. E. Shore and R. W. Johnson. Axiomatic derivation of the principle of maximum entropy and the principle of minimum cross-entropy. *IEEE Trans. Inform. Theory*, IT-26:26–37, Jan. 1980.

[127] P. D. Simic. Statistical mechanics as the underlying theory of elastic and neural optimization. *Network*, 1(1):89–103, 1990.

[128] P. D. Simic. Constrained nets for graph matching and other quadratic assignment problems. *Neural Computation*, 3(2):268–281, 1991.

[129] H. Szu and R. Hartley. Nonconvex optimization by fast simulated annealing. *Proc. of the IEEE*, 75:1538–1540, Nov. 1987.

[130] L. Tarassenko and S. Roberts. Supervised and unsupervised learning in radial basis function classifiers. *Proc. Inst. Elect. Eng. - Vis. Image Sig. Proc.*, 141(4):210–216, 1994.

[131] D. M. Titterington, A. F. M. Smith, and U. E. Makov. *Analysis of Finite Mixture distributions.* John Wiley, 1985.

[132] N. Ueda and R. Nakano. Deterministic annealing variant of the EM algorithm. In *Proc. Neural Inform. Proc. Syst. Conf.*, pages 545–552, 1994.

[133] V. A. Vaishampayan. Design of multiple description scalar quantizers. *IEEE Trans. Inform. Theory*, 39:821–834, May 1993.

[134] S. R. Waterhouse and A. J. Robinson. Non-linear prediction of acoustic vectors using hierarchical mixtures of experts. In *Proc. Neural Inform. Proc. Syst. Conf.*, pages 835–842, 1994.

[135] A. S. Weigend, M. Mangeas, and A. N. Srivastava. Nonlinear gated experts for time series: discovering regimes and avoiding overfitting. *International Journal of Neural Systems*, 6(4):373–399, 1995.

[136] H. Witsenhausen. On source networks with minimal breakdown degradation. *Bell Syst. Tech. Journal*, 59:1083–1087, July-Aug. 1980.

[137] J. Wolf, A. Wyner, and J. Ziv. Source coding for multiple descriptions. *Bell Syst. Tech. Journal*, 59:1417–1426, Oct. 1980.

[138] Y.-F. Wong. Clustering data by melting. *Neural Computation*, 5(1):89–104, 1993.

[139] E. Yair, K. Zeger, and A. Gersho. Competitive learning and soft competition for vector quantizer design. *IEEE Trans. Signal Processing*, 40:294–309, 1992.

[140] A. L. Yuille. Generalized deformable models, statistical physics, and matching problems. *Neural Computation*, 2:1–24, 1990.

[141] K. Zeger and A. Gersho. Vector quantizer design for memoryless noisy channels. In *Proc. IEEE Intl. Conf. Commun.*, pages 1593–1597, 1988.

[142] Z. Zhang and T. Berger. New results in binary multiple descriptions. *IEEE Trans. Inform. Theory*, IT-33:502–521, July 1987.

[143] J. Zhao and J. Shawe-Taylor. Neural network optimization for good generalization performance. In *Proc. Int. Conf. Artificial Neural Networks*, pages 561–564, 1994.

Chapter 6

Local Dynamic Modeling with Self-Organizing Maps and Applications to Nonlinear System Identification and Control

Jose C. Principe, Ludong Wang, Mark A. Motter

Computational NeuroEngineering Laboratory
University of Florida, Gainesville, FL32611

Abstract

The technique of local linear models is appealing for modeling complex time series due to the weak assumptions required and its intrinsic simplicity. Here, instead of deriving the local models from the data, we propose to estimate them directly from the weights of a self organizing map (SOM), which functions as a dynamic-preserving model of the dynamics. We introduce one modification to the Kohonen learning to ensure good representation of the dynamics and use weighted least squares to ensure continuity among the local models. The proposed scheme is tested using synthetic chaotic time series and real world data.

The practicality of the method is illustrated in the identification and control of the NASA Langley wind tunnel during aerodynamic tests of model aircrafts. Modeling the dynamics with a SOM leads to a predictive multiple model control strategy (PMMC). In test runs, a comparison of the new controller against the existing controller shows the superiority of our method.

1. Introduction

System identification and time series prediction are engineering embodiments of the old problem of function approximation. Each seeks to estimate in its way the parameters of the system that created the time series. The traditional approach is *statistical and based on the linear model* [3]: we assume that the time series is produced by a linear system excited by white Gaussian noise. The variability observed in the time series is assigned to the stochastic nature of the excitation, which can not be modeled. Therefore, the goal of linear modeling is limited to the estimation of the parameters of the model which will match the power spectrum of the time series.

More recently, another perspective is surfacing which is called dynamic modeling [21],

[16]. *The time series is considered the output of a deterministic, autonomous (i.e. without input) dynamical system.* The complexity of the time series is linked to the high order and nonlinear nature of the dynamical system and not to the exogenous random excitation, as in the linear case. In this approach the model system must either be *nonlinear or linear with time varying parameters,* because linear time invariant systems have trivial autonomous dynamics (either fixed points of limit cycles).

Perhaps the best example of such a perspective is *chaotic time series modeling* [7], [11], [21]. On the complexity scale, chaotic time series span the gap between periodic signals and random noise. Chaotic time series offer the ultimate difficulty for developing a model from a time series, because the signal time structure is time varying and highly complex with a 1/f spectrum. It is therefore the perfect environment to test new modeling approaches since a minor error in model parameters is amplified by the natural divergence of the trajectories in phase space. Its potential usefulness is also enormous because many real world phenomena are considered chaotic (the weather, sea clutter, lasers, heart beat, some types of brain waves, etc.). Time series produced by chaotic systems that were previously *considered "noise" may in fact have deterministic structure which can be modeled.* Hence, for improved performance, one can develop predictive models whose output can be subtracted from the time series instead of applying filtering techniques to attenuate the noise. Recent results by Haykin [30] show precisely the power of nonlinear dynamic modeling to improve the detectability of targets from sea clutter.

The tools to understand, develop and apply nonlinear dynamic models are very different from linear time series modeling and will be briefly reviewed. We summarize the basic approaches utilized in dynamic modeling, but the main goal of the paper is to develop an *innovative local linear dynamic modeling reminiscent of vector quantization*. Instead of representing the unknown manifolds with state-dependent predictive models which are independently derived from local information [10], here the global dynamics are approximated by a pre-set number of local linear models, which are *concurrently derived through competition* using Kohonen's self organizing map (SOM) [18]. The local linear models are *constructed from the weights of the SOM*, which facilitates training when compared to previous techniques. More significantly, our scheme provides a partial overlap among neighboring subregions with different local models, which effectively alleviates the discontinuity problem. In our method, the SOM is not only employed as a static representation of the signal, *but as a robust dynamic-preserving space*.We discuss first the implementation of our model using the traditional Kohonen learning, but then propose a modified learning strategy for the SOM utilizing the prediction error. Weighted least square estimation of local models is proposed to improve the modeling performance. We show that our method achieves dynamic modeling for both synthetic (Lorenz) and experimental (laser data) chaotic time series.

The applications (and implications) of dynamic modeling to intelligent control are recent and rich. Here the SOM local modeling framework is applied to develop a set point controller for the 16-Foot Transonic Tunnel at NASA Langley Research Center in Hampton, Virginia. The wind tunnel is used for aerodynamic tests of airplane models, and requires precise wind speeds in spite of the large changes in the angle-of-attack of the model which produce different load conditions in the wind tunnel. *The control problem is one of regulating the air speed around a Mach number set point, while the plant has a time varying load which produces very different dynamics.* The control action is very different from set point to set point and is also a function of the disturbance. The approach of modeling the tunnel dynamics with a set of local models seems particularly appropriate for this task, since the winner-take-all-operation of the SOM can switch very fast between very different local dynamic models, tracking the change in the tunnel dynamics.

The fundamental issue in this application is how to go from dynamic modeling (which assumes autonomous dynamics) to a control scheme (which requires a non-autonomous system with forcing inputs). We cluster the control inputs and use them to select one SOM from a set of local dynamic models which are trained with Mach number responses from the full operating range. The winner-take-all operation of the SOM local model creates a switching controller similar to the approach described by Narendra [25]. A set of predefined control inputs are applied to the local model selected by the winning PE with the goal of predicting the tunnel dynamics 50 steps in the future. The control input that best meets the set point specification is applied to the driving turbine.

The experimental system was successfully tested in an operational environment. The Mach number was controlled to within the research requirement at various transonic Mach numbers over a period of approximately 9 hours, while the aircraft model under test was positioned in various attitudes (3 runs of about 3 hours duration). We will present results of this test run and compare them with the operator and the existing controller.

2. Dynamic Modeling

Time series prediction seeks to quantify the system that created the time series. In the linear model framework, the unknown system transfer function is the inverse of the model transfer function which is arrived at through prediction [32]. When the linear methodology is extended to nonlinear systems, the relation between the model system and the unknown system has to be stated in different terms since the nonlinear model system can no longer be described by a transfer function. The work of Takens [45], Casdagli [7] and many others have shown that an equivalent methodology exists for nonlinear dynamic modeling. A Kth order autonomous dynamical system can be represented by a set of Kth ordinary differential equations

$$\frac{d}{dt}\mathbf{s}(t) = \Phi(\mathbf{s}(t)) \tag{1}$$

where $\mathbf{s}(t) = [s_1(t), s_2(t), \ldots, s_K(t)]$ is a vector of system states and Φ is called the vector field. Bold letters will represent vectors. The system state at any time can be specified in a K dimensional space. The vector field maps a manifold M to a tangent space T. If Φ is a nonlinear function of the system state the system is called nonlinear. Assume there is a closed form solution to Eq. 1 $\phi_t\colon M \to M$. For a given initial condition $\mathbf{s}_0$ the function $\phi_t(\mathbf{s}_0)$ represents a state-space trajectory of the system (the mapping ϕ is called the flow). If the vector field is continuously differentiable then the system flow and its inverse exist for any finite time. This implies that the trajectories of an autonomous system never intersect each other.

When working with experimental data produced by dynamical systems, we generally do not know the state equations (Eq.1) [1]. We are restricted to observe the outputs of the dynamical system. So a fundamental issue is what can be inferred about the dynamics from the observation of an output time series. Packard et al [31] showed experimentally and Takens [45] proved that a sampled observable *x(n)* of the dynamical system and its delayed versions

$$\mathbf{x}(n) = [x(n), x(n-\tau), \ldots, x(n-(N-1)\tau))] \tag{2}$$

can be used to create a trajectory in an Euclidean space of size N which preserves the dynamical invariants (correlation dimension and Lyapunov exponents) of the original dynamical system provided N is sufficiently large. In Eq. 2 τ is the normalized delay and it is irrelevant theoretically, but plays an important role when the time series is noisy and or of finite size. The dimension N of the space should be larger than $2D_e$ where D_e is the dimension of the attractor, i.e. the geometric object created by the trajectories after transients die out. This is a remarkable result, since it shows that the system's state information can be recovered from a sufficiently long observation of the output time series, and should be contrasted with the conventional approach of Kalman observables in control theory [46].

In more mathematical terms this statement means that there is a one-to-one smooth map Ψ with a smooth inverse from the Kth dimensional manifold M of the original system to the Euclidean reconstruction space R^N. Such mapping is called an embedding and the theorem is known as Takens embedding theorem [45].

According to Takens' embedding theorem, when $N > 2D_e$ a map $F: R^N \to R^N$ exists that transforms the current reconstructed state $\boldsymbol{x}(n)$ to the next state $\boldsymbol{x}(n+\tau)$, where τ is the normalized delay. For simplicity we will set τ=1, which means

$$\boldsymbol{x}(n+1) = F(\boldsymbol{x}(n)) \tag{3}$$

or

$$\begin{bmatrix} x(n+1) \\ \dots \\ x(n-N+2) \end{bmatrix} = F\left(\begin{bmatrix} x(n) \\ \dots \\ x(n-N+1) \end{bmatrix}\right)$$

Note that Eq. 3 specifies a multiple input, multiple output system *F* built from several (nonlinear) filters and a nonlinear predictor [36]. The predictive mapping is the center piece of modeling since once determined, *F* can be obtained from the predictive mapping by simple matrix operations. The predictive mapping $f: R^N \to R$ can be expressed as

$$x(n+1) = f(\boldsymbol{x}(n)) \tag{4}$$

Eq. 4 defines a deterministic nonlinear autoregressive (NAR) model of the signal. *The existence of this predictive model lays a theoretical basis for dynamic modeling in the sense that it opens the possibility to build from a vector time series a model to approximate the mapping* f. The result and steps in dynamical modeling are depicted in Figure 1.

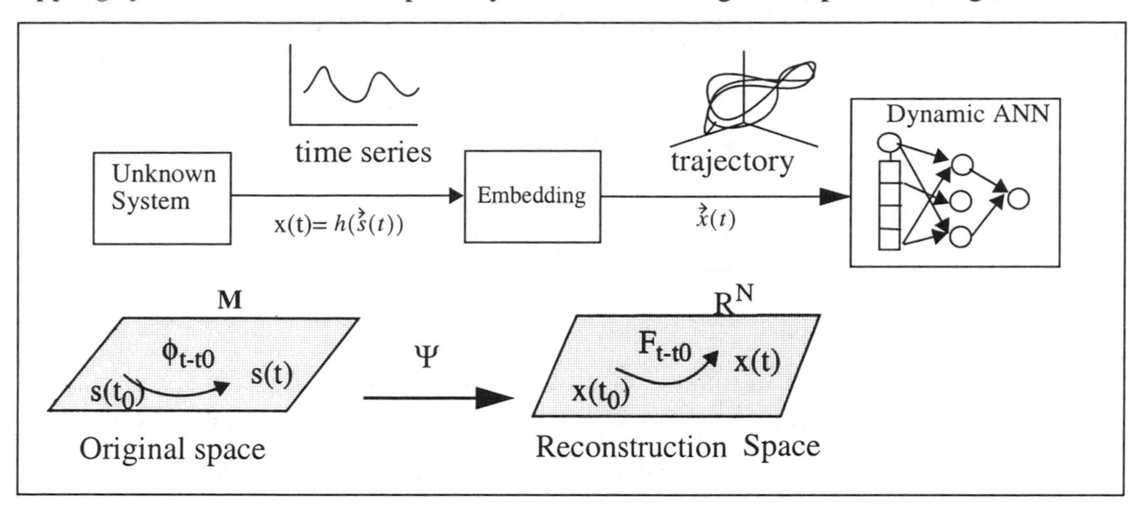

Figure 1. Nonlinear modeling steps and their mathematical translation.

Dynamic modeling implies a two step process [16]. The first step is to transform the observed time series into a trajectory in the reconstruction space by using one of the embedding techniques [52]. The most common is a time delay embedding which can practically be implemented with a delay line (also called a memory structure in neurocomputing) with a size specified by Takens' embedding theorem. The dimension D_e of the attractor can be estimated by the correlation dimension algorithm [13], but other methods exist [1]. The second step in dynamic modeling is to build the predictive model of Eq. 4 from the trajectory in reconstruction space [16]. These theoretical considerations imply that after the embedding step any adaptive nonlinear system trained as a predictor can identify the mapping f, and thus are of no help in deciding which is the best strategy to design the predictive model. So a different approach is required to design predictive models.

3. Global and Local Models

The task of nonlinear dynamic modeling is to approximate the reconstructed flow *F(x)* in Eq. 3 which evolves the trajectory in reconstruction space $\mathbf{x}(n) \rightarrow \mathbf{x}(n+1)$. Unlike the linear case, there is no algorithmic way to determine the functional form for *F(.)*, hence the framework of function approximation has been utilized [37]. The flow *F(x)* is approximated by a functional form $\tilde{F}(.)$ given by

$$\tilde{F}(\boldsymbol{x}(n), \boldsymbol{w}) = \sum_{i=1}^{N} w_i \varphi_i(\boldsymbol{x}(n)) \tag{5}$$

and the error

$$E = \sum_{x \in D} dis(\boldsymbol{x}(n+1) - \tilde{F}(\boldsymbol{x}(n), \boldsymbol{w})) \tag{6}$$

is minimized, where D is the domain of the approximation, and *dis(.)* is an appropriate metric (normally taken as the Euclidean distance). Estimated quantities will be denoted by the superscript ~. Due to the particular form of dynamic modeling (Eq. 4) the function approximation of Eq. 6 is equivalent to prediction in state space, i.e. we can define an instantaneous error

$$\varepsilon(n) = x(n+1) - \tilde{f}(\boldsymbol{x}(n), \boldsymbol{w}) \tag{7}$$

which is minimized with the appropriate error metric over the domain. There are two basic alternatives to proceed with the optimization: either a single map $\tilde{F}(.)$ is used to approximate all the points in the reconstruction space (*global modeling*), i.e. $D=R^N$; or the map $\tilde{F}(.)$ is decomposed in a family of maps $\tilde{F}_r(.)$ with $r=1,...R$, each fitting only the neighbors of the present point in reconstruction space (*local modeling*), i.e. D is a local manifold centered at the present point in reconstruction space. The overall predictive map is then a concatenation of local maps [10]

$$\tilde{F}(\boldsymbol{x}(n)) = \bigcup_{r=1,\ldots R} \tilde{F}_r(\boldsymbol{x}(n)) \tag{8}$$

3.1. Global Dynamic models

Global models based on polynomial fitting of the trajectory have been reported in the literature [12], [43]. Neural networks have also been successfully applied to this problem due to their universal mapping characteristics. After the pioneering work of Lapedes and Farber [21], multilayer perceptrons (MLPs) have been extensively utilized [34][17]. Following the work of Broomhead and Lowe [4], Radial Basis Functions (RBFs) have also become very popular as global models [16]. Note that these two topologies have rather different approximation characteristics since their basis functions are respectively global (sensitive to the entire space) and local (primarily sensitive to a local area). Nevertheless both systems are trained with data from the entire reconstruction space so they belong to the global model category.

The global models have been the most explored for nonlinear dynamic modeling, but will not be reviewed here. Please consult [52] for a broad review, and [36] for a performance comparison with local dynamic models.

3.2. Local Dynamic models

An alternate modeling methodology divides the state space in local regions and models individually the local dynamics in each region. This method is closely related to differential topology [14] and it is more general than the global approach in the sense that fewer statistical and geometric assumptions are required about the data [10]. Each local model $\tilde{F}_r(.)$ ($r = 1, \ldots, R$), can be potentially simpler, eventually linear, but the parameters will have to change across state space. The linear model in Eq. 5 becomes

$$\tilde{F}_r(\boldsymbol{x}(n), \boldsymbol{w}) = \boldsymbol{J}_r\boldsymbol{x}(n) + \boldsymbol{b}_r \tag{9}$$

where $\boldsymbol{J}$ is the Jacobian of F at $\boldsymbol{x}(n)$ and $\boldsymbol{b}$ is a bias vector. The assumption behind this model is that the state dynamics are locally smooth. Assuming the L2 norm for the cost function, the parameters ($\boldsymbol{J},\boldsymbol{b}$) can be estimated by least squares (or with the LMS algorithm [53]) from the data points within the local manifold, by simply substituting Eq. 9 into Eq.6

$$\min_{b,J} E \qquad E = \sum_{k=1}^{N_L} (\boldsymbol{x}(k+1) - \boldsymbol{J}_k\boldsymbol{x}(k) - \boldsymbol{b}_k)^2 \tag{10}$$

where N_L is the number of points in the neighborhood of point $\boldsymbol{x}(k)$, which for stable results must be larger than N, the size of the reconstruction space. Notice that this is a conceptually simple but time consuming operation. Points which are neighbors in reconstruction space may be very far away in the time series (Figure 2).

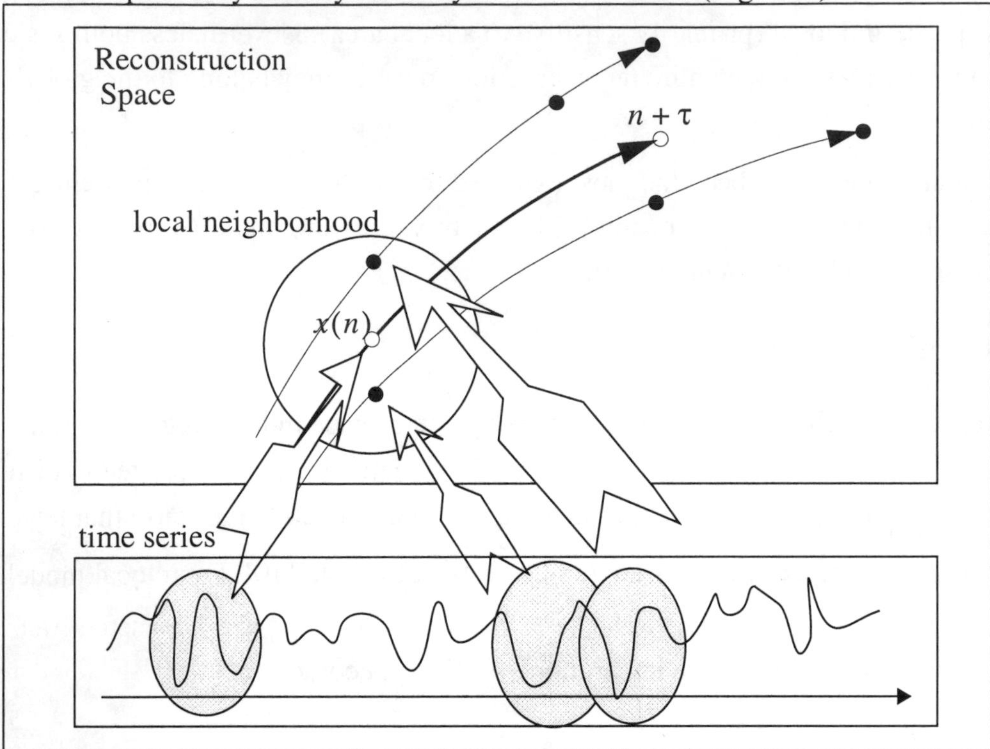

Figure 2. The relationship between time series samples and neighborhoods in reconstruction space.

So, the time series has to be either reconstructed separately for each local model to find the N_L nearest neighbors, or the full trajectory has to be kept in computer memory for a search of the neighbors. Then Eq. 10 has to be computed for each point $\boldsymbol{x}(n)$ in the local neighborhood. The global dynamic model is obtained by piecing together all the local models according to Eq.8.

Another difficulty is how to cover appropriately the trajectories in reconstruction space (placement of centers and neighborhood radius). Normally a uniform coverage of the space is utilized, but the size of the neighborhood is difficult to decide a priori. Applications of this technique have been reported for prediction of chaotic time series [7][11], noise removal of experimental time series [15][20], estimating the largest Lyapunov exponent [5][40], and control of chaotic dynamics [29][42]. For dynamic modeling which seeks to model the long term behavior of the dynamical system, the problem is how to guarantee smoothness at the boundaries among the local models. Crutchfield [10] has experimentally shown that dynamic modeling fails if this condition is not imposed. The extended Kalman filter also develops a local linear model of the trajectory [46], but utilizes a formulation based on a statistical data model which is the main stream of control theory.

3.3. State dependent prediction of nonlinear AR processes.

The approach of locally linear ARMA fitting has been discussed in the time series literature by Priestley [33]. A similar approach was proposed by Farmer and Sidorowich [11] for predicting chaotic time series. They concluded that there was little benefit of using higher order polynomials to fit the local dynamics. Tong also introduced the idea of threshold linear models [47]. Singer et al [44] presented the important idea of *local linear modeling as a state dependent AR scheme*. When the state vectors are constructed from the time series of a single variable by a delay embedding, the locally linear functions are effectively reduced to state dependent AR models, and thus a *codebook prediction scheme reminiscent of vector quantization is formed*. This insight is very valuable because it couples local linear modeling with a combination of vector quantization followed by adaptive linear models.

An Nth order nonlinear AR model can be written as [44]

$$y(n+1) = f(y(n), y(n-1), \ldots, y(n-N+1)) + u(n)$$

which accepts the state space representation

$$\boldsymbol{x}(n+1) = \begin{bmatrix} 0 & 1 & \ldots & 0 \\ \ldots & \ldots & \ldots & \ldots \\ 0 & 0 & \ldots & 1 \\ 0 & 0 & \ldots & 0 \end{bmatrix} \boldsymbol{x}(n) + \begin{bmatrix} 0 \\ \ldots \\ 0 \\ f(\boldsymbol{x}(n)) \end{bmatrix} + \begin{bmatrix} 0 \\ \ldots \\ 0 \\ 1 \end{bmatrix} \boldsymbol{u}(n) \tag{11}$$

$$y(n) = \begin{bmatrix} 0 & \ldots & 0 & 1 \end{bmatrix}^T \boldsymbol{x}(n)$$

where $\boldsymbol{x}(n) \in R^N$, $y(n) \in R$, $\boldsymbol{u}(n) \in R^q$ are respectively the vectors for the state, output and input, and $f: R^N \to R$. Due to the Markovian structure of NAR models of order N, $y(n+1)$ can be estimated from the most recent N values, i.e.

$$P(y(n+1) | y(i), 0 < i < n) = P(y(n+1) | \boldsymbol{x}(n)) \tag{12}$$

as shown by Eq. 11. The minimum mean square error estimate of $y(n+1)$ using the entire signal history is

$$y(n+1) = E[y(n+1)|\boldsymbol{x}(n)] = E[f(\boldsymbol{x}(n)) + u(n)|(\boldsymbol{x}(n))] = f(\boldsymbol{x}(n) \tag{13}$$

i.e. even though $f(\boldsymbol{x}(n))$ is not available, the system state dynamics can be observed in prediction given the recovered state vector $\boldsymbol{x}(n)$.

The time series history $y(n+1) = f(\boldsymbol{x}(n)) + u(n)$ represents a set of noisy state observations nonuniformly distributed in the reconstruction space (see Figure 2). So one option is to create a codebook of state vectors and signal values, and use the present state to lookup the next value of the time series. *The pairs (y(n+1),$\boldsymbol{x}(n)$) contain the information to help us observe local instances of the state, although tainted by noise.* If enough of these observations are available to cover all of the state space and the noise is reduced by local averaging, the nonlinear dynamical system can be identified. With this perspective we can immediately enunciate a procedure to build a dynamic model from a codebook of local linear models [44].

* Form a codebook of pairs *(y(k+1),$\boldsymbol{x}(k)$)* from the signal history;
* Select pairs *(y(k+1),$\boldsymbol{x}(k)$)* from the codebook near the current state $\boldsymbol{x}(n)$;
* Fit a local model $y(k+1) \approx \tilde{f}_r(\boldsymbol{x}(k))$ to the selected pairs in the codebook;
* Apply the local model to obtain $\tilde{x}(n+1) = \tilde{f}_r(\boldsymbol{x}(n))$.

We just have to address the practical aspects of implementing the codebook and estimating the local linear model. Assuming that the underlying state evolution is sufficiently smooth, then the mapping function $f(\boldsymbol{x})$ in the vicinity of $\boldsymbol{x}(n)$ can be approximated by the first few terms of its multidimensional Taylor series expansion,

$$\tilde{f}(\boldsymbol{x}) = f(\boldsymbol{x}(n)) + \nabla f^{\mathrm{T}}(\boldsymbol{x}(n))(\boldsymbol{x} - \boldsymbol{x}(n)) + \ldots \approx \boldsymbol{a}^{\mathrm{T}}\boldsymbol{x} + b \tag{14}$$

which is a local linear predictor. Accordingly, the vector and scalar quantities of $\boldsymbol{a}$ and b are estimated from the selected pairs (*y(n+1),* $\boldsymbol{x}(n)$) in the neighborhood of the present state. The codebook will be developed with a self organizing map (SOM).

4. Kohonen's Self-Organizing Map (SOM)

4.1. SOM Networks and Kohonen learning

Kohonen [18] developed the self-organizing map (SOM) to transform an input signal of arbitrary dimension into a lower (one or two) dimensional discrete representation preserving topological neighborhoods. Let $\Phi: X \rightarrow A$ denote the SOM mapping from an input space X and the discrete output space A. The SOM Φ defines in Kohonen words "an elastic net of points A that are fitted to the input signal space X to approximate its density function in an ordered way" [18]. In order to achieve this goal, the discrete grid A of processing elements (PEs) indexed by $i \in \boldsymbol{A}$ is described by reference vectors $\boldsymbol{w}_i$ which take their values in the input space $\boldsymbol{X}$. The response of a SOM to an input $x \in \boldsymbol{X}$ is determined by the reference vector w_{i° of the PE which produces the best match to the input

$$i^\circ = \arg\min_i \; dist(\boldsymbol{w}_i - \boldsymbol{x}) \qquad i = 1, \ldots, A \ . \tag{15}$$

where the superscript $^\circ$ denotes the winning PE, and *dist(.)* is an appropriate distance metric such as the Euclidean or the dot-product. This means that each PE represents a local neighborhood of the input space also called a Voronoi cell [18]. In this respect a SOM is a vector quantizer, where the weights play the role of the codebook vectors [2]. The nonlinear mapping Φ is obtained by modifying the weight vectors $\boldsymbol{w}_i$ with a learning algorithm. Since the ultimate goal of the SOM is to approximate the input data density, the weights should be "attracted" to the regions of the input space with high sample density which can be accomplished generally with a competitive learning rule as,

$$\boldsymbol{w}_i(n+1) = \begin{cases} \boldsymbol{w}_i(n) + \eta(\boldsymbol{x}(n) - \boldsymbol{w}_i(n)) & i = i^\circ \\ \boldsymbol{w}_i(n) & otherwise \end{cases} \tag{16}$$

The remarkable difference between the SOM and other direct competitive learning schemes is the details of the weight updating. Instead of adjusting the winner exclusively, a scaled adjustment is applied to all the output PEs

$$\Delta \boldsymbol{w}_i(n) = \eta \Lambda_{i^\circ, i}(\boldsymbol{x}(n) - \boldsymbol{w}_i(n)) \tag{17}$$

where $\Lambda_{i^\circ, i}$ is a spatial neighborhood function and η is the learning rate. A typical neighborhood function is

$$\Lambda_{i^\circ, i}(n) = exp\left(-\frac{\left\| r_i - r_{i^\circ} \right\|^2}{2\sigma^2(n)^2} \right) \tag{18}$$

where $\left\| r_i - r_{i^\circ} \right\|$ represents the Euclidean distance in the output space between the ith PE and the winner. The above choice of the neighborhood function adjusts appreciably the output units close to the winner while those further away experience little change. Eq. 16

is a special case of Eq. 17 for

$$\Lambda_{i^\circ,\, i} = \begin{cases} 1 & i = i^\circ \\ 0 & otherwise \end{cases}$$

There are two phases during learning. First, the algorithm should cover the full input space and establish the neighborhood relations that preserve the input data structure. This requires competition among the majority of PEs and a large learning rate such that the PEs can orient themselves to preserve local relationships. The second phase of learning is the convergence phase where the local detail of the input space is preserved. Hence the neighborhood function should cover just one PE and the learning rate should be also small. In order to achieve these properties, both the neighborhood function and the learning rates should be scheduled during learning. So finally, the weight update in SOM training is

$$w_i(n+1) = w_i(n) + \eta(n)\Lambda_{i^\circ,\, i}(n)(x(n) - w_i(n)) \tag{19}$$

$$\text{where } \eta(n) = \frac{1}{a+bn} \qquad \sigma(n) = \frac{1}{c+dn} \tag{20}$$

The SOM has very interesting properties for data modeling. When the network converges to its final stable state following a successful learning process, it displays four major remarkable properties:

1. A discretization and space dimension reduction is attained via the feature map Φ [20]. That is, the continuous input space is mapped onto a discrete output space of lower dimension. This property makes the simple architecture of codebook representation feasible.

2. The SOM map Φ is a good approximation to the input data distribution. Luttrell's analysis of the SOM shows that it approximates the input space data distribution as a vector quantizer [22]. This property is important since it provide a compact representation of the input space data distribution.

3. The feature map Φ embodies a statistical law [20]. In other words, input regions of the same size but with different number of samples occupy different domains in the output space. The larger the number of samples, the larger the domain in the output space $\boldsymbol{A}$. This property helps to make the SOM an optimum codebook of the given input space.

4. The feature map Φ naturally forms a topologically ordered output space, i.e. regions that are adjacent in the input space are mapped onto PEs in the lattice that are spatial neighbors [38]. This feature reduces errors caused by noise because neighborhood PEs tend to be activated in the output space $\boldsymbol{A}$.

4.2. Codebook in Reconstruction Space

The straightforward way to take advantage of the above properties for time series modeling is to create a SOM from the reconstruction space. So the first step is to create an embedding (Eq. 2) that will map the time series to a trajectory in a multidimensional reconstruction space. We will exemplify the use of the SOM when applied to the Lorenz system. The Lorenz equations are

$$\begin{aligned} \dot{x} &= \sigma(y-x) \\ \dot{y} &= x(r-z)-y \\ \dot{z} &= xy-bz \end{aligned} \tag{21}$$

where σ, r, and b are constants. With $\sigma = 10$, $r = 28$, and $b = 8/3$, the system exhibits chaotic dynamics, and can be solved by numerical integration. The time series projected along the x coordinate is shown in Figure 3.

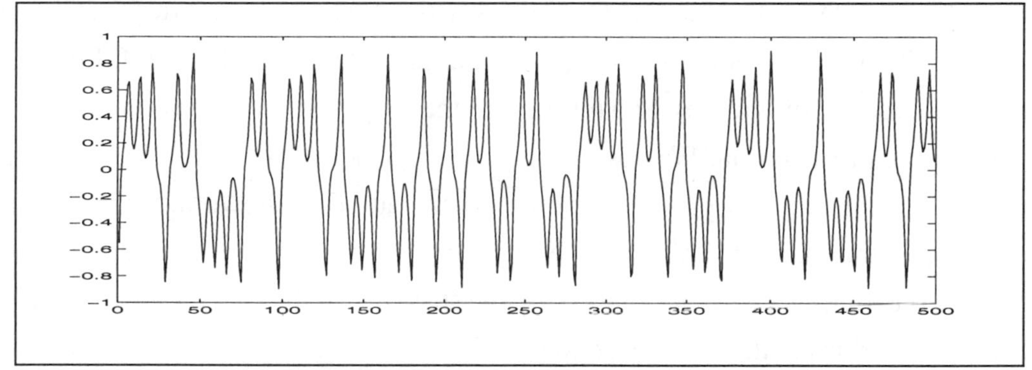

Figure 3. The time series of the Lorenz system (x-coordinate).

Figure 4 shows the 2-D projection of the Lorenz attractor initially reconstructed in 3D space.

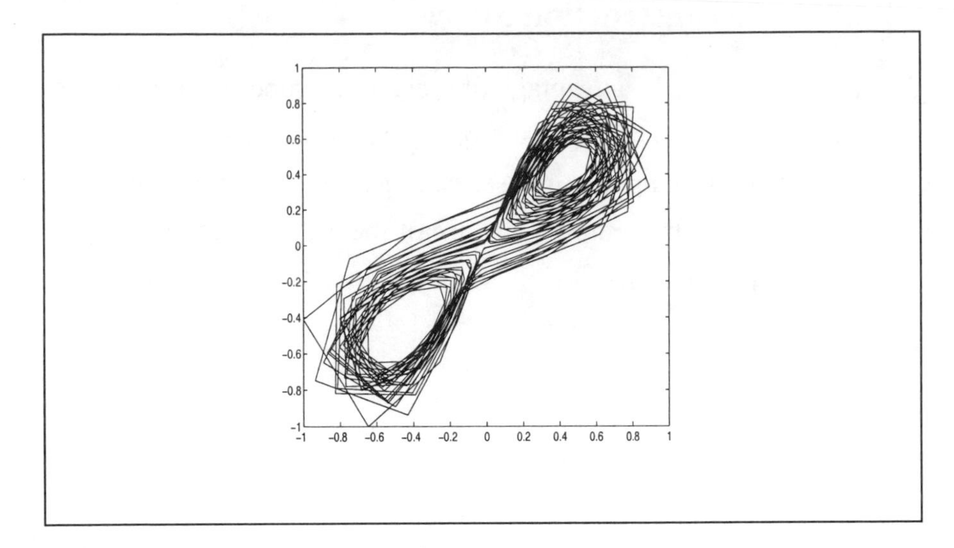

Figure 4. 2-D projection of the Lorenz attractor

We selected an embedding dimension of N=4 (a tap delay line with 3 delays, 4 taps) and trained a 22x22 SOM with a Lorenz time series with 3,000 samples. After training, a new sequence of 500 samples of the Lorenz time series is presented to the SOM. Figure 5 depicts the sequence of winning PEs (circles) in the output SOM space connected with lines during the presentation. We observe that the sequence of winning PEs creates a trajectory that is a faithful 2-D projection of the Lorenz attractor shown in Figure 4. This example shows that the SOM mapped the trajectory at its input (the reconstruction space) into the discrete output 2-D space preserving the structure of the 4-D reconstruction space. Repeatability of this result was very good [51]. Different training runs produced a winning PE evolution which always resembled Figure 5 (apart from rotations and translations)

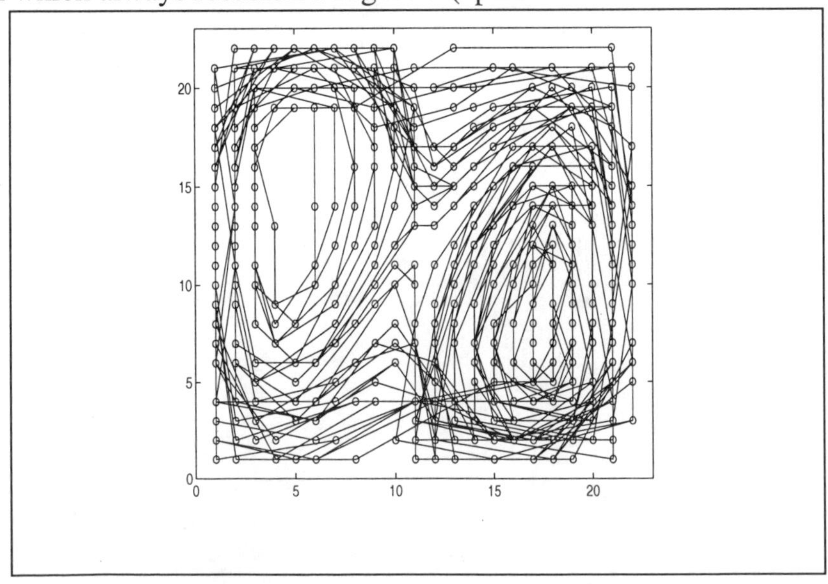

Figure 5. Trajectory of winning PEs in the SOM output space.

Our next step is to seek ways to utilize the SOM for dynamic modeling, harnessing its power of data representation, space discretization, and topological neighborhood preservation.

5. SOM-based Local Linear Models

According to the state dependent theory of nonlinear processes (section 3.3) the three steps needed to identify a nonlinear system are:

* embed the time series in a reconstruction space,
* create a codebook of pairs $(y(n+1), \boldsymbol{x}(n))$,
* fit a local linear model to the selected pairs in the codebook.

These steps naturally lead to our proposed block diagram of SOM based local linear modeling depicted in Figure 6 [35]. There are three function blocks in the topology: the embedding layer, the SOM and the layer of local linear predictors. The first step performs an embedding on the time series using a delay line. But notice that the information for dynamic modeling is contained in the pairs $(y(n+1), \boldsymbol{x}(n))$, so we will augment the embedding space normally of dimension N with one more dimension to represent the functional values, i.e. the embedding space has dimension N+1.

The codebook is implemented by a SOM with a sufficiently large number of PEs (experimentally determined). One and two dimensional neighborhoods have been successfully utilized in our studies using the traditional training procedure suggested by Kohonen. After training, the SOM will represent the system dynamics in the discrete output lattice just like a codebook, but enhanced with neighborhood relationships, i.e. states that are adjacent in the reconstruction space will be represented by PEs that are neighbors in the output space (as illustrated in Figure 5).

We propose to derive the local linear models from the trained SOM by creating an extra layer of linear PEs which receive inputs from the tap delay line (Figure 6). The ith SOM PE has a companion linear model $\tilde{f}_i(\boldsymbol{a}_i, b_i)$ which represents the linear approximation of the local dynamics. The linear PE weights $(\boldsymbol{a}_i, b_i)$ are computed *directly* from the weights of the SOM by a least square fit within a local neighborhood of size N_L (N_L>N+2 where N is the embedding dimension) centered at the current winning PE according to:

$$w_{i,0} = \boldsymbol{a}_i^T \boldsymbol{w}_{ij} + b_i \qquad j = 1, \ldots, N_L \tag{22}$$

where $w_{i,0}$ represents the weight of the ith SOM PE connected to the first tap (most recent sample) of the embedding layer. Obviously, we first train the Kohonen network to find $\mathbf{w}_{ij}$ and only then compute in one sweep the parameters $\mathbf{a}$ and b of the local linear models using least squares. Note that with this training procedure the input data is not used to derive the local models. Comparing Eq. 22 with Eq. 14, we observe that the weights of the SOM are being used instead of the input data samples to derive the local models. The weight $w_{i,0}$ is used as a proxy for $y(n+1)$, while the other weights connected to the same PE represent $\mathbf{x}(n)$. This ordering relation in the SOM weights is a reflection of the flow of time in the delay line. See the Appendix for a summary of the algorithm.

For prediction, the local dynamic model works as follows: The current input $\mathbf{x}(n)$ is placed at the input of the SOM and the winner-take-all operation will select the local PE that best represents the current $(y(n+1), \mathbf{x}(n))$ pair. This winner activates a single linear PE that contains the weights of the local linear model and produces $\tilde{x}(n+1)$ which is a prediction of the next sample of the time series, i.e.

$$\tilde{x}(n+1) = \tilde{f}_{i^\circ}(\mathbf{x}(n)) = \mathbf{a}_{i^\circ}^T \mathbf{x}(n) + b_{i^\circ}. \tag{23}$$

If iterated prediction is required [16], the generated sample is fed back to the input (delayed by one sample) and the procedure repeated as many times as needed.

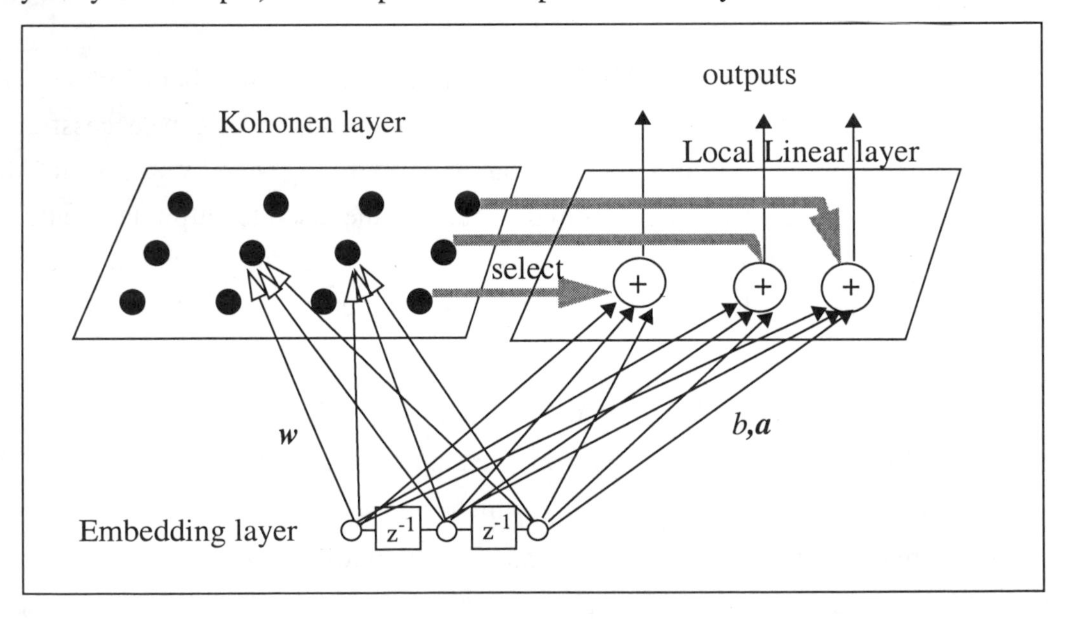

Figure 6. Structure of Kohonen's SOM for dynamic modeling

There are several important advantages of using a SOM based local dynamic model.

1- The SOM mitigates the discontinuity in local modeling addressed in section 3.2. Crutchfield [10] showed experimentally that dynamic modeling fails when this is not enforced. Since the SOM is trained globally ensuring smoothness at the neighborhood boundaries [19], the PE prediction weights at the local linear layer share SOM weights. So, the continuity among the local models has a better chance of being met.

2- The SOM ordered output space allows an easy implementation of linear models. Using the SOM output space for dynamic modeling has two advantages: it positions the local models in state space and identifies the local model for the current input state $(y(n+1), \boldsymbol{x}(n))$. This is crucial to help us build the local linear models from the output lattice when the observations are noisy and we have to perform local averaging.

3- The SOM input space representation simplifies enormously the construction of linear models for dynamic modeling. A major drawback of the local linear method was the need to always reconstruct the neighboring states from the time series. Performing dynamic modeling with the weights of the SOM will avoid this step. The weights of the winning PE of a trained SOM represent the center of the cluster that codes the present state. Due to the neighborhood preserving properties of the SOM, the neighbors of the wining PE in the output space represent the neighboring input states. Hence, *there is no need to go back to the time series to reconstruct the neighboring states of the current state*. The weights of the wining PE and its neighbors contain all the information we need to fit the local linear model to estimate the next sample of the time series. This is a major computational savings.

4- A fourth advantage of the application of the SOM to local dynamic modeling is the discretization of the reconstruction space. Although the reconstruction space and the trajectory are assumed noncountable and continuous subsets of Euclidean space, the feature map is countable and discrete. This is in contrast with the other local linear modeling scenarios.

5- The winner-take-all operation of the SOM makes possible the selection (and training) of a local linear model attached to each output of the network. The method has the advantages of compact state space representation of the original time series, simple state selection (winner-take-all), and state locality. The latter feature is ensured by the neighborhood preserving property which is the direct consequence of Kohonen's SOM training [18].

Hence, the SOM is an explicit quantification of the dynamics and becomes an infra-structure for local model construction. We can use the SOM to identify the system that produced the time series under investigation, or to artificially generate a time series similar to the original one by simply feeding back (through a delay of one sample) the output to the input of the system depicted in Figure 6.

Ritter and co-workers proposed the parametrized self-organizing map (PSOM) for the task of rapid learning multidimensional mappings in robotics and vision. Their goal was to create a continuous mapper based on the SOM which would map the discrete output of the SOM back to the input space using interpolating polynomials [39]. Their impressive results on learning the inverse kinematics of a 3-DOF robot finger with little training data show the power of the technique [49]. Later the same group proposed a Chebyshev PSOM and a local PSOM [50]. The local PSOM includes basically the same components as our work, but both the topology and the training are very different. In our approach the SOM provides simply the selection of the local linear models which work directly with the input instead of the output of the SOM. The training of our local linear models is done directly from the SOM weights while they propose a gradient descent based on the Levenberg-Marquardt algorithm. Moreover, the link to the theory of dynamic modeling was never stated which provides in our opinion new insights and a broader, more principled scope to our topology. In a statistical framework Cherkassky [8] proposes a local linear extension to the SOM for the purpose of improving the approximation to the regression surface, which albeit similar to Ritter's work differs in the global nature of the regression. Previous applications of SOM for dynamic modeling [23],[48] only used the competitive properties of the neural model to create the codebook. Recently, Vesanto [55] proposed a scheme that essentially followed our topology [35].

5.1 Dynamic Learning in the SOM

We propose below an improvement to Kohonen learning for the special task of dynamic modeling by incorporating the model fitting error in the training of the SOM. This modification improves performance but the benefit of independent training of the SOM and of the local models is lost since dynamic learning trains both the SOM and the linear models at the same time. We will also utilize weighted least squares to derive the local linear models for a more precise adaptation of the linear models.

<u>Incorporating the fitting error in SOM training</u>

The SOM obtained with the Kohonen's learning law provides an approximation of the input space with close statistical density matching, i.e. more frequent trajectories will be represented by larger areas in the output space. For the purpose of dynamic modeling where the SOM is employed as a dynamical representation structure, the density matching is not the best criterion.

Regions that are difficult to model (normally the time series segments with large curvature) should be the ones that have larger areas in the output SOM lattice. However, this goal should not interfere with the creation of the map, in particular the topological ordering. Let $m(\boldsymbol{x})$ denote the PE density factor, i.e. the number of PEs representing a small volume of the input space $\boldsymbol{X}$. In the ideal case, $m(\boldsymbol{x})$ should be proportional to the curvature of $f(\boldsymbol{x})$. Although this quantity is not readily available, it can be approximated by the amplitude of the linear prediction error

$$\varepsilon(\boldsymbol{x}(n)) = f(\boldsymbol{x}(n)) - \tilde{f}_{i^\circ}(\boldsymbol{x}(n)) = x(n+1) - (\boldsymbol{a}_{i^\circ}^{\mathrm{T}}\boldsymbol{x}(n) + b_{i^\circ}) \tag{24}$$

Thus the desired property becomes $m(\boldsymbol{x}) \propto \|\varepsilon(\boldsymbol{x}(n))\|$. which implies that $\varepsilon(\boldsymbol{x}(n))$ should be involved in the training process.Two adaptation variables can be considered: the learning rate η and the width of neighborhood function $\Lambda_{i^\circ, i}$. Luttrell [22] shows that the neighborhood function $\Lambda_{i^\circ, i}$ interferes with the power of the noise mismatch, so any disruption to the decreasing trend of the neighborhood function may cause instability during training. Instead, the conventional learning rate η can be modified to be a function of $\varepsilon(\boldsymbol{x}(n))$. A larger value of η_ε will tend to recruit more PEs for the neighborhood function. Considering the constraint that the magnitude of the learning rate is less than unity, a straightforward way to involve ε in the training process is

$$\eta_\varepsilon = \frac{1 - exp(-\mu(\eta + \bar{\varepsilon}))}{1 + exp(-\mu(\eta + \bar{\varepsilon}))} \tag{25}$$

where μ is a constant that controls the slope of the exponential increasing function, and η is the conventional learning rate. A normalized value of $\bar{\varepsilon}$ should be used to offset the amplitude variation,

$$\bar{\varepsilon} = \frac{\|\varepsilon(\boldsymbol{x}(n))\|}{\|x(n+1)\|} \tag{26}$$

When $\bar{\varepsilon}$ is small, the training defaults to the basic Kohonen learning. However, when $\bar{\varepsilon}$ is large, learning is emphasized on the neighborhoods that represent large curvature trajectories. This dynamic learning process will produce a feature map with the density factor $m(\boldsymbol{x})$ consistent with the local complexity of the given dynamics. Similar modifications have been proposed in [28] to increase the accuracy of the SOM in nonparametric regression, but here the estimate of the second derivative is more direct. The authors reported good results.

Weighted Least Square Estimation

The least square algorithm was previous described for the construction of the local linear models [35]. We need to use at least N+1 PEs to derive the local linear model, and we choose the winning PE neighbors for that purpose (the receptive field). In least squares, all the elements of the receptive field are equally weighted for the construction of the local model. Intuitively this may not be an optimal choice, since PEs closer to the center represent dynamics that are closer to the winning PE. A poor weighting of the PE contributions may cause the estimation of the linear models to deviate from the optimal orientation and hinder the reduction of the discontinuity at the boundaries. With this concern in mind, we propose to make the contribution of each PE to the collective response inversely proportional to its metric distance to the center, which yields a weighted least squares solution [41]. Weighted least squares is a very well known technique and it has been proposed to improve the accuracy of the SOM for nonparametric regression [8].

Mathematically, the weighted least square solution is equivalent to inserting a weighting matrix to the optimization process,

$$min(\boldsymbol{y} - \boldsymbol{x}\Theta)^T \mathbf{S}(\boldsymbol{y} - \boldsymbol{x}\Theta) \tag{27}$$

where $\boldsymbol{y} = \boldsymbol{x}\Theta$ and $\mathbf{S}$ is a nonsingular symmetric matrix. The weighted least squares solution satisfies

$$\boldsymbol{x}^T \mathbf{S}(\boldsymbol{y} - \boldsymbol{x}\Theta) = 0 \tag{28}$$

Therefore the solution becomes

$$\tilde{\Theta}_{WLS} = (\boldsymbol{x}^T \mathbf{S}\boldsymbol{x})^{-1}(\boldsymbol{x}^T \mathbf{S}\boldsymbol{y}) \tag{29}$$

and it is unique if and only if $\boldsymbol{x}^T \mathbf{S}\boldsymbol{x}$ is invertible. Using the Euclidean metric for the distance d_i to the center, we construct a diagonal matrix $\mathbf{S}$ defined as

$$\mathbf{S} = \{s_{ij}\}\big|_{1 \le i,\, j \le N_L} \tag{30}$$

where $s_{ij} = 0$ for $i \neq j$, and

$$s_{ii} = 1 - \frac{d_i^m}{\sum_{k=1}^{N} d_k^m} \tag{31}$$

The experimental results show that m ranging from 2 to 4 provides the best performance. In our application,

$$y = \begin{bmatrix} w_{1(0)} \\ \dots \\ w_{N_L(0)} \end{bmatrix}, \quad x = \begin{bmatrix} \begin{bmatrix} \mathbf{w}_1{}^T & 1 \end{bmatrix} \\ \dots \\ \begin{bmatrix} \mathbf{w}_{N_L}{}^T & 1 \end{bmatrix} \end{bmatrix}, \quad \Theta = \begin{bmatrix} a \\ b \end{bmatrix}$$

6. Experimental Results

The proposed topology for local dynamic modeling is tested in the Lorenz system, which exhibits chaotic dynamics with large Lyapunov exponent for $\sigma = 10, b = 8/3, r = 28$. The Lorenz time series is sampled at 10Hz. The SOM system used here is composed of a 2D grid of 22x22 PEs. The embedding dimension is chosen as 4, and so the dimension of the state input during the training process is 5. The parameters for the learning rate and the neighborhood function in Eq. 19 are $a = 1, b = 1/500$, $c = 1/8$, $d = 1/4000$. Further testing of the system is presented in [51]. Here we would like to show the reliability of the method to model the Lorenz system and also the improvements obtained with the new training rule.

The SOM is trained with the conventional Kohonen learning (KL) and with the improvement of section 5.1 which will be called here dynamic learning (DL). The training is performed with a 3,000 sample segment of the time series for 150 epochs. After training, the local models for each PE are constructed using the weighted least algorithm of Eq. 29 with $m = 4$. With all the weights fixed, an initial point in reconstruction space is loaded, and the whole system iterated as an autonomous dynamical system, i.e. its output is fed back to its input to generate a time series. This generation model is called iterated or recursive prediction, and mimics the assumption made for dynamic modeling. Fig. 7 shows the 500 sample trajectory of the winning PE in the output space and the corresponding autonomous prediction.

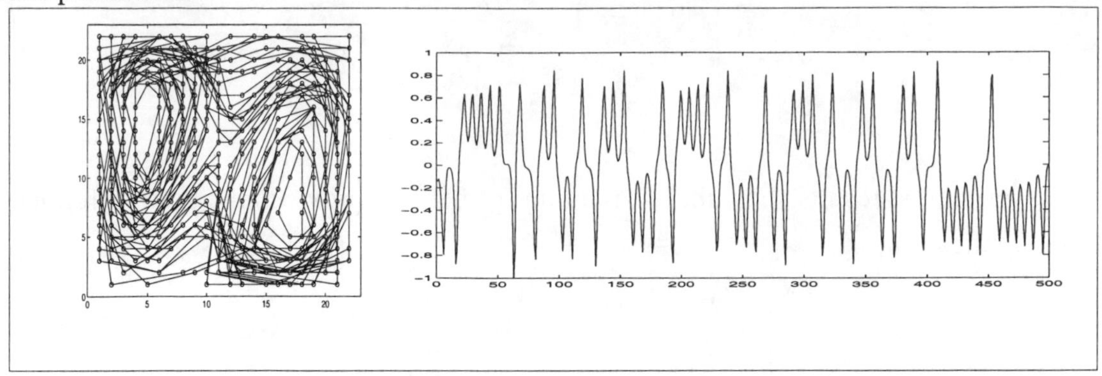

Figure 7. Trajectory of the winning PE and generated Lorenz time series.

Although visually the generated time series resembles the Lorenz signal, the accuracy of dynamic modeling is quantified by comparing the dynamic invariants (correlation dimension and largest Lyapunov exponent (LLexp) between the original time series and the generated time series as suggested in [34]. We use the Grassberger and Procaccia algorithm [13] to estimate the correlation dimension and Wolf's algorithm [54] to estimate the LLexp.

Figure 8 shows that the dynamic learning outperforms the conventional Kohonen learning for the task of dynamic modeling since it produces a smaller normalized MSE during training (0.0011 versus 0.0012).

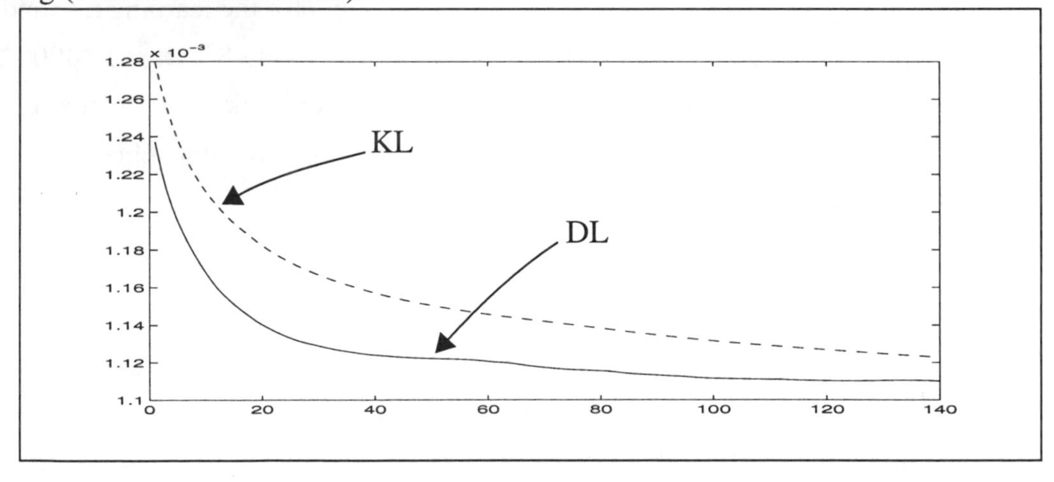

Figure 8. Comparison of the learning curves: KL Kohonen learning and the new rule

An example of the modification of the SOM output neighborhoods produced by the new training is illustrated in Fig. 9. The waveform on top shows an enlarged segment of the Lorenz time series used for training. The segment within the broken lines is selected because it produces one of the largest prediction errors. The sample that is predicted is shown by the cross. Fig. 9 (b) and (c) depicts the contour plot of the SOM activity bubble around the winning PE that corresponds to the selected segment of the time series. Fig. 9b depicts the contour plots obtained with the new dynamic learning, while Fig 9c corresponds to regular Kohonen learning. There are two things to note: first, although these figures correspond to two different training runs of the SOM, and the wining PE actually appears in two distinct places of the output space, both contour plots are rather similar. Second, the area in the output space that codes the peak of the time series is larger for dynamic learning than for regular learning (the first contour line is 0.9 of the maximum). This means that the neighborhood field of this PE was enhanced by dynamic learning which is the expected result (more PEs are recruited enhancing the resolution for the high curvature portion of the trajectory and decreasing the prediction error). Table 1 summarizes the results of the estimated dynamic invariants in the generated time series versus the original time series.

Table 1: Comparison of Dynamic Invariants

Time series	LLexp (bits/sec)	Correlation Dimension
original	2.17	2.07
dynamic learning	2.09	2.08
Kohonen learning	1.83	2.01

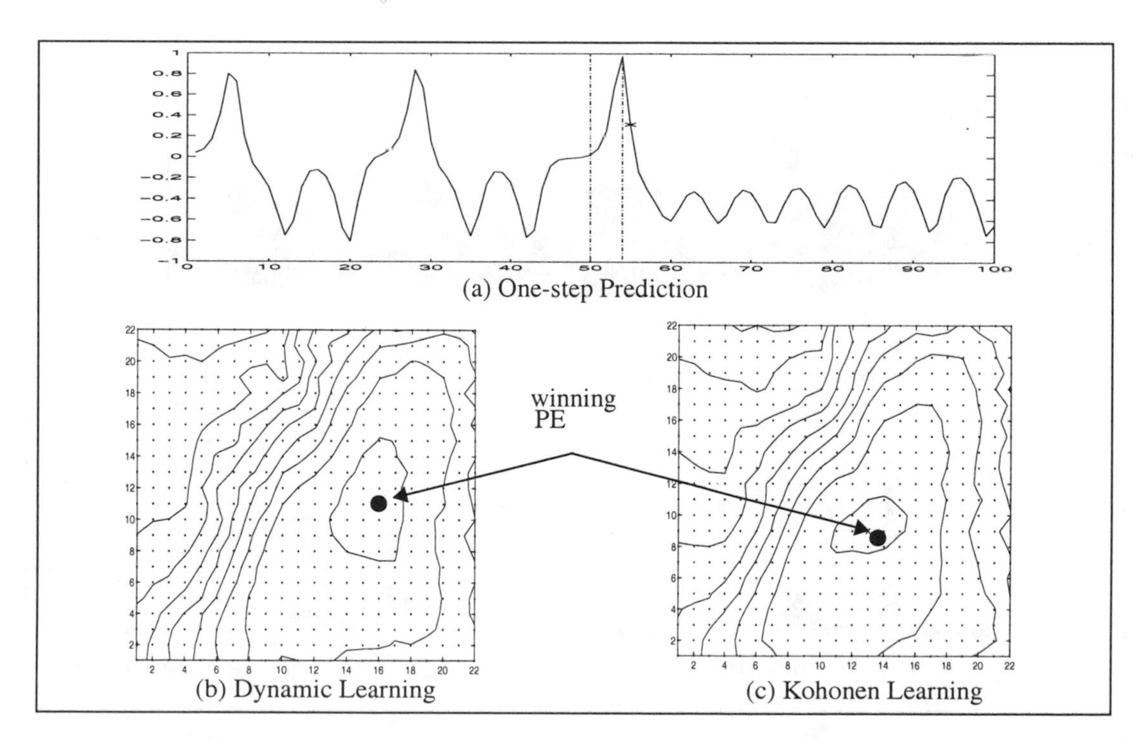

Figure 9. Comparison of the recruiting areas at the SOM output for the two learning rules.

Many more tests of consistency both in time characteristics and on multiple training runs are reported in [51]. We summarize by saying that our proposed SOM based locally linear dynamic model is very robust. It never diverged during our experiments, while both the state dependent model proposed in [10] and a different model created around a simple clustering algorithm [2] showed divergence. Once the system switches to the wrong PE during iterated prediction it will produce a burst of samples that do not conform with the model. The trajectory simply diverges and never recovers. Hence, we conclude that both the neighborhood preservation property of the SOM and our proposed weighted least square estimation of the local modes is preferable for successful local linear dynamic modeling.

Finally, we would like to present results for the modeling of a real-world signal. We selected the laser time series from the Santa Fe Time Series Competition [52]. This time series is difficult to model due to the collapses of the orbits in the attractor. A 28x28 SOM with a=1, b=1/500, c=1/8, d=1/400 with an embedding of 5 and a μ=2, was trained for 130 epochs on a 3,000 sample segment of the laser time series. The output of the autonomous dynamic model obtained through iterated prediction is shown in Fig. 10 (a) which resembles the bursting behavior of the original signal. The spectra of the original and generated time series are shown in Fig. 10 (b) and (c), which demonstrate that the modeling preserved the essential of the deterministic motion on the attractor.

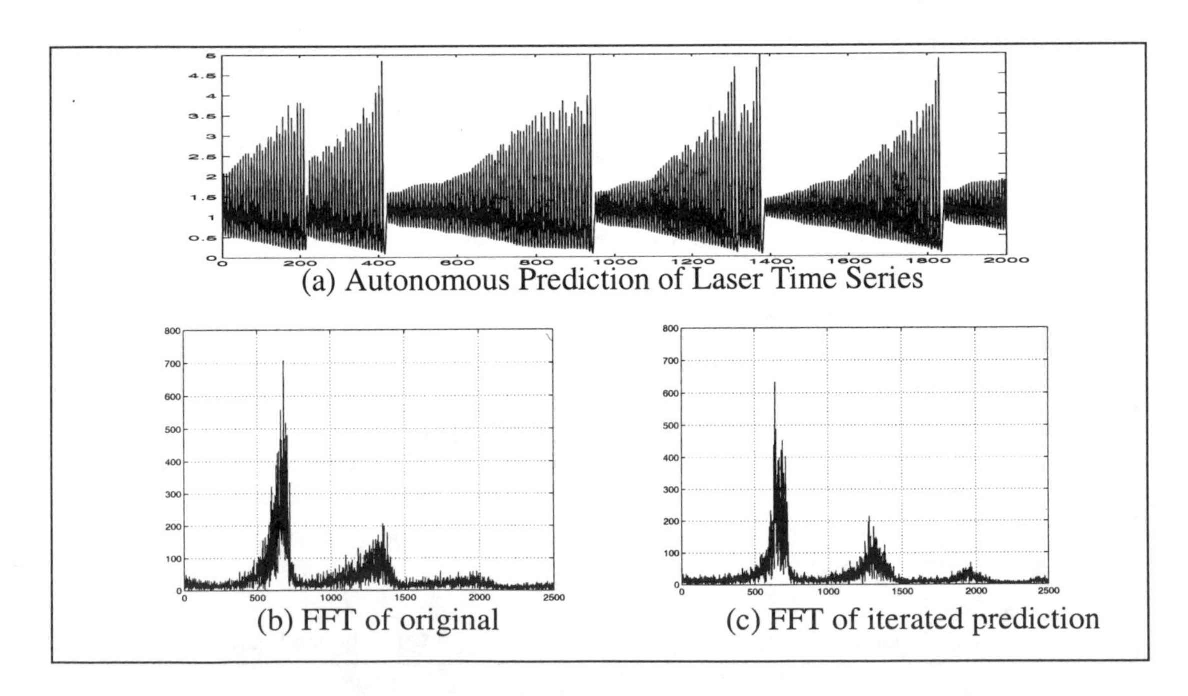

Figure 10. Iterated prediction of the laser time series and FFT spectra.

7. Design of Switching Controllers based on SOM

7.1. Problem Specification

The previous development of local dynamic modeling has many applications, but here we will address its use in system identification and how it can be extended to create a multiple model switching controller. The plant is the 16-Foot Transonic Tunnel at NASA Langley Research Center. The problem is how to ramp up or down and regulate the wind speed (Mach number) in the tunnel such that aerodynamic studies in scaled down models can be conducted with high accuracy. The Mach number accuracy is +/- 0.003 of the set point. The difficulty is that modifying the orientation (angle-of-attack) of the model relative to the airflow produces changes in the effective cross-section of the tunnel test section, changing abruptly the wind speed. This in turn requires quick modification of the main drive motor r.p.m. to stabilize the wind speed at the Mach number set point. The tunnel dynamics vary widely with the Mach number and the angle of attack of the model. Due to the huge power of the main drive system (50 MW) the only available controls are 3 regimes (increase/maintain/decrease the r.p.m. by a nominal amount, which are represented respectively by 1/0/-1), and the time each control is applied. So we can model the possible control inputs as a family of times series with 3 amplitude levels, which differ only in their time structure.

7.2. The Control Strategy

This control application is fundamentally one of regulation around a set point in a nonstationary environment. The leading characteristic of this application is the fast and unpredictable changes in dynamics encountered when the model's angle of attack is modified. The existing automatic controller is a highly tuned but fixed table look-up of drive motor commands based on the error at a given Mach number [6]. An adaptive controller that meets the set point specifications will be very slow to adapt. Alternatively, we developed in advance a set of local dynamic models for the tunnel dynamics and switch them according to the measured Mach number history.

The design of the controller has two phases: first the wind tunnel dynamics have to be identified. Then an appropriate control action has to be decided. The local modeling approach based on a SOM is especially well suited to model the changes in tunnel dynamics when the model position is modified. In fact, the SOM will provide a codebook representation of the tunnel dynamics, and will organize the different dynamic regimes in topological neighborhoods. The difficulty is that the wind tunnel can not be considered as an autonomous system since it is excited by the controller at all times.

We decided to cluster the possible control inputs into a set of representatitive inputs, and design a different SOM per each representative control input. As long as we can create a meaningful and limited catalog of control inputs this is a feasible alternative. The quantized nature of the control input works on our side. Moreover, experience shows that the tunnel operators develop a “style” when dealing with a specific situation. The practical usefulness of a style implies that it is possible to define automatically a set of representative control inputs and also corroborates the opinion that a limited catalog of control inputs suffices to ramp the Mach number up and down and regulate around the set point.

Each SOM will model the *forced tunnel dynamics*, i.e. the combination of the tunnel with one of the representative controls, which can be alternatively thought as a collection of “autonomous” wind tunnels. Each SOM is trained with full range Mach number data under many different experimental conditions (i.e. different model aircrafts and angle-of-attack). When a given SOM PE fires, its corresponding predictor is the best available local linear descriptor for the tunnel dynamics with the applied control input. Hence the system can predict the evolution of the dynamics in the near future. Since the requirement is to regulate the Mach number, the system may choose several possible control commands from the catalog, and see which one best meets the Mach number specification. This reasoning points to the implementation of a predictive multiple model, switching controller

(PMMSC). The block diagram is presented in Figure 11.

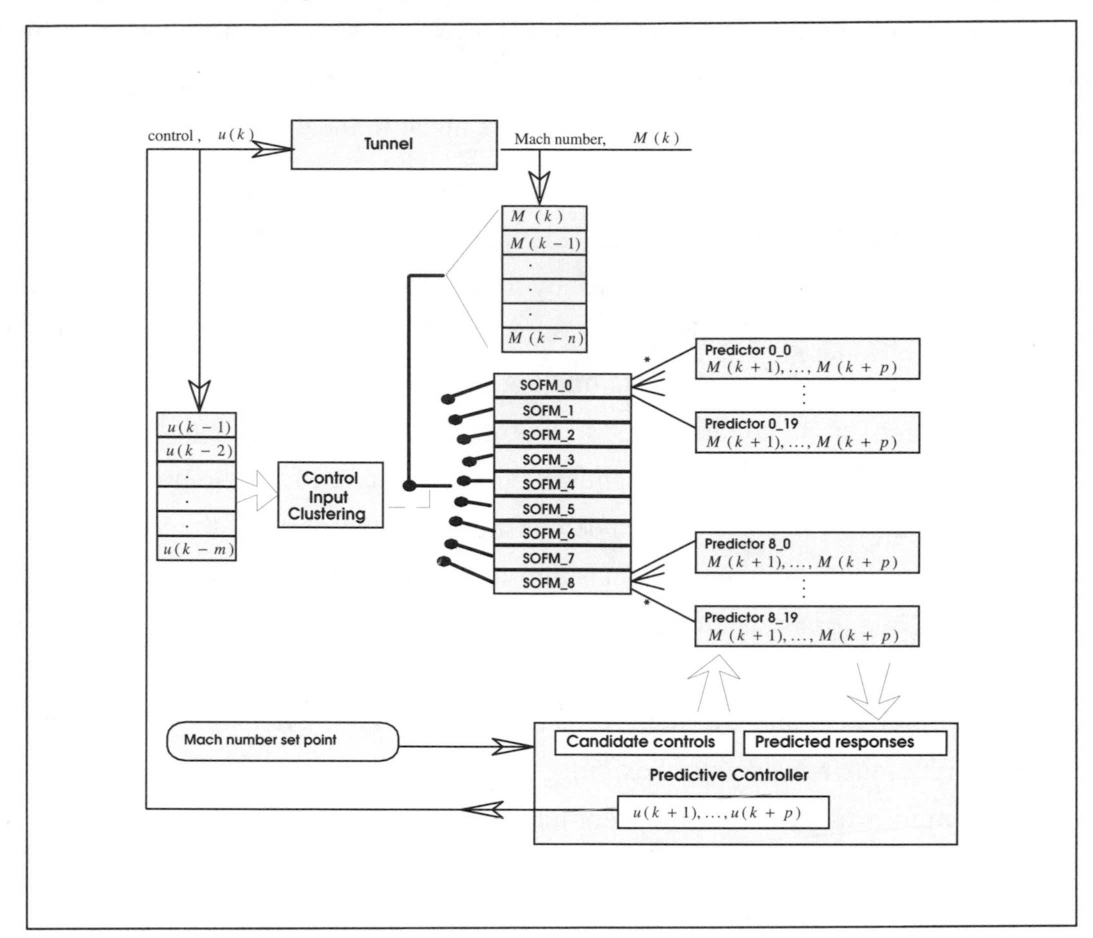

Figure 11. Block diagram of the PMMSC.

We are not the first ones to propose a switching control architecture. A multiple model structure with switching has been proposed by Narendra [25]. The architecture of Narendra's multiple model switching controller also utilizes N predictive models of the plant which are obtained by observing the system over a long period of time. Each has a corresponding controller which is designed off-line. The outputs of the predictive models are compared with the plant output and the best fit to the plant output (according to a performance index) selects one of the models and the corresponding controller at each time instant. This corresponds to the switching part of the architecture. The implementation of the switching scheme employs some hysteresis to prevent arbitrarily fast switching between models. In a more recent paper [26], stability results for an all-fixed models controller were established for linear systems under some mild assumptions. In particular, it is shown that if there is at least one model that is sufficiently close to the actual plant and there is a non-zero waiting time between switches from one model to another, then the overall system is stable, given that each fixed model is stabilized by its corresponding controller. An even more recent paper [27] introduces two classes of approximate non-linear input-output models which reduce the computational complexity of designing a controller based on the fact that the approximate models are linear in the control input.

Our approach can be considered an improvement over this scheme for the following reasons. The SOM approach models all the dynamic regimes observed during training and automatically divides the dynamical space by the number of available PEs (their number specifies the granularity of the Voronoi tessellation). So we are guarantying coverage of the full dynamic space observed during training, which is a difficulty in Narendra's multiple models. Moreover, neighboring SOM PEs represent neighboring regions in the dynamic space. Therefore, wrong assignments in the winning PE due to noise tend to activate similar dynamic models. Kohonen demonstrated this property of the SOM for noisy coding of images, and here the problem is the same [19]. There is no such neighborhood relationship in Narendra's scheme so the selection of a wrong predictive model may temporarily cause a poor control regime. In our case the system tries several control sequences with the winning model and chooses the best one, so even if the dynamic model is not the most appropriate, there is an extra flexibility to match the set point with small error. Finally, our architecture is much more compact, consisting of a SOM and a set of linear predictors obtained directly from the SOM weights. After training, the PMMSC was implemented in an old Intel 486 and controlled the wind tunnel for 9 hours, collecting the data presented in this paper.

7.3 Design of the PMMSC

As seen in Figure 11 both the control input u(k) and the Mach number response M(k) are used as inputs to the PMMSC. The control input is embedded in an m=50 dimensional space to represent the representative control sequences being applied to the tunnel. These control input sequences are clustered in C control classes (C=7). Here we use the available knowledge of operating the wind tunnel to select the representative control inputs to ramp the Mach number up and down and to regulate the set point. The most recent 50 samples of the control input are clustered to the closest representative control input (in a L1 norm sense). Two additional control classes which are only 10 samples long are used for identification purposes only.

The winning cluster selects one of the 7 (C=0,1, … 6) SOM networks. The input to each of these SOM is obtained from the past history of the Mach number responses. An embedding of the Mach number response in a space of N=50 is performed. Each of these SOM is trained in the full operating range of the Mach numbers and works as a codebook for the wind tunnel dynamics under the specified control input. Two additional SOMs, C=7,8, operate essentially in parallel on the m=10 subspace to identify the most recent input-output dynamics while ramping up or down. The size of each SOM here is 20 x 1, i.e. a linear array of PEs. This number of output PEs provided enough precision in the local linear models to achieve the required 0.003 tolerance in the Mach number [24].

Ensembles of Mach number responses resulting from the application of each control prototype were extracted from many hours of wind tunnel test data, sampled at 3 Hz (~2,000,000 samples or 40,000 vectors). Next, each ensemble of responses corresponding to one of the seven control prototypes was clustered using the SOM, trained over 10,000 epochs as explained in section 4.1. Each PE of the SOM is augmented with a local linear model as described in section 5, and the predictor trained as described in the same section. Figure 12 shows the weights of two converged SOM for the Mach number responses belonging to the two regulating control inputs. Notice the smooth organization of the weights across the neural field, from which the local linear models are derived.

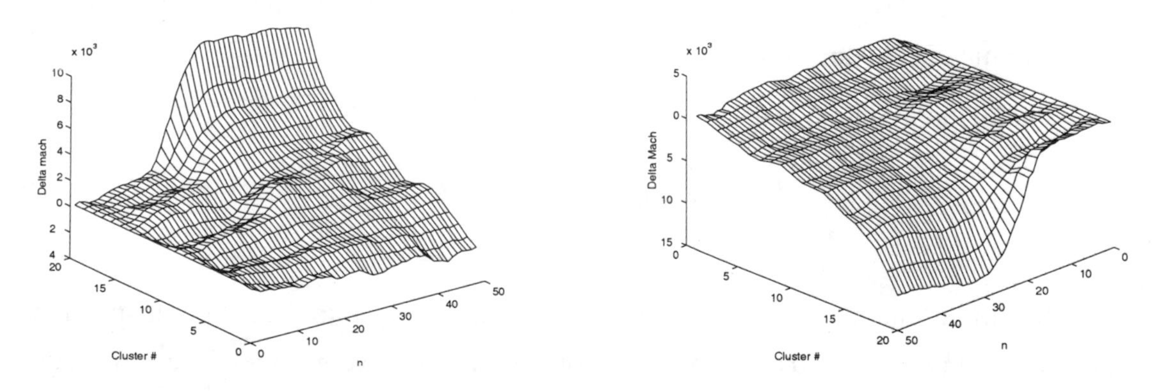

Figure 12. SOM for the Mach number responses of the regulating control prototypes

The local linear model associated with each PE is used to predict the tunnel dynamics for the next 50 samples under the repertoire of our candidate control inputs. We developed a family of 29 control waveforms, shown in Figure 13.

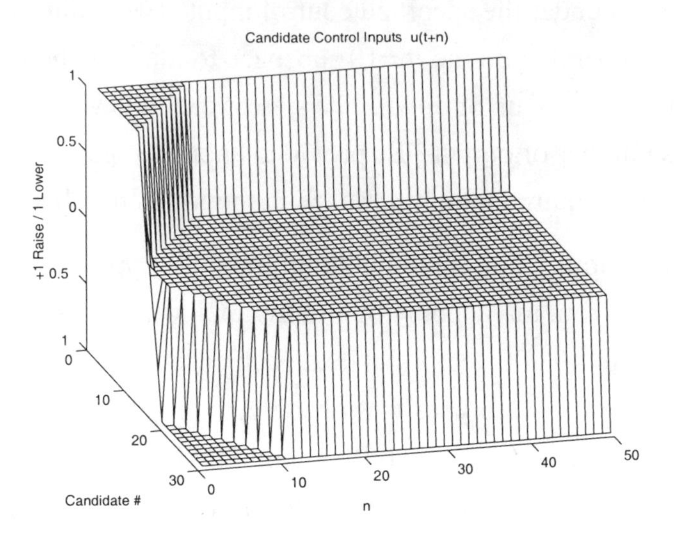

Figure 13. Repertoire of control inputs

These control inputs are developed from experience with the goal of ramping up or down and regulating the Mach number. Each candidate control is fed to the local linear model of the winning PE producing a vector M of Mach number responses which is tested against the desired set point (M_{sp}) within the last 30 points of the 50 point trajectory to emphasize steady state matching, i.e.

$$E = \sum_{k=21}^{50} \left\| M(k) - M_{sp}(k) \right\| \tag{32}$$

The control input that produces the smallest Euclidean distance to the set point specification in steady state is elected as the best control, and sent to the controller. The control u(k) is updated every sample when ramping or steady-state control sequences are selected, but when regulatory control sequences are selected, the entire 50 sample control sequence is applied. Thus, when actively regulating about a set point, the switching between models is performed at most every 50 sample periods.

8. Experimental Control of the Wind Tunnel

8.1 Comparison of PPMSC Control to Existing Automatic Controller and Expert Operator

Figure 14 compares the regulating performance of the existing gain-scheduled control, an expert operator, and the PMMSC under similar conditions over a nominal fifteen minute interval. The first row of figures display the Mach numbers (with the +/-0.003 tolerance lines), followed by the control commands in the second row, and test model angle-of-attack (disturbance) for each of the three control schemes. The control commands are amplitude coded. Control commands whose duration is 0.3 seconds are displayed at amplitude 1. Control pulses of 0.1 second are shown with magnitude 0.33 and 0.2 second pulses are shown with magnitude 0.66. Control pulses of less than 0.1 second in duration are ineffective to produce a change in the fan drive rpm at any condition and are not used.

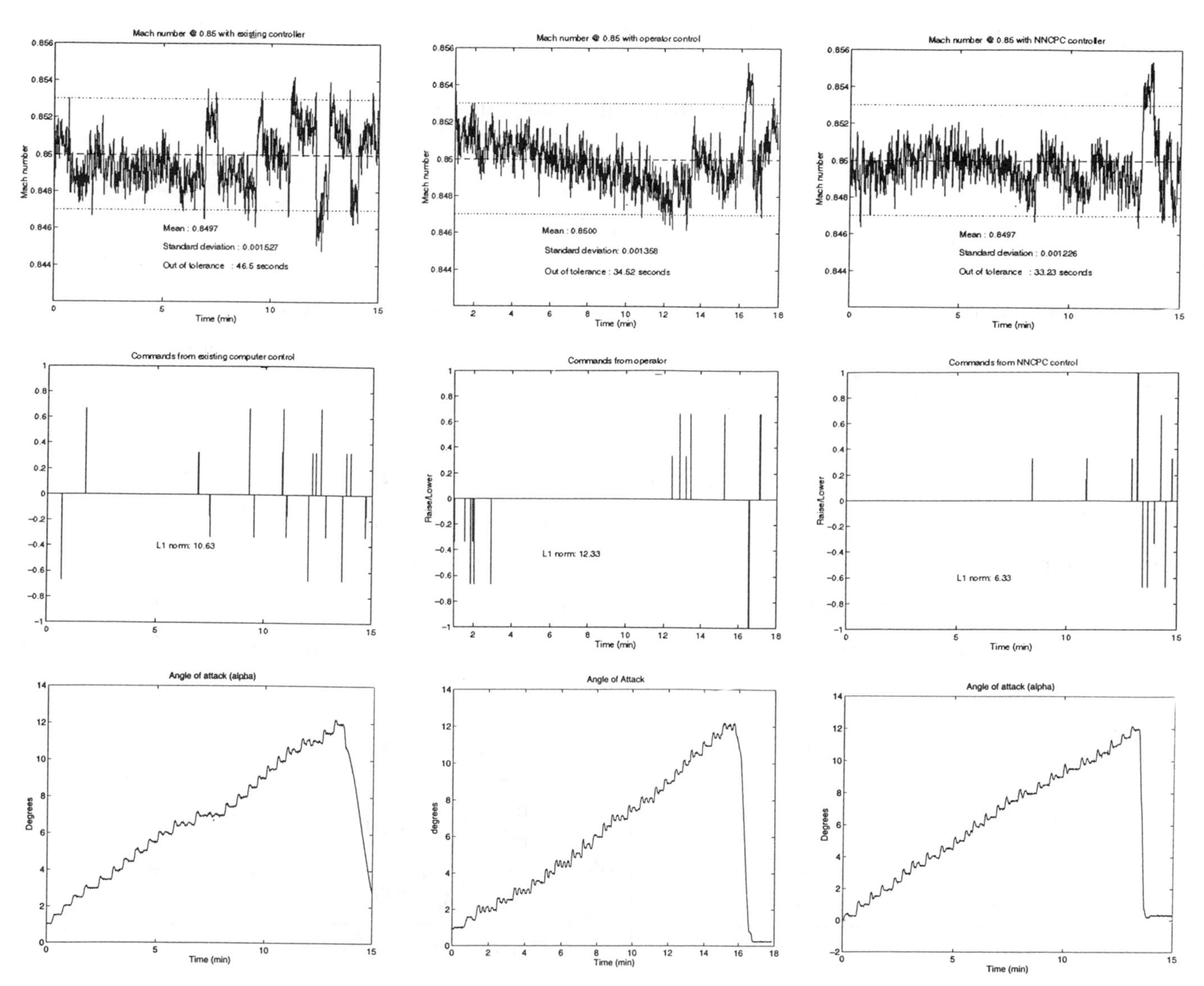

Figure 14. Comparison of the existing controller (left), expert operator(middle) and PMMSC (right).

Derived metrics to quantify the comparisons between the three control strategies are the time out of tolerance and the L1 norm of the control input, u(k). The time out of tolerance is the cumulative sum of time that the measured Mach number deviates beyond the required tolerance of 0.003:

$$T_{out} = \sum_{k=1}^{N} \alpha \Delta t(k) \tag{33}$$

where

$$\begin{cases} \alpha = 0 & |M(k) - M_{sp}(k)| \le 0.003 \\ \alpha = 1 & otherwise \end{cases} \tag{34}$$

and Δt=t(k)-t(k-1). The L1 norm of the control command is

$$L1(u) = \sum_{k=0}^{N} |u(k)| \tag{35}$$

For this particular experiment, the angle-of-attack begins to mildly disturb the Mach number at approximately 5 degrees. Table 2 lists the reduction in the standard deviation of the Mach number, time out of tolerance, control effort, and time required to complete the sweep through the desired range of angle-of-attack while maintaining Mach number steady for this comparison. For this particular test condition, the PMMSC performs slightly better than the expert operator, but with much less control effort and less time to complete the alpha sweep. Compared to the existing automatic control, the PMMSC maintains the Mach number within the desired tolerance with less control effort, completing the alpha sweep in less time, which is the most important figure of merit for the utilization of the wind tunnel facility (relates to the time necessary to conduct the experiment).

Table 2: Comparison of control schemes

	Controller (Cont)	Operator (Oper)	PMMSC	% reduction (Cont/Oper)
Mean	0.8497	0.8500	0.8497	-
SD	0.001527	0.001358	0.001226	20/10
T_{out} (sec)	46.5	34.52	33.2	29/4
L1	10.6	12.33	6.3	40/49
Alpha sweep (sec)	886	930	806	9/13

An additional metric on the control, the control density, was calculated by taking the sum of the absolute value of the control over a 50 sample sliding window:

$$\xi(k) = \sum_{i=0}^{49} |u(k-i)| \tag{36}$$

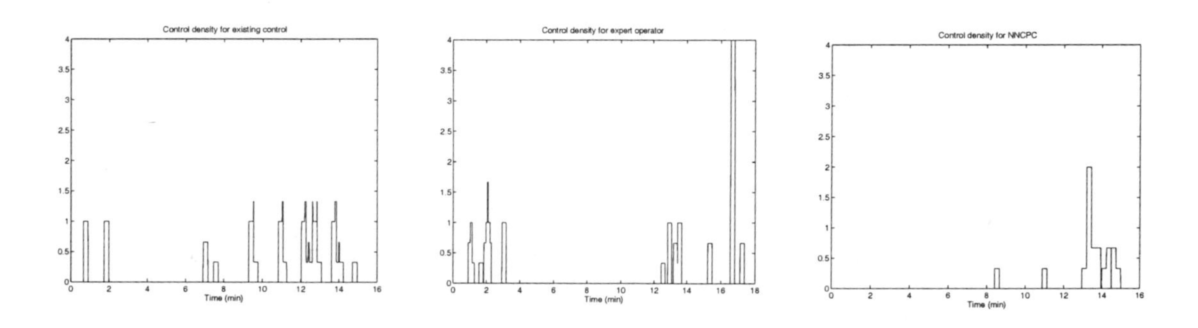

Figure 15. Comparison of Control Densities (controller, operator, PMMSC)

The control density is used to compare the sparseness of the control between the PMMSC, the existing controller, and an expert operator. This quantity measures the accuracy of the present control input so in the PMMSC case it is a measure of the accuracy of the local linear models to predict the tunnel dynamics. As illustrated in Figure 15, the PMMSC is clearly the most sparse, but allows for increased density of control when demanded by the external disturbance. In this respect it is similar to the variation in control density employed by the expert operator. This is in contrast to the existing automatic control, with fixed gains for a particular operating point resulting in a narrow range of control density.

Figure 16 compares the result of controlling the Mach number to several different set points over a nominal 28 minute interval which involved operating point changes and regulation. The goal here is to show how the PMMSC handles ramping up and down the tunnel Mach number. Mach number set points of 0.95, 0.9, and 0.6 are common to all three controllers. The PMMSC controls the Mach number to the intermediate value of 0.85 instead of 0.8 for the operator and existing controller. This difference is minimal so this still provides a reasonable basis for comparison of the controllers. The angle-of-attack was varied extensively during all three runs. Again, the PMMSC maintains the Mach number within tolerance for a higher percentage of the time, with less expenditure of control effort. Table 3 lists the time out of tolerance and the L1 norm for the three runs. The PMMSC reduces the time out of tolerance on the order of 15-20 percent compared to the existing automatic control or an expert operator.

Table 3: Comparison for control schemes

	Controller	Operator	PMMSC	% reduction
T_{out} (sec)	329	310	266	19/16
L1 norm	424	466	374	12/20

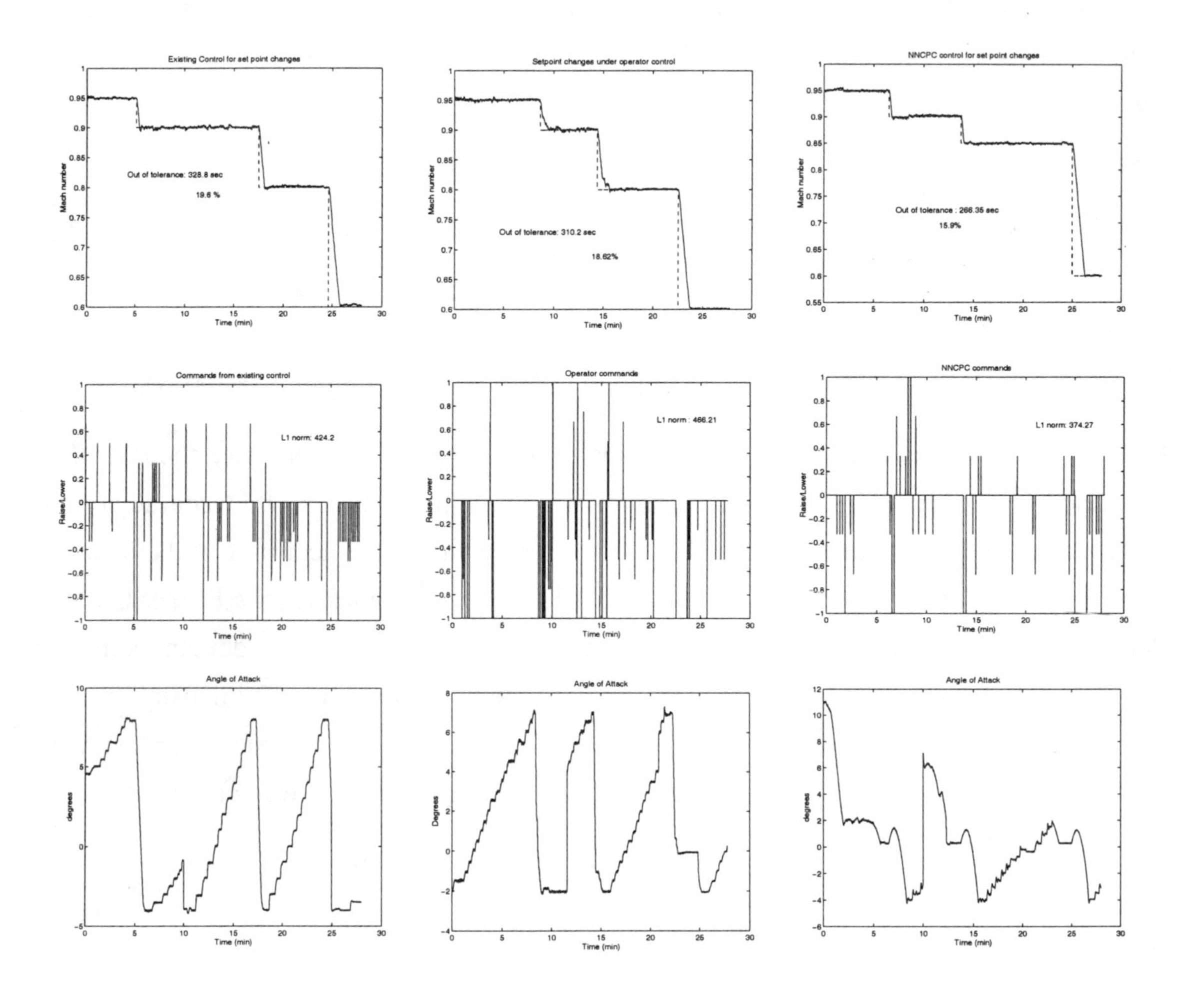

Figure 16. Comparison for controlling to several different set points (controller, operator, PMMSC).

9. Conclusions

Nonlinear dynamical modeling is a rich area with many possible applications. Nonlinear modeling seeks to estimate the nonlinear system that produced the time series under study. Hence it requires a nonlinear global model or a set of local linear models. Local linear modeling is conceptually elegant and requires mild assumptions about the underlying dynamics, but it is hindered by possible discontinuities among the local linear models.

We propose an innovative modeling scheme based on a self-organizing map (SOM) which works not only as an enhanced clustering algorithm which preserves neighborhoods of the reconstruction space dynamics, but also as an estimator of the local linear models. We presented the mathematical principles that support the clustering model and derived a procedure to directly estimate the parameters of the local linear models from the SOM weights. The proposed enhanced SOM architecture implements naturally the necessary steps of identifying a nonlinear dynamical system from a time series.

Our approach has several advantages: First, the SOM based modeling is computationally more efficient than other local models, which makes it very appealing for practical system identification applications. Second, since in our approach the SOM is trained globally to cluster the dynamics, the problem of discontinuities among the local models is reduced. In order to minimize the discontinuities, we applied a weighted least squares estimation to derive the local linear models. Third, we improved Kohonen learning for this task. The conventional Kohonen learning responds only to the density of samples in the input space, while here the fitting error to the trajectory is also important. Therefore we included an extra term in the Kohonen learning which assigns more PEs to regions of the trajectory where the error is larger. This improves the fidelity of the mapping.

It is very interesting to observe that in the appropriately trained SOM for dynamic modeling, the trajectory of the winning PE in the output space describes a complicated path that is a projection of the reconstruction space trajectory. The beauty of the approach is that the complexity of the trajectory is naturally decomposed in the appropriate switching among very simple linear models, which leads to accurate dynamic modeling and very direct engineering applications in system identification and controls.

Our proposed SOM approach is relevant for the switching controllers for two reasons: first it guaranteed that the repertoire of dynamics used for training the SOM are appropriately represented at the output of the SOM. Second, the same system that is performing the mapping is utilized to derive the local linear models, which makes the identification of the plant very compact and computationally efficient. In a sense, the diverse plant dynamics are captured in a compact look-up table of linear models.

We applied this concept to the control of the NASA Langley transonic wind tunnel with very good results. The control problem is one of regulating the set point of the plant under nonstationary loads (produced by the modification of the angle of attack of the aircraft model inside the tunnel). There are several peculiarities to this problem that made this solution so simple and accurate. First, the wind tunnel is stable, which means that the problem is one of regulation. Second, there is plenty of data to train a locally linear dynamic model of the tunnel dynamics under different load conditions. Third, the control input is a time series with 3 levels which enables a meaningful clustering of the control input is a small set of representative control commands.

Hence we can discretize the control space, and simplify the modeling of the tunnel dynamics with the control input as a set of seven "different" wind tunnels. For each we modeled the dynamics with our proposed SOM approach. Once the selected SOM determines the winner PE the system has a local linear model of the plant which can be used to automatically predict the Mach number produced by a small, but effective, set of control commands. Hence the control command that best meets the set point is used as the next command input. Regulating around a set point is distilled into the selection of a local linear model that is used to predict the tunnel dynamics and determines which is the best possible control input (hence the name predictive multiple model switching controller PMMSC). The PMMSC is easily implemented for on-line operation in a 486 PC and was utilized to control the wind tunnel during 9 hours for 3 different aerodynamic modeling experiments. We demonstrated the accuracy of our controller during these runs and showed that it outperformed the existing controller and even trained operators.

We realize that PMMSC is not a general control architecture. The PMMSC is set in advance, so it is not able to adjust to events that are very different from the ones used during training. But what is remarkable is that switching appropriately among a set of 140 local models was able to control to within 0.003 the Mach number of the wind tunnel during normal experimental conditions which create nonstationary loading. Very probably conditions different from the ones used to train the system were encountered during the 9 hour test runs, but as long as the new dynamics could be reasonably approximated locally by the developed dynamic code books the performance remained within bounds. This attests the power of local linear dynamic modeling, and of the proposed SOM based architecture. This reasoning also clearly points the direction of further work. Once we have this simple method of capturing a priori the dynamics about the task, the goal should be to integrate the SOM dynamic infrastructure in more complex control architectures such as the adaptive critics or adaptive controls. These methods tend to work with no a priori information and therefore are very slow to converge. Inclusion of the SOM will speed up the convergence, preserving at the same time their characteristics of adapting to new situations.

Acknowledgments: This work was partially supported by NSF grant ECS-9510715 and ONR grant N00014-94-1-0858.

10. References

[1] Abarbanel, H.D.I., R. Brown, J.J. Sidorowich, and L.S. Tsimring, "The analysis of Observed chaotic data in physical systems," Rev. Mod. Phy., Vol. 65, No. 4, pp. 1331, 1993

[2] Ahalt, S.C., et al., "Competitive learning algorithms for vector quantization," Neural Networks 3, pp. 277, 1990.

[3] Box G. E. P., and Jenkins G. M., "Time series analysis: forecasting and control", Holden-Day, 1976.

[4] Broomhead D.S. and Lowe D., "Multivariate functional interpolation and adaptive networks", Complex Systems 2, 321-355, 1988.

[5] Bryant, P., R. Brown, and H.D.I. Abarbanel, "Lyapunov exponents from observed time series," Phys. Rev. Lett. 65, pp. 1523, 1990.

[6] Capone, F.J., Bangert, L.S., Asbury, S.C., Mills,C.T.L, Bare, E.A., The NASA Langley 16-Foot Tunnel, NASA Technical Paper 3521 , September 1995

[7] Casdagli, M., "Nonlinear prediction of chaotic time series," Physics D 35, pp. 335, 1989.

[8] Cherkassky, V., and H. Lari-Najafi, "Constrained Topological Mapping for Nonparametric Regression Analysis," Neural Networks, Vol. 4, pp. 27, 1991.

[9] Cherkassky V., Gehring D., Mulier F., "Comparison of adaptive methods for function estimation from samples", IEEE Trans. Neural networks, vol 7, #4, 969-984, 1996.

[10] Crutchfield, J.P., and B.S. McNamara, "Equations of motion from a Data Series," Complex Systems 1, pp. 417, 1987.

[11] Farmer, J.D., Sidorowich J.J., "Predicting chaotic time series," Physical Review Letters, vol. 59, No. 8, pp. 845, 1987.

[12] Giona, M., F. Lentini, and V. Cimagalli, "Functional reconstruction and local predic-

tion of chaotic time series," Phys. Rev. A 44, pp. 3496-3502, 1991.

[13] Grassberger, P. and I. Procaccia, "Characterization of strange attractors," Physical Review Letters, Vol. 50, No. 5, pp. 346-349, 1983.

[14] Guilleman, V., and A. Pollack, "Differential Topology", Prentice-Hall, Englewood Cliffs, New Jersey, 1974.

[15] Hammel, S.M., "A noise reduction method for chaotic systems," Physics Letters A, 148, pp. 421, 1990.

[16] Haykin S., and Principe J., "Dynamic modeling of chaotic time series with neural networks", IEEE DSP Magazine, May issue, 1998.

[17] Haykin, S., "Neural Networks: A Comprehensive Foundation", Macmillan College Publishing, 1994.

[18] Kohonen, T., "Self-organizing feature maps", Springer Verlag, 1995.

[19] Kohonen, T., "The self-organizing map," Proceedings of the IEEE, Vol. 78, No. 9, 1990.

[20] Kostelich, E.J., and J.A. Yorke, "Noise reduction: finding the simplest dynamical system consistent with the data," Physica D, 41, pp. 183, 1990.

[21] Lapedes, R., and R. Farber, "Nonlinear signal processing using neural networks: prediction and system modeling," Technical Report LA-UR87-2662, Los Alamos National Laboratory, Los Alamos, New Mexico, 1987.

[22] Luttrell, S.P., "Self-organization: A derivation from first priciples of a class of learning algorithms," IEEE Conference on Neural Networks, pp. 495, Washington, DC, 1989.

[23] Martinetz, T.M., S.G. Berkovich and K.J. Schulten, ""Neural-Gas" network for vector quantization and its application to time-series prediction," IEEE Trans. on Neural Networks, Vol. 4, No. 4, July 1993.

[24] Motter M., "Control of the NASA Transonic wind tunnel with the Self-Organizing Feature Map", Ph.D. dissertation, University of Florida, Dec 1997.

[25] Narendra, K.S. , Balakrishnan, J., and Ciliz, M.K., "Adaptation and Learning Using Multiple Models, Switching, and Tuning", in IEEE Control Systems Magazine, Vol.15, No. 3, 1995.

[26] Narendra, K.S. , Balakrishnan, J., "Adaptive Control Using Multiple Models", in IEEE Transactions on Automatic Control, Vol.42, No. 2, pages 171-187, 1997.

[27] Narendra, K.S. , Mukhopadhyay, S. , (1997). "Adaptive Control Using Neural Net-

works and Approximate Models", in IEEE Transactions on Neural Networks, Vol.8, No.3, pages 475-485.

[28] Najai H., Cherkassky V., “Adaptive knot placement for nonparametric reression”, in Proc. Neural Inf. Proc. Systems, NIPS-6, 247-254, Morgan Kaufman, 1994

[29] Ott, E., C. Grebogi, and J.A. Yorke, “Controlling chaos,” Physical Review Letters, Vol. 64, No. 11, pp. 1196-1199, 1990.

[30] Haykin S. and Li X. B., “Detection of signals in chaos,” Proceedings IEEE 83, 94-122, 1995.

[31] Packard, N.H., J.P. Crutchfield, J.D. Farmer, and R.S. Shaw, “Geometry from a time series,” Phys. Rev. Lett. 45, pp. 712, 1980.

[32] Papoulis, A., “Probability, Random Variables, and Stochastic Processes”, 3rd ed., McGraw-Hill, New York, 1991.

[33] Priestley, M.B., “State-Dependent Models: A general approach to nonlinear time series analysis,” Journal of Time Series Analysis, 1, pp. 47, 1980.

[34] Principe, J.C. and Kuo J-M., “Dynamic modelling of chaotic time series with neural networks”, Proc. of Neural Infor. Proc. Sys, NIPS 7, 311-318, 1995.

[35] Principe, J.C. and Wang L., “Non-linear time series modeling with self-organizing feature maps”, Proc. IEEE Workshop Neural Networks for Signal Proc., 11-20, 1995.

[36] Principe J.C., Wang L., and J-M Kuo, “Chaotic time series modeling with neural networks”, in Signal Analysis and Prediction, Birkhauser, 275–289, 1998.

[37] Rissanen, J., “Stochastic Complexity in Statistical Inquiry”, World Scientific, 1989.

[38] Ritter, H., “Learning with the self-organizing map,” In Artificial Neural Networks (T. Kohonen, K Makisara, O. Simula, and J. Kangas, eds.), Vol. 1, pp. 379-384, Amsterdam: North Holland, 1991.

[39] Ritter H., “Parametrized self-organizing maps for vision tasks”, In Proc. ICANN’94, Italy, pp 803-810, 1994.

[40] Sano, M., and Y. Sawada, “Measurement of the Lyapunov spectrum from a chaotic time series,” Physical Review Letters, Vol. 55, No. 10, pp. 1082, 1985.

[41] Scharf, L.L., “Statistical Signal Processing--Detection, Estimation, and Time Series Analysis”, Addison-Wesley, 1989.

[42] Shinbrot, T., E. Ott, C. Grebogi, and J.A. Yorke, “Using chaos to direct trajectories to targets,” Phys. Rev. Lett. 65, pp. 3250, 1990.

[43] Sidorowich, J.J., "Modeling of chaotic time series for prediction, interpolation, and smoothing," IEEE ICASSP, 4, pp. 121, 1992.

[44] Singer, A.C., Wornell G., Oppenheim A., "Codebook prediction: a nonlinear signal modeling paradigm," IEEE ICASSP, 5, pp. 325, 1992.

[45] Takens, F., "Detecting strange attractors in turbulence," in Dynamical Systems and turbulence, ed. by D.A. Rand and L.-S. Young, Springer Lecture Notes in Mathematics, 898, pp. 365-381, Springer-Verlag, New York, 1980.

[46] Goodwin, G.C., and K.S. Sin, "Adaptive filtering, prediction, and control", Prentice-Hall, 1984.

[47] Tong, H., "Non-linear Time Series Analysis: A Dynamical Systems Approach", Oxford University Press, Oxford, 1990.

[48] Walter, J."Non-linear prediction with self-organization maps," IJCNN, I-589, 1990.

[49] Walter J., Ritter H., "Local PSOMs and Chebyshev PSOMs improving the PSOM", Proc Int. Conf. ANNs (ICANN95), vol 1, 95-102, 1995.

[50] Walter J., Ritter H., "Rapid learning with parametrized self-organizing maps", Neuro-Computing 12, pp 131-153, 1996

[51] Wang, L., "Local linear dynamic modeling with self organizing feature maps", Ph.D. Dissertation, U. of Florida, 1996.

[52] Weigend, A.S., and N.A. Gershenfeld, "Time Series Prediction: Forecasting the future and understanding the past", Addison-Wesley Publishing Company, 1984.

[53] Widrow, B., and S.D. Stearns, "Adaptive Signal Processing", Prentice-Hall, Englewood Cliffs, NJ, 1985.

[54] Wolf, A., J.B. Swift, H.L. Swinney, and J.A. Vastano, "Determining Lyapunov exponents from a time series," Physica D 16, pp. 285, 1985.

[55] Vesanto J., "Using the SOM and local linear models in time series prediction", Workshop on Self-Organizing Maps WSOM97, Helsinky, Finland, Aug 1997.

11. Appendix

Let $\boldsymbol{X}$, $\boldsymbol{A}$ and Φ denote the continuous input space, spatially discrete output space and a competitive projection respectively. The proposed non-linear modeling scenario follows three steps: *a*. Reconstruction of the state space from the input signal; *b*. Training the SOM; *c*. Estimation of the locally linear predictors.

* *Reconstruction of the state space from the training signal.*

Embed the time series in a space of dimension N+1, where N is Takens' embedding dimension ($N \geq 2D_e$). A sequence of N+1 dimensional state vectors$[x(n+\tau), \boldsymbol{x}(n)]$ is created from the time series, where $\boldsymbol{x}(n) = [x(n), x(n-\tau), \ldots, x(n-(N-1)\tau))]^T$ and τ is the appropriate time delay.

* *Training the SOM model.*

This step is accomplished via the Kohonen learning process (or preferably with the improved dynamic learning of section 5.1). With each vector-scalar pair $[x(n+\tau), \boldsymbol{x}(n)]$ presented at the network input, train the system with Eq. 15. The learning process adaptively discretizes the continuous input space $\boldsymbol{X} \subset \mathbf{R}^{N+1}$ into a set of K disjoint cells $\mathbf{A}$ to construct the mapping Φ: $\boldsymbol{X} \to \boldsymbol{A}$. This process continues until the learning rate decreases close to zero and the neighborhood function covers about one output PE. After learning, the output space lattice $\boldsymbol{A}$ represents the input space $\mathbf{X}$. The mapping Φ represents the input space data distribution and preserves local neighborhood relations.

* *Estimate the locally linear predictors.*

Once the SOM is trained, for each PE $u_i \in \boldsymbol{A}$, the corresponding local linear predictor coefficients $[\boldsymbol{a}_i^{\mathrm{T}}, b_i]$ are estimated based on $\alpha_i \subset \boldsymbol{A}$, which is a set of N_L elements in the neighborhood of u_i including u_i itself. Each element u_i, has a corresponding weight vector $[w_{i0}, \boldsymbol{w}_{ij}]^T \in \mathbf{R}^{N+1}$, where $\boldsymbol{w}_{ij} = [w_{i1}, w_{i2}, \ldots, w_{iN}]$. The local prediction model $[\mathbf{a}_i^{\mathrm{T}}, b_i]$ is fitted in the least-square sense to the set of weights in α_i, i.e.

$$w_{i0} = b_i + \boldsymbol{a}_i^T \boldsymbol{w}_{ij} \tag{37}$$

To ensure a stable solution of the above equations, α_i must have more than N+1 elements. Thereafter for each output unit $u_i \subset \boldsymbol{A}$, a unique linearly local model function $\tilde{f}(\boldsymbol{a}_i(.), b_i(.))$ is constructed.

Chapter 7

A Signal Processing Framework Based on Dynamic Neural Networks with Application to Problems in Adaptation, Filtering and Classification

Lee A. Feldkamp and Gintaras V. Puskorius
Ford Research Laboratory, P.O. Box 2053
Dearborn, MI 48121–2053
lfeldkam@ford.com, gpuskori@ford.com

Abstract

We present in this paper a coherent framework, based on the use of time-lagged recurrent neural networks, for solving a variety of difficult signal processing problems. The framework relies on the assertion that time-lagged recurrent networks possess the necessary representational capabilities to act as universal approximators of nonlinear dynamical systems. This property applies to modeling problems posed as system identification, time-series prediction, nonlinear filtering, adaptive filtering, or temporal pattern classification. We address the development of models of nonlinear dynamical systems, in the form of time-lagged recurrent neural networks, which can be used without further training (i.e., as fixed-weight networks). We concentrate here on the recurrent multilayer perceptron (RMLP) architecture, which generalizes the standard MLP with the possibility of single-delay connections within each layer. We have found that a weight update procedure based on the extended Kalman filter (EKF) is far more effective, for both feedforward and recurrent networks, than simple first-order gradient methods. As a solution to the recency effect, *the tendency for a network to forget earlier learning as it processes new examples, we have developed a technique called* multi-stream *training.*

We demonstrate our training framework by applying it to four problems. First, we show that a single time-lagged recurrent neural network can be trained not only to produce excellent one-time-step predictions for two different time series, but also to be robust to severe errors in the input sequence with which it is provided. The second problem involves the modeling of a complex system containing significant process noise, which was shown in [1] to lead to unstable trained models. We illustrate how multi-stream training may be used to enhance the stability of such models.

The remaining two problems are drawn from real-world automotive applications. The first of these involves input-output modeling of a signal that reflects the dynamic behavior of a catalytic converter in the exhaust stream. Finally we consider real-time and continuous detection of engine misfire. This is cast as a signal processing problem that requires a binary decision be made at each time step on the basis of the sequence of available input variables.

1 Introduction

We consider in this paper a practical and quite general framework with which we have approached many different types of problems that involve the temporal processing of signals. Some of these problems have been cast into a more or less standard format, i.e., a single signal, sampled in time, upon which we are expected to perform prediction, filtering, or estimation. More commonly, however, the problems we have encountered have had more of a mixed character. In particular, such problems usually involve a primary signal and one or more additional signals which, taken together, provide context. The problem statement might then involve prediction, estimation (particularly in the form of *virtual sensors*), or classification. Furthermore, we usually find that the systems involved cannot really be regarded as stationary; on the other hand, they are not subject to unlimited variation. Perhaps the best description is a progression among several ill-defined modes of operation. The systems we deal with are usually driven rather than autonomous, though we seldom have access to the underlying drivers.

Our philosophy has been to take a fairly general input-output point of view, in which we assess the available input information sequence and seek to transform it into the required output sequence. Implicit in this view is that we can, in fact, assemble a desired output sequence. (This may be contrasted with problems such as certain types of control, in which desired system outputs are not necessarily available at every moment, and in which the desired output of a neural controller is not provided and must be constructed. The methods described here have been used in such applications as well.)

Our technical approach has been to develop a methodology for effective training of *time-lagged recurrent networks*, usually in the form of *recurrent multilayer perceptrons* (RMLP). The latter are a natural synthesis of feedforward multilayer perceptrons and single-layer fully recurrent networks. For an introduction to neural networks, the reader may wish to consult a textbook, such as [2] or [3].

From a structural point of view, RMLPs are appealing by virtue of subsuming many traditional signal processing structures, including tapped delay lines, FIRs, IIRs, and gamma networks, while retaining the universal approximation property of feedforward networks. Because they have state variables, RMLPs can represent dynamic systems with *strongly hidden state* [4]; indeed, Lo [5] suggests that RMLPs are universal approximators for dynamic systems, just as feedforward MLPs are universal approximators for static mappings (see, for example, [6]). Further, the work of Cotter and Conwell [7] suggests that time-lagged fixed-weight recurrent networks can produce behavior that would usually be called adaptive. In [8, 9], we presented an example of such behavior. Though the referenced work was purely abstract, it was motivated by a contemporaneous practical application involving a system that exhibits nonstationarity over a bounded range.

Recurrent networks have often been regarded as difficult to train. Although we contend that an effective procedure can be carried out relatively routinely, the difficulties presented in training are not mere illusions. In addition to compounding the usual pitfalls encountered in training feedforward networks, including poor local minima, recurrent networks enhance the difficulty of training by the *recency effect*. Simply stated, a tendency always exists for recent weight updates to cause a network to forget what it has learned in the past. Of course, this tendency exists also in the training of feedforward networks, but in that case easy and effective countermeasures exist, viz., scrambling the order of presentation of input-output pairs or employing batch learning (in which an update may be based on all examples). Such methods are cumbersome to employ in training recurrent networks, because the temporal order of the data sequences must be respected. Although recurrent networks can be trained with first- or second-order batch weight update procedures, we prefer to utilize sequential training methods because of frequent weight updates and the benefits of their stochastic characteristics.

In the remainder of this paper, we have organized our presentation as follows. Section 2 presents briefly the RMLP architecture and notes how various signal processing structures can be formed explicitly. Section 3 begins discussion of the training method by describing the calculation of derivatives (gradients). Section 4 describes how these derivatives are used in a second-order weight update procedure, based on the extended Kalman filter. Section 4 also describes our approach to mitigating the recency effect, which we term multi-stream training (or simply multi-streaming). Here we also relate an enabling side benefit of multi-streaming: the ability to coerce desired secondary network behavior at the same time the network is being trained on its primary objective. In Section 5, we discuss two synthetic examples. First, we describe a problem related to the multiple system modeling of [8, 9], with the added requirement of dealing with dropouts in a faulty data sequence. Then we illustrate the modeling of a complex system, presented in [1], together with a practical approach to imposing stability on the neural model. Section 6 is devoted to problems taken from real-world systems. The first of these is a modeling/estimation problem; beyond merely complicated behavior, this application requires that one deal with delays that vary substantially over the operating range of the system. Then we discuss a classification problem in which faults must be detected on the basis of an apparently hopelessly noisy input sequence. Finally, in Section 7 we summarize and make some concluding remarks.

2 Network Architecture and Execution

An RMLP consists of one or more layers of computing nodes, just as in a standard feedforward network or MLP. A simple RMLP architecture is illustrated in Figure 1. In the basic RMLP form we either have full recurrence within a layer, which means that every node is connected through unit time delays to every node in the same layer, or else no recurrence for that layer. (A convenient feature of the RMLP architecture is that it reduces to a simple MLP in the absence of recurrence.) Recurrence can also be present from network outputs to network inputs. It is also possible to

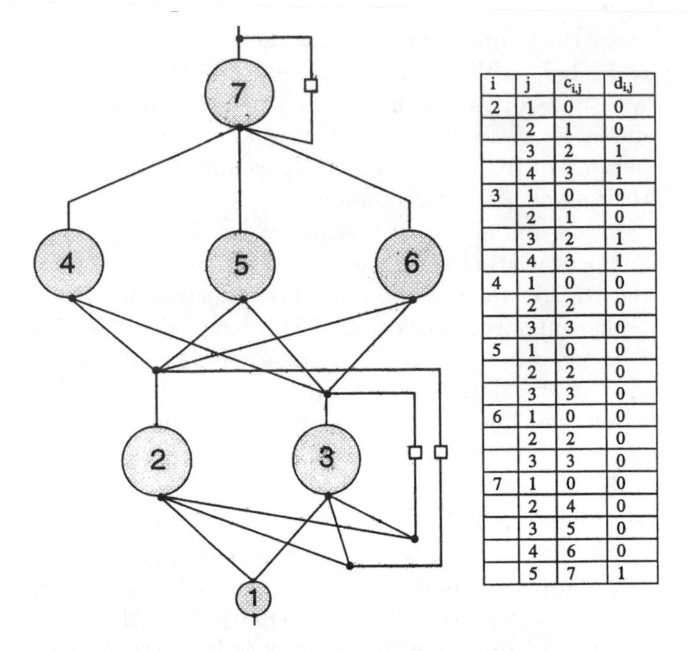

i	j	$c_{i,j}$	$d_{i,j}$
2	1	0	0
	2	1	0
	3	2	1
	4	3	1
3	1	0	0
	2	1	0
	3	2	1
	4	3	1
4	1	0	0
	2	2	0
	3	3	0
5	1	0	0
	2	2	0
	3	3	0
6	1	0	0
	2	2	0
	3	3	0
7	1	0	0
	2	4	0
	3	5	0
	4	6	0
	5	7	1

Figure 1: Schematic illustration of an RMLP, denoted as 1-2R-3-1R. The first hidden layer of two nodes is fully recurrent, the second hidden layer (three nodes) is feedforward, and the output node is recurrent. The small boxes denote unit time delays. The table contains elements of the connection and delay arrays. Entries for index i are absent for the bias (treated as node 0) and the network input (node 1), since neither receives input from any other node. Bias connections are present, as indicated in the connection table, but are not drawn. Note that recurrent connections have unit delays. The elements of the input and output arrays are: $I_1 = 1$ and $O_1 = 7$.

have a sparse connection pattern. Indeed, it is the use of sparse connections together with flexible assignment of node activation functions that permit an RMLP to be structured to act as traditional linear filters (e.g., FIRs, IIRs), gamma networks, time-delay networks, Elman and Jordan networks, and other useful computational forms. This means, of course, that a basic RMLP can, in principle, be trained to subsume any structured network which it contains, although success in training cannot be guaranteed.

It is possible to restructure an RMLP so that it is expressed as a pure feedforward network with external (output-to-input) recurrent connections. An awkward aspect of this transformation is that it is necessary to replicate some of the nodes. Though convenient for some purposes, we feel that such a restructuring contributes little if anything to the training process.

For compactness of presentation, we shall express the operation of an RMLP by treating it as a special case of an *ordered network* [10]. The forward equations for an ordered network with `n_in` inputs and `n_out` outputs may be expressed very compactly, using a pseudocode format similar to that in [11].

Let the network consist of **`n_nodes`** nodes, including `n_in` nodes which serve as receptors for the external inputs, but not including the bias input, which we denote formally as node 0. The latter is set to the constant 1.0. The array **I** contains a list of the input nodes; e.g., I_j is the number of the node that corresponds to the `j`th input, in_j. Similarly, a list of the nodes that correspond to network outputs out_p is contained in the array **O**. We allow for the possibility of network outputs and targets to be advanced or delayed with respect to node outputs by assigning a phase τ_p to each output. In most cases these phases are zero. Node i receives input from `n_con(i)` other nodes and has activation function $f_i(\cdot)$; `n_con(i)` is zero if node `i` is among the nodes listed in the array **I**. The array **c** specifies connections between nodes; $c_{i,j}$ is the node number for the `j`th input for node `i`. Inputs to a given node may originate at the current or past time steps, according to delays contained in the array **d**, and through weights for time step t contained in the array $\mathbf{W}(t)$. Summarizing, the `j`th input to node `i` at time step t is the output of node $c_{i,j}$ at time $t - d_{i,j}$ and is connected through weight $W_{i,j}(t)$.

Prior to beginning network operation, all appropriate memory is initialized to zero. At the beginning of each time step, we execute the following buffer operations on weights and node outputs (in practical implementation, a circular buffer and pointer arithmetic may be employed). Here **`dmax`** is the largest value of delay represented in the array **d**, and **`h`** is the truncation depth of the backpropagation through time gradient calculation that is described in the following section.

for i = 1 **to** n_nodes {
for i_t = t-h-dmax **to** t-1 {

$$\mathbf{W}(i_t) = \mathbf{W}(i_t + 1) \quad (1)$$

$$y_i(i_t) = y_i(i_t + 1) \quad (2)$$

} }

Then the actual network execution is expressed as

for i = 1 **to** n_in {

$$y_{\mathrm{I}_i}(t) = \mathrm{in}_i(t) \quad (3)$$

}
for i = 1 **to** n_nodes {

```
if n_con(i) > 0 {
```

$$y_i(t) = f_i(\sum_{j=1}^{n\_con(i)} W_{i,j}(t) y_{c_{i,j}}(t - d_{i,j})) \quad (4)$$

```
}
}
for p = 1 to n_out {
```

$$\text{out}_p(t + \tau_p) = y_{\text{O}_p}(t) \quad (5)$$

```
}
```

This description of an ordered network does not explicitly involve the concept of *layers*; the layered structure is imposed implicitly by the connection pattern. Pure delays can be described directly, so that tapped delay lines on either external or recurrent inputs are conveniently represented.

3 Gradient Calculation

After the forward propagation at time step t, we compute gradients in preparation for the weight update step. In the past we have made extensive use of forward methods of derivative calculation [12, 13]. Some time ago, however, we replaced the forward method with a form of truncated backpropagation through time (BPTT(h)) [10, 14]. With the truncation depth h suitably chosen, this method produces derivatives that closely approximate those of the forward method with greatly reduced complexity and computational effort.

We describe here the mechanics of the particular form of BPTT(h) we have most commonly employed. We use the Werbos notation in which $\text{F\_}^{q}x$ denotes an ordered derivative of some quantity q with respect to x. In this form of BPTT, $\text{F\_}^{p}x$ denotes the ordered derivative of network output node p with respect to x.

To derive the backpropagation equations, the forward propagation equations are considered in reverse order. From each we derive one or more backpropagation expressions, according to the principle that if $a = g(b, c)$, then $\text{F\_}^{q}b \mathrel{+}= \frac{\partial g}{\partial b}\text{F\_}^{q}a$ and $\text{F\_}^{q}c \mathrel{+}= \frac{\partial g}{\partial c}\text{F\_}^{q}a$. The C-language notation "+ =" indicates that the quantity on the right hand side is to be added to the previously computed value of the left hand side. In this way, the appropriate derivatives are distributed from a given node to all nodes and weights that feed it in the forward direction, with due allowance for any delays that might be present in each connection. The simplicity of the formulation reduces the need for visualizations such as unfolding in time or signal-flow graphs.

```
for p = 1 to n_out {
for i = 1 to n_nodes {
for i_t = t to t-h-1 {
```

$$\text{F\_}^{p}y_i(i_t) = 0 \quad (6)$$

```
}} /* end i and i_t loops */
```

$$\text{F\_}^{p}y_{\text{O}_p}(t) = 1 \quad (7)$$

$$\xi_p(t) = \text{tgt}_p(t + \tau_p) - \text{out}_p(t + \tau_p) \quad (8)$$

```
for i_h = 0 to h {
```

$$i_1 = t - i_h \quad (9)$$

```
for i = n_nodes to 1 {
if n_con(i) > 0 {
for k = n_con(i) to 1 {
```

$$j = c_{i,k} \quad (10)$$

$$i_2 = i_1 - d_{i,k} \quad (11)$$

$$\text{F\_}^{p}y_j(i_2) \mathrel{+}= \gamma^{d_{i,k}} \text{F\_}^{p}y_i(i_1) W_{i,k}(i_1) f'(y_i(i_1)) \quad (12)$$

$$\text{F\_}^{p}W_{i,k} \mathrel{+}= y_j(i_2) \text{F\_}^{p}y_i(i_1) f'(y_i(i_1)) \quad (13)$$

```
} /* end k loop */
}
} /* end i loop */
} /* end i_h loop */
} /* end p loop */
```

Here (6) serves to initialize the derivative array, while (7) expresses the fact that $\frac{\partial y_{\text{O}_p}(t)}{\partial y_{\text{O}_p}(t)} = 1$. The error $\xi_p(t)$ computed in (8) is used in the weight update described in the next section, where the desired value of network output $\text{out}_p(t + \tau_p) = y_{\text{O}_p}(t)$ is denoted as $\text{tgt}_p(t + \tau_p)$. The actual backpropagation occurs in expressions (12) and (13), which derive directly from the forward propagation expression (4). We have included a discount factor γ in expression (12), though it is often set merely to its nominal value of unity.

4 EKF Multi-Stream Training

4.1 The Kalman Recursion

We have made extensive use of training that employs weight updates based on the extended Kalman filter method first proposed by Singhal and Wu [15]. (For background material on the Kalman filter, see [16, 17].) In most of our work, we have made use of a decoupled version of the EKF method [18, 13], which we denote as

DEKF. Decoupling was crucial for early practical use of the method, when speed and memory capabilities of workstations and personal computers were severely limited. At the present time, many problems are small enough to be handled with what we have termed *global* EKF, or GEKF. In many cases, the added coupling brings benefits in terms of quality of solution and overall training time. However, the increased time required for each GEKF update is a potential disadvantage in real-time applications.

For generality, we present the decoupled Kalman recursion; GEKF is recovered in the limit of a single weight group ($g = 1$). The weights in $\mathbf{W}$ are organized into g mutually exclusive weight groups; a convenient and effective choice has been to group together those weights that feed each node. Whatever the chosen grouping, the weights of group i are denoted by $\mathbf{w}_i$. The corresponding derivatives $\overset{p}{\text{F\_}}\mathbf{w}_i$ are placed in `n_out` columns of $\mathbf{H}_i$.

To minimize a cost function $E = \sum_t \frac{1}{2}\boldsymbol{\xi}(t)^T\mathbf{S}(t)\boldsymbol{\xi}(t)$, where $\mathbf{S}(t)$ is a nonnegative definite weighting matrix and $\boldsymbol{\xi}(t)$ is the vector of errors, at time step t, the recursion equations are as follows [11]:

$$\mathbf{A}^*(t) = \left[\frac{1}{\eta(t)}\mathbf{I} + \sum_{j=1}^{g}\mathbf{H}_j^*(t)^T\mathbf{P}_j(t)\mathbf{H}_j^*(t)\right]^{-1}, \tag{14}$$

$$\mathbf{K}_i^*(t) = \mathbf{P}_i(t)\mathbf{H}_i^*(t)\mathbf{A}^*(t), \tag{15}$$

$$\mathbf{w}_i(t+1) = \mathbf{w}_i(t) + \mathbf{K}_i^*(t)\boldsymbol{\xi}^*(t), \tag{16}$$

$$\mathbf{P}_i(t+1) = \mathbf{P}_i(t) - \mathbf{K}_i^*(t)\mathbf{H}_i^*(t)^T\mathbf{P}_i(t) + \mathbf{Q}_i(t). \tag{17}$$

In these equations, the weighting matrix $\mathbf{S}(t)$ is distributed into both the derivative matrices and the error vector: $\mathbf{H}_i^*(t) = \mathbf{H}_i(t)\mathbf{S}(t)^{\frac{1}{2}}$ and $\boldsymbol{\xi}^*(t) = \mathbf{S}(t)^{\frac{1}{2}}\boldsymbol{\xi}(t)$. The matrices $\mathbf{H}_i^*(t)$ thus contain the scaled derivatives of network outputs with respect to the ith group of weights; the concatenation of these matrices forms a global scaled derivative matrix $\mathbf{H}^*(t)$. A common global scaling matrix $\mathbf{A}^*(t)$ is computed with contributions from all g weight groups, through the scaled derivative matrices $\mathbf{H}_j^*(t)$, and from all of the decoupled approximate error covariance matrices, $\mathbf{P}_j(t)$. A user-specified learning rate, $\eta(t)$, appears in this common matrix. For each weight group i, a Kalman gain matrix $\mathbf{K}_i^*(t)$ is computed and is then used in updating the values of the group's weight vector $\mathbf{w}_i(t)$ and in updating the group's approximate error covariance matrix $\mathbf{P}_i(t)$. Each approximate error covariance update is augmented with the addition of a scaled identity matrix, $\mathbf{Q}_i(t)$, that represents the effects of artificial process noise.

In practice, the EKF recursion is typically initialized by setting the approximate error covariance matrices to scaled identity matrices, with a scaling factor of 100 for nonlinear nodes and 1000 for linear nodes. At the beginning of training, we generally set the learning rate low (the actual value depends on characteristics of the problem, but $\eta = 0.1$ is a typical value), and start with a relatively large amount of process noise, e.g., $\mathbf{Q}_i(0) = 10^{-2}\eta\mathbf{I}$. We have previously demonstrated that the artificial process noise extension accelerates training, helps to avoid poor local minima during training, and assists the training procedure in maintaining the necessary property of nonnegative definiteness for the approximate error covariance matrices [18]. As training progresses, we generally decrease the amount of process noise to a limiting value of approximately $\mathbf{Q}_i(t) = 10^{-6}\mathbf{I}$, and increase the learning rate to a limiting value no greater than unity. In addition, we have also found that occasional reinitializations of the error covariance matrices, along with resetting of initial values for the learning rate and process noise terms, may benefit the training process. Finally, one should note that initial choices for the learning rate and error covariance matrices are not independent: a multiplicative increase in the scaling factor for the approximate error covariance matrices can be cancelled by reducing the initial learning rate by the inverse of the scaling factor (the relative scalings of the learning rate and error covariance matrices affect the choice of the artificial process noise term as well).

We wish to emphasize here that the decoupled EKF recursion given by equations (14–17) only performs updates to network weights, and not to the states of the dynamical network model. On the other hand, Matthews [19] and Williams [20] have described *parallel* EKF formulations for recurrent networks in which both network states and weights are updated during training. However, these parallel formulations present certain difficulties. First, the parallel techniques require that gradients be computed by forward methods, and it is not obvious how computationally efficient methods such as BPTT(h) can be used within the parallel formalism. Similarly, we are primarily interested in training recurrent networks to deploy as fixed-weight systems, without having to incur the computational cost of performing the Kalman recursion for state estimation during network use. Finally, we have found, as shown below, that properly trained fixed-weight recurrent networks appear to be capable of performing many filtering, estimation and prediction tasks that are often thought to be suitable candidates for traditional extended Kalman filtering, and it is not obvious that additional filtering of model states would be beneficial.

4.2 Multi-Stream Training

Consider the standard recurrent network training problem: training on a sequence of input-output pairs. If the sequence is in some sense homogeneous, then one or more linear passes through the data may well produce good results. However, in many training problems, especially those in which exogenous inputs are present, the data sequence is heterogeneous. For example, regions of rapid variation of inputs and outputs may be followed by regions of slow change. Or a sequence of outputs that centers about one level may be followed by one that centers about a different level. For any of these cases, in a straightforward training process the tendency always exists for the network weights to be adapted unduly in favor of the currently presented training data. This *recency effect* is analogous to the difficulty that may arise in training feedforward networks because of the order in which the training data are presented.

In this latter case, an effective solution is to scramble the order of presentation; another is to use a batch update algorithm. For recurrent networks, the direct analog of scrambling the presentation order is to present randomly selected sub-sequences, making an update only for the last input-output pair of the subsequence (when the network would be expected to be independent of its initialization at the beginning of the sequence). A full batch update involves running the network through the entire data set, computing the required derivatives that correspond to each input-output pair, and making an update based on the entire set of errors.

The multi-stream procedure largely circumvents the recency effect by combining features of both scrambling and batch updates. Like full batch methods, multi-stream training [21] is based on the principle that each weight update should attempt to satisfy simultaneously the demands from multiple input-output pairs. It retains, however, the useful stochastic aspects of sequential updating and requires much less computation time between updates. We now describe the mechanics of multi-stream training.

In a typical training problem, we deal with one or more files, each of which contains a sequence of data. Breaking the overall data set into multiple files is typical in practical problems, where the data may be acquired in different sessions, for distinct modes of system operation, or under different operating conditions.

In each cycle of training, we choose a specified number N_s of randomly selected starting points in a chosen set of files. Each such starting point is the beginning of a *stream.* The multi-stream procedure consists of progressing in sequence through each stream, carrying out weight updates according to the set of current points. Copies of recurrent node outputs must be maintained separately for each stream. Derivatives are also computed separately for each stream, generally by truncated backpropagation through time (BPTT(h)) as discussed above. Because we generally have no prior information with which to initialize the recurrent network, we typically set all state nodes to values of zero at the start of each stream. Accordingly, the network is executed but updates are suspended for a specified number N_p of time steps, called the *priming length,* at the beginning of each stream. Updates are performed until a specified number N_t of time steps, called the *trajectory length,* have been processed. Hence $N_t - N_p$ updates are performed in each training cycle.

If we take $N_s = 1$ and $N_t - N_p = 1$, we recover the order-scrambling procedure described above; N_t may be identified with the sub-sequence length. On the other hand, we recover the batch procedure if we take N_s equal to the number of time steps for which updates are to be performed, assemble streams systematically to end at the chosen N_s steps, and again take $N_t - N_p = 1$.

Generally speaking, apart from the computational overhead involved (see below), we find that performance tends to improve as the number of streams is increased. Various strategies are possible for file selection. If the number of files is small, it is convenient to choose N_s equal to a multiple of the number of files and to select each file the same number of times. If the number of files is too large to make this practical, then we tend to select files randomly. In this case, each set of $N_t - N_p$ updates is based on only a subset of the files, so it seems reasonable not to make the trajectory length N_t too large.

An important consideration is how to carry out the EKF update procedure. If first-order gradient updates were being used, we would simply average the updates that would have been performed had the streams been treated separately. In the case of EKF training, however, averaging separate updates is incorrect. Instead, we treat this problem as that of training a single shared-weight network with $N_s \times$ `n_out` outputs. From the standpoint of the EKF method, we are simply training a multiple output network in which the number of original outputs is multiplied by the number of streams. The nature of the Kalman recursion is then to produce weight updates which are not a simple average of the weight updates that would be computed separately for each output, as is the case for a simple

$$\mathbf{H} = \left(H_{\substack{p=1\\s=1}} \;\middle|\; H_{\substack{p=2\\s=1}} \;\middle|\; H_{\substack{p=1\\s=2}} \;\middle|\; H_{\substack{p=2\\s=2}} \;\middle|\; H_{\substack{p=1\\s=3}} \;\middle|\; H_{\substack{p=2\\s=3}} \right)$$

$$\xi = \left(\xi_{\substack{p=1\\s=1}} \; \xi_{\substack{p=2\\s=1}} \; \xi_{\substack{p=1\\s=2}} \; \xi_{\substack{p=2\\s=2}} \; \xi_{\substack{p=1\\s=3}} \; \xi_{\substack{p=2\\s=3}} \right)^T$$

Figure 2: Illustration of the augmentation of the derivative matrix and error vector for the case of two outputs and three streams.

gradient descent weight update.

In single-stream EKF training, we place derivatives of network outputs with respect to network weights in the matrix $\mathbf{H}$ constructed from `n_out` column vectors, each of dimension equal to the number of trainable weights, N_w. In multi-stream training, the number of columns is correspondingly increased to $N_s \times$ `n_out`. Similarly, the vector of errors $\boldsymbol{\xi}$ has $N_s \times$ `n_out` elements. Apart from these augmentations of $\mathbf{H}$ and $\boldsymbol{\xi}$, illustrated schematically in Figure 2, the form of the Kalman recursion is unchanged.

Let us consider the computational implications of the multi-stream method. The sizes of the approximate error covariance matrices $\mathbf{P}_i$ and the weight vectors $\mathbf{w}_i$ are independent of the chosen number of streams. The number of columns of the derivative matrices $\mathbf{H}_i^*$, as well as of the Kalman gain matrices $\mathbf{K}_i^*$, increases from `n_out` to $N_s \times$ `n_out`, but the computation required to obtain $\mathbf{H}_i^*$ and to compute updates to $\mathbf{P}_i$ is the same as for N_s separate updates. The major additional computational burden is the inversion required to obtain the $\mathbf{A}^*$ matrix, whose dimension is N_s times larger. Even this tends to be small compared to the cost associated with propagating the $\mathbf{P}_i$ matrices, as long as $N_s \times$ `n_out` is smaller than the number of network weights (GEKF) or the maximum number of weights in a group (DEKF).

If the number of streams chosen is so large as to make the inversion of $\mathbf{A}^*$ impractical, the inversion may be avoided by treating the multiple network outputs with a scalar cost function as described in [22]. The efficacy of multi-stream training performed in this fashion remains to be explored thoroughly, as the presumed advantage of pseudoinverse-like updates, as described below, may be lost.

4.3 Some Insight into the Multi-Stream Technique

A simple means of motivating how multiple training instances can be used simultaneously for a single weight update via the EKF procedure is to consider the training of a single linear node. In this case, the application of EKF training is equivalent to that of recursive least squares (RLS). (A recent discussion of the relationship between RLS and the Kalman filter may be found in [23].) Assume that a training data set is represented by m unique training patterns. The ith training pattern is represented by a d-dimensional input vector $\mathbf{x}(i)$, where we assume that all input vectors include a constant bias component of value equal to 1, and a 1-dimensional output target $y(i)$. The simple linear model for this system is given by

$$\hat{y}(i) = \mathbf{x}(i)^T \mathbf{w}_f \; , \tag{18}$$

where $\mathbf{w}_f$ is the single node's d-dimensional weight vector. The weight vector $\mathbf{w}_f$ can be found by applying m iterations of the RLS procedure as follows:

$$\begin{aligned} a(i) &= \left[1 + \mathbf{x}(i)^T \mathbf{P}(i) \mathbf{x}(i)\right]^{-1} & (19) \\ \mathbf{k}(i) &= \mathbf{P}(i)\mathbf{x}(i)a(i) & (20) \\ \mathbf{w}(i+1) &= \mathbf{w}(i) + \mathbf{k}(i)\left(y(i) - \hat{y}(i)\right) & (21) \\ \mathbf{P}(i+1) &= \mathbf{P}(i) - \mathbf{k}(i)\mathbf{x}(i)^T \mathbf{P}(i) \; , & (22) \end{aligned}$$

where the diagonal elements of $\mathbf{P}$ are initialized to large positive values, here $\mathbf{P}(0) = 1000\mathbf{I}$, and $\mathbf{w}(0)$ to a vector of small random values. Also, $\mathbf{w}_f = \mathbf{w}(m)$ after a single presentation of all training data (i.e., after a single epoch).

We recover a batch, least squares solution to this single node training problem via an extreme application of the multi-stream concept, where we associate m unique streams with each of the m training instances. In this case, we arrange the input vectors into a matrix $\mathbf{X}$ of size $d \times m$, where each column corresponds to a unique training pattern. Similarly, we arrange the target values into a single m-dimensional vector $\mathbf{y}$, where each element of $\mathbf{y}$ is properly aligned with its corresponding feature vector in $\mathbf{X}$. As before, we select the initial weight vector $\mathbf{w}(0)$ to consist of randomly chosen values, and we select $\mathbf{P}(0) = 1000\mathbf{I}$. Given the choice of initial weight vector, we can compute the network output for each training pattern, and arrange all the results using the matrix notation

$$\hat{\mathbf{y}}(0) = \mathbf{X}^T \mathbf{w}(0) \tag{23}$$

A single weight update step of the Kalman filter recursion applied to this m-dimensional output problem at the beginning of training can be written as

$$\mathbf{A}(0) = \left[\mathbf{I} + \mathbf{X}^T\mathbf{P}(0)\mathbf{X}\right]^{-1} \quad (24)$$

$$\mathbf{K}(0) = \mathbf{P}(0)\mathbf{X}\mathbf{A}(0) \quad (25)$$

$$\mathbf{w}(1) = \mathbf{w}(0) + \mathbf{K}(0)\left(\mathbf{y} - \hat{\mathbf{y}}(0)\right) , \quad (26)$$

where we have chosen not to include the error covariance update here for reasons that will soon become obvious. At the beginning of training, we recognize that $\mathbf{P}(0)$ is large, and we assume that the training data set is scaled so that $\mathbf{X}^T\mathbf{P}(0)\mathbf{X} \gg \mathbf{I}$. This allows $\mathbf{A}(0)$ to be approximated by

$$\mathbf{A}(0) \approx \frac{1}{1000}\left(\mathbf{X}^T\mathbf{X}\right)^{-1} , \quad (27)$$

where we have taken advantage of the diagonal nature of $\mathbf{P}(0)$. Given this approximation, we can write the Kalman gain matrix as

$$\mathbf{K}(0) = \mathbf{X}\left(\mathbf{X}^T\mathbf{X}\right)^{-1} . \quad (28)$$

We now substitute equations (23) and (28) into equation (26) to derive the weight vector after one time step of this m-stream Kalman filter procedure:

$$\mathbf{w}(1) = \mathbf{w}(0) + \mathbf{X}\left(\mathbf{X}^T\mathbf{X}\right)^{-1}\left(\mathbf{y} - \mathbf{X}^T\mathbf{w}(0)\right) \quad (29)$$

$$= \mathbf{w}(0) - \mathbf{X}\left(\mathbf{X}^T\mathbf{X}\right)^{-1}\mathbf{X}^T\mathbf{w}(0) + \mathbf{X}\left(\mathbf{X}^T\mathbf{X}\right)^{-1}\mathbf{y} . \quad (30)$$

Applying the matrix equality $\mathbf{X}(\mathbf{X}^T\mathbf{X})^{-1}\mathbf{X}^T = \mathbf{I}$ yields the pseudoinverse solution to this training problem:

$$\mathbf{w}_f = \mathbf{w}(1) = \mathbf{X}\left(\mathbf{X}^T\mathbf{X}\right)^{-1}\mathbf{y} , \quad (31)$$

since $\mathbf{X}(\mathbf{X}^T\mathbf{X})^{-1} = (\mathbf{X}\mathbf{X}^T)^{-1}\mathbf{X}$.

Thus one step of the multi-stream Kalman recursion recovers very closely the least-squares solution. If m is too large to make the inversion operation practical, we could instead divide the problem into subsets and perform the procedure sequentially for each subset, arriving eventually at nearly the same result (in this case, however, the covariance update needs to be performed).

As illustrated in this one-node example, the multi-stream EKF update is not an average of the individual updates, but rather is coordinated through the global scaling matrix $\mathbf{A}^*$. It is intuitively clear that this coordination is most valuable when the various streams place contrasting demands on the network.

4.4 Advantages and Extensions of Multi-Stream Training

Discussions of the training of networks with external recurrence often distinguish between series-parallel and parallel configurations [12]. In the former, target values are substituted for the corresponding network outputs during the training process. This scheme, which is also known as *teacher forcing*, helps the network to get "on track" and stay there during training. Unfortunately it may also compromise the performance of the network when, in use, it must depend on its own output. Hence it is not uncommon to begin with the series-parallel configuration, then switch to the parallel configuration as the network learns the task. Multi-stream training seems to lessen the need for the series-parallel scheme; the response of the training process to the demands of multiple streams tends to keep the network from getting too far off track. In this respect, multi-stream training seems particularly well suited for training RMLPs, where the opportunity to use teacher forcing is limited, because "correct" values for most if not all outputs of recurrent nodes are unknown.

Though our presentation has concentrated on multi-streaming simply as an enhanced training technique, one can also exploit the fact that the streams used to provide input-output data need not arise homogeneously, i.e., from the same training task. Indeed, it is interesting to contemplate teaching a network to do multiple tasks. In contrast to feedforward networks, which can only carry out static input-output mapping, recurrent networks embed the concept of *state*. Hence, it is reasonable to envision that such a network might be trained to exhibit different behavior according to its current region of state space. To one extent or another, all the examples that follow make use of this capability.

A concrete expression of this idea is the use of multi-streaming to coerce a network into desired behavior in conjunction with its primary task. Our first such use was explicit training of controller networks to be robust over a range of systems to be controlled. Section 5.2 uses the technique to induce stability on a model when its primary training objective is modeling performance when driven by a rapidly changing input. In principle, any performance objective or set of objectives that can be cast in a form that produces errors to place in $\boldsymbol{\xi}$ and derivatives that can be placed in the $\mathbf{H}$ matrix can be treated with the multi-stream method.

5 Synthetic Examples

5.1 Multiple Series Prediction and Dropout Rejection

In [8, 9] we demonstrated that a single fixed-weight network could be trained to make single-step predictions for 13 different time series, including versions of the chaotic logistic, Henon, and Ikeda maps, as well as some sinusoids. In testing, the sequence presented as network input could be switched at any time from one series to another. Thus, we required the fixed-weight network to perform steps that are often associated with explicitly adaptive systems: 1) determine the identity of the time series or the parameters of the generating equation from the recent time history of the series; 2) instantiate this information into the prediction system and begin making predictions; and 3) monitor the success of predictions and, as necessary, reevaluate the series identity or its parameters.

In this section, we present an extension of this idea, in which we not only consider multiple series, but also allow the input sequence to be severely disrupted in the form of dropouts, i.e., input values of zero. We use the following two variations of the Henon map. (Note that variation B is a sign-reversed version of variation A.)

`Variation A:`

$$x_a(t+1) = 1 - 1.4x_a^2(t) + 0.3x_a(t-1) \quad (32)$$

`Variation B:`

$$x_b(t+1) = -1 + 1.4x_b^2(t) + 0.3x_b(t-1) \quad (33)$$

5.1.1 Training

We prepared five data sets for training, 1000 points each, for each of the two series; the sequences for two variations were prepared with dropout probabilities 0, 0.1, 0.2, 0.3, and 0.4. No limit on the number of consecutive dropouts was imposed. All sequences are initialized with randomly chosen values for $x(t)$ and $x(t-1)$; transient behavior was minimized by running the generating program for 100 time steps before recording the input-output pairs used for training.

We trained an RMLP with structure 1-5R-5R-1L, i.e., 1 input, two hidden layers with five fully recurrent nodes each (bipolar sigmoid activation), and a linear output node. We employed multi-steam GEKF using 10 streams, a priming length of 10, and a trajectory length of 200. A truncation depth $h = 10$ for BPTT was used. The network was trained for 550×190 updates, based on $550 \times 10 \times 200$ input-output pairs. At termination of training, the RMS error was approximately 0.07, about one-tenth of the standard deviation of each series. The utility of multi-streaming is illustrated by noting that this error is about half that obtained with simple presentation-order scrambling as described in Section 4.2. This was carried out using the same procedure as above except that a single-stream was used and the priming and trajectory lengths are $N_p = 19$ and $N_s = 20$, respectively, and 1,100,000 weight updates were performed (so that training would based on a comparable number of instances).

Considered as a signal processing problem, the challenges include 1) accommodation to the current time series; 2) recognizing the difference between corrupt and good input values; and 3) combining input signals with information stored as network states to form a prediction. Because of the existence of two series, the network must pay attention to the input sequence in order to determine which series is active. Further, because both series are chaotic, even the simpler problem of making predictions for a single series without dropouts requires use of input information to keep the prediction on track.

5.1.2 Testing

To test the performance of the trained network, we generated new data sequences in which we switch between series. We generated sequences with dropout probabilities 0.3 and 0.5. In Figure 3, we show a portion of this sequence with dropout probability 0.3. In spite of corrupt input data, the network output is very close to the desired outputs for both series. At step 400, the desired output is calculated from (33) for series B, using the two previous outputs of series A; having thus been initialized from a state which is not on its attractor, the chaotic time series requires a number of steps to "stabilize." The network suffers a momentary loss of prediction accuracy as it accommodates the new series, but by step 409 it is again making good predictions, even in the face of continuing random dropouts. As we demonstrated in [8], multi-stream training allows one to sacrifice overall accuracy in favor of rapid accommodation by reducing the priming length.

When the dropout probability is set to 0.5, as in Figure 4, the input sequence is substantially more corrupt. (Note the difference between the lower two panels.) The switch between series is handled fairly gracefully, but the chance occurrence of many dropouts between steps 470 and 500 gives rise to substantial prediction errors, which subside as the frequency of dropouts decreases. These prediction errors should not be surpris-

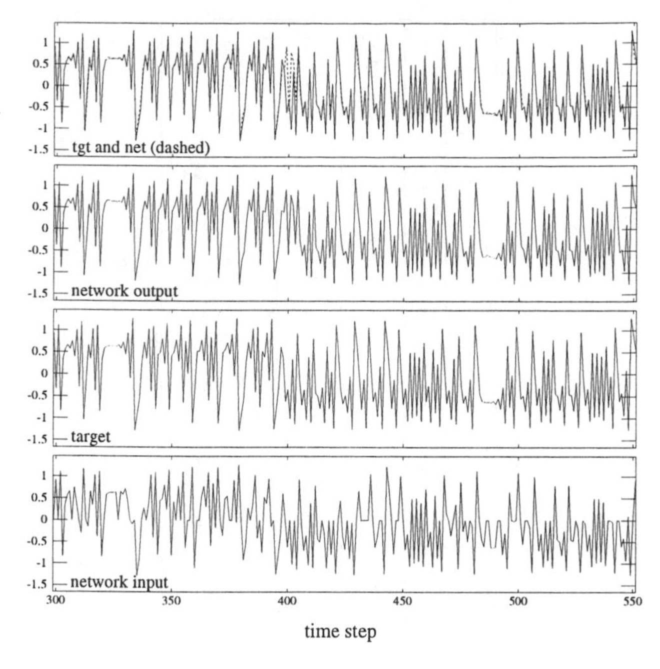

Figure 3: A portion of a testing sequence to illustrate the performance of a network trained to handle dropouts in the input sequence and to make one-step predictions for two variations of the Henon map. For this sequence, the dropout probability is 0.3. The panel labeled *network input* shows the effect of dropouts (input values of 0); in the absence of dropouts, this panel would be identical with the panel labeled *target*, apart from a one-step offset in time. An abrupt switch from variation A to variation B occurs at step 400. Network prediction errors are quite small except just after the switch.

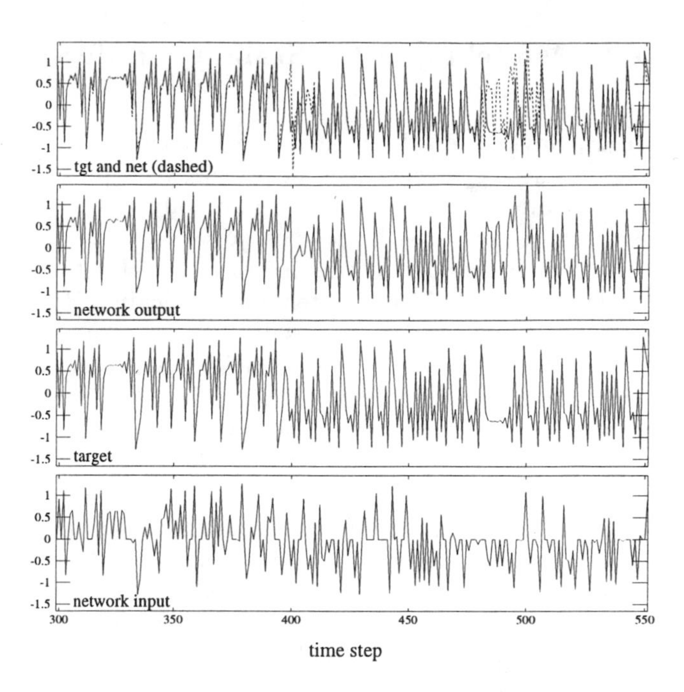

Figure 4: Same as Figure 3 except that the dropout probability has been increased to 0.5. Initialization of the underlying series was identical to that in Figure 3. Again, network prediction errors increase just after the switch from one series variation to the other. In addition, the chance occurrence of a large number of dropouts following step 475 causes a momentary loss in the network's prediction ability.

ing since this testing sequence includes dropouts that are likely to occur more frequently than experienced during training. Of course, for long enough dropout sequences the prediction problem becomes essentially impossible.

5.1.3 Discussion

It is interesting to contemplate how one might write a computer program to solve this problem, given complete information. In general terms, we might proceed as follows: 1) program each of the generating series so as to compute the next sequence value from the proceeding two values; 2) compare the actual observed value with the two generated values to determine which series is active; 3) issue a prediction for the next value based on the determined series.

One might also attempt to make predictions using a feedforward network with inputs based on the present input and a number of previous input values. This approach is based on the assertion that a sufficient number of input values defines a lag space, a point of which can be mapped uniquely to the desired output value. The existence of dropouts complicates matters, since missing values require that a prediction be based on a subspace of the full lag space. Further, the required subspace depends on the positions in which missing values occur in the input sequence.

In order to test whether this problem is easily solved using a modest number of lagged inputs, we carried out training using a feedforward network whose inputs came from a tapped delay line of 8 elements (the current input and 7 previous values). The network structure was 8-10-10-1L. The same procedure as used to train the recurrent network was employed. Training for an equivalent number of updates produced an RMS error of 0.34, substantially inferior to the value of 0.07 reported above. Though this experiment certainly does not prove that mapping from a lag space cannot handle this problem, it confirms our suspicion that even if such a mapping exists, it might be very difficult to learn.

5.2 Modeling with Stability Enforcement

In [1], Suykens et al. consider the problem of training a time-lagged recurrent neural network to identify the model parameters of an internally stable system that is corrupted by process noise. In that work, they demonstrated that training of a neural network model for the considered system via application of dynamic backpropagation can result in a model that exhibits undesirable limit cycle behavior when operated autonomously. Then they established that a neural network model can be trained such that global asymptotic stability is enforced by modifying dynamic backpropagation as in [12] to include stability constraints. Here, we show that the multi-stream training procedure provides an alternative mechanism that allows a neural network model to be developed which satisfies the modeling requirements of the forced system while simultaneously coercing stable autonomous behavior via the use of an auxiliary training data stream.

The single-input/single-output system proposed in [1] is given by

$$x_{k+1} = W_{AB}\tanh(V_A x_k + V_B u_k) + \phi_k \qquad (34)$$

$$y_k = W_{CD}\tanh(V_C x_k + V_D u_k)\,. \qquad (35)$$

In these equations, ϕ_k refers to zero-mean, normally distributed process noise, x_k is a 4-component vector that encodes the state of the system, y_k is the observable system output, u_k is the system input, and W_{AB}, W_{CD}, V_A, V_B, V_C, and V_D encode the model parameters, given by

$$W_{AB} = \begin{bmatrix} 0.4157 & -0.2006 & 0.1260 & -0.0237 \\ 1.1271 & -0.0401 & -0.6084 & 0.4073 \\ -0.2141 & 0.4840 & -0.2966 & -0.0027 \\ -0.1986 & -0.6325 & 0.4208 & -1.0233 \end{bmatrix}$$

$$W_{CD} = \begin{bmatrix} -0.5546 & -0.2603 & 1.3030 & -1.3587 \end{bmatrix}$$

$$V_A = \begin{bmatrix} -0.3152 & -0.8392 & -0.2323 & -0.5119 \\ -0.2872 & -0.7385 & -0.4354 & 0.4126 \\ -0.1009 & -0.6593 & -0.5717 & -0.8109 \\ -0.8850 & -0.7005 & -0.0671 & 0.0592 \end{bmatrix}$$

$$V_C = \begin{bmatrix} 0.3600 & 0.5972 & 1.7870 & -1.4743 \\ 1.3145 & -0.2945 & 0.0347 & 0.2681 \\ -1.2125 & -0.9360 & 0.3255 & 1.7658 \\ 0.5938 & -0.2655 & 0.5651 & -1.7682 \end{bmatrix}$$

$$V_B = \begin{bmatrix} 0.1256 \\ 1.2334 \\ 1.0599 \\ -1.7554 \end{bmatrix} \quad V_D = \begin{bmatrix} 1.6275 \\ -2.3663 \\ -0.8700 \\ -0.7000 \end{bmatrix}.$$

Note that this system can be expressed as a time-lagged recurrent neural network with an ordered network representation, as shown in Table 1. This system can also be expressed as an RMLP, but this representation would require additional nontrainable nodes and weights that encode time-delay connections, as well as connections that pass the control input between layers without modification.

i	j	$c_{i,j}$	$d_{i,j}$	W
2	1	1	0	$V_{B:1}$
	2	6	1	$V_{A:1,1}$
	3	7	1	$V_{A:1,2}$
	4	8	1	$V_{A:1,3}$
	5	9	1	$V_{A:1,4}$
3	1	1	0	$V_{B:2}$
	2	6	1	$V_{A:2,1}$
	3	7	1	$V_{A:2,2}$
	4	8	1	$V_{A:2,3}$
	5	9	1	$V_{A:2,4}$
4	1	1	0	$V_{B:3}$
	2	6	1	$V_{A:3,1}$
	3	7	1	$V_{A:3,2}$
	4	8	1	$V_{A:3,3}$
	5	9	1	$V_{A:3,4}$
5	1	1	0	$V_{B:4}$
	2	6	1	$V_{A:4,1}$
	3	7	1	$V_{A:4,2}$
	4	8	1	$V_{A:4,3}$
	5	9	1	$V_{A:4,4}$
6	1	2	0	$W_{A,B:1,1}$
	2	3	0	$W_{A,B:1,2}$
	3	4	0	$W_{A,B:1,3}$
	4	5	0	$W_{A,B:1,4}$
7	1	2	0	$W_{A,B:2,1}$
	2	3	0	$W_{A,B:2,2}$
	3	4	0	$W_{A,B:2,3}$
	4	5	0	$W_{A,B:2,4}$
8	1	2	0	$W_{A,B:3,1}$
	2	3	0	$W_{A,B:3,2}$
	3	4	0	$W_{A,B:3,3}$
	4	5	0	$W_{A,B:3,4}$
9	1	2	0	$W_{A,B:4,1}$
	2	3	0	$W_{A,B:4,2}$
	3	4	0	$W_{A,B:4,3}$
	4	5	0	$W_{A,B:4,4}$
10	1	1	0	$V_{D:1}$
	2	6	1	$V_{C:1,1}$
	3	7	1	$V_{C:1,2}$
	4	8	1	$V_{C:1,3}$
	5	9	1	$V_{C:1,4}$
11	1	1	0	$V_{D:2}$
	2	6	1	$V_{C:2,1}$
	3	7	1	$V_{C:2,2}$
	4	8	1	$V_{C:2,3}$
	5	9	1	$V_{C:2,4}$
12	1	1	0	$V_{D:3}$
	2	6	1	$V_{C:3,1}$
	3	7	1	$V_{C:3,2}$
	4	8	1	$V_{C:3,3}$
	5	9	1	$V_{C:3,4}$
13	1	1	0	$V_{D:4}$
	2	6	1	$V_{C:4,1}$
	3	7	1	$V_{C:4,2}$
	4	8	1	$V_{C:4,3}$
	5	9	1	$V_{C:4,4}$
14	1	10	0	$W_{C,D:1,1}$
	2	11	0	$W_{C,D:1,2}$
	3	12	0	$W_{C,D:1,3}$
	4	13	0	$W_{C,D:1,4}$

Table 1: Ordered network representation of system equations (34–35). The elements of the input and output arrays are $I_1 = 1$ and $O_1 = 14$. Nodes 2–5 and 10–13 have nonlinear activation functions ($\tanh(\cdot)$), while nodes 6–9 and 14 are linear. It should be noted that this representation is not unique; alternative equivalent ordered network representations are possible.

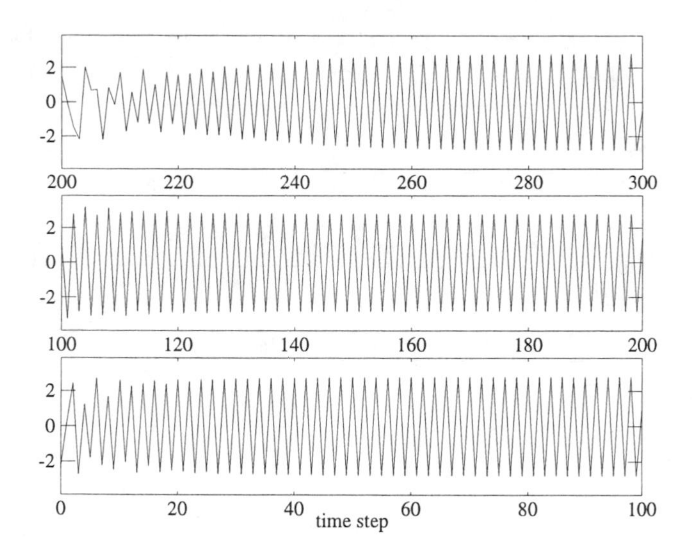

Figure 5: Autonomous behavior of time-lagged network with structure identical to that of the system model trained without stability enforcing constraints. Random state initializations occur at time steps 0, 100 and 200. The unforced network model exhibits limit cycle behavior.

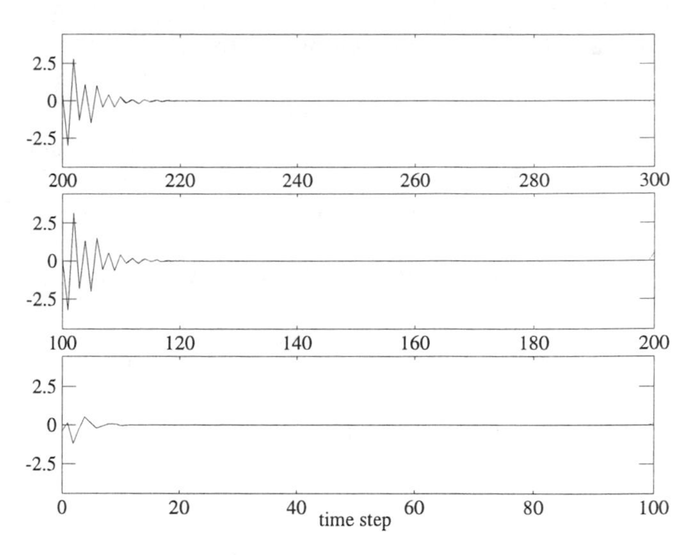

Figure 6: Autonomous behavior of time-lagged network with structure identical to that of the system model trained with multi-stream stability enforcing constraints. Random state initializations occur at time steps 0, 100 and 200. Note that the model quickly reaches stable behavior, unlike the network trained without stability enforcing constraints as shown in Figure 5.

5.2.1 Training without Constraints

We followed closely the problem statement as described in [1]. A training data set is derived by choosing control signals u_k at each time step from a zero-mean, normal distribution with a standard deviation of 5. Similarly, the standard deviation of the process noise is set to 0.1. A set of 1000 data points was generated in this fashion, with the first 500 used for training and the last 500 for independent testing. Values of the initial state variables of the system were set to zero.

We carried out training experiments with two different network architectures. In the first case, we used a network representation of equations (34–35); this network contains 13 trainable nodes (of which 4 are effectively state nodes) with associated trainable weights. The second network considered is a completely trainable RMLP with structure denoted by 1-8R-7R-1L; in this case, we do not assume exact knowledge of the underlying model structure. This network has a substantially greater number of degrees of freedom and states than does the system being modeled.

We first conducted training trials for the two networks using only training data from the forced system with no constraints to enforce stability. In each case, we employed multi-stream global EKF training, with 3 streams, priming length of 5 steps, trajectory length of 100 steps, and a backpropagation truncation depth of $h = 19$. Excellent results were observed for the testing portion of the generated data under conditions of rapidly changing input signal. In [1], the autonomous behavior of the neural model was compared to that of the system, but without the process noise that was present while input-output data were accumulated for network training. (In a real application, it probably would not be possible to examine the actual system behavior in the absence of process noise. For the present system model, the process noise specified is sufficient to drive the unforced ($u_k = 0$) system to substantial output values, over unity in magnitude, making it difficult to observe whether the model exhibits stable behavior.) When the state nodes of the trained network models were initialized to random values in the range [-1,+1] and then executed autonomously (i.e., with zero input), limit cycle output behavior was observed for both networks for many different state initializations. Typical limit cycle behavior of the trained network with structure identical to that of the system model is shown in Figure 5 for three different state initializations. It is noteworthy that limit cycles of different peak-to-peak magnitudes as well as of different periods are observed as the result of different training trials (e.g., different initial conditions and orders of presentation of training data).

5.2.2 Training with Constraints

In order to enforce stability of the autonomous system by training, we must be able to provide the training process with network inputs and target outputs. The most obvious procedure would be to use the observed autonomous behavior of the stable system we are modeling. In the present case, however, we cannot rely on the system itself to provide a reasonable training set, because we cannot use the behavior of the system in the absence of process noise without violating the premise of the problem. Thus we took a different approach, in which we synthesized a secondary training set of 500 points, where each training pattern consists of zero input and zero output. This reflects the goal that the network model should exhibit asymptotic stability. Then we employed multi-stream training to model both the forced system, using the data discussed earlier, and the desired stable behavior of the unforced system, as seen through the synthesized data sequence. We employed 6 training streams altogether, evenly divided between forced and autonomous behaviors. Trajectory lengths of 100 were used, with priming lengths of 5 time steps. Derivatives were computed by BPTT(19) and global multi-stream EKF weight updates were applied. This example is unique among those considered in this paper in that, during training, we instantiate the network with specific values of the state variables, chosen randomly, at the beginning of each training trajectory for both forms of data. A total of 120,000 instances were processed during training of each network.

We show representative autonomous behavior for multiple random state initializations for the two different network architectures in Figures 6 and 7. Note that the outputs of both network models appear to approach zero asymptotically. This behavior is expected for the case of the network model that is structured identically to the original system, since this model has no bias connections. On the other hand, the 1-8R-7R-1L network has a bias connection for each trainable node; it was somewhat surprising to observe stable fixed points of magnitude less than 10^{-4} for this network, given that the bias connections must be trained so that they effectively cancel one another under autonomous operation. Finally, Figure 8 demonstrates that the forced behavior of the 1-8R-7R-1L network that is trained for stable behavior is not significantly compromised relative to the forced behavior of the network that was not trained to be stable; qualitatively similar and comparably accurate behavior is observed for the network that has structure identical to that of the system.

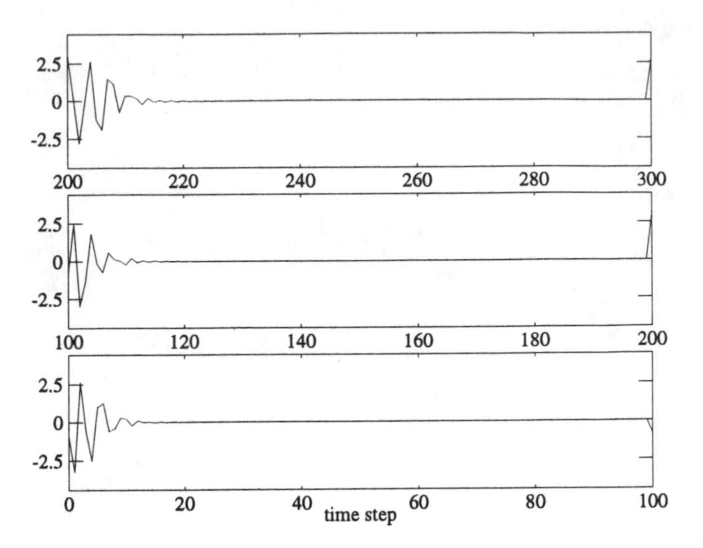

Figure 7: Autonomous behavior of 1-8R-7R-1L RMLP trained with multi-stream stability enforcing constraints. Random state initializations occur at time steps 0, 100 and 200. Even though this network model is of greater complexity than the system model, stable behavior is still easily achieved.

5.2.3 Discussion

It is important to note that use of the multi-stream method to enforce stability affords us some flexibility in tailoring the behavior of the autonomous network as it converges toward an output value of zero. By setting the priming length N_p, we effectively specify a number of steps from the beginning of each training trajectory during which we do not demand a particular network output. In the case of stability enforcement, a larger value of the priming length is expected to lead to less rapid convergence to zero, since the cost function penalizes nonzero output of the autonomous network only for time steps beyond the priming length. In the present case, the convergence of the autonomous model is less rapid than that of networks trained with $N_p = 5$, suggesting that a network model should be trained with a larger priming length. We found that taking $N_p = 25$ and retraining led to autonomous behavior that was closer to that of the actual system. In practice, of course, one would probably not be able to turn off the process noise so as to obtain a convergence standard. Thus it may be possible only to specify a desired convergence characteristic for the model network and select the priming length accordingly.

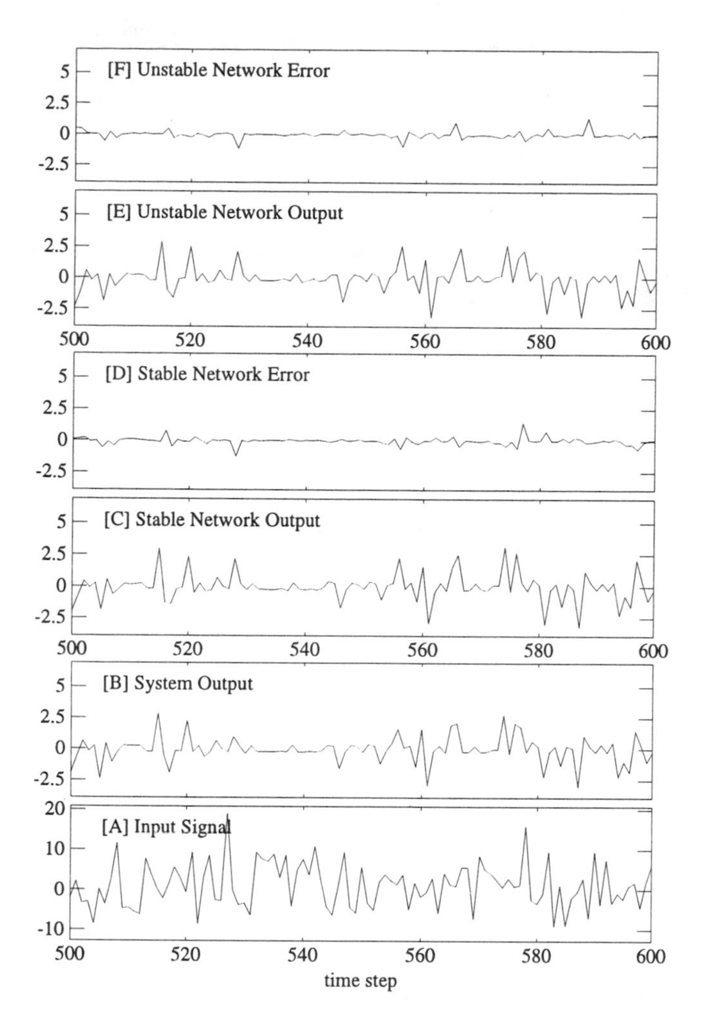

Figure 8: Driven response of the 1-8R-7R-1L RMLP trained with multi-stream stability enforcing constraints compared to that of the same network trained without regard to stability. A sequence of 100 time steps from the noise-corrupted testing set is used. Panel A shows the input signal, while panel B is the noise-corrupted system output. Network responses to the input pattern are shown in panels C (stability-enforced) and E (stability ignored), respectively. Network errors for these two cases are shown in panels D and F, respectively. Note that the networks perform comparably.

6 Automotive Examples

The general area of automotive powertrain control and diagnosis offers substantial opportunity for application of intelligent signal processing methodologies. These opportunities are driven by the steadily increasing demands that are placed on the performance of vehicle control and diagnostic systems as a consequence of global competition and government mandates. Modern automotive powertrain control systems involve several interacting subsystems, any one of which can involve significant engineering challenges. Further, increasingly stringent emissions regulations require that any malfunctioning component or system with the potential to undermine the emissions control system be detected and identified. In this section, we consider two signal processing problems related to automotive diagnostics that are particularly amenable to treatment with recurrent neural networks trained by the multi-stream EKF formalism.

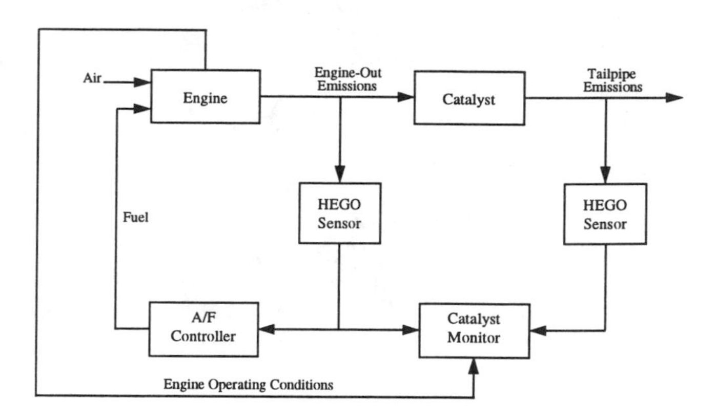

Figure 9: A block diagram of components of a typical catalyst monitoring scheme in the presence of a feedback A/F control system.

6.1 Sensor-Catalyst Modeling

A particularly critical component of a vehicle's emissions control system is the catalytic converter. The role of the catalyst is to chemically transform noxious and environmentally damaging engine-out emissions, which are the by-product of the engine's combustion process, to environmentally benign chemical compounds. In particular, an ideal three-way automotive catalytic converter should completely perform the following three tasks during continuous vehicle operation: (1) oxidation of hydrocarbon (HC) exhaust gases to carbon dioxide (CO_2) and water (H_2O); (2) oxidation of carbon monoxide (CO) to carbon dioxide; and (3) reduction of nitrous oxides (NO_x) to nitrogen (N_2) and oxygen (O_2). In practice, it is possible to achieve high catalytic conversion efficiencies simultaneously for all three types of engine-out exhaust gases only for a very narrow window of operation of air/fuel ratio (A/F); when this occurs, the engine is operating near stoichiometry.

A major role of the engine control system is to regulate A/F about stoichiometry. This is accomplished with an electronic feedback control system that utilizes

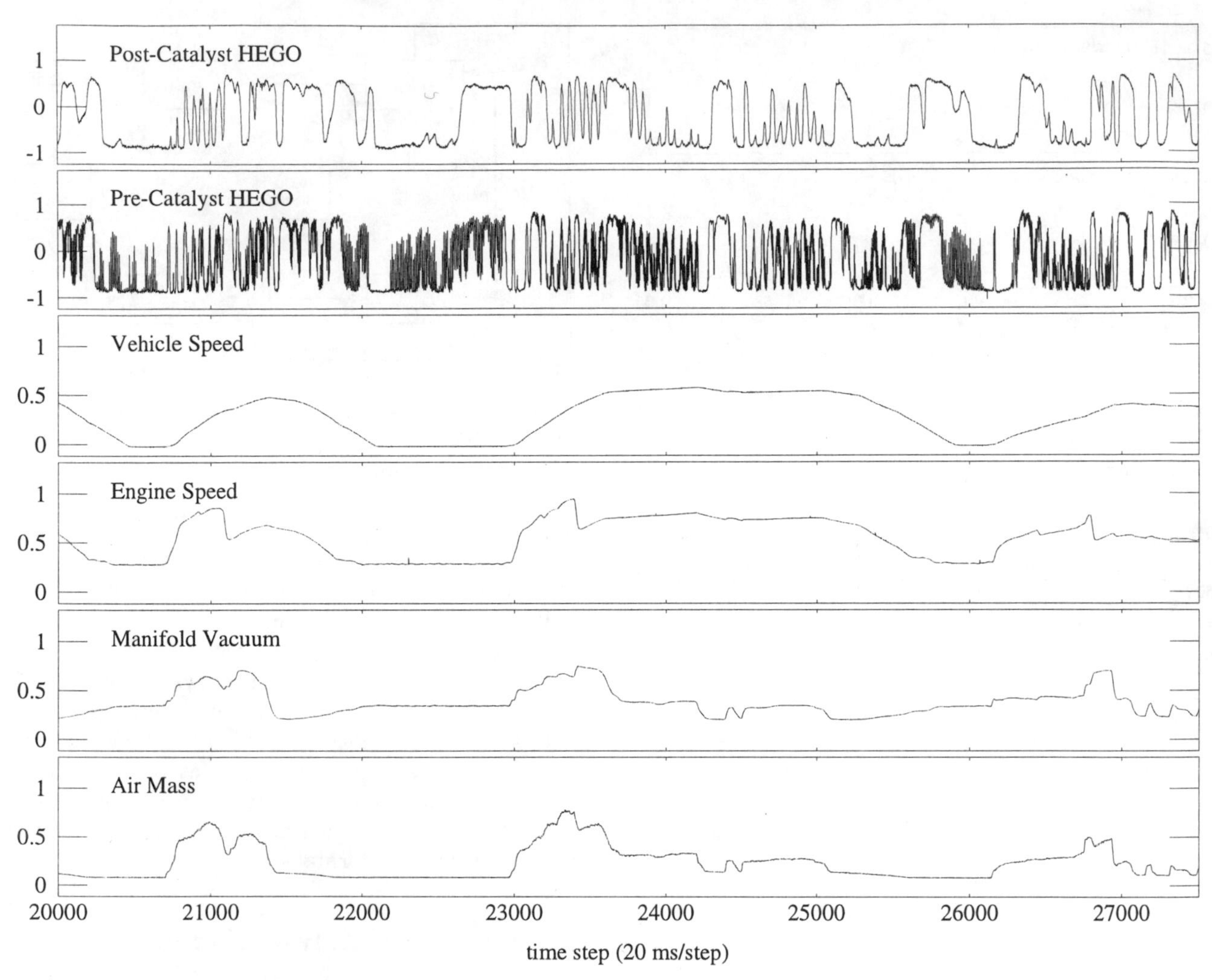

Figure 10: A 150 second segment of training data for the catalyst-sensor model. All variables are shown in scaled units.

a heated exhaust gas oxygen (HEGO) sensor whose role is to indicate whether the engine-out exhaust is rich (i.e., too much fuel) or lean (too much air). Depending on the measured state of the exhaust gases, the A/F control is changed so as to drive the system towards stoichiometry. Since the HEGO sensor is largely considered to be a binary sensor (i.e., it produces high/low voltage levels for lean/rich operations, respectively), and since there are time-varying transport delays, the closed-loop A/F control strategy often takes the form of a jump/ramp strategy, which effectively causes the HEGO output to oscillate between the two voltage levels.

Even in the presence of an effective closed-loop A/F control strategy, vehicle-out (i.e., tailpipe) emissions may be unreasonably high if the catalytic converter has been damaged. For example, the catalytic converter may degrade due to exposure to excessively hot temperatures (as may be the case due to frequent misfires, as we describe below). We would also expect the performance of the catalytic converter to degrade with continued use. Current government regulations require that the performance of a vehicle's catalytic converter be continuously monitored to detect when conversion efficiencies have dropped below some threshold. Unfortunately, it is currently infeasible to equip vehicles with sensors that can measure the various exhaust gas species directly. Instead, current catalytic converter monitors are based on comparing the output of a HEGO sensor that is exposed to engine-out emissions to the output of a second sensor that is mounted downstream of the catalytic converter and is exposed to the tailpipe emissions, as illustrated schematically in Figure 9. This approach is based on the observation that the post-catalyst HEGO sensor switches infrequently, relative to the pre-catalyst HEGO sensor, in the presence of a highly efficient catalyst. Similarly, the average rate of switching of the post-catalyst sensor increases as catalyst efficiency decreases (due to decreasing oxygen storage capability).

As an alternative, a model-based catalyst monitor can be envisioned where the actual output of the post-catalyst HEGO sensor is compared to that of a model of the post-catalyst sensor that assumes a catalytic converter with some nominal converter efficiency. Then, based upon differences between the actual sensor and model outputs, a procedure can be developed to estimate catalytic converter efficiency. However, this modeling is by no means a trivial task. First, the dynamical behavior of catalytic converters, as well as that of HEGO sensors, is not completely understood from a physical and chemical perspective. Second, conditions that affect catalyst and sensor performance are often not observed. Third, the model output must incorporate condition-dependent time delays; e.g., the transport delay associated with the physical placement of the sensors as well as catalyst activity is seen to be a function of engine speed. Here we describe how a time-lagged recurrent neural network can be trained with multi-stream training methods to represent such a system, using only information that could be made available to the vehicle's powertrain control module.

6.1.1 Experimental Data

Data were acquired from a single vehicle equipped with a thermally degraded catalyst. A standard driving cycle was employed to obtain training and testing data of the operating vehicle on a chassis rolls dynamometer facility. Relevant engine variables were sampled at 20 ms intervals. Amongst the variables sampled were ve-

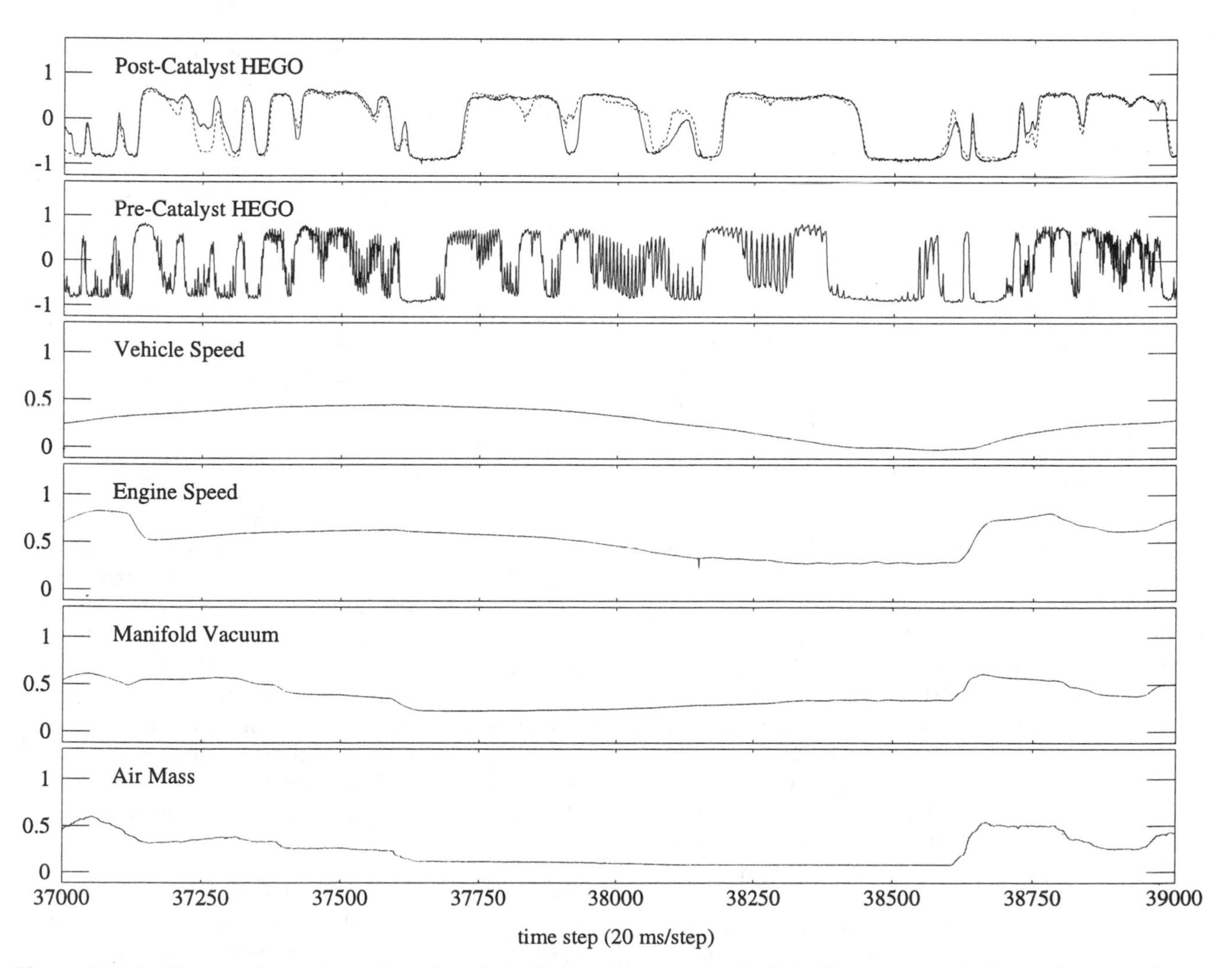

Figure 11: A 40 second segment of testing data for catalyst-sensor model. The top panel shows the actual post-catalyst HEGO sensor output in a solid pattern and the predicted HEGO sensor output in a dashed pattern.

hicle speed, engine speed, air mass, manifold vacuum, pre-catalyst HEGO sensor output, and post-catalyst sensor output. Two data acquisition runs were performed, on different days. Each run had 78,000 data points and required a time of 26 minutes. A representative sample of data is shown in Figure 10. Here we see that there is very slow variation of the air mass, manifold vacuum, engine speed and vehicle speed signals, whereas the pre-catalyst sensor exhibits rapid switching behavior that is somewhat more frequent than that of the post-catalyst sensor. It is noteworthy that although the HEGO sensor output is often considered to be binary in nature, it in fact has an analog nature in which switches do not appear to occur instantaneously and in which the voltage levels reached are not distinctly binary in nature. Analysis of the data discloses that the time delay between the pre- and post-catalyst sensor outputs varies from less than 0.1 s at high engine speed/air mass combinations to more than 1 s at low engine speed/air mass combinations. Finally, depending on vehicle operating conditions, the post-catalyst HEGO sensor output sometimes closely mirrors the switching characteristics of the pre-catalyst sensor, while at other times it appears to be largely independent of the pre-catalyst sensor output.

6.1.2 Training and Testing

A time-lagged recurrent neural network was trained on one of the data sets for which the average HC conversion efficiency was measured to be nearly 80.0%. Only 63,000 of the 78,000 data points gathered were actually used for training; the beginning section corresponding to cold start (5000 points) was ignored, as were data acquired after the engine was turned off. The network inputs at any time step are given by the current air mass, manifold vacuum, engine speed, and vehicle speed. In addition, a sparse tapped delay line representation of the pre-catalyst HEGO sensor output was formed as input to the network, consisting of the current measurement along with 10 additional measurements of the sensor output spaced 5 time steps (0.1 s) apart, spanning a total time of 1 s. The resulting input vector at any time step is comprised of 15 signals plus a bias. The network architecture chosen was an RMLP with structure given by 15-20R-15R-10R-1; the resulting network consisted of 1531 weights. Simpler architectures were found not to be as effective. (We have found the combination of sparse tapped delay line input representations with internal network recurrence to be a particularly effective mechanism for treating the problem highlighted in [24] of learning long-term dependencies and condition dependent time delays.) Multi-stream training was performed utilizing 10 data streams, with trajectory lengths of 1000 instances and priming lengths of 50 time steps. Derivatives were computed by backpropagation through time with a truncation depth of 74 time steps. Due to the complexity and size of the network architecture, node-decoupled multi-stream weight updates were performed. This procedure resulted in the processing of 2.2 million instances during training, with each instance processed an average of 35 times.

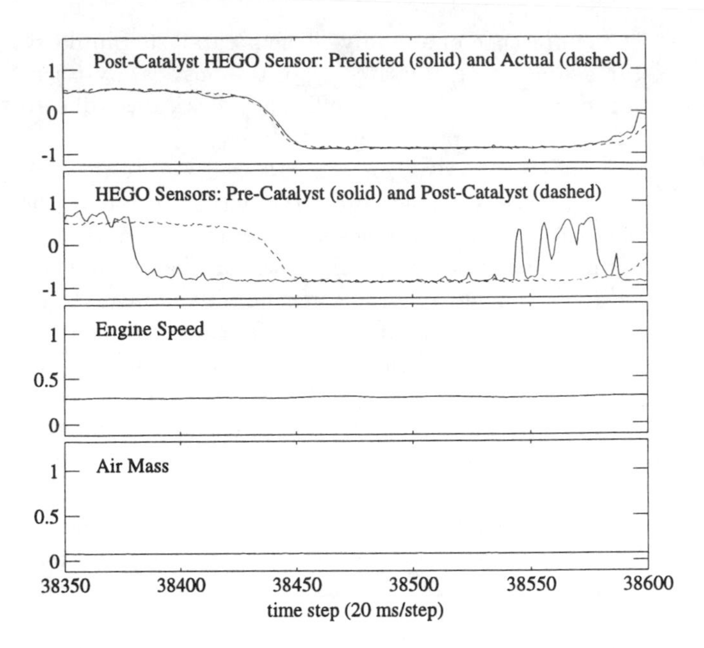

Figure 12: A 5 second segment of network performance for the vehicle operating at idle. For compactness, we have eliminated the manifold vacuum and vehicle speed traces (see Figure 11 for the corresponding segments). In the second panel from the top, the pre- and post-catalyst HEGO sensor outputs are plotted together to provide a visual sense of the context dependent time delay. In the uppermost panel, the post-catalyst HEGO sensor output is repeated as a dashed pattern, and the network output is shown as a solid line. Note that the network output is able to closely capture the dynamic relationship associated with the long time delays between the pre- and post-catalyst sensor outputs.

We performed testing of the trained network with the second data set obtained from the same vehicle/catalyst combination. Interestingly, the average HC conversion efficiency for this data set was measured to be 75%, a 5% difference from the training

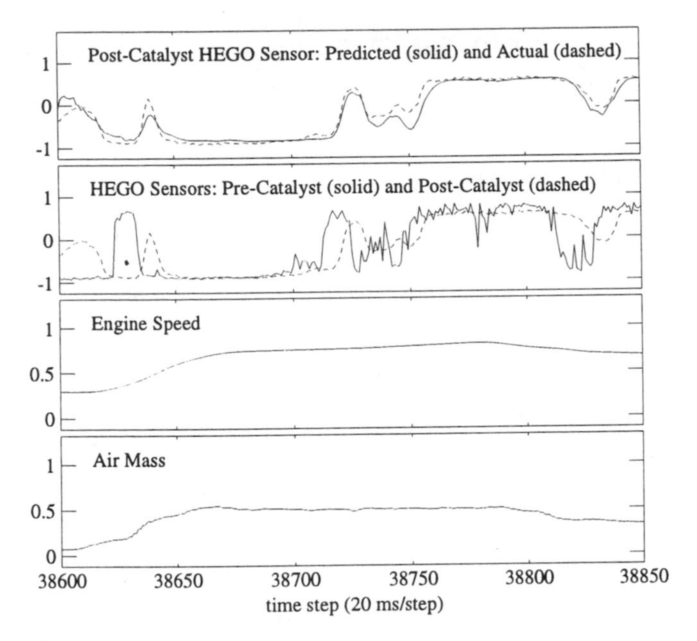

Figure 13: A 5 second segment of network performance for the vehicle accelerating to a high speed, followed by a deceleration. The panels are arranged identically to Figure 12. Note that the network gracefully handles the context transients, and that the the short time delays between the two sensors are properly modeled.

set (it is possible that most of the difference could be attributed to the cold-start portion, which we have ignored for the present purposes). Figure 11 shows a representative sequence of engine operating conditions and network behavior over 40 seconds; this sequence includes both low and high speed operations, as well as some transients due to vehicle accelerations and decelerations. It is evident that the trained network has captured the qualitative behavior of the post-catalyst HEGO sensor output. In Figures 12 through 14, we show 5 second portions of the network performance for different operating conditions. Figure 12 demonstrates network performance for the vehicle operating largely at idle, where the time delay between pre- and post-catalyst sensor outputs is expected to be a maximum. Alternatively, Figure 13 shows results for vehicle acceleration and high engine speed operation, where we expect to see a minimal time delay. Finally, Figure 14 demonstrates 5 seconds of network performance under conditions of medium speed cruise. With the exception of some small phase shifts and amplitude deviations, the network output appears to closely follow that of the actual post-catalyst HEGO sensor.

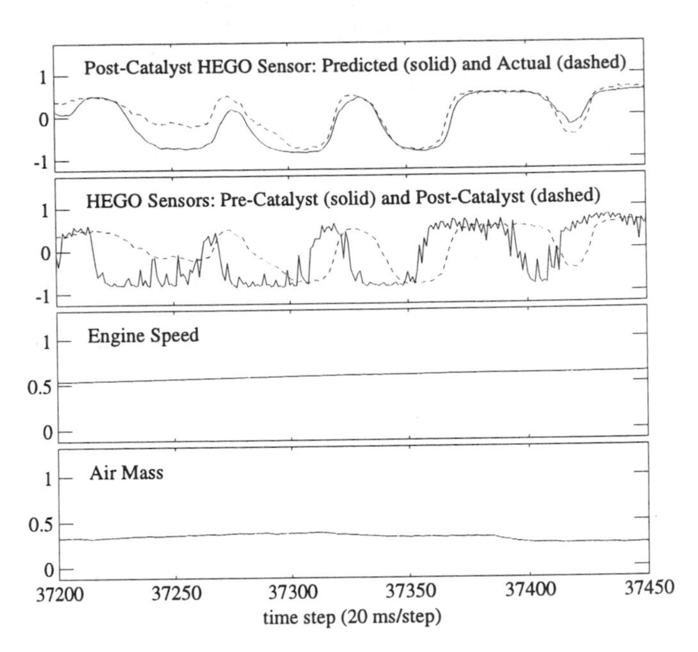

Figure 14: A 5 second segment of network performance for the vehicle operating at a relatively constant speed. The panels are arranged identically to Figure 12. Note that the time delays in this case are of moderate length.

6.2 Engine Misfire Detection

Engine misfire is broadly defined as the condition in which a substantial fraction of a cylinder's air-fuel mixture fails to ignite. Frequent misfires will lead to a deterioration of the catalytic converter, ultimately resulting in unacceptable levels of emitted pollutants and a costly replacement. Consequently, government mandates require that automotive manufacturers provide on-board misfire detection capability under nearly all engine operating conditions for vehicles sold in the United States after 1998.

While there are many ways of detecting engine misfire, all currently practical methods rely on observing engine crankshaft rotational dynamics with a position sensor located at one end of the shaft. Briefly stated, one looks for a crankshaft acceleration deficit following a cylinder firing and attempts to determine whether such a deficit is attributable to a lack of power provided on the most recent firing stroke. (In effect, the momentary unopposed load on the engine causes it to slow down briefly.)

Since every engine firing must be evaluated, the natural "clock" for misfire detection is based on crankshaft rotation, rather than on time. For an n-cylinder engine, there are n engine firings, or *events*, per engine cycle, which requires two engine revolu-

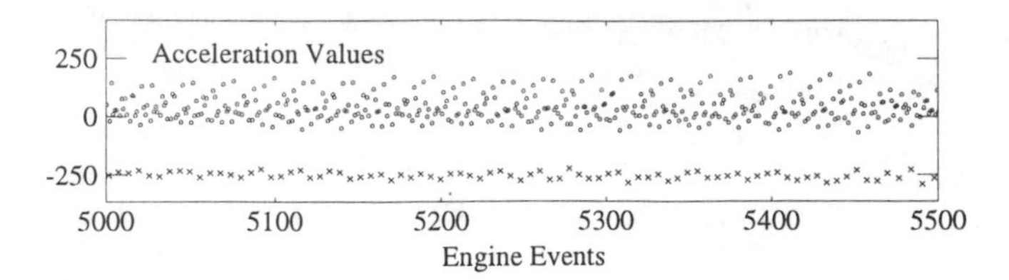

Figure 15: A temporal sequence of acceleration values (scaled units) for low-speed engine operation. Artificially induced misfires are denoted by symbols 'x'.

tions. The actual time interval between events varies considerably; for an 8-cylinder engine, for example, the time interval varies from 20 ms at 750 revolutions per minute (RPM) to 2.5 ms at 6000 RPM. Engine speed, as required for control, is typically derived from measured intervals between marks on a timing wheel. As used in misfire detection, an acceleration value is calculated from the difference between successive intervals. If the timing marks are favorably placed, each such computed acceleration depends sensitively on the firing of just one cylinder.

Figure 15 shows a sequence of acceleration values, taken when the engine is at low speed and lightly loaded. Misfires have been artificially induced (by interrupting the spark), and acceleration values that correspond to misfires have been labeled in the plot. Under these conditions, misfires are easily detected by simple thresholding.

This scheme is complicated by several factors. One of these is that the angular intervals between timing marks may not be precisely equal and may differ from vehicle to vehicle. This is typically handled by an adaptive correction that is carried out on board each vehicle; such a correction has been applied to all data shown here. A more serious problem is that the crankshaft is not infinitely stiff. This causes it to exhibit complex torsional dynamics, even in the absence of misfire. The magnitude of acceleration induced by such torsional vibrations may be large enough to dwarf acceleration deficits from misfire. Further, the torsional vibrations are themselves altered by misfire, so that normal engine firings following a misfire may be misinterpreted.

In the lower panel of Figure 16 we have again plotted an acceleration sequence, but under the more challenging conditions of high speed and moderately high load. It is essentially impossible to pick out misfires visually. In the upper panel we have labeled the misfires; note how they are mixed with normal values.

Although the effect of torsional oscillation is to add what may appear to be random noise to the sequence

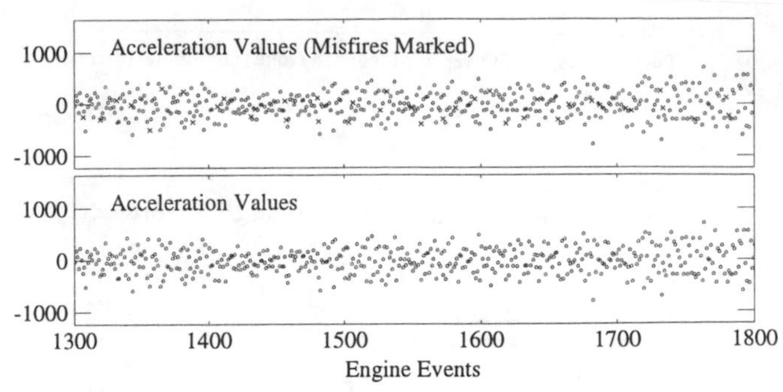

Figure 16: A temporal sequence of acceleration values, illustrating the effects of crankshaft dynamics. In the upper panel misfires are denoted by symbols 'x'.

of acceleration values, we knew from prior work of colleagues that the pattern, in the absence of misfire, was fairly reproducible at a given operating condition, although it varied as the operating condition changed. Hence, we speculated that the problem might yield to a suitable unraveling of the acceleration signal. In effect, our approach is to use a training process to form a context-dependent nonlinear filter.

6.2.1 Applying Recurrent Networks

We have approached the misfire detection problem with recurrent networks in two related ways. In the first of these, we attempt to convert the observed acceleration sequence into a good approximation of an idealized sensor; here the latter amounts to laboratory quality time measurement on a timing wheel which is placed in a crankshaft location less subject to torsional vibration. Details of the first use of this method are presented in [25]. Performing misfire classification with this approach requires the same final steps employed in current production misfire detection. Typically, one forms the difference between each sensor value and a central measure of its temporal neighbors and normalizes the result according to the acceleration deficit expected for that engine state. The resulting value is then compared to a chosen threshold to effect the classification.

In the second approach [26] the final steps may be bypassed by training a network to perform the classification directly. Here we choose the network architecture to be 4-15R-7R-1. The inputs are engine speed, engine load (a derived quantity that is based primarily on air mass), acceleration (as described above), and a binary flag to identify the beginning of the cylinder firing sequence. The target is either +1 or -1, according to whether a misfire had been artificially induced for the current cylinder during the *previous* engine cycle.

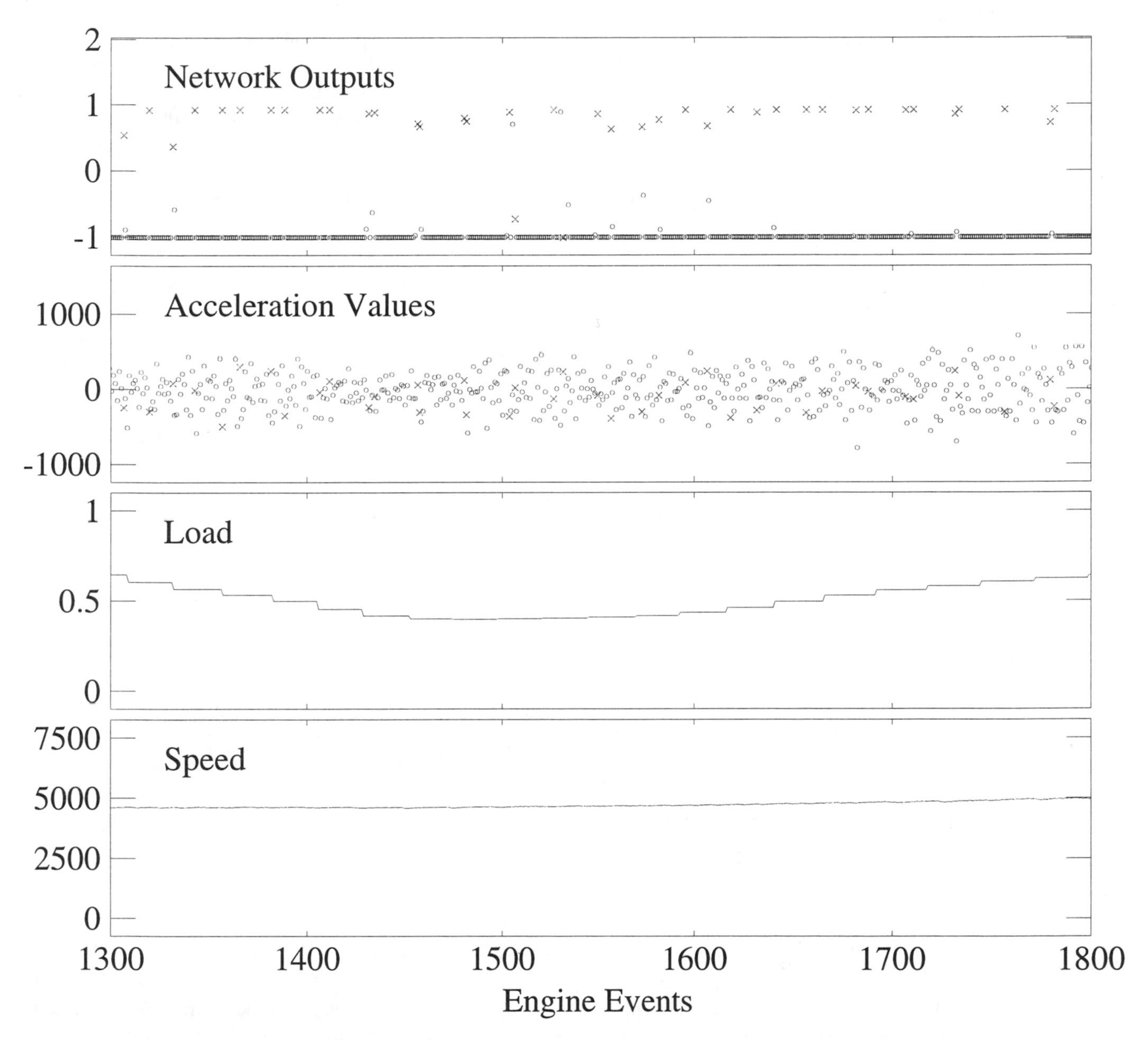

Figure 17: The segment of acceleration values of Figure 16 is plotted, together with inputs to a recurrent classification network and the output of the trained network. An additional network input assumes a value of 1 every 8 steps and is 0 otherwise. Misfires are again denoted by symbols 'x'.

This phasing, $\tau_1 = -8$ in the notation of Section 2, enables the network to make use of information contained in measured accelerations that follow the engine event being classified. The consequently *noncausal* nature of the nonlinear filter presents no practical problem, since the classifications are used statistically, rather than for immediate action.

An important practical aspect is that, unlike the method which depends on targets based on an idealized sensor, the acquisition of data for training does not require special equipment beyond that used to extract information from the engine computer.

In the example shown here, the training database was acquired on a production vehicle with an 8-cylinder engine over a wide range of operation, including engine speed and load combinations beyond those encountered in normal driving. Misfires were deliberately induced at various intervals. Although the database consists of more than 600,000 examples (one per cylinder event), it only approximates full coverage of the space of operating conditions and possible misfire patterns. It is therefore important to test the network's generalization by applying it to independent data.

In Figure 17, we display the same segment of data as in Figure 16, but here we have also plotted network inputs corresponding to engine speed and load and have shown the output of the trained recurrent network. Both the acceleration values and the network output values have been labeled according to whether misfire had been induced. It should be noted that this segment is not part of the training set, but rather is part of a test set acquired *after* the network had been trained. The network is making remarkably few classification errors. Most network errors occur during moments of rapid positive or negative acceleration, not shown here. (Actually, misfire detection is not required during conditions of negative engine torque, such as during deceleration, when the engine is not producing power.)

7 Summary and Conclusions

In this paper we have presented a set of techniques that, used together, have allowed us to address a range of interesting and potentially useful training problems. We have used substantially the same procedure for each of the examples presented as well as for many others. The RMLP has served us very well, with architectural details chosen largely according to how hard a given problem is thought to be, with due regard for the amount of available data. (We have no solution to the occasionally encountered problem of a difficult mapping supported by little data.) For most problems, the networks we train are probably higher order than would minimally be required. More study is required to determine under what conditions this may be problematical. In any case, if the minimal architecture for a given problem is known, then our framework may be applied to it.

Though we have concentrated on off-line applications, it is entirely possible to employ EKF updates in real time, as we demonstrated for controller training in [11]. With sufficient computational power, multi-stream methods can also be incorporated into real-time applications. A quite promising scenario is that of combining a primary data stream based on real-time data with one or more secondary streams synthesized as in Section 5.2 to encourage desired secondary behavior, such as stability or steady-state accuracy.

Some of the networks we have discussed here require significant computation merely to execute. As this could be a stumbling block to production applications, a collaboration was initiated between Ford Research and the Jet Propulsion Laboratory, resulting in an elegant and inexpensive hardware implementation [27] that allows fast recurrent network execution.

Acknowledgments

It is a pleasure to acknowledge the various important contributions of our colleagues to this effort. In particular, we are indebted to Ken Marko, John James and Tim Feldkamp of Ford's Advanced Diagnostics Group for inviting our participation in and sharing with us their critical insights on the misfire detection problem. We are also grateful to Jerry Jesion for his many technical contributions to the experimental portions of this work, including training of the misfire classification network whose results are shown here.

References

[1] J. A. K. Suykens, J. Vandewalle, and B. L. R. De Moor, "NL$_q$ theory: checking and imposing stability of recurrent neural networks for nonlinear modeling," *IEEE Transactions on Signal Processing,* vol. 45, no. 11, pp. 2682–2691, 1997.

[2] S. Haykin. *Neural Networks: A Comprehensive Foundation*, 2nd edition. Upper Saddle River, NJ: Prentice Hall, 1999.

[3] C. M. Bishop. *Neural Networks for Pattern Recognition.* Oxford: Oxford University Press, 1995.

[4] R. J. Williams, "Adaptive state representation and estimation using recurrent connectionist networks," in W. T. Miller III, R. S. Sutton, and P. J. Werbos (eds), *Neural Networks for Control*, pp. 97–114. Cambridge, MA: MIT Press, 1990.

[5] J. T.-H. Lo, "System identification by recurrent multilayer perceptrons," in *Proceedings of the World Congress on Neural Networks*, Portland, OR, 1993, pp. IV-589 - IV-600.

[6] K. Hornik, M. Stinchcombe, and H. White, "Multilayer feedforward networks are universal approximators," *Neural Networks*, vol. 2, pp. 359–366, 1989.

[7] N. E. Cotter and P. R. Conwell, "Fixed-weight networks can learn," in *Proceedings of the International Joint Conference on Neural Networks*, San Diego, 1990, vol. III, pp. 553–559; N. E. Cotter and P. R. Conwell, "Learning algorithms and fixed dynamics," in *Proceedings of the International Joint Conference on Neural Networks*, Seattle, 1991, vol. I, pp. 799–804.

[8] L. A. Feldkamp, G. V. Puskorius, and P. C. Moore, "Adaptive behavior from fixed weight networks," *Information Sciences*, vol. 98, pp. 217–235, 1997.

[9] L. A. Feldkamp, G. V. Puskorius, and P. C. Moore, "Adaptation from Fixed Weight Dynamic Networks," in *Proceedings of the IEEE International Conference on Neural Networks*, Washington D.C., 1996, pp. 155–160.

[10] P. J. Werbos, "Backpropagation through time: What it does and how to do it," *Proceedings of the IEEE*, vol. 78, no. 10, pp. 1550–1560, 1990.

[11] G. V. Puskorius, L. A. Feldkamp, and L. I. Davis, Jr., "Dynamic neural network methods applied to on-vehicle idle speed control," *Proceedings of the IEEE*, vol. 84, no. 10, pp. 1407–1420, 1996.

[12] K. S. Narendra and K. Parthasarathy, "Identification and control of dynamical systems containing neural networks," *IEEE Transactions on Neural Networks*, vol. 1, no.1, pp. 4–27, 1990.

[13] G. V. Puskorius and L. A. Feldkamp, "Neurocontrol of nonlinear dynamical systems with Kalman filter trained recurrent networks," *IEEE Transactions on Neural Networks*, vol. 5, no. 2, pp. 279–297, 1994.

[14] R. J. Williams and J. Peng, "An efficient gradient-based algorithm for on-line training of recurrent network trajectories," *Neural Computation*, vol. 2, pp. 490–501, 1990.

[15] S. Singhal and L. Wu, "Training multilayer perceptrons with the extended Kalman algorithm," in D. S. Touretzky (ed), *Advances in Neural Information Processing Systems 1*, pp. 133–140. San Mateo, CA: Morgan Kaufmann, 1989.

[16] B. D. O. Anderson and J. B. Moore. *Optimal Filtering.* Englewood Cliffs, NJ: Prentice Hall, 1979.

[17] S. Haykin. *Adaptive Filter Theory.* Upper Saddle River, NJ: Prentice-Hall, 1996.

[18] G. V. Puskorius and L. A. Feldkamp, "Decoupled extended Kalman filter training of feedforward layered networks," in *Proceedings of the International Joint Conference on Neural Networks*, Seattle, 1991, vol. I, pp. 771–777.

[19] M. B. Matthews, "Neural network nonlinear adaptive filtering using the extended Kalman filter algorithm," in *Proceedings of the International Neural Networks Conference*, Paris, 1990, vol. I, pp. 115–119.

[20] R. J. Williams, "Training recurrent networks using the extended Kalman filter," in *International Joint Conference on Neural Networks*, Baltimore, 1992, vol. IV, pp. 241–246.

[21] L. A. Feldkamp and G. V. Puskorius, "Training of robust neural controllers," in *Proceedings of the 33rd IEEE International Conference on Decision and Control*, Orlando, 1994, vol. III, pp. 2754–2760.

[22] G. V. Puskorius and L. A. Feldkamp, "Extensions and enhancements of decoupled extended Kalman filter training," in *Proceedings of the 1997 International Conference on Neural Networks*, Houston TX, 1997, vol. 3, pp. 1879–1883.

[23] A. H. Sayed and T. Kailath, "A state-space approach to adaptive RLS filtering," *IEEE Signal Processing Magazine*, vol. 11, no. 3, pp. 18–60, 1994.

[24] Y. Bengio, P. Simard, and P. Frasconi, "Learning long-term dependencies with gradient descent is difficult," *IEEE Transaction on Neural Networks*, vol. 5, no. 2, pp. 157–166, 1994.

[25] G. V. Puskorius and L. A. Feldkamp, "Signal processing by dynamic neural networks with application to automotive misfire detection," in *Proceedings of the World Congress on Neural Networks*, San Diego, 1996, pp. 585–590.

[26] K. A. Marko, J. V. James, T. M. Feldkamp, G. V. Puskorius, L. A. Feldkamp, and D. Prokhorov, "Training recurrent networks for classification: realization of automotive engine diagnostics," in *Proceedings of the World Congress on Neural Networks*, San Diego, 1996, pp. 845–850.

[27] R. Tawel, N. Aranki, G. V. Puskorius, K. A. Marko, L. A. Feldkamp, J. V. James, G. Jesion, and T. M. Feldkamp, "Custom VLSI ASIC for automotive applications with recurrent networks," *Proceedings of the 1998 International Joint Conference on Neural Networks*, Anchorage AK, 1998, vol. 3, pp. 598–602.

Chapter 8

Semiparametric Support Vector Machines for Nonlinear Model Estimation

DAVIDE MATTERA AND FRANCESCO PALMIERI
DIPARTIMENTO DI INGEGNERIA ELETTRONICA E DELLE TELECOMUNICAZIONI
UNIVERSITÀ DEGLI STUDI DI NAPOLI FEDERICO II, VIA CLAUDIO, 21, 80125, NAPOLI, ITALY

SIMON HAYKIN
COMMUNICATIONS RESEARCH LABORATORY. MCMASTER UNIVERSITY, 1280 MAIN STREET WEST, HAMILTON, ONTARIO, CANADA, L8S 4K1

***Abstract*—The semiparametric Support Vector Machine (SVM) constitutes a powerful method for measurement-based modeling of nonlinear systems. It generalizes the classical nonparametric SVM by adding to the model a parametric component which can account for the *a priori* knowledge about the system to be estimated. The existing approach for the application of a semiparametric SVM is based on the utilization of general-purpose software packages; this limits the control on the memory requirements and on the computational complexity of the algorithm.**

In this chapter a comprehensive set of explicit algorithms are proposed for training the semiparametric SVM, which include simple coordinate descent algorithms for estimating the nonparametric component and an augmented Lagrangian approach for the parametric component. Various versions of the general method described herein are considered, including the classical SVM and the most general form augmented with an arbitrary kernel.

Key Words—Regression estimation, Support Vector Machines, Radial Basis Functions, Nonlinear system estimation

1. Introduction

Support Vector Machines have proved to be one of the most powerful tools for solving nonlinear regression and classification problems from empirical data [1]. The success of this paradigm is due to the careful formulation of the problem in terms of cost function and regularization constraints [2] in such a way that SVM can be considered an important method for measurement-based system modeling, given a limited amount of data.

Unfortunately, problems still arise when we try to apply the method in high dimensions or utilize it for improving the solution of a problem previously approached, for example, with a linear regressor. Think of the many applications in signal processing [3], in which linear estimators already provide good performance but may be improved by the addition of a nonlinear capability. In such cases, classical SVM [2] provides a poor trade-off between complexity and generality because it is a purely nonparametric method and a large number of support vectors have to be included in the representation to outperform linear estimators.

Semiparametric SVM [4] overcomes this limitation by introducing in the system representation an unregularized parametric component, which models *a priori* knowledge about the system of interest. The parametric component can be utilized to include in the model a linear term, possibly present in the unknown system, and, therefore, to combine some practical advantages of a linear model with the generality of a nonparametric approach. The semiparametric SVM determines jointly the parametric and the nonparametric components and, therefore, differs from the simple approach in which we first apply a parametric method and then use the nonparametric approach to model the residue term. This provides a significant advantage in terms of modeling error and complexity of the nonparametric component when there is a large residue term resulting from the parametric approach.

An important limitation to wide application of semiparametric SVM is the lack of transparency in the algorithm that is often based on black-box optimization routines. In many applications, it would be desirable to utilize an explicit algorithm directly tuned to the SVM problem to allow designers to have more direct control of memory requirements and computational complexity. Although some methods have been proposed [5–10] to overcome this limitation with reference to the classical SVM, to the best of our knowledge no explicit algorithm for the application of the semiparametric SVM has been proposed in the literature.

In this chapter, we propose an analytical method for the line search procedure in the optimization problem required for SVM training and utilize it to derive a set of explicit algorithms for classical and semiparametric SVM training. These algorithms become difficult to implement when the dimension of the parametric component grows; therefore, we also propose a general method, based on an augmented Lagrangian approach, for training semiparametric SVM by the iterative application of any method for solving classical SVM. This directly extends all the explicit algorithms for solving classical SVM to the semiparametric case.

The chapter is organized as follows. In Section 2, we introduce the semiparametric SVM and the optimization problem to be solved for training it. In Section 3, we show an analytical method for the line search procedure in any optimization algorithm that determines the optimum by moving along directions that, at each step, satisfy the equality constraints. In Sections 4 and 5, we utilize this analytical method for determining explicit algorithms with reference to the classical SVM. In Section 6, we generalize the previous algorithms to the case of semiparametric SVM training. In Section 7, we propose an explicit algorithm for training parametric and regularized SVM such as linear SVM [10] and in Section 8 we show, by computer simulations, the effectiveness of the proposed algorithms in nonlinear system estimation.

2. Problem Setting

Model estimation is a fundamental problem in signal processing. When the underlying physical mechanism responsible for the operation of a system of interest is nonlinear, the model could naturally have to be nonlinear. In practice, however, we often find that there is a preference for the adoption of a linear parametric model because of the simplicity involved in the construction of such a model and the physical insight it provides into the operation of the system. In situations of this kind, the use of a semiparametric approach offers the following distinct advantages:

- Compared to the linear parametric model, the nonlinear semiparametric model provides improved accuracy in representing the input-output behavior of the system.
- Compared to a purely nonparametric nonlinear model, the semiparametric model provides a significant reduction in memory requirement by avoiding one to work in a higher dimensional feature space.

In the problem setting described herein, the incorporation of the linear parametric model in the design of the nonlinear semiparametric model may be viewed as a method of exploiting prior knowledge.

Consider then the general set of nonlinear functions

$$f(\boldsymbol{x}) = \boldsymbol{w}^T\boldsymbol{\phi}(\boldsymbol{x}) + \boldsymbol{w}_1^T\boldsymbol{\psi}(\boldsymbol{x}) \qquad [1]$$

where $\boldsymbol{\phi}(\boldsymbol{x})$ is a fixed nonlinear d-dimensional expansion of the n-dimensional vector $\boldsymbol{x}$ and $\boldsymbol{\psi}(\boldsymbol{x})$ is a fixed m-dimensional expansion of the vector $\boldsymbol{x}$.

Let us assume that a set of examples $(\boldsymbol{x}_i, yi)$ are given and we are required to search for an approximation $\hat{f}(\boldsymbol{x})$ of the unknown desired function $\bar{f}(\boldsymbol{x})$ by minimizing, with respect to the weight vectors $\boldsymbol{w}$ and $\boldsymbol{w}_1$, the following regularized function:

$$J(\boldsymbol{w}, \boldsymbol{w}_1) \triangleq \frac{1}{2}\|\boldsymbol{w}\|^2 + \sum_{i=1}^{\ell} C_i c_{\varepsilon_i}(y_i - \boldsymbol{w}^T\boldsymbol{\phi}(\boldsymbol{x}_i) - \boldsymbol{w}_1^T\boldsymbol{\psi}(\boldsymbol{x}_i)), \qquad [2]$$

where $c_{\varepsilon_i}(\xi) \triangleq \max(0, |\xi| - \varepsilon_i)$ is the cost function and ε_i is a small positive constant, which specifies the required accuracy, usually set to the same value for all i. The constants C_i, usually set to a constant value C, control the amount of regularization; the smaller C is, the stronger is the regularization. Note that we have split the expansion into two parts and we have regularized only the first one. This approach can be useful when a rough parametric approximation $\boldsymbol{w}_1^T\boldsymbol{\psi}(\boldsymbol{x})$ is available and we want to improve it with the nonparametric component[1] $\boldsymbol{w}\boldsymbol{\phi}(\boldsymbol{x})$. Note that some, or all, components of $\boldsymbol{\psi}(\boldsymbol{x})$ can be set equal to the input vector $\boldsymbol{x}$, obtaining general extensions of classical linear regression techniques. The cost function of (2) is well suited to signal processing problems where a linear estimator has already been shown to be effective [3] but its approximation performance needs to be improved with some nonlinear extension.

It can be shown [6, 4, 2] that the optimal semiparametric function can be written in the form:

$$\hat{f}(\boldsymbol{x}) = \sum_{i=1}^{\ell} u_i \mathcal{K}(\boldsymbol{x}, \boldsymbol{x}_i) + \boldsymbol{w}_1^T\boldsymbol{\psi}(\boldsymbol{x}) \qquad [3]$$

where the kernel $\mathcal{K}(\boldsymbol{x}_1, \boldsymbol{x}_2) \triangleq \boldsymbol{\phi}^T(\boldsymbol{x}_1)\boldsymbol{\phi}(\boldsymbol{x}_2)$ and the coefficients u_i are the solution of the following optimization problem:

$$\begin{cases} \min\limits_{u} \frac{1}{2}\boldsymbol{u}^T\mathcal{K}\boldsymbol{u} - \boldsymbol{y}^T\boldsymbol{u} + \varepsilon^T|\boldsymbol{u}| \\ \boldsymbol{A}_2^T\boldsymbol{u} = 0 \\ |\boldsymbol{u}| \leq \boldsymbol{C}, \end{cases} \qquad [4]$$

[1] The term $\boldsymbol{w}^T\boldsymbol{\phi}(\boldsymbol{x})$ is called nonparametric because we assume that the number of components of $\boldsymbol{\phi}(\boldsymbol{x})$ is so high (possibly infinite) that we cannot work with them directly.

where $\boldsymbol{u} \triangleq [u_1, \ldots, u_\ell]^T$, $|\boldsymbol{u}| \triangleq [|u_1|, \ldots, |u_\ell|]^T$, $\boldsymbol{y} \triangleq [y_1, \ldots, y_\ell]^T$, $\boldsymbol{\varepsilon} \triangleq [\varepsilon_1, \ldots, \varepsilon_\ell]^T$, $\boldsymbol{C} \triangleq [C_1, \ldots, C_\ell]^T$, $\boldsymbol{A}_2 \triangleq [\boldsymbol{\psi}(\boldsymbol{x}_1), \ldots, \boldsymbol{\psi}(\boldsymbol{x}_\ell)]^T$, the generic element of the matrix $\mathcal{K}$ is equal to $\mathcal{K}(\boldsymbol{x}_i, \boldsymbol{x}_j)$, and the condition $|\boldsymbol{u}| \leq \boldsymbol{C}$ is defined componentwise (i.e., it is equivalent to the conditions $|u_i| \leq C_i$, $i = 1, \ldots, \ell$). Note that, differently from (1) and (2), the function complexity of the representation in (3) and of the optimization problem (4) does not depend on the dimension d of the nonlinear expansion $\boldsymbol{\phi}(\boldsymbol{x})$, but only on the kernel $\mathcal{K}(\cdot, \cdot)$ and on the number of examples ℓ just as in a nonparametric approach.

At optimum, the following conditions have to be satisfied:

$$u_i = 0 \Rightarrow |e_i| \leq \varepsilon_i \qquad [5]$$

$$0 < |u_i| < C_i \Rightarrow e_i = \varepsilon_i \operatorname{sign}(u_i) \qquad [6]$$

$$u_i = C_i \operatorname{sign}(e_i) \Rightarrow |e_i| \geq \varepsilon_i \qquad [7]$$

where $[e_1, e_2, \ldots, e_\ell]^T \triangleq \boldsymbol{y} - \mathcal{K}\boldsymbol{u} - \boldsymbol{A}_2\boldsymbol{w}_1$ and $\operatorname{sign}(\cdot)$ is the classical signum function (i.e., $\operatorname{sign}(\xi) = 1$ if $\xi > 0$; $\operatorname{sign}(\xi) = 0$ if $\xi = 0$; $\operatorname{sign}(\xi) = -1$ if $\xi < 0$). Once we have solved (4), conditions (5)–(7) can be used to determine $\boldsymbol{w}_1$; otherwise, we can determine it as the vector of the Lagrangian multipliers of (4). Moreover, from (5) it follows that at the optimum points $\boldsymbol{u}$ will be sparse because typically a large number of points $\boldsymbol{x}_i$ will have an error $e_i < \varepsilon_i$; therefore, a few values of $\boldsymbol{x}_i$ (support vectors) are utilized for representing the solution. The complexity of the function representation (3) is, therefore, dependent on the number of support vectors, which, in turn, depends on the required accuracy ε.

The semiparametric SVM has been derived as a generalization of three special cases:

1. The *purely nonparametric approach*, when (1) reduces to $f(\boldsymbol{x}) = \boldsymbol{w}^T\boldsymbol{\phi}(\boldsymbol{x})$ and, therefore, the equality constraint in (4) is not imposed.
2. The *classical SVM*,[2] where (1) reduces to $f(\boldsymbol{x}) = \boldsymbol{w}^T\boldsymbol{\phi}(\boldsymbol{x}) + w_1$, with $m = 1$, $\psi(\boldsymbol{x}) = 1\ \forall \boldsymbol{x}$, and $\boldsymbol{A}_2^T = [1, \ldots, 1]$.
3. The *regularized parametric SVM approach*[3] (e.g., linear SVM), where (1) reduces again to $f(\boldsymbol{x}) = \boldsymbol{w}^T\boldsymbol{\phi}(\boldsymbol{x})$ but with dimension d of $\boldsymbol{\phi}(\boldsymbol{x})$ so small that it is no longer efficient to rewrite the approximating function as in (3) introducing the kernel $\mathcal{K}$.

The solution of (4), required for applying semiparametric SVM, is usually determined by a black-box software package; simple and explicit algorithms have been proposed only for nonparametric SVM [5–10]. We first propose an analytical method for optimum line search in (4) and then utilize it, in conjunction with an augmented Lagrangian approach, to derive simple and explicit algorithms for training semiparametric SVM.

[2] It is still a nonparametric approach in practice because only a constant term is introduced as parametric component.

[3] Let us note that the regularization is useful only if the number of examples is not much larger than the number of parametric components.

3. Optimum Line Search for a Generic Direction

An iterative search for the optimum solution of (4) can be usually written as $\boldsymbol{u}_{k+1} = \boldsymbol{u}_k + \mu_k\boldsymbol{r}_k$, where $\boldsymbol{u}_k$ denotes the value of $\boldsymbol{u}$ at a generic step k of the procedure assumed here to satisfy the constraints in (4), i.e., $|\boldsymbol{u}_k| \leq \boldsymbol{C}$ and $\boldsymbol{A}_2^T\boldsymbol{u}_k = 0$. Any iterative procedure consists in determining at each step a direction of movement $\boldsymbol{r}_k$ and the value of μ_k that achieves the optimum of the cost function along that direction. The choice of the direction $\boldsymbol{r}_k$ can be a classical one such as the gradient or its modification. Later on we will suggest some particular choices. For now, we concentrate on determining the value of μ_k such that $\boldsymbol{u}_{k+1} = \boldsymbol{u}_k + \mu_k\boldsymbol{r}_k$ satisfies the bound and minimizes the cost function in (4). We derive here a closed-form solution for μ_k suited to the specific objective function in (4). The problem to be solved at step k is

$$\begin{cases} \min\limits_{\mu_k} \dfrac{1}{2}(\boldsymbol{u}_k + \mu_k\boldsymbol{r}_k)^T\mathcal{K}(\boldsymbol{u}_k + \mu_k\boldsymbol{r}_k) \\ \quad - \boldsymbol{y}^T(\boldsymbol{u}_k + \mu_k\boldsymbol{r}_k) + \boldsymbol{\varepsilon}^T|\boldsymbol{u}_k + \mu_k\boldsymbol{r}_k|, \\ \boldsymbol{A}_2^T(\boldsymbol{u}_k + \mu_k\boldsymbol{r}_k) = \boldsymbol{0}, \\ |\boldsymbol{u}_k + \mu_k\boldsymbol{r}_k| \leq \boldsymbol{C}. \end{cases} \qquad [8]$$

Assuming that the value of $\boldsymbol{u}_k$, determined in the previous step, satisfies all the constraints, we can rewrite the problem as

$$\begin{cases} \min\limits_{\mu_k} \dfrac{1}{2}(\boldsymbol{r}_k^T\mathcal{K}\boldsymbol{r}_k)\mu_k^2 - [(\boldsymbol{y} - \mathcal{K}\boldsymbol{u}_k)^T\boldsymbol{r}_k]\mu_k \\ \quad + \left[\sum\limits_{i=1}^{\ell} \varepsilon_i r_{ki} \operatorname{sign}(u_{ki} + \mu_k r_{ki})\right]\mu_k \\ \quad + \sum\limits_{i=1}^{\ell} \varepsilon_i u_{ki} \operatorname{sign}(u_{ki} + \mu_k r_{ki}), \\ \mu_k\boldsymbol{A}_2^T\boldsymbol{r}_k = \boldsymbol{0} \\ |\boldsymbol{u}_k + \mu_k\boldsymbol{r}_k| \leq \boldsymbol{C}, \end{cases} \qquad [9]$$

We need here to assume that the search direction $\boldsymbol{r}_k$ is such that $\boldsymbol{A}_2^T\boldsymbol{r}_k = \boldsymbol{0}$ (we will discuss this condition later). Let us suppose that the vector $\boldsymbol{r}_k$ contains ℓ_k nonnull components and define the set $\mathcal{V}_k$ containing the ℓ_k indices i such that $r_{ki} \neq 0$. Define also a vector $\boldsymbol{v}_k$, with ℓ_k components, such that $v_{ki} = -\dfrac{u_{ki}}{r_{ki}}$ for each $i \in \mathcal{V}_k$. The problem then becomes:

$$\begin{cases} \min\limits_{\mu_k} \dfrac{1}{2}(\boldsymbol{r}_k^T \mathcal{K} \boldsymbol{r}_k)\mu_k^2 - [(\boldsymbol{y} - \mathcal{K}\boldsymbol{u}_k)^T \boldsymbol{r}_k]\,\mu_k \\ \quad + \left[\sum_{i \in \mathcal{V}_k} \varepsilon_i r_{ki}\,\mathrm{sign}(u_{ki} + \mu_k r_{ki})\right]\mu_k \\ \quad + \sum_{i=1}^{\ell} \varepsilon_i u_{ki}\,\mathrm{sign}(u_{ki} + \mu_k r_{ki}), \\ |u_{ki} + \mu_k r_{ki}| \le C_i \qquad \forall i \in \mathcal{V}_k. \end{cases} \quad [10]$$

Let us note that problem (10) is a simple one-dimensional optimization problem in μ_k with a piecewise quadratic objective function on the following intervals:

$$-\infty < v_{k(1)} < \cdots < v_{k(\ell_k)} < \infty, \quad [11]$$

where the vector $[v_{k(1)}, \ldots, v_{k(\ell_k)}]^T$ is the ranked version of $\boldsymbol{v}_k$. Let us denote with $I_{k0}, \ldots, I_{k\ell_k}$ the corresponding intervals, that is, $I_{k0} \triangleq (-\infty, v_{k(1)})$, $I_{kj} \triangleq (v_{k(j)}, v_{k(j+1)})$ for $j \in \{1, \ldots, \ell_k - 1\}$, and $I_{k\ell_k} \triangleq (v_{k(\ell_k)}, +\infty)$. We also denote with

$$[1], \ldots, [\ell_k] \quad [12]$$

the indices from the set $\mathcal{V}_k \subseteq \{1, \ldots, \ell\}$ of the elements in the ranked version of $\boldsymbol{v}_k$.

The objective function $f(\mu_k)$ is continuous and it can be written as

$$f(\mu_k) = f_i(\mu_k), \qquad \mu_k \in I_{ki}, \qquad i \in \{0, 1, \ldots, \ell_k\} \quad [13]$$

where $f_i(\mu_k)$, $\mu_k \in (-\infty, +\infty)$, is a quadratic concave function that attains its minimum value in μ_{ki} with

$$\mu_{k0} \triangleq \frac{1}{2}\gamma_k[(\boldsymbol{y} - \mathcal{K}\boldsymbol{u}_k)^T \boldsymbol{r}_k + \boldsymbol{\varepsilon}^T |\boldsymbol{r}_k|], \quad \gamma_k \triangleq \frac{2}{\boldsymbol{r}_k^T \mathcal{K} \boldsymbol{r}_k}, \quad [14]$$

$$\mu_{ki} = \mu_{k,i-1} - \gamma_k \varepsilon_{[i]} |r_{k[i]}|, \qquad \forall i \in \{1, \ldots, \ell_k\}. \quad [15]$$

By noting[4] that $\mu_{ki} \le \mu_{k,i-1}$ $\forall i \in \{0, 1, \ldots, \ell_k\}$, it follows that the subset of $\{0, 1, \ldots, \ell_k\}$ satisfying the property

$$\mu_{ki} \in I_{ki} \quad [16]$$

cannot[5] contain more than one element. Moreover, the minimum of f over I_{ki} is μ_{ki} if $\mu_{ki} \in I_{ki}$; if $\mu_{ki} < v_{k(i)}$, the minimum of f over I_{ki} is the lower limit $v_{k(i)}$ of I_{ki}; if $\mu_{ki} > v_{k(i+1)}$, the minimum of f over I_{ki} is the upper limit $v_{k(i+1)}$ of I_{ki}.

If an element, say m, satisfies property (16), then the global solution is[6] just μ_{km}. If no element satisfies (16), then a value, say m, exists[7] such that

$$\mu_{k,m+1} < v_{k(m+1)} < \mu_{km} \quad [17]$$

and $v_{k(m+1)}$ represents[8] the global solution.

The following pseudocode can be used for determining the global solution:

```
i = 0; term=0;
while term == 0
    calculate μ_ki from (14) and (15)
    if μ_ki ∈ I_ki
        global solution = μ_ki;
        term=1;
    end if;
    if μ_ki < v_k(i)
        global solution = v_k(i);
        term=1;
    end if;
    i=i+1:
end while
```

When the solution to the unconstrained problem is determined, it is easy to show that the satisfaction of the constraints can be imposed by clipping it with α_1 and α_2, where $\alpha_1 \triangleq -\min\limits_{i \in \mathcal{V}_k}\left[\dfrac{C_i}{|r_{ki}|} + \dfrac{u_{ki}}{r_{ki}}\right]$ and $\alpha_2 = \min\limits_{i \in \mathcal{V}_k}\left[\dfrac{C_i}{|r_{ki}|} - \dfrac{u_{ki}}{r_{ki}}\right]$.

This procedure allows us to calculate analytically[9] the optimum step size in the optimization process, thereby avoiding a possibly expensive line search, and it can therefore be applied to any possible choice of the vector $\boldsymbol{r}_k$ such that $\boldsymbol{A}_2^T \boldsymbol{r}_k = \boldsymbol{0}$. Next, we give some particular examples of a possible procedure for choosing $\boldsymbol{r}_k$.

4. Simple Algorithms for Purely Nonparametric Approach

The rule for determining at each step the vector $\boldsymbol{r}_k$ constitutes an important part of the whole algorithm. There is a natural trade-off between the complexity of the procedure for determining $\boldsymbol{r}_k$ and the number of iterations that are needed to obtain a certain accuracy. We start by considering the purely nonparametric approach. Then, the direc-

[4] This condition holds since $\mathcal{K}$ is positive semidefinite and, therefore, $\gamma_k \ge 0$; moreover, $\varepsilon_i > 0$, by definition, $\forall_i \in \{1, \ldots, \ell\}$.

[5] In fact, let us denote with m a value that satisfies (16). Then, all the elements of I_{kj} with $j > m$ are larger than the maximum value in I_{km} while $\mu_{kj} \le \mu_{km} \in I_{km}$; on the other hand, all the elements of I_{kj} with $j < m$ are smaller than the minimum value in I_{km} while $\mu_{kj} \ge \mu_{km} \in I_{km}$.

[6] The minimum of f over I_{ki} is its upper limit if $i < m$ and it is its lower limit if $i > m$. Then, from continuity of f, it follows that the global minimum is given by the minimum over I_{km} that is μ_{km}.

[7] From the condition $\mu_{ki} \le \mu_{k,i-1}$ $\forall_i \in \{0, 1, \ldots, \ell_k\}$, it follows that

$$\mu_{ki} \notin I_{ki} \qquad \forall i \in \{0, 1, \ldots, \ell_k\}$$
$$\equiv \exists m : \begin{cases} \mu_{ki} > v_{k(i+1)} & \forall i \in \{0, \ldots, m\} \\ \mu_{ki} < v_{k(i)} & \forall i \in \{m+1, \ldots, \ell_k\} \end{cases}$$

which implies (17).

[8] From (17) it follows that the minimum of f over I_{ki} is its upper limit if $i \le m$ and it is its lower limit if $i > m$. Then, from the continuity of f, it follows that the optimum is given by the upper limit of I_{km} which coincides with the lower limit of $I_{k,m+1}$.

[9] In practice, it is useful to set to zero the components of $\boldsymbol{u}_{k+1} = \boldsymbol{u}_k + \mu_k \boldsymbol{r}_k$ whose magnitude is smaller than a fixed threshold (e.g., 10^{-6}). Moreover, it is useful to set to C the components of $\boldsymbol{u}_k$ very close to C.

tion $\boldsymbol{r}_k$ is not required to satisfy the constraint $\boldsymbol{A}_2^T\boldsymbol{r}_k = \boldsymbol{0}$ and, therefore, a simpler direction finding procedure can be used. Examples are included to illustrate the procedure.

4.1. A Gradient-Based Approach

The gradient of the cost function in (4) is $\boldsymbol{g}_k = -(\boldsymbol{y} - \mathcal{K}\boldsymbol{u}_k) + \boldsymbol{\varepsilon}^T \operatorname{sign}(\boldsymbol{u}_k)$ where $\operatorname{sign}(\boldsymbol{u}_k) \triangleq [\operatorname{sign}(u_{k1}), \ldots, \operatorname{sign}(u_{k\ell})]^T$. We cannot choose $\boldsymbol{r}_k = \boldsymbol{g}_k$ because of the bound constraint and because the gradient exists only if all the components of $\boldsymbol{u}$ are nonnull. Therefore, we can choose the direction $\boldsymbol{r}_k$ in the following way:

$$r_{ki} = \begin{cases} 0, & u_{ki} = C, \quad g_{ki} < 0 \\ 0, & u_{ki} = -C, \quad g_{ki} > 0 \\ 0, & u_{ki} = 0, \quad |g_{ki}| < \varepsilon_i \\ g_{ki}, & \text{otherwise} \end{cases} \qquad [18]$$

4.2. A Coordinate Descent Approach

The simplest choice of search direction is to run along the space coordinates, changing $\boldsymbol{r}_k$ cyclically over all the vectors $[\underbrace{0, \ldots, 0}_{i-1}, 1, 0, \ldots, 0]^T$ with $i \in \{1, \ldots, \ell\}$. In this case, the search for the optimum value of μ_k reduces to

$$\begin{cases} \ell_k = 1\ \forall k, \qquad \gamma_k = \dfrac{2}{\mathcal{K}(\boldsymbol{x}_i, \boldsymbol{x}_i)}, \\ \mu_{k0} = \dfrac{1}{\mathcal{K}(\boldsymbol{x}_i, \boldsymbol{x}_i)}[y_i - [\mathcal{K}\boldsymbol{u}_k]_i + \varepsilon_i] \\ \quad = \dfrac{1}{\mathcal{K}(\boldsymbol{x}_i, \boldsymbol{x}_i)}[y_i - \hat{g}_i + \varepsilon_i] - u_{ki} \\ \quad = \dfrac{1}{\mathcal{K}(\boldsymbol{x}_i, \boldsymbol{x}_i)}[e_i + \varepsilon_i] - u_{ki} \\ \mu_{k1} = \dfrac{1}{\mathcal{K}(\boldsymbol{x}_i, \boldsymbol{x}_i)}[e_i - \varepsilon_i] - u_{ki} \\ \hat{g}_i \triangleq \sum_{j=1, j\neq i}^{\ell} u_{kj}\mathcal{K}(\boldsymbol{x}_i, \boldsymbol{x}_j) \\ e_i \triangleq y_i - \hat{g}_i \end{cases} \qquad [19]$$

and, therefore,

$$[\boldsymbol{u}_{k+1}]_i = u_{ki} + \mu_k = \begin{cases} \dfrac{1}{\mathcal{K}(\boldsymbol{x}_i, \boldsymbol{x}_i)}[e_i + \varepsilon_i] & e_i \le -\varepsilon_i \\ \dfrac{1}{\mathcal{K}(\boldsymbol{x}_i, \boldsymbol{x}_i)}[e_i - \varepsilon_i] & e_i \ge \varepsilon_i \\ 0 & |e_i| \le \varepsilon_i \end{cases} \qquad [20]$$

or, equivalently,

$$[\boldsymbol{u}_{k+1}]_i = \frac{|e_i| - \varepsilon_i}{\mathcal{K}(\boldsymbol{x}_i, \boldsymbol{x}_i)} \operatorname{sign}(e_i)\Theta(|e_i| - \varepsilon_i) \qquad [21]$$

where $\Theta(\cdot)$ is the classical step function (i.e., $\Theta(\xi) = 0$ if $\xi \le 0$ and $\Theta(\xi) = 1$ if $\xi > 0$). If we consider the bound constraint, we need to clip between $-C_i$ and C_i. In fact, μ_k needs to be clipped between $\alpha_1 \triangleq -\min_{i \in \mathcal{Y}_k} \left[\frac{C_i}{|r_{ki}|} + \frac{u_{ki}}{r_{ki}}\right] = -C_i - u_{ki}$ and $\alpha_2 = \min_{i \in \mathcal{Y}_k} \left[\frac{C_i}{|r_{ki}|} - \frac{u_{ki}}{r_{ki}}\right] = C_i - u_{ki}$. Therefore, we can write the next value $\boldsymbol{u}_{k+1}$ in the following form[10]:

$$u_{k+1,i} = \min\left\{C_i\Theta(|e_i| - \varepsilon_i), \frac{\| e_i| - \varepsilon_i|}{\mathcal{K}(\boldsymbol{x}_i, \boldsymbol{x}_i)}\right\} \operatorname{sign}(e_i), \qquad [22]$$

which is the simple coordinate descent procedure proposed in [5].

This approach consists of starting from a point that satisfies the constraint $\boldsymbol{u} \le \boldsymbol{C}$ and in cycling on all the variables, utilizing (22), until the variations in the components of $\boldsymbol{u}$ become negligible (i.e, smaller than a small constant δ) for an entire cycle or for some cycles.

Two modifications of the basic algorithm allow us to reduce the computational burden:

1. At each step, the largest part of the computation is required by the calculation of $\hat{g}_i$; we could calculate $\hat{g}_j$ for each j once and store the results to avoid the calculation of $\hat{g}_i$ at each step. This requires, however, that when the value of $[\boldsymbol{u}_{k+1}]_i$ calculated by means of (22) is different from the previous one, we need to update all the values of $\hat{g}_j$ for $j \neq i$. However, if, during a cycle, many variables do not change, the computational burden is strongly reduced.
2. At the end of each cycle on all the variables, we can subcycle simply on the variables for which a modification occurred during the last cycle or the last few cycles. We stop subcycling on these variables when we have optimized the problem with respect to these variables. Care needs to be taken to avoid imposing a strong accuracy in the subcycling when we are far from the final solution. The advantage of subcycling is that we need to update the values of $\hat{g}_j$ not considered for subcycling just once at the end of the subcycle. Alternative methods for choosing the subset of variables for subcycling can be considered (see, for example, [11, 12]). A proof of convergence of this strategy in a finite number of steps may strongly depend on the subset selection procedure and, to the best of our knowledge, it is still an open problem [10]. First results on this topic can be found in [13].

In a large-scale problem, we cannot store in memory the matrix $\mathcal{K}$. A first direct approach (that is valid for any algorithm) consists in evaluating each element of the matrix when needed. The computational complexity of this choice depends mainly on how complex the calculation of the generic elements of the matrix is. In particular, our simple iterative algorithm can also be applied in this case,

[10] Obviously, the other components are not modified, i.e., $u_{k+1,j} = u_{kj}$ $\forall_j \neq i$.

provided that, when updating $\hat{g}_j$ $(j \neq i)$, the value of k_{ij} is calculated for each $j \neq i$.

With this modification, we pass from a memory-intensive algorithm to minimum memory requirements. If some RAM memory is available (not sufficient, however, to store the entire matrix), we can utilize it to cache, the most often utilized elements of the matrix.

It is important to note that a relation should exist between the dimension of the subproblem in subcycling and the number of matrix elements that can be maintained in the cache memory. In particular, a wise choice of the subproblem dimension should allow one to store, in the RAM memory utilized as cache, the matrix elements needed to solve a subproblem.

5. The Classic SVM Problem

In the classic SVM problem, we need to satisfy the equality constraint $\mathbf{1}^T \boldsymbol{r}_k = \mathbf{0}$, where $\mathbf{1} \triangleq [1, \ldots, 1]^T$. In such a case, we can choose the vector $\boldsymbol{r}_k$ by cyclically choosing the following vectors $\boldsymbol{r}_k$: $[\underbrace{\underbrace{0, \ldots, 0}_{i_1 - 1}, 1, 0, \ldots, 0}_{i_2 - 1}, -1, 0, \ldots, 0]^T$ over all the couples (i_1, i_2) with $i_1 < i_2$. Then,

$$\begin{cases} \mathcal{V}_k = \{i_1, i_2\}, \quad \boldsymbol{v}_k = [u_{ki_1}, u_{ki_2}]^T, \quad \ell_k = 2\, \forall k, \quad \gamma_k = \dfrac{2}{\beta_2}, \\ \mu_{k0} = \dfrac{\beta_1 + \varepsilon_{i_1} + \varepsilon_{i_2}}{\beta_2} - u_{ki_1} \\ \beta_1 \triangleq y_{i_1} - \hat{f}_{i_1} - y_{i_2} + \hat{f}_{i_2} + K(\mathcal{K}(\boldsymbol{x}_{i_2}, \boldsymbol{x}_{i_2}) - \mathcal{K}(\boldsymbol{x}_{i_1}, \boldsymbol{x}_{i_2})) \\ \beta_2 \triangleq \mathcal{K}(\boldsymbol{x}_{i_1}, \boldsymbol{x}_{i_1}) + \mathcal{K}(\boldsymbol{x}_{i_2}, \boldsymbol{x}_{i_2}) - 2\,\mathcal{K}(\boldsymbol{x}_{i_1}, \boldsymbol{x}_{i_2}) \\ \quad = \|\phi(\boldsymbol{x}_{i_1}) - \phi(\boldsymbol{x}_{i_2})\|^2 \geq 0 \\ \hat{f}_{i_1} \triangleq \sum_{j=1, j \neq i_1, j \neq i_2} u_{kj}\, \mathcal{K}(\boldsymbol{x}_{i_1}, \boldsymbol{x}_j), \quad K \triangleq u_{ki_1} + u_{ki_2}. \end{cases} \qquad [23]$$

It follows that

$$\begin{cases} \mu_{k1} \triangleq \begin{cases} \dfrac{\beta_1 - \varepsilon_{i_1} + \varepsilon_{i_2}}{\beta_2} - u_{ki_1} & K \geq 0 \\ \dfrac{\beta_1 + \varepsilon_{i_1} - \varepsilon_{i_2}}{\beta_2} - u_{ki_1} & K \leq 0 \end{cases} \\ \mu_{k2} \triangleq \dfrac{\beta_1 - \varepsilon_{i_1} - \varepsilon_{i_2}}{\beta_2} - u_{ki_1}. \end{cases} \qquad [24]$$

If we use the same bound constraint for all variables: $C_i = C\, \forall_i$, we can implement the analytical procedure with the following pseudocode:

$u_{i_1} = \beta_1 + \varepsilon_{i_1} + \varepsilon_{i_2}$;
if $K \geq C$ then $u_{i_1} = \beta_1 - \varepsilon_{i_1} + \varepsilon_{i_2}$;
if $K \leq -C$ then $u_{i_1} = \beta_1 + \varepsilon_{i_1} - \varepsilon_{i_2}$;
$u_{i_1} = \dfrac{u_{i_1}}{\beta_2}$;
if $K < max(-C, K - C)$ then $u_{i_1} = max(-C, K - C)$;
if $0 \leq K < C$ then
  call rou $\left(u_{i_1}, 0, \dfrac{2\varepsilon_{i_1}}{\beta_2}\right)$;
  call rou $\left(u_{i_1}, K, \dfrac{2\varepsilon_{i_2}}{\beta_2}\right)$;
end if
if $-C \leq K < 0$ then
  call rou $\left(u_{i_1}, K, \dfrac{2\varepsilon_{i_2}}{\beta_2}\right)$;
  call rou $\left(u_{i_1}, 0, \dfrac{2\varepsilon_{i_1}}{\beta_2}\right)$;
end if
$u_{i_1} > min(C, K - C)$ then $u_{i_1} = \min(C, K - C)$;
$u_{i_2} = K - u_{i_1}$;

subroutine rou(a,b,c)
if $a > b$ then
  $a = a - c$;
  if $a < b$ then $a = b$
end if

The pseudocode can be be written in the following analytical form, obtaining the algorithm introduced in [5]:

$$u_{k+1, i_1} = \min\{C, K + C, \max[-C, K - C, z_1]\}, \qquad [25]$$

where

$$z_1 \triangleq g_{\max(0,K)}\left[z_2; \frac{2}{K\beta_1}(\varepsilon_{i_1} \min(0, K) + \varepsilon_{i_2} \max(0, K))\right], \qquad [26]$$

$g_{\xi_1}(a; \xi_2) \triangleq \xi_1 + (a - \xi_1 - \xi_2)\Theta(a - \xi_1 - \xi_2) + (a - \xi_1)$ $\Theta(\xi_1 - a)$, $\Theta(\cdot)$ denotes the usual step function and

$$z_2 \triangleq g_{\min(0,K)}\left[z_3; \frac{2}{K\beta_1}(\varepsilon_{i_1} \max(0, K) + \varepsilon_{i_2} \min(0, K))\right], \qquad [27]$$

with

$$z_3 \triangleq \frac{\beta_2 + \varepsilon_{i_1} + \varepsilon_{i_2}}{\beta_1}. \qquad [28]$$

The algorithm can be summarized in the following steps:

(a) For $k = 1$ and $i = 1, \ldots, \ell$ initialize u_{ki} to generic values satisfying the constraints in (4) (e.g., $u_{ki} = 0$) and the quantities $\hat{g}_i \triangleq \sum_{j=1, j \neq i}^{\ell} u_{kj}\, \mathcal{K}(\boldsymbol{x}_i, \boldsymbol{x}_j)$;
(b) Define a couple (i_1, i_2);
(c) Compute $K \triangleq -\sum_{i=1, i \neq i_1, i \neq i_2}^{\ell} u_{ki} = u_{ki_1} + u_{ki_2}$;
(d) Compute β_1 and β_2 from (23) with $\hat{f}_{i_1} = \hat{g}_{i_1} - u_{ki_2}\, \mathcal{K}(\boldsymbol{x}_{i_1}, \boldsymbol{x}_{i_2})$ and $\hat{f}_{i_2} = \hat{g}_{i_2} - u_{ki_1}\, \mathcal{K}(\boldsymbol{x}_{i_1}, \boldsymbol{x}_{i_2})$;
(e) Compute u_{k+1, i_1} from (25) or, better, from the pseudocode;

(f) If $|u_{k+1,i_1} - u_{ki_1}| < 0.1 \cdot \delta_i$, where δ_i is the desired accuracy on $\boldsymbol{u}$, then $u_{k+1,i_1} = u_{k,i_1}$ and $u_{k+1,i_2} = u_{k,i_2}$; otherwise, keep u_{k+1,i_1} computed in (e), set $u_{k+1,i_2} = K - u_{k+1,i_1}$, and update the $\hat{g}_i$;

(g) Go back to step (b) by considering a new couple (i_1, i_2). All the $\frac{\ell(\ell-1)}{2}$ couples should be explored in cycle.

The calculation ends when, in the last cycle, all the couples satisfy the condition $|u_{k+1,i_1} - u_{ki_1}| < \delta_i$.

Storing in memory the quantities $\hat{g}_i$ avoids calculating $\hat{g}_{i_1}$ and $\hat{g}_{i_2}$ at each step; their updating is the most intensive part of the algorithm. The steps in which $\boldsymbol{u}$ is not updated are therefore much faster than the others. This strongly reduces the need for efficiently choosing at each step the couple (i_1, i_2) between all the possible ones.

Slight modification of the basic algorithm can be useful for improving the training speed. For example, at the end of each cycle on all the couples, one can form a subset $S \subseteq L$ (with $L \triangleq \{1, \ldots, \ell\}$) of the indices of the couples for which the updating occurred during the last cycle (or during the last cycles), and start to cycle on all the couples in S until all of them satisfy the accuracy condition (or also a less accurate one) before starting the new cycle on L. The advantage of this solution is that cycling on the couples in S requires updating at each step only the values of $\hat{g}_i$ in S while the $\hat{g}_i$ in $L - S$ can be updated once before starting the new cycle on L.

The proposed algorithm has minimal memory requirements. The whole memory can be dedicated to the specification of the matrix $\mathcal{K}(\boldsymbol{x}_i, \boldsymbol{x}_j)$. However, if ℓ is large so that it cannot be stored, one can explicitly evaluate the values $\mathcal{K}(\boldsymbol{x}_i, \boldsymbol{x}_j)$ when they are needed, shifting memory requirements to computational complexity. Note that in the partitioned version previously mentioned, one can choose to store in memory only the matrix needed for subcycling.

6. General Semiparametric SVM

When the number of parametric components m increases, it is not easy to satisfy at each step the equality constraint: $\boldsymbol{A}_2^T \boldsymbol{r}_k = \boldsymbol{0}$.

The generalization to this case of the simple iterative algorithm considered in the previous section would consist in cycling over vectors $\boldsymbol{r}_k$ with $m + 1$ components different from zero, say $i_1, i_2, \ldots, i_{m+1}$, obtaining the vector with the nonnull components of $\boldsymbol{r}_k$ in the following form: $[\boldsymbol{B}^{-1}\boldsymbol{s}, -1]^T$, where $\boldsymbol{B}$ (assumed to be invertible) is the submatrix of $\boldsymbol{A}_2^T$ containing the columns $i_1, \ldots, i_m$ and $\boldsymbol{s}$ is the i_{m+1}-th column. The loop should be done over all m-tuples of indices in $\{1, \ldots, \ell\}$. Note that the simple iterative algorithm for the classic SVM represents a special case of this general algorithm when $m = 1$ and $\boldsymbol{A}_2 = \boldsymbol{1}$.

The most significant difficulty of the considered method is represented by the procedure for determining, at each step, the vector $\boldsymbol{r}_k$. Also in the simplest case, the inversion of the matrix $\boldsymbol{B}$ is required when the set of indices $i, \ldots, i_m$ is modified. An efficient algorithm should be provided with a procedure for updating its inverse or, when introduced, its factorization. This can make the implementation of this algorithm difficult, especially when the number m of parametric components is high (say $m = 100$).[11]

When m is large, it is preferable to approach the problem by other means such as the augmented Lagrangian approach that will be discussed next.

6.1. The Augmented Lagrangian Approach

Given the following optimization problem:

$$\min_{\boldsymbol{u}} J(\boldsymbol{u}) \qquad [29]$$

with the constraints $|\boldsymbol{u}| \leq \boldsymbol{C}$ (defined componentwise) and $\boldsymbol{h}(\boldsymbol{u}) = \boldsymbol{0}$, the augmented Lagrangian approach consists [14] in solving iteratively the problem:

$$\boldsymbol{u}_j = arg \min_{\boldsymbol{u}} J(\boldsymbol{u}) + \boldsymbol{b}_j^T \boldsymbol{h}(\boldsymbol{u}) + \frac{1}{2} \sum_{i=1}^{m} c_{ji} |h_i(\boldsymbol{u})|^2 \qquad [30]$$

subject to the constraint $|\boldsymbol{u}| \leq \boldsymbol{C}$ where $\boldsymbol{h}(\boldsymbol{u}) \triangleq [h_1(\boldsymbol{u}), \ldots, h_m(\boldsymbol{u})]^T$ and $\boldsymbol{c}_j \triangleq [c_{j1}, \ldots, c_{jm}]^T$. The updating rules for $\boldsymbol{b}_j$ and $\boldsymbol{c}_j$ are the following: $b_{j+1,i} = b_{ji} + c_{ji} h_i(\boldsymbol{u}_j)$, $c_{j+1,i} = \beta c_{ji}$ if $|h_i(\boldsymbol{u}_j)| > \gamma |h_i(\boldsymbol{u}_{j-1})|$ and $c_{j+1,i} = c_{ji}$ otherwise; the updating rule refers to each $i \in \{1, \ldots, m\}$. At each step j, problem (30) needs to be solved with accuracy equal to $\delta \leq \min(1, |h_1(\boldsymbol{u}_j)|, \ldots, |h_m(\boldsymbol{u}_j)|)$ and the iteration on j can be stopped if $|h_i(\boldsymbol{u}_j)| \leq \delta_1 \; \forall i \in \{1, \ldots, m\}$. Choices such as $\beta = 5$ or 10, $\boldsymbol{b}_0 = 0$, $c_0^i = 1 \; \forall i$, $\delta_1 = 10^{-8}$, and $\gamma = 0.25$ are typically recommended [14].

6.2. Training Semiparametric SVM

We now describe the application of the augmented Lagrangian approach to problem (4) in the generic semiparametric form (1). Let us assume that a starting point $\boldsymbol{w}_1(0)$ has been chosen.

Since the optimum parameter $\boldsymbol{w}_1$ represents the Lagrangian multiplier $\boldsymbol{b}_j$, it follows from (30) and (4) that we can find $\boldsymbol{u}_j$ by solving the nonparametric problem with $\boldsymbol{y}$ replaced by $\boldsymbol{y} - \boldsymbol{A}_2 \boldsymbol{w}_1(j)$ and the matrix $\mathcal{K}$ replaced by $\mathcal{K} + \boldsymbol{A}_2 \Sigma \boldsymbol{A}_2^T$ with $\Sigma \triangleq \mathrm{diag}(\boldsymbol{c}_j)$. The vector $\boldsymbol{w}_1(j+1)$ is determined as $\boldsymbol{w}_1(j) + \boldsymbol{c}_j \odot \boldsymbol{A}_2^T \boldsymbol{u}_j$ and $c_{j+1,i} = \beta c_{ji}$ if $|\boldsymbol{a}_i^T \boldsymbol{u}_j| > \gamma |\boldsymbol{a}_i^T \boldsymbol{u}_{j-1}|$ and $c_{j+1,i} = c_{ji}$ otherwise, where $\boldsymbol{a}_i$ is the ith column of $\boldsymbol{A}_2$ and $\odot$ denotes the Hadamard product (i.e., the element-by-element multiplication).

[11] This can be the case in the presence of a large number of input components with small influence on the complete function. In such a case, they can managed only linearly in the parametric part while only a subset of them can be considered in the nonparametric part.

The proposed approach solves optimization problem (4) by virtue of its capability to solve the problem in the nonparametric case (i.e., when the equality constraint is not imposed). We propose the algorithm presented in the previous subsection for solving it; nevertheless, any other algorithm can be utilized. Since the optimization problem relative to the classical SVM represents a particular case of (4), the proposed approach allows us to train the classical and semiparametric SVM by utilizing a software package for solving the problem in the nonparametric case. In particular, if we solve the problem, at each step of iteration, by means of the coordinate descent algorithm (22), we obtain another simple and iterative algorithm for solving the classical SVM and the semiparametric SVM.

6.2.1. A Possible Simplification. It may appear strange at first sight that we would prefer solving a sequence of minimization problems rather than a single problem. However, in practice, we need to approximately solve a simpler problem for a finite number of time instants. Moreover, we may utilize efficiently the solution of the previous problem as starting point for the next problem.

Nevertheless, the complexity of the considered method depends on how much the chosen $\boldsymbol{w}_1(0)$ is far from its optimum value. A simple choice consists in setting it to zero; a better choice consists in utilizing as starting value for $\boldsymbol{w}_1$ the optimum value with reference to the simple quadratic cost function; in fact, by choosing as cost function in (2) $c(\xi) = \frac{1}{2}\xi^2$, the optimization problem (4) changes in such a way that its solution can be written in closed form.[12] Specifically, we need to solve the following linear system:

$$\left[(\boldsymbol{I}^k - \boldsymbol{A}_2\boldsymbol{A}_2^+)\mathcal{K} + \frac{1}{C}\boldsymbol{I}^k\right]\boldsymbol{u} = (\boldsymbol{I}^k - \boldsymbol{A}_2\boldsymbol{A}_2^+)\boldsymbol{y}, \qquad [31]$$

and then calculate $\boldsymbol{w}_1 = \boldsymbol{A}_2^+(\boldsymbol{b} - \mathcal{K}\boldsymbol{u})$, where $\boldsymbol{A}_2^+$ denotes the pseudoinverse of $\boldsymbol{A}_2$. Therefore, this initialization procedure requires that we calculate the pseudoinverse of an $\ell \times m$ matrix and that we solve an ℓ-dimensional linear system. This procedure is obviously inefficient when ℓ is very large but the number of support vectors in the nonparametric part is much smaller than ℓ. With a good initial choice, as the proposed one, it is possible to reach a good solution by solving only a few times the optimization problem without the equality constraint. As an extreme simplification we can solve it only once utilizing the $\boldsymbol{w}_1$ obtained with the quadratic cost function.

7. Regularized Parametric SVM Approach

Let us consider the SV approximation for the case $m = 0$: $f(\boldsymbol{x}) = \boldsymbol{w}^T\Phi(\boldsymbol{x})$ where $\boldsymbol{w} = \sum_{i=1}^{\ell} u_i\Phi(\boldsymbol{x}_i)$. If the dimension d of the vector $\boldsymbol{w}$ and $\Phi(\boldsymbol{x})$ is sufficiently small, one can work directly in this space avoiding the use of a kernel approach. A simple case in which this is verified is given by the linear SVM; however, when the dimension[13] n of the input vector $\boldsymbol{x}$ is small (e.g., not larger than 10), it is not difficult to construct a sufficiently rich nonlinear expansion $\phi(\boldsymbol{x})$ with a dimension d manageable in practice.

When d is sufficiently small, we can try to store directly in memory the quantities $\boldsymbol{w}$ and $\{\Phi(\boldsymbol{x}_1) \ldots, \Phi(\boldsymbol{x}_\ell)\}$. By applying the iterative algorithm (22) with $e_i = y_i - \boldsymbol{w}_k^T\Phi(\boldsymbol{x}_i) + u_{ki}\|\Phi(\boldsymbol{x}_i)\|^2$, at each step $u_{k+1,i}$ is calculated; if $|u_{k+1,i} - u_{k,i}| \geq \delta$, then the vector $\boldsymbol{w}$ is updated by means of the following relation:

$$\boldsymbol{w}_{k+1} = \boldsymbol{w}_k + (u_{k+1,i} - u_{ki})\Phi(\boldsymbol{x}_i). \qquad [32]$$

Moreover, in an adaptive environment, it may be that the generic kth example becomes available at step k but cannot be stored. In such a case, we can utilize each example for a single step before forgetting it; then, the adaptation formula (32) is modified by setting $u_{ki} = 0$:

$$\boldsymbol{w}_{k+1} = \boldsymbol{w}_k + \min\left(C\Theta(|e_k| - \varepsilon), \frac{\|e_k| - \varepsilon|}{\|\Phi(\boldsymbol{x}_k)\|^2}\right)\operatorname{sign}(e_k)\Phi(\boldsymbol{x}_k), \qquad [33]$$

where $e_k \triangleq y_k - \boldsymbol{w}_k^T\Phi(\boldsymbol{x}_k)$ and $(\boldsymbol{x}_k, y_k)$ represents the kth example. The updating rule (33) closely resembles the classical least mean square (LMS) techniques for adaptive filtering that, however, are not regularized and refer to the choice of a quadratic cost function that is not ε insensitive.

8. Experimental Results

We apply the semiparametric techniques introduced in the previous sections to the problem of modeling a nonlinear time-invariant system and analyze their performances by computer simulations.

Consider the following system with input $x(n)$ and output $y(n)$

$$y(n) = b + a\exp[-\alpha(x(n) - \mu)^2] + \sum_{i=0}^{4} h(i)x(n-i) \qquad [34]$$

with $b = a = 1000$, $\alpha = 0.01$, $\mu = 60$, and $[h(0), h(1), h(2), h(3), h(4)] = 0.01\ [10\ \ -8\ \ 15\ \ 6\ \ 1] \triangleq \boldsymbol{h}$.

We choose such a system because we want to show the effectiveness of the proposed methods in modeling systems for which there is an additive linear component with a long memory and a nonlinear component with a smaller memory. In such a case, the classical nonparametric SVM as well as a classical linear method would be obviously ineffective. Moreover, we choose a system with a relevant nonlinear component to show the superiority of the semiparametric SVM over a residue-based approach that separates the estimation of the parametric component from the nonparametric one.

The input time series can be embedded to form the vector $\boldsymbol{x}_n \triangleq [x(n)\ x(n-1)\ x(n-2)\ x(n-3)\ x(n-4)]^T$;

[12] Unfortunately, the resulting solution for $\boldsymbol{u}$ is not sparse and, therefore, the number of support vectors is equal to ℓ.

[13] The vector $\boldsymbol{x}$ here can be considered as the output of a supervised feature extraction stage which has achieved the dimensionality reduction [15].

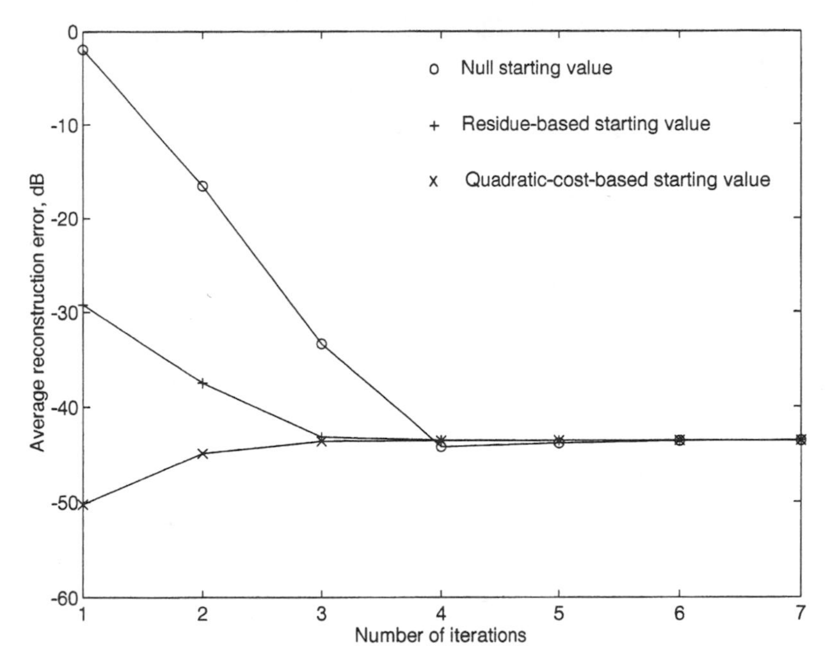

Fig. 1. The average value of the reconstruction error in dB versus the number of iterations for the three considered initialization procedures.

from (34) it follows that $y(n) = \bar{f}(\boldsymbol{x}_n)$ where $\bar{f}(\boldsymbol{x}_n) = b + \boldsymbol{h}^T\boldsymbol{x}_n + a \exp[-\alpha(x_{n,1} - \mu)^2]$. We generate the input time series as independent and identically distributed (i.i.d.) values uniformly distributed in (10,110) and embed it to form $\ell = 200$ noisy examples $\{(\boldsymbol{x}_i, \bar{f}(\boldsymbol{x}_i) + \xi_i), i = 1, \ldots, \ell\}$, for training and $\ell_{te} = 600$ uncorrupted test examples $\{(\boldsymbol{x}_i, \bar{f}(\boldsymbol{x}_i)), i = \ell + 1, \ldots, \ell + \ell_{te}\}$, where $\{\xi_i\}$ are ℓ i.i.d. samples of a contaminated Gaussian random variable with the following probability density function:

$$0.9\mathcal{N}(0; \chi_1) + 0.1\mathcal{N}(0; 100\chi_1), \qquad [35]$$

where $\mathcal{N}(0; \chi_1)$ is the zero-mean Gaussian density function with variance χ_1. The value of χ_1 is chosen to set the signal-to-noise ratio

$$10 \log_{10}\left(\frac{\sum_{i=1}^{\ell} (\bar{f}(x_i) - b)^2}{\sum_{i=1}^{\ell} \xi_i^2}\right) \qquad [36]$$

to 60 dB. We choose the following kernel function $\mathcal{K}(\boldsymbol{x}_1, \boldsymbol{x}_2) \triangleq \exp[-\alpha(x_{1,1} - x_{2,1})^2]$ for the nonparametric component and the following set of $m = 6$ functions $\boldsymbol{\psi}(\boldsymbol{x}) \triangleq [1\ \boldsymbol{x}^T]^T$ for the nonparametric components.

We test three different choices for initializing the value of $\boldsymbol{w}_1$ in the general semiparametric approach: (a) $\boldsymbol{w}_1 = \boldsymbol{0}$ (null starting value); (b) $\boldsymbol{w}_1 = \boldsymbol{A}_2^+\boldsymbol{y}$ (residue-based starting value); (c) $\boldsymbol{w}_1 = \boldsymbol{A}_2^+(\boldsymbol{y} - \mathcal{K}\boldsymbol{u}_d)$ where $\boldsymbol{u}_d$ is the solution of (31) (quadratic cost function–based starting value).

We solve the semiparametric training problem by adopting the augmented Lagrangian approach[14] introduced in Section 6 with parameters $\beta = 10$, $\gamma = 0.9$. At each step of the iterative procedure we solve the unconstrained optimization problem (30) by the coordinate descent algorithm (22) with $C = 100$, $\varepsilon = 10$, $\delta = 10^{-8}$.

We measure the reconstruction error E by adopting as cost function the absolute value of the difference e_i between the desired value $y(n)$ and its approximation:

$$E \triangleq 20 \log_{10}\left(\left\langle \frac{\sum_{i=1}^{\ell_{te}} |e_i|}{\sum_{i=1}^{\ell_{te}} |y_i|} \right\rangle\right). \qquad [37]$$

where $\langle\cdot\rangle$ denotes the average over 50 independent trials of the experiment. We also measure the estimation accuracy of the linear component of our system by means of the following parameter L:

$$L \triangleq 20 \log_{10}\left(\left\langle \sum_{i=0}^{4} |h(i) - w_{1,i+1}| \right\rangle\right). \qquad [38]$$

The average reconstruction error E, the average approximation error L of the linear component of the system (34), and the average amount of violation, at iteration j, of the equality constraint in (4) $D \triangleq \left\langle \frac{\|\boldsymbol{A}_2^T\boldsymbol{u}_j\|_1}{\|\boldsymbol{A}_2\|_1} \right\rangle$ are represented with respect to the number of iterations of the augmented Lagrangian approach in Figures 1, 2, and 3, respectively.

The smaller number of iterations for convergence[15] is

[14] We multiply all the columns of the matrix $\boldsymbol{A}_2$, except the first (and, obviously, also the corresponding components of the multiplier $\boldsymbol{w}_1$) by a constant so that their norm is similar to that of the first column.

[15] Figure 3 shows that the value of D can be reduced at each step of the iterative procedure provided that a sufficiently small value of δ is utilized.

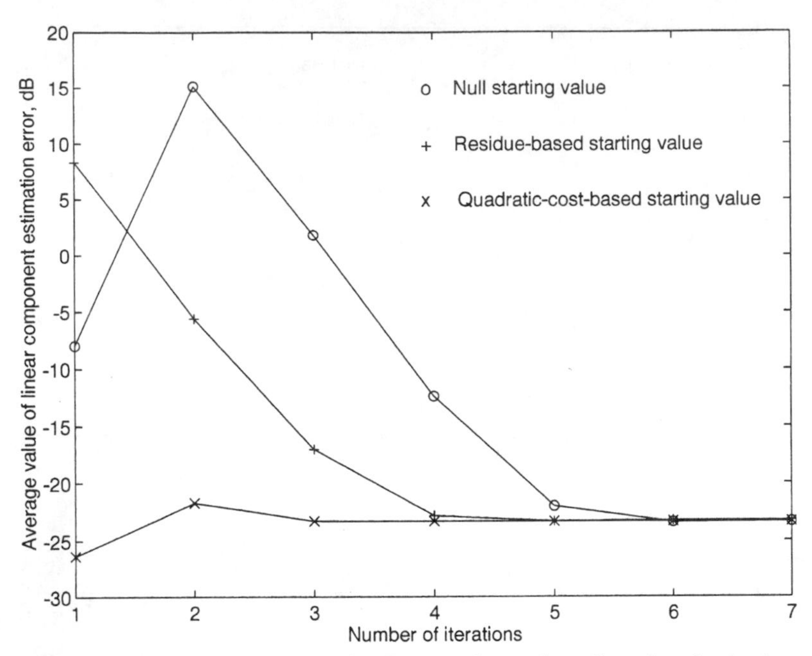

Fig. 2. The average value of the linear-component estimation error in dB versus the number of iterations for the three considered initialization procedures.

obtained for the initialization procedure (c), which, however, requires the solution of the linear system (31) of dimension ℓ. The solution based on the residue approach can be a good compromise; this is especially true when, differently from the considered example, the linear component is comparable to or larger in magnitude than the nonlinear one; however, only a few more iterations are needed with the simplest choice $\boldsymbol{w}_1 = \mathbf{0}$.

We notice that the choice of a quadratic cost function determines, after a few iterations, a smaller reconstruction error with respect to the chosen ε-insensitive cost function ($\varepsilon = 10$). However, it requires ℓ support vectors for representing the nonparametric component while the ε-insensitive solution requires an average value of 27 support vectors. The average value of support vectors is practically equal to this value at any iteration step for any of the three procedures, except for the value, equal to 53, after the first iteration in the procedure (a). This clearly shows that the

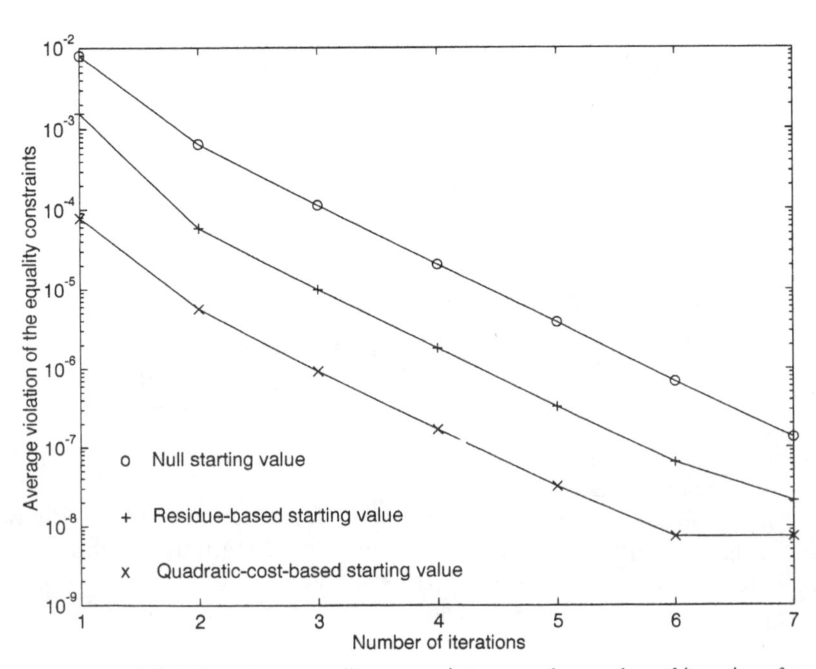

Fig. 3. The average value of the amount of violation of the equality constraint versus the number of iterations for the three considered initialization procedures.

approach consisting of estimating w_1 as in (c) and, then, solving only once the unconstrained optimization problem (30) to force the sparsity of the nonparametric representation gives very good results. This is the preferred approach when the value of ℓ is sufficiently low so that we can solve, with reasonable complexity, the linear system (31); when ℓ is too large, an alternative procedure would consist of estimating w_1 by the same procedure by utilizing a smaller number of training examples.

9. Conclusions

The problem of nonlinear model estimation by means of measured input-output examples is central to the study of intelligent signal processing, especially in applications where no model for the data is available. When a traditional parametric technique is inadequate for accurate nonlinear system modeling, a nonparametric approach is usually adopted (e.g., the classical SVM). The inclusion in the model of a parametric component can help to improve the accuracy in system representation and, most important, simplify the structure of the remaining nonparametric component, hence the motivation for a semiparametric approach.

Semiparametric SVM is a powerful generalization of the classical SVM that jointly determines both the parametric and the nonparametric components in model representation. Such an approach provides superior performance over simpler approaches that separately determine the parametric and nonparametric components.

One of the difficulties for practical application of the semiparametric SVM is the unavailability of simple and explicit algorithms for determining the system model. In this chapter, we have proposed new explicit algorithms for training semiparametric SVM; their principal advantages are the simplicity of design and minimal memory requirement. The trade-off for this simplicity, which is a necessary requirement in real-time adaptive algorithms, is an increase in the number of iterations required to reach convergence. However, the time to converge depends on the starting value of the iterative method. Simulation experiments presented herein confirm the effectiveness of the proposed algorithms and show that a smaller number of iterations are sufficient if we utilize more complex initialization procedures. These algorithms pave the way to solving challenging signal-processing problems in which the system model needs to be adaptively modified in order to account for possible nonstationarities of both the incoming time series and system under test.

Acknowledgments

The authors express their gratitude to Dr. Bernhard Schölkopf, Microsoft, Cambridge, United Kingdom, for many useful comments on an earlier version of the manuscript of this chapter.

References

[1] B. Schölkopf, K. Sung, C. Burges, F. Girosi, P. Niyogi, T. Poggio, and V. Vapnik, "Comparing support vector machines with gaussian kernels to radial basis function classifiers," *IEEE Trans. Signal Processing* 45:2758–2765, 1997.

[2] V. Vapnik, *Statistical Learning Theory.* New York: Wiley, 1998.

[3] S. Haykin, *Adaptive Filter Theory,* 3rd ed. New York: Prentice-Hall, 1996.

[4] A. J. Smola, T. Friess, and B. Schölkopf, "Semiparametric support vector and linear programming machines," in *Advances in Neural Information Processing Systems,* pp. 585–591. Cambridge, MA: MIT Press, 1999.

[5] D. Mattera, F. Palmieri, and S. Haykin, "An explicit algorithm for training support vector machines," *IEEE Signal Processing Letters* 6:243–245, September 1999.

[6] D. Mattera, F. Palmieri, and S. Haykin, "Simple and robust methods for support vector expansions," *IEEE Trans. Neural Networks* 10:1038–1047, September 1999.

[7] O. L. Mangasarian and D. R. Musicant, "Successive overrelaxation for support vector machines," Tech. Rep. 98-12, Computer Sciences Department, University of Wisconsin, USA, 1998. Available at http://www.cs.wisc.edu/~ olvi/olvi.html.

[8] J. Platt, "Fast training of support vector machines using sequential minimal optimization," in *Advances in Kernel Methods—Support Vector Learning* (B. Schölkopf, C. Burges, and A. Smola, eds.), pp. 185–208. Cambridge, MA: MIT Press, 1998.

[9] T. Friess, N. Cristianini, and C. Campbell, "The kernel adatron algorithm: a fast and simple learning procedure for support vector machines," in *15th Intl. Conf. Machine Learning,* pp. 188–196. Morgan Kaufman Publishers, 1998.

[10] A. Smola and B. Schölkopf, "A tutorial on support vector regression," Tech. Rep. TR-1998-030, NEURO-COLT, Royal Holloway College, 1998.

[11] T. Joachims, "Making large-scale support vector machine learning practical," in *Advances in Kernel Methods—Support Vector Learning* (B. Schölkopf, C. Burges, and A. Smola, eds.), pp. 169–184. Cambridge, MA: MIT Press, 1998.

[12] E. Osuna, R. Freund, and F. Girosi, "An improved training algorithm for support vector machines," in *Neural Networks for Signal Processing VII—Proceedings of the 1997 IEEE Workshop,* pp. 276–285, 1997.

[13] C.-C. Chang, C.-W. Hsu, and C.-J. Lin, "The analysis of decomposition methods for support vector machines," in *Proceedings of the SVM Workshop in IJCAI99,* 1999. Available at http://www.csie.ntu.edu.tw/~cjlin/papers.html.

[14] D. P. Bertsekas, *Constrained Optimization and Lagrange Multiplier Methods.* San Diego, CA: Academic Press, 1982.

[15] K. Fukunaga, *Introduction to Statistical Pattern Recognition,* 2nd ed. San Diego, CA: Academic Press, 1990.

Chapter 9

Gradient-Based Learning Applied to Document Recognition

Yann LeCun, Léon Bottou, Yoshua Bengio, and Patrick Haffner

Abstract—

Multilayer Neural Networks trained with the backpropagation algorithm constitute the best example of a successful Gradient-Based Learning technique. Given an appropriate network architecture, Gradient-Based Learning algorithms can be used to synthesize a complex decision surface that can classify high-dimensional patterns such as handwritten characters, with minimal preprocessing. This paper reviews various methods applied to handwritten character recognition and compares them on a standard handwritten digit recognition task. Convolutional Neural Networks, that are specifically designed to deal with the variability of 2D shapes, are shown to outperform all other techniques.

Real-life document recognition systems are composed of multiple modules including field extraction, segmentation, recognition, and language modeling. A new learning paradigm, called Graph Transformer Networks (GTN), allows such multi-module systems to be trained globally using Gradient-Based methods so as to minimize an overall performance measure.

Two systems for on-line handwriting recognition are described. Experiments demonstrate the advantage of global training, and the flexibility of Graph Transformer Networks.

A Graph Transformer Network for reading bank check is also described. It uses Convolutional Neural Network character recognizers combined with global training techniques to provides record accuracy on business and personal checks. It is deployed commercially and reads several million checks per day.

Keywords— **Neural Networks, OCR, Document Recognition, Machine Learning, Gradient-Based Learning, Convolutional Neural Networks, Graph Transformer Networks, Finite State Transducers.**

Nomenclature

- GT Graph transformer.
- GTN Graph transformer network.
- HMM Hidden Markov model.
- HOS Heuristic oversegmentation.
- K-NN K-nearest neighbor.
- NN Neural network.
- OCR Optical character recognition.
- PCA Principal component analysis.
- RBF Radial basis function.
- RS-SVM Reduced-set support vector method.
- SDNN Space displacement neural network.
- SVM Support vector method.
- TDNN Time delay neural network.
- V-SVM Virtual support vector method.

The authors are with the Speech and Image Processing Services Research Laboratory, AT&T Labs-Research, 200 Laurel Avenue, Middletown, NJ 07748. E-mail: {yann,leonb,yoshua,haffner}@research.att.com. Yoshua Bengio is also with the Département d'Informatique et de Recherche Opérationelle, Université de Montréal, C.P. 6128 Succ. Centre-Ville, 2920 Chemin de la Tour, Montréal, Québec, Canada H3C 3J7.

I. Introduction

Over the last several years, machine learning techniques, particularly when applied to neural networks, have played an increasingly important role in the design of pattern recognition systems. In fact, it could be argued that the availability of learning techniques has been a crucial factor in the recent success of pattern recognition applications such as continuous speech recognition and handwriting recognition.

The main message of this paper is that better pattern recognition systems can be built by relying more on automatic learning, and less on hand-designed heuristics. This is made possible by recent progress in machine learning and computer technology. Using character recognition as a case study, we show that hand-crafted feature extraction can be advantageously replaced by carefully designed learning machines that operate directly on pixel images. Using document understanding as a case study, we show that the traditional way of building recognition systems by manually integrating individually designed modules can be replaced by a unified and well-principled design paradigm, called *Graph Transformer Networks*, that allows training all the modules to optimize a global performance criterion.

Since the early days of pattern recognition it has been known that the variability and richness of natural data, be it speech, glyphs, or other types of patterns, make it almost impossible to build an accurate recognition system entirely by hand. Consequently, most pattern recognition systems are built using a combination of automatic learning techniques and hand-crafted algorithms. The usual method of recognizing individual patterns consists in dividing the system into two main modules shown in figure 1. The first module, called the feature extractor, transforms the input patterns so that they can be represented by low-dimensional vectors or short strings of symbols that (a) can be easily matched or compared, and (b) are relatively invariant with respect to transformations and distortions of the input patterns that do not change their nature. The feature extractor contains most of the prior knowledge and is rather specific to the task. It is also the focus of most of the design effort, because it is often entirely hand-crafted. The classifier, on the other hand, is often general-purpose and trainable. One of the main problems with this approach is that the recognition accuracy is largely determined by the ability of the designer to come up with an appropriate set of features. This turns out to be a daunting task which, unfortunately, must be redone for each new problem. A large amount of the pattern recognition literature is devoted to describing and comparing the relative

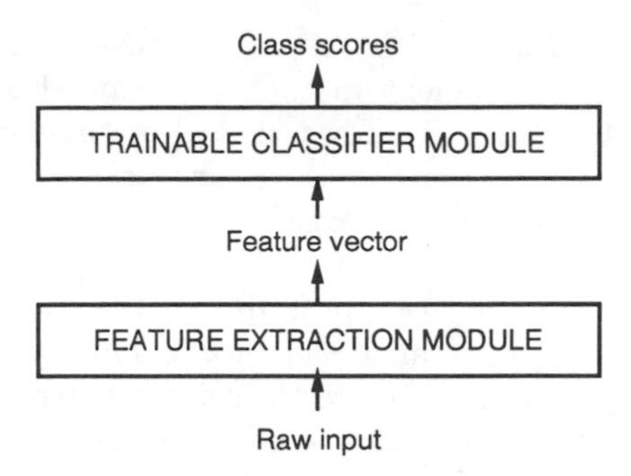

Fig. 1. Traditional pattern recognition is performed with two modules: a fixed feature extractor, and a trainable classifier.

merits of different feature sets for particular tasks.

Historically, the need for appropriate feature extractors was due to the fact that the learning techniques used by the classifiers were limited to low-dimensional spaces with easily separable classes [1]. A combination of three factors has changed this vision over the last decade. First, the availability of low-cost machines with fast arithmetic units allows to rely more on brute-force "numerical" methods than on algorithmic refinements. Second, the availability of large databases for problems with a large market and wide interest, such as handwriting recognition, has enabled designers to rely more on real data and less on hand-crafted feature extraction to build recognition systems. The third and very important factor is the availability of powerful machine learning techniques that can handle high-dimensional inputs and can generate intricate decision functions when fed with these large data sets. It can be argued that the recent progress in the accuracy of speech and handwriting recognition systems can be attributed in large part to an increased reliance on learning techniques and large training data sets. As evidence to this fact, a large proportion of modern commercial OCR systems use some form of multi-layer Neural Network trained with back-propagation.

In this study, we consider the tasks of handwritten character recognition (Sections I and II) and compare the performance of several learning techniques on a benchmark data set for handwritten digit recognition (Section III). While more automatic learning is beneficial, no learning technique can succeed without a minimal amount of prior knowledge about the task. In the case of multi-layer neural networks, a good way to incorporate knowledge is to tailor its architecture to the task. Convolutional Neural Networks [2] introduced in Section II are an example of specialized neural network architectures which incorporate knowledge about the invariances of 2D shapes by using local connection patterns, and by imposing constraints on the weights. A comparison of several methods for isolated handwritten digit recognition is presented in section III. To go from the recognition of individual characters to the recognition of words and sentences in documents, the idea of combining multiple modules trained to reduce the overall error is introduced in Section IV. Recognizing variable-length objects such as handwritten words using multi-module systems is best done if the modules manipulate directed graphs. This leads to the concept of trainable *Graph Transformer Network* (GTN) also introduced in Section IV. Section V describes the now classical method of heuristic over-segmentation for recognizing words or other character strings. Discriminative and non-discriminative gradient-based techniques for training a recognizer at the word level without requiring manual segmentation and labeling are presented in Section VI. Section VII presents the promising Space-Displacement Neural Network approach that eliminates the need for segmentation heuristics by scanning a recognizer at all possible locations on the input. In section VIII, it is shown that trainable Graph Transformer Networks can be formulated as multiple generalized transductions, based on a general graph composition algorithm. The connections between GTNs and Hidden Markov Models, commonly used in speech recognition is also treated. Section IX describes a globally trained GTN system for recognizing handwriting entered in a pen computer. This problem is known as "on-line" handwriting recognition, since the machine must produce immediate feedback as the user writes. The core of the system is a Convolutional Neural Network. The results clearly demonstrate the advantages of training a recognizer at the word level, rather than training it on pre-segmented, hand-labeled, isolated characters. Section X describes a complete GTN-based system for reading handwritten and machine-printed bank checks. The core of the system is the Convolutional Neural Network called LeNet-5 described in Section II. This system is in commercial use in the NCR Corporation line of check recognition systems for the banking industry. It is reading millions of checks per month in several banks across the United States.

A. Learning from Data

There are several approaches to automatic machine learning, but one of the most successful approaches, popularized in recent years by the neural network community, can be called "numerical" or *gradient-based learning.* The learning machine computes a function $Y^p = F(Z^p, W)$ where Z^p is the p-th input pattern, and W represents the collection of adjustable parameters in the system. In a pattern recognition setting, the output Y^p may be interpreted as the recognized class label of pattern Z^p, or as scores or probabilities associated with each class. A loss function $E^p = \mathcal{D}(D^p, F(W, Z^p))$, measures the discrepancy between D^p, the "correct" or desired output for pattern Z^p, and the output produced by the system. The average loss function $E_{train}(W)$ is the average of the errors E^p over a set of labeled examples called the training set $\{(Z^1, D^1), ...(Z^P, D^P)\}$. In the simplest setting, the learning problem consists in finding the value of W that minimizes $E_{train}(W)$. In practice, the performance of the system on a training set is of little interest. The more relevant measure is the error rate of the system in the field, where it would be used in practice. This performance is estimated by measuring the accuracy on a set of samples disjoint from the training set, called the test set. Much theoretical and experimental work [3], [4], [5] has shown that the gap between the expected error rate on the test set E_{test} and the error rate on the training set E_{train} decreases with the number of training samples approximately as

$$E_{test} - E_{train} = k(h/P)^{\alpha} \quad (1)$$

where P is the number of training samples, h is a measure of "effective capacity" or complexity of the machine [6], [7], α is a number between 0.5 and 1.0, and k is a constant. This gap always decreases when the number of training samples increases. Furthermore, as the capacity h increases, E_{train} decreases. Therefore, when increasing the capacity h, there is a trade-off between the decrease of E_{train} and the increase of the gap, with an optimal value of the capacity h that achieves the lowest generalization error E_{test}. Most learning algorithms attempt to minimize E_{train} as well as some estimate of the gap. A formal version of this is called structural risk minimization [6], [7], and is based on defining a sequence of learning machines of increasing capacity, corresponding to a sequence of subsets of the parameter space such that each subset is a superset of the previous subset. In practical terms, Structural Risk Minimization is implemented by minimizing $E_{train} + \beta H(W)$, where the function $H(W)$ is called a regularization function, and β is a constant. $H(W)$ is chosen such that it takes large values on parameters W that belong to high-capacity subsets of the parameter space. Minimizing $H(W)$ in effect limits the capacity of the accessible subset of the parameter space, thereby controlling the tradeoff between minimizing the training error and minimizing the expected gap between the training error and test error.

B. Gradient-Based Learning

The general problem of minimizing a function with respect to a set of parameters is at the root of many issues in computer science. Gradient-Based Learning draws on the fact that it is generally much easier to minimize a reasonably smooth, continuous function than a discrete (combinatorial) function. The loss function can be minimized by estimating the impact of small variations of the parameter values on the loss function. This is measured by the gradient of the loss function with respect to the parameters. Efficient learning algorithms can be devised when the gradient vector can be computed analytically (as opposed to numerically through perturbations). This is the basis of numerous gradient-based learning algorithms with continuous-valued parameters. In the procedures described in this article, the set of parameters W is a real-valued vector, with respect to which $E(W)$ is continuous, as well as differentiable almost everywhere. The simplest minimization procedure in such a setting is the gradient descent algorithm where W is iteratively adjusted as follows:

$$W_k = W_{k-1} - \epsilon \frac{\partial E(W)}{\partial W}. \quad (2)$$

In the simplest case, ϵ is a scalar constant. More sophisticated procedures use variable ϵ, or substitute it for a diagonal matrix, or substitute it for an estimate of the inverse Hessian matrix as in Newton or Quasi-Newton methods. The Conjugate Gradient method [8] can also be used. However, Appendix B shows that despite many claims to the contrary in the literature, the usefulness of these second-order methods to large learning machines is very limited.

A popular minimization procedure is the stochastic gradient algorithm, also called the on-line update. It consists in updating the parameter vector using a noisy, or approximated, version of the average gradient. In the most common instance of it, W is updated on the basis of a single sample:

$$W_k = W_{k-1} - \epsilon \frac{\partial E^{p_k}(W)}{\partial W} \quad (3)$$

With this procedure the parameter vector fluctuates around an average trajectory, but usually converges considerably faster than regular gradient descent and second order methods on large training sets with redundant samples (such as those encountered in speech or character recognition). The reasons for this are explained in Appendix B. The properties of such algorithms applied to learning have been studied theoretically since the 1960's [9], [10], [11], but practical successes for non-trivial tasks did not occur until the mid eighties.

C. Gradient Back-Propagation

Gradient-Based Learning procedures have been used since the late 1950's, but they were mostly limited to linear systems [1]. The surprising usefulness of such simple gradient descent techniques for complex machine learning tasks was not widely realized until the following three events occurred. The first event was the realization that, despite early warnings to the contrary [12], the presence of local minima in the loss function does not seem to be a major problem in practice. This became apparent when it was noticed that local minima did not seem to be a major impediment to the success of early non-linear gradient-based Learning techniques such as Boltzmann machines [13], [14]. The second event was the popularization by Rumelhart, Hinton and Williams [15] and others of a simple and efficient procedure, the back-propagation algorithm, to compute the gradient in a non-linear system composed of several layers of processing. The third event was the demonstration that the back-propagation procedure applied to multi-layer neural networks with sigmoidal units can solve complicated learning tasks. The basic idea of back-propagation is that gradients can be computed efficiently by propagation from the output to the input. This idea was described in the control theory literature of the early sixties [16], but its application to machine learning was not generally realized then. Interestingly, the early derivations of back-propagation in the context of neural network learning did not use gradients, but "virtual targets" for units in intermediate layers [17], [18], or minimal disturbance arguments [19]. The Lagrange formalism used in the control theory literature provides perhaps the best rigorous method for deriving back-propagation [20], and for deriving generalizations of back-propagation to recurrent networks [21], and networks of heterogeneous modules [22]. A simple derivation for generic multi-layer systems is given in Section I-E.

The fact that local minima do not seem to be a problem for multi-layer neural networks is somewhat of a theoretical mystery. It is conjectured that if the network is oversized for the task (as is usually the case in practice), the presence of "extra dimensions" in parameter space reduces the risk of unattainable regions. Back-propagation is by far the most widely used neural-network learning algorithm, and probably the most widely used learning algorithm of any form.

D. *Learning in Real Handwriting Recognition Systems*

Isolated handwritten character recognition has been extensively studied in the literature (see [23], [24] for reviews), and was one of the early successful applications of neural networks [25]. Comparative experiments on recognition of individual handwritten digits are reported in Section III. They show that neural networks trained with Gradient-Based Learning perform better than all other methods tested here on the same data. The best neural networks, called Convolutional Networks, are designed to learn to extract relevant features directly from pixel images (see Section II).

One of the most difficult problems in handwriting recognition, however, is not only to recognize individual characters, but also to separate out characters from their neighbors within the word or sentence, a process known as segmentation. The technique for doing this that has become the "standard" is called *Heuristic Over-Segmentation.* It consists in generating a large number of potential cuts between characters using heuristic image processing techniques, and subsequently selecting the best combination of cuts based on scores given for each candidate character by the recognizer. In such a model, the accuracy of the system depends upon the quality of the cuts generated by the heuristics, and on the ability of the recognizer to distinguish correctly segmented characters from pieces of characters, multiple characters, or otherwise incorrectly segmented characters. Training a recognizer to perform this task poses a major challenge because of the difficulty in creating a labeled database of incorrectly segmented characters. The simplest solution consists in running the images of character strings through the segmenter, and then manually labeling all the character hypotheses. Unfortunately, not only is this an extremely tedious and costly task, it is also difficult to do the labeling consistently. For example, should the right half of a cut up 4 be labeled as a 1 or as a non-character? should the right half of a cut up 8 be labeled as a 3?

The first solution, described in Section V consists in training the system at the level of whole strings of characters, rather than at the character level. The notion of Gradient-Based Learning can be used for this purpose. The system is trained to minimize an overall loss function which measures the probability of an erroneous answer. Section V explores various ways to ensure that the loss function is differentiable, and therefore lends itself to the use of Gradient-Based Learning methods. Section V introduces the use of directed acyclic graphs whose arcs carry numerical information as a way to represent the alternative hypotheses, and introduces the idea of GTN.

The second solution described in Section VII is to eliminate segmentation altogether. The idea is to sweep the recognizer over every possible location on the input image, and to rely on the "character spotting" property of the recognizer, i.e. its ability to correctly recognize a well-centered character in its input field, even in the presence of other characters besides it, while rejecting images containing no centered characters [26], [27]. The sequence of recognizer outputs obtained by sweeping the recognizer over the input is then fed to a Graph Transformer Network that takes linguistic constraints into account and finally extracts the most likely interpretation. This GTN is somewhat similar to Hidden Markov Models (HMM), which makes the approach reminiscent of the classical speech recognition [28], [29]. While this technique would be quite expensive in the general case, the use of Convolutional Neural Networks makes it particularly attractive because it allows significant savings in computational cost.

E. *Globally Trainable Systems*

As stated earlier, most practical pattern recognition systems are composed of multiple modules. For example, a document recognition system is composed of a field locator, which extracts regions of interest, a field segmenter, which cuts the input image into images of candidate characters, a recognizer, which classifies and scores each candidate character, and a contextual post-processor, generally based on a stochastic grammar, which selects the best grammatically correct answer from the hypotheses generated by the recognizer. In most cases, the information carried from module to module is best represented as graphs with numerical information attached to the arcs. For example, the output of the recognizer module can be represented as an acyclic graph where each arc contains the label and the score of a candidate character, and where each path represent a alternative interpretation of the input string. Typically, each module is manually optimized, or sometimes trained, outside of its context. For example, the character recognizer would be trained on labeled images of pre-segmented characters. Then the complete system is assembled, and a subset of the parameters of the modules is manually adjusted to maximize the overall performance. This last step is extremely tedious, time-consuming, and almost certainly suboptimal.

A better alternative would be to somehow train the entire system so as to minimize a global error measure such as the probability of character misclassifications at the document level. Ideally, we would want to find a good minimum of this global loss function with respect to all the parameters in the system. If the loss function E measuring the performance can be made differentiable with respect to the system's tunable parameters W, we can find a local minimum of E using Gradient-Based Learning. However, at first glance, it appears that the sheer size and complexity of the system would make this intractable.

To ensure that the global loss function $E^p(Z^p, W)$ is differentiable, the overall system is built as a feed-forward network of differentiable modules. The function implemented by each module must be continuous and differentiable *almost everywhere* with respect to the internal parameters of the module (e.g. the weights of a Neural Net character recognizer in the case of a character recognition module), and with respect to the module's inputs. If this is the case, a simple generalization of the well-known back-propagation procedure can be used to efficiently compute the gradients of the loss function with respect to all the parameters in the system [22]. For example, let us consider a system built as a cascade of modules, each of which implements a function $X_n = F_n(W_n, X_{n-1})$, where X_n is a vector representing the output of the module, W_n is the vector of tunable parameters in the module (a subset of W), and X_{n-1} is the module's input vector (as well as the previous module's output vector). The input X_0 to the first module is the input pattern Z^p. If the partial derivative of E^p with respect to X_n is known, then the partial derivatives of E^p with respect to W_n and X_{n-1} can be computed using the backward recurrence

$$\begin{aligned} \frac{\partial E^p}{\partial W_n} &= \frac{\partial F}{\partial W}(W_n, X_{n-1})\frac{\partial E^p}{\partial X_n} \\ \frac{\partial E^p}{\partial X_{n-1}} &= \frac{\partial F}{\partial X}(W_n, X_{n-1})\frac{\partial E^p}{\partial X_n} \end{aligned} \quad (4)$$

where $\frac{\partial F}{\partial W}(W_n, X_{n-1})$ is the Jacobian of F with respect to W evaluated at the point (W_n, X_{n-1}), and $\frac{\partial F}{\partial X}(W_n, X_{n-1})$ is the Jacobian of F with respect to X. The Jacobian of a vector function is a matrix containing the partial derivatives of all the outputs with respect to all the inputs. The first equation computes some terms of the gradient of $E^p(W)$, while the second equation generates a backward recurrence, as in the well-known back-propagation procedure for neural networks. We can average the gradients over the training patterns to obtain the full gradient. It is interesting to note that in many instances there is no need to explicitly compute the Jacobian matrix. The above formula uses the product of the Jacobian with a vector of partial derivatives, and it is often easier to compute this product directly without computing the Jacobian beforehand. By analogy with ordinary multi-layer neural networks, all but the last module are called hidden layers because their outputs are not observable from the outside. more complex situations than the simple cascade of modules described above, the partial derivative notation becomes somewhat ambiguous and awkward. A completely rigorous derivation in more general cases can be done using Lagrange functions [20], [21], [22].

Traditional multi-layer neural networks are a special case of the above where the state information X_n is represented with fixed-sized vectors, and where the modules are alternated layers of matrix multiplications (the weights) and component-wise sigmoid functions (the neurons). However, as stated earlier, the state information in complex recognition systems is best represented by graphs with numerical information attached to the arcs. In this case, each module, called a Graph Transformer, takes one or more graphs as input, and produces a graph as output. Networks of such modules are called Graph Transformer Networks (GTN). Sections IV, VI and VIII develop the concept of GTNs, and show that Gradient-Based Learning can be used to train all the parameters in all the modules so as to minimize a global loss function. It may seem paradoxical that gradients can be computed when the state information is represented by essentially discrete objects such as graphs, but that difficulty can be circumvented, as shown later.

II. Convolutional Neural Networks for Isolated Character Recognition

The ability of multi-layer networks trained with gradient descent to learn complex, high-dimensional, non-linear mappings from large collections of examples makes them obvious candidates for image recognition tasks. In the traditional model of pattern recognition, a hand-designed feature extractor gathers relevant information from the input and eliminates irrelevant variabilities. A trainable classifier then categorizes the resulting feature vectors into classes. In this scheme, standard, fully-connected multi-layer networks can be used as classifiers. A potentially more interesting scheme is to rely on as much as possible on learning in the feature extractor itself. In the case of character recognition, a network could be fed with almost raw inputs (e.g. size-normalized images). While this can be done with an ordinary fully connected feed-forward network with some success for tasks such as character recognition, there are problems.

Firstly, typical images are large, often with several hundred variables (pixels). A fully-connected first layer with, say one hundred hidden units in the first layer, would already contain several tens of thousands of weights. Such a large number of parameters increases the capacity of the system and therefore requires a larger training set. In addition, the memory requirement to store so many weights may rule out certain hardware implementations. But, the main deficiency of unstructured nets for image or speech applications is that they have no built-in invariance with respect to translations, or local distortions of the inputs. Before being sent to the fixed-size input layer of a neural net, character images, or other 2D or 1D signals, must be approximately size-normalized and centered in the input field. Unfortunately, no such preprocessing can be perfect: handwriting is often normalized at the word level, which can cause size, slant, and position variations for individual characters. This, combined with variability in writing style, will cause variations in the position of distinctive features in input objects. In principle, a fully-connected network of sufficient size could learn to produce outputs that are invariant with respect to such variations. However, learning such a task would probably result in multiple units with similar weight patterns positioned at various locations in the input so as to detect distinctive features wherever they appear on the input. Learning these weight configurations requires a very large number of training instances to cover the space of possible variations. In convolutional networks, described below, shift invariance is automatically obtained by forcing the replication of weight configurations across space.

Secondly, a deficiency of fully-connected architectures is that the topology of the input is entirely ignored. The input variables can be presented in any (fixed) order without affecting the outcome of the training. On the contrary, images (or time-frequency representations of speech) have a strong 2D local structure: variables (or pixels) that are spatially or temporally nearby are highly correlated. Local correlations are the reasons for the well-known advantages of extracting and combining *local* features before recognizing spatial or temporal objects, because configurations of neighboring variables can be classified into a small number of categories (e.g. edges, corners...). *Convolutional Networks* force the extraction of local features by restricting the receptive fields of hidden units to be local.

A. Convolutional Networks

Convolutional Networks combine three architectural ideas to ensure some degree of shift, scale, and distortion invariance: *local receptive fields*, *shared weights* (or weight replication), and spatial or temporal *sub-sampling*. A typical convolutional network for recognizing characters, dubbed LeNet-5, is shown in figure 2. The input plane receives images of characters that are approximately size-normalized and centered. Each unit in a layer receives inputs from a set of units located in a small neighborhood in the previous layer. The idea of connecting units to local receptive fields on the input goes back to the Perceptron in the early 60s, and was almost simultaneous with Hubel and Wiesel's discovery of locally-sensitive, orientation-selective neurons in the cat's visual system [30]. Local connections have been used many times in neural models of visual learning [31], [32], [18], [33], [34], [2]. With local receptive fields, neurons can extract elementary visual features such as oriented edges, end-points, corners (or similar features in other signals such as speech spectrograms). These features are then combined by the subsequent layers in order to detect higher-order features. As stated earlier, distortions or shifts of the input can cause the position of salient features to vary. In addition, elementary feature detectors that are useful on one part of the image are likely to be useful across the entire image. This knowledge can be applied by forcing a set of units, whose receptive fields are located at different places on the image, to have identical weight vectors [32], [15], [34]. Units in a layer are organized in planes within which all the units share the same set of weights. The set of outputs of the units in such a plane is called a *feature map*. Units in a feature map are all constrained to perform the same operation on different parts of the image. A complete convolutional layer is composed of several feature maps (with different weight vectors), so that multiple features can be extracted at each location. A concrete example of this is the first layer of LeNet-5 shown in Figure 2. Units in the first hidden layer of LeNet-5 are organized in 6 planes, each of which is a feature map. A unit in a feature map has 25 inputs connected to a 5 by 5 area in the input, called the *receptive field* of the unit. Each unit has 25 inputs, and therefore 25 trainable coefficients plus a trainable bias. The receptive fields of contiguous units in a feature map are centered on correspondingly contiguous units in the previous layer. Therefore receptive fields of neighboring units overlap. For example, in the first hidden layer of LeNet-5, the receptive fields of horizontally contiguous units overlap by 4 columns and 5 rows. As stated earlier, all the units in a feature map share the same set of 25 weights and the same bias so they detect the same feature at all possible locations on the input. The other feature maps in the layer use different sets of weights and biases, thereby extracting different types of local features. In the case of LeNet-5, at each input location six different types of features are extracted by six units in identical locations in the six feature maps. A sequential implementation of a feature map would scan the input image with a single unit that has a local receptive field, and store the states of this unit at corresponding locations in the feature map. This operation is equivalent to a convolution, followed by an additive bias and squashing function, hence the name *convolutional network*. The kernel of the convolution is the set of connection weights used by the units in the feature map. An interesting property of convolutional layers is that if the input image is shifted, the feature map output will be shifted by the same amount, but will be left unchanged otherwise. This property is at the basis of the robustness of convolutional networks to shifts and distortions of the input.

Once a feature has been detected, its exact location becomes less important. Only its approximate position relative to other features is relevant. For example, once we know that the input image contains the endpoint of a roughly horizontal segment in the upper left area, a corner in the upper right area, and the endpoint of a roughly vertical segment in the lower portion of the image, we can tell the input image is a 7. Not only is the precise position of each of those features irrelevant for identifying the pattern, it is potentially harmful because the positions are likely to vary for different instances of the character. A simple way to reduce the precision with which the position of distinctive features are encoded in a feature map is to reduce the spatial resolution of the feature map. This can be achieved with a so-called *sub-sampling layers* which performs a local averaging and a sub-sampling, reducing the resolution of the feature map, and reducing the sensitivity of the output to shifts and distortions. The second hidden layer of LeNet-5 is a sub-sampling layer. This layer comprises six feature maps, one for each feature map in the previous layer. The receptive field of each unit is a 2 by 2 area in the previous layer's corresponding feature map. Each unit computes the *average* of its four inputs, multiplies it by a trainable coefficient, adds a trainable bias, and passes the result through a sigmoid function. Contiguous units have non-overlapping contiguous receptive fields. Consequently, a sub-sampling layer feature map has half the number of rows and columns

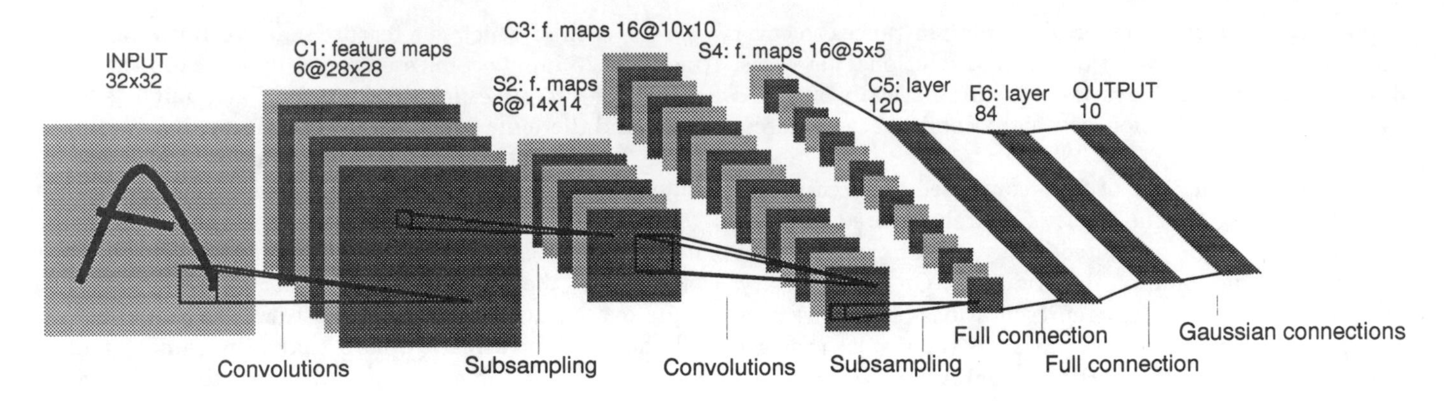

Fig. 2. Architecture of LeNet-5, a Convolutional Neural Network, here for digits recognition. Each plane is a feature map, i.e. a set of units whose weights are constrained to be identical.

as the feature maps in the previous layer. The trainable coefficient and bias control the effect of the sigmoid non-linearity. If the coefficient is small, then the unit operates in a quasi-linear mode, and the sub-sampling layer merely blurs the input. If the coefficient is large, sub-sampling units can be seen as performing a "noisy OR" or a "noisy AND" function depending on the value of the bias. Successive layers of convolutions and sub-sampling are typically alternated, resulting in a "bi-pyramid": at each layer, the number of feature maps is increased as the spatial resolution is decreased. Each unit in the third hidden layer in figure 2 may have input connections from several feature maps in the previous layer. The convolution/sub-sampling combination, inspired by Hubel and Wiesel's notions of "simple" and "complex" cells, was implemented in Fukushima's Neocognitron [32], though no globally supervised learning procedure such as back-propagation was available then. A large degree of invariance to geometric transformations of the input can be achieved with this progressive reduction of spatial resolution compensated by a progressive increase of the richness of the representation (the number of feature maps).

Since all the weights are learned with back-propagation, convolutional networks can be seen as synthesizing their own feature extractor. The weight sharing technique has the interesting side effect of reducing the number of free parameters, thereby reducing the "capacity" of the machine and reducing the gap between test error and training error [34]. The network in figure 2 contains 340,908 connections, but only 60,000 trainable free parameters because of the weight sharing.

Fixed-size Convolutional Networks have been applied to many applications, among other handwriting recognition [35], [36], machine-printed character recognition [37], on-line handwriting recognition [38], and face recognition [39]. Fixed-size convolutional networks that share weights along a single temporal dimension are known as Time-Delay Neural Networks (TDNNs). TDNNs have been used in phoneme recognition (without sub-sampling) [40], [41], spoken word recognition (with sub-sampling) [42], [43], on-line recognition of isolated handwritten characters [44], and signature verification [45].

B. LeNet-5

This section describes in more detail the architecture of LeNet-5, the Convolutional Neural Network used in the experiments. LeNet-5 comprises 7 layers, not counting the input, all of which contain trainable parameters (weights). The input is a 32x32 pixel image. This is significantly larger than the largest character in the database (at most 20x20 pixels centered in a 28x28 field). The reason is that it is desirable that potential distinctive features such as stroke end-points or corner can appear *in the center* of the receptive field of the highest-level feature detectors. In LeNet-5 the set of centers of the receptive fields of the last convolutional layer (C3, see below) form a 20x20 area in the center of the 32x32 input. The values of the input pixels are normalized so that the background level (white) corresponds to a value of -0.1 and the foreground (black) corresponds to 1.175. This makes the mean input roughly 0, and the variance roughly 1 which accelerates learning [46].

In the following, convolutional layers are labeled Cx, sub-sampling layers are labeled Sx, and fully-connected layers are labeled Fx, where x is the layer index.

Layer C1 is a convolutional layer with 6 feature maps. Each unit in each feature map is connected to a 5x5 neighborhood in the input. The size of the feature maps is 28x28 which prevents connection from the input from falling off the boundary. C1 contains 156 trainable parameters, and 122,304 connections.

Layer S2 is a sub-sampling layer with 6 feature maps of size 14x14. Each unit in each feature map is connected to a 2x2 neighborhood in the corresponding feature map in C1. The four inputs to a unit in S2 are added, then multiplied by a trainable coefficient, and added to a trainable bias. The result is passed through a sigmoidal function. The 2x2 receptive fields are non-overlapping, therefore feature maps in S2 have half the number of rows and column as feature maps in C1. Layer S2 has 12 trainable parameters and 5,880 connections.

Layer C3 is a convolutional layer with 16 feature maps. Each unit in each feature map is connected to several 5x5 neighborhoods at identical locations in a subset of S2's feature maps. Table I shows the set of S2 feature maps

	0	1	2	3	4	5	6	7	8	9	10	11	12	13	14	15
0	X				X	X	X			X	X	X	X		X	X
1	X	X				X	X	X			X	X	X	X		X
2	X	X	X				X	X	X			X		X	X	X
3		X	X	X			X	X	X	X			X		X	X
4			X	X	X			X	X	X	X		X	X		X
5				X	X	X			X	X	X	X		X	X	X

TABLE I

EACH COLUMN INDICATES WHICH FEATURE MAP IN S2 ARE COMBINED BY THE UNITS IN A PARTICULAR FEATURE MAP OF C3.

combined by each C3 feature map. Why not connect every S2 feature map to every C3 feature map? The reason is twofold. First, a non-complete connection scheme keeps the number of connections within reasonable bounds. More importantly, it forces a break of symmetry in the network. Different feature maps are forced to extract different (hopefully complementary) features because they get different sets of inputs. The rationale behind the connection scheme in table I is the following. The first six C3 feature maps take inputs from every contiguous subsets of three feature maps in S2. The next six take input from every contiguous subset of four. The next three take input from some discontinuous subsets of four. Finally the last one takes input from all S2 feature maps. Layer C3 has 1,516 trainable parameters and 151,600 connections.

Layer S4 is a sub-sampling layer with 16 feature maps of size 5x5. Each unit in each feature map is connected to a 2x2 neighborhood in the corresponding feature map in C3, in a similar way as C1 and S2. Layer S4 has 32 trainable parameters and 2,000 connections.

Layer C5 is a convolutional layer with 120 feature maps. Each unit is connected to a 5x5 neighborhood on all 16 of S4's feature maps. Here, because the size of S4 is also 5x5, the size of C5's feature maps is 1x1: this amounts to a full connection between S4 and C5. C5 is labeled as a convolutional layer, instead of a fully-connected layer, because if LeNet-5 input were made bigger with everything else kept constant, the feature map dimension would be larger than 1x1. This process of dynamically increasing the size of a convolutional network is described in the section Section VII. Layer C5 has 48,120 trainable connections.

Layer F6, contains 84 units (the reason for this number comes from the design of the output layer, explained below) and is fully connected to C5. It has 10,164 trainable parameters.

As in classical neural networks, units in layers up to F6 compute a dot product between their input vector and their weight vector, to which a bias is added. This weighted sum, denoted a_i for unit i, is then passed through a sigmoid squashing function to produce the state of unit i, denoted by x_i:

$$x_i = f(a_i) \tag{5}$$

The squashing function is a scaled hyperbolic tangent:

$$f(a) = A \tanh(Sa) \tag{6}$$

where A is the amplitude of the function and S determines its slope at the origin. The function f is odd, with horizontal asymptotes at $+A$ and $-A$. The constant A is chosen to be 1.7159. The rationale for this choice of a squashing function is given in Appendix A.

Finally, the output layer is composed of Euclidean Radial Basis Function units (RBF), one for each class, with 84 inputs each. The outputs of each RBF unit y_i is computed as follows:

$$y_i = \sum_j (x_j - w_{ij})^2. \tag{7}$$

In other words, each output RBF unit computes the Euclidean distance between its input vector and its parameter vector. The further away is the input from the parameter vector, the larger is the RBF output. The output of a particular RBF can be interpreted as a penalty term measuring the fit between the input pattern and a model of the class associated with the RBF. In probabilistic terms, the RBF output can be interpreted as the unnormalized negative log-likelihood of a Gaussian distribution in the space of configurations of layer F6. Given an input pattern, the loss function should be designed so as to get the configuration of F6 as close as possible to the parameter vector of the RBF that corresponds to the pattern's desired class. The parameter vectors of these units were chosen by hand and kept fixed (at least initially). The components of those parameters vectors were set to -1 or +1. While they could have been chosen at random with equal probabilities for -1 and +1, or even chosen to form an error correcting code as suggested by [47], they were instead designed to represent a stylized image of the corresponding character class drawn on a 7x12 bitmap (hence the number 84). Such a representation is not particularly useful for recognizing isolated digits, but it is quite useful for recognizing strings of characters taken from the full printable ASCII set. The rationale is that characters that are similar, and therefore confusable, such as uppercase O, lowercase O, and zero, or lowercase l, digit 1, square brackets, and uppercase I, will have similar output codes. This is particularly useful if the system is combined with a linguistic post-processor that can correct such confusions. Because the codes for confusable classes are similar, the output of the corresponding RBFs for an ambiguous character will be similar, and the post-processor will be able to pick the appropriate interpretation. Figure 3 gives the output codes for the full ASCII set.

Another reason for using such distributed codes, rather than the more common "1 of N" code (also called place code, or grand-mother cell code) for the outputs is that non distributed codes tend to behave badly when the number of classes is larger than a few dozens. The reason is that output units in a non-distributed code must be off most of the time. This is quite difficult to achieve with sigmoid units. Yet another reason is that the classifiers are often used to not only recognize characters, but also to reject non-characters. RBFs with distributed codes are more appropriate for that purpose because unlike sigmoids, they are activated within a well circumscribed region of their in-

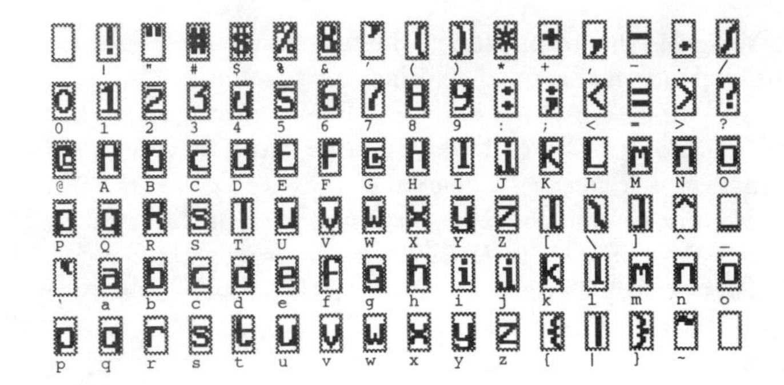

Fig. 3. Initial parameters of the output RBFs for recognizing the full ASCII set.

put space that non-typical patterns are more likely to fall outside of.

The parameter vectors of the RBFs play the role of target vectors for layer F6. It is worth pointing out that the components of those vectors are +1 or -1, which is well within the range of the sigmoid of F6, and therefore prevents those sigmoids from getting saturated. In fact, +1 and -1 are the points of maximum curvature of the sigmoids. This forces the F6 units to operate in their maximally non-linear range. Saturation of the sigmoids must be avoided because it is known to lead to slow convergence and ill-conditioning of the loss function.

C. Loss Function

The simplest output loss function that can be used with the above network is the Maximum Likelihood Estimation criterion (MLE), which in our case is equivalent to the Minimum Mean Squared Error (MSE). The criterion for a set of training samples is simply:

$$E(W) = \frac{1}{P}\sum_{p=1}^{P} y_{D^p}(Z^p, W) \tag{8}$$

where y_{D^p} is the output of the D_p-th RBF unit, i.e. the one that corresponds to the correct class of input pattern Z^p. While this cost function is appropriate for most cases, it lacks three important properties. First, if we allow the parameters of the RBF to adapt, $E(W)$ has a trivial, but totally unacceptable, solution. In this solution, all the RBF parameter vectors are equal, and the state of F6 is constant and equal to that parameter vector. In this case the network happily ignores the input, and all the RBF outputs are equal to zero. This collapsing phenomenon does not occur if the RBF weights are not allowed to adapt. The second problem is that there is no competition between the classes. Such a competition can be obtained by using a more discriminative training criterion, dubbed the MAP (maximum a posteriori) criterion, similar to Maximum Mutual Information criterion sometimes used to train HMMs [48], [49], [50]. It corresponds to maximizing the posterior probability of the correct class D_p (or minimizing the logarithm of the probability of the correct class), given that the input image can come from one of the classes or from a background "rubbish" class label. In terms of penalties, it means that in addition to pushing down the penalty of the correct class like the MSE criterion, this criterion also pulls up the penalties of the incorrect classes:

$$E(W) = \frac{1}{P}\sum_{p=1}^{P}(y_{D^p}(Z^p, W) + \log(e^{-j} + \sum_i e^{-y_i(Z^p,W)})) \tag{9}$$

The negative of the second term plays a "competitive" role. It is necessarily smaller than (or equal to) the first term, therefore this loss function is positive. The constant j is positive, and prevents the penalties of classes that are already very large from being pushed further up. The posterior probability of this rubbish class label would be the ratio of e^{-j} and $e^{-j} + \sum_i e^{-y_i(Z^p,W)}$. This discriminative criterion prevents the previously mentioned "collapsing effect" when the RBF parameters are learned because it keeps the RBF centers apart from each other. In Section VI, we present a generalization of this criterion for systems that learn to classify multiple objects in the input (e.g., characters in words or in documents).

Computing the gradient of the loss function with respect to all the weights in all the layers of the convolutional network is done with back-propagation. The standard algorithm must be slightly modified to take account of the weight sharing. An easy way to implement it is to first compute the partial derivatives of the loss function with respect to each *connection*, as if the network were a conventional multi-layer network without weight sharing. Then the partial derivatives of all the connections that share a same parameter are added to form the derivative with respect to that parameter.

Such a large architecture can be trained very efficiently, but doing so requires the use of a few techniques that are described in the appendix. Section A of the appendix describes details such as the particular sigmoid used, and the weight initialization. Section B and C describe the minimization procedure used, which is a stochastic version of a diagonal approximation to the Levenberg-Marquardt procedure.

III. Results and Comparison with Other Methods

While recognizing individual digits is only one of many problems involved in designing a practical recognition system, it is an excellent benchmark for comparing shape recognition methods. Though many existing methods combine a hand-crafted feature extractor and a trainable classifier, this study concentrates on adaptive methods that operate directly on size-normalized images.

A. Database: the Modified NIST set

The database used to train and test the systems described in this paper was constructed from the NIST's Special Database 3 and Special Database 1 containing binary images of handwritten digits. NIST originally designated SD-3 as their training set and SD-1 as their test set. However, SD-3 is much cleaner and easier to recognize than SD-1. The reason for this can be found on the fact that SD-3 was collected among Census Bureau employees, while SD-1 was collected among high-school students. Drawing sensible conclusions from learning experiments requires that the result be independent of the choice of training set and test among the complete set of samples. Therefore it was necessary to build a new database by mixing NIST's datasets.

SD-1 contains 58,527 digit images written by 500 different writers. In contrast to SD-3, where blocks of data from each writer appeared in sequence, the data in SD-1 is scrambled. Writer identities for SD-1 are available and we used this information to unscramble the writers. We then split SD-1 in two: characters written by the first 250 writers went into our new training set. The remaining 250 writers were placed in our test set. Thus we had two sets with nearly 30,000 examples each. The new training set was completed with enough examples from SD-3, starting at pattern # 0, to make a full set of 60,000 training patterns. Similarly, the new test set was completed with SD-3 examples starting at pattern # 35,000 to make a full set with 60,000 test patterns. In the experiments described here, we only used a subset of 10,000 test images (5,000 from SD-1 and 5,000 from SD-3), but we used the full 60,000 training samples. The resulting database was called the Modified NIST, or MNIST, dataset.

The original black and white (bilevel) images were size normalized to fit in a 20x20 pixel box while preserving their aspect ratio. The resulting images contain grey levels as result of the anti-aliasing (image interpolation) technique used by the normalization algorithm. Three versions of the database were used. In the first version, the images were centered in a 28x28 image by computing the center of mass of the pixels, and translating the image so as to position this point at the center of the 28x28 field. In some instances, this 28x28 field was extended to 32x32 with background pixels. This version of the database will be referred to as the *regular* database. In the second version of the database, the character images were deslanted and cropped down to 20x20 pixels images. The deslanting computes the second moments of inertia of the pixels (counting a foreground pixel as 1 and a background pixel as 0), and shears the image by horizontally shifting the lines so that the principal axis is vertical. This version of the database will be referred to as the *deslanted* database. In the third version of the database, used in some early experiments, the images were reduced to 16x16 pixels. The regular database (60,000 training examples, 10,000 test examples size-normalized to 20x20, and centered by center of mass in 28x28 fields) is available at `http://www.research.att.com/~yann/ocr/mnist`. Figure 4 shows examples randomly picked from the test set.

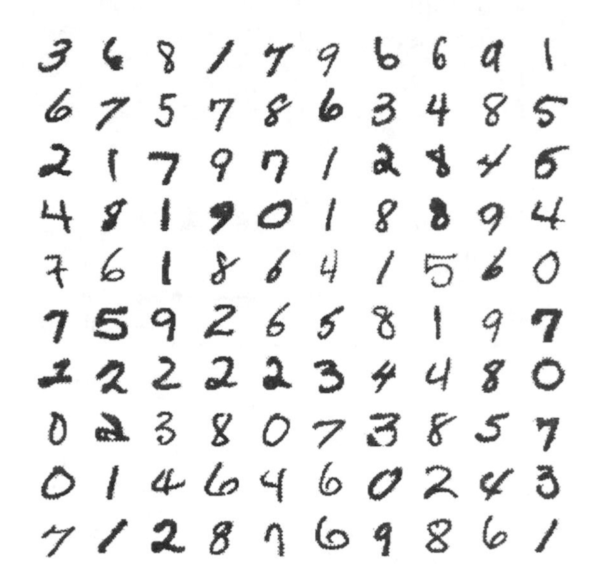

Fig. 4. Size-normalized examples from the MNIST database.

B. Results

Several versions of LeNet-5 were trained on the regular MNIST database. 20 iterations through the entire training data were performed for each session. The values of the global learning rate η (see Equation 21 in Appendix C for a definition) was decreased using the following schedule: 0.0005 for the first two passes, 0.0002 for the next three, 0.0001 for the next three, 0.00005 for the next 4, and 0.00001 thereafter. Before each iteration, the diagonal Hessian approximation was reevaluated on 500 samples, as described in Appendix C and kept fixed during the entire iteration. The parameter μ was set to 0.02. The resulting effective learning rates during the first pass varied between approximately 7×10^{-5} and 0.016 over the set of parameters. The test error rate stabilizes after around 10 passes through the training set at 0.95%. The error rate on the training set reaches 0.35% after 19 passes. Many authors have reported observing the common phenomenon of over-training when training neural networks or other adaptive algorithms on various tasks. When over-training occurs, the training error keeps decreasing over time, but the test error goes through a minimum and starts increasing after a certain number of iterations. While this phenomenon is very common, it was not observed in our case as the learning curves in figure 5 show. A possible reason is that the learning rate was kept relatively large. The effect of this is that the weights never settle down in the local minimum but keep oscillating randomly. Because of those fluctuations, the average cost will be lower in a broader minimum. Therefore, stochastic gradient will have a similar effect as a regularization term that favors broader minima. Broader minima correspond to solutions with large entropy of the parameter distribution, which is beneficial to the generalization error.

The influence of the training set size was measured by training the network with 15,000, 30,000, and 60,000 examples. The resulting training error and test error are shown in figure 6. It is clear that, even with specialized architectures such as LeNet-5, more training data would improve the accuracy.

To verify this hypothesis, we artificially generated more training examples by randomly distorting the original training images. The increased training set was composed of the 60,000 original patterns plus 540,000 instances of

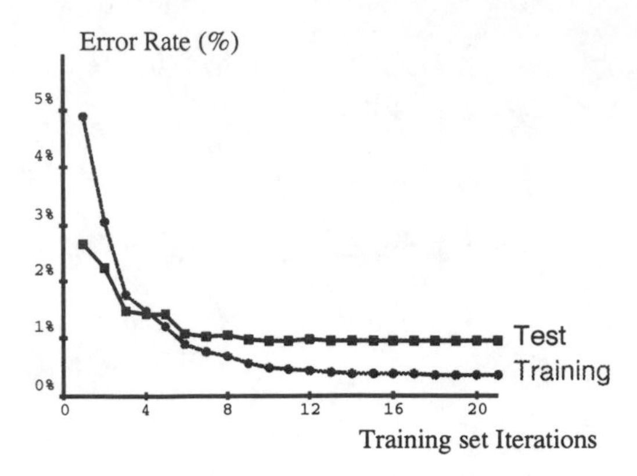

Fig. 5. Training and test error of LeNet-5 as a function of the number of passes through the 60,000 pattern training set (without distortions). The average training error is measured on-the-fly as training proceeds. This explains why the training error appears to be larger than the test error. Convergence is attained after 10 to 12 passes through the training set.

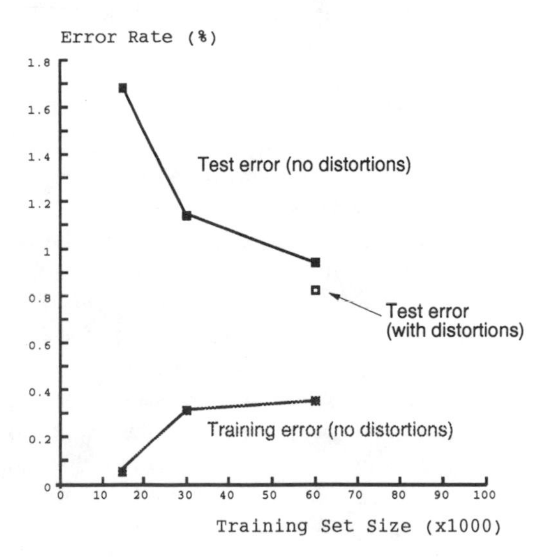

Fig. 6. Training and test errors of LeNet-5 achieved using training sets of various sizes. This graph suggests that a larger training set could improve the performance of LeNet-5. The hollow square show the test error when more training patterns are artificially generated using random distortions. The test patterns are not distorted.

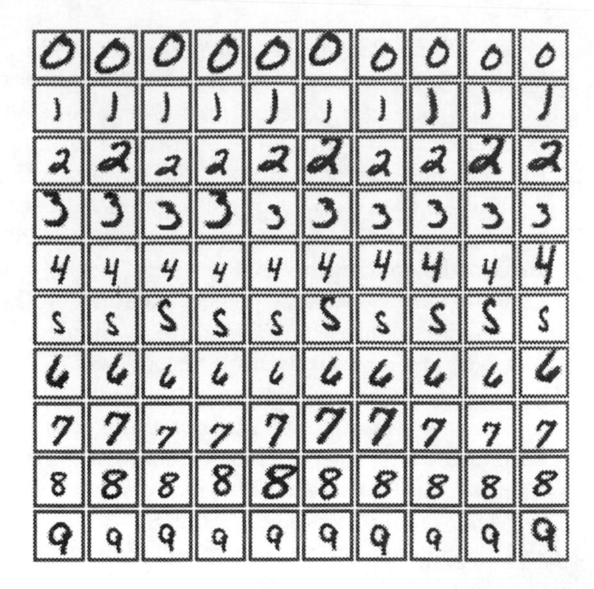

Fig. 7. Examples of distortions of ten training patterns.

Fig. 8. The 82 test patterns misclassified by LeNet-5. Below each image is displayed the correct answers (left) and the network answer (right). These errors are mostly caused either by genuinely ambiguous patterns, or by digits written in a style that are under-represented in the training set.

distorted patterns with randomly picked distortion parameters. The distortions were combinations of the following planar affine transformations: horizontal and vertical translations, scaling, squeezing (simultaneous horizontal compression and vertical elongation, or the reverse), and horizontal shearing. Figure 7 shows examples of distorted patterns used for training. When distorted data was used for training, the test error rate dropped to 0.8% (from 0.95% without deformation). The same training parameters were used as without deformations. The total length of the training session was left unchanged (20 passes of 60,000 patterns each). It is interesting to note that the network effectively sees each individual sample only twice over the course of these 20 passes.

Figure 8 shows all 82 misclassified test examples. some of those examples are genuinely ambiguous, but several are perfectly identifiable by humans, although they are written in an under-represented style. This shows that further improvements are to be expected with more training data.

C. Comparison with Other Classifiers

For the sake of comparison, a variety of other trainable classifiers was trained and tested on the same database. An early subset of these results was presented in [51]. The error rates on the test set for the various methods are shown in figure 9.

C.1 Linear Classifier, and Pairwise Linear Classifier

Possibly the simplest classifier that one might consider is a linear classifier. Each input pixel value contributes to a weighted sum for each output unit. The output unit with the highest sum (including the contribution of a bias con-

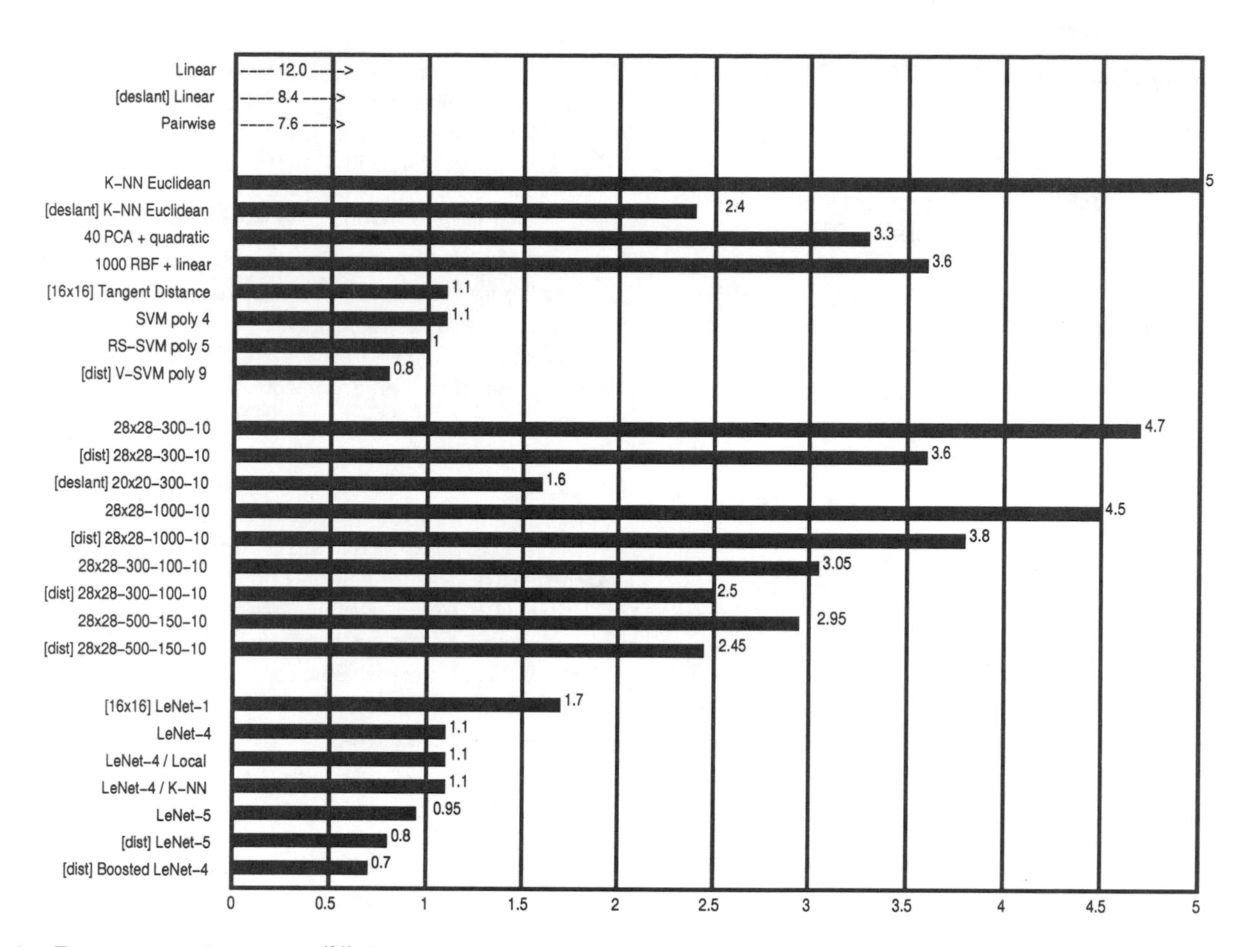

Fig. 9. Error rate on the test set (%) for various classification methods. [deslant] indicates that the classifier was trained and tested on the deslanted version of the database. [dist] indicates that the training set was augmented with artificially distorted examples. [16x16] indicates that the system used the 16x16 pixel images. The uncertainty in the quoted error rates is about 0.1%.

stant) indicates the class of the input character. On the regular data, the error rate is 12%. The network has 7850 free parameters. On the deslanted images, the test error rate is 8.4% The network has 4010 free parameters. The deficiencies of the linear classifier are well documented [1] and it is included here simply to form a basis of comparison for more sophisticated classifiers. Various combinations of sigmoid units, linear units, gradient descent learning, and learning by directly solving linear systems gave similar results.

A simple improvement of the basic linear classifier was tested [52]. The idea is to train each unit of a single-layer network to separate each class from each other class. In our case this layer comprises 45 units labeled 0/1, 0/2,...0/9, 1/2....8/9. Unit i/j is trained to produce +1 on patterns of class i, -1 on patterns of class j, and is not trained on other patterns. The final score for class i is the sum of the outputs all the units labeled i/x minus the sum of the output of all the units labeled y/i, for all x and y. The error rate on the regular test set was 7.6%.

C.2 Baseline Nearest Neighbor Classifier

Another simple classifier is a K-nearest neighbor classifier with a Euclidean distance measure between input images. This classifier has the advantage that no training time, and no brain on the part of the designer, are required. However, the memory requirement and recognition time are large: the complete 60,000 twenty by twenty pixel training images (about 24 Megabytes at one byte per pixel) must be available at run time. Much more compact representations could be devised with modest increase in error rate. On the regular test set the error rate was 5.0%. On the deslanted data, the error rate was 2.4%, with $k = 3$. Naturally, a realistic Euclidean distance nearest-neighbor system would operate on feature vectors rather than directly on the pixels, but since all of the other systems presented in this study operate directly on the pixels, this result is useful for a baseline comparison.

C.3 Principal Component Analysis (PCA) and Polynomial Classifier

Following [53], [54], a preprocessing stage was constructed which computes the projection of the input pattern on the 40 principal components of the set of training vectors. To compute the principal components, the mean of each input component was first computed and subtracted from the training vectors. The covariance matrix of the resulting vectors was then computed and diagonalized using Singular Value Decomposition. The 40-dimensional feature vector was used as the input of a second degree polynomial classifier. This classifier can be seen as a linear classifier with 821 inputs, preceded by a module that computes all products of pairs of input variables. The error on the regular test set was 3.3%.

C.4 Radial Basis Function Network

Following [55], an RBF network was constructed. The first layer was composed of 1,000 Gaussian RBF units with 28x28 inputs, and the second layer was a simple 1000 inputs / 10 outputs linear classifier. The RBF units were divided into 10 groups of 100. Each group of units was trained on all the training examples of one of the 10 classes using the adaptive K-means algorithm. The second layer weights were computed using a regularized pseudo-inverse method. The error rate on the regular test set was 3.6%

C.5 One-Hidden Layer Fully Connected Multilayer Neural Network

Another classifier that we tested was a fully connected multi-layer neural network with two layers of weights (one hidden layer) trained with the version of back-propagation described in Appendix C. Error on the regular test set was 4.7% for a network with 300 hidden units, and 4.5% for a network with 1000 hidden units. Using artificial distortions to generate more training data brought only marginal improvement: 3.6% for 300 hidden units, and 3.8% for 1000 hidden units. When deslanted images were used, the test error jumped down to 1.6% for a network with 300 hidden units.

It remains somewhat of a mystery that networks with such a large number of free parameters manage to achieve reasonably low testing errors. We conjecture that the dynamics of gradient descent learning in multilayer nets has a "self-regularization" effect. Because the origin of weight space is a saddle point that is attractive in almost every direction, the weights invariably shrink during the first few epochs (recent theoretical analysis seem to confirm this [56]). Small weights cause the sigmoids to operate in the quasi-linear region, making the network essentially equivalent to a low-capacity, single-layer network. As the learning proceeds, the weights grow, which progressively increases the effective capacity of the network. This seems to be an almost perfect, if fortuitous, implementation of Vapnik's "Structural Risk Minimization" principle [6]. A better theoretical understanding of these phenomena, and more empirical evidence, are definitely needed.

C.6 Two-Hidden Layer Fully Connected Multilayer Neural Network

To see the effect of the architecture, several two-hidden layer multilayer neural networks were trained. Theoretical results have shown that any function can be approximated by a one-hidden layer neural network [57]. However, several authors have observed that two-hidden layer architectures sometimes yield better performance in practical situations. This phenomenon was also observed here. The test error rate of a 28x28-300-100-10 network was 3.05%, a much better result than the one-hidden layer network, obtained using marginally more weights and connections. Increasing the network size to 28x28-1000-150-10 yielded only marginally improved error rates: 2.95%. Training with distorted patterns improved the performance somewhat: 2.50% error for the 28x28-300-100-10 network, and 2.45% for the 28x28-1000-150-10 network.

C.7 A Small Convolutional Network: LeNet-1

Convolutional Networks are an attempt to solve the dilemma between small networks that cannot learn the training set, and large networks that seem overparameterized. LeNet-1 was an early embodiment of the Convolutional Network architecture which is included here for comparison purposes. The images were down-sampled to 16x16 pixels and centered in the 28x28 input layer. Although about 100,000 multiply/add steps are required to evaluate LeNet-1, its convolutional nature keeps the number of free parameters to only about 2600. The LeNet-1 architecture was developed using our own version of the USPS (US Postal Service zip codes) database and its size was tuned to match the available data [35]. LeNet-1 achieved 1.7% test error. The fact that a network with such a small number of parameters can attain such a good error rate is an indication that the architecture is appropriate for the task.

C.8 LeNet-4

Experiments with LeNet-1 made it clear that a larger convolutional network was needed to make optimal use of the large size of the training set. LeNet-4 and later LeNet-5 were designed to address this problem. LeNet-4 is very similar to LeNet-5, except for the details of the architecture. It contains 4 first-level feature maps, followed by 8 subsampling maps connected in pairs to each first-layer feature maps, then 16 feature maps, followed by 16 subsampling map, followed by a fully connected layer with 120 units, followed by the output layer (10 units). LeNet-4 contains about 260,000 connections and has about 17,000 free parameters. Test error was 1.1%. In a series of experiments, we replaced the last layer of LeNet-4 with a Euclidean Nearest Neighbor classifier, and with the "local learning" method of Bottou and Vapnik [58], in which a local linear classifier is retrained each time a new test pattern is shown. Neither of those methods improved the raw error rate, although they did improve the rejection performance.

C.9 Boosted LeNet-4

Following theoretical work by R. Schapire [59], Drucker et al. [60] developed the "boosting" method for combining multiple classifiers. Three LeNet-4s are combined: the first one is trained the usual way. the second one is trained on patterns that are filtered by the first net so that the second machine sees a mix of patterns, 50% of which the first net got right, and 50% of which it got wrong. Finally, the third net is trained on new patterns on which the first and the second nets disagree. During testing, the outputs of the three nets are simply added. Because the error rate of LeNet-4 is very low, it was necessary to use the artificially distorted images (as with LeNet-5) in order to get enough samples to train the second and third nets. The test error rate was 0.7%, the best of any of our classifiers. At first glance, boosting appears to be three times more expensive as a single net. In fact, when the first net produces a high confidence answer, the other nets are not called. The average computational cost is about 1.75 times that of a single net.

C.10 Tangent Distance Classifier (TDC)

The Tangent Distance classifier (TDC) is a nearest-neighbor method where the distance function is made insensitive to small distortions and translations of the input image [61]. If we consider an image as a point in a high dimensional pixel space (where the dimensionality equals the number of pixels), then an evolving distortion of a character traces out a curve in pixel space. Taken together, all these distortions define a low-dimensional manifold in pixel space. For small distortions, in the vicinity of the original image, this manifold can be approximated by a plane, known as the tangent plane. An excellent measure of "closeness" for character images is the distance between their tangent planes, where the set of distortions used to generate the planes includes translations, scaling, skewing, squeezing, rotation, and line thickness variations. A test error rate of 1.1% was achieved using 16x16 pixel images. Prefiltering techniques using simple Euclidean distance at multiple resolutions allowed to reduce the number of necessary Tangent Distance calculations.

C.11 Support Vector Machine (SVM)

Polynomial classifiers are well-studied methods for generating complex decision surfaces. Unfortunately, they are impractical for high-dimensional problems, because the number of product terms is prohibitive. The Support Vector technique is an extremely economical way of representing complex surfaces in high-dimensional spaces, including polynomials and many other types of surfaces [6].

A particularly interesting subset of decision surfaces is the ones that correspond to hyperplanes that are at a maximum distance from the convex hulls of the two classes in the high-dimensional space of the product terms. Boser, Guyon, and Vapnik [62] realized that any polynomial of degree k in this "maximum margin" set can be computed by first computing the dot product of the input image with a subset of the training samples (called the "support vectors"), elevating the result to the k-th power, and linearly combining the numbers thereby obtained. Finding the support vectors and the coefficients amounts to solving a high-dimensional quadratic minimization problem with linear inequality constraints. For the sake of comparison, we include here the results obtained by Burges and Schölkopf reported in [63]. With a regular SVM, their error rate on the regular test set was 1.4%. Cortes and Vapnik had reported an error rate of 1.1% with SVM on the same data using a slightly different technique. The computational cost of this technique is very high: about 14 million multiply-adds per recognition. Using Schölkopf's Virtual Support Vectors technique (V-SVM), 1.0% error was attained. More recently, Schölkopf (personal communication) has reached 0.8% using a modified version of the V-SVM. Unfortunately, V-SVM is extremely expensive: about twice as much as regular SVM. To alleviate this problem, Burges has proposed the Reduced Set Support Vector technique (RS-SVM), which attained 1.1% on the regular test set [63], with a computational cost of only 650,000 multiply-adds per recognition, i.e. only about 60% more expensive than LeNet-5.

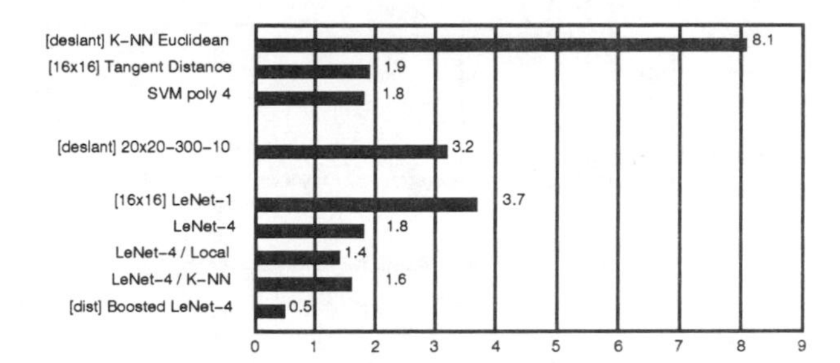

Fig. 10. Rejection Performance: percentage of test patterns that must be rejected to achieve 0.5% error for some of the systems.

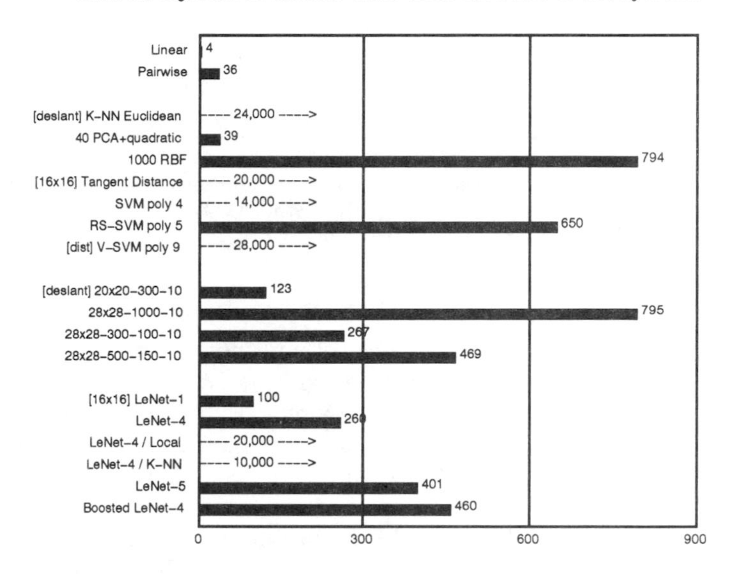

Fig. 11. Number of multiply-accumulate operations for the recognition of a single character starting with a size-normalized image.

D. Discussion

A summary of the performance of the classifiers is shown in Figures 9 to 12. Figure 9 shows the raw error rate of the classifiers on the 10,000 example test set. Boosted LeNet-4 performed best, achieving a score of 0.7%, closely followed by LeNet-5 at 0.8%.

Figure 10 shows the number of patterns in the test set that must be rejected to attain a 0.5% error for some of the methods. Patterns are rejected when the value of corresponding output is smaller than a predefined threshold. In many applications, rejection performance is more significant than raw error rate. The score used to decide upon the rejection of a pattern was the difference between the scores of the top two classes. Again, Boosted LeNet-4 has the best performance. The enhanced versions of LeNet-4 did better than the original LeNet-4, even though the raw

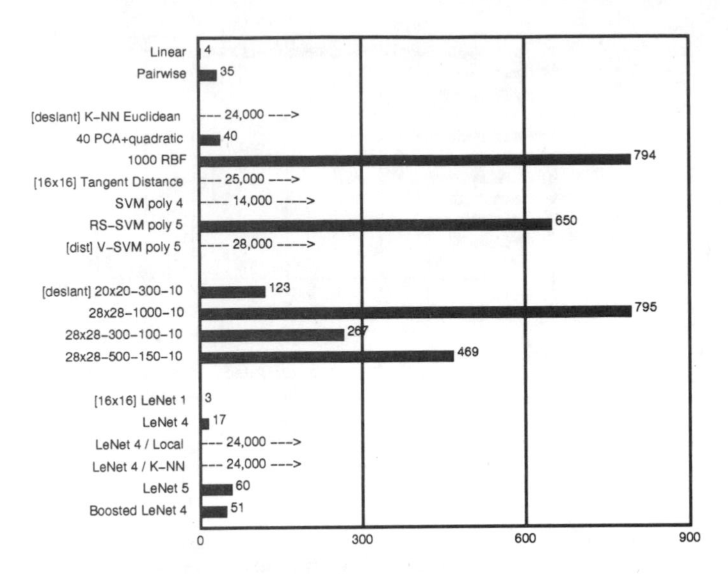

Fig. 12. Memory requirements, measured in number of variables, for each of the methods. Most of the methods only require one byte per variable for adequate performance.

accuracies were identical.

Figure 11 shows the number of multiply-accumulate operations necessary for the recognition of a single size-normalized image for each method. Expectedly, neural networks are much less demanding than memory-based methods. Convolutional Neural Networks are particularly well suited to hardware implementations because of their regular structure and their low memory requirements for the weights. Single chip mixed analog-digital implementations of LeNet-5's predecessors have been shown to operate at speeds in excess of 1000 characters per second [64]. However, the rapid progress of mainstream computer technology renders those exotic technologies quickly obsolete. Cost-effective implementations of memory-based techniques are more elusive, due to their enormous memory requirements, and computational requirements.

Training time was also measured. K-nearest neighbors and TDC have essentially zero training time. While the single-layer net, the pairwise net, and PCA+quadratic net could be trained in less than an hour, the multilayer net training times were expectedly much longer, but only required 10 to 20 passes through the training set. This amounts to 2 to 3 days of CPU to train LeNet-5 on a Silicon Graphics Origin 2000 server, using a single 200MHz R10000 processor. It is important to note that while the training time is somewhat relevant to the designer, it is of little interest to the final user of the system. Given the choice between an existing technique, and a new technique that brings marginal accuracy improvements at the price of considerable training time, any final user would chose the latter.

Figure 12 shows the memory requirements, and therefore the number of free parameters, of the various classifiers measured in terms of the number of variables that need to be stored. Most methods require only about one byte per variable for adequate performance. However, Nearest-Neighbor methods may get by with 4 bits per pixel for storing the template images. Not surprisingly, neural networks require much less memory than memory-based methods.

The Overall performance depends on many factors including accuracy, running time, and memory requirements. As computer technology improves, larger-capacity recognizers become feasible. Larger recognizers in turn require larger training sets. LeNet-1 was appropriate to the available technology in 1989, just as LeNet-5 is appropriate now. In 1989 a recognizer as complex as LeNet-5 would have required several weeks' training, and more data than was available, and was therefore not even considered. For quite a long time, LeNet-1 was considered the state of the art. The local learning classifier, the optimal margin classifier, and the tangent distance classifier were developed to improve upon LeNet-1 – and they succeeded at that. However, they in turn motivated a search for improved neural network architectures. This search was guided in part by estimates of the capacity of various learning machines, derived from measurements of the training and test error as a function of the number of training examples. We discovered that more capacity was needed. Through a series of experiments in architecture, combined with an analysis of the characteristics of recognition errors, LeNet-4 and LeNet-5 were crafted.

We find that boosting gives a substantial improvement in accuracy, with a relatively modest penalty in memory and computing expense. Also, distortion models can be used to increase the effective size of a data set without actually requiring to collect more data.

The Support Vector Machine has excellent accuracy, which is most remarkable, because unlike the other high performance classifiers, it does not include *a priori* knowledge about the problem. In fact, this classifier would do just as well if the image pixels were permuted with a fixed mapping and lost their pictorial structure. However, reaching levels of performance comparable to the Convolutional Neural Networks can only be done at considerable expense in memory and computational requirements. The reduced-set SVM requirements are within a factor of two of the Convolutional Networks, and the error rate is very close. Improvements of those results are expected, as the technique is relatively new.

When plenty of data is available, many methods can attain respectable accuracy. The neural-net methods run much faster and require much less space than memory-based techniques. The neural nets' advantage will become more striking as training databases continue to increase in size.

E. Invariance and Noise Resistance

Convolutional networks are particularly well suited for recognizing or rejecting shapes with widely varying size, position, and orientation, such as the ones typically produced by heuristic segmenters in real-world string recognition systems.

In an experiment like the one described above, the importance of noise resistance and distortion invariance is not obvious. The situation in most real applications is quite different. Characters must generally be segmented out of their context prior to recognition. Segmentation algorithms are rarely perfect and often leave extraneous marks in character images (noise, underlines, neighboring characters), or sometimes cut characters too much and produce incomplete characters. Those images cannot be reliably size-normalized and centered. Normalizing incomplete characters can be very dangerous. For example, an enlarged stray mark can look like a genuine 1. Therefore many systems have resorted to normalizing the images at the level of fields or words. In our case, the upper and lower profiles of entire fields (amounts in a check) are detected and used to normalize the image to a fixed height. While this guarantees that stray marks will not be blown up into character-looking images, this also creates wide variations of the size and vertical position of characters after segmentation. Therefore it is preferable to use a recognizer that is robust to such variations. Figure 13 shows several examples of distorted characters that are correctly recognized by LeNet-5. It is estimated that accurate recognition occurs for scale variations up to about a factor of 2, vertical shift variations of plus or minus about half the height of the character, and rotations up to plus or minus 30 degrees. While fully invariant recognition of complex shapes is still an elusive goal, it seems that Convolutional Networks offer a partial answer to the problem of invariance or robustness with respect to geometrical distortions.

Figure 13 includes examples of the robustness of LeNet-5 under extremely noisy conditions. Processing those images would pose unsurmountable problems of segmentation and feature extraction to many methods, but LeNet-5 seems able to robustly extract salient features from these cluttered images. The training set used for the network shown here was the MNIST training set with salt and pepper noise added. Each pixel was randomly inverted with probability 0.1. More examples of LeNet-5 in action are available on the Internet at `http://www.research.att.com/~yann/ocr`.

IV. Multi-Module Systems and Graph Transformer Networks

The classical back-propagation algorithm, as described and used in the previous sections, is a simple form of Gradient-Based Learning. However, it is clear that the gradient back-propagation algorithm given by Equation 4 describes a more general situation than simple multi-layer feed-forward networks composed of alternated linear transformations and sigmoidal functions. In principle, derivatives can be back-propagated through any arrangement of functional modules, as long as we can compute the product of the Jacobians of those modules by any vector. Why would we want to train systems composed of multiple heterogeneous modules? The answer is that large and complex trainable systems need to be built out of simple, specialized modules. The simplest example is LeNet-5, which mixes convolutional layers, sub-sampling layers, fully-connected layers, and RBF layers. Another less trivial example, described in the next two sections, is a system for recognizing words, that can be trained to simultaneously segment and recognize words, without ever being given the correct segmentation.

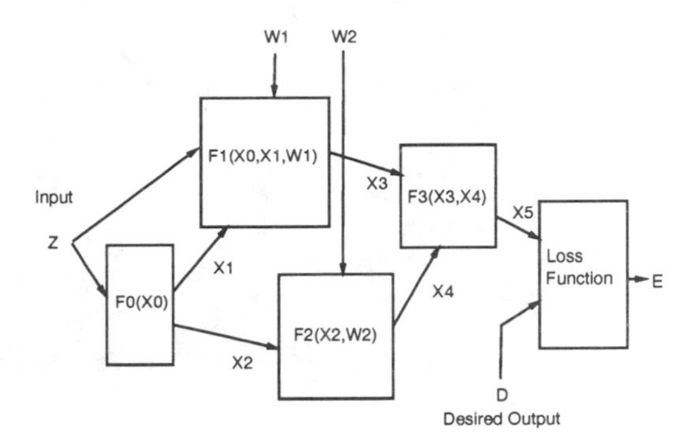

Fig. 14. A trainable system composed of heterogeneous modules.

Figure 14 shows an example of a trainable multi-modular system. A multi-module system is defined by the function implemented by each of the modules, and by the graph of interconnection of the modules to each other. The graph implicitly defines a partial order according to which the modules must be updated in the forward pass. For example in Figure 14, module 0 is first updated, then modules 1 and 2 are updated (possibly in parallel), and finally module 3. Modules may or may not have trainable parameters. Loss functions, which measure the performance of the system, are implemented as module 4. In the simplest case, the loss function module receives an external input that carries the desired output. In this framework, there is no qualitative difference between trainable parameters (`W1,W2` in the figure), external inputs and outputs (`Z,D,E`), and intermediate state variables(`X1,X2,X3,X4,X5`).

A. An Object-Oriented Approach

Object-Oriented programming offers a particularly convenient way of implementing multi-module systems. Each module is an instance of a class. Module classes have a "forward propagation" method (or member function) called `fprop` whose arguments are the inputs and outputs of the module. For example, computing the output of module 3 in Figure 14 can be done by calling the method `fprop` on module 3 with the arguments `X3,X4,X5`. Complex modules can be constructed from simpler modules by simply defining a new class whose slots will contain the member modules and the intermediate state variables between those modules. The `fprop` method for the class simply calls the `fprop` methods of the member modules, with the appropriate intermediate state variables or external input and outputs as arguments. Although the algorithms are easily generalizable to any network of such modules, including those whose influence graph has cycles, we will limit the discussion to the case of directed acyclic graphs (feed-forward networks).

Computing derivatives in a multi-module system is just as simple. A "backward propagation" method, called `bprop`, for each module class can be defined for that purpose. The `bprop` method of a module takes the same ar-

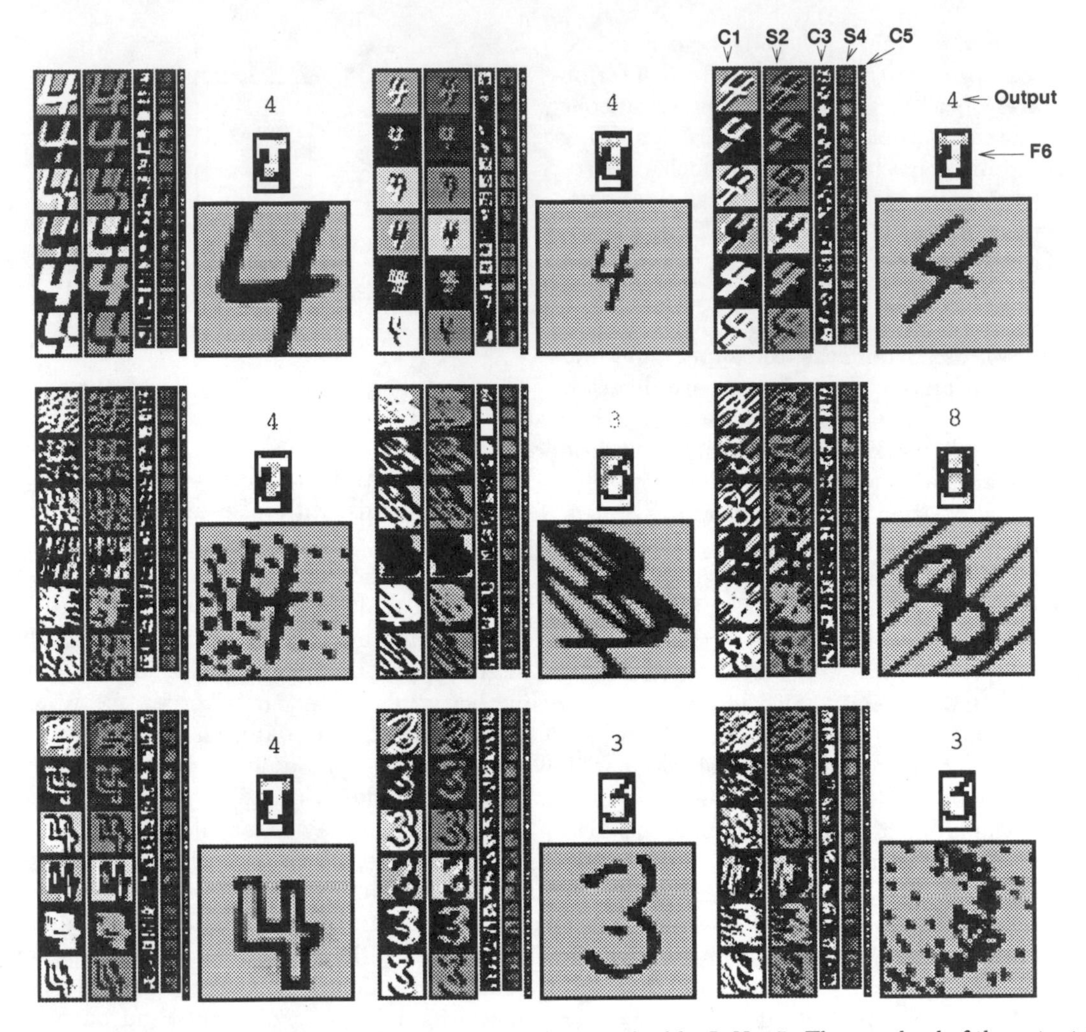

Fig. 13. Examples of unusual, distorted, and noisy characters correctly recognized by LeNet-5. The grey-level of the output label represents the penalty (lighter for higher penalties).

guments as the `fprop` method. All the derivatives in the system can be computed by calling the `bprop` method on all the modules in reverse order compared to the forward propagation phase. The state variables are assumed to contain slots for storing the gradients computed during the backward pass, in addition to storage for the states computed in the forward pass. The backward pass effectively computes the partial derivatives of the loss E with respect to all the state variables and all the parameters in the system. There is an interesting duality property between the forward and backward functions of certain modules. For example, a sum of several variables in the forward direction is transformed into a simple fan-out (replication) in the backward direction. Conversely, a fan-out in the forward direction is transformed into a sum in the backward direction. The software environment used to obtain the results described in this paper, called SN3.1, uses the above concepts. It is based on a home-grown object-oriented dialect of Lisp with a compiler to C.

The fact that derivatives can be computed by propagation in the reverse graph is easy to understand intuitively. The best way to justify it theoretically is through the use of Lagrange functions [21], [22]. The same formalism can be used to extend the procedures to networks with recurrent connections.

B. Special Modules

Neural networks and many other standard pattern recognition techniques can be formulated in terms of multimodular systems trained with Gradient-Based Learning. Commonly used modules include matrix multiplications and sigmoidal modules, the combination of which can be used to build conventional neural networks. Other modules include convolutional layers, sub-sampling layers, RBF layers, and "softmax" layers [65]. Loss functions are also represented as modules whose single output produces the value of the loss. Commonly used modules have simple `bprop` methods. In general, the `bprop` method of a function F is a multiplication by the Jacobian of F. Here are a few commonly used examples. The `bprop` method of a fanout (a "Y" connection) is a sum, and vice versa. The `bprop` method of a multiplication by a coefficient is a multiplication by the same coefficient. The `bprop` method of a multiplication by a matrix is a multiplication by the transpose of that matrix. The `bprop` method of an addition with a constant is the identity.

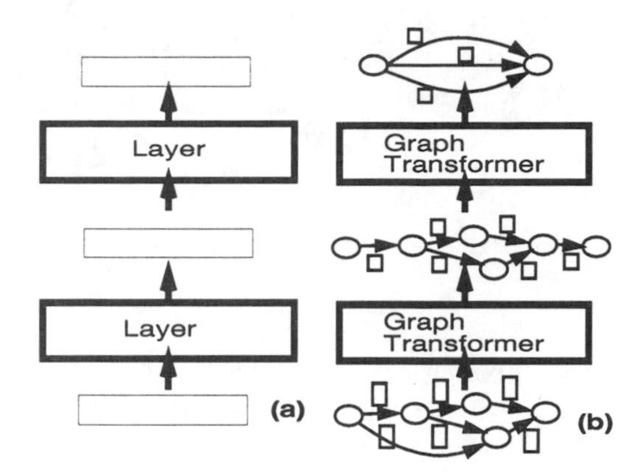

Fig. 15. Traditional neural networks, and multi-module systems communicate fixed-size vectors between layer. Multi-Layer Graph Transformer Networks are composed of trainable modules that operate on and produce graphs whose arcs carry numerical information.

Interestingly, certain non-differentiable modules can be inserted in a multi-module system without adverse effect. An interesting example of that is the multiplexer module. It has two (or more) regular inputs, one switching input, and one output. The module selects one of its inputs, depending upon the (discrete) value of the switching input, and copies it on its output. While this module is not differentiable with respect to the switching input, it is differentiable with respect to the regular inputs. Therefore the overall function of a system that includes such modules will be differentiable with respect to its parameters as long as the switching input does not depend upon the parameters. For example, the switching input can be an external input.

Another interesting case is the `min` module. This module has two (or more) inputs and one output. The output of the module is the minimum of the inputs. The function of this module is differentiable everywhere, except on the switching surface which is a set of measure zero. Interestingly, this function is continuous and reasonably regular, and that is sufficient to ensure the convergence of a Gradient-Based Learning algorithm.

The object-oriented implementation of the multi-module idea can easily be extended to include a `bbprop` method that propagates Gauss-Newton approximations of the second derivatives. This leads to a direct generalization for modular systems of the second-derivative back-propagation Equation 22 given in the Appendix.

The multiplexer module is a special case of a much more general situation, described at length in Section VIII, where the architecture of the system changes dynamically with the input data. Multiplexer modules can be used to dynamically rewire (or reconfigure) the architecture of the system for each new input pattern.

C. Graph Transformer Networks

Multi-module systems are a very flexible tool for building large trainable system. However, the descriptions in the previous sections implicitly assumed that the set of parameters, and the state information communicated between the modules, are all fixed-size vectors. The limited flexibility of fixed-size vectors for data representation is a serious deficiency for many applications, notably for tasks that deal with variable length inputs (e.g continuous speech recognition and handwritten word recognition), or for tasks that require encoding relationships between objects or features whose number and nature can vary (invariant perception, scene analysis, recognition of composite objects). An important special case is the recognition of strings of characters or words.

More generally, fixed-size vectors lack flexibility for tasks in which the state must encode probability distributions over sequences of vectors or symbols as is the case in linguistic processing. Such distributions over sequences are best represented by stochastic grammars, or, in the more general case, *directed graphs in which each arc contains a vector* (stochastic grammars are special cases in which the vector contains probabilities and symbolic information). Each path in the graph represents a different sequence of vectors. Distributions over sequences can be represented by interpreting elements of the data associated with each arc as parameters of a probability distribution or simply as a penalty. Distributions over sequences are particularly handy for modeling linguistic knowledge in speech or handwriting recognition systems: each sequence, i.e., each path in the graph, represents an alternative interpretation of the input. Successive processing modules progressively refine the interpretation. For example, a speech recognition system might start with a single sequence of acoustic vectors, transform it into a lattice of phonemes (distribution over phoneme sequences), then into a lattice of words (distribution over word sequences), and then into a single sequence of words representing the best interpretation.

In our work on building large-scale handwriting recognition systems, we have found that these systems could much more easily and quickly be developed and designed by viewing the system as a networks of modules that take one or several graphs as input and produce graphs as output. Such modules are called *Graph Transformers*, and the complete systems are called *Graph Transformer Networks*, or GTN. Modules in a GTN communicate their states and gradients in the form of directed graphs whose arcs carry numerical information (scalars or vectors) [66].

From the statistical point of view, the fixed-size state vectors of conventional networks can be seen as representing the means of distributions in state space. In variable-size networks such as the Space-Displacement Neural Networks described in section VII, the states are variable-length sequences of fixed size vectors. They can be seen as representing the mean of a probability distribution over variable-length *sequences* of fixed-size vectors. In GTNs, the states are represented as graphs, which can be seen as representing mixtures of probability distributions over structured collections (possibly sequences) of vectors (Figure 15).

One of the main points of the next several sections is to show that Gradient-Based Learning procedures are not limited to networks of simple modules that communicate through fixed-size vectors, but can be generalized to GTNs. Gradient back-propagation through a Graph Transformer takes gradients with respect to the numerical information in the output graph, and computes gradients with respect to the numerical information attached to the input graphs, and with respect to the module's internal parameters. Gradient-Based Learning can be applied as long as differentiable functions are used to produce the *numerical data* in the output graph from the *numerical data* in the input graph and the functions parameters.

The second point of the next several sections is to show that the functions implemented by many of the modules used in typical document processing systems (and other image recognition systems), though commonly thought to be combinatorial in nature, are indeed differentiable with respect to their internal parameters as well as with respect to their inputs, and are therefore usable as part of a globally trainable system.

In most of the following, we will purposely avoid making references to probability theory. All the quantities manipulated are viewed as penalties, or costs, which if necessary can be transformed into probabilities by taking exponentials and normalizing.

V. Multiple Object Recognition: Heuristic Over-Segmentation

One of the most difficult problems of handwriting recognition is to recognize not just isolated characters, but strings of characters, such as zip codes, check amounts, or words. Since most recognizers can only deal with one character at a time, we must first *segment* the string into individual character images. However, it is almost impossible to devise image analysis techniques that will infallibly segment naturally written sequences of characters into well formed characters.

The recent history of automatic speech recognition [28], [67] is here to remind us that training a recognizer by optimizing a global criterion (at the word or sentence level) is much preferable to merely training it on hand-segmented phonemes or other units. Several recent works have shown that the same is true for handwriting recognition [38]: optimizing a word-level criterion is preferable to solely training a recognizer on pre-segmented characters because the recognizer can learn not only to recognize individual characters, but also to reject mis-segmented characters thereby minimizing the overall word error.

This section and the next describe in detail a simple example of GTN to address the problem of reading strings of characters, such as words or check amounts. The method avoids the expensive and unreliable task of hand-truthing the result of the segmentation often required in more traditional systems trained on individually labeled character images.

A. Segmentation Graph

A now classical method for word segmentation and recognition is called Heuristic Over-Segmentation [68], [69]. Its main advantages over other approaches to segmentation are

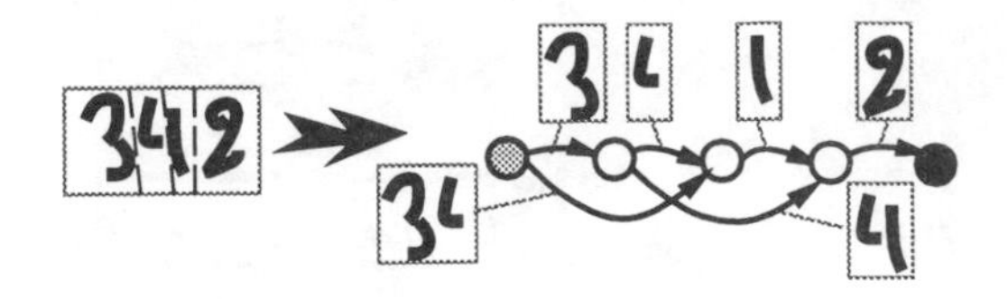

Fig. 16. Building a segmentation graph with Heuristic Over-Segmentation.

that it avoids making hard decisions about the segmentation by taking a large number of different segmentations into consideration. The idea is to use heuristic image processing techniques to find candidate cuts of the word or string, and then to use the recognizer to score the alternative segmentations thereby generated. The process is depicted in Figure 16. First, a number of candidate cuts are generated. Good candidate locations for cuts can be found by locating minima in the vertical projection profile, or minima of the distance between the upper and lower contours of the word. Better segmentation heuristics are described in section X. The cut generation heuristic is designed so as to generate more cuts than necessary, in the hope that the "correct" set of cuts will be included. Once the cuts have been generated, alternative segmentations are best represented by a graph, called the *segmentation graph*. The segmentation graph is a *Directed Acyclic Graph* (DAG) with a start node and an end node. Each internal node is associated with a candidate cut produced by the segmentation algorithm. Each arc between a source node and a destination node is associated with an image that contains all the ink between the cut associated with the source node and the cut associated with the destination node. An arc is created between two nodes if the segmentor decided that the ink between the corresponding cuts could form a candidate character. Typically, each individual piece of ink would be associated with an arc. Pairs of successive pieces of ink would also be included, unless they are separated by a wide gap, which is a clear indication that they belong to different characters. Each complete path through the graph contains each piece of ink once and only once. Each path corresponds to a different way of associating pieces of ink together so as to form characters.

B. Recognition Transformer and Viterbi Transformer

A simple GTN to recognize character strings is shown in Figure 17. It is composed of two graph transformers called the *recognition transformer* T_{rec}, and the *Viterbi transformer* T_{vit}. The goal of the recognition transformer is to generate a graph, called the *interpretation graph* or *recognition graph* G_{int}, that contains all the possible interpretations for all the possible segmentations of the input. Each path in G_{int} represents one possible interpretation of one particular segmentation of the input. The role of the Viterbi transformer is to extract the best interpretation from the interpretation graph.

The recognition transformer T_{rec} takes the segmentation graph G_{seg} as input, and applies the recognizer for single characters to the images associated with each of the arcs

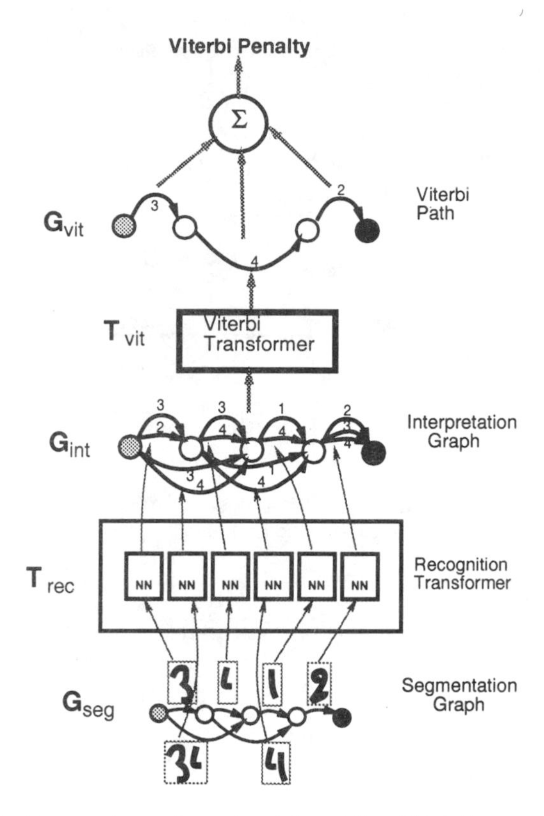

Fig. 17. Recognizing a character string with a GTN. For readability, only the arcs with low penalties are shown.

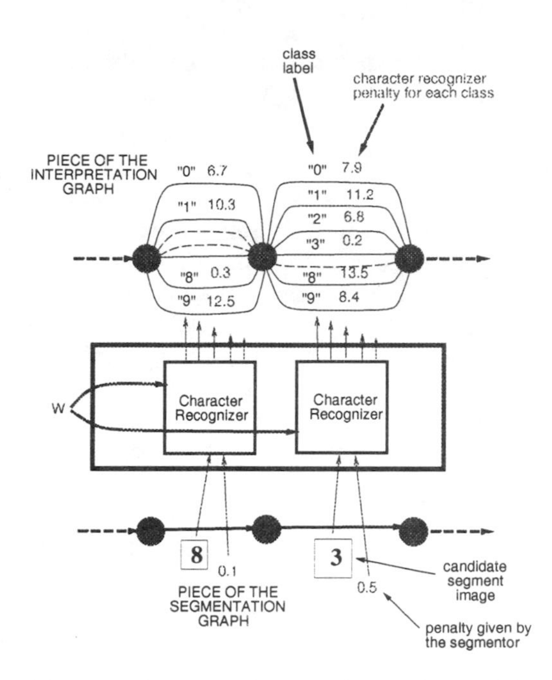

Fig. 18. The recognition transformer refines each arc of the segmentation arc into a set of arcs in the interpretation graph, one per character class, with attached penalties and labels.

in the segmentation graph. The interpretation graph G_{int} has almost the same structure as the segmentation graph, except that each arc is replaced by a set of arcs from and to the same node. In this set of arcs, there is one arc for each possible class for the image associated with the corresponding arc in G_{seg}. As shown in Figure 18, to each arc is attached a class label, and the penalty that the image belongs to this class as produced by the recognizer. If the segmentor has computed penalties for the candidate segments, these penalties are combined with the penalties computed by the character recognizer, to obtain the penalties on the arcs of the interpretation graph. Although combining penalties of different nature seems highly heuristic, the GTN training procedure will tune the penalties and take advantage of this combination anyway. Each path in the interpretation graph corresponds to a possible interpretation of the input word. The penalty of a particular interpretation for a particular segmentation is given by the sum of the arc penalties along the corresponding path in the interpretation graph. Computing the penalty of an interpretation independently of the segmentation requires to combine the penalties of all the paths with that interpretation. An appropriate rule for combining the penalties of parallel paths is given in section VI-C.

The Viterbi transformer produces a graph G_{vit} with a single path. This path is the path of least cumulated penalty in the Interpretation graph. The result of the recognition can be produced by reading off the labels of the arcs along the graph G_{vit} extracted by the Viterbi transformer. The Viterbi transformer owes its name to the famous *Viterbi algorithm* [70], an application of the principle of dynamic programming to find the shortest path in a graph efficiently. Let c_i be the penalty associated to arc i, with source node s_i, and destination node d_i (note that there can be multiple arcs between two nodes). In the interpretation graph, arcs also have a label l_i. The Viterbi algorithm proceeds as follows. Each node n is associated with a cumulated Viterbi penalty v_n. Those cumulated penalties are computed in any order that satisfies the partial order defined by the interpretation graph (which is directed and acyclic). The start node is initialized with the cumulated penalty $v_{\text{start}} = 0$. The other nodes cumulated penalties v_n are computed recursively from the v values of their parent nodes, through the upstream arcs $U_n = \{\text{arc } i \text{ with destination } d_i = n\}$:

$$v_n = \min_{i \in U_n}(c_i + v_{s_i}). \tag{10}$$

Furthermore, the value of i for each node n which minimizes the right hand side is noted m_n, the minimizing entering arc. When the end node is reached we obtain in v_{end} the total penalty of the path with the smallest total penalty. We call this penalty the *Viterbi penalty*, and this sequence of arcs and nodes the *Viterbi path*. To obtain the Viterbi path with nodes $n_1 \ldots n_T$ and arcs $i_1 \ldots i_{T-1}$, we trace back these nodes and arcs as follows, starting with n_T = the end node, and recursively using the minimizing entering arc: $i_t = m_{n_{t+1}}$, and $n_t = s_{i_t}$ until the start node is reached. The label sequence can then be read off the arcs of the Viterbi path.

VI. Global Training for Graph Transformer Networks

The previous section describes the process of recognizing a string using Heuristic Over-Segmentation, assuming that the recognizer is trained so as to give low penalties for the correct class label of correctly segmented characters, high penalties for erroneous categories of correctly segmented characters, and high penalties for all categories for badly formed characters. This section explains how to train the system at the string level to do the above without requiring manual labeling of character segments. This training will be performed with a GTN whose architecture is slightly different from the recognition architecture described in the previous section.

In many applications, there is enough a priori knowledge about what is expected from each of the modules in order to train them separately. For example, with Heuristic Over-Segmentation one could individually label single-character images and train a character recognizer on them, but it might be difficult to obtain an appropriate set of non-character images to train the model to reject wrongly segmented candidates. Although separate training is simple, it requires additional supervision information that is often lacking or incomplete (the correct segmentation and the labels of incorrect candidate segments). Furthermore it can be shown that separate training is sub-optimal [67].

The following section describes three different gradient-based methods for training GTN-based handwriting recognizers at the string level: Viterbi training, discriminative Viterbi training, forward training, and discriminative forward training. The last one is a generalization to graph-based systems of the MAP criterion introduced in Section II-C. Discriminative forward training is somewhat similar to the so-called Maximum Mutual Information criterion used to train HMM in speech recognition. However, our rationale differs from the classical one. We make no recourse to a probabilistic interpretation, but show that, within the Gradient-Based Learning approach, discriminative training is a simple instance of the pervasive principle of error correcting learning.

Training methods for graph-based sequence recognition systems such as HMMs have been extensively studied in the context of speech recognition [28]. Those methods require that the system be based on probabilistic *generative* models of the data, which provide normalized likelihoods over the space of possible input sequences. Popular HMM learning methods, such as the the Baum-Welsh algorithm, rely on this normalization. The normalization cannot be preserved when non-generative models such as neural networks are integrated into the system. Other techniques, such as discriminative training methods, must be used in this case. Several authors have proposed such methods to train neural network/HMM speech recognizers at the word or sentence level [71], [72], [73], [74], [75], [76], [77], [78], [29], [67].

Other globally trainable sequence recognition systems avoid the difficulties of statistical modeling by not resorting to graph-based techniques. The best example is Recurrent Neural Networks (RNN). Unfortunately, despite early enthusiasm, the training of RNNs with gradient-based techniques has proved very difficult in practice [79].

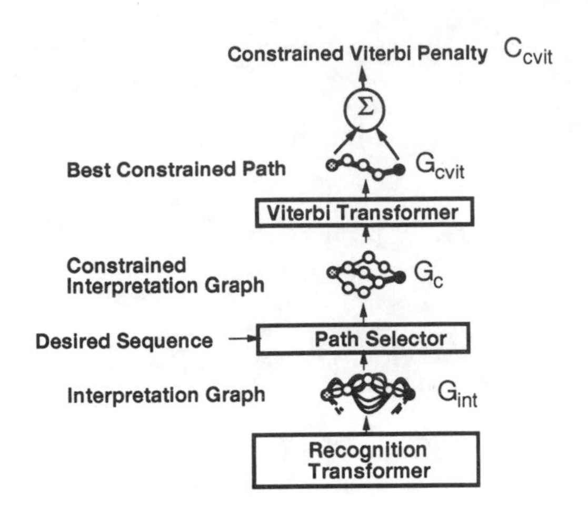

Fig. 19. Viterbi Training GTN Architecture for a character string recognizer based on Heuristic Over-Segmentation.

The GTN techniques presented below simplify and generalize the global training methods developed for speech recognition.

A. Viterbi Training

During recognition, we select the path in the Interpretation Graph that has the lowest penalty with the Viterbi algorithm. Ideally, we would like this path of lowest penalty to be associated with the correct label sequence as often as possible. An obvious loss function to minimize is therefore the average over the training set of the penalty of the path *associated with the correct label sequence* that has the lowest penalty. The goal of training will be to find the set of recognizer parameters (the weights, if the recognizer is a neural network) that minimize the average penalty of this "correct" lowest penalty path. The gradient of this loss function can be computed by back-propagation through the GTN architecture shown in figure 19. This training architecture is almost identical to the recognition architecture described in the previous section, except that an extra graph transformer called a *path selector* is inserted between the Interpretation Graph and the Viterbi Transformer. This transformer takes the interpretation graph and the desired label sequence as input. It extracts from the interpretation graph those paths that contain the correct (desired) label sequence. Its output graph G_c is called the *constrained interpretation graph* (also known as *forced alignment* in the HMM literature), and contains all the paths that correspond to the correct label sequence. The constrained interpretation graph is then sent to the Viterbi transformer which produces a graph G_{cvit} with a single path. This path is the "correct" path with the lowest penalty. Finally, a path scorer transformer takes G_{cvit}, and simply computes its cumulated penalty C_{cvit} by adding up the penalties along the path. The output of this GTN is the loss function for the current pattern:

$$E_{\text{vit}} = C_{\text{cvit}} \tag{11}$$

The only label information that is required by the above system is the sequence of desired character labels. No knowledge of the correct segmentation is required on the part of the supervisor, since it chooses among the segmentations in the interpretation graph the one that yields the lowest penalty.

The process of back-propagating gradients through the Viterbi training GTN is now described. As explained in section IV, the gradients must be propagated backwards through all modules of the GTN, in order to compute gradients in preceding modules and thereafter tune their parameters. Back-propagating gradients through the path scorer is quite straightforward. The partial derivatives of the loss function with respect to the individual penalties on the constrained Viterbi path G_{cvit} are equal to 1, since the loss function is simply the sum of those penalties. Back-propagating through the Viterbi Transformer is equally simple. The partial derivatives of E_{vit} with respect to the penalties on the arcs of the constrained graph G_{c} are 1 for those arcs that appear in the constrained Viterbi path G_{cvit}, and 0 for those that do not. Why is it legitimate to back-propagate through an essentially discrete function such as the Viterbi Transformer? The answer is that the Viterbi Transformer is nothing more than a collection of `min` functions and adders put together. It was shown in Section IV that gradients can be back-propagated through `min` functions without adverse effects. Back-propagation through the path selector transformer is similar to back-propagation through the Viterbi transformer. Arcs in G_{int} that appear in G_{c} have the same gradient as the corresponding arc in G_{c}, i.e. 1 or 0, depending on whether the arc appear in G_{cvit}. The other arcs, i.e. those that do not have an *alter ego* in G_{c} because they do not contain the right label have a gradient of 0. During the forward propagation through the recognition transformer, one instance of the recognizer for single character was created for each arc in the segmentation graph. The state of recognizer instances was stored. Since each arc penalty in G_{int} is produced by an individual output of a recognizer instance, we now have a gradient (1 or 0) for each output of each instance of the recognizer. Recognizer outputs that have a non zero gradient are part of the correct answer, and will therefore have their value pushed down. The gradients present on the recognizer outputs can be back-propagated through each recognizer instance. For each recognizer instance, we obtain a vector of partial derivatives of the loss function with respect to the recognizer instance parameters. All the recognizer instances share the same parameter vector, since they are merely clones of each other, therefore the full gradient of the loss function with respect to the recognizer's parameter vector is simply the sum of the gradient vectors produced by each recognizer instance. Viterbi training, though formulated differently, is often use in HMM-based speech recognition systems [28]. Similar algorithms have been applied to speech recognition systems that integrate neural networks with time alignment [71], [72], [76] or hybrid neural-network/HMM systems [29], [74], [75].

While it seems simple and satisfying, this training architecture has a flaw that can potentially be fatal. The problem was already mentioned in Section II-C. If the recognizer is a simple neural network with sigmoid output units, the minimum of the loss function is attained, not when the recognizer always gives the right answer, but when it ignores the input, and sets its output to a constant vector with small values for all the components. This is known as *the collapse problem.* The collapse only occurs if the recognizer outputs can simultaneously take their minimum value. If on the other hand the recognizer's output layer contains RBF units with fixed parameters, then there is no such trivial solution. This is due to the fact that a set of RBF with fixed distinct parameter vectors cannot simultaneously take their minimum value. In this case, the complete collapse described above does not occur. However, this does not totally prevent the occurrence of a milder collapse because the loss function still has a "flat spot" for a trivial solution with constant recognizer output. This flat spot is a saddle point, but it is attractive in almost all directions and is very difficult to get out of using gradient-based minimization procedures. If the parameters of the RBFs are allowed to adapt, then the collapse problems reappears because the RBF centers can all converge to a single vector, and the underlying neural network can learn to produce that vector, and ignore the input. A different kind of collapse occurs if the width of the RBFs are also allowed to adapt. The collapse only occurs if a trainable module such as a neural network feeds the RBFs. The collapse does not occur in HMM-based speech recognition systems because they are generative systems that produce normalized likelihoods for the input data (more on this later). Another way to avoid the collapse is to train the whole system with respect to a discriminative training criterion, such as maximizing the conditional probability of the correct interpretations (correct sequence of class labels) given the input image.

Another problem with Viterbi training is that the penalty of the answer cannot be used reliably as a measure of confidence because it does not take low-penalty (or high-scoring) competing answers into account.

B. Discriminative Viterbi Training

A modification of the training criterion can circumvent the collapse problem described above and at the same time produce more reliable confidence values. The idea is to not only minimize the cumulated penalty of the lowest penalty path with the correct interpretation, but also to somehow increase the penalty of competing and possibly incorrect paths that have a dangerously low penalty. This type of criterion is called *discriminative*, because it plays the good answers against the bad ones. Discriminative training procedures can be seen as attempting to build appropriate separating surfaces between classes rather than to model individual classes independently of each other. For exam-

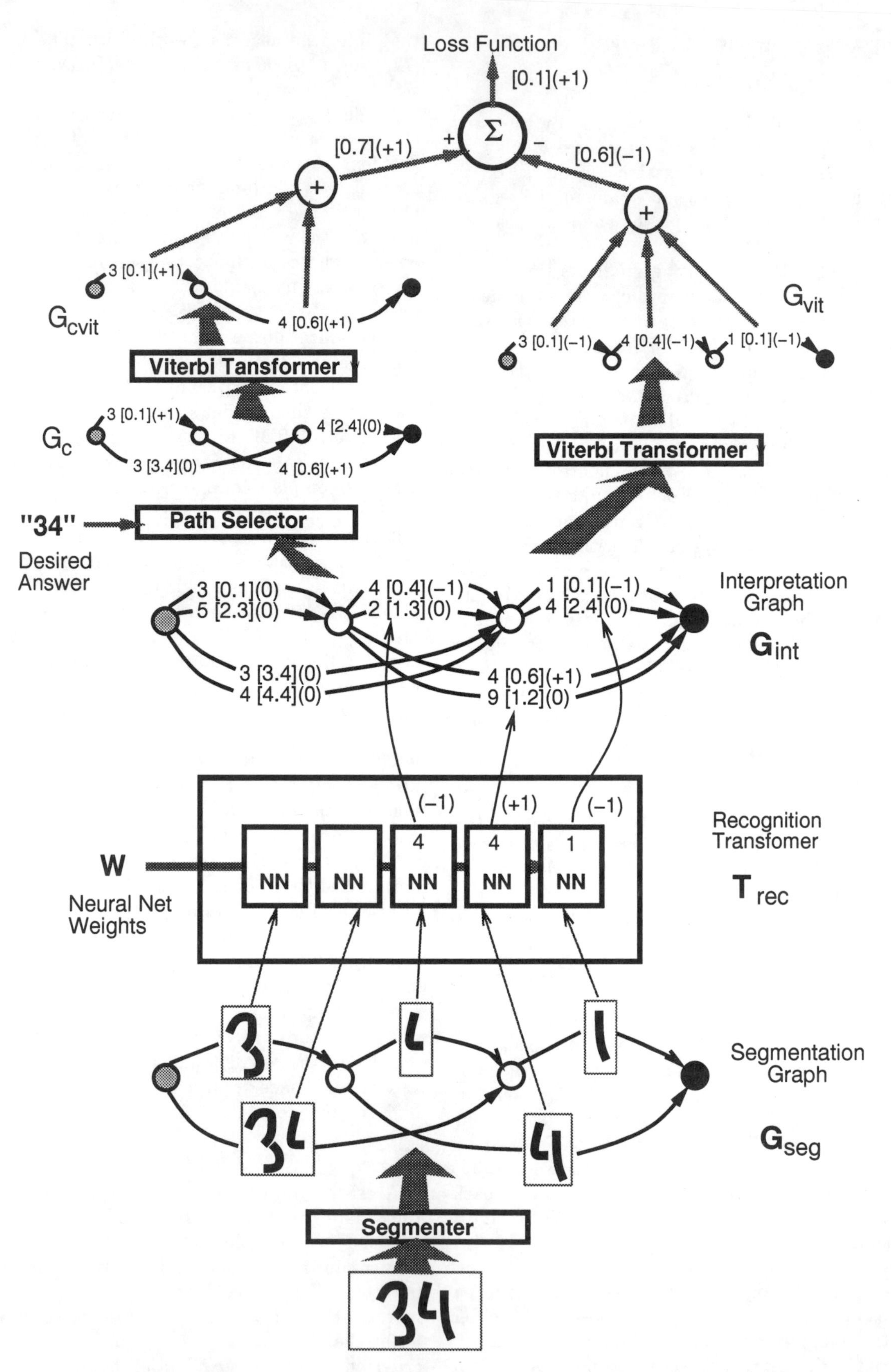

Fig. 20. Discriminative Viterbi Training GTN Architecture for a character string recognizer based on Heuristic Over-Segmentation. Quantities in square brackets are penalties computed during the forward propagation. Quantities in parentheses are partial derivatives computed during the backward propagation.

ple, modeling the conditional distribution of the classes given the input image is more discriminative (focus-sing more on the classification surface) than having a separate generative model of the input data associated to each class (which, with class priors, yields the whole joint distribution of classes and inputs). This is because the conditional approach does not need to assume a particular form for the distribution of the input data.

One example of discriminative criterion is the difference between the penalty of the Viterbi path in the constrained graph, and the penalty of the Viterbi path in the (unconstrained) interpretation graph, i.e. the difference between the penalty of the best correct path, and the penalty of the best path (correct or incorrect). The corresponding GTN training architecture is shown in figure 20. The left side of the diagram is identical to the GTN used for non-discriminative Viterbi training. This loss function reduces the risk of collapse because it forces the recognizer to *increases* the penalty of wrongly recognized objects. Discriminative training can also be seen as another example of *error correction procedure*, which tends to minimize the difference between the desired output computed in the left half of the GTN in figure 20 and the actual output computed in the right half of figure 20.

Let the discriminative Viterbi loss function be denoted E_{dvit}, and let us call C_{cvit} the penalty of the Viterbi path in the constrained graph, and C_{vit} the penalty of the Viterbi path in the unconstrained interpretation graph:

$$E_{\text{dvit}} = C_{\text{cvit}} - C_{\text{vit}} \tag{12}$$

E_{dvit} is always positive since the constrained graph is a subset of the paths in the interpretation graph, and the Viterbi algorithm selects the path with the lowest total penalty. In the ideal case, the two paths C_{cvit} and C_{vit} coincide, and E_{dvit} is zero.

Back-propagating gradients through the discriminative Viterbi GTN adds some "negative" training to the previously described non-discriminative training. Figure 20 shows how the gradients are back-propagated. The left half is identical to the non-discriminative Viterbi training GTN, therefore the back-propagation is identical. The gradients back-propagated through the right half of the GTN are multiplied by -1, since C_{vit} contributes to the loss with a negative sign. Otherwise the process is similar to the left half. The gradients on arcs of G_{int} get positive contributions from the left half and negative contributions from the right half. The two contributions must be added, since the penalties on G_{int} arcs are sent to the two halves through a "Y" connection in the forward pass. Arcs in G_{int} that appear neither in G_{vit} nor in G_{cvit} have a gradient of zero. They do not contribute to the cost. Arcs that appear in both G_{vit} and G_{cvit} also have zero gradient. The -1 contribution from the right half cancels the the +1 contribution from the left half. In other words, when an arc is rightfully part of the answer, there is no gradient. If an arc appears in G_{cvit} but not in G_{vit}, the gradient is +1. The arc should have had a lower penalty to make it to G_{vit}. If an arc is in G_{vit} but not in G_{cvit}, the gradient is -1. The arc had a low penalty, but should have had a higher penalty since it is not part of the desired answer.

Variations of this technique have been used for the speech recognition. Driancourt and Bottou [76] used a version of it where the loss function is saturated to a fixed value. This can be seen as a generalization of the Learning Vector Quantization 2 (LVQ-2) loss function [80]. Other variations of this method use not only the Viterbi path, but the K-best paths. The Discriminative Viterbi algorithm does not have the flaws of the non-discriminative version, but there are problems nonetheless. The main problem is that the criterion does not build a margin between the classes. The gradient is zero as soon as the penalty of the constrained Viterbi path is *equal* to that of the Viterbi path. It would be desirable to push up the penalties of the wrong paths when they are dangerously close to the good one. The following section presents a solution to this problem.

C. Forward Scoring, and Forward Training

While the penalty of the Viterbi path is perfectly appropriate for the purpose of recognition, it gives only a partial picture of the situation. Imagine the lowest penalty paths corresponding to several *different* segmentations produced the same answer (the same label sequence). Then it could be argued that the overall penalty for the interpretation should be smaller than the penalty obtained when only one path produced that interpretation, because multiple paths with identical label sequences are more evidence that the label sequence is correct. Several rules can be used compute the penalty associated to a graph that contains several parallel paths. We use a combination rule borrowed from a probabilistic interpretation of the penalties as negative log posteriors. In a probabilistic framework, the posterior probability for the interpretation should be the sum of the posteriors for all the paths that produce that interpretation. Translated in terms of penalties, the penalty of an interpretation should be the negative logarithm of the sum of the negative exponentials of the penalties of the individual paths. The overall penalty will be smaller than all the penalties of the individual paths.

Given an interpretation, there is a well known method, called the *forward algorithm* for computing the above quantity efficiently [28]. The penalty computed with this procedure for a particular interpretation is called the *forward penalty*. Consider again the concept of constrained graph, the subgraph of the interpretation graph which contains only the paths that are consistent with a particular label sequence. There is one constrained graph for each possible label sequence (some may be empty graphs, which have infinite penalties). Given an interpretation, running the forward algorithm on the corresponding constrained graph gives the forward penalty for that interpretation. The forward algorithm proceeds in a way very similar to the Viterbi algorithm, except that the operation used at each node to combine the incoming cumulated penalties, instead of being the `min` function is the so-called `logadd` operation, which can be seen as a "soft" version of the `min` function:

$$f_n = \text{logadd}_{i \in U_n}(c_i + f_{s_i}). \tag{13}$$

where $f_{\text{start}} = 0$, U_n is the set of upstream arcs of node n, $c_i$is the penalty on arc i, and

$$\text{logadd}(x_1, x_2, \ldots, x_n) = -\log(\sum_{i=1}^{n} e^{-x_i}) \tag{14}$$

Note that because of numerical inaccuracies, it is better to factorize the largest e^{-x_i} (corresponding to the smallest penalty) out of the logarithm.

An interesting analogy can be drawn if we consider that a graph on which we apply the forward algorithm is equivalent to a neural network on which we run a forward propagation, except that multiplications are replaced by additions, the additions are replaced by logadds, and there are no sigmoids.

One way to understand the forward algorithm is to think about multiplicative scores (e.g., probabilities) instead of additive penalties on the arcs: score = exp(− penalty). In that case the Viterbi algorithm selects the path with the largest cumulative score (with scores multiplied along the path), whereas the forward score is the sum of the cumulative scores associated to each of the possible paths from the start to the end node. The forward penalty is always lower than the cumulated penalty on any of the paths, but if one path "dominates" (with a much lower penalty), its penalty is almost equal to the forward penalty. The forward algorithm gets its name from the forward pass of the well-known Baum-Welsh algorithm for training Hidden Markov Models [28]. Section VIII-E gives more details on the relation between this work and HMMs.

The advantage of the forward penalty with respect to the Viterbi penalty is that it takes into account all the different ways to produce an answer, and not just the one with the lowest penalty. This is important if there is some ambiguity in the segmentation, since the combined forward penalty of two paths C_1 and C_2 associated with the same label sequence may be less than the penalty of a path C_3 associated with another label sequence, even though the penalty of C_3 might be less than any one of C_1 or C_2.

The Forward training GTN is only a slight modification of the previously introduced Viterbi training GTN. It suffices to turn the Viterbi transformers in Figure 19 into *Forward Scorers* that take an interpretation graph as input an produce the forward penalty of that graph on output. Then the penalties of all the paths that contain the correct answer are lowered, instead of just that of the best one.

Back-propagating through the forward penalty computation (the forward transformer) is quite different from back-propagating through a Viterbi transformer. All the penalties of the input graph have an influence on the forward penalty, but penalties that belong to low-penalty paths have a stronger influence. Computing derivatives with respect to the forward penalties f_n computed at each n node of a graph is done by back-propagation through the graph

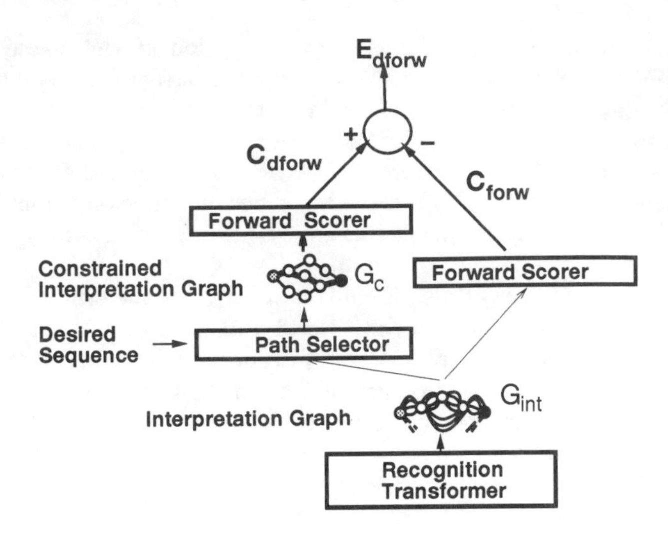

Fig. 21. Discriminative Forward Training GTN Architecture for a character string recognizer based on Heuristic Over-Segmentation.

G_c

$$\frac{\partial E}{\partial f_n} = e^{-f_n} \sum_{i \in D_n} \frac{\partial E}{\partial f_{d_i}} e^{f_{d_i} - c_i} \tag{15}$$

where $D_n = \{\text{arc } i \text{ with source } s_i = n\}$ is the set of downstream arcs from node n. From the above derivatives, the derivatives with respect to the arc penalties are obtained:

$$\frac{\partial E}{\partial c_i} = \frac{\partial E}{\partial f_{d_i}} e^{-c_i - f_{s_i} + f_{d_i}} \tag{16}$$

This can be seen as a "soft" version of the back-propagation through a Viterbi scorer and transformer. All the arcs in G_c have an influence on the loss function. The arcs that belong to low penalty paths have a larger influence. Back-propagation through the path selector is the same as before. The derivative with respect to G_{int} arcs that have an *alter ego* in G_c are simply copied from the corresponding arc in G_c. The derivatives with respect to the other arcs are 0.

Several authors have applied the idea of back-propagating gradients through a forward scorer to train speech recognition systems, including Bridle and his α-net model [73] and Haffner and his $\alpha\beta$-TDNN model [81], but these authors recommended discriminative training as described in the next section.

D. Discriminative Forward Training

The information contained in the forward penalty can be used in another discriminative training criterion which we will call the *discriminative forward criterion*. This criterion corresponds to *maximization of the posterior probability of choosing the paths associated with the correct interpretation*. This posterior probability is defined as the exponential of minus the constrained forward penalty, normalized by the exponential of minus the unconstrained forward penalty. Note that the forward penalty of the constrained graph is always larger or equal to the forward penalty of the unconstrained interpretation graph. Ideally, we would like the forward penalty of the constrained graph to be equal to the forward penalty of the complete interpretation graph. Equality between those two quantities is achieved when the combined penalties of the paths with the correct label sequence is negligibly small compared to the penalties of all the other paths, or that the posterior probability associated to the paths with the correct interpretation is almost 1, which is precisely what we want. The corresponding GTN training architecture is shown in figure 21.

Let the difference be denoted E_{dforw}, and let us call C_{cforw} the forward penalty of the constrained graph, and C_{forw} the forward penalty of the complete interpretation graph:

$$E_{\text{dforw}} = C_{\text{cforw}} - C_{\text{forw}} \tag{17}$$

E_{dforw} is always positive since the constrained graph is a subset of the paths in the interpretation graph, and the forward penalty of a graph is always larger than the forward penalty of a subgraph of this graph. In the ideal case, the penalties of incorrect paths are infinitely large, therefore the two penalties coincide and E_{dforw} is zero. Readers familiar with the Boltzmann machine connectionist model might recognize the constrained and unconstrained graphs as analogous to the "clamped" (constrained by the observed values of the output variable) and "free" (unconstrained) phases of the Boltzmann machine algorithm [13].

Back-propagating derivatives through the discriminative Forward GTN distributes gradients more evenly than in the Viterbi case. Derivatives are back-propagated through the left half of the the GTN in Figure 21 down to the interpretation graph. Derivatives are negated and back-propagated through the right-half, and the result for each arc is added to the contribution from the left half. Each arc in G_{int} now has a derivative. Arcs that are part of a correct path have a positive derivative. This derivative is very large if an incorrect path has a lower penalty than all the correct paths. Similarly, the derivatives with respect to arcs that are part of a low-penalty incorrect path have a large negative derivative. On the other hand, if the penalty of a path associated with the correct interpretation is much smaller than all other paths, the loss function is very close to 0 and almost no gradient is back-propagated. The training therefore concentrates on examples of images which yield a classification error, and furthermore, it concentrates on the pieces of the image which cause that error. Discriminative forward training is an elegant and efficient way of solving the infamous *credit assignment problem* for learning machines that manipulate "dynamic" data structures such as graphs. More generally, the same idea can be used in all situations where a learning machine must choose between discrete alternative interpretations.

As previously, the derivatives on the interpretation graph penalties can then be back-propagated into the character recognizer instances. Back-propagation through the character recognizer gives derivatives on its parameters. All the gradient contributions for the different candidate segments are added up to obtain the total gradient associated to one pair (input image, correct label sequence), that is, one example in the training set. A step of stochastic gradient descent can then be applied to update the parameters.

E. Remarks on Discriminative Training

In the above discussion, the global training criterion was given a probabilistic interpretation, but the individual penalties on the arcs of the graphs were not. There are good reasons for that. For example, if some penalties are associated to the different class labels, they would (1) have to sum to 1 (class posteriors), or (2) integrate to 1 over the input domain (likelihoods).

Let us first discuss the first case (class posteriors normalization). This local normalization of penalties may eliminate information that is important for locally rejecting all the classes [82], e.g., when a piece of image does not correspond to a valid character class, because some of the segmentation candidates may be wrong. Although an explicit "garbage class" can be introduced in a probabilistic framework to address that question, some problems remain because it is difficult to characterize such a class probabilistically and to train a system in this way (it would require a density model of unseen or unlabeled samples).

The probabilistic interpretation of individual variables plays an important role in the Baum-Welsh algorithm in combination with the Expectation-Maximization procedure. Unfortunately, those methods cannot be applied to discriminative training criteria, and one is reduced to using gradient-based methods. Enforcing the normalization of the probabilistic quantities while performing gradient-based learning is complex, inefficient, time consuming, and creates ill-conditioning of the loss-function.

Following [82], we therefore prefer to postpone normalization as far as possible (in fact, until the final decision stage of the system). Without normalization, the quantities manipulated in the system do not have a direct probabilistic interpretation.

Let us now discuss the second case (using a generative model of the input). Generative models build the boundary indirectly, by first building an independent density model for each class, and then performing classification decisions on the basis of these models. This is not a discriminative approach in that it does not focus on the ultimate goal of learning, which in this case is to learn the classification decision surface. Theoretical arguments [6], [7] suggest that estimating input densities when the real goal is to obtain a discriminant function for classification is a suboptimal strategy. In theory, the problem of estimating densities in high-dimensional spaces is much more ill-posed than finding decision boundaries.

Even though the internal variables of the system do not have a direct probabilistic interpretation, the overall system can still be viewed as producing posterior probabilities for the classes. In fact, assuming that a particular label sequence is given as the "desired sequence" to the GTN in figure 21, the exponential of minus E_{dforw} can be interpreted as an estimate of the posterior probability of that label sequence given the input. The sum of those posteriors for all the possible label sequences is 1. Another approach would consists of directly minimizing an approximation of the number of misclassifications [83] [76]. We prefer to use the discriminative forward loss function because it causes

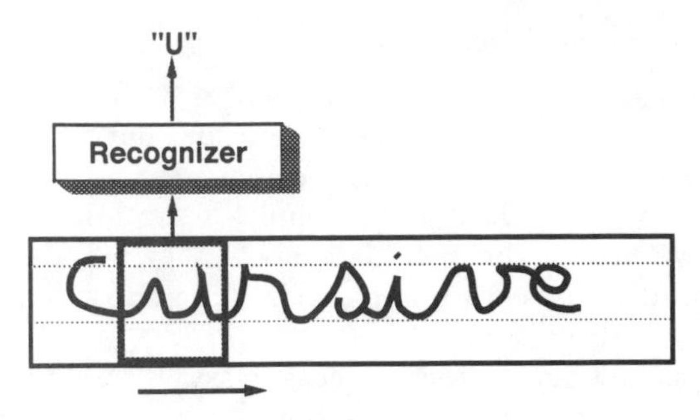

Fig. 22. Explicit segmentation can be avoided by sweeping a recognizer at every possible location in the input field.

less numerical problems during the optimization. We will see in Section X-C that this is a good way to obtain scores on which to base a rejection strategy. The important point being made here is that one is free to choose *any* parameterization deemed appropriate for a classification model. The fact that a particular parameterization uses internal variables with no clear probabilistic interpretation does not make the model any less legitimate than models that manipulate normalized quantities.

An important advantage of global and discriminative training is that learning focuses on the most important errors, and the system learns to integrate the ambiguities from the segmentation algorithm with the ambiguities of the character recognizer. In Section IX we present experimental results with an on-line handwriting recognition system that confirm the advantages of using global training versus separate training. Experiments in speech recognition with hybrids of neural networks and HMMs also showed marked improvements brought by global training [77], [29], [67], [84].

VII. Multiple Object Recognition: Space Displacement Neural Network

There is a simple alternative to explicitly segmenting images of character strings using heuristics. The idea is to sweep a recognizer at all possible locations across a normalized image of the entire word or string as shown in Figure 22. With this technique, no segmentation heuristics are required since the system essentially examines *all* the possible segmentations of the input. However, there are problems with this approach. First, the method is in general quite expensive. The recognizer must be applied at every possible location on the input, or at least at a large enough subset of locations so that misalignments of characters in the field of view of the recognizers are small enough to have no effect on the error rate. Second, when the recognizer is centered on a character to be recognized, the neighbors of the center character will be present in the field of view of the recognizer, possibly touching the center character. Therefore the recognizer must be able to correctly recognize the character in the center of its input field, even if neighboring characters are very close to, or touching the central character. Third, a word or character string cannot be perfectly size normalized. Individual characters within a string may have widely varying sizes and baseline positions. Therefore the recognizer must be very robust to shifts and size variations.

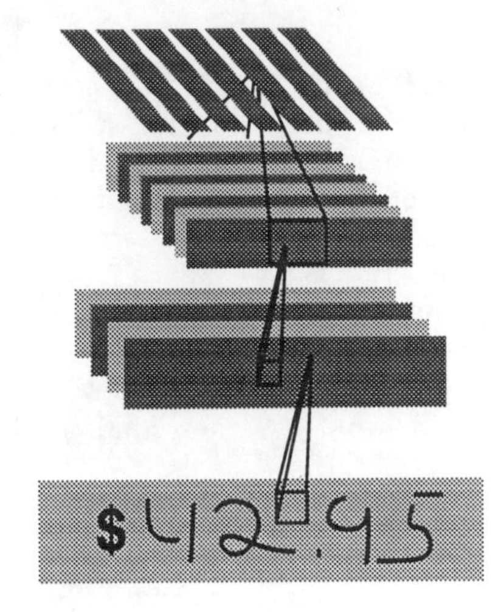

Fig. 23. A Space Displacement Neural Network is a convolutional network that has been replicated over a wide input field.

These three problems are elegantly circumvented if a convolutional network is replicated over the input field. First of all, as shown in section III, convolutional neural networks are very robust to shifts and scale variations of the input image, as well as to noise and extraneous marks in the input. These properties take care of the latter two problems mentioned in the previous paragraph. Second, convolutional networks provide a drastic saving in computational requirement when replicated over large input fields. A replicated convolutional network, also called a *Space Displacement Neural Network* or SDNN [27], is shown in Figure 23. While scanning a recognizer can be prohibitively expensive in general, convolutional networks can be scanned or replicated very efficiently over large, variable-size input fields. Consider one instance of a convolutional net and its *alter ego* at a nearby location. Because of the convolutional nature of the network, units in the two instances that look at identical locations on the input have identical outputs, therefore their states do not need to be computed twice. Only a thin "slice" of new states that are not shared by the two network instances needs to be recomputed. When all the slices are put together, the result is simply a larger convolutional network whose structure is identical to the original network, except that the feature maps are larger in the horizontal dimension. In other words, replicating a convolutional network can be done simply by increasing the size of the fields over which the convolutions are performed, and by replicating the output layer accordingly. The output layer effectively becomes a convolutional layer. An output whose receptive field is centered on an elementary object will produce the class of this object, while an in-between output may indicate no character or contain rubbish. The outputs can be interpreted as evidences for the presence of objects at all possible positions in the input field.

The SDNN architecture seems particularly attractive for

recognizing cursive handwriting where no reliable segmentation heuristic exists. Although the idea of SDNN is quite old, and very attractive by its simplicity, it has not generated wide interest until recently because, as stated above, it puts enormous demands on the recognizer [26], [27]. In speech recognition, where the recognizer is at least one order of magnitude smaller, replicated convolutional networks are easier to implement, for instance in Haffner's Multi-State TDNN model [78], [85].

A. Interpreting the Output of an SDNN with a GTN

The output of an SDNN is a sequence of vectors which encode the likelihoods, penalties, or scores of finding character of a particular class label at the corresponding location in the input. A post-processor is required to pull out the best possible label sequence from this vector sequence. An example of SDNN output is shown in Figure 25. Very often, individual characters are spotted by several neighboring instances of the recognizer, a consequence of the robustness of the recognizer to horizontal translations. Also quite often, characters are erroneously detected by recognizer instances that see only a piece of a character. For example a recognizer instance that only sees the right third of a "4" might output the label 1. How can we eliminate those extraneous characters from the output sequence and pull-out the best interpretation? This can be done using a new type of Graph Transformer with two input graphs as shown in Figure 24. The sequence of vectors produced by the SDNN is first coded into a linear graph with multiple arcs between pairs of successive nodes. Each arc between a particular pair of nodes contains the label of one of the possible categories, together with the penalty produced by the SDNN for that class label at that location. This graph is called the *SDNN Output Graph.* The second input graph to the transformer is a *grammar transducer*, more specifically a *finite-state transducer* [86], that encodes the relationship between input strings of class labels and corresponding output strings of recognized characters.The transducer is a weighted finite state machine (a graph) where each arc contains a pair of labels and possibly a penalty. Like a finite-state machine, a transducer is in a state and follows an arc to a new state when an observed input symbol matches the first symbol in the symbol pair attached to the arc. At this point the transducer emits the second symbol in the pair together with a penalty that combines the penalty of the input symbol and the penalty of the arc. A transducer therefore transforms a weighted symbol sequence into another weighted symbol sequence. The graph transformer shown in figure 24 performs a *composition* between the recognition graph and the grammar transducer. This operation takes every possible sequence corresponding to every possible path in the recognition graph and matches them with the paths in the grammar transducer. The composition produces the interpretation graph, which contains a path for each corresponding output label sequence. This composition operation may seem combinatorially intractable, but it turns out there exists an efficient algorithm for it described in more details in Section VIII.

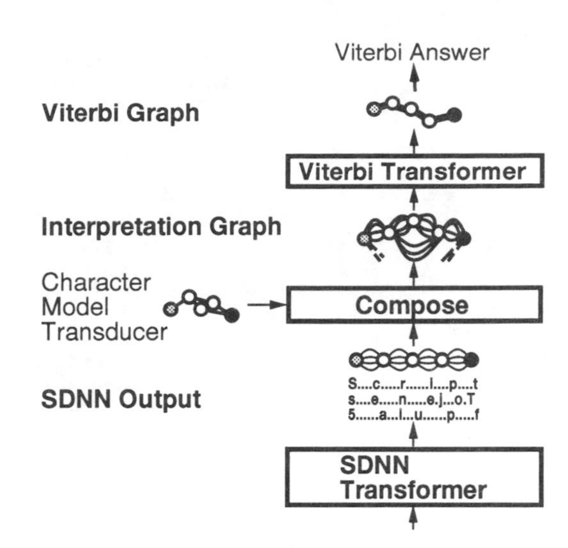

Fig. 24. A Graph Transformer pulls out the best interpretation from the output of the SDNN.

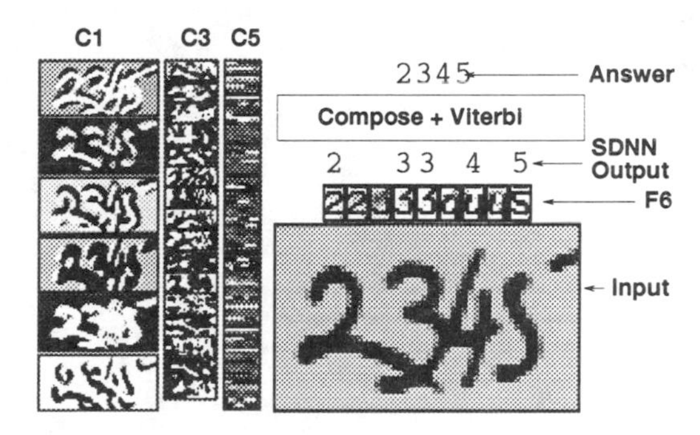

Fig. 25. An example of multiple character recognition with SDNN. With SDNN, no explicit segmentation is performed.

B. Experiments with SDNN

In a series of experiments, LeNet-5 was trained with the goal of being replicated so as to recognize multiple characters without segmentations. The data was generated from the previously described Modified NIST set as follows. Training images were composed of a central character, flanked by two side characters picked at random in the training set. The separation between the bounding boxes of the characters were chosen at random between -1 and 4 pixels. In other instances, no central character was present, in which case the desired output of the network was the blank space class. In addition, training images were degraded with 10% salt and pepper noise (random pixel inversions).

Figures 25 and 26 show a few examples of successful recognitions of multiple characters by the LeNet-5 SDNN. Standard techniques based on Heuristic Over-Segmentation would fail miserably on many of those examples. As can be seen on these examples, the network exhibits striking invariance and noise resistance properties. While some authors have argued that invariance requires more sophisticated models than feed-forward neural networks [87], LeNet-5 exhibits these properties to a large extent.

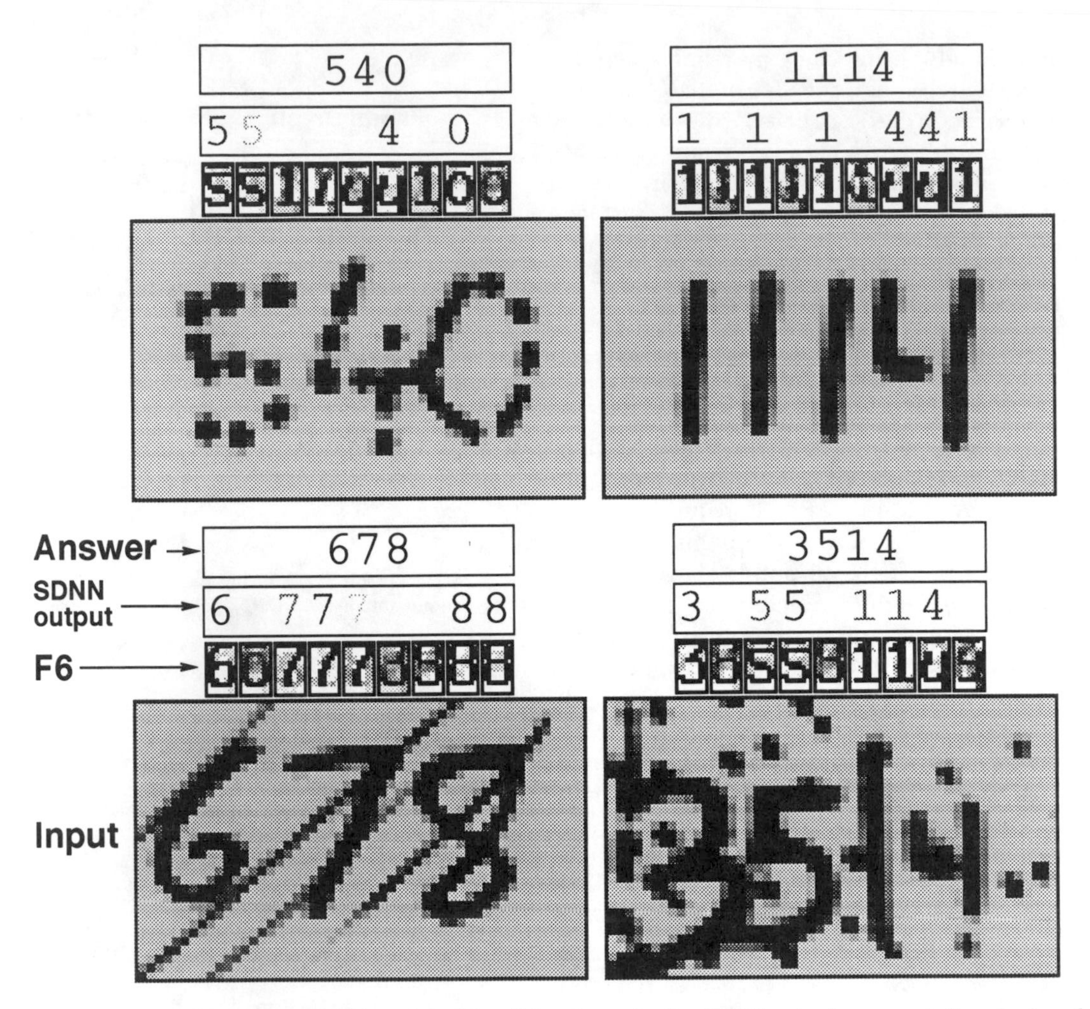

Fig. 26. An SDNN applied to a noisy image of digit string. The digits shown in the SDNN output represent the winning class labels, with a lighter grey level for high-penalty answers.

Similarly, it has been suggested that accurate recognition of multiple overlapping objects require explicit mechanisms that would solve the so-called *feature binding* problem [87]. As can be seen on Figures 25 and 26, the network is able to tell the characters apart, even when they are closely intertwined, a task that would be impossible to achieve with the more classical Heuristic Over-Segmentation technique. The SDNN is also able to correctly group disconnected pieces of ink that form characters. Good examples of that are shown in the upper half of figure 26. In the top left example, the 4 and the 0 are more connected to each other than they are connected with themselves, yet the system correctly identifies the 4 and the 0 as separate objects. The top right example is interesting for several reasons. First the system correctly identifies the three individual ones. Second, the left half and right half of disconnected 4 are correctly grouped, even though no geometrical information could decide to associate the left half to the vertical bar on its left or on its right. The right half of the 4 does cause the appearance of an erroneous 1 on the SDNN output, but this one is removed by the character model transducer which prevents characters from appearing on contiguous outputs.

Another important advantage of SDNN is the ease with which they can be implemented on parallel hardware. Specialized analog/digital chips have been designed and used in character recognition, and in image preprocessing applications [88]. However the rapid progress of conventional processor technology with reduced-precision vector arithmetic instructions (such as Intel's MMX) make the success of specialized hardware hypothetical at best.

Short video clips of the LeNet-5 SDNN can be viewed at `http://www.research.att.com/~yann/ocr`.

C. Global Training of SDNN

In the above experiments, the string image were artificially generated from individual character. The advantage is that we know in advance the location and the label of the important character. With real training data, the correct sequence of labels for a string is generally available, but the precise locations of each corresponding character in the input image are unknown.

In the experiments described in the previous section, the best interpretation was extracted from the SDNN output using a very simple graph transformer. Global training of an SDNN can be performed by back-propagating gradients through such graph transformers arranged in architectures similar to the ones described in section VI.

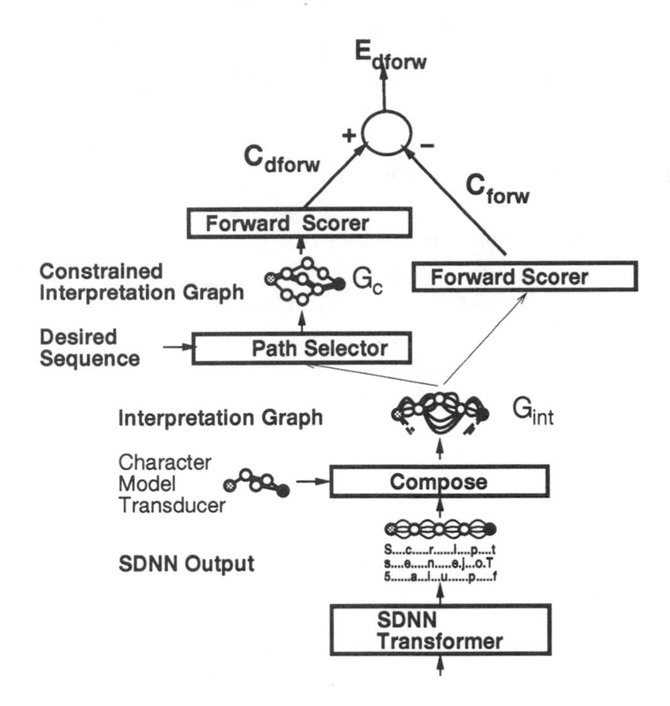

Fig. 27. A globally trainable SDNN/HMM hybrid system expressed as a GTN.

This is somewhat equivalent to modeling the output of an SDNN with a Hidden Markov Model. Globally trained, variable-size TDNN/HMM hybrids have been used for speech recognition and on-line handwriting recognition [77], [89], [90], [67]. Space Displacement Neural Networks have been used in combination with HMMs or other elastic matching methods for handwritten word recognition [91], [92].

Figure 27 shows the graph transformer architecture for training an SDNN/HMM hybrid with the Discriminative Forward Criterion. The top part is comparable to the top part of figure 21. On the right side the composition of the recognition graph with the grammar gives the interpretation graph with all the possible legal interpretations. On the left side the composition is performed with a grammar that only contains paths with the desired sequence of labels. This has a somewhat similar function to the path selector used in the previous section. Like in Section VI-D the loss function is the difference between the forward score obtained from the left half and the forward score obtained from the right half. To back-propagate through the composition transformer, we need to keep a record of which arc in the recognition graph originated which arcs in the interpretation graph. The derivative with respect to an arc in the recognition graph is equal to the sum of the derivatives with respect to all the arcs in the interpretation graph that originated from it. Derivative can also be computed for the penalties on the grammar graph, allowing to learn them as well. As in the previous example, a discriminative criterion must be used, because using a non-discriminative criterion could result in a collapse effect if the network's output RBF are adaptive. The above training procedure can be equivalently formulated in term of HMM. Early experiments in zip code recognition [91], and more recent experiments in on-line handwriting recognition [38] have demonstrated the idea of globally-trained SDNN/HMM hybrids. SDNN is an extremely promising and attractive technique for OCR, but so far it has not yielded better results than Heuristic Over-Segmentation. We hope that these results will improve as more experience is gained with these models.

D. Object Detection and Spotting with SDNN

An interesting application of SDNNs is object detection and spotting. The invariance properties of Convolutional Networks, combined with the efficiency with which they can be replicated over large fields suggest that they can be used for "brute force" object spotting and detection in large images. The main idea is to train a single Convolutional Network to distinguish images of the object of interest from images present in the background. In utilization mode, the network is replicated so as to cover the entire image to be analyzed, thereby forming a two-dimensional Space Displacement Neural Network. The output of the SDNN is a two-dimensional plane in which activated units indicate the presence of the object of interest in the corresponding receptive field. Since the sizes of the objects to be detected within the image are unknown, the image can be presented to the network at multiple resolutions, and the results at multiple resolutions combined. The idea has been applied to face location, [93], address block location on envelopes [94], and hand tracking in video [95].

To illustrate the method, we will consider the case of face detection in images as described in [93]. First, images containing faces at various scales are collected. Those images are filtered through a zero-mean Laplacian filter so as to remove variations in global illumination and low spatial frequency illumination gradients. Then, training samples of faces and non-faces are manually extracted from those images. The face sub-images are then size normalized so that the height of the entire face is approximately 20 pixels while keeping fairly large variations (within a factor of two). The scale of background sub-images are picked at random. A single convolutional network is trained on those samples to classify face sub-images from non-face sub-images.

When a scene image is to be analyzed, it is first filtered through the Laplacian filter, and sub-sampled at powers-of-two resolutions. The network is replicated over each of multiple resolution images. A simple voting technique is used to combine the results from multiple resolutions.

A two-dimensional version of the global training method described in the previous section can be used to alleviate the need to manually locate faces when building the training sample [93]. Each possible location is seen as an alternative interpretation, i.e. one of several parallel arcs in a simple graph that only contains a start node and an end node.

Other authors have used Neural Networks, or other classifiers such as Support Vector Machines for face detection with great success [96], [97]. Their systems are very similar to the one described above, including the idea of presenting the image to the network at multiple scales. But since those systems do not use Convolutional Networks, they cannot take advantage of the speedup described here, and have to rely on other techniques, such as pre-filtering and real-time tracking, to keep the computational requirement within reasonable limits. In addition, because those classifiers are much less invariant to scale variations than Convolutional Networks, it is necessary to multiply the number of scales at which the images are presented to the classifier.

VIII. Graph Transformer Networks and Transducers

In Section IV, Graph Transformer Networks (GTN) were introduced as a generalization of multi-layer, multi-module networks where the state information is represented as graphs instead of fixed-size vectors. This section re-interprets the GTNs in the framework of *Generalized Transduction*, and proposes a powerful *Graph Composition* algorithm.

A. Previous Work

Numerous authors in speech recognition have used Gradient-Based Learning methods that integrate graph-based statistical models (notably HMM) with acoustic recognition modules, mainly Gaussian mixture models, but also neural networks [98], [78], [99], [67]. Similar ideas have been applied to handwriting recognition (see [38] for a review). However, there has been no proposal for a systematic approach to multi-layer graph-based trainable systems. The idea of transforming graphs into other graphs has received considerable interest in computer science, through the concept of *weighted finite-state transducers* [86]. Transducers have been applied to speech recognition [100] and language translation [101], and proposals have been made for handwriting recognition [102]. This line of work has been mainly focused on efficient search algorithms [103] and on the algebraic aspects of combining transducers and graphs (called acceptors in this context), but very little effort has been devoted to building globally trainable systems out of transducers. What is proposed in the following sections is a systematic approach to automatic training in graph-manipulating systems. A different approach to graph-based trainable systems, called Input-Output HMM, was proposed in [104], [105].

B. Standard Transduction

In the established framework of finite-state transducers [86], discrete symbols are attached to arcs in the graphs. Acceptor graphs have a single symbol attached to each arc whereas transducer graphs have two symbols (an input symbol and an output symbol). A special null symbol is absorbed by any other symbol (when concatenating symbols to build a symbol sequence). Weighted transducers and acceptors also have a scalar quantity attached to each arc. In this framework, the composition operation takes as input an acceptor graph and a transducer graph and builds an output acceptor graph. Each path in this output graph (with symbol sequence S_{out}) corresponds to one path (with symbol sequence S_{in}) in the input acceptor graph and one path and a corresponding pair of input/output sequences (S_{out},S_{in}) in the transducer graph. The weights on the arcs of the output graph are obtained by adding the weights from the matching arcs in the input acceptor and transducer graphs. In the rest of the paper, we will call this graph composition operation using transducers the *(standard) transduction operation.*

A simple example of transduction is shown in Figure 28. In this simple example, the input and output symbols on the transducer arcs are always identical. This type of transducer graph is called a grammar graph. To better understand the transduction operation, imagine two tokens sitting each on the start nodes of the input acceptor graph and the transducer graph. The tokens can freely follow any arc labeled with a null input symbol. A token can follow an arc labeled with a non-null input symbol if the other token also follows an arc labeled with the *same* input symbol. We have an *acceptable trajectory* when both tokens reach the end nodes of their graphs (i.e. the tokens have reached the terminal configuration). This trajectory represents a sequence of input symbols that complies with both the acceptor and the transducer. We can then collect the corresponding sequence of output symbols along the trajectory of the transducer token. The above procedure produces a tree, but a simple technique described in Section VIII-C can be used to avoid generating multiple copies of certain subgraphs by detecting when a particular output state has already been seen.

The transduction operation can be performed very efficiently [106], but presents complex book-keeping problems concerning the handling of all combinations of null and non null symbols. If the weights are interpreted as probabilities (normalized appropriately) then an acceptor graph represents a probability distribution over the language defined by the set of label sequences associated to all possible paths (from the start to the end node) in the graph.

An example of application of the transduction operation is the incorporation of linguistic constraints (a lexicon or a grammar) when recognizing words or other character strings. The recognition transformer produces the recognition graph (an acceptor graph) by applying the neural network recognizer to each candidate segment. This acceptor graph is composed with a transducer graph for the grammar. The grammar transducer contains a path for each legal sequence of symbol, possibly augmented with penalties to indicate the relative likelihoods of the possible sequences. The arcs contain identical input and output symbols. Another example of transduction was mentioned in Section V: the path selector used in the heuristic over-segmentation training GTN is implementable by a composition. The transducer graph is linear graph which contains the correct label sequence. The composition of the interpretation graph with this linear graph yields the constrained graph.

C. Generalized Transduction

If the data structures associated to each arc took only a finite number of values, composing the input graph and an appropriate transducer would be a sound solution. For our applications however, the data structures attached to the arcs of the graphs may be vectors, images or other high-dimensional objects that are not readily enumerated. We present a new composition operation that solves this problem.

Instead of only handling graphs with discrete symbols and penalties on the arcs, we are interested in considering graphs whose arcs may carry complex data structures, including continuous-valued data structures such as vectors and images. Composing such graphs requires additional information:

- When examining a pair of arcs (one from each input graph), we need a criterion to decide whether to create corresponding arc(s) and node(s) in the output graph, based on the information attached to the input arcs. We can decide to build an arc, several arcs, or an entire sub-graph with several nodes and arcs.
- When that criterion is met, we must build the corresponding arc(s) and node(s) in the output graph and compute the information attached to the newly created arc(s) as a function the the information attached to the input arcs.

These functions are encapsulated in an object called a *Composition Transformer.* An instance of Composition Transformer implements three methods:

- `check(arc1, arc2)`

compares the data structures pointed to by arcs `arc1` (from the first graph) and `arc2` (from the second graph) and returns a boolean indicating whether corresponding arc(s) should be created in the output graph.

- `fprop(ngraph, upnode, downnode, arc1, arc2)`

is called when `check(arc1, arc2)` returns true. This method creates new arcs and nodes between nodes `upnode` and `downnode` in the output graph `ngraph`, and computes the information attached to these newly created arcs as a function of the attached information of the input arcs `arc1` and `arc2`.

- `bprop(ngraph, upnode, downnode, arc1, arc2)`

is called during training in order to propagate gradient information from the output sub-graph between `upnode` and `downnode` into the data structures on the `arc1` and `arc2`, as well as with respect to the parameters that were used in the `fprop` call with the same arguments. This assumes that the function used by `fprop` to compute the values attached to its output arcs is differentiable.

The `check` method can be seen as constructing a dynamic *architecture* of functional dependencies, while the `fprop` method performs a forward propagation through that architecture to compute the numerical information attached to the arcs. The `bprop` method performs a backward propagation through the same architecture to compute the partial derivatives of the loss function with respect to the information attached to the arcs. This is illustrated in Figure 28.

Figure 29 shows a simplified generalized graph composition algorithm. This simplified algorithm does not handle null transitions, and does not check whether the tokens tra-

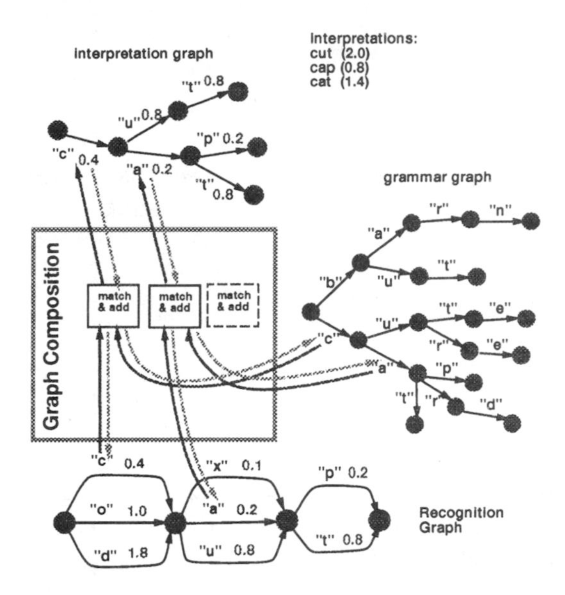

Fig. 28. Example of composition of the recognition graph with the grammar graph in order to build an interpretation that is consistent with both of them. During the forward propagation (dark arrows), the methods **check** and **fprop** are used. Gradients (dashed arrows) are back-propagated with the application of the method **bprop**.

jectory is acceptable (i.e. both tokens simultaneously reach the end nodes of their graphs). The management of null transitions is a straightforward modification of the token simulation function. Before enumerating the possible non null joint token transitions, we loop on the possible null transitions of each token, recursively call the token simulation function, and finally call the method `fprop`. The safest way for identifying acceptable trajectories consists in running a preliminary pass for identifying the token configurations from which we can reach the terminal configuration (i.e. both tokens on the end nodes). This is easily achieved by enumerating the trajectories in the opposite direction. We start on the end nodes and follow the arcs upstream. During the main pass, we only build the nodes that allow the tokens to reach the terminal configuration.

Graph composition using transducers (i.e. standard transduction) is easily and efficiently implemented as a generalized transduction. The method `check` simply tests the equality of the input symbols on the two arcs, and the method `fprop` creates a single arc whose symbol is the output symbol on the transducer's arc.

The composition between pairs of graphs is particularly useful for incorporating linguistic constraints in a handwriting recognizer. Examples of its use are given in the on-line handwriting recognition system described in Section IX) and in the check reading system described in Section X).

In the rest of the paper, the term *Composition Transformer* will denote a Graph Transformer based on the generalized transductions of multiple graphs. The concept of generalized transduction is a very general one. In fact, many of the graph transformers described earlier in this paper, such as the segmenter and the recognizer, can be

```
Function generalized_composition(PGRAPH graph1,
                                 PGRAPH graph2,
                                 PTRANS trans)
Returns PGRAPH
{
  // Create new graph
  PGRAPH ngraph = new_graph()

  // Create map between token positions
  // and nodes of the new graph
  PNODE map[PNODE,PNODE] =  new_empty_map()
  map[endnode(graph1), endnode(graph2)] =
    endnode(newgraph)

  // Recursive subroutine for simulating tokens
  Function simtokens(PNODE node1, PNODE node2)
  Returns PNODE
  {
    PNODE currentnode = map[node1, node2]
    // Check if already visited
    If (currentnode == nil)
      // Record new configuration
      currentnode = ngraph->create_node()
      map[node1, node2] = currentnode
      // Enumerate the possible non-null
      // joint token transitions
      For ARC arc1 in down_arcs(node1)
        For ARC arc2 in down_arcs(node2)
          If (trans->check(arc1, arc2))
            PNODE newnode =
             simtokens(down_node(arc1),
                       down_node(arc2))
            trans->fprop(ngraph, currentnode,
                         newnode, arc1, arc2)
    // Return node in composed graph
    Return currentnode
  }

  // Perform token simulation
  simtokens(startnode(graph1), startnode(graph2))
  Delete map
  Return ngraph
}
```

Fig. 29. Pseudo-code for a simplified generalized composition algorithm. For simplifying the presentation, we do not handle null transitions nor implement dead end avoidance. The two main component of the composition appear clearly here: (a) the recursive function `simtoken()` enumerating the token trajectories, and, (b) the associative array `map` used for remembering which nodes of the composed graph have been visited.

formulated in terms of generalized transduction. In this case the, the generalized transduction does not take two input graphs but a single input graph. The method `fprop` of the transformer may create several arcs or even a complete subgraph for each arc of the initial graph. In fact the pair (`check`, `fprop`) itself can be seen as procedurally defining a transducer.

In addition, It can be shown that the generalized transduction of a single graph is theoretically equivalent to the standard composition of this graph with a particular transducer graph. However, implementing the operation this way may be very inefficient since the transducer can be very complicated.

In practice, the graph produced by a generalized transduction is represented procedurally, in order to avoid building the whole output graph (which may be huge when for example the interpretation graph is composed with the grammar graph). We only instantiate the nodes which are visited by the search algorithm during recognition (e.g. Viterbi). This strategy propagates the benefits of pruning algorithms (e.g. Beam Search) in all the Graph Transformer Network.

D. Notes on the Graph Structures

Section VI has discussed the idea of global training by back-propagating gradient through simple graph transformers. The `bprop` method is the basis of the back-propagation algorithm for generic graph transformers. A generalized composition transformer can be seen as dynamically establishing functional relationships between the numerical quantities on the input and output arcs. Once the `check` function has decided that a relationship should be established, the `fprop` function implements the numerical relationship. The `check` function establishes the structure of the ephemeral network inside the composition transformer.

Since `fprop` is assumed to be differentiable, gradients can be back-propagated through that structure. Most parameters affect the scores stored on the arcs of the successive graphs of the system. A few threshold parameters may determine whether an arc appears or not in the graph. Since non existing arcs are equivalent to arcs with very large penalties, we only consider the case of parameters affecting the penalties.

In the kind of systems we have discussed until now (and the application described in Section X), much of the knowledge about the structure of the graph that is produced by a Graph Transformer is determined by the nature of the Graph Transformer, but it may also depend on the value of the parameters and on the input. It may also be interesting to consider Graph Transformer modules which attempt to learn the structure of the output graph. This might be considered a combinatorial problem and not amenable to Gradient-Based Learning, but a solution to this problem is to generate a large graph that contains the graph candidates as sub-graphs, and then select the appropriate sub-graph.

E. GTN and Hidden Markov Models

GTNs can be seen as a generalization and an extension of HMMs. On the one hand, the probabilistic interpretation can be either kept (with penalties being log-probabilities), pushed to the final decision stage (with the difference of the constrained forward penalty and the unconstrained forward penalty being interpreted as negative log-probabilities of label sequences), or dropped altogether (the network just represents a decision surface for label sequences in input space). On the other hand, Graph Transformer Networks extend HMMs by allowing to combine in a well-principled framework multiple levels of processing, or multiple models (e.g., Pereira et al. have been using the transducer framework for stacking HMMs representing different levels of processing in automatic speech recognition [86]).

Unfolding a HMM in time yields a graph that is very similar to our interpretation graph (at the final stage of processing of the Graph Transformer Network, before Viterbi recognition). It has nodes $n(t, i)$ associated to each time step t and state i in the model. The penalty c_i for an arc from $n(t-1, j)$ to $n(t, i)$ then corresponds to the negative log-probability of emitting observed data o_t at position t and going from state j to state i in the time interval $(t-1, t)$. With this probabilistic interpretation, the forward penalty is the negative logarithm of the likelihood of whole observed data sequence (given the model).

In Section VI we mentioned that the collapsing phenomenon can occur when non-discriminative loss functions are used to train neural networks/HMM hybrid systems. With classical HMMs with fixed preprocessing, this problem does not occur because the parameters of the emission and transition probability models are forced to satisfy certain probabilistic constraints: the sum or the integral of the probabilities of a random variable over its possible values must be 1. Therefore, when the probability of certain events is increased, the probability of other events must automatically be decreased. On the other hand, if the probabilistic assumptions in an HMM (or other probabilistic model) are not realistic, discriminative training, discussed in Section VI, can improve performance as this has been clearly shown for speech recognition systems [48], [49], [50], [107], [108].

The Input-Output HMM model (IOHMM) [105], [109], is strongly related to graph transformers. Viewed as a probabilistic model, an IOHMM represents the conditional distribution of output sequences given input sequences (of the same or a different length). It is parameterized from an emission probability module and a transition probability module. The emission probability module computes the conditional emission probability of an output variable (given an input value and the value of discrete "state" variable). The transition probability module computes conditional transition probabilities of a change in the value of the "state" variable, given the an input value. Viewed as a graph transformer, it assigns an output graph (representing a probability distribution over the sequences of the output variable) to each path in the input graph. All these output graphs have the same structure, and the penalties on their arcs are simply added in order to obtain the complete output graph. The input values of the emission and transition modules are read off the data structure on the input arcs of the IOHMM Graph Transformer. In practice, the output graph may be very large, and needs not be completely instantiated (i.e., it is pruned: only the low penalty paths are created).

IX. An On-Line Handwriting Recognition System

Natural handwriting is often a mixture of different "styles", lower case printed, upper case, and cursive. A reliable recognizer for such handwriting would greatly improve interaction with pen-based devices, but its implementation presents new technical challenges. Characters taken in isolation can be very ambiguous, but considerable information is available from the context of the whole word. We have built a word recognition system for pen-based devices based on four main modules: a preprocessor that normalizes a word, or word group, by fitting a geometrical model to the word structure; a module that produces an "annotated image" from the normalized pen trajectory; a replicated convolutional neural network that spots and recognizes characters; and a GTN that interprets the networks output by taking word-level constraints into account. The network and the GTN are *jointly* trained to minimize an error measure defined at the word level.

In this work, we have compared a system based on SDNNs (such as described in Section VII), and a system based on Heuristic Over-Segmentation (such as described in Section V). Because of the sequential nature of the information in the pen trajectory (which reveals more information than the purely optical input from in image), Heuristic Over-Segmentation can be very efficient in proposing candidate character cuts, especially for non-cursive script.

A. Preprocessing

Input normalization reduces intra-character variability, simplifying character recognition. We have used a word normalization scheme [92] based on fitting a geometrical model of the word structure. Our model has four "flexible" lines representing respectively the ascenders line, the core line, the base line and the descenders line. The lines are fitted to local minima or maxima of the pen trajectory. The parameters of the lines are estimated with a modified version of the EM algorithm to maximize the joint probability of observed points and parameter values, using a prior on parameters that prevents the lines from collapsing on each other.

The recognition of handwritten characters from a pen trajectory on a digitizing surface is often done in the time domain [110], [44], [111]. Typically, trajectories are normalized, and local geometrical or dynamical features are extracted. The recognition may then be performed using curve matching [110], or other classification techniques such as TDNNs [44], [111]. While these representations have several advantages, their dependence on stroke ordering and individual writing styles makes them difficult to

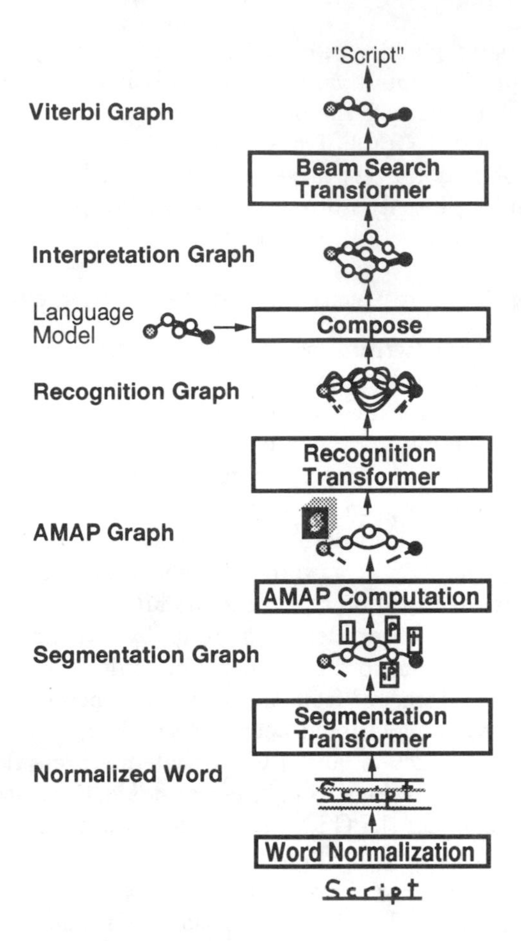

Fig. 30. An on-line handwriting recognition GTN based on heuristic over-segmentation

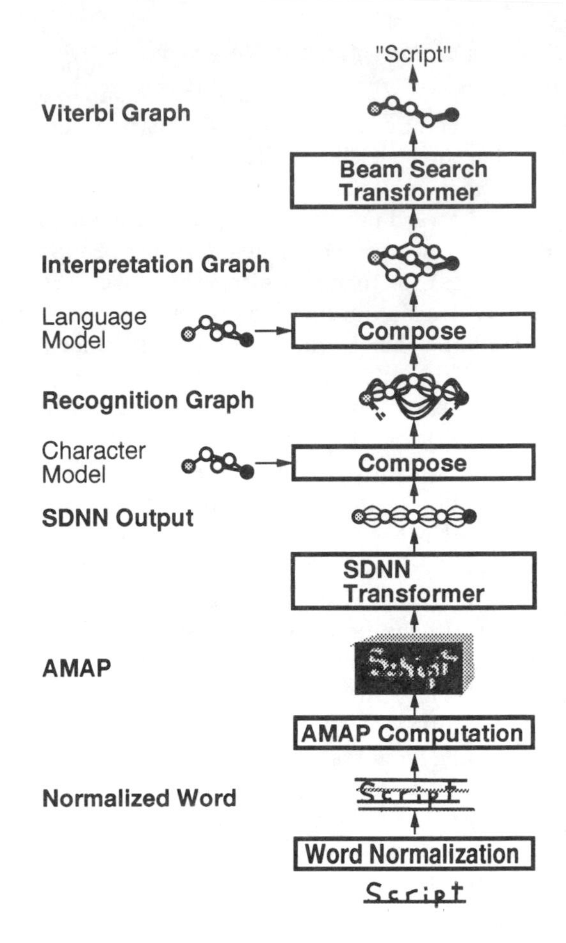

Fig. 31. An on-line handwriting recognition GTN based on Space-Displacement Neural Network

use in high accuracy, writer independent systems that integrate the segmentation with the recognition.

Since the intent of the writer is to produce a legible *image*, it seems natural to preserve as much of the pictorial nature of the signal as possible, while at the same time exploit the sequential information in the trajectory. For this purpose we have designed a representation scheme, called AMAP [38], where pen trajectories are represented by low-resolution images in which each picture element contains information about the local properties of the trajectory. An AMAP can be viewed as an "annotated image" in which each pixel is a 5-element feature vector: 4 features are associated to four orientations of the pen trajectory in the area around the pixel, and the fifth one is associated to local curvature in the area around the pixel. A particularly useful feature of the AMAP representation is that it makes very few assumptions about the nature of the input trajectory. It does not depend on stroke ordering or writing speed, and it can be used with all types of handwriting (capital, lower case, cursive, punctuation, symbols). Unlike many other representations (such as global features), AMAPs can be computed for complete words without requiring segmentation.

B. Network Architecture

One of the best networks we found for both online and offline character recognition is a 5-layer convolutional network somewhat similar to LeNet-5 (Figure 2), but with multiple input planes and different numbers of units on the last two layers; layer 1: convolution with 8 kernels of size 3x3, layer 2: 2x2 sub-sampling, layer 3: convolution with 25 kernels of size 5x5, layer 4 convolution with 84 kernels of size 4x4, layer 5: 2x1 sub-sampling, classification layer: 95 RBF units (one per class in the full printable ASCII set). The distributed codes on the output are the same as for LeNet-5, except they are adaptive unlike with LeNet-5. When used in the heuristic over-segmentation system, the input to above network consisted of an AMAP with five planes, 20 rows and 18 columns. It was determined that this resolution was sufficient for representing handwritten characters. In the SDNN version, the number of columns was varied according to the width of the input word. Once the number of sub-sampling layers and the sizes of the kernels are chosen, the sizes of all the layers, including the input, are determined unambiguously. The only architectural parameters that remain to be selected are the number of feature maps in each layer, and the information as to what feature map is connected to what other feature map. In our case, the sub-sampling rates were chosen as small as possible (2x2), and the kernels as small as possible in the first layer (3x3) to limit the total number of connections. Kernel sizes in the upper layers are chosen to be as small as possible while satisfying the size constraints mentioned above. Larger architectures did not necessarily perform better and required considerably more time to be trained. A very small architecture with half the input field also performed worse, because of insufficient input resolution. Note that the input resolution is nonetheless much less than for optical character recognition, because the angle and curvature provide more information than would a single grey level at each pixel.

C. Network Training

Training proceeded in two phases. First, we kept the centers of the RBFs fixed, and trained the network weights so as to minimize the output distance of the RBF unit corresponding to the correct class. This is equivalent to minimizing the mean-squared error between the previous layer and the center of the correct-class RBF. This bootstrap phase was performed on isolated characters. In the second phase, all the parameters, network weights and RBF centers were trained globally to minimize a discriminative criterion at the word level.

With the Heuristic Over-Segmentation approach, the GTN was composed of four main Graph Transformers:

1. The **Segmentation Transformer** performs the Heuristic Over-Segmentation, and outputs the segmentation graph. An AMAP is then computed for each image attached to the arcs of this graph.
2. The **Character Recognition Transformer** applies the the convolutional network character recognizer to each candidate segment, and outputs the recognition graph, with penalties and classes on each arc.
3. The **Composition Transformer** composes the recognition graph with a grammar graph representing a language model incorporating lexical constraints.
4. The **Beam Search Transformer** extracts a good interpretation from the interpretation graph. This task could have been achieved with the usual Viterbi Transformer. The Beam Search algorithm however implements pruning strategies which are appropriate for large interpretation graphs.

With the SDNN approach, the main Graph Transformers are the following:

1. The **SDNN Transformer** replicates the convolutional network over the a whole word image, and outputs a recognition graph that is a linear graph with class penalties for every window centered at regular intervals on the input image.
2. The **Character-Level Composition Transformer** composes the recognition graph with a left-to-right HMM for each character class (as in Figure 27).
3. The **Word-Level Composition Transformer** composes the output of the previous transformer with a language model incorporating lexical constraints, and outputs the interpretation graph.
4. The **Beam Search Transformer** extracts a good interpretation from the interpretation graph.

In this application, the language model simply constrains the final output graph to represent sequences of character labels from a given dictionary. Furthermore, the interpretation graph is not actually completely instantiated: the only nodes created are those that are needed by the Beam Search module. The interpretation graph is therefore represented procedurally rather than explicitly.

A crucial contribution of this research was the joint training of all graph transformer modules within the network with respect to a single criterion, as explained in Sections VI and VIII. We used the Discriminative Forward loss function on the final output graph: minimize the forward penalty of the constrained interpretation (i.e., along all the "correct" paths) while maximizing the forward penalty of the whole interpretation graph (i.e., along all the paths).

During global training, the loss function was optimized with the stochastic diagonal Levenberg-Marquardt procedure described in Appendix C, that uses second derivatives to compute optimal learning rates. This optimization operates on *all* the parameters in the system, most notably the network weights and the RBF centers.

D. Experimental Results

In the first set of experiments, we evaluated the generalization ability of the neural network classifier coupled with the word normalization preprocessing and AMAP input representation. All results are in *writer independent* mode (different writers in training and testing). Initial training on isolated characters was performed on a database of approximately 100,000 hand printed characters (95 classes of upper case, lower case, digits, and punctuation). Tests on a database of isolated characters were performed separately on the four types of characters: upper case (2.99% error on 9122 patterns), lower case (4.15% error on 8201 patterns), digits (1.4% error on 2938 patterns), and punctuation (4.3% error on 881 patterns). Experiments were performed with the network architecture described above. To enhance the robustness of the recognizer to variations in position, size, orientation, and other distortions, additional training data was generated by applying local affine transformations to the original characters.

The second and third set of experiments concerned the recognition of lower case words (writer independent). The tests were performed on a database of 881 words. First we evaluated the improvements brought by the word normalization to the system. For the SDNN/HMM system we *have* to use word-level normalization since the network sees one whole word at a time. With the Heuristic Over-Segmentation system, and before doing any word-level training, we obtained with character-level normalization 7.3% and 3.5% word and character errors (adding insertions, deletions and substitutions) when the search was constrained within a 25461-word dictionary. When using the word normalization preprocessing instead of a character level normalization, error rates dropped to 4.6% and 2.0% for word and character errors respectively, i.e., a relative drop of 37% and 43% in word and character error respectively. This suggests that normalizing the word in its entirety is better than first segmenting it and then normalizing and processing each of the segments.

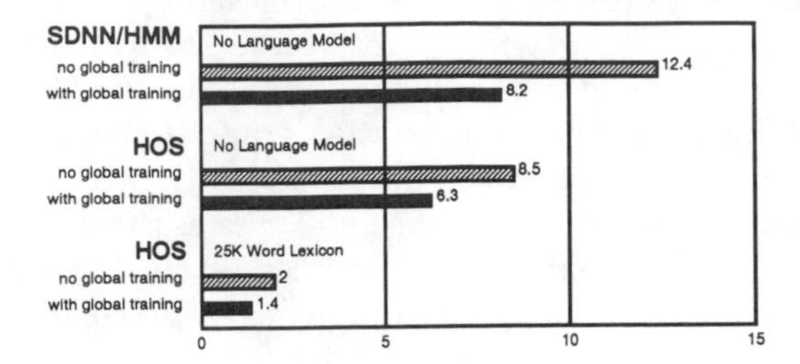

Fig. 32. Comparative results (character error rates) showing the improvement brought by global training on the SDNN/HMM hybrid, and on the Heuristic Over-Segmentation system (HOS), without and with a 25461 words dictionary.

In the third set of experiments, we measured the improvements obtained with the joint training of the neural network and the post-processor with the word-level criterion, in comparison to training based only on the errors performed at the character level. After initial training on individual characters as above, global word-level discriminative training was performed with a database of 3500 lower case words. For the SDNN/HMM system, without any dictionary constraints, the error rates dropped from 38% and 12.4% word and character error to 26% and 8.2% respectively after word-level training, i.e., a relative drop of 32% and 34%. For the Heuristic Over-Segmentation system and a slightly improved architecture, without any dictionary constraints, the error rates dropped from 22.5% and 8.5% word and character error to 17% and 6.3% respectively, i.e., a relative drop of 24.4% and 25.6%. With a 25461-word dictionary, errors dropped from 4.6% and 2.0% word and character errors to 3.2% and 1.4% respectively after word-level training, i.e., a relative drop of 30.4% and 30.0%. Even lower error rates can be obtained by drastically reducing the size of the dictionary to 350 words, yielding 1.6% and 0.94% word and character errors.

These results clearly demonstrate the usefulness of globally trained Neural-Net/HMM hybrids for handwriting recognition. This confirms similar results obtained earlier in speech recognition [77].

X. A Check Reading System

This section describes a GTN based Check Reading System, intended for immediate industrial deployment. It also shows how the use of Gradient Based-Learning and GTNs make this deployment fast and cost-effective while yielding an accurate and reliable solution.

The verification of the amount on a check is a task that is extremely time and money consuming for banks. As a consequence, there is a very high interest in automating the process as much as possible (see for example [112], [113], [114]). Even a partial automation would result in considerable cost reductions. The threshold of economic viability for automatic check readers, as set by the bank, is when 50% of the checks are read with less than 1% error. The other 50% of the check being rejected and sent to human operators. In such a case, we describe the performance of the system as *50% correct / 49% reject / 1% error*. The system presented here was one of the first to cross that threshold on representative mixtures of business and personal checks.

Checks contain at least two versions of the amount. The *Courtesy amount* is written with numerals, while the *Legal amount* is written with letters. On business checks, which are generally machine-printed, these amounts are relatively easy to read, but quite difficult to find due to the lack of standard for business check layout. On the other hand, these amounts on personal checks are easy to find but much harder to read.

For simplicity (and speed requirements), our initial task is to read the Courtesy amount only. This task consists of two main steps:

- The system has to find, among all the fields (lines of text), the candidates that are the most likely to contain the courtesy amount. This is obvious for many personal checks, where the position of the amount is standardized. However, as already noted, finding the amount can be rather difficult in business checks, even for the human eye. There are many strings of digits, such as the check number, the date, or even "not to exceed" amounts, that can be confused with the actual amount. In many cases, it is very difficult to decide which candidate is the courtesy amount before performing a full recognition.
- In order to read (and choose) some Courtesy amount candidates, the system has to segment the fields into characters, read and score the candidate characters, and finally find the best interpretation of the amount using contextual knowledge represented by a stochastic grammar for check amounts.

The GTN methodology was used to build a check amount reading system that handles both personal checks and business checks.

A. A GTN for Check Amount Recognition

We now describe the successive graph transformations that allow this network to read the check amount (cf. Figure 33). Each Graph Transformer produces a graph whose paths encode and score the current hypotheses considered at this stage of the system.

The input to the system is a trivial graph with a single arc that carries the image of the whole check (cf. Figure 33).

The field location transformer T_{field} first performs classical image analysis (including connected component analysis, ink density histograms, layout analysis, etc...) and heuristically extracts rectangular zones that may contain the check amount. T_{field} produces an output graph, called the *field graph* (cf. Figure 33) such that each candidate zone is associated with one arc that links the start node to the end node. Each arc contains the image of the zone, and a penalty term computed from simple features extracted from the zone (absolute position, size, aspect ratio, etc...). The penalty term is close to zero if the features suggest that the field is a likely candidate, and is large if the field is deemed less likely to be an amount. The penalty

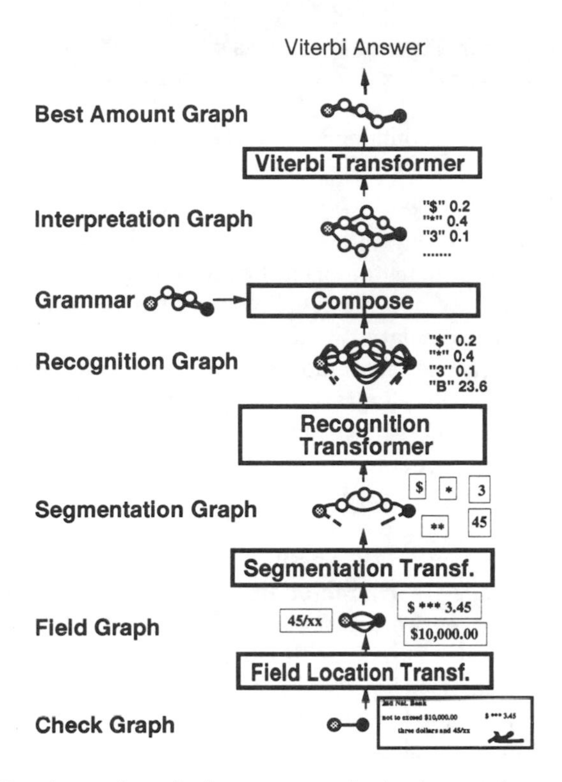

Fig. 33. A complete check amount reader implemented as a single cascade of Graph Transformer modules. Successive graph transformations progressively extract higher level information.

function is differentiable, therefore its parameter are globally tunable.

An arc may represent separate dollar and cent amounts as a sequence of fields. In fact, in handwritten checks, the cent amount may be written over a *fractional bar*, and not aligned at all with the dollar amount. In the worst case, one may find several cent amount candidates (above and below the fraction bar) for the same dollar amount.

The segmentation transformer T_{seg}, similar to the one described in Section VIII examines each zone contained in the field graph, and cuts each image into pieces of ink using heuristic image processing techniques. Each piece of ink may be a whole character or a piece of character. Each arc in the field graph is replaced by its corresponding segmentation graph that represents all possible groupings of pieces of ink. Each field segmentation graph is appended to an arc that contains the penalty of the field in the field graph. Each arc carries the segment image, together with a penalty that provides a first evaluation of the likelihood that the segment actually contains a character. This penalty is obtained with a differentiable function that combines a few simple features such as the space between the pieces of ink or the compliance of the segment image with a global baseline, and a few tunable parameters. The segmentation graph represents *all* the possible segmentations of *all* the field images. We can compute the penalty for one segmented field by adding the arc penalties along the corresponding path. As before using a differentiable function for computing the penalties will ensure that the parameters can be optimized globally.

The segmenter uses a variety of heuristics to find candidate cut. One of the most important ones is called "hit and deflect" [115]. The idea is to cast lines downward from the top of the field image. When a line hits a black pixel, it is deflected so as to follow the contour of the object. When a line hits a local minimum of the upper profile, i.e. when it cannot continue downward without crossing a black pixel, it is just propagated vertically downward through the ink. When two such lines meet each other, they are merged into a single cut. The procedure can be repeated from the bottom up. This strategy allows the separation of touching characters such as double zeros.

The recognition transformer T_{rec} iterates over all segment arcs in the segmentation graph and runs a character recognizer on the corresponding segment image. In our case, the recognizer is LeNet-5, the Convolutional Neural Network described in Section II, whose weights constitute the largest and most important subset of tunable parameters. The recognizer classifies segment images into one of 95 classes (full printable ASCII set) plus a rubbish class for unknown symbols or badly-formed characters. Each arc in the input graph T_{rec} is replaced by 96 arcs in the output graph. Each of those 96 arcs contains the label of one of the classes, and a penalty that is the sum of the penalty of the corresponding arc in the input (segmentation) graph and the penalty associated with classifying the image in the corresponding class, as computed by the recognizer. In other words, the recognition graph represents a weighted trellis of scored character classes. Each path in this graph represents a possible character string for the corresponding field. We can compute a penalty for this interpretation by adding the penalties along the path. This sequence of characters may or may not be a valid check amount.

The composition transformer T_{gram} selects the paths of the recognition graph that represent valid character sequences for check amounts. This transformer takes two graphs as input: the recognition graph, and the grammar graph. The grammar graph contains all possible sequences of symbols that constitute a well-formed amount. The output of the composition transformer, called the interpretation graph, contains all the paths in the recognition graph that are compatible with the grammar. The operation that combines the two input graphs to produce the output is a *generalized transduction* (see Section VIII).A differentiable function is used to compute the data attached to the output arc from the data attached to the input arcs. In our case, the output arc receives the class label of the two arcs, and a penalty computed by simply summing the penalties of the two input arcs (the recognizer penalty, and the arc penalty in the grammar graph). Each path in the interpretation graph represents one interpretation of one segmentation of one field on the check. The sum of the penalties along the path represents the "badness" of the corresponding interpretation and combines evidence from each of the modules along the process, as well as from the grammar.

The Viterbi transformer finally selects the path with the lowest accumulated penalty, corresponding to the best

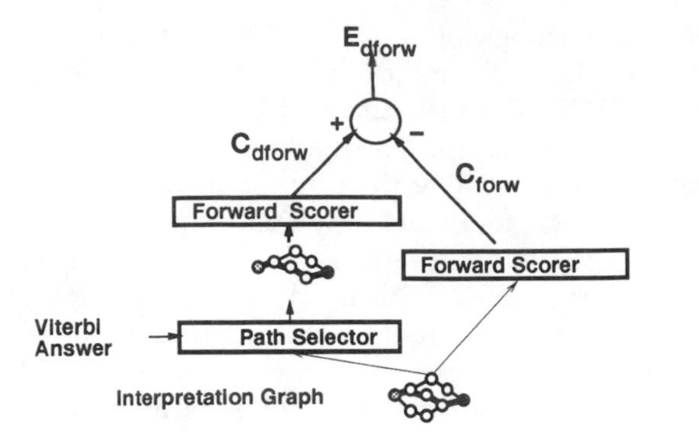

Fig. 34. Additional processing required to compute the confidence.

grammatically correct interpretations.

B. Gradient-Based Learning

Each stage of this check reading system contains tunable parameters. While some of these parameters could be manually adjusted, for example the parameters of the field locator and segmenter, the vast majority of them *must* be learned, particularly the weights of the neural net recognizer.

Prior to globally optimizing the system, each module parameters are initialized with reasonable values. The parameters of the field locator and the segmenter are initialized by hand, while the parameters of the neural net character recognizer are initialized by training on a database of pre-segmented and labeled characters. Then, the entire system is trained globally from whole check images labeled with the correct amount. No explicit segmentation of the amounts is needed to train the system: it is trained at the check level.

The loss function E minimized by our global training procedure is the Discriminative Forward criterion described in Section VI: the difference between (a) the forward penalty of the constrained interpretation graph (constrained by the correct label sequence), and (b) the forward penalty of the unconstrained interpretation graph. Derivatives can be back-propagated through the entire structure, although it only practical to do it down to the segmenter.

C. Rejecting Low Confidence Checks

In order to be able to reject checks which are the most likely to carry erroneous Viterbi answers, we must rate them with a *confidence*, and reject the check if this confidence is below a given threshold. To compare the unnormalized Viterbi Penalties of two different checks would be meaningless when it comes to decide which answer we trust the most.

The optimal measure of confidence is the probability of the Viterbi answer given the input image. As seen in Section VI-E, given a target sequence (which, in this case, would be the Viterbi answer), the *discriminative forward* loss function is an estimate of the logarithm of this probability. Therefore, a simple solution to obtain a good estimate of the confidence is to reuse the interpretation graph (see Figure 33) to compute the discriminative forward loss as described in Figure 21, using as our desired sequence the Viterbi answer. This is summarized in Figure 34, with:

$$\text{confidence} = \exp(E_{\text{dforw}})$$

D. Results

A version of the above system was fully implemented and tested on machine-print business checks. This system is basically a generic GTN engine with task specific heuristics encapsulated in the `check` and `fprop` method. As a consequence, the amount of code to write was minimal: mostly the adaptation of an earlier segmenter into the segmentation transformer. The system that deals with hand-written or personal checks was based on earlier implementations that used the GTN concept in a restricted way.

The neural network classifier was initially trained on 500,000 images of character images from various origins spanning the entire printable ASCII set. This contained both handwritten and machine-printed characters that had been previously size normalized at the string level. Additional images were generated by randomly distorting the original images using simple affine transformations of the images. The network was then further trained on character images that had been automatically segmented from check images and manually truthed. The network was also initially trained to reject non-characters that resulted from segmentation errors. The recognizer was then inserted in the check reading system and a small subset of the parameters were trained globally (at the field level) on whole check images.

On 646 business checks that were automatically categorized as *machine printed* the performance was 82% correctly recognized checks, 1% errors, and 17% rejects. This can be compared to the performance of the previous system on the same test set: 68% correct, 1% errors, and 31% rejects. A check is categorized as machine-printed when characters that are near a standard position Dollar sign are detected as machine printed, or when, if nothing is found in the standard position, at least one courtesy amount candidate is found somewhere else. The improvement is attributed to three main causes. First the neural network recognizer was bigger, and trained on more data. Second, because of the GTN architecture, the new system could take advantage of grammatical constraints in a much more efficient way than the previous system. Third, the GTN architecture provided extreme flexibility for testing heuristics, adjusting parameters, and tuning the system. This last point is more important than it seems. The GTN framework separates the "algorithmic" part of the system from the "knowledge-based" part of the system, allowing easy adjustments of the latter. The importance of global training was only minor in this task because the global training only concerned a small subset of the parameters.

An independent test performed by systems integrators in 1995 showed the superiority of this system over other commercial Courtesy amount reading systems. The system was integrated in NCR's line of check reading systems. It has been fielded in several banks across the US since June 1996, and has been reading millions of checks per day since then.

XI. Conclusions

During the short history of automatic pattern recognition, increasing the role of learning seems to have invariably improved the overall performance of recognition systems. The systems described in this paper are more evidence to this fact. Convolutional Neural Networks have been shown to eliminate the need for hand-crafted feature extractors. Graph Transformer Networks have been shown to reduce the need for hand-crafted heuristics, manual labeling, and manual parameter tuning in document recognition systems. As training data becomes plentiful, as computers get faster, as our understanding of learning algorithms improves, recognition systems will rely more and more of learning, and their performance will improve.

Just as the back-propagation algorithm elegantly solved the credit assignment problem in multi-layer neural networks, the gradient-based learning procedure for Graph Transformer Networks introduced in this paper solves the credit assignment problem in systems whose functional architecture dynamically changes with each new input. The learning algorithms presented here are in a sense nothing more than unusual forms of gradient descent in complex, dynamic architectures, with efficient back-propagation algorithms to compute the gradient. The results in this paper help establish the usefulness and relevance of gradient-based minimization methods as a general organizing principle for learning in large systems.

It was shown that all the steps of a document analysis system can be formulated as graph transformers through which gradients can be back-propagated. Even in the non-trainable parts of the system, the design philosophy in terms of graph transformation provides a clear separation between domain-specific heuristics (e.g. segmentation heuristics) and generic, procedural knowledge (the generalized transduction algorithm)

It is worth pointing out that *data generating models* (such as HMMs) and the *Maximum Likelihood Principle* were *not* called upon to justify most of the architectures and the training criteria described in this paper. Gradient based learning applied to global discriminative loss functions guarantees optimal classification and rejection without the use of "hard to justify" principles that put strong constraints on the system architecture, often at the expense of performances.

More specifically, the methods and architectures presented in this paper offer generic solutions to a large number of problems encountered in pattern recognition systems:

1. Feature extraction is traditionally a fixed transform, generally derived from some expert prior knowledge about the task. This relies on the probably incorrect assumption that the human designer is able to capture all the relevant information in the input. We have shown that the application of Gradient-Based Learning to Convolutional Neural Networks allows to learn appropriate features from examples. The success of this approach was demonstrated in extensive comparative digit recognition experiments on the NIST database.
2. Segmentation and recognition of objects in images cannot be completely decoupled. Instead of taking hard segmentation decisions too early, we have used Heuristic Over-Segmentation to generate and evaluate a large number of hypotheses in parallel, postponing any decision until the overall criterion is minimized.
3. Hand truthing images to obtain segmented characters for training a character recognizer is expensive and does not take into account the way in which a whole document or sequence of characters will be recognized (in particular the fact that some segmentation candidates may be wrong, even though they may look like true characters). Instead we train multi-module systems to optimize a global measure of performance, which does not require time consuming detailed hand-truthing, and yields significantly better recognition performance, because it allows to train these modules to cooperate towards a common goal.
4. Ambiguities inherent in the segmentation, character recognition, and linguistic model should be integrated optimally. Instead of using a sequence of task-dependent heuristics to combine these sources of information, we have proposed a unified framework in which generalized transduction methods are applied to graphs representing a weighted set of hypotheses about the input. The success of this approach was demonstrated with a commercially deployed check reading system that reads millions of business and personal checks per day: the generalized transduction engine resides in only a few hundred lines of code.
5. Traditional recognition systems rely on many hand-crafted heuristics to isolate individually recognizable objects. The promising Space Displacement Neural Network approach draws on the robustness and efficiency of Convolutional Neural Networks to avoid explicit segmentation altogether. Simultaneous automatic learning of segmentation and recognition can be achieved with Gradient-Based Learning methods.

This paper presents a small number of examples of graph transformer modules, but it is clear that the concept can be applied to many situations where the domain knowledge or the state information can be represented by graphs. This is the case in many audio signal recognition tasks, and visual scene analysis applications. Future work will attempt to apply Graph Transformer Networks to such problems, with the hope of allowing more reliance on automatic learning, and less on detailed engineering.

Appendices

A. Pre-conditions for faster convergence

As seen before, the squashing function used in our Convolutional Networks is $f(a) = A \tanh(Sa)$. Symmetric functions are believed to yield faster convergence, although the learning can become extremely slow if the weights are too small. The cause of this problem is that in weight space the origin is a fixed point of the learning dynamics, and, although it is a saddle point, it is attractive in almost all directions [116]. For our simulations, we use $A = 1.7159$ and $S = \frac{2}{3}$ (see [20], [34]). With this choice of parameters, the equalities $f(1) = 1$ and $f(-1) = -1$ are satisfied. The rationale behind this is that the overall gain of the squashing transformation is around 1 in normal operating conditions, and the interpretation of the state of the network is simplified. Moreover, the absolute value of the second derivative of f is a maximum at $+1$ and -1, which improves the convergence towards the end of the learning session. This particular choice of parameters is merely a convenience, and does not affect the result.

Before training, the weights are initialized with random values using a uniform distribution between $-2.4/F_i$ and $2.4/F_i$ where F_i is the number of inputs (fan-in) of the unit which the connection belongs to. Since several connections share a weight, this rule could be difficult to apply, but in our case, all connections sharing a same weight belong to units with identical fan-ins. The reason for dividing by the fan-in is that we would like the initial standard deviation of the weighted sums to be in the same range for each unit, and to fall within the normal operating region of the sigmoid. If the initial weights are too small, the gradients are very small and the learning is slow. If they are too large, the sigmoids are saturated and the gradient is also very small. The standard deviation of the weighted sum scales like the square root of the number of inputs when the inputs are independent, and it scales linearly with the number of inputs if the inputs are highly correlated. We chose to assume the second hypothesis since some units receive highly correlated signals.

B. Stochastic Gradient vs Batch Gradient

Gradient-Based Learning algorithms can use one of two classes of methods to update the parameters. The first method, dubbed "Batch Gradient", is the classical one: the gradients are accumulated over the entire training set, and the parameters are updated after the exact gradient has been so computed. In the second method, called "Stochastic Gradient", a partial, or noisy, gradient is evaluated on the basis of one single training sample (or a small number of samples), and the parameters are updated using this approximate gradient. The training samples can be selected randomly or according to a properly randomized sequence. In the stochastic version, the gradient estimates are noisy, but the parameters are updated much more often than with the batch version. An empirical result of considerable practical importance is that on tasks with large, redundant data sets, the stochastic version is considerably faster than the batch version, sometimes by orders of magnitude [117]. Although the reasons for this are not totally understood theoretically, an intuitive explanation can be found in the following extreme example. Let us take an example where the training database is composed of two copies of the same subset. Then accumulating the gradient over the whole set would cause redundant computations to be performed. On the other hand, running Stochastic Gradient once on this training set would amount to performing two complete learning iterations over the small subset. This idea can be generalized to training sets where there exist no precise repetition of the same pattern but where some redundancy is present. In fact stochastic update *must* be better when there is redundancy, i.e., when a certain level of generalization is expected.

Many authors have claimed that second-order methods should be used in lieu of gradient descent for neural net training. The literature abounds with recommendations [118] for classical second-order methods such as the Gauss-Newton or Levenberg-Marquardt algorithms, for Quasi-Newton methods such as the Broyden-Fletcher-Goldfarb-Shanno method (BFGS), Limited-storage BFGS, or for various versions of the Conjugate Gradients (CG) method. Unfortunately, all of the above methods are unsuitable for training large neural networks on large data sets. The Gauss-Newton and Levenberg-Marquardt methods require $O(N^3)$ operations per update, where N is the number of parameters, which makes them impractical for even moderate size networks. Quasi-Newton methods require "only" $O(N^2)$ operations per update, but that still makes them impractical for large networks. Limited-Storage BFGS and Conjugate Gradient require only $O(N)$ operations per update so they would appear appropriate. Unfortunately, their convergence speed relies on an accurate evaluation of successive "conjugate descent directions" which only makes sense in "batch" mode. For large data sets, the speed-up brought by these methods over regular batch gradient descent cannot match the enormous speed up brought by the use of stochastic gradient. Several authors have attempted to use Conjugate Gradient with small batches, or batches of increasing sizes [119], [120], but those attempts have not yet been demonstrated to surpass a carefully tuned stochastic gradient. Our experiments were performed with a stochastic method that scales the parameter axes so as to minimize the eccentricity of the error surface.

C. Stochastic Diagonal Levenberg-Marquardt

Owing to the reasons given in Appendix B, we prefer to update the weights after each presentation of a single pattern in accordance with stochastic update methods. The patterns are presented in a constant random order, and the training set is typically repeated 20 times.

Our update algorithm is dubbed the Stochastic Diagonal Levenberg-Marquardt method where an individual learning rate (step size) is computed for each parameter (weight) before each pass through the training set [20], [121], [34]. These learning rates are computed using the diagonal terms of an estimate of the Gauss-Newton approximation to the Hessian (second derivative) matrix. This algorithm is not believed to bring a tremendous increase in learning speed but it converges reliably without requiring extensive adjustments of the learning parameters. It corrects major ill-conditioning of the loss function that are due to the peculiarities of the network architecture and the training data. The additional cost of using this procedure over standard stochastic gradient descent is negligible.

At each learning iteration a particular parameter w_k is

updated according to the following stochastic update rule

$$w_k \leftarrow w_k - \epsilon_k \frac{\partial E^p}{\partial w_k}. \tag{18}$$

where E^p is the instantaneous loss function for pattern p. In Convolutional Neural Networks, because of the weight sharing, the partial derivative $\frac{\partial E^p}{\partial w_k}$ is the sum of the partial derivatives with respect to the connections that share the parameter w_k:

$$\frac{\partial E^p}{\partial w_k} = \sum_{(i,j)\in V_k} \frac{\partial E^p}{\partial u_{ij}} \tag{19}$$

where u_{ij} is the connection weight from unit j to unit i, V_k is the set of unit index pairs (i, j) such that the connection between i and j share the parameter w_k, i.e.:

$$u_{ij} = w_k \quad \forall (i,j) \in V_k \tag{20}$$

As stated previously, the step sizes ϵ_k are not constant but are function of the second derivative of the loss function along the axis w_k:

$$\epsilon_k = \frac{\eta}{\mu + h_{kk}} \tag{21}$$

where μ is a hand-picked constant and h_{kk} is an estimate of the second derivative of the loss function E with respect to w_k. The larger h_{kk}, the smaller the weight update. The parameter μ prevents the step size from becoming too large when the second derivative is small, very much like the "model-trust" methods, and the Levenberg-Marquardt methods in non-linear optimization [8]. The exact formula to compute h_{kk} from the second derivatives with respect to the connection weights is:

$$h_{kk} = \sum_{(i,j)\in V_k} \sum_{(k,l)\in V_k} \frac{\partial^2 E}{\partial u_{ij} \partial u_{kl}} \tag{22}$$

However, we make three approximations. The first approximation is to drop the off-diagonal terms of the Hessian with respect to the connection weights in the above equation:

$$h_{kk} = \sum_{(i,j)\in V_k} \frac{\partial^2 E}{\partial u_{ij}^2} \tag{23}$$

Naturally, the terms $\frac{\partial^2 E}{\partial u_{ij}^2}$ are the average over the training set of the local second derivatives:

$$\frac{\partial^2 E}{\partial u_{ij}^2} = \frac{1}{P} \sum_{p=1}^{P} \frac{\partial^2 E^p}{\partial u_{ij}^2} \tag{24}$$

Those local second derivatives with respect to connection weights can be computed from local second derivatives with respect to the total input of the downstream unit:

$$\frac{\partial^2 E^p}{\partial u_{ij}^2} = \frac{\partial^2 E^p}{\partial a_i^2} x_j^2 \tag{25}$$

where x_j is the state of unit j and $\frac{\partial^2 E^p}{\partial a_i^2}$ is the second derivative of the instantaneous loss function with respect to the total input to unit i (denoted a_i). Interestingly, there is an efficient algorithm to compute those second derivatives which is very similar to the back-propagation procedure used to compute the first derivatives [20], [121]:

$$\frac{\partial^2 E^p}{\partial a_i^2} = f'(a_i)^2 \sum_k u_{ki}^2 \frac{\partial^2 E^p}{\partial a_k^2} + f''(a_i) \frac{\partial E^p}{\partial x_i} \tag{26}$$

Unfortunately, using those derivatives leads to well-known problems associated with every Newton-like algorithm: these terms can be negative, and can cause the gradient algorithm to move uphill instead of downhill. Therefore, our second approximation is a well-known trick, called the Gauss-Newton approximation, which guarantees that the second derivative estimates are non-negative. The Gauss-Newton approximation essentially ignores the non-linearity of the estimated function (the Neural Network in our case), but not that of the loss function. The back-propagation equation for Gauss-Newton approximations of the second derivatives is:

$$\frac{\partial^2 E^p}{\partial a_i^2} = f'(a_i)^2 \sum_k u_{ki}^2 \frac{\partial^2 E^p}{\partial a_k^2} \tag{27}$$

This is very similar to the formula for back-propagating the first derivatives, except that the sigmoid's derivative and the weight values are squared. The right-hand side is a sum of products of non-negative terms, therefore the left-hand side term is non-negative.

The third approximation we make is that we do not run the average in Equation 24 over the entire training set, but run it on a small subset of the training set instead. In addition the re-estimation does not need to be done often since the second order properties of the error surface change rather slowly. In the experiments described in this paper, we re-estimate the h_{kk} on 500 patterns before each training pass through the training set. Since the size of the training set is 60,000, the additional cost of re-estimating the h_{kk} is negligible. The estimates are not particularly sensitive to the particular subset of the training set used in the averaging. This seems to suggest that the second-order properties of the error surface are mainly determined by the structure of the network, rather than by the detailed statistics of the samples. This algorithm is particularly useful for shared-weight networks because the weight sharing creates ill-conditionning of the error surface. Because of the sharing, one single parameter in the first few layers can have an enormous influence on the output. Consequently, the second derivative of the error with respect to this parameter may be very large, while it can be quite small for other parameters elsewhere in the network. The above algorithm compensates for that phenomenon.

Unlike most other second-order acceleration methods for back-propagation, the above method works in stochastic mode. It uses a diagonal approximation of the Hessian. Like the classical Levenberg-Marquardt algorithm, it uses a "safety" factor μ to prevent the step sizes from getting too large if the second derivative estimates are small. Hence the method is called the Stochastic Diagonal Levenberg-Marquardt method.

Acknowledgments

Some of the systems described in this paper is the work of many researchers now at AT&T, and Lucent Technologies. In particular, Christopher Burges, Craig Nohl, Troy Cauble and Jane Bromley contributed much to the check reading system. Experimental results described in section III include contributions by Chris Burges, Aymeric Brunot, Corinna Cortes, Harris Drucker, Larry Jackel, Urs Müller, Bernhard Schölkopf, and Patrice Simard. The authors wish to thank Fernando Pereira, Vladimir Vapnik, John Denker, and Isabelle Guyon for helpful discussions, Charles Stenard and Ray Higgins for providing the applications that motivated some of this work, and Lawrence R. Rabiner and Lawrence D. Jackel for relentless support and encouragements.

References

[1] R. O. Duda and P. E. Hart, *Pattern Classification And Scene Analysis*, Wiley and Son, 1973.

[2] Y. LeCun, B. Boser, J. S. Denker, D. Henderson, R. E. Howard, W. Hubbard, and L. D. Jackel, "Backpropagation applied to handwritten zip code recognition," *Neural Computation*, vol. 1, no. 4, pp. 541–551, Winter 1989.

[3] S. Seung, H. Sompolinsky, and N. Tishby, "Statistical mechanics of learning from examples," *Physical Review A*, vol. 45, pp. 6056–6091, 1992.

[4] V. N. Vapnik, E. Levin, and Y. LeCun, "Measuring the vc-dimension of a learning machine," *Neural Computation*, vol. 6, no. 5, pp. 851–876, 1994.

[5] C. Cortes, L. Jackel, S. Solla, V. N. Vapnik, and J. Denker, "Learning curves: asymptotic values and rate of convergence," in *Advances in Neural Information Processing Systems 6*, J. D. Cowan, G. Tesauro, and J. Alspector, Eds., San Mateo, CA, 1994, pp. 327–334, Morgan Kaufmann.

[6] V. N. Vapnik, *The Nature of Statistical Learning Theory*, Springer, New-York, 1995.

[7] V. N. Vapnik, *Statistical Learning Theory*, John Wiley & Sons, New-York, 1998.

[8] W. H. Press, B. P. Flannery, S. A. Teukolsky, and W. T. Vetterling, *Numerical Recipes: The Art of Scientific Computing*, Cambridge University Press, Cambridge, 1986.

[9] S. I. Amari, "A theory of adaptive pattern classifiers," *IEEE Transactions on Electronic Computers*, vol. EC-16, pp. 299–307, 1967.

[10] Ya. Tsypkin, *Adaptation and Learning in automatic systems*, Academic Press, 1971.

[11] Ya. Tsypkin, *Foundations of the theory of learning systems*, Academic Press, 1973.

[12] M. Minsky and O. Selfridge, "Learning in random nets," in *4th London symposium on Information Theory*, London, 1961, pp. 335–347.

[13] D. H. Ackley, G. E. Hinton, and T. J. Sejnowski, "A learning algorithm for boltzmann machines," *Cognitive Science*, vol. 9, pp. 147–169, 1985.

[14] G. E. Hinton and T. J. Sejnowski, "Learning and relearning in Boltzmann machines," in *Parallel Distributed Processing: Explorations in the Microstructure of Cognition. Volume 1: Foundations*, D. E. Rumelhart and J. L. McClelland, Eds. MIT Press, Cambridge, MA, 1986.

[15] D. E. Rumelhart, G. E. Hinton, and R. J. Williams, "Learning internal representations by error propagation," in *Parallel distributed processing: Explorations in the microstructure of cognition*, vol. I, pp. 318–362. Bradford Books, Cambridge, MA, 1986.

[16] A. E. Jr. Bryson and Yu-Chi Ho, *Applied Optimal Control*, Blaisdell Publishing Co., 1969.

[17] Y. LeCun, "A learning scheme for asymmetric threshold networks," in *Proceedings of Cognitiva 85*, Paris, France, 1985, pp. 599–604.

[18] Y. LeCun, "Learning processes in an asymmetric threshold network," in *Disordered systems and biological organization*, E. Bienenstock, F. Fogelman-Soulié, and G. Weisbuch, Eds., Les Houches, France, 1986, pp. 233–240, Springer-Verlag.

[19] D. B. Parker, "Learning-logic," Tech. Rep., TR-47, Sloan School of Management, MIT, Cambridge, Mass., April 1985.

[20] Y. LeCun, *Modèles connexionnistes de l'apprentissage (connectionist learning models)*, Ph.D. thesis, Université P. et M. Curie (Paris 6), June 1987.

[21] Y. LeCun, "A theoretical framework for back-propagation," in *Proceedings of the 1988 Connectionist Models Summer School*, D. Touretzky, G. Hinton, and T. Sejnowski, Eds., CMU, Pittsburgh, Pa, 1988, pp. 21–28, Morgan Kaufmann.

[22] L. Bottou and P. Gallinari, "A framework for the cooperation of learning algorithms," in *Advances in Neural Information Processing Systems*, D. Touretzky and R. Lippmann, Eds., Denver, 1991, vol. 3, Morgan Kaufmann.

[23] C. Y. Suen, C. Nadal, R. Legault, T. A. Mai, and L. Lam, "Computer recognition of unconstrained handwritten numerals," *Proceedings of the IEEE, Special issue on Optical Character Recognition*, vol. 80, no. 7, pp. 1162–1180, July 1992.

[24] S. N. Srihari, "High-performance reading machines," *Proceedings of the IEEE, Special issue on Optical Character Recognition*, vol. 80, no. 7, pp. 1120–1132, July 1992.

[25] Y. LeCun, L. D. Jackel, B. Boser, J. S. Denker, H. P. Graf, I. Guyon, D. Henderson, R. E. Howard, and W. Hubbard, "Handwritten digit recognition: Applications of neural net chips and automatic learning," *IEEE Communication*, pp. 41–46, November 1989, invited paper.

[26] J. Keeler, D. Rumelhart, and W. K. Leow, "Integrated segmentation and recognition of hand-printed numerals," in *Neural Information Processing Systems*, R. P. Lippmann, J. M. Moody, and D. S. Touretzky, Eds., vol. 3, pp. 557–563. Morgan Kaufmann Publishers, San Mateo, CA, 1991.

[27] Ofer Matan, Christopher J. C. Burges, Yann LeCun, and John S. Denker, "Multi-digit recognition using a space displacement neural network," in *Neural Information Processing Systems*, J. M. Moody, S. J. Hanson, and R. P. Lippman, Eds. 1992, vol. 4, Morgan Kaufmann Publishers, San Mateo, CA.

[28] L. R. Rabiner, "A tutorial on hidden Markov models and selected applications in speech recognition," *Proceedings of the IEEE*, vol. 77, no. 2, pp. 257–286, February 1989.

[29] H. A. Bourlard and N. Morgan, *CONNECTIONIST SPEECH RECOGNITION: A Hybrid Approach*, Kluwer Academic Publisher, Boston, 1994.

[30] D. H. Hubel and T. N. Wiesel, "Receptive fields, binocular interaction, and functional architecture in the cat's visual cortex," *Journal of Physiology (London)*, vol. 160, pp. 106–154, 1962.

[31] K. Fukushima, "Cognitron: A self-organizing multilayered neural network," *Biological Cybernetics*, vol. 20, no. 6, pp. 121–136, November 1975.

[32] K. Fukushima and S. Miyake, "Neocognitron: A new algorithm for pattern recognition tolerant of deformations and shifts in position," *Pattern Recognition*, vol. 15, pp. 455–469, 1982.

[33] M. C. Mozer, *The perception of multiple objects: A connectionist approach*, MIT Press-Bradford Books, Cambridge, MA, 1991.

[34] Y. LeCun, "Generalization and network design strategies," in *Connectionism in Perspective*, R. Pfeifer, Z. Schreter, F. Fogelman, and L. Steels, Eds., Zurich, Switzerland, 1989, Elsevier, an extended version was published as a technical report of the University of Toronto.

[35] Y. LeCun, B. Boser, J. S. Denker, D. Henderson, R. E. Howard, W. Hubbard, and L. D. Jackel, "Handwritten digit recognition with a back-propagation network," in *Advances in Neural Information Processing Systems 2 (NIPS*89)*, David Touretzky, Ed., Denver, CO, 1990, Morgan Kaufmann.

[36] G. L. Martin, "Centered-object integrated segmentation and recognition of overlapping hand-printed characters," *Neural Computation*, vol. 5, no. 3, pp. 419–429, 1993.

[37] J. Wang and J Jean, "Multi-resolution neural networks for omnifont character recognition," in *Proceedings of International Conference on Neural Networks*, 1993, vol. III, pp. 1588–1593.

[38] Y. Bengio, Y. LeCun, C. Nohl, and C. Burges, "Lerec: A NN/HMM hybrid for on-line handwriting recognition," *Neural Computation*, vol. 7, no. 5, 1995.

[39] S. Lawrence, C. Lee Giles, A. C. Tsoi, and A. D. Back, "Face recognition: A convolutional neural network approach," *IEEE*

Transactions on Neural Networks, vol. 8, no. 1, pp. 98–113, 1997.

[40] K. J. Lang and G. E. Hinton, "A time delay neural network architecture for speech recognition," Tech. Rep. CMU-CS-88-152, Carnegie-Mellon University, Pittsburgh PA, 1988.

[41] A. H. Waibel, T. Hanazawa, G. Hinton, K. Shikano, and K. Lang, "Phoneme recognition using time-delay neural networks," *IEEE Transactions on Acoustics, Speech and Signal Processing*, vol. 37, pp. 328–339, March 1989.

[42] L. Bottou, F. Fogelman, P. Blanchet, and J. S. Lienard, "Speaker independent isolated digit recognition: Multilayer perceptron vs dynamic time warping," *Neural Networks*, vol. 3, pp. 453–465, 1990.

[43] P. Haffner and A. H. Waibel, "Time-delay neural networks embedding time alignment: a performance analysis," in *EUROSPEECH'91, 2nd European Conference on Speech Communication and Technology*, Genova, Italy, Sept. 1991.

[44] I. Guyon, P. Albrecht, Y. LeCun, J. S. Denker, and W. Hubbard, "Design of a neural network character recognizer for a touch terminal," *Pattern Recognition*, vol. 24, no. 2, pp. 105–119, 1991.

[45] J. Bromley, J. W. Bentz, L. Bottou, I. Guyon, Y. LeCun, C. Moore, E. Säckinger, and R. Shah, "Signature verification using a siamese time delay neural network," *International Journal of Pattern Recognition and Artificial Intelligence*, vol. 7, no. 4, pp. 669–687, August 1993.

[46] Y. LeCun, I. Kanter, and S. Solla, "Eigenvalues of covariance matrices: application to neural-network learning," *Physical Review Letters*, vol. 66, no. 18, pp. 2396–2399, May 1991.

[47] T. G. Dietterich and G. Bakiri, "Solving multiclass learning problems via error-correcting output codes.," *Journal of Artificial Intelligence Research*, vol. 2, pp. 263–286, 1995.

[48] L. R. Bahl, P. F. Brown, P. V. de Souza, and R. L. Mercer, "Maximum mutual information of hidden Markov model parameters for speech recognition," in *Proc. Int. Conf. Acoust., Speech, Signal Processing*, 1986, pp. 49–52.

[49] L. R. Bahl, P. F. Brown, P. V. de Souza, and R. L. Mercer, "Speech recognition with continuous-parameter hidden Markov models," *Computer, Speech and Language*, vol. 2, pp. 219–234, 1987.

[50] B. H. Juang and S. Katagiri, "Discriminative learning for minimum error classification," *IEEE Trans. on Acoustics, Speech, and Signal Processing*, vol. 40, no. 12, pp. 3043–3054, December 1992.

[51] Y. LeCun, L. D. Jackel, L. Bottou, A. Brunot, C. Cortes, J. S. Denker, H. Drucker, I. Guyon, U. A. Muller, E. Säckinger, P. Simard, and V. N. Vapnik, "Comparison of learning algorithms for handwritten digit recognition," in *International Conference on Artificial Neural Networks*, F. Fogelman and P. Gallinari, Eds., Paris, 1995, pp. 53–60, EC2 & Cie.

[52] I Guyon, I. Poujaud, L. Personnaz, G. Dreyfus, J. Denker, and Y. LeCun, "Comparing different neural net architectures for classifying handwritten digits," in *Proc. of IJCNN, Washington DC*. 1989, vol. II, pp. 127–132, IEEE.

[53] R. Ott, "construction of quadratic polynomial classifiers," in *Proc. of International Conference on Pattern Recognition*. 1976, pp. 161–165, IEEE.

[54] J. Schürmann, "A multi-font word recognition system for postal address reading," *IEEE Transactions on Computers*, vol. C-27, no. 8, pp. 721–732, August 1978.

[55] Y. Lee, "Handwritten digit recognition using k-nearest neighbor, radial-basis functions, and backpropagation neural networks," *Neural Computation*, vol. 3, no. 3, pp. 440–449, 1991.

[56] D. Saad and S. A. Solla, "Dynamics of on-line gradient descent learning for multilayer neural networks," in *Advances in Neural Information Processing Systems*, David S. Touretzky, Michael C. Mozer, and Michael E. Hasselmo, Eds. 1996, vol. 8, pp. 302–308, The MIT Press, Cambridge.

[57] G. Cybenko, "Approximation by superpositions of sigmoidal functions," *Mathematics of Control, Signals, and Systems*, vol. 2, no. 4, pp. 303–314, 1989.

[58] L. Bottou and V. N. Vapnik, "Local learning algorithms," *Neural Computation*, vol. 4, no. 6, pp. 888–900, 1992.

[59] R. E. Schapire, "The strength of weak learnability," *Machine Learning*, vol. 5, no. 2, pp. 197–227, 1990.

[60] H. Drucker, R. Schapire, and P. Simard, "Improving performance in neural networks using a boosting algorithm," in *Advances in Neural Information Processing Systems 5*, S. J. Hanson, J. D. Cowan, and C. L. Giles, Eds., San Mateo, CA, 1993, pp. 42–49, Morgan Kaufmann.

[61] P. Simard, Y. LeCun, and Denker J., "Efficient pattern recognition using a new transformation distance," in *Advances in Neural Information Processing Systems*, S. Hanson, J. Cowan, and L. Giles, Eds., vol. 5. Morgan Kaufmann, 1993.

[62] B. Boser, I. Guyon, and V. Vapnik, "A training algorithm for optimal margin classifiers," in *Proceedings of the Fifth Annual Workshop on Computational Learning Theory*, 1992, vol. 5, pp. 144–152.

[63] C. J. C. Burges and B. Schoelkopf, "Improving the accuracy and speed of support vector machines," in *Advances in Neural Information Processing Systems 9*, M. Jordan M. Mozer and T. Petsche, Eds. 1997, The MIT Press, Cambridge.

[64] Eduard Säckinger, Bernhard Boser, Jane Bromley, Yann LeCun, and Lawrence D. Jackel, "Application of the ANNA neural network chip to high-speed character recognition," *IEEE Transaction on Neural Networks*, vol. 3, no. 2, pp. 498–505, March 1992.

[65] J. S. Bridle, "Probabilistic interpretation of feedforward classification networks outputs, with relationship to statistical pattern recognition," in *Neurocomputing, Algorithms, Architectures and Applications*, F. Fogelman, J. Herault, and Y. Burnod, Eds., Les Arcs, France, 1989, Springer.

[66] Y. LeCun, L. Bottou, and Y. Bengio, "Reading checks with graph transformer networks," in *International Conference on Acoustics, Speech, and Signal Processing*, Munich, 1997, vol. 1, pp. 151–154, IEEE.

[67] Y. Bengio, *Neural Networks for Speech and Sequence Recognition*, International Thompson Computer Press, London, UK, 1996.

[68] C. Burges, O. Matan, Y. LeCun, J. Denker, L. Jackel, C. Stenard, C. Nohl, and J. Ben, "Shortest path segmentation: A method for training a neural network to recognize character strings," in *International Joint Conference on Neural Networks*, Baltimore, 1992, vol. 3, pp. 165–172.

[69] T. M. Breuel, "A system for the off-line recognition of handwritten text," in *ICPR'94*, IEEE, Ed., Jerusalem 1994, 1994, pp. 129–134.

[70] A. Viterbi, "Error bounds for convolutional codes and an asymptotically optimum decoding algorithm," *IEEE Transactions on Information Theory*, pp. 260–269, April 1967.

[71] Lippmann R. P. and Gold B., "Neural-net classifiers useful for speech recognition," in *Proceedings of the IEEE First International Conference on Neural Networks*, San Diego, June 1987, pp. 417–422.

[72] H. Sakoe, R. Isotani, K. Yoshida, K. Iso, and T. Watanabe, "Speaker-independent word recognition using dynamic programming neural networks," in *International Conference on Acoustics, Speech, and Signal Processing*, Glasgow, 1989, pp. 29–32.

[73] J. S. Bridle, "Alphanets: a recurrent 'neural' network architecture with a hidden markov model interpretation," *Speech Communication*, vol. 9, no. 1, pp. 815–819, 1990.

[74] M. A. Franzini, K. F. Lee, and A. H. Waibel, "Connectionist viterbi training: a new hybrid method for continuous speech recognition," in *International Conference on Acoustics, Speech, and Signal Processing*, Albuquerque, NM, 1990, pp. 425–428.

[75] L. T. Niles and H. F. Silverman, "Combining hidden markov models and neural network classifiers," in *International Conference on Acoustics, Speech, and Signal Processing*, Albuquerque, NM, 1990, pp. 417–420.

[76] X. Driancourt and L. Bottou, "MLP, LVQ and DP: Comparison & cooperation," in *Proceedings of the International Joint Conference on Neural Networks*, Seattle, 1991, vol. 2, pp. 815–819.

[77] Y. Bengio, R. De Mori, G. Flammia, and R. Kompe, "Global optimization of a neural network-hidden Markov model hybrid," *IEEE Transactions on Neural Networks*, vol. 3, no. 2, pp. 252–259, 1992.

[78] P. Haffner and A. H. Waibel, "Multi-state time-delay neural networks for continuous speech recognition," in *Advances in Neural Information Processing Systems*. 1992, vol. 4, pp. 579–588, Morgan Kaufmann, San Mateo.

[79] Y. Bengio, , P. Simard, and P. Frasconi, "Learning long-term dependencies with gradient descent is difficult," *IEEE Transactions on Neural Networks*, vol. 5, no. 2, pp. 157–166, March 1994, Special Issue on Recurrent Neural Network.

[80] T. Kohonen, G. Barna, and R. Chrisley, "Statistical pattern recognition with neural network: Benchmarking studies," in *Proceedings of the IEEE Second International Conference on Neural Networks*, San Diego, 1988, vol. 1, pp. 61–68.

[81] P. Haffner, "Connectionist speech recognition with a global MMI algorithm," in *EUROSPEECH'93, 3rd European Conference on Speech Communication and Technology*, Berlin, Sept. 1993.

[82] J. S. Denker and C. J. Burges, "Image segmentation and recognition," in *The Mathematics of Induction*. 1995, Addison Wesley.

[83] L. Bottou, *Une Approche théorique de l'Apprentissage Connexionniste: Applications à la Reconnaissance de la Parole*, Ph.D. thesis, Université de Paris XI, 91405 Orsay cedex, France, 1991.

[84] M. Rahim, Y. Bengio, and Y. LeCun, "Discriminative feature and model design for automatic speech recognition," in *Proc. of Eurospeech*, Rhodes, Greece, 1997.

[85] U. Bodenhausen, S. Manke, and A. Waibel, "Connectionist architectural learning for high performance character and speech recognition," in *International Conference on Acoustics, Speech, and Signal Processing*, Minneapolis, 1993, vol. 1, pp. 625–628.

[86] F. Pereira, M. Riley, and R. Sproat, "Weighted rational transductions and their application to human language processing," in *ARPA Natural Language Processing workshop*, 1994.

[87] M. Lades, J. C. Vorbrüggen, J. Buhmann, and C. von der Malsburg, "Distortion invariant object recognition in the dynamic link architecture," *IEEE Trans. Comp.*, vol. 42, no. 3, pp. 300–311, 1993.

[88] B. Boser, E. Säckinger, J. Bromley, Y. LeCun, and L. Jackel, "An analog neural network processor with programmable topology," *IEEE Journal of Solid-State Circuits*, vol. 26, no. 12, pp. 2017–2025, December 1991.

[89] M. Schenkel, H. Weissman, I. Guyon, C. Nohl, and D. Henderson, "Recognition-based segmentation of on-line hand-printed words," in *Advances in Neural Information Processing Systems 5*, S. J. Hanson, J. D. Cowan, and C. L. Giles, Eds., Denver, CO, 1993, pp. 723–730.

[90] C. Dugast, L. Devillers, and X. Aubert, "Combining TDNN and HMM in a hybrid system for improved continuous-speech recognition," *IEEE Transactions on Speech and Audio Processing*, vol. 2, no. 1, pp. 217–224, 1994.

[91] Ofer Matan, Henry S. Baird, Jane Bromley, Christopher J. C. Burges, John S. Denker, Lawrence D. Jackel, Yann Le Cun, Edwin P. D. Pednault, William D. Satterfield, Charles E. Stenard, and Timothy J. Thompson, "Reading handwritten digits: A ZIP code recognition system," *Computer*, vol. 25, no. 7, pp. 59–62, July 1992.

[92] Y. Bengio and Y. Le Cun, "Word normalization for on-line handwritten word recognition," in *Proc. of the International Conference on Pattern Recognition*, IAPR, Ed., Jerusalem, 1994, IEEE.

[93] R. Vaillant, C. Monrocq, and Y. LeCun, "Original approach for the localization of objects in images," *IEE Proc on Vision, Image, and Signal Processing*, vol. 141, no. 4, pp. 245–250, August 1994.

[94] R. Wolf and J. Platt, "Postal address block location using a convolutional locator network," in *Advances in Neural Information Processing Systems 6*, J. D. Cowan, G. Tesauro, and J. Alspector, Eds. 1994, pp. 745–752, Morgan Kaufmann Publishers, San Mateo, CA.

[95] S. Nowlan and J. Platt, "A convolutional neural network hand tracker," in *Advances in Neural Information Processing Systems 7*, G. Tesauro, D. Touretzky, and T. Leen, Eds., San Mateo, CA, 1995, pp. 901–908, Morgan Kaufmann.

[96] H. A. Rowley, S. Baluja, and T. Kanade, "Neural network-based face detection," in *Proceedings of CVPR'96*. 1996, pp. 203–208, IEEE Computer Society Press.

[97] E. Osuna, R. Freund, and F. Girosi, "Training support vector machines: an application to face detection," in *Proceedings of CVPR'96*. 1997, pp. 130–136, IEEE Computer Society Press.

[98] H. Bourlard and C. J. Wellekens, "Links between Markov models and multilayer perceptrons," in *Advances in Neural Information Processing Systems*, D. Touretzky, Ed., Denver, 1989, vol. 1, pp. 186–187, Morgan-Kaufmann.

[99] Y. Bengio, R. De Mori, G. Flammia, and R. Kompe, "Neural network - gaussian mixture hybrid for speech recognition or density estimation," in *Advances in Neural Information Processing Systems 4*, J. E. Moody, S. J. Hanson, and R. P. Lippmann, Eds., Denver, CO, 1992, pp. 175–182, Morgan Kaufmann.

[100] F. C. N. Pereira and M. Riley, "Speech recognition by composition of weighted finite automata," in *Finite-State Devices for Natural Langue Processing*, Cambridge, Massachusetts, 1997, MIT Press.

[101] M. Mohri, "Finite-state transducers in language and speech processing," *Computational Linguistics*, vol. 23, no. 2, pp. 269–311, 1997.

[102] I. Guyon, M. Schenkel, and J. Denker, "Overview and synthesis of on-line cursive handwriting recognition techniques," in *Handbook on Optical Character Recognition and Document Image Analysis*, P. S. P. Wang and Bunke H., Eds. 1996, World Scientific.

[103] M. Mohri and M. Riley, "Weighted determinization and minimization for large vocabulary recognition," in *Proceedings of Eurospeech '97*, Rhodes, Greece, September 1997, pp. 131–134.

[104] Y. Bengio and P. Frasconi, "An input/output HMM architecture," in *Advances in Neural Information Processing Systems*, G. Tesauro, D Touretzky, and T. Leen, Eds. 1996, vol. 7, pp. 427–434, MIT Press, Cambridge, MA.

[105] Y. Bengio and P. Frasconi, "Input/Output HMMs for sequence processing," *IEEE Transactions on Neural Networks*, vol. 7, no. 5, pp. 1231–1249, 1996.

[106] M. Mohri, F. C. N. Pereira, and M. Riley, *A rational design for a weighted finite-state transducer library*, Lecture Notes in Computer Science. Springer Verlag, 1997.

[107] M. Rahim, C. H. Lee, and B. H. Juang, "Discriminative utterance verification for connected digits recognition," *IEEE Trans. on Speech & Audio Proc.*, vol. 5, pp. 266–277, 1997.

[108] M. Rahim, Y. Bengio, and Y. LeCun, "Discriminative feature and model design for automatic speech recognition," in *Eurospeech '97*, Rhodes, Greece, 1997, pp. 75–78.

[109] S. Bengio and Y. Bengio, "An EM algorithm for asynchronous input/output hidden Markov models," in *International Conference On Neural Information Processing*, L. Xu, Ed., Hong-Kong, 1996, pp. 328–334.

[110] C. Tappert, C. Suen, and T. Wakahara, "The state of the art in on-line handwriting recognition," *IEEE Transactions on Pattern Analysis and Machine Intelligence*, vol. 8, no. 12, pp. 787–808, 1990.

[111] S. Manke and U. Bodenhausen, "A connectionist recognizer for on-line cursive handwriting recognition," in *International Conference on Acoustics, Speech, and Signal Processing*, Adelaide, 1994, vol. 2, pp. 633–636.

[112] M. Gilloux and M. Leroux, "Recognition of cursive script amounts on postal checks," in *European Conference dedicated to Postal Technologies*, Nantes, France, June 1993, pp. 705–712.

[113] D. Guillevic and C. Y. Suen, "Cursive script recognition applied to the processing of bank checks," in *Int. Conf. on Document Analysis and Recognition*, Montreal, Canada, August 1995, pp. 11–14.

[114] L. Lam, C. Y. Suen, D. Guillevic, N. W. Strathy, M. Cheriet, K. Liu, and J. N. Said, "Automatic processing of information on checks," in *Int. Conf. on Systems, Man & Cybernetics*, Vancouver, Canada, October 1995, pp. 2353–2358.

[115] C. J. C. Burges, J. I. Ben, J. S. Denker, Y. LeCun, and C. R. Nohl, "Off line recognition of handwritten postal words using neural networks," *Int. Journal of Pattern Recognition and Artificial Intelligence*, vol. 7, no. 4, pp. 689, 1993, Special Issue on Applications of Neural Networks to Pattern Recognition (I. Guyon Ed.).

[116] Y. LeCun, Y. Bengio, D. Henderson, A. Weisbuch, H. Weissman, and Jackel. L., "On-line handwriting recognition with neural networks: spatial representation versus temporal representation.," in *Proc. International Conference on handwriting and drawing*. 1993, Ecole Nationale Superieure des Telecommunications.

[117] U. Müller, A. Gunzinger, and W. Guggenbühl, "Fast neural net simulation with a DSP processor array," *IEEE Trans. on Neural Networks*, vol. 6, no. 1, pp. 203–213, 1995.

[118] R. Battiti, "First- and second-order methods for learning: Between steepest descent and newton's method.," *Neural Computation*, vol. 4, no. 2, pp. 141–166, 1992.

[119] A. H. Kramer and A. Sangiovanni-Vincentelli, "Efficient parallel learning algorithms for neural networks," in *Advances in Neural Information Processing Systems*, D.S. Touretzky, Ed.,

Denver 1988, 1989, vol. 1, pp. 40–48, Morgan Kaufmann, San Mateo.

[120] M. Moller, *Efficient Training of Feed-Forward Neural Networks*, Ph.D. thesis, Aarhus University, Aarhus, Denmark, 1993.

[121] S. Becker and Y. LeCun, "Improving the convergence of back-propagation learning with second-order methods," Tech. Rep. CRG-TR-88-5, University of Toronto Connectionist Research Group, September 1988.

Chapter 10

PATTERN RECOGNITION USING A FAMILY OF DESIGN ALGORITHMS BASED UPON THE GENERALIZED PROBABILISTIC DESCENT METHOD

SHIGERU KATAGIRI

ATR Human Information Processing Research Laboratories
2-2 Hikaridai, Seika-cho, Soraku-gun, Kyoto 619-02, Janan
Email: katagiri@hip.atr.co.jp

BIING-HWANG JUANG

Bell Laboratories, Lucent Technologies
600-700 Mountain Avenue, Murray Hill, NJ 07974-0636, USA
Email: bjuang@lucent.com

and

CHIN-HUI LEE

Bell Laboratories, Lucent Technologies
600-700 Mountain Avenue, Murray Hill, NJ 07974-0636, USA
Email: chl@research.bell-labs.com

Abstract

This paper provides a comprehensive introduction to a novel approach to pattern recognition, which is based on the Generalized Probabilistic Descent method (GPD) and its related design algorithms. The paper contains a survey of recent recognizer design techniques, the formulation of GPD, the concept of Minimum Classification Error learning that is closely related to the GPD formalization, a relational analysis between GPD and other important design methods, and various embodiments of GPD-based design, including Segmental-GPD, Minimum Spotting Error training, Discriminative Utterance Verification, and Discriminative Feature Extraction. GPD development has its origins in basic pattern recognition and Bayes decision theory. It represents a simple but careful reinvestigation of the classical theory and successfully leads to an innovative framework. For clarity of presentation, detailed discussions about its embodiments are provided **for examples of speech pattern recognition tasks that use** a distance-based classifier. Experimental results in speech pattern recognition tasks clearly demonstrate the remarkable utility of the family of GPD-based design algorithms.

Keywords: Bayes decision theory, Discriminant function approach, Discriminative feature extraction, Discriminative training, Generalized probabilistic descent method, Minimum classification error learning, Pattern recognition, Speech recognition

1. Introduction

Pattern recognition has long been a topic of fundamental importance in a wide range of science and technology. In these days of rapidly-growing information-oriented societies, improvement in its performance is an urgent technological issue. In particular, a mathematically-proven, effective and efficient method is desired for designing highly-accurate recognizers. As one of the solutions for meeting this need, a discriminative training method called the Generalized Probabilistic Descent method (GPD) was developed for classifier design [54]. This method has been shown very useful in various speech pattern classification tasks. Since its development it has been deeply analyzed and further extended to a more general methodological framework for pattern recognition (the terminological difference between "classification" and "recognition" will be shown later.). This paper is therefore devoted to providing a comprehensive review of the GPD-based approach to pattern recognition.

GPD is a general pattern recognition framework. For clarity of presentation, we consider here the problem of speech pattern recognition, which is one of the crucial research areas in the development of multimedia and artificial intelligence technologies. In the following paragraphs of this section, we shall summarize the motivations for GPD's development, addressing problems in speech recognizer design.

1.1. Speech pattern recognition using modular systems

We refer to the acoustic output of the human speech production system as a speech instantiation, and it can be considered as a sequence of linguistic units, such as phonemes and words. The goal of speech *recognition* then is to map a speech instantiation to its corresponding correct sequence of linguistic units.

As can be easily observed, the duration of each linguistic unit is highly variable, mainly due to speaking-rate changes. This indicates that a speech instantiation is a *dynamic* (variable-durational) temporal sample. In addition, the acoustic properties of speech waves are highly variable due to various factors such as the speakers themselves and the speaking-fluency. The size of a vocabulary, i.e., the number of words used, often exceeds several tens of thousands of words. From this, it is obvious that coping with various issues appropriately and comprehensively is an indispensable requirement in the design of the recognizer. However, it is not necessarily recommended to start introductory discussions with such a large-scale complicated design framework. Let us therefore start preparations on discussions by using the following basic statement: speech recognition involves a process of mapping a dynamic speech instantiation, which is *a priori* correctly extracted from its surrounding acoustic signal and belongs to one of a given set of M speech classes, to a class index; C_j $(j = 1, \cdots, M)$. We specially consider the design problem of training the adjustable parameter set Ψ of a recognizer (see below for a more precise description), aiming at achieving the *optimal* (best in recognition accuracy for all future instantiations) recognition decision performance.

One of the fundamental approaches to this problem is the Bayes approach using the following Bayes decision rule, rigorous execution of which is well known to lead to the minimum recognition error rate:

$$C(u_1^{T_o}) = C_i, \qquad \text{iff } i = \arg\max_j p(C_j \mid u_1^{T_o}), \tag{1}$$

where $u_1^{T_o}$ is a dynamic speech instantiation with length T_o, $C(\cdot)$ represents the recognition operation, and it is assumed that the *a posteriori* probability for the dynamic instantiation, $p(C_j \mid u_1^{T_o})$, exists and is known. A training goal in this approach is to find a state of Ψ that enables the corresponding estimate $p_\Psi(C_j \mid u_1^{T_o})$, which is a function of Ψ, to precisely approximate the true *a posteriori* probability (density) $p(C_j \mid u_1^{T_o})$, or in other words, to adjust Ψ so that $p_\Psi(C_j \mid u_1^{T_o})$ can approximate $p(C_j \mid u_1^{T_o})$ as precisely as possible. For example, one could attempt a direct estimate of this *a posteriori* probability for the dynamic instantiation by using a system having sufficiently-large approximation capability. However, to the best of the authors' knowledge, there has not been a successful design example based on such an optimistic strategy. Therefore, an alternative to this ideal but simple-minded attempt is obviously needed.

Actually, most speech recognizers are *transparent* (meaning that an internal process is explicitly described), modular systems. Such a recognizer consists of several observable modules, each carefully designed based on scientific experiences. As illustrated in Fig. 1, a typical recognizer consists of 1) a feature extractor (*feature extraction* module) and 2) a classifier (*classification* module) that is further divided into a language model and an acoustic model. Let us represent the adjustable parameter sets of the feature extractor, the acoustic model, and the language model by Φ, Λ, and Ξ, respectively: $\Psi = \Phi \cup \Lambda \cup \Xi$.

The feature extraction module converts a speech wave sample $u_1^{T_0}$ to a dynamic pattern $x_1^T = (\mathbf{z}_1, \mathbf{z}_2, \cdots, \mathbf{z}_t, \cdots, \mathbf{z}_T)$ that is a sequence of *static* (fixed-dimensional) F-dimensional acoustic feature vectors, where T is the duration of the dynamic pattern and $\mathbf{z}_t$ is the t-th feature vector of the sequence. The feature vector is generally made up of cepstrum or bank-of-filters output coefficients. Φ is then the designable parameter set, such as lifter and bank-of-filters functions, that controls the nature of the feature vectors.

Next, the classification module assigns a class index to this converted feature pattern. This assignment is generally performed by using the classification rule,

$$C(x_1^T) = C_i, \qquad \text{iff } i = \arg\max_j p(x_1^T \mid C_j) p(C_j), \tag{2}$$

which is conceptually equivalent to (1). Note in (2) that in accordance with the Bayes rule of probability, the *a posteriori* probability is replaced by the conditional probability (density) and the *a priori* probability, which are both suited for the estimation based on the well-analyzed Maximum Likelihood (ML) method. In fact, the conditional probability [density] $p(x_1^T \mid C_j)$ is often computed as the estimate $p_\Lambda(x_1^T \mid C_j)$ using Hidden Markov Models (HMMs) for

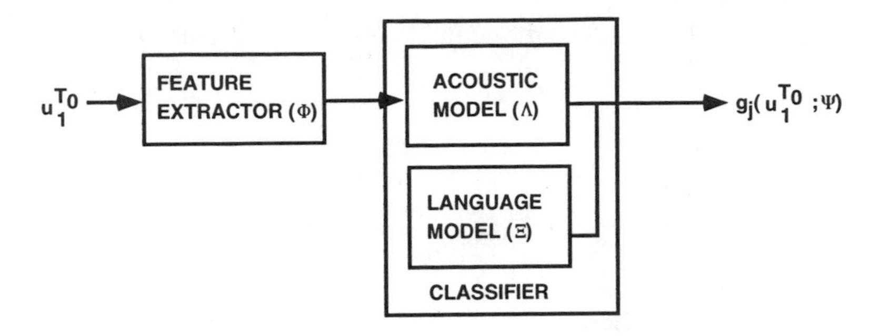

Fig. 1: Typical structure of a modular speech recognizer.

the acoustic models; Λ corresponds to a set of model parameters, such as the HMM state transition probabilities and the mean vectors of the Gaussian continuous HMMs. Also, the *a priori* probability $p(C_j)$ is often computed as the estimate $p_\Xi(C_j)$ using a language model such as a N-gram and a probabilistic context-free grammar; Ξ then corresponds to the probability of the N-gram and the parameters that determine the grammatical rules.

1.2. Classifier design based on maximum-likelihood method

In the same sense as (1), substituting accurate estimates for the probabilities in (2) enables one to fundamentally achieve the optimal, minimum classification error status. Thus, it seems plausible to consider the accurate estimation of these conditional and *a priori* probabilities to be a desirable design objective. Actually, the classifiers of most existing recognizers have been designed based on the design principle of this ML approach (to classification); that is, the Expectation-Maximization method, which is an extended ML estimation method for incomplete data[1] [9, 10, 28], and segmental k-means clustering [50] are used for training the HMM acoustic model, and a simple computation of the relative frequency of occurance of symbols (e.g., phonemes or words) is used for

[1] The incompleteness here corresponds to the fact that one cannot observe the state transition behavior of HMM.

designing the N-gram language model [49]. Note here that the minimum distortion principle, which underlies the design of the reference pattern based models of a distance classifier (widely used prior to HMM classifiers) is fundamentally equivalent to this ML principle.

However, this conventional ML-based approach actually has a basic problem in that the functional form of the class distribution (the conditional probability density) function to be estimated is in practice rarely known and the likelihood maximization of these estimated functions, performed to model each entire class distribution individually, is not direct with regard to the minimization of classification errors (the accurate estimate of class boundaries). Also, the ML-based approach covers only the classifier design; it does not optimize the overall recognizer, in other words, its design target is too far from emulating the original decision strategy (2).

The techniques of feature extraction and probability estimation are described in detail in textbooks such as [85].

1.3. Classifier design based on a discriminant function approach

Recently, an alternative to the common ML approach, based on the concept of discriminative training, has been vigorously investigated to especially improve the acoustic model parameter Λ. Discriminative training has been called many different names, because it has various backgrounds, including multi-variate analysis, artificial intelligence, and ANN. In fact, it is sometimes called competitive learning and discriminant analysis. In this paper, we refer to it as the Discriminant Function Approach (DFA) that has been widely used in pattern recognition.

In DFA, a discriminant function $g_j(x_1^T; \Lambda)$ is introduced (for C_j) to measure the class membership of the input x_1^T (the degree to which x_1^T belongs to one class), where one should note that the discriminant function is a function of the classifier parameters Λ. This discriminant function does not need to be a probability function; it can be any reasonable type of measure, such as distance or similarity. In the approach, the following decision rule is used in place of (2);

$$C(x_1^T) = C_i, \qquad \text{iff } i = \arg\max_j g_j(x_1^T; \Lambda). \tag{3}$$

In this approach, Λ is trained in order to reduce a loss that reflects a classification error in a certain manner. Since the classification result is evaluated in the design stage, this approach is fundamentally more direct with regard to the minimization of classification errors than the ML-based approach where class model parameters are designed independently of each other. In fact, designs using this approach have successfully improved the classification accuracy of ML-based baseline systems in various speech classification tasks (See later paragraphs.). However, there was plenty of room left for improvement in this apparently powerful DFA: Each of these designs had a mathematical or procedural inadequacy, as summarized in the following.

1. Execution of rule (3) using an arbitrary measure as the discriminant func-

tion does not necessarily lead to the minimum Bayes error probability situation.

2. The design scope stays within the acoustic modeling and does not cover the overall recognizer.

3. Most of the existing training procedures are empirical or heuristic, and their mathematical optimality is thus unclear.

The following is a review of the research situation concerning DFA-based acoustic model training in the years around 1990, which was actually a direct motivation for GPD development. In the early stages of investigation, attempts were made to improve HMM acoustic models. The concept of maximum mutual (inter-class) information was incorporated in the model design [7], and corrective training similar to traditional error-correction training was developed [8]. In the next stage, ANN concepts, such as Feed-Forward Network (FFN) [91] and Learning Vector Quantization (LVQ) [61], were applied to the acoustic modeling. Typical examples of such applications are categorized as follows, based on system structure and training methods:

1. Discriminative (of DFA) ANN with a time-delay structure

   (a) FFN- and LVQ-based systems with a time-delay structure were proposed, aiming to accurately classify a short speech fragment [70, 101].

   (b) An analog ANN with a time-delay structure was developed for continuous speech pattern classification [99].

2. A hybrid of an ANN and nonlinear Dynamic Time Warping (DTW)

   (a) ANN classifiers with a time-delay structure were used for front-end processing of DTW [74].

   (b) FFN estimation of local probabilities of discriminative HMMs were used for front-end processing of DTW [16].

   (c) DTW was used for front-end processing of ANN static pattern classifiers [34, 43, 57, 92].

3. A discriminative HMM based on ANN design concepts

   (a) LVQ was used for designing the codebook of a discrete HMM [47, 48, 60, 109].

   (b) Empirical training rules similar to LVQ were applied to the design of the mean vectors of the Gaussian distributions of a continuous (Gaussian) HMM [75].

   (c) The concept of a recurrent network was applied to an HMM, and discriminative HMM classifiers were designed by using a training objective, which is equivalent in its fundamentals to the maximization of mutual information [18, 78].

Actually, these methods led to successful results to some extent. However, as cited before, the resulting recognizers were not necessarily satisfactory. There are two likely causes. The first is that, as summarized in the following, the discriminative training procedures used therein were mathematically inadequate.

1. Empirical rules such as error correction learning and LVQ did not have enough of a mathematical basis to guarantee design optimality.

2. The mathematical properties, such as training convergence, of an adaptive version of the Error-Back Propagation (EBP) used for designing the FFNs were unclear. Note that here the term "adaptive" means a procedure in which learning (adjustment) was performed every time one design sample was presented.

3. As is well known, the minimization of the squared error loss between a classifier output and its corresponding supervising signal is not necessarily equivalent to the minimization of misclassifications [29, 39].

4. The maximization of mutual information does not necessarily imply the minimization of misclassifications (see Section 4).

The second possibility is that improvement efforts were too limited to acoustic modeling and lacked the global scope of designing an overall recognizer. With this in mind, we reexamine the classifier examples introduced above, recategorizing them according to the location of the DFA-based design execution in Fig. 2. Fig. 2 (A) illustrates a method of using DFA to increase the classification accuracy of short speech fragments, each being merely a part of the input

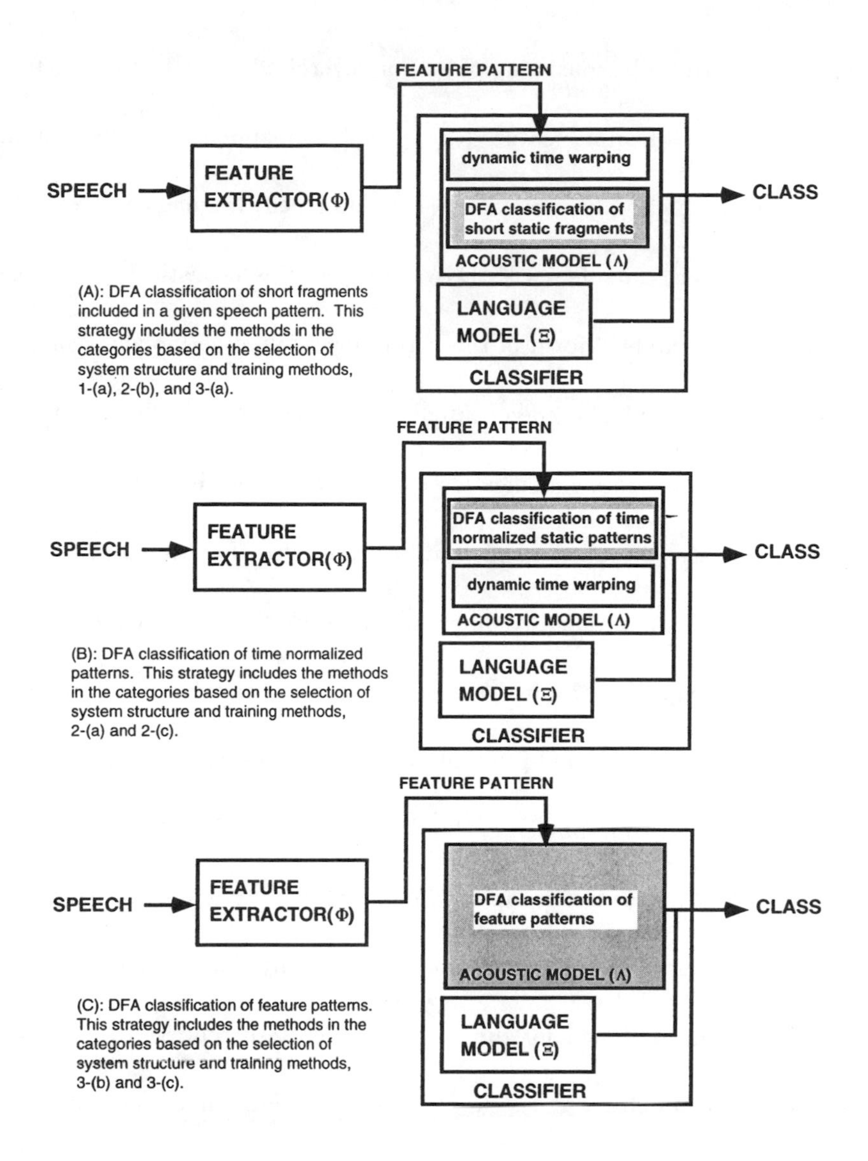

Fig. 2:Schematic explanation of the role of the discriminant function approach classification in recent ANN-based approaches to speech pattern recognition.

feature pattern (usually consisting of one or several adjacent feature vectors). The original input dynamic feature pattern is first mapped to a new dynamic pattern by using highly discriminative (DFA-designed) fragment models, then this new pattern is classified with DTW; that is, this acoustical modeling is performed using a hybrid form of the highly discriminative modeling of short fragments and the DTW-based concatenation of these models. Note here that the effort made to increase classification capability does not bear directly on the classification of the overall dynamic pattern, and also that the DTW process is based on the minimum distortion or maximum likelihood principle, which is not necessarily relevant to an increase of classification accuracy. It is therefore obvious that this improvement attempt is not consistent with the optimal classification of the dynamic pattern. On the other hand, Fig. 2 (B) illustrates a method of first mapping the input dynamic pattern to a static pattern with DTW and then performing classification with highly discriminative acoustic models, which are designed for this new static pattern representation. Here, the static pattern models are designed based on a minimum loss principle (e.g., minimum squared error loss) that is different from the minimization of classification error counts. It is thus obvious that these hybrid methods are also inconsistent with the optimal structure for classifying the dynamic pattern. The method shown in Fig. 2 (C) trains HMM parameters by using a discriminative training method, aiming at a direct increase of dynamic pattern classification accuracy. This is clearly a more advanced approach than before, because it attempts to directly improve the HMM acoustical models corresponding to the dynamic pattern representation. However, the figure obviously shows that even this method suffers from the problem of narrow design scope, as had the others.

1.4. Motivation of GPD development and paper organization

From the above survey, one may recognize the necessity of a novel design method for pursuing the overall optimality of a recognizer that covers the acoustic modeling process as well as the feature extraction and language modeling processes. Note here that although we have discussed the problems in speech recognition, most of the arguments hold true in many other cases of pattern recognition. The development of GPD was motivated by such circumstances in the problems of pattern recognizer design.

GPD relies on traditional adaptive discriminative training, called the Probabilistic Descent Method (PDM) [1, 2], for pattern classification. In the early stage of development, GPD was formalized as a training algorithm for classifiers [54]. It was then extended to a more general method for designing an entire recognizer, e.g., [58]; it was also extended to particular tasks such as keyword spotting, e.g., [63]; it was applied to speaker recognition as well, e.g., [66]. Its extension through such continued research efforts has resulted in a novel family of discriminative training methods, members of which are based upon the original formalization concept of GPD.

In this paper, we shall comprehensively present the GPD-based approach to pattern recognizer design. The paper is organized as follows. Following the development history of the original formalization of GPD and its extensions, we start our description by focusing on design problems in pattern classification. In Section 2, we provide the fundamentals of the conventional, DFA-based pattern classification, and discuss the background of GPD development. In Section 3, we introduce GPD. Its formalization and mathematical nature, such as the training optimality, are described in detail. So as to maintain the concreteness of the formulation as well, we focus in this section on classifier design problems using a multiple reference distance classifier, which has long been used for speech pattern classification and is the structure for the LVQ pattern quantizer. In Section 3, we also summarize the concept of Minimum Classification Error learning (MCE), which is closely related to the GPD formalization and has a significant theoretical contribution to pattern classification study [52]. In Section 4, we discuss the relationship between GPD and existing design methods. In Section 5, we introduce derivatives of GPD, or in other words, algorithms based on extensions of the GPD concept. The paper is then concluded in Section 6. Several additional issues related to GPD implementation are finally provided in the Appendixes.

2. Discriminative Pattern Classification

2.1. Bayes decision theory

We first summarize the Bayes decision theory, which underlies most approaches to pattern classification, including the DFA approach. We assume for simplicity of discussion that given an observed feature pattern sample, we aim to classify it *accurately*. What is "accuracy" here? The meaning of this common term is not necessarily clear for a particular operation of classification. A significant contribution of the Bayes decision theory is to give this accuracy a general statistics-based definition, i.e., the minimum expected loss (risk) situation, and to show that this situation can be achieved by observing the Bayes decision rule.

For the general task of classifying M-class patterns, the Bayes theory is formulated as follows (see [29] for details). According to the conventional approach in this theoretical framework, we assume a sample to be static. A static feature pattern $\mathbf{x}(\in C_k)$ is given. To measure the accuracy, the theory first introduces an (individual) loss $\ell_k(C(\mathbf{x}))$ that is incurred by judging the given pattern's class C_k to be one of the M possible classes, where $C(\cdot)$ denotes the classification operation as in (1). It is obvious that the accuracy of the task should be evaluated over all of the possible samples. Thus, based on statistics, the theory next introduces an expected loss incurred by classifying $\mathbf{x}$, called the conditional risk, as

$$L(C(\mathbf{x}) \mid \mathbf{x}) = \sum_k \ell_k(C(\mathbf{x}))1(\mathbf{x} \in C_k)P(C_k \mid \mathbf{x}), \tag{4}$$

and also introduces an expected loss associated with $C(\cdot)$, called the overall risk, as

$$\mathcal{L} = \int L(C(\mathbf{x}) \mid \mathbf{x}) p(\mathbf{x}) d\mathbf{x}, \tag{5}$$

where $1(\mathcal{A})$ is the following indicator function

$$1(\mathcal{A}) = \begin{cases} 1 & \text{(if } \mathcal{A} \text{ is true),} \\ 0 & \text{(otherwise).} \end{cases} \tag{6}$$

The accuracy is accordingly defined by the overall risk. The smaller the overall risk, the better its corresponding classification result. A desirable decision is thus the very one that minimizes the overall risk. Consequently, the theory leads to the following well-known Bayes decision rule that is justified by (5).

Rule 1 (Bayes decision rule) *To minimize the overall risk, compute all of the (M) possible conditional risks $L(C_i \mid \mathbf{x})$ $(i = 1, \cdots, M)$ and select the C_j for which $L(C_j \mid \mathbf{x})$ is minimum.*

2.2. Minimum error rate classification

Using the loss enables one to evaluate classification results flexibly. This is one of the big attractions of the Bayes decision theory. However, in practice, a loss that evaluates results uniformly for all of the possible classes (i.e., class-independent loss) has been used widely, due to the difficulty of setting losses in a reasonable class-by-class manner. A natural loss in this simple case is the following error count loss

$$\ell_k(C(\mathbf{x})) = \begin{cases} 0 & (C(\mathbf{x}) = k), \\ 1 & \text{(otherwise),} \end{cases} \tag{7}$$

and then its corresponding conditional risk becomes

$$L(C(\mathbf{x}) \mid \mathbf{x}) = 1 - P(C(\mathbf{x}) \mid \mathbf{x}) \tag{8}$$

and the overall risk becomes the average probability of error. Thus, the minimization of the overall risk using (7) leads to the minimization of the average probability of error, in other words, the minimum error rate classification (e.g., [29]).

According to the above Bayes decision rule, it is obvious that the desirable classification decision is the one that minimizes the conditional risk (8), or maximizes the *a posteriori* probability $P(C(\mathbf{x}) \mid \mathbf{x})$. As far as minimum error classification is concerned, the best classifier design is theoretically achieved by simulating (8) as accurately as possible. Hence this way of thinking justifies the Bayes approach of aiming at a correct estimation of the *a posteriori* probability, or its corresponding *a priori* probability and conditional probability. Note here that we assume the Bayes approach includes the ML method and the so-called Bayesian approach [29]. The principal attraction of the Bayes approach may be that the classifier design is primarily based on a mathematically well-analyzed probability computation. However, in realistic situations where only limited resources are available, an accurate estimation of these probability functions is difficult, and generally this approach does not achieve satisfactory design results.

Originally, the probability functions are given for the task at hand. In the Bayes approach, the classifier design is replaced by the estimation of these functions. Taking into account the fact that these given functions are unobservable and essentially difficult to estimate accurately, one may have to find an alternative approach to the design; that is, one can consider that the design is to execute an accurate evaluation based on (7) for every sample. This is the very concept of DFA design. However, unfortunately, this point of view has not been explicitly indicated in most conventional DFA design attempts. Usually, the design formalization has started with (3), merely selecting arbitrary discriminant functions.

2.3. Discriminant function approach

Let us discuss in more detail the DFA design for a task of classifying the static pattern $\mathbf{x}$. The decision rule is

$$C(\mathbf{x}) = C_i, \qquad \text{iff } i = \arg\max_j g_j(\mathbf{x}; \Lambda). \tag{9}$$

As cited before, DFA design is fundamentally performed in a competitive way (with regard to classes), aimed at realizing the discriminant function set (consequently Λ) that achieves the least classification error over the design sample set. Its implementation is basically characterized by the following factors:

1. functional form of the discriminant function (classifier structure or measure),
2. design (training) objective,
3. optimization method,

4. consistency with unknown samples (robustness, generalization capability, or estimation sensitivity).

Classifier structure, which determines the type of Λ, is generally selected based on the nature of the patterns, and this selection often determines the selection of the measure or the selection of $g_j(\mathbf{x};\Lambda)$. The linear discriminant function is a typical example of a classical classifier structure, and the measure used therein is a linearly-weighted sum of input vector components. A distance classifier that uses the distance between an input and a reference vector as the measure is another widely used example.

The design objective is a function used for evaluating a classification result in the design stage and is equivalent to the concept of risk in Bayes decision theory. Usually, an individual loss that is a design criterion for an individual design sample is first introduced; the individual loss for $\mathbf{x}$ ($\in C_k$) is denoted by $\ell_k(\mathbf{x};\Lambda)$. As discussed in the previous section, a natural loss form is the classification error count, as:

$$\ell_k(\mathbf{x};\Lambda) = \begin{cases} 0 & (C(\mathbf{x}) = k), \\ 1 & (\text{otherwise}), \end{cases} \tag{10}$$

where the fact that Λ is included in the loss definition means that the individual loss is a function of Λ. Next, similar to the overall risk, an ideal overall loss, i.e., the expected loss, is defined using the individual loss, as:

$$L(\Lambda) = \sum_k P(C_k) \int \ell_k(\mathbf{x};\Lambda) p(\mathbf{x}|C_k) d\mathbf{x}. \tag{11}$$

However, since sample distributions are essentially unknown, it is impossible to use this expected loss in practice. For this reason, an empirical average loss, defined in the following, is usually used:

$$L_0(\Lambda) = \frac{1}{N} \sum_k \sum_n \ell_k(\mathbf{x}_n; \Lambda) 1(\mathbf{x}_n \in C_k), \tag{12}$$

where n of $\mathbf{x}_n$ explicitly means that the sample $\mathbf{x}_n$ is the n-th sample of the finite (consisting of N samples) design sample set $X = \{\mathbf{x}_1, \cdots, \mathbf{x}_n, \cdots, \mathbf{x}_N\}$. In the case of using the error count loss of (10), this empirical average loss becomes the total count of classification errors measured over X.

In addition to (10), several different forms of loss have been used; e.g., perceptron loss, squared error loss (between a discriminant function and its corresponding supervising signal), and mutual information. However, these are merely temporary expedients, and it is known that the design results obtained by using them are not necessarily consistent with minimum classification error probability condition (e.g., see [29]; also see Section 4.2 for the use of mutual information).

Optimization is a method of finding the state of Λ that minimizes loss over X, and its embodiments are grouped into a heuristic method and a mathematically-proven algorithm.

Among the many examples of heuristic optimization, error correction training has been widely used (e.g., see [79]). Most of these heuristic methods do not guarantee, however, the achievement of a true optimal situation due to the lack of a theoretical justification.

The mathematically-proven methods are further grouped into ones which search for locally-optimal conditions and ones which search for a globally-optimal condition. Traditionally, a practical, gradient search-based, local optimization method such as the steepest descent method has been used. In recent ANN frameworks, global optimization methods, such as simulated annealing and its special case, called the Boltzmann machine, have also been extensively studied [35, 91]. Their fundamental capability of globally optimizing the objective function is fascinating, even though the optimization is done in a probabilistic sense. Moreover, these ANN-related methods do not require the continuity, or the *smoothness* (being at least the first differentiable in system parameters), of the loss function (See the following paragraphs in this section and Section 2.4.3.). In fact, simulated annealing was applied to the minimization of the unsmooth, average empirical loss based on (10) [4]. However, these global optimization methods are usually impractical due to their time-consuming nature.

Optimization methods are categorized from another point of view; i.e., the batch type search vs. the adaptive search. The batch type search aims to minimize (or locally minimize) the average empirical loss. The steepest descent method is a typical example of the batch type local minimization method. The adaptive search minimizes the individual loss $\ell_j(\cdot;\Lambda)$ for a small set of design samples (usually one sample) randomly selected from X; the adjustment of Λ is repeated while changing this set of samples. Methods based on stochastic approximation are traditional examples of adaptive local minimization (e.g., [29, 32]).

Ideally, the purpose of classifier design is to realize a state of Λ that leads to accurate classification over all of the samples of a task at hand instead of the given design samples. Remember that the expected loss was defined as an ideal overall loss. Thus it is desirable that the design result obtained by using design samples be consistent with unknown samples too. Obviously, the pursuit of such a result for unknown data requires some assumptions concerning unobservable, unknown samples or about the entire sample distribution of the task. In contrast with the Bayes approach, measures to increase the consistency in DFA are generally moderate: the loss is individually defined for every design sample, and the design is essentially formed using the given design samples, as shown in the use of the average empirical loss (12).

Naturally, the most *practical* method of DFA is to search for the state of Λ that corresponds to the minimum of the average empirical loss consisting of (10). However, (10) is not a smooth function (not differentiable in Λ). Thus, the corresponding average empirical loss may be difficult to minimize directly. Moreover, this average empirical loss is sensitive only to design samples (see Section 3.3), which means that a design using this loss may not clearly contribute to increasing the consistency with unknown samples. Consequently, a fundamental improvement for DFA is desired.

We have considered the classification of static patterns in this section. The issues discussed above also hold true in the classification of dynamic speech patterns: the discussion does not depend upon the static nature of patterns. Hence, the most effective design method for dynamic pattern classification can again be the local minimization of the average empirical, classification error count loss. Detailed discussions about dynamic pattern classification will be given in Section 2.4.3.

2.4. Probabilistic descent method

2.4.1. Formalization

Almost 30 years ago, PDM was developed as a practical method for locally minimizing the expected loss based on the gradient search [1, 2]. In particular, its minimization mechanism is based on stochastic approximation. Clearly, this method can be applied to locally minimizing the practical, average empirical loss consisting of individual classification error count losses. The use of stochastic approximation for pattern classification has a long history and, in fact, many studies have been done, as shown in [29, 32]. However, PDM is clearly distinguished from other stochastic approximation-based methods by the following two points:

1. It proves that the reduction of individual losses leads to a reduction of the expected loss in the probabilistic sense.

2. It proposes a design concept that enables the classifier design problem to

be handled systematically by incorporating the classification process in a three-step functional formalization.

Our purpose in this section is to show the background of GPD development by reviewing classical PDM. Let us consider the previous M-class pattern classification task. It is assumed in PDM that a pattern is static. As cited in Section 1.4, we use a distance classifier consisting of multiple reference vectors as our formalization framework: $\Lambda = \left\{ \{\mathbf{r}_j^b\}_{b=1}^{B_j} \right\}_{j=1}^{j=M}$, where $\mathbf{r}_j^b$ is the b-th closest reference vector of C_j (in the sense of squared Euclidean distance to an input pattern $\mathbf{x}$, described later), defined in the same vector space as $\mathbf{x}$, and B_j is the number of C_j's reference vectors. The classification rule that we consider is:

$$C(\mathbf{x}) = C_i, \qquad \text{iff } i = \arg\min_j g_j(\mathbf{x}; \Lambda), \tag{13}$$

which is essentially the same as (3).

Suppose our design is to be done adaptively; that is, we update the classifier parameters Λ every time one design sample is presented, and we pursue the optimal status of the parameters by repeating this updating procedure. Let us assume that $\mathbf{x}(t)$ of C_k is presented at the design stage time index t (natural number). PDM is then formalized in the following three-step manner.

The first step is to choose the discriminant function $g_j(\mathbf{x}(t); \Lambda)$. Naturally, for every class, we define the function as the squared Euclidean distance between the input and its closest reference vector:

$$g_j(\mathbf{x}(t); \Lambda) \;\; = \;\; d_E(\mathbf{x}(t), \mathbf{r}_j^{c_j}) = \parallel \mathbf{x}(t) - \mathbf{r}_j^{c_j} \parallel^2,$$

$$\text{where } c_j = \arg\min_b d_E(\mathbf{x}(t), \mathbf{r}_j^b). \tag{14}$$

In the second step, a misclassification measure $d_k(\mathbf{x}(t);\Lambda)$ is introduced so as to emulate the decision process of (13), i.e., the comparison/decision among the competing classes:

$$d_k(\mathbf{x}(t);\Lambda) = \frac{1}{N_t}\sum_{C_j\in\Im}\{g_k(\mathbf{x}(t);\Lambda) - g_j(\mathbf{x}(t);\Lambda)\}, \tag{15}$$

where $\Im$ is a set of confusing classes defined as

$$\Im = \{C_j; g_j(\mathbf{x}(t);\Lambda) > g_k(\mathbf{x}(t);\Lambda)\}, \tag{16}$$

and N_t is the number of $\Im$ classes. Note in (15) that (13) is represented by a scalar value decision. A positive value of $d_k(\;)$ means a misclassification, and a negative value means a correct classification.

The third step conforms to the same loss evaluation as DFA. The misclassification measure is embedded in a loss in order to evaluate the corresponding classification result. A general form of the loss is represented as

$$\ell_k(\mathbf{x}(t);\Lambda) = l(d_k(\mathbf{x}(t);\Lambda)), \tag{17}$$

where $l(\;)$ is a monotonically-increasing function. Various definitions of the loss are possible; for example,

$$\ell_k(\mathbf{x}(t);\Lambda) = \begin{cases} \{d_k(\mathbf{x}(t);\Lambda)\}^{\varrho_1} & (d_k(\mathbf{x}(t);\Lambda) > 0), \\ 0 & (\text{otherwise}), \end{cases} \tag{18}$$

and

$$\ell_k(\mathbf{x}(t);\Lambda) = \begin{cases} 1 & (\varrho_2 < d_k(\mathbf{x}(t);\Lambda)), \\ \frac{d_k(\mathbf{x}(t);\Lambda)}{\varrho_2} & (0 < d_k(\mathbf{x}(t);\Lambda) \le \varrho_2), \\ 0 & (d_k(\mathbf{x}(t);\Lambda) \le 0), \end{cases} \tag{19}$$

where ϱ_1 and ϱ_2 are positive numbers. The loss in (19) is a linear approximation of the classification error count. In particular, the loss in (18) approximates the classification error count when $\varrho_1 \rightarrow 0$; similarly, the loss in (19) approximates the classification error count when $\varrho_2 \rightarrow 0$.

2.4.2. Probabilistic descent theorem

The PDM training by which the parameters are adjusted through the above three-step formalization is summarized in the following probabilistic descent theorem.

Theorem 1 (Probabilistic descent theorem) *Assume that a given design sample* $\mathbf{x}(t)$ *belongs to* C_k. *If the classifier parameter adjustment* $\delta\Lambda(\mathbf{x}(t), C_k, \Lambda)$ *is specified by*

$$\delta\Lambda(\mathbf{x}(t), C_k, \Lambda) = -\epsilon \mathbf{U} \nabla \ell_k(\mathbf{x}(t); \Lambda), \tag{20}$$

where $\mathbf{U}$ *is a positive-definite matrix and* ϵ *is a small positive real number, then*

$$E[\delta L(\Lambda)] \leq 0. \tag{21}$$

Furthermore, if an infinite sequence of randomly selected samples $\mathbf{x}_t$ *is used for learning (designing) and the adjustment rule of (20) is utilized with a corresponding [learning] weight sequence* $\epsilon(t)$ *which satisfies*

$$\sum_{t=1}^{\infty} \epsilon(t) \rightarrow \infty, \tag{22}$$

$$\sum_{t=1}^{\infty} \epsilon(t)^2 < \infty, \tag{23}$$

then the parameter sequence $\Lambda(t)$ *(the state of* Λ *at* t*) according to*

$$\Lambda(t+1) = \Lambda(t) + \delta\Lambda(\mathbf{x}(t), C_k, \Lambda(t)) \tag{24}$$

converges with probability one (1) to Λ^* *which is at least a local minimum of* $L(\Lambda)$.

The nature of the adjustment convergence, such as accuracy and speed, is analyzed in detail in [2].

It is obviously unrealistic to rigidly observe the infinitely-repeated probabilistic descent adjustments. In practice, the learning coefficient $\epsilon(t)$ is often approximated by a finite monotonically-decreasing function such as

$$\epsilon(t) = \epsilon(0)\left(1 - \frac{t}{N_{max}}\right), \tag{25}$$

where N_{max} is a preset number of adjustment repetitions. In particular, when the individual loss of (19) is used and $\mathbf{U}$ is assumed for simplicity to be a unit matrix, the adjustment rule for our multi-reference distance classifier is given "speciously" (The reason for using this particular "critical" term will be shown later) as follows: If and only if $0 < d_k(\mathbf{x}(t);\Lambda) \leq \varrho_2$ holds,

$$\mathbf{r}_j^q(t+1) = \begin{cases} \mathbf{r}_j^q(t) + \frac{2\epsilon(t)(\mathbf{X}(t) - \mathbf{r}_j^q(t))}{\varrho_2} & \text{(for } j = k \text{ and } q = \arg\min_b d_E(\mathbf{x}(t), \mathbf{r}_j^b)), \\ \mathbf{r}_j^q(t) - \frac{2\epsilon(t)(\mathbf{X}(t) - \mathbf{r}_j^q(t))}{\varrho_2} & \text{(for } C_j \in \Im \text{ and } q = \arg\min_b d_E(\mathbf{x}(t), \mathbf{r}_j^b)), \\ \mathbf{r}_j^q(t) & \text{(otherwise)}. \end{cases} \tag{26}$$

The probabilistic descent theorem shows that an infinite repetition of the adjustment (20) based on the gradient computation of the individual loss for one training sample leads to a local minimum of the expected loss. Using an infinite number of samples is essentially equivalent to a situation in which the corresponding sample distribution is accessible. Therefore, criticizing that the minimum classification error probability status is known in this ideal situation, one may doubt the technical significance of PDM (or the methods based on the stochastic approximation).

However, the contribution of PDM is clear. In reality, design samples are finite and it is impossible to know the minimum classification error probability situation. In the Bayes approach, the estimation of the class distributions obviously suffers from estimation errors that may spread over the entire sample space instead of the region near the class boundaries. Note here that accurate classification relies on an accurate estimation of the class boundary. In contrast with this approach, PDM aims to minimize the average empirical loss in a manner that corresponds directly to the classification task (the classification rule (13)); in particular, PDM design using (19) will minimize the actual classification error count, resulting in an accurate class boundary estimation. This point is clearly distinct from other DFA methods as well as the Bayes approach when the exact distribution form is not known to the designer. If this minimization (even the local minimization) is successfully done, the resulting classification performance should be superior to that of the Bayes approach. Moreover, though finite, a huge number of design samples are often used. In fact, it is often difficult to store all of the design samples for training. Given a finite set of design samples X, the PDM adjustment will be repeated by extracting a sample $\mathbf{x}_t$ randomly from X. Furthermore, it is often desired that a classifier be adaptable to new circumstances in a realistic situation where future samples cannot be observed in the design stage. PDM provided a fundamental solution based on simple gradient computation to this adaptive design. Therefore, the true fundamental value of PDM should be widely recognized, even though the practical aspects of its adaptive learning mechanism must be further investigated.

Recall the adaptive training version of EBP and LVQ. Both are useful but have only intuitive validity (meaning that the convergence mechanism has not been mathematically elaborated). One should note that the probabilistic descent adjustment is quite similar to adaptive EBP [3, 5]. In fact, the adjustment principle of adaptive EBP is to locally minimize the loss for a given design sample according to its gradient; also, the mechanism of error back-propagation is simply a special case of differential calculus for a functional embedded in a smooth functional form. Moreover, one should note that (26) closely resembles LVQ, especially an improved version of LVQ2 [69]. Furthermore, the monotonically-decreasing learning weight function used in these learning methods is essentially the same as (25). Thus, it appears that PDM may be useful for analyzing these recent implementations of DFA.

2.4.3. Problems

Encouraged by the advantages of PDM, i.e., the direct minimization of the classification error count and the use of an adaptive training mechanism, which were described in the last few paragraphs of Section 2.4.2, one might attempt to classify dynamic patterns with PDM. However, PDM is a method for the classification of static patterns. Some additional procedures will thus be needed for applying this method to dynamic pattern classification. Furthermore, the formulation of PDM actually suffers from a mathematical difficulty. Solving this inadequacy was the motivation for the development of the more general GPD. In this section, we clarify the problems of PDM.

Recall that we used the critical term "speciously" in the derivation of (26). Actually, the derivation is mathematically impossible.

The probabilistic descent adjustment rule is based on gradient computation. The overall process included in the formulation must thus be differentiable. However, the "min" operation selecting the closest reference vector, in (14), is not a differentiable function; $\Im$ can change discretely when Λ changes, with the result that the misclassification measure is also discontinuous. It then becomes apparent that the PDM formalization is inadequate. To apply the probabilistic descent adjustment in the right way, this lack of smoothness must therefore be overcome.

One should note that the above smoothness difficulty does not appear in some special cases. In fact, the problem concerning (14) does not occur for a classifier that represents each class by only one reference pattern. Moreover, the problem concerning (15) does not occur in a two-class task ($M = 2$). However, these are very limited and unrealistic situations. A satisfactory method should be able to handle general and realistic tasks.

Let us next consider the task of classifying spoken English letter syllables, such as {b} and {c}, by using a distance classifier. We assume here that each syllable is uttered in an isolated mode and is converted to an acoustic feature vector sequence with some preset feature extractor. It is also assumed that speech signal is correctly extracted from its surrounding signal and input to the system. A dynamic pattern to classify is then represented as

$$x_1^T = (\mathbf{z}_1, \ldots, \mathbf{z}_\tau, \ldots, \mathbf{z}_T) \in \underbrace{\Re^F \times \cdots \times \Re^F}_{T}, \tag{27}$$

where $\mathbf{z}_\tau$ is the fixed F-dimensional component vector (usually corresponding to an acoustic feature vector or a short sequence of such acoustic feature vectors) at the time τ ($\mathbf{z}_\tau \in \Re^F$, where F is a fixed natural number), and T is a variable but finite natural number that represents the duration of the component vector sequence.

A simple application of PDM for classifying this dynamic speech pattern would be similar to the structural hybrid system described in the Introduction; i.e., PDM would be used to increase the discriminative power of a distance classifier consisting of reference vectors defined in the same F-dimensional component vector space, and DTW would be incorporated with this classifier in order to handle the dynamics of the entire input pattern. However, this local improvement of classification capability is not necessarily consistent with improvement of classification accuracy for the dynamic pattern. Thus, a natural solution to this problem must be to improve the overall classification process, including the dynamics normalization of a DTW distance classifier (which has long been used in speech pattern classification) by using PDM.

A DTW distance classifier is generally given as

$$\Lambda = \{\lambda_j\} = \left\{ \{r_j^b\}_{b=1}^{b=B_j} \right\}_{j=1}^{j=M}, \tag{28}$$

where r_j^b is C_j's b-th finite-length dynamic reference pattern[2]. The distance computation based on the DTW procedure between dynamic patterns is used for the classification decision. First, a DTW matching path and its corresponding *path distance* is introduced between an input pattern and each reference pattern. The path distance is given as

$$\begin{aligned} D_\theta(x_1^T, r_j^b) &= \sum_{\tau=1}^{T} w_{j,\tau}^b \delta_{\varpi(j,b,\tau,\theta)} \\ &= \| \mathbf{z}_{\varpi(j,b,\tau,\theta)} - \mathbf{r}_{j,\tau}^b \|^2, \end{aligned} \tag{29}$$

where $\mathbf{z}_{\varpi(j,b,\tau,\theta)}$ is the $\varpi(j,b,\tau,\theta)$-th component vector of x_1^T, to which the τ-th component vector of r_j^b, $\mathbf{r}_{j,\tau}^b$ corresponds along the θ-th matching path of r_j^b; $\delta_{\varpi(j,b,\tau,\theta)}$ is a *local distance*, defined by the squared Euclidean distance between these corresponding component vectors (Note that the local distance can be defined by using any other reasonable distance measure.); $w_{j,\tau}^b$ is a weight coefficient. Next, using the path distances, the distance between an input and each reference pattern, i.e., a *reference pattern distance*, is defined as the smallest

[2]Note that in this simple case for introductory discussions, we consider only the acoustic model of the classifier, assuming that the classifier has no language model.

(best) path distance for each reference pattern:

$$d_A(x_1^T, r_j^b) = \min_\theta \{D_\theta(x_1^T, r_j^b)\}. \tag{30}$$

Usually, the search for the smallest path distance is performed by Dynamic Programming (DP). Lastly, the discriminant function that represents the degree to which an input belongs to each class is defined as the reference pattern distance of the closest (in the sense of (30)) reference to the input:

$$\begin{aligned} g_j(x_1^T; \Lambda) &= d_A(x_1^T, r_j^{c_j}), \\ \text{where } c_j &= \arg\min_b d_A(x_1^T, r_j^b). \end{aligned} \tag{31}$$

The most direct use of PDM for this classifier is to adjust Λ by using the discriminant function (31). However, (31) is doubly unsmooth: This new discriminant function includes the unsmooth operation of "min" in both the best path search and the closest reference search. Obviously, there is a fundamental difficulty in designing the DTW distance classifier with PDM in its original form.

However, the reason for the difficulty in applying PDM to dynamic pattern classification is the same as the problem included in the formulation of the PDM training method. The fundamental problem to be solved is that the functions used are discontinuous, i.e., not differentiable in Λ. We come, therefore, naturally to the motivation for the development of GPD, which is to re-formulate PDM using some smoothing function.

3. Generalized Probabilistic Descent Method

3.1. Formalization concept

The fundamental concept of the GPD formalization is to directly embed the overall process of classifying a dynamic pattern of speech acoustic vectors x_1^T in a smooth functional form (at least first order differentiable with respect to classifier parameters) that is suited for the use of a practical optimization method, especially gradient search optimization [52, 54, 56]. As can be easily noticed from the previous discussions, a key point for achieving this formalization is to overcome the discontinuity of PDM. For this purpose, GPD uses the L_p norm form and a sigmoidal function that is widely used in ANN implementations.

Similar to PDM, GPD uses an adaptive update mechanism. However, the above formalization concept can be achieved, without any lack of formal rigor, even using a batch-type optimization such as the steepest descent method. Clearly, the selection of optimization methods is distinct from the formalization of the emulation of the classification decision process.

In the following subsection, we present in detail an embodiment of GPD for the distance classifier, defined by (28), for the task used in Section 2.4.3, i.e., the classification of M class feature vector sequence patterns, each corresponding to an isolated spoken syllable.

3.2. GPD Embodiment for distance classifier

3.2.1. Discriminant function for dynamic patterns

The problem with PDM in handling dynamic patterns is shown in (30) and (31), i.e., the discriminant function includes a discontinuous search for the closest reference and for the best matching path. To solve this, GPD defines the discriminant function as

$$g_j(x_1^T;\Lambda) = \left[\sum_{b=1}^{B_j} \{D(x_1^T, r_j^b)\}^{-\zeta}\right]^{-1/\zeta}, \tag{32}$$

where $D(x_1^T, r_j^b)$ is a *generalized reference pattern distance* between x_1^T and r_j^b, and ζ is a positive constant; also it defines this generalized reference pattern distance as

$$D(x_1^T, r_j^b) = \left[\sum_{\theta=1}^{\Theta_j^b} \{D_\theta(x_1^T, r_j^b)\}^{-\xi}\right]^{-1/\xi}, \tag{33}$$

where $D_\theta(\cdot)$ is the path distance of (29) and ξ is a positive constant. Accordingly, the method achieves a smooth discriminant function for the dynamic pattern by replacing the two "min" operations by the L_p norm form functions.

The use of the L_p norm form affords us an interesting flexibility in the implementation. Let ζ and ξ approach ∞. Then, clearly, (32) approximates the operation of searching for the closest reference pattern, and (33) for operation of searching for the best matching path. Thus, these smooth definitions are generalized versions of the existing search operation.

3.2.2. Formulation

GPD uses the three-step formalization of PDM in order to represent the task at hand (classification in our case) in a smooth functional form. The classification rule here is

$$C(x_1^T) = C_i, \quad \text{iff } i = \arg\min_j g_j(x_1^T; \Lambda), \tag{34}$$

which is essentially the same as (13), except that the pattern is dynamic.

We already described the first step of defining the discriminant function; the function is given as (32).

The second step defines a smooth misclassification measure. Among many possibilities, the following is a typical definition for x_1^T $(\in C_k)$:

$$d_k(x_1^T; \Lambda) = g_k(x_1^T; \Lambda) - \left[\frac{1}{M-1} \sum_{j, j \neq k} \{g_j(x_1^T; \Lambda)\}^{-\mu} \right]^{-1/\mu}, \tag{35}$$

where μ is a positive constant. Similar to (15), $d_k(\ ) > 0$ indicates a misclassification, and $d_k(\ ) < 0$ indicates a correct classification. Similar to ζ and ξ, controlling μ enables one to simulate various decision rules. In particular, when μ approaches ∞, (35) resembles rule (34).

The third step of defining the loss is almost the same as in PDM. GPD defines an individual loss as a smooth, monotonically-increasing function of the misclassification:

$$\ell_k(x_1^T; \Lambda) = l_k\left(d_k(x_1^T; \Lambda)\right). \tag{36}$$

There are various possibilities for defining the loss. Among them, the GPD method usually uses a smooth function as

$$\ell_k(x_1^T, \Lambda) = l_k(d_k(x_1^T; \Lambda)) = \frac{1}{1 + e^{-(\alpha d_k(x_1^T; \Lambda) + \beta)}} \quad (\alpha > 0), \tag{37}$$

where α and β are constants. We shall mention the selection of the loss function again later.

The embodiment is completed by deriving an adjustment rule such as (26), according to the probabilistic descent theorem. Given a dynamic input pattern at the design stage time index t, i.e., $x_1^T(t) = (\mathbf{z}_1(t), \ldots, \mathbf{z}_T(t))$ ($\in C_k$), the resulting adjustment rule using loss (37) for the reference patterns is given as

$$\mathbf{r}_{j,\tau}^b(t+1) = \begin{cases} \mathbf{r}_{j,\tau}^b(t) + 2\epsilon(t)\nu_k w_{j,\tau}^b \phi_j \rho_j \varphi_j & (\text{for } j = k), \\ \mathbf{r}_{j,\tau}^b(t) - \frac{2}{M-1}\epsilon(t)\nu_k w_{j,\tau}^b \sigma_j \phi_j \rho_j \varphi_j & (\text{for } j \neq k), \end{cases} \tag{38}$$

where

$$\nu_k = \alpha l_k(d_k(x_1^T(t);\Lambda))\{1 - l_k(d_k(x_1^T(t);\Lambda))\}, \tag{39}$$

$$\phi_j = \left[\sum_{b'=1}^{B_j} \left\{ \frac{D(x_1^T(t), r_j^b)}{D(x_1^T(t), r_j^{b'})} \right\}^{\zeta}\right]^{-(1+\zeta)/\zeta}, \tag{40}$$

$$\rho_j = \left[\sum_{\theta=1}^{\Theta_1^b} \{D_\theta(x_1^T(t), r_j^b)\}^{-\xi}\right]^{-(1+\xi)/\xi}, \tag{41}$$

$$\varphi = \sum_{\theta=1}^{\Theta_j^b} \frac{\mathbf{z}_{\varpi(j,b,\tau,\theta)}(t) - r_{j,\tau}^b}{\{D_\theta(x_1^T(t), r_j^b)\}^{\xi+1}}, \tag{42}$$

$$\sigma_j = \left[\frac{1}{M-1}\sum_{j,j\neq k}^{M} \left\{ \frac{g_k(x_1^T(t);\Lambda)}{g_j(x_1^T(t);\Lambda)} \right\}^{\mu}\right]^{-(1+\mu)/\mu}, \tag{43}$$

and $\varpi(j,b,\tau,\theta)$ indicates the component vector index of $x_1^T(t)$, to which the τ-th component vector of r_j^b, $r_{j,\tau}^b$, corresponds along the θ-th matching path of r_j^b. In (38), the adjustment is done for all of the possible paths and all of the possible reference patterns. This is a remarkable distinction from (26), in which the adjustment is selectively done. Furthermore, treating $w_{j,\tau}^b$'s as adjustable parameters, one can achieve an adjustment rule similar to (38), though we omit the result. See this point in [20].

The rule of (38) is rather complicated. In the last paragraph of the implementation, we present for informational purposes an extremely simplified version of this rule. That is, letting ζ, ξ, and μ approach ∞, we can rewrite (38) as

$$\mathbf{r}_{j,\tau}^b(t+1) = \begin{cases} \mathbf{r}_{j,\tau}^b(t) + 2\epsilon(t)\nu_k w_{j,\tau}^b(\mathbf{z}_{1(j,b,\tau)}(t) - \mathbf{r}_{j,\tau}^b(t)) \\ \qquad \text{(for } j = k \text{ and } b = 1\text{)}, \\ \mathbf{r}_{j,\tau}^b(t) - 2\epsilon(t)\nu_k w_{j,\tau}^b(\mathbf{z}_{1(j,b,\tau)}(t) - \mathbf{r}_{j,\tau}^b(t)) \\ \qquad \text{(for } j = i \text{ and } b = 1\text{)}, \\ \mathbf{r}_{j,\tau}^b(t) \qquad \text{(otherwise)}, \end{cases} \tag{44}$$

where $1(j,b,\tau) = \theta(j,b,\tau)\mid_{\theta=1}$, and C_i is the class having the smallest discriminant function value among classes other than C_k, i.e., the most likely competing class. This simple result weakens the smoothness policy of the GPD formalization, but increases its feasibility by focusing on the most likely rival class instead of all rival classes. The rule (44) will be referred to later in Section 4.1.

3.3. Design optimality

In a realistic situation where only finite design samples are available, the state of Λ that can be achieved by probabilistic descent training is at most a local optimum over a set of design samples. Morever, in the case of finite learning repetitions, Λ does not necessarily achieve even the local optimum solution. However, GPD addresses this problem in a more advanced way than the previous DFA-based methods.

To discuss this point, we consider an illustrative task. The task is to classify one-dimensional static patterns as one of two classes. The distribution of samples is shown on the horizontal axis in Fig. 3. In the same figure, the classification error count, which is a function of the estimated class boundary position, is also shown. This error count is equivalent to the average empirical loss using (10), and is a discontinuous step function. Note that the classification error count graph changes only at the locations of the design samples. Moreover, in Fig. 3, several continuous curves, each corresponding to the average empirical loss using the smooth classification error count loss (37), are shown; the difference among these continuous loss curves relies on the value setting of α in (37). These curves change smoothly, even at positions other than the sample positions; which demonstrates that controlling α in (37) can produce an interesting change in the smoothness of the loss.

As shown in the figure, conventional use of the discontinuous loss does not take into account the sample distribution at locations other than the given sample positions. In constrast, GPD substantially incorporates the region around the given samples in the design process by using the smooth representation [51]. Methods in the Bayes approach usually attempt to solve this sparse sample problem by introducing parametric distribution models. Compared with them, GPD assumes only the continuity of the sample distribution. This modeling is moderate but quite natural. Thus GPD aims to increase consistency with unknown samples for the estimate of the proper decision boundary simply by making use of the sample distribution's continuity.

Fig. 3 also suggests a way to alleviate the problems of local optima during

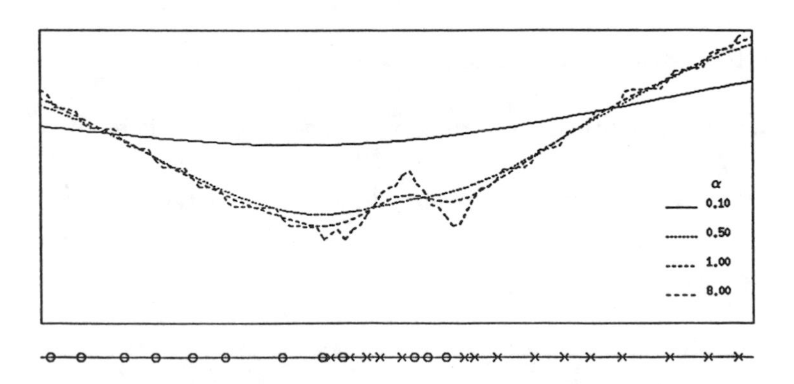

Fig. 3: The relation in smoothness between the smooth classification error count loss and its corresponding empirical average loss.

training; that is, one sets the smoothing larger in the early stage of training and gradually decreases it as training proceeds [59]. Learning is first done in a coarse but global search mode and gradually changes to a fine-adjustment search mode. This is similar to the idea in [42], and furthermore to simulated annealing global search optimization [35].

Consequently, we can argue that GPD has a larger fundamental possibility of achieving the [global] minimum of the loss than the traditional PDM. This characteristic of GPD can be a useful mathematical framework for advanced analysis, such as for training robustness to unknown samples. It is also worthwhile noting that this characteristic can be applied to the entire recognition problem as well as for the classification discussed in this section.

3.4. Minimum classification error learning

The ultimate goal of recognizer/classifier design is to find the parameter set that achieves the condition of minimum error. In this section, we summarize the results of [52], which successfully showed that a GPD-formalization based minimization of the expected smooth classification error count loss, i.e., MCE, possesses a fundamental capability for achieving this minimum error condition. Let us indicate here that this discussion holds true in general approaches to minimum-Bayes-risk recognition as well as in classification.

To maintain the mathematical rigor of our discussions for handling dynamic patterns, we begin our summary by providing a new probability measure $p(x_1^T)$

for dynamic patterns. Let us assume that this probability can be computed by an HMM, which has been widely used in speech recognition.

First, let us consider an [unrealistic] case where the true functional form of the probability is known. We assume that a parameter set determining the functional form is $\dot{\Lambda}$. We also consider a discriminant function that is computed by the HMM probability,

$$g_j(x_1^T;\dot{\Lambda}) = p_{\dot{\Lambda}}(C_j \mid x_1^T). \tag{45}$$

The classification rule used here is (2). Then, for example, defining the misclassification measure as

$$d_k(x_1^T;\Lambda) = -g_k(x_1^T;\Lambda) + \left[\frac{1}{M-1}\sum_{j,j\neq k}\{g_j(x_1^T;\Lambda)\}^{\mu}\right]^{1/\mu}, \tag{46}$$

we can rewrite the expected loss, defined by using the smooth classification error count loss (37) in the GPD formalization, as follows:

$$\begin{aligned} L(\dot{\Lambda}) &= \sum_k \int_{\Omega} p(x_1^T, C_k)\ell_k(x_1^T;\dot{\Lambda})1(x_1^T \in C_k)dx_1^T \\ &\simeq \sum_k \int_{\Omega} p(x_1^T, C_k)1(x_1^T \in C_k) \\ &\quad 1\left(p_{\dot{\Lambda}}(C_k \mid x_1^T) \neq \max_j p_{\dot{\Lambda}}(C_k \mid x_1^T)\right) dx_1^T, \end{aligned} \tag{47}$$

where Ω is the entire sample space of the dynamic patterns x_1^T's, and it is assumed that $dp(x_1^T, C_k) = p(x_1^T, C_k)dx_1^T$. Controlling the smoothness of functions such as L_p norm and the sigmoidal function, we can arbitrarily make $L(\dot{\Lambda})$ closer to the last equation in (47). Note here that we use $\dot{\Lambda}$. Based on this fact, the status of $\dot{\Lambda}$ that corresponds to the minimum of $L(\dot{\Lambda})$ in (47) (which is achieved by adjusting $\dot{\Lambda}$) is clearly equal to the $\dot{\Lambda}^*$ that corresponds to a true probability, or in other words, achieves the maximum *a posteriori* probability condition. Accordingly, the minimum condition of $L(\dot{\Lambda})$ can become arbitrarily close to the minimum classification error probability

$$\mathcal{E} = \sum_k \int_{\Omega_k} p_{\dot{\Lambda}^*}(x_1^T, C_k) 1(x_1^T \in C_k) dx_1^T, \tag{48}$$

where Ω_k is a partial space of Ω that causes a classification error according to the maximum *a posteriori* probability rule, i.e.,

$$\Omega_k = \left\{ x_1^T \in \Omega \,|\, p_{\dot{\Lambda}^*}(C_k \,|\, x_1^T) \neq \max_j p_{\dot{\Lambda}^*}(C_k \,|\, x_1^T) \right\}. \tag{49}$$

The above result is quite interesting, since it shows that one can achieve the minimum classification error probability situation with DFA. However, the assumption that $\dot{\Lambda}$ is known is obviously unsatisfactory, similar to the criticism of the Bayes approach. Recent results concerning the ANN's approximation capability have provided useful suggestions for studying this inadequacy. In the literature, e.g., [33], it was shown that the three-layer perceptron has the fundamental capability of approximating an arbitrary function. In [40, 84], it was also shown that the Radial-Basis Function (RBF) network had universal approximation capability. Furthermore, a mixtured Gaussian distribution function was shown to approximate an arbitrary function in the sense of the minimum L_p norm distortion [95]. Based on these results, one could argue that an HMM with sufficient adjustable parameters has the fundamental capability of modeling an [unknown] true probability function. If $L(\Lambda)$ (and $\mathcal{E}$) has a unique minimum, and if Λ and $L(\Lambda)$ (and $\mathcal{E}$) are monotonic to each other, then the minimum corresponds to the case in which $g_j(x_1^T; \Lambda)$ is equal to the true probability function. Thus, it follows that under this assumption, which is softer (more realistic) than that in the preceding paragraph, DFA enables one to fundamentally achieve minimum classification error probability, even if the true parametric form of the probability function is unknown.

The preceding discussions assumed the impractical condition that sufficient design samples and classifier parameters are given, allowing the remarkable results concerning the fundamental capability of DFA to be shown. However, one should notice that MCE shows a high degree of utility and originality in practical circumstances in which only finite resources are available. In fact, the more limited the design resources are, the more distinctive from others the MCE result is: MCE always directly pursues the minimum classification error situation, conditioned by given circumstances, while the Bayes approach aims at estimating the entire probability function, and moreover, the conventional DFA methods aim at minimizing the average empirical loss, which is not necessarily consistent with the classification error count. It may be appropriate in such practical circumstances to treat the probability form discriminant function as a class membership function that expresses the possibility that the input belongs to its corresponding class, instead of as an estimate of probability concerning sample distribution [83]. This type of understanding would greatly help to extend the application range of MCE.

The above argument is independent of the selection of practical minimization methods, such as the probabilistic update of GPD. Instead, an essential point included therein is the use of the three-step formalization of GPD and the smooth classification error count loss such as (37). This interpretation is the reason why we refer to the result in [52] as MCE and distinguish it from GPD.

Although MCE is a somewhat separate concept from GPD, as shown above, it is quite natural that an implementation of GPD design uses the smooth classification error count loss. Actually, most GPD applications use this smooth classification-oriented loss, and such combined usage is often referred to as an MCE/GPD training, e.g., [58]. In the rest of this section, Section 3, and in Section 4, we use the combined naming, MCE/GPD, so as to make clear the fact that the GPD implementations introduced therein each use a smooth classification error count loss such as (37).

3.5. Speech recognition using MCE/GPD-trained distance classifiers

Let us examine the power of MCE/GPD by summarizing speech pattern classification results, produced by the multi-reference distance classifier represented in (28) [19, 20, 62]. The adjustment rules used were basically the same as (38) and (44). The task was to classify 9-class spoken American E-rhyme letter syllables. In most of the experimental settings of [19], [20], and [62], for a step-by-step analysis, only the reference patterns were considered to be adjustable, while the weights $w_{j,\tau}^{b}$'s were fixed to a constant value of one (1). [19] and [20] additionally reported results for the training of the weights. For simplicity, the smoothing of the sigmoidal error count loss was fixed, and the learning factor, $\epsilon(t)$, was set to a finite sequence of small, monotonically-decreasing numbers, as in (25); i.e., the training did not use the global optimum search strategy described in Section 3.3. Instead, the reference patterns were initialized by using the conventional, modified k-means clustering, which basically relies on the minimum distortion design principle: It was expected that this initialization reduced the risk that the GPD-based gradient search would fall into a local optimum.

The data of this task consisted of the 9-class E-rhyme letter syllables: {b, c, d, e, g, p, t, v, z}. Each sample was recorded over dial-up telephone lines from one hundred (50 male and 50 female) untrained speakers. The sampling rate was 6.67 kHz, and each sample was converted to a sequence of 24-dimensional acoustic feature vectors (12 cepstral coefficients and 12 delta-cepstral coefficients) by shifting a time window of 300 samples with a window overlap of 200 samples. Speaking was done in isolated mode. Since all of the samples included the common phoneme {e}, and were recorded over telephone lines, this task was essentially rather difficult. Recognition experiments were done in multi-speaker mode; i.e., each speaker uttered each of the E-set syllables twice, once for designing and once for testing. Thus, for every class, the design and testing data sets consisted of 100 samples, respectively.

Let us first observe the classification results with the conventional, modified

k-means clustering that was used for initialization. The following classification rates are all ones obtained over the unknown, testing data. The comparison in resultant accuracy (classification rates) between the conventional method and MCE/GPD will illustrate the effectiveness of MCE/GPD.

According to [62], the modified k-means clustering results ranged from 55.0% to 59.8% in the case of using one reference pattern for every class. MCE/GPD achieved rates ranging from 74.2% to 75.4% for the same classifier. In the case of using three reference patterns for every class, the modified k-means clustering resulted in the range of 64.1% to 64.9%; with 74.0% to 77.2% for MCE/GPD[3].

[19] and [20] investigated an implementation somewhat different from (38) and (44), using an exponential-form distance measure and also considering the weights, $w^b_{j,\tau}$'s, to be adjustable. Consequently, in these further experiments, the classification accuracy increased to 79.4% with the use of only one reference pattern per class and reached a remarkable 84.4% by using four references for every class.

For comparison, let us summarize the classification results of HMM classifiers, each designed by using the very same acoustic feature extraction module. Each of the HMM systems had a left-to-right, no skip, mixtured Gaussian structure and was designed by the conventional segmental k-means clustering. Its accuracy was 61.7% for the 5-state and 5-component mixtured Gaussian struc-

[3]Additionally, it was reported that a larger size distance classifier having 12 reference patterns for every class produced only 67.6%. For this classifier, no result based on MCE/GPD was reported.

ture, 66.7% for the 10-state and 5-component mixtured Gaussian structure, and 69.0% for the 15-state and 5-component mixtured Gaussian structure. These results illustrate that the task was quite difficult and that the achievable classification rate of conventional design methods was 70% at most.

The results clearly demonstrate the high utility of MCE/GPD. In particular, it is worthwhile noting that the method achieved a higher accuracy with the smaller size Λ. For example, the MCE/GPD-designed classifier that used only one reference per class had an error rate about three fourths that of the conventionally-designed, twelve-times larger classifier, which used twelve references per class (see the footnote of this subsection).

3.6. Remarks

Although we have used the task of classifying dynamic speech patterns as our basis for discussing the GPD formalism, one may notice that the resultant MCE/GPD training rules can be used for the classification of various kinds of (static and dynamic) patterns other than speech sounds. In fact, MCE/GPD applications to signals other than speech are being reported in current literature, e.g., [76, 93, 107]. However, every class of signals has its own special nature, and there are also a variety of possible task goals, in other words, a variety of recognition criteria (the Bayes decision rule being one among many). Therefore, we should state here that applying GPD to a new task always requires some (usually small) effort to adapt the basic GPD formulations so that they adequately reflect the unique nature of the signals at hand.

4. Relationships between MCE/GPD and Others

4.1. Relation with LVQ

As seen in [61] and [67], several implementations of LVQ have been reported. However, the following fundamental concept of LVQ underlies all of the implementations; that is, *only if a given design sample is misclassified and is located near the actual class boundary, are the reference patterns of the correct class (of the given sample) and those of some competing classes adjusted so as to increase the possibility of classifying the sample correctly*. Moreover, all of them have in common a heuristic window function for restricting the adjustment to the class boundary region.

Recall the simplified MCE/GPD adjustment rule for the distance classifier, i.e., (44). One can easily see that in the circumstances of handling static patterns, this rule is quite similar to an improved version of LVQ, which is characterized as follows [69]: *If and only if a given sample is misclassified and is located in a small region (window) near the actual class boundary, then 1) the closest reference (to the sample) of the correct class is pushed closer to the sample, and 2) the closest reference of the best (most probable) competing class is pulled farther from the sample; otherwise no adjustment is incurred*. Clearly, the adjustment mechanism of this LVQ training is realized in (44). In particular, it is worthwhile noting that the window set in LVQ corresponds to a derivative function of the MCE/GPD's smooth classification error count loss. Consequently, this correspondence proves that in substance, LVQ uses the design objective of classification error count.

In LVQ, the adjustment was generally performed only when a sample was misclassified. In contrast with this (44), incorporating (37), performs the adjustment when a sample is correctly classified but the corresponding misclassification measure is small (close to zero), in other words, when the certainty of the corresponding classification decision is small. This is a natural result of the MCE/GPD formalization using the smooth decision process. Consequently, this "learning-in-correct" contributes to increasing design robustness [62]. The effect of "learning-in-correct" is also reported in [1], [4], [57], and [97].

The close relation between MCE/GPD and LVQ enables us to use many application results of LVQ as circumstantial evidence of the effectiveness of MCE/GPD [57, 61, 70]. The inadequacy of LVQ, which was reported in the literature, e.g., [48], may be due to a lack of consistency with regard to the design objective (loss) formulation between a given task and the LVQ training. In most cases, LVQ has been used for reducing the misclassification of static patterns, such as the elemental acoustic feature vectors of the dynamic speech pattern to be classified originally. Solving this inconsistency was one of the motivations for the development of MCE/GPD.

4.2. Relation with maximization of mutual information

As befits the Bayes decision theory, we have considered classification error count loss in this paper. On the other hand, various other objective functions have also been extensively studied for the sake of increasing classification accuracy. Among them, the use of mutual information has attracted much research interest in speech pattern classification [7].

This approach aims to select the state of Λ so as to increase as much as possible the mutual information between class C_j and a sample x_1^T ($\in C_k$), which is defined in the following:

$$I_k(x_1^T;\Lambda) = \ln \frac{p_\Lambda(x_1^T \mid C_k)}{\sum_j^M p_\Lambda(x_1^T \mid C_j)P(C_j)}. \tag{50}$$

Let us consider the effect of maximizing the mutual information in the GPD (instead of MCE/GPD) framework. For convenience, we use a negative value of mutual information. Then, the goal of GPD design is to minimize this negative measure. The negative mutual information is rewritten as:

$$\begin{aligned} -I_k(x_1^T;\Lambda) &= \ln \frac{\sum_j^M p_\Lambda(x_1^T \mid C_j)P(C_j)}{p_\Lambda(x_1^T \mid C_k)} \\ &= \ln \left\{ P(C_k) + \frac{\sum_{j,j\neq k}^M p_\Lambda(x_1^T \mid C_j)P(C_j)}{p_\Lambda(x_1^T \mid C_k)} \right\} \\ &\geq \ln \frac{\sum_{j,j\neq k}^M p_\Lambda(x_1^T \mid C_j)P(C_j)}{p_\Lambda(x_1^T \mid C_k)} \\ &= -\ln p_\Lambda(x_1^T \mid C_k) + \ln \left\{ \sum_{j,j\neq k}^M P(C_j) e^{\ln p_\Lambda(x_1^T \mid C_j)} \right\}. \end{aligned} \tag{51}$$

Here, defining the logarithmic likelihood, $\ln p_\Lambda(x_1^T \mid C_k)$, as the discriminant function, one can treat the bottom line expression of (51) as a kind of misclassification measure:

$$d_k(x_1^T;\Lambda) = -g_k(x_1^T;\Lambda) + \ln\left\{\sum_{j,j\neq k}^{M} P(C_j)e^{g_j(x_1^T;\Lambda)}\right\}. \tag{52}$$

Then, the inequality,

$$- I_k(x_1^T;\Lambda) \geq d_k(x_1^T;\Lambda) \tag{53}$$

holds true, and therefore maximizing the mutual information leads at least to minimizing the misclassification measure (52). Consequently, it turns out that a classifier design based on maximizing mutual information has the same effect as a GPD design that uses the misclassification measure (52) and the linear loss function. Note here that the loss used is not a smoothed error count but a simple linear function of the misclassification measure. Obviously, this method is not guaranteed to be consistent with the minimum classification error condition unless a $0-1$ nonlinear function is imposed.

4.3. Relation with minimization of squared error loss

Most ANN classifiers have used the squared error between the teaching signal and the classifier output (the value of the discriminant function) as the loss; the design employed therein aims to minimize the expected loss or the average empirical loss, computed by using this individual squared error loss. Similar to the mutual information maximization method, let us consider the nature of this squared error loss minimization method in the GPD framework. First, the squared error loss is represented as

$$\ell_k(x_1^T;\Lambda) = \frac{1}{2}\sum_{j}^{M}\left\{g_j(x_1^T;\Lambda) - \varepsilon_j\right\}^2, \tag{54}$$

where $\{\varepsilon_j\}$ is a teaching signal which is usually set, for a design sample x_1^T ($\in C_k$), to

$$\varepsilon_j = \begin{cases} 1 & (j = k), \\ 0 & \text{(otherwise)}. \end{cases} \tag{55}$$

The loss is thus rewritten as

$$\begin{aligned} \ell_k(x_1^T;\Lambda) &= -g_k(x_1^T;\Lambda) + \frac{1}{2}\left\{(g_k(x_1^T;\Lambda))\right\}^2 + \frac{1}{2} + \frac{1}{2}\sum_{j,j\neq k}^{M}\left\{(g_j(x_1^T;\Lambda))\right\}^2 \\ &> -g_k(x_1^T;\Lambda) + \frac{1}{2}\sum_{j,j\neq k}^{M}\left\{(g_j(x_1^T;\Lambda))\right\}^2. \end{aligned} \tag{56}$$

Then, we can treat the bottom line expression of (56) as a kind of misclassification measure:

$$d_k(x_1^T;\Lambda) = -g_k(x_1^T;\Lambda) + \frac{1}{2}\sum_{j,j\neq k}^{M}\left\{(g_j(x_1^T;\Lambda))\right\}^2. \tag{57}$$

It turns out that the reduction of the squared error loss results in the reduction of this misclassification measure. Hence, we can conclude that the squared error loss minimization leads at least to an optimal situation, in the GPD sense, which is based on the linear loss using the misclassification measure (57). Note here again that the loss used is not a smoothed classification error count but a simple linear function of the misclassification measure. Obviously, it is not guaranteed that the resulting status of this method is consistent with that of the minimum classification error condition.

(57) is difficult to use as the misclassification measure. Since the second term (of the right hand side) of (57), i.e., the average discriminant function of the competing classes, is not normalized by the number of possible classes (M), the design result using this measure is fundamentally affected by the number of classes. Furthermore, since the discriminant function of C_k is compared with the squared values of the other competing class discriminant functions, the scalar decision based on (57) does not emulate the classification appropriately.

5. Derivatives of GPD

5.1. Overview

The key concept of GPD formalization is to emulate the overall process of a task at hand in a tractable functional form. Actually, the implementation of MCE/GPD presented in Section 3 properly realized this concept for the task of classifying a dynamic speech pattern. However, in the example task, the pattern to be classified was *a priori* extracted from its surrounding input signal and was represented in the preset form of an acoustic feature vector sequence. The task defined therein is one simplified and limited case among many. Clearly, the design concept of GPD should be applied directly to more complicated and realistic task situations. For example, the concept of GPD development should be applied to either a connected word recognition task or an open-vocabulary speech recognition task. It should also be applied to the entire process of recognition, which includes the spotting (detection) of target speech sounds, the design of the feature extraction parameters, and the design of the language model. In this light GPD has actually been quite extensively studied, and its original formalization for classification has been dramatically extended to a new family of discriminative design methods, which are more suitable for handling complex real-world tasks. In the following pages of this section, we shall summarize several important members of the GPD family, such as Segmental-GPD, Minimum SPotting Error learning (MSPE) [63], Discriminative Utterance Verification [89], Discriminative Feature Extraction (DFE), e.g., [15, 58], and also introduce application examples of GPD to speaker recognition [66].

5.2. Segmental-GPD for continuous speech recognition

Most definitions used in the simple example task of classifying spoken syllables, in Sections 2 and 3, can be applied to a more realistic task of classifying continuous speech utterances. The most important issue in this advanced application is still to embed the entire process of classification in an appropriate GPD-based functional form. For example, for a connected word recognition task, a discriminant function should be defined so as to directly measure the degree to which a connected word sample belongs to its corresponding class. Due to resource restrictions, most continuous speech recognizers employ subword models, such as phonemes, as the basic unit of acoustic modelling, and represent words and connected words by concatenating the subword models.

The classification process in this task consequently includes the mechanism of dividing a long input of continuous speech into short subword segments. Also, due to the need for tackling the statistical variation of speech sounds, most present continuous speech recognizers use HMM-based acoustic models. Efforts of applying GPD to continuous speech recognition must appropriately take into account these two requirements, i.e. the segmentation mechanism and the probabilistic modelling. Indeed, the development of Segmental-GPD was motivated by these concerns [24, 53].

Assume the use of a modular connected word recognizer that consists of a feature extraction module and a classification module. Then, consider $x_1^T = (\mathbf{z}_1, \mathbf{z}_2, \cdots, \mathbf{z}_T)$ to be speech input to the classifier, where $\mathbf{z}_t$ is an acoustic feature vector (observation) at time index t. Also let $W = (w_1, w_2, \cdots, w_S)$ be a word sequence that usually constitutes a sentence. Generally, an HMM classifier for x_1^T is defined by using a first-order S-state Markov chain governed by the following manifolds:

1. a state transition probability matrix $A = [a_{\iota\kappa}]$, where $a_{\iota\kappa}$ is the probability of making a transition from state ι to state κ.

2. an initial state probability $\pi_\iota = P(q_0 = \iota)$, which specifies the state of the system q_0 at time index $t = 0$.

3. an observation emission probability according to a distribution $b_{q_t}(\mathbf{z}_t) = P(\mathbf{z}_t \mid q_t), q_t = 1, 2, \cdots, S$, which is usually defined with a mixtured Gaus-

sian distribution.

Accordingly, assuming that Λ is a set of the above probability parameters, i.e., the state transition probabilities, the initial state probabilities, and the observation emission probabilities, the discriminant function for x_1^T is given as the following likelihood measure

$$g_{W_r}(x_1^T;\Lambda) = \log P(x_1^T, \mathbf{q}_{W_r}, W_r \mid \Lambda), \tag{58}$$

where

$$\begin{aligned} W_r &= \arg\max_{W \neq W_1, \cdots, W_{r-1}} P(x_1^T, \mathbf{q}_{W_r}, W_r \mid \Lambda) \\ &= r\text{th best word sequence,} \\ \mathbf{q}_{W_r} &= \text{best state sequence corresponding to } W_r \end{aligned} \tag{59}$$

and $P(x_1^T, \mathbf{q}_{W_r}, W_r \mid \Lambda)$ is the joint state-word sequence likelihood.

The goal of training is to find an optimal Λ that leads to accurate classification of word sequences. Given $x_1^T \in W_k$ for training, Segmental-GPD defines the misclassification measure as

$$d_{W_r}(x_1^T;\Lambda) = -g_{W_k}(x_1^T;\Lambda) + \log \left\{ \frac{1}{r_m} \sum_{r=1}^{r_m} e^{g_{W_r}(x_1^T;\Lambda)\cdot\eta} \right\}^{1/\eta} \tag{60}$$

where r_m is the total number of the competing word sequences, different from W_k, that will be taken into consideration in training. The training procedure of Segmental-GPD is then completed by embedding this misclassification measure in the smooth error count loss, as in Section 3. Note here that by setting r_m to a small number in (60), in other words, by focusing only on a limited number of likely rival classes, one can make simpler the adjustment.

Clearly, the Markov modelling satisfies the above two requirements for continuous speech recognition: It performs the segmentation process with its state transition mechanism, and it also provides a probabilistic representation framework. A point to specially note here is that an implementation with probabilistic models should maintain the probabilistic constraint of the model parameters, i.e., the sum of the probabilistic parameters, such as the state transition probability, over all the possible cases should be equal to one (1). Segmental-GPD thus uses parameter transformation as

$$a_{ij} \rightarrow \tilde{a}_{ij} \quad \text{where } a_{ij} = \frac{e^{\tilde{a}_{ij}}}{\sum_m e^{\tilde{a}_{im}}} \tag{61}$$

in the parameter adaptation.

The power of Segmental-GPD has been demonstrated in several experimental tasks. For example, [24] reported quite successful recognition results in the 9-class American English E-ryhme task, used in Section3; i.e., an HMM recognizer with 10-state, 5-component mixtured Gaussian(/state) models scored 99% over the training data and 88% over the testing data. For the connected word case, [25] and [26] reported results in the TI connected-digit recognition task. In particular, in [26] the best results reported so far on the database (a string error rate of 0.72% and a word error rate of 0.24% on testing data) were successfully achieved by a Segmental-GPD-trained, context-dependent subword model system.

The same design concept employed in Segmental-GPD, i.e., optimizing to increase recognition accuracy for concatenated word inputs, has been tested in several slightly different ways and has further demonstrated its high utility [72, 73, 87].

5.3. Minimum error training for open-vocabulary speech recognition

5.3.1. Open-vocabulary speech recognition

Usually, spontaneous conversation utterances contain speech segments that are not directly relevant to the tasks at hand, such as false-starts, interjections, and repairs, and also suffer from various acoustic problems such as increased coarticulation. Many large-vocabulary continuous-speech recognizers have been successfully developed for dictation-style [recognition of read speech] tasks, but the recognition of such casual conversational utterances is still an important research issue [108]. Actually, it is costly to fully model the highly variable acoustic phenomena of large-vocabulary conversational utterances. Therefore, research interest has recently been increasing on a practical alternative, i.e., an open-vocabulary paradigm where only selected keywords are to be recognized. Fig. 4 illustrates the mechanism of open-vocabulary speech recognition.

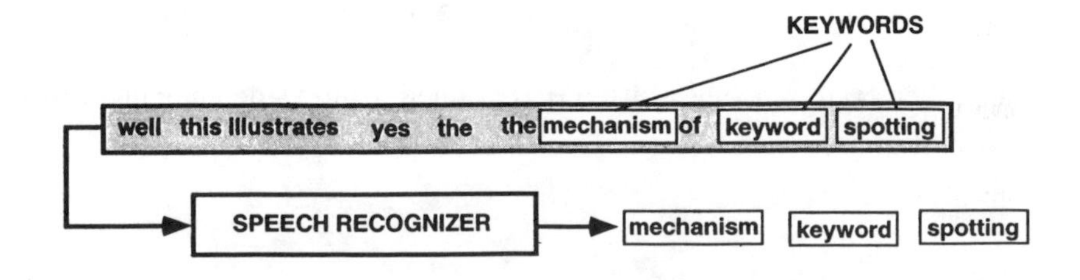

Fig.4: Concept of open-vocabulary speech recognition.

There are two main approaches to open-vocabulary recognition [63, 90]. The first is keyword spotting based on a threshold comparison. A system prepares keyword models, computes the similarity between a segment of input utterance and each keyword model, and decides whether the segment contains the keyword or not by comparing a similarity value and a preset threshold. The second is to use a continuous-speech recognizer having a "filler" model. In this case, the recognizer attempts to segment and classify an input utterance in the same manner as conventional speech recognizers for isolated or concatenated spoken words do, but where a non-keyword segment is classified as the class corresponding to the filler model. Note here that the filler model can fundamentally cover an infinite number of non-keyword segments, and accordingly, a continuous-speech recognizer with this filler model can run as an open-vocabulary system. There are clear differences between the two approaches, especially in terms of segmentation strategy and implementation approach. However, one can also note that the approaches still have a close link to each other because the similarity comparison between the filler class and keyword classes works as a kind of thresholding operation. Evaluation of these approaches is one of the important on-going research issues in the speech recognition field.

5.3.2. Minimum spotting error learning

Depending upon the decision strategy used for spotting, one can clearly note that spotting performance relies heavily upon the adequacy of the model and the threshold. Conventionally, the model has been designed using ML training, as is done for standard continuous speech recognition, and the threshold has been simply determined in a trial-and-error fashion. Obviously, such an unintegrated design method does not guarantee optimal spotting accuracy. One natural solution to this problem is to design both the model and the threshold jointly in the sense of minimizing spotting errors. In this light, GPD was reformulated as Minimum SPotting Error learning (MSPE) for keyword spotting in [63].

Since spotting can be done in keyword-by-keyword mode, or in other words independently for each keyword class, in the algorithm described here we will consider a simple task with only one keyword class and thus a spotter (spotting system) that contains a single keyword model λ (such as reference patterns or HMMs) and its threshold h. The goal of MSPE design is to optimize $\Lambda = \{\lambda, h\}$. For clarity of description, we assume λ to be a reference-pattern-based keyword model, i.e., a sequence of reference patterns, each defined in the acoustic feature vector space. A spotting decision is fundamentally done at every time index. Therefore, a discriminant function $g_t(x_1^T, S_t; \lambda)$ is defined as a function that measures a distance between a selected segment S_t of input utterance x_1^T and the model λ, and the spotting decision rule is formalized as: *If the discriminant function meets*

$$g_t(x_1^T, S_t; \lambda) < h, \tag{62}$$

then the spotter judges at "t" that a keyword exists in the segment S_t*;no keyword is spotted otherwise.* Importantly here, this type of decision could produce in principle two types of spotting errors: *false-detection* (the spotter decides that

S_t does not include the keyword when S_t actually includes.) and *false-alarm* (the spotter decides that S_t includes the keyword even when S_t does not include.).

Similar to original GPD, the MSPE formalization aims to embed the above-cited spotting decision process in an optimizable functional form and provide a concrete algorithm of optimization so that one can consequently reduce the spotting errors. The spotting decision process is then emulated as *spotting measure* $d_t(X;\Lambda)$ which is defined as

$$d_t(X;\Lambda) = h - \ln \left\{ \frac{1}{|I_t|} \sum_{\varsigma \in I_t} \exp(-\xi g_\varsigma(X;\Lambda)) \right\}^{-1/\xi}, \tag{63}$$

where I_t is a short segment, which is set for increasing the reliability and stability of the decision, around the time position t, $|I_t|$ is the size of I_t (the number of acoustic feature frames in I_t), and ξ is a positive constant. A positive value of $d_t(X;\Lambda)$ implies that at least one keyword exists in I_t and a negative value implies that no keyword exists in I_t. Decision results are evaluated by using a loss function that is defined by using two types of smoothed $0-1$ functions, illustrated in Fig. 5, as:

$$\ell_t(X;\Lambda) = \begin{cases} \acute{\ell}(d_t(X;\Lambda)), & \text{if } I_t \text{ includes a training keyword,} \\ \gamma\hat{\ell}(d_t(X;\Lambda)), & \text{if } I_t \text{ includes no training keyword.} \end{cases} \tag{64}$$

$\ell_t(\cdot)$ approximates 1) one (1) for one false-detection (the left side of $\acute{\ell}$, Fig. 5 (a)), 2) γ for one false-alarm (the right side of $\gamma\hat{\ell}$, Fig. 5 (b)), and 3) zero (0) for correct spotting (the right side of $\acute{\ell}$ and the left side of $\gamma\hat{\ell}$), where γ is a parameter controlling the characteristics of the spotting decision. By setting γ to one (1), one can treat all of the errors evenly. By letting γ be less than

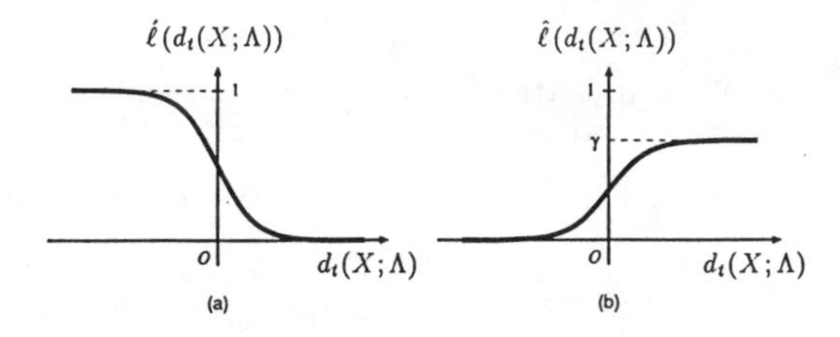

Fig. 5: An example of loss functions for keyword spotting (left side: for false-detection, right side: for false-alarm) (after [52]).

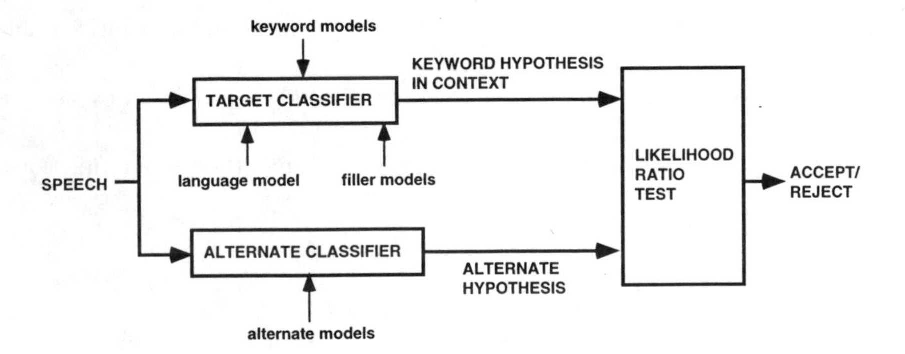

Fig. 6: A speech recognizer with the capability for verifying the word hypotheses produced by a continuous speech recognizer (after [74] with some simplification).

one (1), one can tune the system toward reduction of false-detection errors, which are often more troublesome than false-alarm errors in many application situations.

The training procedure of MSPE is accordingly obtained by applying the adjustment rule (24) of the probabilistic descent theorem to the above-defined functions, i.e., the loss, the spotting measure, and the discriminant function. In [63], readers can find its promising aspects, demonstrated in a task of spotting Japanese consonants.

Even in the open-vocabulary situation, a system is often required to spot a sequence (or set) of keywords instead of only one single keyword. In order to meet this requirement, MSPE was reformulated as the Minimum Error Classification of Keyword-sequences method (MECK). See [64] for details.

5.3.3. Discriminative utterance verification

One can easily notice that the keyword spotting procedure is considered a hypothesis testing problem, based especially on Neyman-Pearson testing and a Likelihood Ratio Test (LRT). The design concept of GPD was also used to increase the LRT-based keyword spotting capability of a continuous speech recognizer employing a filler model.

Fig. 6 illustrates a recognizer used for the above-cited spotting purpose. The system consists of two sub-classifiers: 1) the target classifier and 2) the alternate classifier. The target classifier labels an input as a sequence of the hypothesized keyword and the out-of-vocabulary filler segments. The alternate classifier then generates an alternate hypotheses for the LRT test at the segment where the keyword is hypothesized by the target classifier. An LRT tests the hypothesis that an input x_1^T is generated by the model λ_C corresponding to the keyword W_C versus x_1^T having been generated by an imposter model λ_I corresponding to the alternate hypothesis W_I, according to the likelihood ratio $p(x_1^T \mid \lambda_C)/p(x_1^T \mid \lambda_I)$. Accordingly, if the ratio exceeds a preset threshold, the system accepts the hypothesized keyword; otherwise it rejects the hypothesis.

Obviously, the quality of hypothesis testing relies on the adequacy of the estimated likelihood values. However, similar to most cases of the ML-based speech recognition, it is quite difficult to achieve accurate estimates due to the fact that both the probability density functions and their parameterization forms are often unknown. To alleviate this problem, GPD was used as a design algorithm for directly minimizing the false-detection errors and the false-alarm errors [89]. Fig. 7 illustrates the GPD-based training scheme of the keyword hypothesis verification, i.e., Discriminative Utterance Verification (DUV). The training first defines a distance based on the likelihood ratio as:

$$d_C(x_1^T; \Lambda) = \log p(x_1^T \mid \lambda_I) - \log p(x_1^T \mid \lambda_C). \tag{65}$$

One should notice here that a positive value of the distance implies hypothesis rejection and a negative value implies hypothesis acceptance. Then, a loss is defined so as to embed the two types of errors, i.e., the false-detection and the false-alarm, directly as

$$\ell(x_1^T;\Lambda) = \ell^*(d_C(x_1^T;\Lambda))1(x_1^T \in W_C) + \ell^*(-d_C(x_1^T;\Lambda))1(x_1^T \in W_I), \quad (66)$$

where ℓ^* is a smooth monotonic $0-1$ function. Clearly, a training target here is to minimize the loss over possible design samples in the same manner as the other GPD training cases. Refer to [89] for details.

5.4. Discriminative feature extraction

5.4.1. Fundamentals

In the previous sections, we have considered only the GPD applications to the post-end decision modules, such as the classification (or identification) module, the spotting decision module, and the verification module. Actually several more advanced applications of GPD have been reported where the GPD training was applied to both the acoustic modelling and the language modelling [45, 96], and the training scope was extended to the front-end feature extraction stage [12, 13, 15]. In particular, in recent years, GPD based overall design of the feature extraction and classification modules has attracted the attention of speech researchers who recognize the importance of feature space design (on which the quality of the post-end classification decision indispensably relies), and who also seek solutions to unsatisfactory aspects of current feature extractor design techniques. DFE, which is one of the most straightforward embodiments of GPD's concept of global optimization, is becoming a novel paradigm for

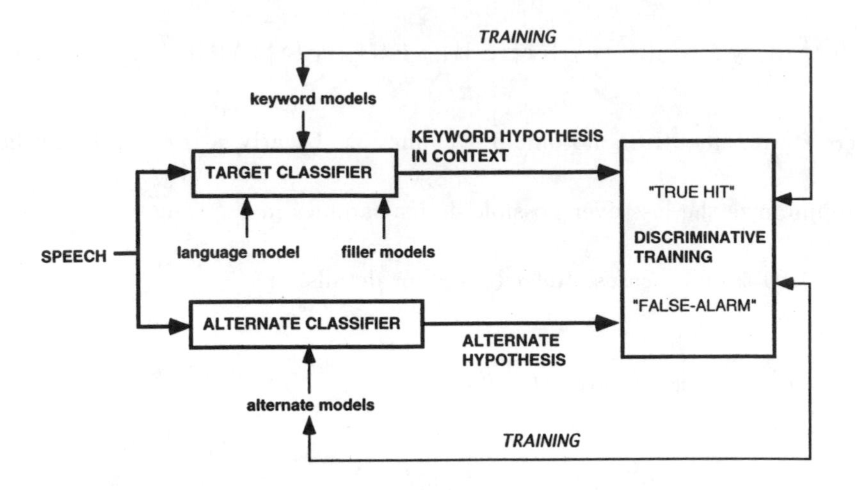

Fig. 7: Block diagram illustraing a GPD-based discriminative training for a speech recognizer verifying hypothesized vocabulary words (after [74] with some simplification).

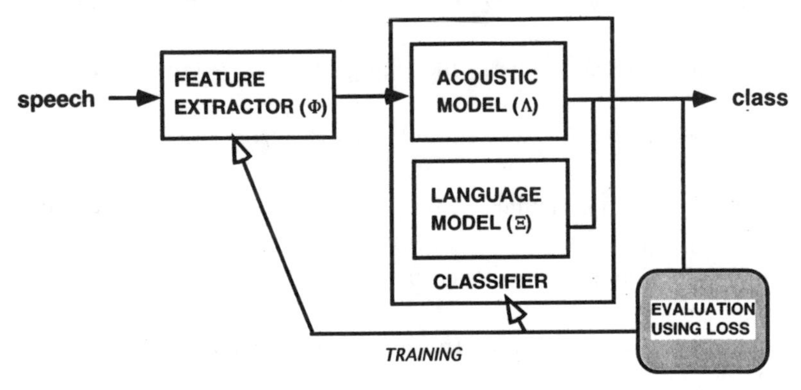

Fig. 8: Training strategy of discriminative feature extraction.

design algorithm study [14, 22, 23, 27, 30, 41, 82, 86].

The formalization of DFE is provided by simply replacing the classification rule (3) with the following recognition rule for an input instantiation $u_1^{T_0}$ in the GPD paradigm:

$$C(u_1^{T_0}) = C_i, \quad \text{iff } i = \arg\max_j g_j(\mathcal{T}_\Phi(u_1^{T_0}); \Lambda), \tag{67}$$

where $\mathcal{T}_\Phi(\ )$ is a feature extraction operation with the designable parameters Φ. One should notice that the feature extraction process is embedded in the discriminant function, and that by assuming the feature extraction process to be differentiable in Φ, both the feature extractor parameters Φ and the classifier parameters Λ now become the targets of the GPD optimization. Λ is trained in the very same manner as that of GPD training for classification, and Φ is trained by using the chain rule of differential calculus that is additionally used for back-propagating the derivative of the loss to the feature extraction module. Fig. 8 illustrates the DFE training strategy. One should clearly notice here that the DFE design jointly optimizes the overall recognizer with the single objective of minimizing recognition errors.

In (67), we used the speech wave sample $u_1^{T_0}$. However, one can use any reasonable form of input, which may be interpreted as an intermediate feature representation of the original utterance input, such as a sequence of FFT-computed power spectrum vectors. In such cases, the DFE training scope is slightly limited but the training can be performed without any degradation of formalization rigor.

In the DFE framework, any reasonable differentiable (in Φ) function, e.g., linear transformation [82], MLP-based nonlinear transformation, or filtering [14], can be applied to the design of the feature extraction module. In the rest of this subsection, Section 5.4, we particularly focus on one example of a direct DFE application, in which a linear transformation called *liftering* is used, and on two [slightly advanced] topics related to DFE training.

5.4.2. An example implementation for cepstrum-based speech recognition

Among many implementation possibilities, [15] studies DFE applied to the design of a cepstrum-based speech recognizer. The cepstrum is one of the most widely-used parameter selections for the acoustic feature vector. It has been shown that its low-quefrency components contain salient information for speech recognition and various lifter shapes have been investigated with the aim of more accurate recognition. However, since there were no methods to design the lifter shape under the criterion of minimizing recognition errors, the shape has usually been determined in an trial-and-error fashion. Obviously, such conventional lifter shapes are not guaranteed to be optimal. An expectation of DFE here is therefore that DFE training can lead to more accurate recognition through the realization of an optimal design for both the lifter feature extractor and the post-end classifier.

Fig. 9 shows a typical DFE-trained lifter shape, which was obtained in the Japanese 5 vowel pattern recognition task [15]. One can find that this lifter de-emphasizes 1) the high quefrency region that corresponds to pitch harmonics and minute spectral structure and 2) the lower quefrency region (0-2 quefrency region) that is dominated by the bias and slant of the overall spectrum while

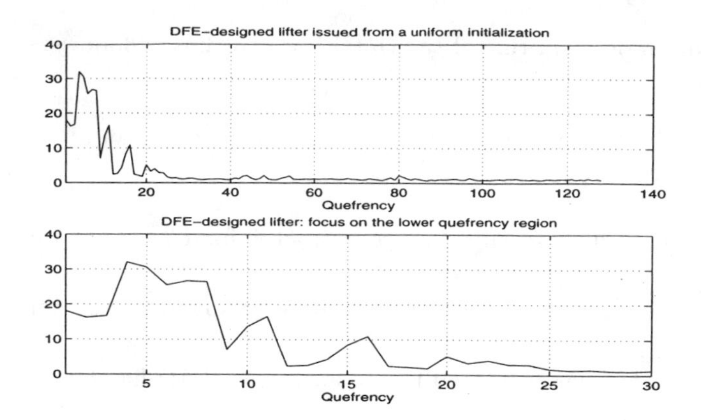

Fig. 9 A typical DFE-trained lifter shape (Upper: the entire view issued from the uniform initialization that set the lifter values at a constant before DFE training. Bottom: the same lifter with a focus on the lower quefrency region) (after [11]).

enhancing the region of 3-20 quefrency that corresponds primarily to the formant structure. The observed shape suggests that the DFE training successfully extracted the salient features for vowel pattern recognition. The training actually achieved a clear error reduction, from 14.5% (best by a baseline system) to 11.3% (DFE-trained system), over unknown testing data using the very same size of recognizer.

5.4.3. Discriminative metric design

Feature extraction can be viewed as a process that forms a metric for measuring the class membership of an input pattern. In general, however, the definition of class identity can be different from class to class. Thus, the feature representation framework, or metric, does not have to be common to all of the classes; i.e. each class could have its own metric that is suitable for representing its identity as effectively as possible. From this point of view, DFE, whose formalization originally assumed a feature extractor common to all sample classes, is not quite sufficient. Motivated by this point, DFE was extended to the Discriminative Metric Design method (DMD) by using the Subspace Method (SM) paradigm [103].

A recognition decision rule of the DMD framework is as follows:

$$C(u_1^{T_0}) = C_i, \qquad \text{iff } i = \arg\max_j g_j(T_\Phi^j(u_1^{T_0}); \Lambda). \tag{68}$$

where $T_\Phi^j(\ )$ is a class-specific feature extractor for C_j. Note that for each class the feature extractor is embedded in its corresponding discriminant function, i.e. $g_j(\cdot)$. Similar to the DFE formalization, DMD has many implementation possibilities. Among them, [103] in particular studied implementation using a quadratic discriminant function, which is shown to be closely related to MCE/GPD training for Gaussian kernel functions and accordingly to recent MCE/GPD applications in HMM speech recognizers.

Readers are referred to [103] and [105] for details.

5.4.4. Minimum error learning subspace method

Recognition-oriented feature design has been most extensively studied in the Subspace Method (SM) paradigm, especially for problems in character and image recognition [81]. SM originated from the similarity method where the recognition (classification) decision was made by measuring the angle between a pattern (vector) to be recognized and a class model for every class. In particular, a recognizer based on this method is defined by (lower-dimensional) *class subspaces*, each assumed to represent the salient features of its corresponding class. There is no distinction between the feature extraction process and the classification process in this system framework. Given an input sample, the recognizer computes the orthogonal projection of the input onto each class subspace and then classifies the pattern to the class giving the maximum projection value.

The performance of SM-based recognizers relies on the quality of the class subspaces. The most fundamental algorithms for designing the subspaces were the CLAFIC [102] and the Multiple Similarity Method (MSM) [46], where each class subspace was designed by running the Karhunen-Loève Transformation or Principal Component Analysis over the design data of its corresponding class. Obviously, this class-by-class design does not directly guarantee recognition error reduction: the recognition error of design samples are not reflected in the subspace design. These methods were later improved, aiming at increasing recognition accuracy. The CLAFIC was extended to the Learning Subspace Method (LSM) (e.g., [81]); the MSM was reformed as the Compound Similarity Method [46]. In these new versions, subspaces are trained iteratively (adaptively) depending upon the recognition result of each design sample; i.e., when an input design pattern is misrecognized, the subspace of the true class and that of the most likely but incorrect class are adjusted so that projection onto the true class subspace increases and projection onto the competing incorrect class subspace decreases. This discriminative iteration actually contributed to improvements in accuracy. However, the training mechanism had only an intuitive validity, and its mathematical optimality in terms of error minimization has long remained unclear.

To alleviate the above problem, [104] studied a DFE formalization for the

SM framework and proposed the Minimum Error Learning Subspace method (MELS) that guarantees the optimal subspace design in the MCE/GPD sense. Detailed discussions including the training convergence proof is included in [104] and [106].

5.5. Speaker recognition using GPD

Speaker recognition is usually classified into two major categories, i.e., *speaker identification*, which is the process of identifying an unknown speaker from a known population, and *speaker verification*, which is the process of verifying the identity of a claimed speaker from a known population. Given a test utterance x_1^T, which is assumed to be represented as a sequence of feature vectors, the likelihood of observing x_1^T being generated by speaker k is denoted as $g_k(x_1^T;\Lambda)$ where Λ is a set of trainable recognizer parameters. Then, the decision operation of speaker identification is formalized as

$$\hat{k} = \arg\max_k g_k(x_1^T;\Lambda), \tag{69}$$

with $\hat{k}$ being the identified speaker, attaining the highest likelihood score among all competing speakers; the decision operation of speaker verification is formalized as: *If the likelihood meets*

$$g_k(x_1^T;\Lambda) > h_k, \tag{70}$$

then the recognizer accepts the claimed speaker identity k; the recognizer rejects it otherwise. One can easily notice here that the formalization for speaker identification is equivalent to that for speech recognition and also that the formalization for speaker verification is essentially the same as that for keyword spotting. One may also notice that the recognizer training here incurs the same difficulty in the likelihood estimation as does the conventional training of continuous speech recognizers and keyword spotters. To alleviate the problem, in [66], GPD training was successfully applied to the design of both an HMM-based speaker identification system and an HMM-based speaker verification system. The derivation of GPD-based training is fundamentally the same as that for speech recognition. Note that whereas GPD training was applied to threshold training in MECK [63], the threshold h_k was determined experimentally in [66]. Refer to [66] for details of the implementation and experimental results.

For speaker verification, [66] also proposed the use of a GPD-trained normalized likelihood score. The decision rule in this case is formalized as *If the likelihood meets*

$$\log \frac{g_k(x_1^T; \Lambda)}{g_{k'}(x_1^T; \Lambda)} > h_k, \tag{71}$$

then the recognizer accepts the claimed speaker identity k; the recognizer rejects it otherwise, where k is the claimed speaker and k' is the antispeaker (the set of speakers other than k). Clearly, the likelihood ratio of (71) is quite similar to the misclassification measure of GPD, and one can naturally expect that the discriminative power of the normalized score is increased by GPD training. In [66], a remarkable improvement due to GPD training was actually demonstrated.

To see other GPD applications to the task of speaker recognition, readers are referred to recent literature such as [31], [68], [94], and [98].

6. Concluding Remarks

6.1. Recent progress of DFA-based speech recognition

In this paper, we have reviewed a recent approach to pattern recognizer design, based on the Generalized Probabilistic Descent method. Since our main purpose is to introduce this recent design approach comprehensively, we have focused our discussions on issues rather closely related to the GPD family of methods. This type of close focus may give readers a somewhat biased understanding however. Thus, for the purpose of being informative and to avoid misunderstanding, we provide a brief review on the current situation of discriminative training methods, other than GPD, in the speech recognition field.

As cited in the Introduction, the development of GPD was motivated by the insufficiency of DFA-based speech pattern recognition in the years of around 1990 and the classical formulation of PDM. However, in parallel with the progress of the family of GPD methods, several other DFA-based methods, such as hybrid HMM/ANN methods (e.g., [77]), have also been greatly extended in this decade.

We showed, in Section 4, that the Minimization of the Squared Error (MSE) loss is not directly tied to the minimization of classification errors. On the other hand, it has been shown that a discriminant function based on this minimization method can be an estimate of the corresponding class *a posteriori* probability function in the statistical sense (e.g., see [6] and [36]). Actually, several hybrid HMM/ANN speech recognizers, in which the ANN works as an *a posteriori* probability estimator, have successfully utilized the discriminative power of MSE-based training in advanced frameworks suited for making classification decisions for the entire speech input instead of its elemental acoustic feature vectors (e.g., [17] and [77]).

Concerning MSE-based training, it seems that there is still some gap between the theoretical criticisms being made and some practical successes; actually, it has also been pointed out that fundamentally, the MSE-based estimation of *a posteriori* probabilities is not accurate in the class boundary region that is usually represented by small values in the sample distribution function [29] but which is important for classification. This point may be a good future research issue.

Efforts to integrate classification decision results, one for every acoustic feature vector, into an overall classification result for an entire speech input have also been made for the framework of Time-Delay Neural Network (TDNN) [101]. That is, TDNN has been successfully extended to Multiple-State TDNN (MSTDNN), in which a loss function is defined over the entire input and explicitly reflects the classification decision at the level of the final category outputs of the problem [37, 38]. In the MSTDNN framework, most parts of the network classifier are adjusted under a single training objective, defined for the entire speech input, i.e., the MSE objective in [37] and the Maximization of Mutual Information (MMI) objective in [38].

Moreover, a similar effort has been made for another type of ANN/HMM hybrid [11], in which the MMI objective is used for training. Clearly, the global design scope of these advanced methods, which attempts to reflect the classification decision of the entire speech input in the training step, is in the same spirit as that of the GPD formalism, though the MMI criterion used therein is different from the MCE training target.

It may be worthwhile mentioning the fact that the MMI-based discriminative training has attracted research interest in recent years, though its mathematical rigor is not sufficiently guaranteed (e.g., see Section 4 of this paper and [77]). Among the many accessible reports, [88] specially evaluates MMI-based training for recurrent networks, and [80] and [100] do the same for HMM classifiers.

In this decade, there has been remarkable progress not only of research in DFA-based design algorithms (e.g., the one based on GPD methods), but also of speech recognition technology in general. A survey of this advancement would be beyond the scope of this paper, and we refer reader to recent reviews such as [65], [77], and [108].

6.2. Advantages of GPD formalization

The most important point of the GPD concept is to embed the entire process of a given recognition task into a smooth functional form so that one can optimize all of the adjustable system parameters in a manner consistent with the design objective of minimizing recognition errors.

There is the natural intuition that discriminative design methods such as GPD can achieve a higher recognition accuracy using fewer trainable system parameters than methods based on the ML principle: the discriminative design method focuses on classification results near the class borders. In fact, we demonstrated this efficiency in terms of the number of trainable system parameters in many experimental results in literature, e.g., [73], a GPD-trained prototype (reference pattern)-based classifier achieved higher recognition scores than an ML-trained hidden Markov network having about twenty times more trainable parameters than that of the GPD-trained system. Unfortunately, this "efficiency" advantage has often been criticized: the high discriminative power is considered to lead to a less robust design result, i.e., an overlearning of the given design samples. However, such criticism is simplistic and misses an essential point. The overlearning phenomenon is commonly observed in any system with a surplus of adjustable parameters; on the other hand, it is rarely observed in discriminative designs using fewer parameters. Any overlearning in a discriminative design has its basis in the directness by which such designs are linked to a given classification task and therefore demonstrates the power of the approach.

The significance of GPD is that it has both mathematical rigor and a great degree of practicality. GPD was shown to provide attractive solutions to three of the four major DFA issues, i.e., 1) the design objective, 2) optimization method, and 3) design consistency with unknown samples. In addition, the high utility of GPD has been demonstrated in various experiments, as were introduced in the paper. The remaining DFA issue, i.e., the selection of the discriminant function form, has not been fully studied to date. Basically, this point should be investigated in a task-by-task basis, and GPD, which gives a sound mathematical framework for the other design issues, can be quite useful in this future study.

6.3. Future issues in GPD-based pattern recognition

As introduced in this paper, GPD provides useful solutions to many of the existing problems in speech pattern recognition. However, there are still issues needing further investigation. A point of utmost importance is the discovery (for any given task) of a desirable form for the discriminant function, as cited in the last paragraph of the previous subsection. Solving this problem will dramatically advance speech recognition technology, but it is obviously difficult and needs significant research effort. Another important point is to find a reasonable way of controling the smoothness of the functions, such as the smooth classification error count loss, defined in the GPD training. This point is closely linked to the issue of training sensitivity to unknown samples, and study on the point may necessitate advanced analyses from the viewpoint of statistics. The following points are rather minor (compared with the issues of the discriminant function form selection and smoothness control) but clearly important for the effective implementation of GPD methods:

1. DFA-based training suffers essentially from a scaling problem; that is, in a large-scale task such as large-vocabulary continuous speech recognition, extensive computation is involved in evaluating the inter-class competition over the tremendous number of possible classes (of connected words). Obviously, the misclassification measure of GPD possesses the same problem. In addition to the simplification based on the use of the L_∞ norm function, further simplification, such as the use of N-best hypotheses [21, 53], may be needed to reduce the adjustment computation in the training stage.

2. The gradient search optimization used in GPD is usually slower than the common EM method, especially in a large-scale task. Incorporating a faster optimization mechanism in the GPD formalism will clearly be useful.

3. The success of the gradient-based adjustment relies on a good selection of several controllable parameters such as the learning factor. The selection often controls the balance of interaction among the modules (such as the feature extractor and the classifier) as well as the speed of training convergence. Since the selection is usually performed experimentally due to a lack of theory, a more systematic and theoretically grounded selection method is needed.

The most important concept of the GPD formalization is, in short, to *formalize the overall procedure of the task at hand into an optimizable design process.* As the original form of GPD for classification problems has been greatly extended so as to cope with various aspects of pattern recognition, such as the open vocabulary problem and speaker verification, we could further extend the family to other types of tasks as well. The simple but significant concept of GPD, i.e., formalizing the overall procedure of the task at hand into an optimizable design process, can be quite useful in general fields such as system design and information processing.

Acknowledgments

We take this opportunity to express our thanks to our many colleagues at the Bell Laboratories and the ATR Laboratories for their significant contributions, over the years, to the development of the GPD family. In particular, our development work would have been impossible without the contributions of Alain Biem, Wu Chou, Erik McDermott, Hideyuki Watanabe, and Eric Woudenberg. We would also like to give special thanks to Eric Woudenberg for his patience and assistance during the preparation of this paper. Finally, we want to acknowledge the careful review work of our two reviewers. Their detailed comments greatly helped us to substantially improve the quality of this paper.

REFERENCES

1. S. Amari; "A Theory of Adaptive Pattern Classifiers", IEEE Trans. Electronic Computers, Vol. EC-16, No. 3, pp. 299-307 (1967 6).

2. S. Amari; "Information Theory II - Geometrical Theory of Information", Kyoritsu (1968) (in Japanese).

3. S. Amari; "Neural Network Model and Connectionism", Tokyo University Press (1989) (in Japanese).

4. A. Ando, and K. Ozeki; "A Clustering Algorithm to Minimize Recognition Error Function", IEICE Trans. A, Vol. J74-A, No. 3, pp. 360-367 (1991 3).

5. H. Asou; "Neural Network Information Processing - Introduction to Connectionism, or Toward Soft Symbolic Representation", Sangyo Tosho (1988) (in Japanese).

6. H. Asou, and N. Otsu; "Nonlinear Data Analysis and Multilayer Perceptrons", IEEE, Proc. IJCNN, Vol. 2, pp. 411-415 (1989 6).

7. L. Bahl, P. Brown, P. de Souza, and R. Mercer; "Maximum Mutual Information Estimation of Hidden Markov Model Parameters for Speech Recognition", IEEE, Proc. ICASSP86, Vol. 1, pp. 49-52 (1986 4).

8. L. Bahl, P. Brown, P. de Souza, and R. Mercer; "A New Algorithm for the Estimation of Hidden Markov Model Parameters", IEEE, Proc. ICASSP88, Vol. 1, pp. 493-496 (1988 4).

9. L. E. Baum, and T. Petrie; "Statistical Inference for Probabilistic Functions of Finite State Markov Chains," Ann. Math. Stat., Vol. 37, pp. 1554-1563 (1966).

10. L. E. Baum, and G. R. Sell; "Growth Functions for Transformations on Manifolds," Pac. J. Math., Vol. 27, No. 2, pp. 211-227 (1968).

11. Y. Bengio, R. De Mori, G. Flammia, and R. Kompe; "Global Optimization of a Neural Network-Hidden Markov Model Hybrid", IEEE Trans. NN, Vol. 3., No. 2, pp. 252-259 (1992 3).

12. A. Biem, and S. Katagiri; "Feature Extraction Based on Minimum Classification Error/Generalized Probabilistic Descent Method", IEEE, Proc. ICASSP93, Vol. 2, pp. 275-278 (1993 4).

13. A. Biem, S. Katagiri, and B.-H. Juang; "Discriminative Feature Extraction for Speech Recognition", IEEE, Neural Networks for Signal Processing III - Proc. of the 1993 IEEE Workshop, pp. 392-401 (1993 9).

14. A. Biem, E. McDermott, and S. Katagiri; "Discriminative Feature Extraction to Filter Bank Design", IEEE, Neural Networks for Signal Processing VI, pp. 273-282 (1996 9).

15. A. Biem, S. Katagiri, and B.-H. Juang; "Pattern Recognition Using Discriminative Feature Extraction", IEEE Trans. Signal Processing, Vol. 45, No. 2 pp. 500-504 (1997 2).

16. H. Bourlard, and C. Wellekens; "Links Between Markov Models and Multilayer Perceptrons", IEEE Trans. PAMI, Vol. 12, No. 12, pp. 1167-1178 (1990).

17. H. Bourlard, and N. Morgan; "Connectionist Speech Recognition -A Hybrid Approach", Kluwer Academic Publishers (1994).

18. J. Bridle; "Alpha-Nets: A Recurrent 'Neural' Network Architecture with a Hidden Markov Model Interpretation", Speech Communication, Vol. 9, pp. 83-92 (1990).

19. P.-C. Chang, and B.-H. Juang; "Discriminative Template Training for Dynamic Programming Speech Recognition", IEEE, Proc. ICASSP92, Vol. 1, pp. 493-496 (1992 3).

20. P.-C. Chang, and B.-H. Juang; "Discriminative Training of Dynamic Programming Based Speech Recognizers", IEEE Trans. SAP, Vol. 1, No. 2, pp. 135-143 (1993 4).

21. J.-K. Chen and F. K. Soong; "An N-best Candidates-Based Discriminative Training for Speech Recognition Applications", IEEE Trans. SAP, Vol. 2, No. 1, pp. 206-216 (1994 1).

22. R. Chengalvarayan and L. Deng; "Use of Generalized Dynamic Feature Parameters for Speech Recognition", IEEE Trans. SAP, Vol. 5, No. 3, pp. 232-242 (1997 5).

23. R. Chengalvarayan and L. Deng; "HMM-Based Speech Recognition Using State-Dependent, Discriminatively Derived Transforms on Mel-Warped DFT Features", IEEE Trans. SAP, Vol. 5, No. 3, pp. 243-256 (1997 5).

24. W. Chou, B.-H. Juang, and C.-H. Lee; "Segmental GPD Training of HMM

Based Speech Recognition", IEEE, Proc. ICASSP92, Vol. 1, pp. 473-476 (1992 3).

25. W. Chou, C.-H. Lee, and B.-H. Juang; "Minimum Error Rate Training Based on N-Best String Models", IEEE, Proc. ICASSP93, Vol. 2, pp. 652-655 (1993 4).

26. W. Chou, C.-H. Lee, and B.-H. Juang; "Minimum Error Rate Training of Inter-Word Context Dependent Acoustic Model Units in Speech Recognition", Proc. ICSLP94, pp. 439-442 (1994 9).

27. W. Chou, M. G. Rahim, and E. Buhrke; "Signal Conditioned Minimum Error Rate Training", Proc. EUROSPEECH95, pp. 495-498 (1995 9).

28. A. P. Dempster, N. M. Laird, and D. B. Rubin; "Maximum Likelihood from Incomplete Data via the EM Algorithm", J. Roy. Stat. Soc., Vol. 39, No. 1, pp. 1-38 (1977).

29. R. Duda, and P. Hart; "Pattern Classification and Scene Analysis", John Wiley and Sons (1973).

30. S. Euler; "Integrated Optimization of Feature Transformation for Speech Recognition", Proc. EUROSPEECH95, pp. 109-112 (1995 9).

31. K. R. Farrell, R. J. Mammone, and K. T. Assaleh; "Speaker Recognition Using Neural Networks and Conventional Classifiers", IEEE Trans. SAP, Vol. 2, No. 1, PART II, pp. 194-205 (1994 1).

32. K. Fukunaga; "Introduction to Statistical Pattern Recognition", Academic Press

(1972).

33. K. Funahashi; "On the Approximate Realization of Continuous Mappings by Neural Networks", Neural Networks, Vol. 2, No. 3, pp. 183-191 (1989).

34. Y.-Q. Gao, T.-Y. Huang, D.-W. Chen; "HMM-Based Warping in Neural Networks", IEEE, Proc. ICASSP90, Vol. 1, pp. 501-504 (1990 4).

35. S. Geman, and D. Geman; "Stochastic Relaxation, Gibbs Distributions, and the Bayesian Restoration of Images", IEEE Trans. on PAMI, Vol. PAMI-6, No. 6, pp. 721-741 (1984 11).

36. H. Gish; "A Probabilistic Approach to the Understanding and Training of Neural Network Classifiers", IEEE, Proc. ICASSP90, Vol. 3, pp. 1361-1364 (1990 4).

37. P. Haffner, M. Franzini, and A. Waibel; "Integrating Time Alignment and Neural Networks for High Performance Continuous Speech Recognition", IEEE, Proc. ICASSP91, pp. 105-108 (1991 5).

38. P. Haffner; "A New Probabilistic Framework for Connectionist Time Alignment", Proc. ICSLP94, pp. 1559-1562 (1994 9).

39. J. Hampshire II, and A. Waibel; "A Novel Objective Function for Improved Phoneme Recognition Using Time-Delay Neural Networks", IEEE Tras. NN, Vol. 1, No. 2, pp. 216-228 (1990 6).

40. E. Hartman, J. Keeler, and J. Kowalski; "Layered Neural Networks with Gaussian Hidden Units as Universal Approximations", Neural Computation, Vol. 2, pp. 210-215 (1990).

41. J. Hernando, J. Ayarte, and E. Monte; "Optimization of Speech Parameter Weighting for CDHMM word recognition", Proc. EUROSPEECH95, pp. 105-108 (1995 9).

42. J. Hopfield, and D. Tank; ""Neural" Computation of Decisions in Optimization Problems", Springer-Verlag, Biological Cybernetics, Vol. 52, pp. 141-152 (1985).

43. D. Howell; "The Multi-Layer Perceptron as a Discriminating Post Processor for Hidden Markov Networks", FASE, Proc. 7th FASE Symposium -Speech-, pp. 1389-1396 (1988).

44. X. Huang, Y. Ariki, and M. Jack; "Hidden Markov Models for Speech Recognition", Edinburg University Press (1990).

45. X. Huang, M. Belin, F. Alleva, and M. Hwang; "Unified Stochastic Engine (USE) for Speech Recognition", IEEE, Proc. ICASSP93, Vol. 2, pp. 636-639 (1993 4).

46. T. Iijima; "Pattern Recognition Theory", Morikita Publisher (1989) (in Japanese).

47. H. Iwamida, S. Katagiri, E. McDermott, and Y. Tohkura; "A Hybrid Speech Recognition System Using HMMs with an LVQ-Trained Codebook", ASJ, J. Acoust. Soc. Jpn. (E), Vol. 11, No. 5, pp. 277-286 (1990 9).

48. H. Iwamida, S. Katagiri, and E. McDermott; "Re-Evaluation of LVQ-HMM Hybrid Algorithm", ASJ, J. Acoust. Soc. Jpn. (E), Vol. 14, No. 4, pp. 267-274

(1993 7).

49. F. Jelinek; "Self-Organized Language Modeling for Speech Recognition", IBM, T. J. Watson Research Center Report (1985).

50. B.-H. Juang, and L. Rabiner; "The Segmental K-Means Algorithm for Estimating Parameters of Hidden Markov Models", IEEE Trans. ASSP, Vol. 38, No. 9, pp. 1639-1641 (1990 9).

51. B.-H. Juang, and S. Katagiri; "Discriminative Training", ASJ, J. Acoust. Soc. Jpn. (E), Vol. 13, No. 6, pp. 333-339 (1992 11).

52. B.-H. Juang, and S. Katagiri; "Discriminative Learning for Minimum Error Classification", IEEE Trans. SP., Vol. 40, No. 12, pp. 3043-3054 (1992 12).

53. B.-H. Juang, W. Chou, and C.-H. Lee; "Minimum Classification Error Rate Methods for Speech Recognition", IEEE Trans. SAP., Vol. 5, No. 3, pp. 257-265 (1997 3).

54. S. Katagiri, C. H. Lee, and B.-H. Juang; "A Generalized Probabilistic Descent Method", ASJ, Proc. Conf., pp. 141-142 (1990 9).

55. S. Katagiri, C.-H. Lee, and B.-H. Juang; "Discriminative Multi-Layer Feed-Forward Networks", IEEE, Neural Networks for Signal Processing, pp. 11-20 (1991 9).

56. S. Katagiri, C.-H. Lee, and B.-H. Juang; "New Discriminative Training Algorithms Based on the Generalized Probabilistic Descent Method", IEEE, Neural Networks for Signal Processing, pp. 299-308 (1991 9).

57. S. Katagiri, and C.-H. Lee; "A New Hybrid Algorithm for Speech Recognition Based on HMM Segmentation and Learning Vector Quantization", IEEE Trans. SAP, Vol. 1, No. 4, pp. 421-430 (1993 10).

58. S. Katagiri, B.-H. Juang, and A. Biem; "Discriminative Feature Extraction", in "Artificial Neural Networks for Speech and Vision (ed. R. Mammone)", Chapman and Hall, pp. 278-293 (1994).

59. S. Katagiri; "A Unified Approach to Pattern Recognition", Proc. ISANN94, pp. 561-570 (1994 12).

60. D. Kimber, M. Bush, and G. Tajchman; "Speaker-Independent Vowel Classification Using Hidden Markov Models and LVQ2", IEEE, Proc. ICASSP90, Vol. 1, pp. 497-500 (1990 4).

61. T. Kohonen; "The Self-Organizing Map", Proc. IEEE, Vol. 78, No. 9, pp. 1464-1480 (1990 9).

62. T. Komori, and S. Katagiri; "GPD Training of Dynamic Programming -Based Speech Recognizers", ASJ, J. Acoust. Soc. Jpn. (E), Vol. 13, No. 6, pp. 341-349 (1992 11).

63. T. Komori, and S. Katagiri; "A Minimum Error Approach to Spotting-Based Pattern Recognition", IEICE Trans. Inf. & Syst., Vol. E78-D, No. 8, pp. 1032-1043 (1995 8).

64. T. Komori, and S. Katagiri; "A Novel Spotting-Based Approach to Continuous Speech Recognition: Minimum Error Classification of Keyword-Sequences", J.

Acous. Soc. Jpn. (E), Vol. 16, No. 3, pp. 147-157 (1995 5).

65. C.-H. Lee, F. K. Soong, and K. K. Paliwal (eds.); "Automatic Speech and Speaker Recognition", Kluwer Academic Publishers (1996).

66. C.-S. Liu, C.-H. Lee, W. Chou, B.-H. Juang, and A. Rosenberg; "A Study on Minimum Error Discriminative Training for Speaker Recognition", J. Acoust. Soc. Am., Vol. 97, No. 1, pp. 637-648 (1995 1).

67. S. Makino, M. Endo, T. Sone, and K. Kido; "Recognition of Phonemes in Continuous Speech Using a Modified LVQ2 Method", J. Acoust. Soc. Jpn. (E), Vol. 13, No. 6, pp. 351-360 (1992 11).

68. T. Matsui and S. Furui; "A Study of Speaker Adaptation Based on Minimum Classification Error Training", Proc. EUROSPEECH95, pp. 81-84 (1995 9).

69. E. McDermott; "LVQ3 for Phoneme Recognition", ASJ, Proc. Spring Conf., pp. 151-152 (1990 3).

70. E. McDermott, and S. Katagiri; "LVQ-Based Shift-Tolerant Phoneme Recognition", IEEE Trans. SP, Vol. 39, No. 6, pp. 1398-1411 (1991 6).

71. E. McDermott, and S. Katagiri; "Prototype-Based MCE/GPD Training for Word Spotting and Connected Word Recognition", IEEE, Proc. ICASSP93, Vol. 2, pp. 291-294 (1993 4).

72. E. McDermott, and S. Katagiri; "Prototype-Based MCE/GPD Training for Various Speech Units", Computer Speech and Language, Vol. 8, pp. 351-368 (1994).

73. E. McDermott, and S. Katagiri; "String-Level MCE for Continuous Phoneme Recognition", Proc. EUROSPEECH97, Vol. 1, pp. 123-126 (1997 9).

74. M. Miyatake, H. Sawai, Y. Minami, and K. Shikano; "Integrated Training for Spotting Japanese Phonemes Using Large Phonemic Time-Delay Neural Networks", IEEE, Proc. ICASSP90, Vol. 1, pp. 449-452 (1990 4).

75. S. Mizuta, and K. Nakajima; "A Discriminative Training Method for Continuous Mixture Density HMMs and its Implementation to Recognize Noisy Speech", J. Acoust. Soc. Jpn. (E), Vol. 13, No. 6, pp. 389-393 (1992 11).

76. H. Mizutani; "Discriminative Learning for Minimum Error Classification with Reject Options", IEICE, Technical Report PRMU97-245 (1998 2) (in Japanese).

77. N. Morgan and H. A. Bourlard; "Neural Networks for Statistical Recognition of Continuous Speech", Proc. of IEEE, Vol. 83, No. 5, pp. 742-770 (1995 5).

78. L. Niles, and H. Silverman; "Combining Hidden Markov Model and Neural Network Classifier", IEEE, Proc. ICASSP90, Vol. 1, pp. 417-420 (1990 4).

79. N. Nilsson; "The Mathematical Foundations of Learning Machines", Morgan Kaufmann Publishers (1990).

80. Y. Normandin, R. Cardin, and R. De Mori; "High-Performance Connected Digit Recognition Using Maximum Mutual Information Estimation", IEEE Trans. SAP, Vol. 2, No. 2, pp. 299-311 (1994 4).

81. E. Oja; "Subspace Methods of Pattern Recognition", Research Studies Press (1983).

82. K. K. Paliwal, M. Bacchiani, and Y. Sagisaka; "Minimum Classification Error Training Algorithm for Feature Extractor and Pattern Classifier in Speech Recognition", Proc. EUROSPEECH95, pp. 541-544 (1995 9).

83. Y.-H. Pao: "Adaptive Pattern Recognition and Neural Networks", Addison-Wesley Publishing Company (1989).

84. J. Park, and I. Sandberg; "Universal Approximation Using Radial-Basis-Function Networks", Neural Computation, Vol. 3, pp. 246-257 (1991).

85. L. Rabiner, and B.-H. Juang; "Fundamentals of Speech Recognition", Prentice Hall (1993).

86. M. G. Rahim and C.-H. Lee; "Simultaneous ANN Feature and HMM Recognizer Design Using String-Based Minimum Classification Error (MCE) Training", Proc. ICSLP96, pp. 1824-1827 (1995 9).

87. D. Rainton, and S. Sagayama; "Minimum Error Classification Training of HMMs -Implementation Details and Experimental Results", ASJ, J. Acoust. Soc. Jpn. (E), Vol. 13, No. 6, pp. 379-387 (1992 11).

88. A. J. Robinson; "An Application of Recurrent Nets to Phone Probability Estimation", IEEE Trans. NN, Vol. 5, No. 2, pp. 298-305 (1994 3).

89. R. C. Rose, B.-H. Juang, and C.-H. Lee; "A Training Procedure for verifying string hypothesis in continuous speech recognition," Proc. ICASSP95, pp. 281-284 (1995 4).

90. R. C. Rose; "Word Spotting from Continuous Speech Utterances", in "Auto-

matic Speech and Speaker Recognition: Advanced Topics (ed. C.-H. Lee, F. K. Soong, and K. K. Paliwal)", Kluwer Academic Publishers, pp. 303-329 (1996).

91. D. E. Rumelhart, J. L. McClelland, and the PDP Research Group; "Parallel Distributed Processing", MIT Press (1986).

92. H. Sakoe, R. Isotani, K. Yoshida, K. Iso, and T. Watanabe; "Speaker-Independent Word Recognition Using Dynamic Programming Neural Networks", IEEE, Proc. ICASSP89, Vol. 1, pp. 29-32 (1989 5).

93. A. Sato and K. Yamada; "A Learning Method for Definite Canonicalization Based on Minimum Classification Error", IEICE, Technical Report PRMU97-244 (1998 2) (in Japanese).

94. A. Setlur and T. Jacobs; "Results of a Speaker Verification Service Trial Using HMM Models", Proc. EUROSPEECH95, pp. 639-642 (1995 9).

95. H. Sorenson, and D. Alspach; "Recursive Baysian Estimation Using Gaussian Sums", Automatica, Vol. 7, pp. 465-479 (1971).

96. K.-Y. Su, T.-H. Chiang, and Y.-C. Lin; "A Unified Framework to Incorporate Speech and Language Information in Spoken Language Processing", Proc. ICASSP92, Vol. 1, pp. 185-188 (1992 3).

97. K.-Y. Su, and C.-H. Lee; "Speech Recognition Using Weighted HMM and Subspace Projection Approaches", IEEE Trans. SAP, Vol. 2, No. 1, pp. 69-79 (1994 1).

98. M. Sugiyama, and K. Kurinami; "Minimal Classification Error Optimization

for a Speaker Mapping Neural Network", IEEE, Neural Networks for Signal Processing II, pp. 233-242 (1992 10).

99. K. Unnikrishnan, J. Hopfield, and D. Tank; "Connected-Digit Speaker-Dependent Speech Recognition Using a Neural Network with Time-Delayed Connections", IEEE Trans. SP, Vol. 39, No. 3, pp. 698-713 (1991 3).

100. V. Valtchev, J. J. Odell, P. C. Woodland, and S. J. Young; "Lattice-Based Discriminative Training for Large Vocabulary Speech Recognition", IEEE, Proc. ICASSP96, pp. 605-608 (1996 5).

101. A. Waibel, T. Hanazawa, G. Hinton, K. Shikano, K. Lang; "Phoneme Recognition Using Time-Delay Neural Networks", IEEE Trans. ASSP, Vol. 37, No. 3, pp. 328-339 (1989 3).

102. S. Watanabe, P. F. Lambert, C. A. Kulikowski, J. L. Buxton, and R. Walker; "Evaluation and Selection of Variables in Pattern Recognition", in Computer and Information Sciences II (J. T. Tou Ed.), pp. 91-122, Academic Press (1967).

103. H. Watanabe, T. Yamaguchi, and S. Katagiri; "A Novel Approach to Pattern Recognition Based on Discriminative Metric Design", IEEE, Neural Networks for Signal Processing V, pp. 48-57 (1995 8).

104. H. Watanabe and S. Katagiri; "Discriminative Subspace Method for Minimum Error Pattern Recognition", IEEE, Neural Networks for Signal Processing V, pp. 77-86 (1995 8).

105. H. Watanabe, T. Yamaguchi, and S. Katagiri; "Discriminative Metric Design for

Robust Pattern Recognition", IEEE Trans. SP, Vol. 45, No. 11, pp. 2655-2662 (1997 11).

106. H. Watanabe and S. Katagiri; "Subspace Method for Minimum Error Pattern Recognition", IEICE Trans. Info. & Sys., Vol. E80-D, No. 12, pp. 1195-1204 (1997 12).

107. H. Watanabe, Y. Matsumoto, and S. Katagiri; "A Novel Approach to Signal Detection Based on the Generalized Probabilistic Descent Method", IEEE, Proc. ICASSP98 (1998 5) (to appear).

108. S. Young; "A Review of Large-Vocabulary Continuous-Speech Recognition", IEEE SP Magazine, Vol. 13, No. 5, pp. 45-57 (1996 9).

109. G. Yu, W. Russel, R. Schwartz, and J. Makhoul; "Discriminant Continuous Speech Recognition", IEEE, Proc. ICASSP90, Vol. 2, pp. 685-688 (1990 4).

APPENDIX

A-1. Probabilistic descent theorem for probability-based discriminant functions

In the case of using a probability function as the discriminant function, the adjustment convergence of (20) no longer holds true: The probabilistic descent theorem assumed the classifier parameter Λ to be arbitrary scalar variables and/or arbitrary vector variables, and did not anticipate that each class parameter vector $\lambda_j = (\lambda_j[1], \cdots, \lambda_j[D])^T$ had to meet the following probability constraint:

$$\sum_d \lambda_j[d] = 1 \quad \text{and} \quad \lambda_j[d] \geq 0 \text{ (for } \forall d), \tag{72}$$

where we assume that λ_j is of D-dimension. The **HMM** state transition probability and the mixture coefficients of the mixtured Gaussian HMM are typical examples of this type of constrained parameter vector.

One solution to this problem is to use the transformation of (61). In addition, [54] and [56] provids a more general solution, i.e., the following, constrained probabilistic descent theorem for applying the GPD-based adjustment to the constrained parameters.

Theorem 2 (Constrained probabilistic descent theorem) *Assume that a design sample $x_1^T(t)(\in C_k)$ is given and a classifier parameter set λ_j for C_j includes the parameter vector set $\{\rho_j\}$ that satisfies the constraint of (72). Then, the adjustment using $(\delta\Lambda(x_1^T(t), C_k, \Lambda))_\Phi$, which is obtained by project-*

ing $\delta\Lambda(x_1^T(t), C_k, \Lambda)$ *of (20) to the* ρ_j*'s subspace that is spanned according to the constraint, reduces* $\delta L(\Lambda)$ *in the sense of expectation; here* $(\ )_\Phi$ *is the orthogonal projection of the parenthesized parameter vector onto the (72)-based subspace of the parameter vector space.*

Proof. Considering that the gradient computation is independently done for every dimensional element of the parameter vector, we can prove the theorem by showing that the adjustment in terms of one parameter vector ρ_j, which satisfies the constraint, reduces $\delta L(\Lambda)$ in the expectation sense. In fact, the following inequality clearly proves that the adjustment can reduce $\delta L(\Lambda)$ in the expectation sense:

$$\begin{aligned} E[\delta L(\Lambda)] &= E[(\delta\Lambda(x_n, C_k, \Lambda))_\Phi \cdot \nabla L(\Lambda)] \\ &= -\epsilon\,\{\mathbf{U}E[\nabla\ell_k(x_n;\Lambda)]\} \cdot (\nabla L(\Lambda))_\Phi \\ &= -\epsilon\,\{\mathbf{U}\nabla L(\Lambda)\} \cdot (\nabla L(\Lambda))_\Phi \\ &= -\epsilon\mathbf{U}||\nabla L(\Lambda)||^2 cos\vartheta \leq 0, \end{aligned} \tag{73}$$

where ϑ is the angle between $\nabla L(\Lambda)$ and $(\nabla L(\Lambda))_\Phi$, and $\vartheta \leq \pi/2$ always holds true. ∎

Now it is clear that the characteristics of the original theorem, such as the stochastic approximation-based convergence to a locally optimal situation, hold true.

Let us point out that even in the case of using the probability-based discriminant function, all the parameters do not need to observe the constrained probabilistic descent adjustment. For example, the mean vector of the component Gaussian distribution of the mixtured Gaussian HMM system can be updated according to (20). Moreover, the covariance matrix of the component distribution can be updated in the same way. The adjustment of the diagonal covariance matrix, which is widely used in speech pattern recognition, is actually easy. The adjustment of the full covariance matrix is somewhat complicated and difficult: It should satisfy the positive-definiteness constraint of the matrix. However, it can be done fundamentally based on (20). A GPD implementation for this case is presented in detail in [103].

A-2. Selection of discriminant function forms

In this paper, we have especially used the distance classifier for explanations. Thus, the form of the discriminant function used has been distance. Although briefly, we have also mentioned the case of the probability function form of the discriminant function. Obviously, the distance is closely related to the probability function, and thus it is unreasonable to consider these two forms to be separate. In order to embody GPD effectively, we should comprehensively understand the nature of the discriminant functions selected. In light of this, we summarize here the nature of several important discriminant function forms (or the structure of classifiers).

Discussions focusing on the fixed-dimensional sample space can not necessarily be applied directly to the classification of dynamic patterns. However, it should be of some help to summarize various structures of the static pattern classifiers.

As shown in the literature, such as [33], it is known that the Multi-Layer Perceptron (MLP), having a sigmoidal output function at its hidden nodes, possesses quite a large potential for function approximation. [52] introduced the GPD implementation for this type of nonlinear network classifier and demonstrated the high efficiency and effectiveness of its training.

There are several other types of feedforward networks that are similar to MLP in terms of structure but that compute different measures of distance. Among them, let us focus on a three-layer RBF network. Interestingly, it is shown that this type of network achieves universal function approximation under mild conditions, even though linear (e.g., see [40]). Thus, similar to the above MLP case [52], an RBF network trained with GPD would thus greatly contribute to achieving the minimum classification error probability condition.

If the three-layer RBF network uses a continuous probabilistic basis function and, furthermore, the upper connection coefficients satisfy the probability constraint (72), the network works as a likelihood network that uses probability or likelihood as the discriminant function [55]. In particular, a likelihood network using the Gaussian-form probability function as the discriminant function, i.e., a mixtured Gaussian likelihood network, would be quite useful due to its simple computation. Due to the probability constraint, a likelihood network would not have the universal approximation capability of the original RBF network in the sense of [40]. However, it was shown that the mixtured Gaussian density function could approximate an arbitrary continuous probability density function in the sense of minimizing the L_p norm error between the two corresponding density functions [95]. These results would therefore enable us to expect highly effective applications of GPD to this Gaussian-basis likelihood network.

The mixtured Gaussian likelihood network is obviously equivalent to the one state case of the mixtured Gaussian-component, continuous HMM. The discussions of the above paragraphs thus suggest that the GPD-trained mixtured Gaussian-component HMM classifier would have a large potential for achieving the minimum error probability condition of dynamic pattern classification.

Similar to the Mahalanobis distance, several likelihood-based distance measures can be defined based on the Gaussian distribution probability density function. The squared Euclidean distance is the most limited but simplest version among these likelihood-based distances. Let us replace the probability computation with a computation using one of these distances in a Gaussian-basis likelihood network. We assume here that the lower connection weights correspond to the mean vectors of the Gaussian-basis function. Then, the resulting network is obviously equivalent to a distance network using the distance between an input vector and a reference vector as the discriminant function [55]. One may note here that this distance network can also be seen as a generalized version of a distance classifier. Fundamentally, in a distance network, the output nodes are fully connected to the hidden nodes. On the other hand, as described in Section 2.4, a distance classifier usually assigns one reference vector to one of all of the possible classes exclusively. The distance network does not take such *a priori* assignment of the reference vectors; instead, this network attempts to share the reference vectors (inner models) among the classes. The same concept of sharing is also implemented for a continuous HMM [44].

In practical applications of RBF networks and distance networks for classification, the network classifiers usually use a much smaller number of hidden nodes than given design samples. That is, any clustering concept for the given samples underlies the setting of the hidden nodes. Let us consider the situation in which as we increase the number of hidden nodes, we decrease the size of the cluster corresponding to each hidden node. In the extreme case in which each hidden-node cluster corresponds to one sample, the RBF network performs a generalized Parzen approximation and the distance network works as a generalized k-nearest neighbor classifier. The generalization here relies on flexibility in determining the upper layer connection weights. The Parzen estimate of probability distribution is an unbiased and consistent estimate of the true distribution function. Moreover, the error probability of the k-nearest neighbor method is bounded to be below twice the Bayes error probability. These well known characteristics suggest that the distance network is also worth further study.

Fundamentally, one should select classifier structures, discussed above, that determine the discriminant function forms, based on ease of hardware implementation, computational efficiency, and the nature of a given feature representation.

A-3. Various definitions of discriminant functions, misclassification measures, and losses

There is a large variety in the defining functions used in GPD, such as the discriminant functions and the misclassification measures. Assuming that the discriminant function represents the degree to which an input sample belongs to a class, we present several examples of definitions which have not been used in the previous sections. Note that we assume a dynamic pattern $x_1^T (\in C_k)$ to be an input.

- **discriminant function**

$$g_j(x_1^T;\Lambda) = \ln p_\Lambda(x_1^T \mid C_j) \tag{74}$$

- **misclassification measure**

$$d_k(x_1^T;\Lambda) = -1 + \frac{\left[\frac{1}{M-1}\sum_{j,j\neq k}\{g_j(x_1^T;\Lambda)\}^\mu\right]^{1/\mu}}{g_k(x_1^T;\Lambda)} \tag{75}$$

$$d_k(x_1^T;\Lambda) = 1 - \frac{g_k(x_1^T;\Lambda)}{\left[\frac{1}{M-1}\sum_{j,j\neq k}\{g_j(x_1^T;\Lambda)\}^\mu\right]^{1/\mu}} \tag{76}$$

$$d_k(x_1^T;\Lambda) = -g_k(x_1^T;\Lambda) + \ln\left\{\frac{1}{M-1}\sum_{j,j\neq k} e^{g_j(x_1^T;\Lambda)\mu}\right\}^{1/\mu} \tag{77}$$

The form of (77) was given in [24].

• **loss**

$$\tilde{\ell}_k(x_1^T;\Lambda) = \frac{\sum_{j,j\neq k}^{M} \{g_j(x_1^T;\Lambda)\}^{\gamma}\,\delta_{jk}}{\sum_{j,j\neq k}^{M} \{g_j(x_1^T;\Lambda)\}^{\gamma}} \ell_k(x_1^T;\Lambda), \tag{78}$$

where $\ell_k(x_1^T;\Lambda)$ is such a usual loss as (37), and δ_{jk} is a symbolic distance between C_j and C_k, which is often determined by parsing in speech recognition. (78) was proposed in [71]. Using this loss, one can attempt to pursue the minimum of a specific risk to a given task instead of the general misclassification account.

Chapter 11

An Approach to Adaptive Classification

LEE A. FELDKAMP, TIMOTHY M. FELDKAMP, AND DANIL V. PROKHOROV
POWERTRAIN CONTROL SYSTEMS DEPARTMENT
FORD RESEARCH LABORATORY
FORD MOTOR COMPANY
DEARBORN, MI 48121-2053

***Abstract*—We present an on-line learning system that is capable of analyzing an input-output data sequence to construct a sequence of binary classifications without being provided correct class information as part of the training process. The system employs a combination of supervised and unsupervised learning techniques to form two or more behavior models. By examining these models for consistency with the sequence of observed data, an estimate of the class at each time step may be constructed. The learning system has been formulated by considering the general characteristics of the automotive misfire detection problem. The present chapter, however, concentrates on the general aspects of the approach and uses synthetic problems as illustrative examples.**

1. Introduction

By *adaptive classification* we have in mind the process of developing a classifier according to the following set of assumptions. Suppose that we have a (usually physical) system for which measurements are available at intervals. We assume that these intervals are evenly spaced according to a natural "clock" for the system. Whether or not this clock is based strictly on time, we shall refer to such intervals as "time steps." We assume that, at every time step, the system behavior falls into one of two mutually exclusive classes, which we term "normal" or "fault." (These labels are arbitrary and for other types of applications would be changed appropriately. For example, in a communication application the classes might be denoted merely by 0 and 1.) We assume that the output of the physical system is measured and available. We assume further that this observed output depends on the measurable inputs to the system and on the current fault class and possibly the fault classes of a finite and known number of previous time steps. We make the weak assumption that overall the probability of normal is greater than that of fault, i.e., summed over many time steps the number of normals exceeds the number of faults. Knowledge of whether a particular time step corresponds to normal or fault is assumed *not* to be known. Further, we do not assume knowledge of the form of the dependence of the observed output variable on the measured input variable(s) or on the existence of occurrence of fault during the current or previous time steps.

Of course, in developing an adaptive classifier for a specific system, we would naturally employ the known characteristics of the system, and we would use knowledge of the existence of fault conditions to verify the performance of a proposed learning system. However, our intention is to develop the learning system such that it can bootstrap itself to good performance with little or no preloading.

It is clear that any ability to develop a classifier depends on observing the system for a sufficient number of time steps in the presence of both normal and fault conditions. It is also implicitly assumed that dependence of the observed output on the existence of fault is not overwhelmed by unmeasured disturbances, such as noise.

The available information in the problem statement just described is less than that employed in the training of recurrent networks for misfire detection, as described in [1, 2], because fault-labeled data are not assumed. We attempt to make up for this missing information by imposing structure on the learning system to reflect our prior understanding of the physical system. Even with this structure, the performance of a classifier trained by powerful supervised methods may be difficult to equal for a given *fixed* problem. However, use of an adaptive learning system may be advantageous if the physical system is expected to change significantly over the lifetime of the classifier.

The following features of the automotive misfire problem drive the present method. 1) The presence of fault (misfire) affects the primary observed variable (acceleration). 2) The observed variable is a function of context variables (engine speed and load). 3) The observed variable is affected by the presence/absence of fault for several previous steps. 4) The probability of fault is known to be (much) less than the probability of normal. All of these features have been incorporated into the synthetic problems used here for illustration. However, none of these problems is meant in any way to model the misfire problem, and no attempt was made to achieve the same balance of complications presented by the misfire problem.

The next section presents our learning method and discusses its various elements. We also relate the method to several other techniques and approaches (most notably, hidden Markov models (HMMs)). Section 3 provides examples to illustrate our method. To calibrate the difficulty of the examples, we compare the classification performance of the adaptive methods with that attained by supervised training of a standard RMLP in a setting that is equivalent (apart from the availability of labeled data). Section 4 concludes the chapter with a summary of the results and a mention of future directions.

2. The Proposed Learning Method

2.1. Problem Statement

By assumption, the observed input-output behavior of the physical system depends on the recent pattern of faults/normals. A portion of a typical fault sequence, with faults denoted by 1 and normals by 0, might look like

01001011000011100101001000001 . . .

The essence of the present method is the training of a set of neural networks to model the input-output behavior. Each network is expected to model the system for one specific fault pattern. The number of networks required thus depends on the number of relevant fault patterns expected to be present in significant proportions. If the networks can be trained such that each specializes to a specific fault pattern, we might expect classification to be accomplished easily by identifying (at each time step) the network that best models the current output behavior and then declaring fault or normal accordingly. In practice, because of noise and the possible coincidence of the input-output behavior at a given time step, such a simple scheme may not be adequate. Hence, as discussed below, we perform a more complete class inference by enforcing a measure of consistency.

We regard the input-output behavior that generates the training data as that of a dynamical system

$$\mathbf{x}(k) = f(\mathbf{x}(k-1), \mathbf{u}(k)) \tag{1}$$

$$y(k) = g(s(k), \mathbf{x}(k)) \tag{2}$$

Here, $\mathbf{x}(k)$ is a state vector of the system. The vector $\mathbf{u}(k)$ is the current measured input to the system, and $y(k)$ is the current measured output of the system. (For simplicity we assume a single output variable.) We use $s(k)$ as a state variable to denote the *fault state* of the system. We adopt a compact notation of m bits to denote the pattern of fault for the current step and the $m-1$ previous steps. In this chapter, we concentrate on the case $m = 3$, so that the fault state comprises fault categories from three consecutive steps. We use the obvious m bit notation such that, for example, $s(k) = (110)$ denotes fault at step $k-2$, fault at step $k-1$, and normal at step k. It is sometimes convenient to treat the bit string as a binary number and convert it to a decimal number. For the previous example we would have $s(k) = 6$. By virtue of its definition, a particular state $s(k)$ can evolve to one of two possible *successor* states. For example, if $s(k) = (110)$, $s(k+1)$ must be either (100) or (101), i.e., the two least significant bits of $s(k)$ shift left, and the least significant bit (LSB) of $s(k+1)$ can be either 0 or 1. Thus s evolves according to a state transition diagram or state transition matrix. The state transition diagram for $m = 3$ is shown in Figure 1. As an example, the fault/normal sequence

01001011000011100101001000001 . . .

gives rise to the fault state sequence (in decimal notation)

01241253640013754125241000001 . . .

If the current output of the system is a function only of the current input and the state s, it may be simplified to

$$y(k) = h(s(k), \mathbf{u}(k)) \tag{3}$$

Our synthetic examples, for simplicity, will assume this form. The more general system in (1) and (2) may be

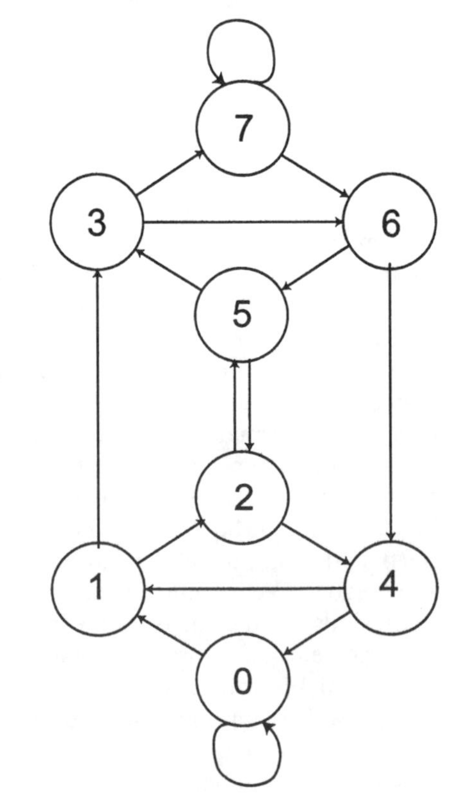

Fig. 1. State transition diagram for 3 bits.

approximated by augmenting the inputs with l previous input vectors:

$$y(k) = h(s(k), \mathbf{u}(k), \mathbf{u}(k-1), \ldots \mathbf{u}(k-l)) \quad (4)$$

2.2. Model Inference and Updates

We model the system (in either form) using a set of neural network models, each of which is assigned to one of the possible states s. Let the networks be numbered consecutively from 0 to $2^m - 1$, where 2^m is the number of networks. Let the *label* assigned to network l be denoted by p_l. These labels, such as (000), (001), (010), . . . (or, equivalently, 0, 1, 2, . . .) are used to specify to which state s the network applies. If the actual state $s(k)$ were to be known for time step k, we would merely estimate the system output by the output of the appropriate network, whose output we denote as $y_{s(k)}(k)$. In reality, as a consequence of not having labeled data, we do not know *a priori* the state $s(k)$. Hence, it is termed a *hidden state*; if each hidden state can be inferred from the available data, the original classification problem is solved. From here on, for notational simplicity, we denote networks by labels p rather than by network numbers l. Of course, in practice we must maintain a table which specifies p_l for each l.

Let us assume for the moment that we were given trained networks that accurately model the physical system for each $s(k)$. In this case, the correct state may be determined by merely executing each network in turn and selecting the one which is most consistent with the observed system output. This determination may involve more than just choosing the network with the smallest error. As an example, suppose we wish to conclude that the network labeled (101) (decimal label 5) best describes the input and output observations at time step k. Then, as illustrated in Figure 2, we should ensure that either label (010) or label (110) best describes the observations at $k-1$ and that either (010) or (011) is reasonable for step $k+1$.

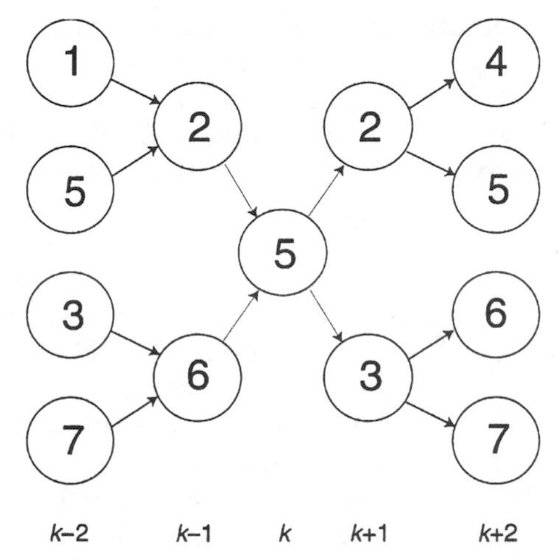

Fig. 2. Illustration of the states from time step $k-2$ to $k+2$ that are consistent with and include state (101) (decimal 5) at time step k.

At time step k, each network p is provided with the measured system input vector $\mathbf{u}(k)$ (or a time-lagged sequence of such vectors) and produces an estimate of the system output $\hat{y}_p(k)$ (here assumed to be a scalar). Each network p has its own error distribution, characterized by a variance σ_p^2. The calculation of this variance will be discussed shortly. A measure of the probability of network p being appropriate to describe the system is given by the quantity

$$d_p(k) = \frac{1}{\sqrt{2\pi\sigma_p^2}} \exp\left(-\frac{(y(k) - \hat{y}_p(k))^2}{2\sigma_p^2}\right) \quad (5)$$

By itself, this quantity does not reflect the requirement that the states s evolve according to the transition diagram. Simply choosing the network p with the largest value of $d_p(k)$ may not be consistent with the network chosen by the same method at the previous time step. In principle, to infer the sequence of p values we should optimize over the data sequence as a whole.

We may utilize the classical results of the HMM theory [3]. The *forward-backward* procedure is used in situations in which one needs to estimate the probability $P_p(k)$ of being in state p at time k, given the entire sequence of T steps

$$P_p(k) = \frac{\alpha_p(k)\beta_p(k)}{\sum_{j=0}^{2^m-1} \alpha_j(k)\beta_j(k)} \quad (6)$$

Here $\alpha_p(k)$ is a probability measure related to observing the sequence $y(1), y(2), \ldots, y(k)$ and label p at time k, and $\beta_p(k)$ is a probability measure related to observing the sequence $y(k+1), y(k+2), \ldots, y(T)$ starting with label p at time k.

Each $\alpha_j(k)$ can be determined from the forward part of the procedure. First, we need to initialize:

$$\alpha_j(1) = \pi_j d_j(1), j = 0, 1, \ldots, 2^m - 1, \quad (7)$$

where π_j is the prior probability of label j. For all α_j at the next time step we carry out the recursion

$$\alpha_j(k) = \frac{d_j(k) \sum_i \alpha_i(k-1)a_{ij}}{\sum_l d_l(k) \sum_i \alpha_i(k-1)a_{il}}, k = 2, 3, \ldots, T, \quad (8)$$

where a_{ij} is an element of the transition matrix A and represents the probability of transition from label i to label j. The index l in the denominator of (8) runs across all labels.

Each $\beta_j(k)$ is determined from the backward part of the procedure. We begin by initializing

$$\beta_j(T) = 1, j = 0, 1, \ldots, 2^m - 1. \quad (9)$$

For all β at the previous time step we carry out the recursion

$$\beta_j(k'-1) = \frac{\sum_i \beta_i(k')d_i(k')a_{ji}}{\sum_l \sum_i \beta_i(k')d_i(k')a_{li}}, \quad (10)$$

$$k' = T, T-1, \ldots, k+1.$$

In our adaptive setting, the recursion (8) poses no practical problem. However, we cannot make use of data arbitrarily far beyond the current step k. Hence, for the calculation of $\beta_p(k)$ we take T to be given by $T = k + T_h$, where we term T_h the *forward horizon*. The choice of T_h determines the latency associated with classifications available from the learning system, i.e., how long we must wait after a given step for a classification to be available. A larger T_h also increases the storage required to carry out the calculation of $\beta_p(k)$.

We may treat $P_p(k)$ as the probability that the state $s(k)$ is given by p. Then we may estimate the probability of fault at step k by summing over networks whose labels have LSB equal to 1 (i.e., are odd):

$$\text{prob}_{fault}(k) = \sum_{p=\text{odd}} P_p(k) \tag{11}$$

In analogy with the procedure of mixture models, the composite output of the learning system may be computed as the weighted average of the individual networks' outputs:

$$\hat{y}_{ave}(k) = \sum_{p} P_p(k)\hat{y}_p(k) \tag{12}$$

This quantity, if compared with the measured output, provides a qualitative indication of how well the learning system is modeling the input-output behavior of the physical system. However, it plays no direct role in the classification procedure.

At the beginning of the learning process, we start with naive networks and no knowledge of the correct states $s(k)$. Hence, we must employ a bootstrap procedure, in which we attempt to train networks and identify the states at the same time.

Our procedure is executed in a sequential fashion with only short-term data caching (over the forward horizon), since we have a continual supply of new data. At each time step we execute all networks with the same inputs and compute the probabilities $P_p(k)$ for each network label. For well-trained networks, only one label should have significant probability at any given time step and the probabilities $P_p(k)$ should be consistent with the correct state $s(k)$ at each step. From the $P_p(k)$ we also compute $\text{prob}_{fault}(k)$ from (11), which provides a classification for time step k.

Variances of all networks and an estimate of the overall probability of fault are computed every N_t steps, based on information accumulated at each step:

$$\sigma_p^2 = \frac{\sum_k (y(k) - \hat{y}_p(k))^2 P_p(k)}{\sum_k P_p(k)} \tag{13}$$

$$\text{prob}_{fault} = (1/N_t) \sum_k \text{prob}_{fault}(k) \tag{14}$$

The sums are taken over all the N_t steps. We could estimate the elements of the transition matrix, as described in [3]. For the examples presented here, we instead derive a_{ij} directly from prob_{fault}, i.e., $a_{ij} = \text{prob}_{fault}$ if the LSB of state j is 1, $a_{ij} = 1 - \text{prob}_{fault}$ otherwise.

At each step, we update the networks in proportion to their probabilities $P_p(k)$. We employ global EKF training as described in [2] and apply a scale factor of $P_p(k)/\sigma_p^2$. This scaling normalizes the squared error of each network to its variance.

When updating weights of the network using the EKF formalism, it is important to make use of artificial process noise, which has been recognized as helping to avoid poor local minima (see [2]). For efficient use with our method, it is crucial to scale not only the error of each network but also the matrix $\mathbf{Q}$ that reflects process noise. To understand why, consider the situation that $P_p(k)$ is nearly zero for a long period of time while σ_p^2 is much larger than $P_p(k)$. If $\mathbf{Q}$ is not modified appropriately when the network for p is being trained, the EKF approximate error covariance matrix will be subject to repeated additions to its diagonal elements, even when vanishingly small weight changes are being made. To avoid this, we scale the matrix $\mathbf{Q}$ by $P_p(k)/\sigma_p^2$ before invoking the EKF recursion for each network.

We assume for simplicity that the error variance for each model is uniform over the space of model inputs $\mathbf{u}(k)$. If required, this assumption can be relaxed by training a network to map the variance as a function of the input variables [4].

2.3. Restarting

The learning system we have described consists of several interacting elements. The inference procedure, in addition to providing a classification, operates in an unsupervised fashion to divide up the space of behavior and allocate portions thereof to the individual neural network models. Having thus been provided a target output and a scale factor that acts as a learning rate, the neural networks are updated. The difficulty of any given problem is affected by the degree to which the behavior of the various fault states acts to differentiate them. The presence of noise naturally increases the difficulty, because it serves to reduce the distinction between input-output behaviors that are locally similar, thereby misleading the inference procedure and possibly causing the network training to be inappropriate. Not surprisingly, we have observed that successful learning is subject to statistical fluctuation, just as occurs in conventional supervised learning. With a probability p_s that depends on the inherent difficulty of the problem, we observe that after some number of steps N_{rt}, the learning process is successful, meaning that the misclassification percentage has been reduced to a reasonable number (in the presence of noise this may not be zero) and is relatively stable as new data are processed. (Naturally, the number of steps N_{rt} at which success is judged must be chosen large enough to give the process a reasonable chance to take hold.) When the learning process has not been successful after N_{rt} steps, the failure may persist indefinitely. Without implying rigor, we regard this as a local minimum. In supervised learning, common practice, whether formalized or not, is merely to abandon an apparently unsuccessful training trial and start again, at least with a new set of initial weights and frequently with the change of some other parameter of the process. In our adaptive setting, we wish to be able to use a similar restarting strategy, so that after no more than n sets of N_{rt} steps the probability of success will be increased to

$$p_{success} = 1 - (1 - p_s)^n. \tag{15}$$

Unless p_s is very small, such a scheme provides a high probability of eventual success. A possible snag, of course, is that in our assumed adaptive setting we do not *know* whether our classifications are correct or not. Hence, we seek measures that can be computed on the basis of available information and which individually or collectively can be shown to correlate with success in classification. For example, on the basis of experiments with test problems, we have noted that the following probability-weighted averages are useful:

$$\overline{\alpha}_p = \frac{\sum_k P_p(k)\alpha_p(k)}{\sum_k P_p(k)} \tag{16}$$

For a given 3-bit problem, it is frequently possible to find a threshold value of $(\overline{\alpha}_{001} + \overline{\alpha}_{010} + \overline{\alpha}_{100})/3$, computed after N_{rt} steps, that reliably separates classification success from failure. In addition, the quantity $\overline{\alpha}_{000}$ appears to be useful in certain cases.

The best way to make use of such measures will differ from problem to problem. However, in a specific application it may be possible to place bounds on the extent to which the underlying problem can vary, such that it becomes possible to select appropriate restart criteria.

2.4. The Viterbi Algorithm

An alternative to basing our classification on the quantities $P_p(k)$, as just described, is the well-known *Viterbi algorithm* [3]. This algorithm produces the state *sequence* that is the most likely, given an entire sequence of observations.

The Viterbi algorithm is similar to the forward procedure already described. For a sequence with index k in the range $[1, T]$, we begin with initialization:

$$\delta_j(1) = \pi_j d_j(1), j = 0, 1, \ldots, 2^m - 1, \tag{17}$$

$$\psi_1(j) = 0 \tag{18}$$

The next step is the the recursion

$$\delta_j(k) = \max[\delta_i(k-1)a_{ij}]d_j(k) \tag{19}$$

$$\psi_k(j) = \text{argmax}[\delta_i(k-1)a_{ij}]$$

where max and argmax are taken over all labels. The algorithm terminates at the final time step T with

$$q^*(T) = \text{argmax}[\delta_i(T)] \tag{20}$$

To determine the optimal sequence of labels, one needs to perform backtracking:

$$q^*(k-1) = \psi_k(q^*(k)) \text{ for } k = T, T-1, \ldots, 2. \tag{21}$$

To prevent scaling difficulties when multiplying many small numbers in the recursion (19), it is helpful to implement it using logarithms.

For the current problem, we need to divide the data stream into subsequences, to which the Viterbi algorithm is applied. The optimal label at each point in the sequence determines the classification to be reported and is used to determine which model network to train. Our experience is that an unsatisfactory local minimum results more frequently than when we use the "softer" probabilities from the forward-backward procedure. On the other hand, we found that using the Viterbi algorithm only as a diagnostic tool is valuable, although the classification decisions from the forward-backward and the Viterbi procedures are usually little different, especially at convergence.

2.5. Synopsis of Our Learning Method

We summarize the mechanics of our learning method as follows. The implementation consists of two stages. The first stage is initialization. We initialize nonzero entries a_{ij} of the transition matrix A with our prior assumption about fault probability. We choose scaling parameters such that the system output used as neural network targets will have approximately zero mean and unit variance. We set initial variances of all the networks to chosen values (e.g., unity for each of them) and we randomly initialize network weights.

The next stage of our implementation is execution. For each point k of the data block, we carry out the forward recursion (8). At the same time we maintain the history buffer up to a certain point $k + T_h$ determined by the specified forward horizon T_h. We perform the backward recursion (10) beginning from the point $k + T_h$. Further, we compute $P_p(k)$ of (6) for every label p. We carry out the classification and accumulate all the quantities necessary to compute the statistics of the method (i.e., variances of (13) and estimated prob_{fault} of (14)). We also update each of the model networks. Every N_t steps ($N_t = 2000$ in the examples later) we update the network variances and the nonzero elements of the transition matrix. If we have decided to invoke a restart strategy, we evaluate the situation every N_r blocks of N_t steps ($N_{rt} = N_r N_t$). In this case we determine whether the chosen restarting criteria are satisfied and, if so, reinitialize the learning system.

2.6. Relation to Other Learning Methods

The learning system described here bears resemblance to various other well-known forms of learning or training but also has certain critical differences. A primary difference from many other methods is that we have a different underlying goal. We are not trying to estimate the observed output from the observed input, which would be impossible to do accurately without knowing the sequence of faults, but rather to infer whether a fault is present.

The present system shares with the Kohonen self-organizing map (SOM) and various other clustering schemes the mechanism of units competing to describe data examples (instances). A key difference is that the present system partitions input-output behavior among the various networks, whereas clustering schemes effectively partition the input space.

The present system makes use of the same mechanics used in such systems as *mixture models* (see [5] for an excellent discussion), *gated experts* [4], and *cluster-weighted modeling* (CWM) [6]. In contrast to the first two of these, we do not partition the input space. Like CWM, we deal with the joint space of inputs and outputs, but with a different goal. It seems likely, nevertheless, that the present method could be reformulated to make good use of elements of CWM. Further, a synthesis of a direct classifier and the present approach could be accomplished by treating the classifier as a gating network.

The concept of a structured transition relationship among the various models or experts is not a part of the other methods just mentioned. The sequence of labels in our system may be regarded as a Markov chain: from each state two successor states are possible, with the actual successor determined by a random process. Our learning system may be regarded as a hidden Markov model (HMM) [3] in that we attempt to determine at each step the particular hidden state of the actual system. As with HMMs, we employ state transition diagrams and we attempt to learn models concurrently with determining the state. Basic HMMs do not treat input-output relationships. Hence, it might be natural to regard our system as an input-output HMM. However, this name has already be appropriated [7] to describe an HMM variant that might be described as a combination of an HMM and a gated-expert system. This type of HMM bears only a superficial resemblance to our approach.

3. Examples

Our examples employ systems with 3-bit fault states, i.e., eight hidden states. In each case the input-output relationship that corresponds to each fault state is relatively simple. The challenge in each case is to sort out this structure from the available input-output data stream, which usually is not obviously separated into normal instances and fault instances.

3.1. Fixed Linear and Nonlinear Generating Functions

In our first example, we assign a different function to each fault state:

$$y_0(k) = \sin(3\pi u(k)) \tag{22}$$

$$y_1(k) = 1 - u(k) \tag{23}$$

$$y_2(k) = -1 + u(k) \tag{24}$$

$$y_3(k) = u^3(k) \tag{25}$$

$$y_4(k) = u^2(k) \tag{26}$$

$$y_5(k) = u(k) \tag{27}$$

$$y_6(k) = \cos(3\pi u(k)) + \sin(0.8\pi u(k)) \tag{28}$$

$$y_7(k) = -1.5u(k) \tag{29}$$

For clarification, let us consider a fragment of a sequence of faults and normals given by

0110101000100000 . . .

with fault state state sequence

0136525240124000 . . .

We assume that the first element of the sequence from the left is at the current time step k. It is generated by the function $y_0(k)$. The second element is generated by the function $y_1(k+1)$, the third one is created by $y_3(k+2)$, etc.

We choose the input $u(k)$ to be a uniform random variable from the interval $[-1, 1]$. As an additional complication, we corrupt the measurement of both the input u and the output y with additive Gaussian noise with a standard deviation between 0 (the no-noise case) and 0.07.

The preceding system generates a data sequence according to a specified true probability of fault. For our experiments, we create sequences with true probabilities of fault ranging from 0.01 to 0.3. As the initial prob_{fault}, we choose values different from true probabilities of fault.

Figure 3 shows the input-out relationship when noise of standard deviation 0.07 is present. Different symbols are used according to whether the function corresponds to fault or normal. The observed input and output for a small segment of sequential data are shown in Figure 4. The correct classification of any given input-output pair is not obvious.

To apply our learning method to this system, we generate a continuous sequence of data. A typical training trial lasts

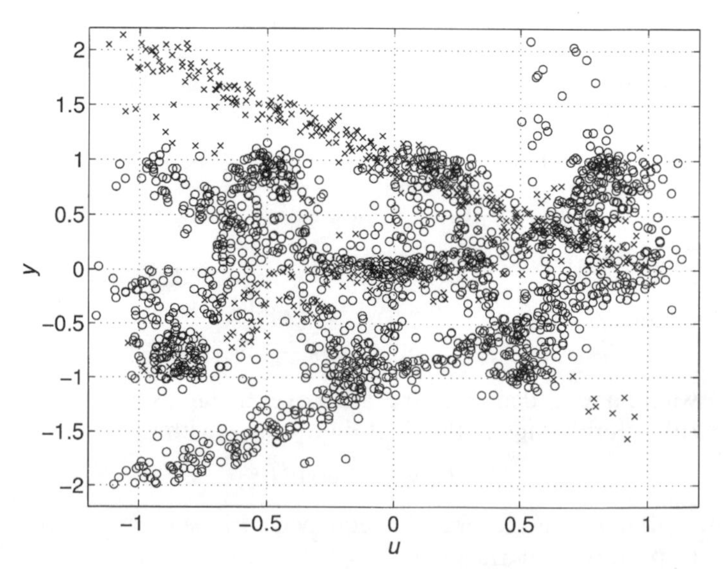

Fig. 3. System input-output map corrupted by Gaussian measurement noise with standard deviation 0.07. Here, for clarity, approximately 25% of the points correspond to fault, rather than the 5% used in the examples. Faults are drawn as crosses, normals as circles.

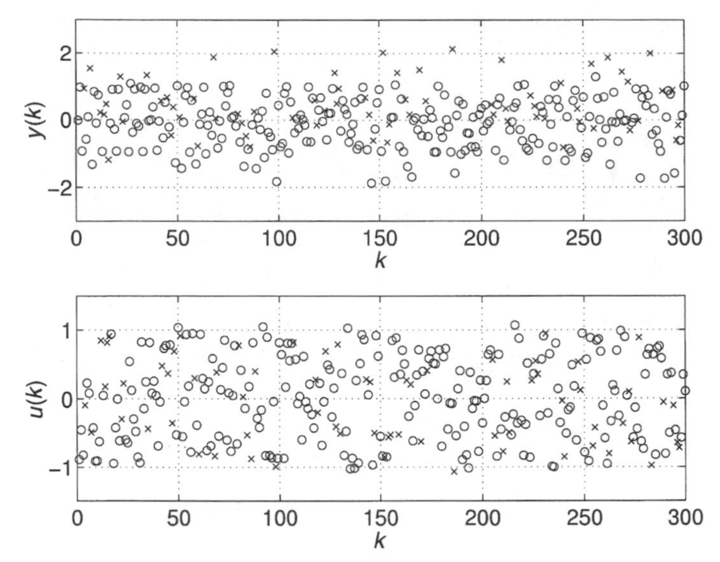

Fig. 4. Portion of the sequence of inputs $u(k)$ and corresponding outputs $y(k)$ (top panel) for the system of the first example with Gaussian noise of 0.07. Faults are again drawn as crosses, normals as circles.

until we process 200,000 pairs $(u(k), y(k))$. (Restarts are not considered here but will be discussed later.) We use eight feedforward networks, each with architecture 1-5-1L, i.e., one input, five nonlinear nodes in a single hidden layer, and a linear output node. For data generated with noise in the range [0.0, 0.07] and initial $prob_{fault}$ in the range [0.01, 0.3], the learning process has a reasonable probability of success, typically 0.2 to 0.9, judged by whether a low percentage of classification errors is achieved by the end of the trial data sequence and maintained thereafter. (We also term this convergence of the learning process.) Depending on the noise level, successful runs range from no errors to about a 2% error rate, the larger rates being common for the larger noise levels. If convergence is achieved, it usually happens within the first 100,000 points processed.

To assess the difficulty of such classification problems, we trained a recurrent multilayer perceptron (RMLP) using our usual EKF-based methods in the supervised setting. The problem is then stated as follows: train an RMLP which inputs the pair $(u(k), y(k))$ and outputs a number as close to $+1(-1)$ as possible when fault (normal) is present at time step k, provided the sequence of faults and normals is known. This formulation of the problem turned out to be nontrivial for the RMLP to handle, even in the case of very little measurement noise. In a typical training session consisting of a fixed data sequence of 20,000 points, we trained a 2-10R-8R-1 RMLP using five-stream NDEKF (the length of each stream was 2000 data points). We trained for 300 epochs with the parameter $\eta = 0.01$, 200 more epochs with $\eta = 0.1$, and 200 additional epochs with $\eta = 1$. We kept the matrix $\mathbf{Q} = 10^{-4}\mathbf{I}$ throughout the entire training session. We observed overfitting if we continued training longer (as indicated on an independently generated validation set). Although our training data set was limited in size, the *effective* length of the training sequence is about 7 million points (five streams by 2000 points each by 700 total epochs), since the order in which fragments of the data sequence were assembled for training with streams was always different from one epoch to another. The reader is referred to [2] for more information about the EKF algorithm and training strategies.

Our best results achieved with the trained RMLP can be summarized as follows. For the no-noise case, testing on an independent data sequence results in around 20 errors per 2000 points. This number grows to about 80 errors (4%) for noise of 0.07. Although it may be possible to improve the results when using a larger network and continuously generated data sequences (to counter the tendency to overfitting), we are confident that, in terms of total number of points used for training, the adaptive learning scheme proposed in this chapter is quite competitive.

3.2. *Time-Varying Functions*

Our second example deals with time-varying functions to illustrate the ability of our learning method to adapt in a slowly varying environment. We also show a simple restarting strategy. The system of generating functions is given as

$$y_0(k) = \gamma\sin(3\pi u(k)) + (1-\gamma)\cos(2\pi u(k)) \quad (30)$$

$$y_1(k) = \gamma(1 - u(k)) + (1-\gamma)(1 - u^4(k)) \quad (31)$$

$$y_2(k) = \gamma(-1 + u(k)) + (1-\gamma)\sin(0.5\pi u(k)) \quad (32)$$

$$y_3(k) = \gamma u^3(k) + (1-\gamma)(0.5u(k) - 1) \quad (33)$$

$$y_4(k) = \gamma u^2(k) + (1-\gamma)[\sin(\pi u(k)) + \cos(\pi u(k))] \quad (34)$$

$$y_5(k) = \gamma u(k) + (1-\gamma)(0.5 + 1.5u(k)) \quad (35)$$

$$y_6(k) = \gamma[\cos(3\pi u(k)) + \sin(0.8\pi u(k))] - (1-\gamma)u^2(k) \quad (36)$$

$$y_7(k) = -1.5\gamma u(k) + (1-\gamma)(u^3(k) - 0.5) \quad (37)$$

where $\gamma \geq 0$ is a time-varying parameter. Note that this system is identical to that of the previous example (Section 3.1) if $\gamma = 1.0$. Through the course of the experiment the true probability of fault was 0.05 and Gaussian measurement noise of 0.07 was present. Under these conditions, multiple independent runs, each of which consisted of 100 blocks of 2000 pairs, converged successfully about 40% of the time. Increasing the run length did not appear to result in significantly more successes.

We maintained $\gamma = 1.0$ for the first 500 blocks of 2000 pairs $(u(k), y(k))$. We then decreased γ by 1% for each of the next 100 blocks. By the end of this phase of the experiment we have processed 1,200,000 pairs and have transformed from the original system with $\gamma = 1.0$ to an entirely different system with $\gamma = 0.0$. Finally, we ran for 600 additional blocks without changing γ. As in the previous example, the input $u(k)$ was drawn from a uniform random

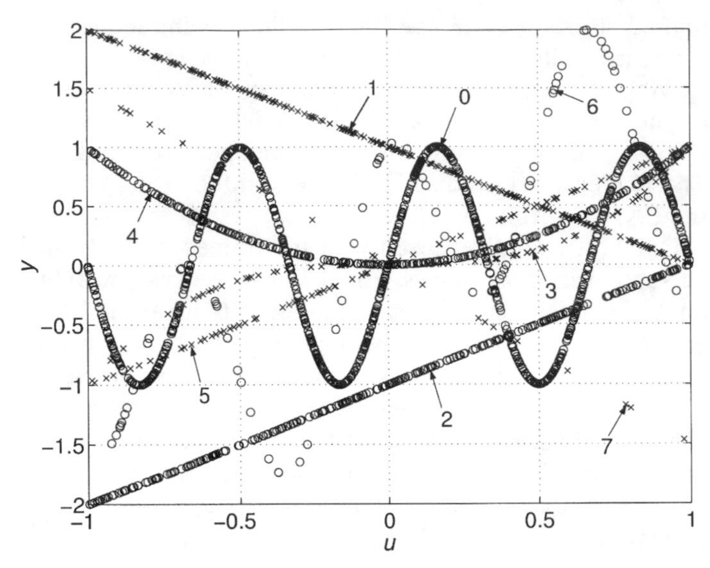

Fig. 5. System input-output map for $\gamma = 1$ before noise is added. The various functions are labeled by the states (decimal notation) to which they are assigned.

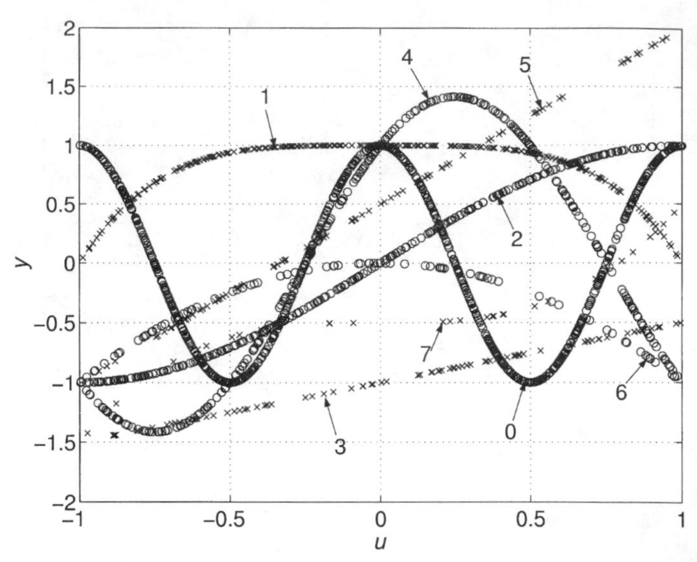

Fig. 7. System input-output map after transition is complete ($\gamma = 0.0$). The hidden states are completely different from those of the original system.

distribution in the interval $[-1, 1]$ and we used feedforward networks of the same architecture. Figures 5, 6, and 7 show the input-output relationships at the beginning, after 500 blocks, and after 600 blocks, respectively. For clarity, they are shown in the absence of noise. As already mentioned, the specific run to be illustrated here was actually performed with Gaussian measurement noise of 0.07. The noisy equivalent to Figure 5 is Figure 3.

We also used a simple restart strategy to recover from local minima. Every $N_r = 100$ blocks of $N_t = 2000$ steps we evaluated whether $(\overline{\alpha}_{001} + \overline{\alpha}_{010} + \overline{\alpha}_{100})/3 < 0.6$. If so, we restarted the training process. In Figure 8 we illustrate

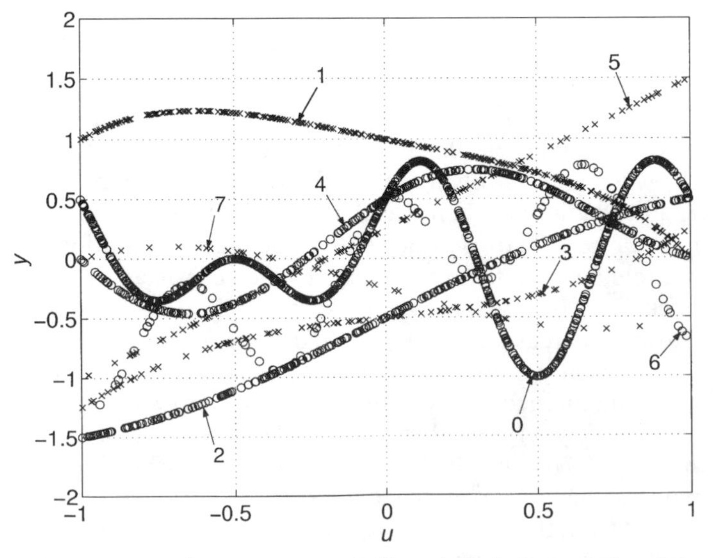

Fig. 6. System input-output map halfway through the transition phase ($\gamma = 0.5$).

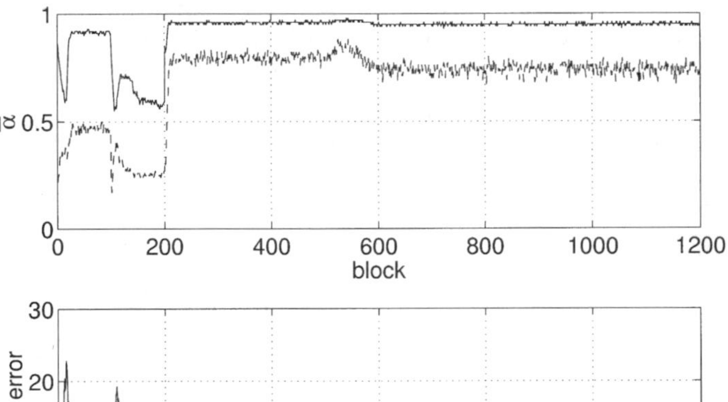

Fig. 8. The upper panel contains the plot of the probability-weighted average $\overline{\alpha}_{000}$ (see (16)) (solid line) and the plot of the sum $(\overline{\alpha}_{001} + \overline{\alpha}_{010} + \overline{\alpha}_{100})/3$ (dashed line). Each point of the plot represents values averaged over blocks of 2000 points of the data sequence generated with true probability of fault 0.05 and noise level 0.07. The lower panel shows the corresponding percentage of errors. For the first 200 blocks, learning is unsatisfactory. To initiate restart, we monitor the values shown on the upper panel for 100 consecutive blocks. Restarting is carried out if $(\overline{\alpha}_{001} + \overline{\alpha}_{010} + \overline{\alpha}_{100})/3 < 0.6$ at the end of 100 blocks of 2000 points. The restart (after the 100th block) fails to lead to convergence. The second attempt (at the 200th block) is successful. Furthermore, between blocks 500 and 600 we perform a slow transition between the original system of generating functions ($\gamma = 1.0$), and the final system ($\gamma = 0.0$), as described in Section 3.2.

one particular run. Convergence did not occur in the first 100 blocks, triggering the first restart. Similarly, another restart occurred after the next 100 blocks. Convergence soon followed, i.e., $(\overline{\alpha}_{001} + \overline{\alpha}_{010} + \overline{\alpha}_{100})/3$ climbed above its threshold. Correspondingly, the error rate dropped to a small value. Once convergence took place, the learning system gracefully adapted to the system change between blocks 500 and 600 and remained in the converged state until the end of the run.

The threshold was determined by inspection of many training trials under similar conditions and would not be expected to be appropriate if the fault probability or the noise level were to change significantly.

This example demonstrates the ability of the adaptive classifier to bootstrap itself both to learn the correct fault class and to map the hidden states of the system. It also shows that the method can track sufficiently gradual changes in the system. This example is similar to how we envision our learning method being used in a real application. A real system could change slowly (probably much more slowly than we have simulated here) and our learning system would adapt to the changes. Of course, in practice we would try to have the initially deployed system at least partially adapted, so that an initial period of poor performance and restarts would not be likely.

4. Conclusions and Future Directions

We have presented here a model-learning approach to the problem of detecting faults from a continuous sequence of data. This method combines supervised and unsupervised elements. Considerable room remains for both improved understanding and refinement of the technique. A crucial practical question involves understanding the level of complexity of underlying input-output behavior (e.g., as in Figure 5) that can be learned when the labels are not provided. Intuitively, the learning process would seem more likely to be successful when the functions for states that differ only in their LSB are well differentiated, as such states share the same predecessor state. (However, rearranging the generating functions assigned to the various states to make those assigned to (000) and (001) more similar did not prevent the system from learning.) By observing the early stages of the learning process in detail, we have noted that one network usually begins to model the overall input-output behavior, which for small fault probability tends to be dominated by the state (000). It is then crucial for another network to start to model the state (001). If this happens, the process usually converges satisfactorily. However, this does not always take place; in one such failure mechanism P_{000} remains higher than P_{001} for the (001) instances, i.e., the (000) network does not "release" them. Analogous behavior is not uncommon in other types of unsupervised learning, such as clustering, so it is not particularly surprising here.

In a practical application, partial label information, such as might be available from an existing classifier, would expedite the learning process. Perhaps more importantly, it may be possible to have the system preadapted to the physical system in its original form, so that the initial bootstrap portion of the learning process may not be necessary. In this regard, our experiment with time-varying systems is very encouraging, as it suggests that once the learning system "locks on," it is quite reliable at following slow changes. We have also observed that a converged learning system seems to be able to adapt to slow changes in the probability of fault and the level of measurement noise. As noise is increased, the error rate increases but the degradation in performance is graceful.

Although we have treated them without distinction, the regimes of small and large prob_{fault} pose distinct challenges. For small prob_{fault}, states with two or more bits equal to 1 occur very seldom. At some point, it may become expedient not to represent them in the learning system, because a poorly trained network may interfere with the inference procedure. When prob_{fault} is higher, states such as (111) play a more important role and must be represented well for errors to be minimized.

The problem statement we have addressed may also be viewed in terms of clustering. The simplest clustering problem statement would be to assign the points in the two-dimensional space of points $(u(k), y(k))$ to two sets of clusters, one set for normal instances and another for fault. It appears very difficult to accomplish this for the points in Figure 3. More appropriate, but difficult to visualize, would be to cluster the points in a multidimensional lag space, e.g., the points $(u(k-2), y(k-2), u(k-1), y(k-1), u(k), y(k))$.

If class labels are known, we have the very different problem of supervised learning. One possibility is to use a recurrent network, as we did here in order to assess the difficulty of the classification task. Another is to use a feedforward network with time-lagged inputs. A third is to employ the mechanics of the present method, modified to make use of label information.

Finally, we briefly recall the misfire detection problem that motivated this work. For the most part, this problem falls into the small prob_{fault} regime. The observed acceleration attributed to the current cylinder firing depends on the presence or absence of misfire during at least the most recent eight firings (for an eight-cylinder engine). A further complication is that the acceleration map for each cylinder in terms of engine speed and load is different from that for all other cylinders, thereby increasing the number of models required to describe the input-output behavior. Application to this problem is under way.

References

[1] G. V. Puskorius, L. A. Feldkamp, K. A. Marko, J. V. James, and T. M. Feldkamp. Method for Identifying Misfire Events of an Internal Combustion Engine. U.S. Patent 5,732,382, 1998.

[2] L. A. Feldkamp and G. V. Puskorius. A Signal Processing Framework Based on Dynamic Neural Networks with Application to Problems in Adaptation, Filtering and Classification, *Proceedings of the IEEE* 86(11):2259–2277, 1998.
[3] L. R. Rabiner. A Tutorial on Hidden Markov Models and Selected Applications in Speech Recognition, *Proceedings of the IEEE* 77(2):257–286, 1989.
[4] A. S. Weigend, M. Mangeas, and A. N. Srivastava. Nonlinear Gated Experts for Time Series: Discovering Regimes and Avoiding Overfitting, *International Journal of Neural Systems* 6:373–399, 1995.
[5] C. M. Bishop. *Neural Networks for Pattern Recognition.* Oxford: Oxford University Press, 1995.
[6] N. A. Gershenfeld. *The Nature of Mathematical Modeling.* Cambridge University Press, 1999.
[7] Y. Bengio and P. Frasconi. Input/Output HMMs for Sequence Processing, *IEEE Transactions on Neural Networks* 7(5):1231–1249, 1996.

Chapter 12

Reduced-Rank Intelligent Signal Processing with Application to Radar

J.S. GOLDSTEIN, J.R. GUERCI, AND I.S. REED
SAIC, ADAPTIVE SIGNAL EXPLOITATION,
4001 FAIRFAX DRIVE, SUITE 675,
ARLINGTON, VA 22203

1. Introduction

The technologies associated with radar signal processing have developed and advanced at a tremendous rate over the past 60 years. This evolution is driven by the desire to detect more stealthy targets in increasingly challenging noise environments. Two fundamental requirements on signal processing have developed as advanced radar systems strive to achieve these detection goals: 1) The dimensionality of the signal space is increased in order to find subspaces in which the targets can be discriminated from the noise; and 2) the bandwidth of each of these dimensions is increased to provide the degrees of freedom and resolution that are needed to accomplish this discrimination when the competing noise and the target are in close proximity. To be more precise, radar has developed from having only a spatial dimension to the utilization of a Doppler frequency (or slow-time) dimension to combat monostatic clutter, to a signal frequency (or fast-time) dimension to defeat terrain scattered interference, to multiple polarization dimensions for target discrimination, etc. The number of degrees of freedom required to separate the target from nearby competing noise within any one of these signal dimensions (i.e., the resolution or bandwidth in that dimension) grows as the target becomes smaller and the noise becomes more challenging. The total number of degrees of freedom for the radar is then given by the Cartesian product of the degrees of freedom for each individual dimension.

As radar signal processing evolved, it quickly became clear that one needed an estimate of the noise environment in order to realize detectors that worked well in the real world. This concept led simultaneously to the development of adaptive radar signal processing and adaptive constant false-alarm rate (CFAR) detectors. The theory of adaptive arrays [1, 2] was developed at a time when the spatial dimension was predominantly used alone. The theory of adaptive radar was next advanced to apply adaptivity simultaneously to the spatial and Doppler dimensions [3], introducing the popular field called space-time adaptive processing (STAP). These adaptive techniques for radar signal processing were based upon second-order statistics for wide-sense stationary (WSS) random processes. The ideas that the noise was Gaussian as well as independent and identically distributed (IID) over range served as fundamental assumptions which were embedded in these theoretical developments. The same assumptions were used in the tools which evaluated the performance of radar systems.

The estimation of the noise environment, with the preceding assumptions, requires a training region composed of a number of samples which are at least on the order of twice the number of the radar's degrees of freedom[1] [4]. This famous "$2N$" rule means that, for N degrees of freedom, the Gaussian noise field estimation requires a minimum of $2N$ IID samples. Where does this training data, or sample support for noise field estimation, come from? In radar, one tries to detect a target within some specified range cell. If there are N total degrees of freedom, then a data cube consisting of at least $2N + 1$ range cells is processed coherently. The group of $2N$ or more range cells which exclude the test cell are termed auxiliary range cells. The auxiliary range cells are used to estimate the statistics

[1] More formally, a training region consisting of at least twice the number of the radar's degrees of freedom is required to obtain a statistical estimate of the noise which results in an output signal-to-interference plus noise ratio within 3 dB of that obtained with the true statistics.

of the noise within the test range cell, thereby providing a target-free training set. The statistical estimate of interest here is the covariance matrix, which contains all of the second-order information that is needed when the underlying assumptions are satisfied. This estimate of the noise statistics is then used to compute an adaptive weight vector, or Wiener filter, which maximizes the output signal-to-interference plus noise ratio (SINR). This Wiener filter is equivalent to a normalized colored-noise matched filter.

The advancement of sensor technology easily allows radars to be constructed with large numbers of degrees of freedom in the spatial, Doppler, fast-time, and polarization dimensions. The new fundamental problem for radar signal processing is that this large number of degrees of freedom makes it impossible for the IID, WSS, second-order assumptions embedded in adaptive signal processing to be valid. A moderate STAP radar design includes a minimum of a few hundred degrees of freedom. For example, the DARPA Mountain Top radar uses 16 Doppler pulses and 14 receive channels for a total of $N = 224$ degrees of freedom. This implies a requirement for at least 448 range cells that contain noise which is IID with respect to that present in the range cell of interest. The IID assumption accompanies a spatial ergodicity argument with respect to the stationarity of the noise. However, with the parameters given above, the range extent of the sample support would cover approximately 338 kilometers. Given the topography of the earth, it is unreasonable to assume homogeneity over a region of this size or larger, and therefore the stationarity assumption on the data received at the radar cannot, in general, be valid for advanced radars.

The proposed solution to this problem requires intelligent signal processing for both subspace compression and training region selection. It is necessary to enlarge the signal space greatly in order to find that subspace which permits target detection. This drives the requirement for large numbers of antenna elements, large sets of tapped delay lines (at the relevant frequency spacing), many polarization channels, etc. However, this signal space enlargement also drives the need for large sample support and, as a consequence, stresses or breaks the underlying statistical assumptions. Since the space is enlarged, the true noise subspace is generally overmodeled in order to guarantee that a subspace for detection can be found. It is therefore necessary to perform intelligent signal representation so that the smallest possible subspace that contains the majority (if not all) of the noise power can be determined. This representation must take into account some information about the manifold of the noise subspace that it is desirable to estimate rather than to blindly estimate the entire noise subspace. This representation would of necessity allow optimal compression of the noise subspace, permit both rapid convergence and tracking of the noise statistics, and reduce the size of the required data region for statistical sample support.

It is also necessary to introduce intelligent training methods in order to determine which data in the auxiliary training set is most similar, in some appropriate statistical measure, to that present in the range gate of interest. Finally, an intelligent signal processing approach should also be capable of utilizing prior or additional knowledge to incorporate information about the structure of roads or other man-made objects and geospatial information such as the U.S. National Imagery and Mapping Agency's digital terrain elevation database (DTED).

This chapter introduces a new method of intelligent signal representation and compression. This theory is presented in an application-independent manner since the concepts are valid in nearly every statistical signal processing problem. The radar application is then revisited to demonstrate the principles developed herein.

2. Background

Consider the representation of discrete-time, wide-sense stationary signals, which is fundamental in the many applications of statistical detection and estimation theory. For the purposes of this work, the efficiency of a signal representation is evaluated by its ability to compact useful signal energy as a function of rank. This criterion is equivalent to optimal signal compression.

The multiple-signal problem is considered herein, where a nonwhite signal of interest is observed only in the presence of at least one other generally nonwhite process.[2] Signal processing for multiple signals, under these conditions, is described within the general framework of the discrete-time, finite-impulse response Wiener filter. The Wiener filter is a fundamental component in the solution to virtually every problem that is concerned with the optimality of linear filtering, detection, estimation, classification, smoothing, and prediction in the framework of statistical signal processing in the presence of stationary random processes. This same approach provides the least-squares solution for the processing of collected data either in a deterministic framework or by invoking some form of ergodicity.

The fundamental issue in signal representation and compression is the determination of an optimal coordinate system. It is well known that the eigenvectors associated with the covariance matrix of an N-dimensional WSS signal provide the basis set for the Karhunen-Lóeve expansion of that signal. The minimax theorem establishes that this set of eigenvectors represent the particular basis for an N-dimensional space which is most efficient in the energy sense. This autocorrelation-based energy maximization for a single process satisfies the stated representation criterion, and the eigenvectors form the best basis representation for this single process. If the eigenvectors are ordered to correspond with the magnitude of the eigenvalues in a descending manner, then this enumeration is termed order-

[2] This scenario is very common; even if a true white noise field is observed by a sensor, for example a radar or a sonar receiver, the output from the sensor is a filtered process which, in general, will not be white.

ing by principal components. Optimal signal representation and compression as a function of rank (or dimension) are then obtained by a truncation of the principal components. In other words, the rule for optimal rank M basis selection ($M < N$) and compression is to choose those M eigenvectors which correspond to the largest M eigenvalues of the observed signal covariance matrix.

However, there are many statistical decision problems where the criterion of interest is more general. This fact is readily verified by considering the popular problems of detection, estimation, or any of the many other statistical signal processing applications. Here, for the problem to be nontrivial, there are a minimum of two additive nonwhite signal processes: the signal process of interest and a process of colored noise. If one now speaks of signal representation or compression of one process, the solution must take both processes into account in order to determine an optimal basis.

The goal of signal representation and compression for detection and estimation is to find an optimal basis *without prior knowledge* of the inverse of the covariance matrix. Optimal basis selection allows for signal compression, rank reduction, and a lower computational complexity in order to obtain the Wiener solution. Reduced sample support for estimating statistics and faster convergence of filter parameters are also obtained with rank reduction if the Wiener filter is implemented in an adaptive manner. Note that there is a subtle information-theoretic idea embedded in this goal; if the covariance matrix inverse is known *a priori,* then so is the Wiener filter, and signal representation is irrelevant.

Previous work in the area of signal representation has centered around principal component analysis for the single process case [5, 6] or canonical correlation analysis for the multiple process case [7–10]. The vector Wiener filter is unique because, while it is among the most common filtering structures, neither of these analysis tools applies in this case to the optimization of performance as a function of rank. The solution to the canonical correlation analysis degenerates into the vector Wiener filter itself and therefore provides no insight into the selection of an optimal basis. It is demonstrated explicitly in this chapter that the principal-components approach is not the correct enumeration of the eigenvectors to achieve optimal representation for the Wiener filtering problem. Once this fact is established, it is shown herein that the standard Karhunen-Lóeve decomposition no longer provides a solution to optimal basis selection and that a new basis set must be derived that takes into account the presence of other signals; it is called here a generalized joint-process Karhunen-Lóeve transformation.

Previous attempts to solve the rank reduction problem for the vector Wiener filter only result in solutions which, at best, dictate the computation of the Wiener filter itself [11–13]; they do not provide a basis set for the vector Wiener filter problem. A new approach is now presented that provides an optimal basis set through the natural extension of the Karhunen-Lóeve transform (KLT) for the Wiener filter.

Classical Karhunen-Lóeve analysis is briefly reviewed next in Section 3. The vector Wiener filtering model is then introduced in Section 4. The necessary modifications to the KLT are addressed in Section 5 to obtain an optimal Wiener filter with an eigenvector basis. In Section 6 a new method is developed to obtain an optimal basis without having the need for knowledge of the covariance matrix, its inverse, or the Wiener filter. The radar application is then discussed in Section 7, where the use of space time adaptive processing with intelligent signal processing is examined. Concluding remarks and a summary are presented in Section 8.

3. Karhunen-Lóeve Analysis

This section presents a review of the Karhunen-Lóeve transformation and signal expansion. These preliminaries set the stage for an analysis of signal representation and compression when multiple signals are present and the goal is to perform detection, estimation, classification or prediction of a signal of interest.

3.1. The Karhunen-Lóeve Transformation

Consider an N-dimensional, complex, WSS signal $\mathbf{x}_0$ with an $N \times 1$ mean-vector $\boldsymbol{\mu}_x$ (assumed without loss in generality to be the zero vector) and a nonnegative definite, Hermitian, $N \times N$ covariance matrix $\mathbf{R}_{x_0}$. Let the covariance matrix $\mathbf{R}_{x_0}$ be represented by its Karhunen-Lóeve transformation,

$$\mathbf{R}_{x_0} = \mathbf{V}\boldsymbol{\Lambda}\mathbf{V}^H = \sum_{i=1}^{N} \lambda_i \mathbf{v}_i \mathbf{v}_i^H, \tag{1}$$

where $(\cdot)^H$ denotes the complex Hermitian transpose operator, the $N \times N$ matrix $\mathbf{V}$ is composed of the N-unitary eigenvectors $\{\mathbf{v}_i\}_{i=1}^N$, and the diagonal matrix $\boldsymbol{\Lambda}$ is composed of the corresponding eigenvalues $\{\lambda_i\}_{i=1}^N$. It is assumed that the eigenvectors are ordered in a descending manner in accordance with the magnitude of the corresponding eigenvalues and, for convenience, that all of the eigenvalues are distinct.

The KLT of the covariance matrix $\mathbf{R}_{x_0}$, with the assumptions presented above, yields N orthonormal eigenvectors $\mathbf{v}_i$. Now denote the complex N-dimensional space spanned by the columns of $\mathbf{R}_{x_0}$ as $\mathscr{C}^N$. Then these eigenvectors form a basis for the space $\mathscr{C}^N$, and any vector $\mathbf{x}_0 \in \mathscr{C}^N$ can be represented by a linear combination of any basis vectors for $\mathscr{C}^N$.

3.2. The Karhunen-Lóeve Expansion

The Karhunen-Lóeve expansion of the N-vector $\mathbf{x}_0$ is obtained by its representation in terms of the basis generated by the eigenvectors of $\mathbf{R}_{x_0}$; that is,

$$\mathbf{x}_0 = \sum_{i=1}^{N} \alpha_i \mathbf{v}_i, \text{ where } \alpha_i = \mathbf{v}_i^H \mathbf{x}_0. \tag{2}$$

It is easily verified that

$$\mathbf{E}[\alpha_i] = 0 \ \forall i, \tag{3}$$

and

$$\mathbf{E}[\alpha_i \alpha_j^*] = \begin{cases} \lambda_i, & i = j \\ 0 & i \neq j \end{cases}, \tag{4}$$

where $(\cdot)^*$ is the complex conjugation operator.

A k-dimensional subspace $\mathscr{C}^k \subset \mathscr{C}^N$ is formed by any arbitrary $\binom{N}{k}$ collection of basis vectors for the space spanned by the columns of $\mathbf{R}_{x_0}$. The k-dimensional principal-components subspace is defined to be that subspace spanned by the k principal eigenvectors and denoted $\mathscr{C}^k_{pc}$. A new reduced-rank N-dimensional vector, denoted $\mathbf{z}_0$, is given by the truncated series representation of $\mathbf{x}_0$ in (2) using only the k principal components,

$$\mathbf{z}_0 = \sum_{j=1}^{k} \alpha_j \mathbf{v}_j. \tag{5}$$

Note that in the k-dimensional subspace $\mathscr{C}^k_{pc}$, the same vector has the k-dimensional representation,

$$\mathbf{z}_0 = \mathbf{V}_{pc}^H \mathbf{x}_0, \tag{6}$$

where $\mathbf{V}_{pc}$ is the $N \times k$ matrix composed of the k principal eigenvectors.[3] Finally, from (5) and (6), it is seen that z_0 is the projection of $\mathbf{x}_0$ onto the k-dimensional principal-components subspace $\mathscr{C}^k_{pc} \subset \mathscr{C}^N$ and therefore represents the compression of the N-dimensional vector to k coefficients. This geometrical relationship is depicted in Figure 1 for $\mathscr{R}^3$, the three-dimensional vector space of real numbers. The principal-components method of compression and representation is optimal for a single signal in the sense that it provides the best representation of the full-rank space in terms of autocorrelation energy retained as a function of rank [5, 6].

3.3. Implementing the KLT

The KLT of the covariance matrix associated with discrete-time random processes is most often calculated by a two-step process which yields the eigendecomposition of the covariance matrix $\mathbf{R}_{x_0}$. The first step is a tridiagonalization, achieved through the use of Householder reduction. The second step is a QR, zero-chasing, iterative method which completes the diagonalization upon convergence [14, 15].

To visualize this process, consider the following $N \times N$

[3] For this reason, the terms dimension and rank are used interchangeably in the rest of this chapter.

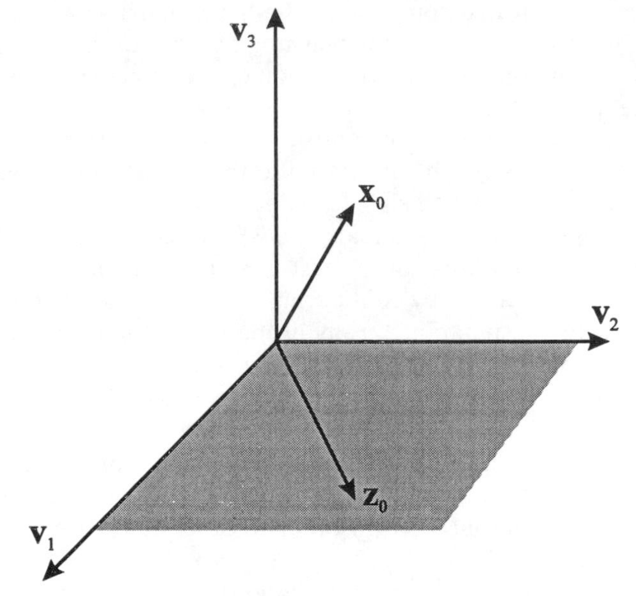

Fig. 1. The representation of the vector x_0 in $\mathscr{R}^3$ and its projection $\mathbf{z}_0$ in $\mathscr{R}^2$.

Hermitian covariance matrix:

$$\mathbf{R}_{x_0} = \begin{bmatrix} r_{1,1} & r_{2,1}^* & r_{3,1}^* & \cdots & r_{N,1}^* \\ r_{2,1} & r_{2,2} & r_{3,2}^* & \cdots & r_{N,2}^* \\ \vdots & & & \ddots & \vdots \\ r_{N,1} & r_{N,2} & r_{N,3} & \cdots & r_{N,N} \end{bmatrix}. \tag{7}$$

Define the $(N-1)$-dimensional vector $\mathbf{s}_1$ to be the first column of $\mathbf{R}_{x_0}$ excluding the top element,

$$\mathbf{s}_1 = [r_{2,1} \ \ r_{3,1} \cdots r_{N,1}]^T. \tag{8}$$

Next the vector $\mathbf{h}_1$ is given by

$$\mathbf{h}_1 = \frac{\mathbf{s}_1}{\|\mathbf{s}_1\|}, \tag{9}$$

where $\|(\cdot)\|$ represents the Euclidean norm and an orthonormal $(N-2) \times (N-1)$ matrix $\mathbf{B}_1$ is computed such that its nullspace is $\mathbf{h}_1$. Then the first Householder reflection is given by the $N \times N$ unitary matrix operator $\mathbf{T}_1$,

$$\mathbf{T}_1 = \begin{bmatrix} \mathbf{I}_1 & \mathbf{0}_{N-1,1}^H \\ \mathbf{0}_{N-1,1} & \begin{bmatrix} \mathbf{h}_1^H \\ \mathbf{B}_1 \end{bmatrix} \end{bmatrix}, \tag{10}$$

which initializes the tridiagonalization. The notation $\mathbf{I}_k$ is used to represent the rank-k identity matrix and $\mathbf{0}_{p,k}$ is the $p \times k$ zero matrix. The first-stage quadratic transformation of the covariance is then given by

$$\mathbf{T}_1 \mathbf{R}_{x_0} \mathbf{T}_1^H = \begin{bmatrix} r_{1,1} & \tilde{r}_{2,1}^* & 0 & \cdots & 0 \\ \tilde{r}_{2,1} & \tilde{r}_{2,2} & \tilde{r}_{3,2}^* & \cdots & \tilde{r}_{N,2}^* \\ 0 & \tilde{r}_{3,2} & \tilde{r}_{3,3}^* & \cdots & \tilde{r}_{N,3}^* \\ \vdots & & & \ddots & \vdots \\ 0 & \tilde{r}_{N,2} & \tilde{r}_{N,3} & \cdots & \tilde{r}_{N,N} \end{bmatrix}. \tag{11}$$

and it is easily verified from (7)–(11) that $\tilde{r}_{2,1} = \|r_{2,1}\|$.

TABLE 1. UNITARY KLT DECOMPOSITION RECURSION

$$\mathbf{R}_t = \mathbf{R}_{x_0}$$

$$\mathbf{h} = \frac{\mathbf{R}_t(2{:}N,1)}{\|\mathbf{R}_t(2{:}N,1)\|}$$

for $k = 1{:}N - 2$

$$\mathbf{B} = null(\mathbf{h})$$

$$\mathbf{T}_k = \begin{bmatrix} \mathbf{I}_k & \mathbf{O}_{k,N-k} \\ \mathbf{O}_{N-k,k} & \begin{bmatrix} \mathbf{h}^H \\ \mathbf{B} \end{bmatrix} \end{bmatrix}$$

$$\mathbf{R}_t = \mathbf{T}_k \mathbf{R}_t \mathbf{T}_k^H$$

$$\mathbf{h} = \frac{\mathbf{R}_t(k+2{:}N,k+1)}{\|\mathbf{R}_t(k+2{:}N,k+1)\|}$$

end

This process is continued by defining the $(N - 2)$-dimensional vector $\mathbf{s}_2$ to be the second column of the covariance matrix $\mathbf{T}_1\mathbf{R}_{x_0}\mathbf{T}_1^H$ excluding the top two elements. The vector $\mathbf{h}_2$, the orthonormal matrix $\mathbf{B}_2$, and the unitary matrix $\mathbf{T}_2$ are then calculated in a manner analogous to (9) and (10). This second stage replaces all but the first element of $\mathbf{s}_2$ with zeros. This iteration is repeated and at stage $N - 1$ the covariance matrix is tridiagonalized by the product of the $N - 1$ unitary operators $\mathbf{T}_k$. The algorithm which generates the tridiagonal covariance matrix $\mathbf{R}_t$ is described in Table 1.

Note that this operation produces a representation of the signal powers at different lags on the main diagonal and the cross-correlations between the lags, compressed into positive scalars, on the upper and lower diagonals. The tridiagonal matrix $\mathbf{R}_t$ is diagonalized by an iterative procedure using the QR algorithm which guarantees a unitary transfer function and results in the KLT due to the uniqueness theorem for unitary matrices. One popular version of the QR algorithm is shown in Table 2. The QR factorization is first applied to the tridiagonal covariance matrix to yield a unitary matrix $\mathbf{Q}_1$ and an upper triangular matrix $\mathbf{R}_1$. These resulting factors are then multiplied in reverse order to update the covariance matrix. It is easily verified that, under suitable conditions, the QR algorithm converges and the tridiagonal covariance matrix is diagonalized via a sequence of unitary similarity transformations,

$$\Lambda_k = \mathbf{Q}_k^H \Lambda_{k-1} \mathbf{Q}_k. \tag{12}$$

TABLE 2. UNITARY KLT QR SYNTHESIS RECURSION

$\Lambda_0 = \mathbf{R}_t$

for $k = 1,2,\ldots$

$$\Lambda_{k-1} = \mathbf{Q}_k \mathbf{R}_k$$

$$\Lambda_k = \mathbf{R}_k \mathbf{Q}_k$$

end

Convergence is declared when the magnitude of the off-diagonal elements is within some acceptable tolerance of zero. Finally, let $\mathbf{T}$ represent the product of the $\mathbf{T}_k$ generated in Table 1 and $\mathbf{Q}$ represent the product of the $\mathbf{Q}_k$ generated in Table 2. Then the KLT is computed as follows,

$$\begin{aligned} \Lambda &= \mathbf{Q}_N^H \cdots \mathbf{Q}_1^H \mathbf{T}_N \cdots \mathbf{T}_1 \mathbf{R}_{x_0} \mathbf{T}_1^H \cdots \mathbf{T}_N^H \mathbf{Q}_1 \cdots \mathbf{Q}_N \\ &= \mathbf{Q}^H \mathbf{T} \mathbf{R}_{x_0} \mathbf{T}^H \mathbf{Q} \\ &= \mathbf{V}^H \mathbf{R}_{x_0} \mathbf{V}, \end{aligned} \tag{13}$$

where $\mathbf{V} = \mathbf{T}^H\mathbf{Q}$ is the unitary eigenvector matrix and the sequence Λ_k converges to the diagonal matrix of eigenvalues Λ.

4. The Multiple Signal Model and Wiener Filtering

In a more general setting, the classical problems of statistical signal processing are concerned with joint signal representation and compression. These problems are often characterized by the Wiener filter, depicted in Figure 2, where there are two processes present. The N-dimensional process $\mathbf{x}_0$ is now considered to be the composite of potentially many processes, while d_0 is a scalar process which is correlated with $\mathbf{x}_0$.

The process d_0, normally termed a desired process, is representative of a signal of interest in some way, and the goal is to estimate d_0 from $\mathbf{x}_0$. For example, in the radar and sonar detection problem the "desired" signal is usually the output of a beamformer or matched-field processor, and the observed data vector $\mathbf{x}_0$ consists of data different from this signal that is received at a sensor array. In the communications application of multiuser detection and demodulation, the process d_0 may be generated by a known correlation with the signal of interest such as the code of a user in a CDMA wireless network. As a final example, in classification for automatic target recognition the desired signal may be a template image from training data, while the differing observed data is an image received by the fielded sensor. In general, the mechanism which generates the reference signal is application specific; however nearly every problem in linear statistical signal processing may be represented by the use of this model.

The problem at hand is the determination of optimal

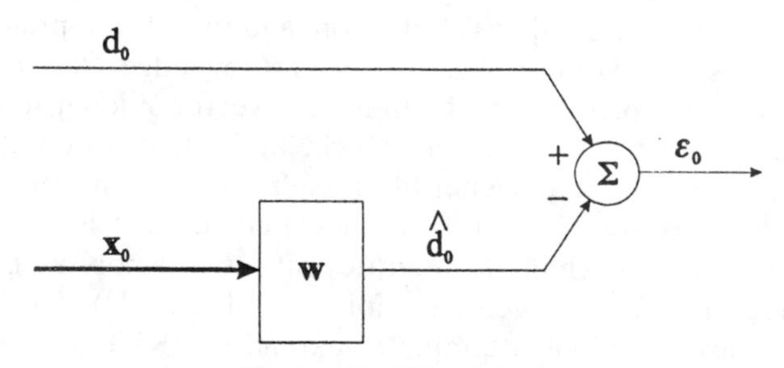

Fig. 2. The Wiener filter.

signal representation and compression for the observed data vector $\mathbf{x}_0$. The resulting basis will still span $\mathscr{C}^N$, the space spanned by the columns of $\mathbf{R}_{x_0}$, but the energy, which must be compactly represented, is now the estimation energy $\mathbf{E}[|\hat{d}_0|^2]$ and not the autocorrelation energy described by $\mathbf{R}_{x_0}$.

The signal model, introduced in Section 3 is extended here to two jointly stationary zero-mean processes. The process d_0 is a zero-mean scalar process with variance $\sigma_{d_0}^2$ and $\mathbf{x}_0$ is an observed N-dimensional signal, which itself may be a composite random process, with covariance $\mathbf{R}_{x_0}$. The filter to be defined, $\mathbf{w}$, processes the observed data to form an estimate of the desired signal $\hat{d}_0 = \mathbf{w}^H\mathbf{x}_0$. The error process ε_0,

$$\varepsilon_0 = d_0 - \mathbf{w}^H\mathbf{x}_0, \tag{14}$$

is the signal which characterizes the performance of the filter, and the optimal Wiener filter minimizes the mean-square value of this error signal.

The minimum mean-square error (MMSE) optimization criterion is formally stated as follows:

$$\min_w \mathbf{E}[|\varepsilon_0|^2] = \min_w \{\sigma_{d_0}^2 - \mathbf{w}^H\mathbf{r}_{x_0d_0} - \mathbf{r}_{x_0d_0}^H\mathbf{w} + \mathbf{w}^H\mathbf{R}_{x_0}\mathbf{w}\}, \tag{15}$$

where the N-vector $\mathbf{r}_{x_0d_0} = \mathbf{E}[\mathbf{x}_0 d_0^*]$ is the cross-correlation between the processes d_0 and $\mathbf{x}_0$. The well-known solution to (15) is the Wiener filter, which is computed as follows:

$$\mathbf{w} = \mathbf{R}_{x_0}^{-1}\mathbf{r}_{x_0d_0}. \tag{16}$$

The MMSE is calculated by substituting (16) into the expression for the mean-square value of the error,

$$\xi_0 = \sigma_{d_0}^2 - \mathbf{w}^H\mathbf{R}_{x_0}\mathbf{w} \tag{17}$$

where the far right-hand term in (17) is the optimal value of the estimation energy, given by

$$\mathbf{E}[|\hat{d}_0|^2] = \mathbf{w}^H\mathbf{R}_{x_0}\mathbf{w} = \mathbf{r}_{x_0d_0}^H\mathbf{R}_{x_0}^{-1}\mathbf{r}_{x_0d_0}. \tag{18}$$

It is now apparent that the optimal basis vector to select is the Wiener filter. However, this result requires the solution to the full-rank problem and provides no insight with respect to the selection of an optimal basis set when the Wiener filter (or the inverse of the covariance matrix) is unknown *a priori*. The desired information-theoretic goal is to achieve optimal basis selection, and therefore optimal compression, without complete prior knowledge. Previous attempts to solve this problem for the vector Wiener filter resulted only in a solution which, at best, dictated the computation of the Wiener filter itself [11–13]. This previous work is also related to canonical correlation analysis, the solution to which degenerates into the vector Wiener filter in this case as well (see, for example [7–10]). Therefore these previous attempts to extend the KLT and the related canonical correlation analysis are not discussed further. Instead, in Section 5, the necessary modifications to the KLT are addressed to obtain optimality with an eigenvector basis, which is the natural extension to the KLT previously sought by other researchers. A method to obtain an optimal basis without complete knowledge of the covariance matrix, its inverse, or the Wiener filter is then developed in Section 6.

5. The Signal-Dependent KLT for Statistical Signal Processing

The role of the KLT and the principal components in signal representation and compression is now examined within the framework of Wiener filtering for statistical signal processing, detection, and estimation. The result developed in this section demonstrates that the principal-components procedure is a suboptimal basis selection rule for detection and estimation problems. This fact motivates the question of whether the KLT and its eigenvector basis are optimal for these statistical signal processing applications, and the answer to this question serves as the topic of Section 6.

5.1. The KLT and Principal Components

A low-dimensional numerical example serves as a valuable tool for understanding the behavior of the KLT and the principal components in the statistical signal processing framework. Consider a simple example in $\mathscr{R}^2$, where there is a two-dimensional observed-data vector $\mathbf{x}_0$ with covariance matrix $\mathbf{R}_{x_0}$,

$$\mathbf{R}_{x_0} = \begin{bmatrix} 10 & 4 \\ 4 & 10 \end{bmatrix}, \tag{19}$$

and a desired signal d_0 with a variance $\sigma_{d_0}^2 = 10$. The two processes, d_0 and $\mathbf{x}_0$, are assumed to be zero mean, jointly stationary, and correlated. The cross-correlation between the two processes is given by the vector $\mathbf{r}_{x_0d_0}$,

$$\mathbf{r}_{x_0d_0} = \begin{bmatrix} 9 \\ 1 \end{bmatrix}. \tag{20}$$

The Wiener-Hopf equation in (16) may also be expressed in the form

$$\mathbf{R}_{x_0}\mathbf{w} = \mathbf{r}_{x_0d_0}, \tag{21}$$

which explicitly demonstrates that the optimal filter $\mathbf{w}$ is that particular linear combination of the columns of R_{x_0} which yields $\mathbf{r}_{x_0d_0}$. It also means that $\mathbf{w}$ is in the space spanned by the columns of $\mathbf{R}_{x_0}$ and therefore that efficient signal representation is still equivalent to optimal basis selection in $\mathscr{C}^N$. These facts indicate that the KLT be considered.

The KLT provides the eigendecomposition of the matrix $\mathbf{R}_{x_0}$:

$$\mathbf{R}_{x_0} = \begin{bmatrix} 10 & 4 \\ 4 & 10 \end{bmatrix} = \mathbf{V}\mathbf{\Lambda}\mathbf{V}^H \tag{22}$$
$$= \frac{1}{2}\begin{bmatrix} 1 & 1 \\ 1 & -1 \end{bmatrix}\begin{bmatrix} 14 & 0 \\ 0 & 6 \end{bmatrix}\begin{bmatrix} 1 & 1 \\ 1 & -1 \end{bmatrix}$$
$$= \frac{1}{2}\left(14\begin{bmatrix} 1 \\ 1 \end{bmatrix}[1 \quad 1] + 6\begin{bmatrix} 1 \\ -1 \end{bmatrix}[1 \quad -1]\right),$$

which demonstrates that one eigenvalue is significantly greater than the other. The KLT takes into account this self-directional preference of the signal. The Rayleigh quotient [16],

$$\Psi = \mathbf{e}^H\mathbf{R}_{x_0}\mathbf{e} = \sum_{i=1}^{N} \lambda_i|\mathbf{v}_i^H\mathbf{e}|^2, \tag{23}$$

mathematically describes this self-directional preference, where $\mathbf{e}$ is a unit-norm direction-varying vector. The Rayleigh quotient is maximized when $\mathbf{e} = \mathbf{v}_{max}$, the eigenvector corresponding to the largest eigenvalue. The Rayleigh quotient values form an ellipsoid, where the eigenvectors serve as the principal axes and the length of the ellipsoid along each eigenvector is proportional to the magnitude of the corresponding eigenvalue. For the example under consideration, this ellipsoid is depicted in Figure 3. It is evident that the KLT provides the most efficient representation of the autocorrelation energy in the signal.

The question at hand, however, is the determination of the *best basis representation for* $\mathbf{x}_0$ *in terms of estimating* d_0. To explore whether the KLT and the principal components are the best basis choice, consider preprocessing the observed process $\mathbf{x}_0$ with a filter composed of the eigenvectors of $\mathbf{R}_{x_0}$. This situation is depicted in Figure 4, where the new process $\mathbf{z}_0$,

$$\mathbf{z}_0 = \mathbf{V}^H\mathbf{x}_0, \tag{24}$$

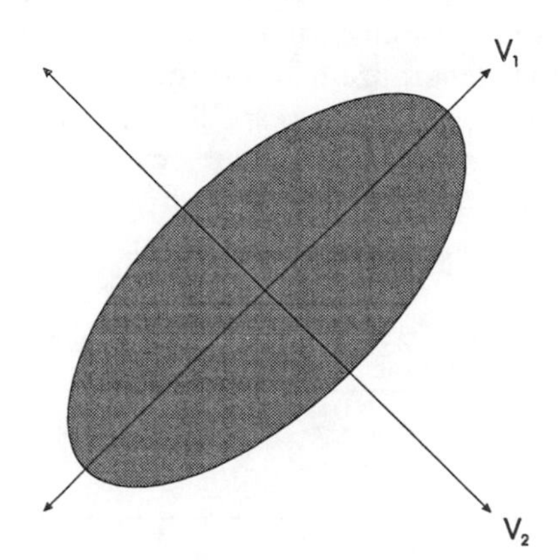

Fig. 3. The Rayleigh quotient ellipsoid.

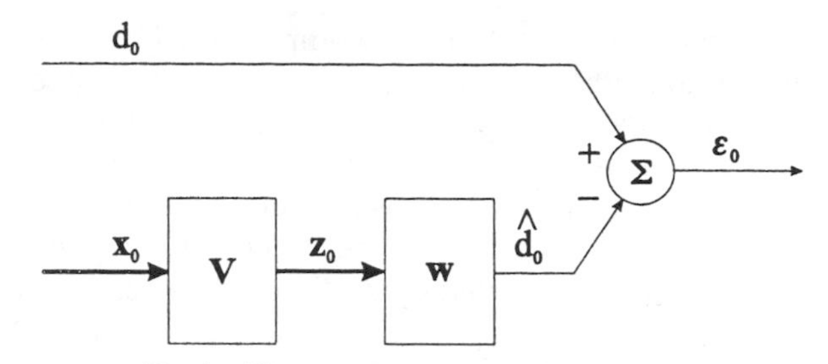

Fig. 4. The transform domain Wiener filter.

has a diagonal covariance matrix given by $\mathbf{\Lambda}$. Also, the cross-correlation between $\mathbf{z}_0$ and d_0 is now given by

$$\mathbf{r}_{z_0d_0} = \mathbf{V}^H\mathbf{r}_{x_0d_0}. \tag{25}$$

The MMSE performance of the eigenvector basis as a function of rank is now evaluated. The full-rank solution is identical regardless of basis representation since the performance measure is invariant to invertible transformations of the observed signal $\mathbf{x}_0$,

$$\xi_0 = \sigma_{d_0}^2 - \mathbf{r}_{z_0d_0}^H\mathbf{R}_{z_0}^{-1}\mathbf{r}_{z_0d_0} = \sigma_{d_0}^2 - \mathbf{r}_{x_0d_0}^H\mathbf{R}_{x_0}^{-1}\mathbf{r}_{x_0d_0}, \tag{26}$$

where $\mathbf{R}_{z_0} = \mathbf{V}^H\mathbf{R}_{x_0}\mathbf{V}$. However, the results are different for this example in the rank-1 case, where one of the two eigenvectors which compose $\mathbf{V}$ is discarded. The principal-components algorithm states that the eigenvector corresponding to the largest eigenvalue should be retained and that the eigenvector corresponding to the smaller eigenvalue should be discarded. Here, the largest eigenvalue magnitude corresponds to the first eigenvector, and the MMSE for this case is given by

$$\xi_{PC} = \sigma_{d_0}^2 - \mathbf{r}_{x_0d_0}^H\mathbf{v}_1\lambda_1^{-1}\mathbf{v}_1^H\mathbf{r}_{x_0d_0}. \tag{27}$$

The MMSE performance, converted to decibels, for the full-rank Wiener filter is 0.3951 dB and that for the rank-1 principal-components Wiener filter is 8.0811 dB. Thus, there is a loss of 7.6860 dB in reducing the rank from 2 to 1.

To answer the question of whether the principal-components approach is optimal for signal representation in this problem, first evaluate the MMSE which results if the smaller eigenvector is retained in this example:

$$\xi_{test} = \sigma_{d_0}^2 - \mathbf{r}_{x_0d_0}^H\mathbf{v}_2\lambda_2^{-1}\mathbf{v}_2^H\mathbf{r}_{x_0d_0}, \tag{28}$$

yielding an MMSE of 6.6901 dB. Here the MMSE loss is approximately 1.4 dB less than that experienced by the principal-components selection; that is, a performance enhancement is obtained by selecting a different eigenvector than that indicated by the principal components.

5.2. The Cross-Spectral Metric: An Intelligent and Signal-Dependent KLT

It is necessary to explain what, at first exposure, appears to be a breakdown in theory. Thus far it may seem strange that the selection of the principal components is not optimal for signal representation and compression in detection and estimation problems. An insight is now developed to demonstrate why the principal components are the wrong performance measure for these problems. It is also shown that the result of this example is not an anomaly. Finally, the necessary modification to the principal-components algorithm is provided so that optimal signal representation and compression is again gained if one restricts signal representation to an eigenvector basis: an intelligent KLT which takes all available statistical information into account to select the best eigenvectors.

The optimal solution for the performance measure of interest[4] in (17) is given by

$$\xi_0 = \sigma_{d_0}^2 - \mathbf{r}_{x_0d_0}^H \mathbf{R}_{x_0}^{-1} \mathbf{r}_{x_0d_0} = \sigma_{d_0}^2 - \sum_{i=1}^{N} \frac{|\mathbf{v}_i^H \mathbf{r}_{x_0d_0}|^2}{\lambda_i}. \tag{29}$$

The expression in (29) demonstrates that the presence of d_0 induces another directional preference through $\mathbf{r}_{x_0d_0}$ in the spectrum of $\mathbf{x}_0$. This induced directional preference must be taken into account to achieve the optimal subspace selection for detection and estimation. Here it is evident that minimizing the performance measure requires maximizing the summation on the right-hand side of (29). This is accomplished by evaluating the square of a normalized projection of the cross-correlation vector upon the KLT basis vectors and, by (18), retaining those eigenvectors which maximize the estimation energy. Note that the result in the middle term of (29) may be interpreted as the difference between the desired signal variance and the mean-squared value of the whitened cross-correlation (or whitened matched-filter replica vector), $|\mathbf{R}_{x_0}^{-1/2}\mathbf{r}_{x_0d_0}|^2$.

This new metric chooses those eigenvectors which correspond to the largest values of a ratio which takes into account both the directional preference of the process $\mathbf{x}_0$ and the impact of the correlated portion of the spectrum due to the process d_0, namely the projection of the cross-correlation along the ith basis vector,

$$\frac{|\mathbf{v}_i^H \mathbf{r}_{x_0d_0}|^2}{\lambda_i}. \tag{30}$$

Accordingly, consider a different enumeration of the eigenvectors, in a descending order based upon the largest magnitude of those N terms in (30). This ranking is called the cross-spectral metric [17], and keeping those k eigenvectors which maximize this cross-spectral metric provide the lowest mean-square error as a function of k, the dimension of the eigenvector basis. The subspace spanned by the eigenvectors selected by the cross-spectral metric is denoted $\mathscr{C}_{csm}^k$ and is in general different from the subspace $\mathscr{C}_{pc}^k$, especially for small k. The cross-spectral metric is the magnitude-squared value of a direction cosine in $\mathscr{C}^N$ which measures the "closeness" of the basis vector $\mathbf{v}_i$ to the cross-correlation vector [17]. Recall from (21) that the Wiener-Hopf equation yields the optimal coefficient vector for linearly combining the basis vectors to obtain this cross-correlation vector.

An interesting fact is realized by constraining the vector $\mathbf{w}$ in (18) to be unit norm and allowing it to vary over all possible values. The quadratic form of the Rayleigh's quotient in (23) is again obtained, with the optimal solution not being an eigenvector but the Wiener filter in (16). This is another interpretation of the change in directional preference induced by the observation of multiple signals.

The example concerned with the selection of rank-1 Wiener filters is now revisited in order to complete the KLT analysis. The magnitude of the cross-spectral metric for the first eigenvector, corresponding to the larger eigenvalue of magnitude 14, is 3.5714. The magnitude of the cross-spectral metric for the second eigenvector, corresponding to the smaller eigenvalue of magnitude 6, is 5.3333. Thus, $\xi_{test} = \xi_{CSM}$ in (28), and it is verified that the cross-spectral metric provides the better solution. A summary of the KLT analysis for this example is provided in Table 3.

6. Intelligent Signal Representation for Statistical Signal Processing

The preceding results demonstrate that the principal-components method is not the optimal metric for signal representation and compression in Wiener filtering problems. From a pedagogical perspective, it has now been demonstrated that the principal components must be modified to take into account the directional preference induced by other signals to retain optimality (with respect to the KLT basis) for signal representation in detection and estimation problems. However, it is reasonable to argue that knowledge of the eigenstructure of $\mathbf{R}_{x_0}$ is equivalent to knowledge of $\mathbf{R}_{x_0}^{-1}$, and therefore a different solution is desired.

This section derives a generalization of the KLT which naturally includes this cross-spectral information to generate a new basis for signal representation with optimal properties when the covariance matrix inverse is unknown. In particular, the generalized joint-process KLT is demonstrated to be optimal with respect to a maximization of the "cross-correlation" energy, subject to a unity response filter gain. This may be directly interpreted as a matched filter criterion which maximizes the signal response energy.

TABLE 3. THE EFFICIENCY OF KLT-BASED SIGNAL REPRESENTATION

	$\lvert\lambda_i\rvert$	$\frac{\lvert\mathbf{v}_i^H \mathbf{r}_{x_0d_0}\rvert^2}{\lambda_i}$	ξ_{PC}	ξ_{CSM}
i = 1	14	3.5714	8.0811 dB	
i = 2	6	5.3333		6.6901 dB

[4] The MMSE performance measure is actually very general; for example, under many conditions it is also equivalent to the maximum likelihood, maximum output signal-to-interference plus noise ratio and maximum mutual information performance measures, among others.

6.1. A New Criterion for Signal Representation and Its Implementation

A multistage Wiener filter, which derives a signal-dependent basis for compression [18–25], is seen in this section to achieve the joint-process KLT from a "whitening" or innovations perspective [26]. The objective here is to optimize sequentially each rank-one basis selection so as to both further whiten the error residue and compress (compactly represent) the colored portion of the observed-data subspace, thereby achieving the desired signal-dependent rank reduction.

Consider the selection of the first stage of the multistage Wiener filter depicted in Figure 5a. At each stage, a rank-one basis is chosen to both reduce the mean-square error and compactly represent the correlated portion of the observed random process. In so doing, however, it is desired to avoid the obvious Wiener solution, which requires knowledge of the full-rank covariance matrix associated with the observed process $\mathbf{x}_0$. Note that, in general, this requirement also eliminates the usual KLT-based methods. However, in order to contribute to a reduction of the mean-square error, the basis selection process should produce signals which are, in some appropriate measure, maximally correlated with the desired process d_0. Therefore consider the following basis selection procedure: At stage 1 (see Figure 5a), select that rank-one subspace $\mathbf{h}_1 \in \mathscr{C}^N$ which is maximally correlated with the desired signal d_0. More specifically, since $d_1 = \mathbf{h}_1^H\mathbf{x}_0$, the following optimization problem is obtained:

$$\max_{\mathbf{h}_1} \mathbf{E}[|\mathbf{h}_1^H\mathbf{x}_0 d_0^*|^2] \text{ subject to } \mathbf{h}_1^H\mathbf{h}_1 = 1. \quad (31)$$

A direct application of Schwarz's inequality with $\mathbf{R}_{x_0}$ unknown readily yields the selection,

$$\mathbf{h}_1 = \frac{\mathbf{r}_{\mathbf{x}_0 d_0}}{\|\mathbf{r}_{\mathbf{x}_0 d_0}\|}. \quad (32)$$

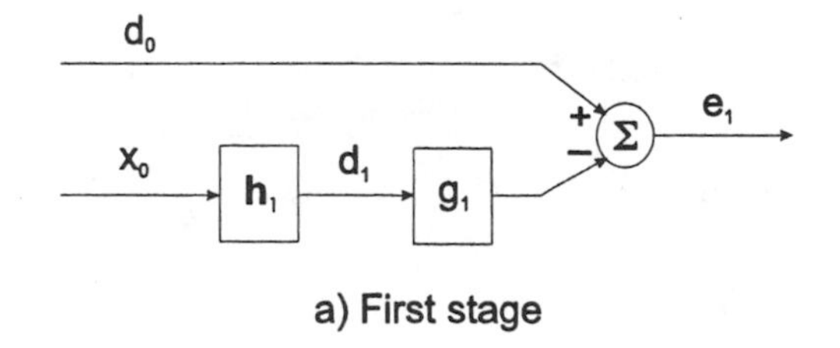

a) First stage

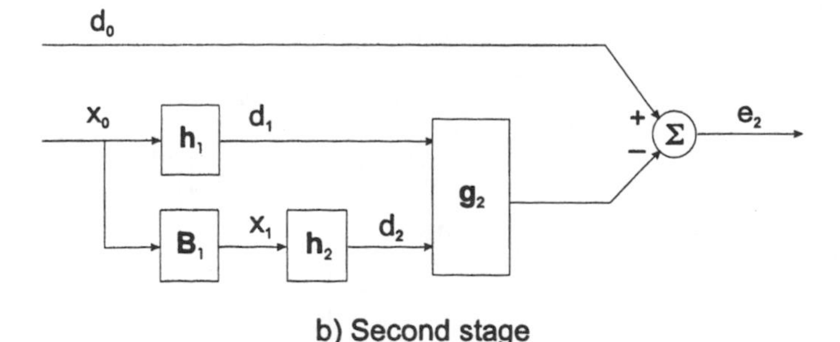

b) Second stage

Fig. 5. The first two stages of a matched filter decomposition.

Note that, as desired, the solution does not depend on full knowledge of $\mathbf{R}_{x_0}$. Since $\mathbf{h}_1$ is not, in general, colinear with the Wiener solution (unless $\mathbf{R}_{x_0}$ is white), further mean-square error reduction (whitening) is possible by adding additional stages. Thus, consider the error residue, e_1, that results from the first-stage basis selection:

$$e_1 = d_0 - g_1^* d_1 = d_0 - g_1^* \mathbf{h}_1^H \mathbf{x}_0 \quad (33)$$

where $g_1 \in \mathscr{C}$, is the optimal scalar Wiener weight (filter) for linearly estimating d_0 from d_1.

All of the information required for estimating d_0 from $\mathbf{x}_0$ which is not in the direction of $\mathbf{r}_{x_0 d_0}$ is contained in its $(N - 1)$-dimensional nullspace. The second stage, therefore, is introduced by first filtering $\mathbf{x}_0$ with the operator $\mathbf{B}_1$, an $(N - 1) \times N$ matrix whose rows form an orthonormal basis for the nullspace of $\mathbf{r}_{x_0 d_0}$. This operator is again easily realized by one stage of a Householder reflection. Now, $\mathbf{h}_2$ is chosen by an optimization which is identical in form to (31), thereby resulting in the selection

$$\mathbf{h}_2 = \frac{\mathbf{r}_{x_1 e_1}}{\|\mathbf{r}_{x_1 e_1}\|}, \quad (34)$$

which is the matched filter for estimating the error residual from the remaining information available in $\mathbf{x}_0$ (see Figure 5b). This decomposition is continued and a new basis is constructed that is based on maximal correlation.

In general, the objective at the ith stage, $2 \leq i \leq N - 1$, is to select that $\mathbf{h}_i \in \mathscr{C}^{N-i+1}$ which is maximally correlated with the residue e_{i-1} from all previous stages, i.e.,

$$\max_{\mathbf{h}_i} \mathbf{E}[|\mathbf{h}_i^H\mathbf{x}_{i-1} e_{i-1}^*|^2] \text{ subject to } \mathbf{h}_i^H\mathbf{h}_i = 1, \quad (35)$$

without knowledge of $\mathbf{R}_{x_0}$, where

$$e_i = d_0 - \mathbf{g}_i^H \mathbf{d}_i, \quad (36)$$

$\mathbf{d}_i = [d_1\ d_2 \ldots d_i]^T \in \mathscr{C}^i$, and $\mathbf{g}^i \in \mathscr{C}^i$ is the optimal weight vector in the transformed coordinates. For example, Figure 5b shows filtering structure for the case where $i = 2$.

A significant simplification in solving for $\mathbf{h}_i$, which eliminates the need for explicitly computing the weight vector $\mathbf{g}_i$ in the basis selection process, results by recognizing that $\mathbf{r}_{x_i e_i}$ is colinear with $\mathbf{r}_{x_i d_i}$. Thus,

$$\mathbf{h}_i = \frac{\mathbf{r}_{x_i e_i}}{\|\mathbf{r}_{x_i e_i}\|} = \frac{\mathbf{r}_{x_i d_i}}{\|\mathbf{r}_{x_i d_i}\|}, \quad (37)$$

and the equivalent filter structure can be represented in the form of a multistage Wiener filter [18, 19, 21]. This filter, depicted in Figure 6 for the $N = 4$ case, is readily interpreted as a nested chain of *scalar* Wiener filters following a bank of white-noise matched filters whose composite form solves the colored-noise matched filtering problem. Define $\xi_i = \mathbf{E}[|\varepsilon_i|^2]$ to be the error variance for the ith stage and $\delta_N = r_{x_N d_N}$. Then the scalar weights in Figure 6

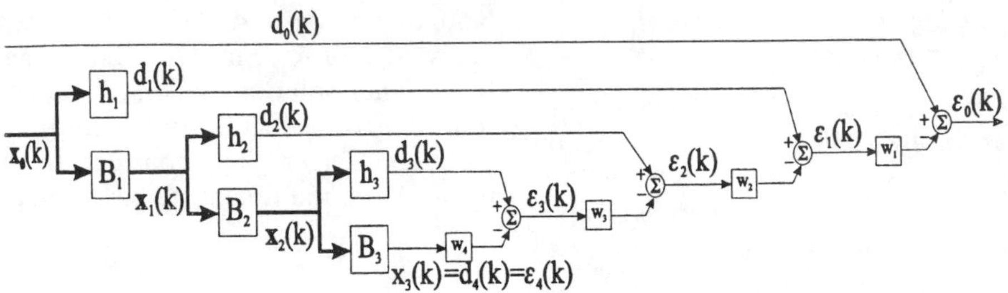

Fig. 6. The multistage Wiener filters for $N = 4$.

are computed as follows:

$$w_i = \xi_i^{-1}\delta_i, \quad 1 \le i \le N, \tag{38}$$

where

$$\delta_i = \mathbf{h}_i^H \mathbf{r}_{x_{i-1}d_{i-1}}, \quad 1 \le i < N. \tag{39}$$

Note that the error signal for the last stage ε_N is defined by

$$\varepsilon_N = d_N = x_{N-1}, \tag{40}$$

as depicted in Figure 6.

It is now emphasized that the filterbank-based multistage Wiener filter in Figure 6 does not require that the covariance matrix be estimated, inverted, or decomposed. This structure and the vector Wiener filter yield identical solutions, implying that the multistage Wiener filter solves the colored-noise match filtering problem via a sequence of white noise match filters which do not require a matrix whitening operator or a matrix inversion. These white noise matched filters are determined by a sequence of correlation vectors, which may be estimated directly. A significant advantage of this structure, emphasized in Section 6.2, is that in general all stages of this filter need not be computed to obtain excellent performance.

6.2. *A Generalized Joint-Process KLT: Nonunitary Diagonalization of the Covariance*

The implementation of the KLT is now modified to motivate its generalization for the induced directional preference which occurs in detection and estimation. The Householder reduction of a covariance matrix to tridiagonal form [14, 15] is the first step for most algorithms which solve the eigenvalue problem, as noted in Section 3.3. The second step is the diagonalization of the tridiagonal covariance matrix in a unitary manner, yielding the eigenvalues and eigenvectors. The use of Householder reflectors to tridiagonalize the $N \times N$ covariance matrix $\mathbf{R}_x$ is again considered here, where a modification to the presentation in Section 3.3 is introduced to provide the desired directional preference. It is seen that this modification to the KLT results directly in the multistage Wiener filter.

The $N \times N$ reflector $\mathbf{T}_1$ is now defined to satisfy the following relation [21],

$$\mathbf{T}_1 \mathbf{r}_{x_0 d_0} = [\delta_1 \quad 0 \cdots 0]^T, \tag{41}$$

where $\delta_1 = \|\mathbf{r}_{x_0 d_0}\|$. This reflector is found by normalizing the cross-correlation vector

$$\mathbf{h}_1 = \frac{\mathbf{r}_{x_0 d_0}}{\|\mathbf{r}_{x_0 d_0}\|}. \tag{42}$$

and solving the relation

$$\mathbf{B}_1 = null(\mathbf{h}_1). \tag{43}$$

The matrix $\mathbf{T}_1$ in (41) is given by

$$\mathbf{T}_1 = \begin{bmatrix} \mathbf{h}_1^H \\ \mathbf{B}_1 \end{bmatrix}, \tag{44}$$

and the operation of $\mathbf{T}_1$ on $\mathbf{x}_0$ yields a covariance matrix $\mathbf{R}_{x_1}$,

$$\mathbf{R}_{x_1} = \mathbf{T}_1 \mathbf{R}_{x_0} \mathbf{T}_1^H = \begin{bmatrix} \sigma_{d_1}^2 & \mathbf{r}_{x_1 d_1}^H \\ \mathbf{r}_{x_1 d_1} & \mathbf{R}_{x_2} \end{bmatrix}, \tag{45}$$

where the scalar $\sigma_{d_1}^2 = \mathbf{h}_1^H \mathbf{R}_{x_0} \mathbf{h}_1$, the $(N-1) \times (N-1)$-dimensional matrix $\mathbf{R}_{x_1} = \mathbf{B}_1 \mathbf{R}_{x_0} \mathbf{B}_1^H$, and the $(N-1) \times 1$ vector $\mathbf{r}_{x_1 d_1} = \mathbf{B}_1 \mathbf{R}_{x_0} \mathbf{h}_1$. This operation is then repeated as described in Table 1 of Section 3.3 and shown here in Table 4. Note that the operations in Table 4 are identical to those in Section 3.3, with the exception that the first reflection (or pivot) uses the directional information from

TABLE 4. MODIFIED UNITARY DECOMPOSITION RECURSION

$\mathbf{R}_t = \mathbf{R}_{x_0}$

$\mathbf{h} = \dfrac{\mathbf{r}_{x_0 d_0}}{\|\mathbf{r}_{x_0 d_0}\|}$

for k = 1:N − 1

$\quad \mathbf{B} = null(\mathbf{h})$

$$\quad \mathbf{T}_k = \begin{bmatrix} \mathbf{I}_{k-1} & \mathbf{O}_{k-1,N-k+1} \\ \mathbf{O}_{N-k+1,k-1} & \begin{bmatrix} \mathbf{h}^H \\ \mathbf{B} \end{bmatrix} \end{bmatrix}$$

$\quad \mathbf{R}_t = \mathbf{T}_k \mathbf{R}_t \mathbf{T}_k^H$

$\quad \mathbf{h} = \dfrac{\mathbf{R}_t(k+1{:}N,k)}{\|\mathbf{R}_t(k+1{:}N,k)\|}$

end

the desired process. Also note that available algorithms for Householder reflections can solve the tridiagonalization, initialized by the matrix multiplications in (11) and (45), without explicitly forming the product of the matrices $\mathbf{T}_i$.

This unitary reduction of the covariance matrix $\mathbf{R}_{x_0}$ results in the matrix $\mathbf{R}_t$ having a tridiagonal form. For example, the following decomposition is realized for $N = 3$:

$$\mathbf{R}_{x_0} \overset{\mathbf{T}_1}{\Rightarrow} \begin{bmatrix} \sigma^2_{d_1} & \mathbf{r}^H_{x_1 d_1} \\ \mathbf{r}_{x_1 d_1} & \mathbf{R}_{x_1} \end{bmatrix} \overset{\mathbf{T}_2}{\Rightarrow} \begin{bmatrix} \sigma^2_{d_1} & \delta^*_2 & 0 \\ \delta_2 & \sigma^2_{d_2} & \delta^*_3 \\ 0 & \delta_3 & \sigma^2_{d_3} \end{bmatrix} = \mathbf{R}_t, \quad (46)$$

where $\mathbf{h}_1 = \dfrac{\mathbf{r}_{x_0 d_0}}{\|\mathbf{r}_{x_0 d_0}\|}$, $\mathbf{h}_2 = \dfrac{\mathbf{r}_{x_1 d_1}}{\|\mathbf{r}_{x_1 d_1}\|}$, $\mathbf{R}_{x_0} \in \mathscr{C}^{3\times 3}$, $\mathbf{R}_{x_1} \in \mathscr{C}^{2\times 2}$, $\mathbf{r}_{x_1 d_1} \in \mathscr{C}^{2\times 1}$, and the remaining variables are all scalars with $\delta_2 = \|\mathbf{r}_{x_1 d_1}\|$, $\delta_3 = r_{x_2 d_2}$, and $\sigma^2_{d_3} = \mathbf{R}_{x_2}$.

The decomposition is equivalently represented as an analysis filterbank, shown in Figure 6, which transforms the observed-data N-vector $\mathbf{x}_0$ to a new N-vector $\mathbf{d}_N$,

$$\mathbf{d}_N = \mathbf{T}\mathbf{x}_0 = [d_1 \quad d_2 \cdots d_N]^T, \quad (47)$$

where $\mathbf{T}$ is the product of the matrices $\mathbf{T}_i$ from each stage [21]. The covariance matrix associated with vector $\mathbf{d}_N$ is tridiagonal, as shown in (46). Because each of the $\mathbf{T}_i$ is unitary, the matrix $\mathbf{T}$ is unitary, and the MMSE is not modified. However, the basis vectors are now ordered in a manner based upon the maximum correlation between the desired and observed processes. If this matrix is diagonalized using a unitary operator, then the KLT is obtained, the ordering is lost, and the basis is altered. It is therefore necessary to consider the diagonalization of the covariance matrix *in a manner which is not restricted to be unitary.*

In particular, to minimize the residual and generate an innovations process in cross-correlation, let the error-synthesis filterbank minimize the mean-square error recursively at each stage of the analysis filterbank. This nonunitary error-synthesis filterbank generates the recursion described in Table 5 and depicted in Figure 6. The nonunitary diagonalization of $\mathbf{R}_{x_0}$ may be expressed in the following matrix equation:

$$\xi = \begin{bmatrix} \xi_1 & & & \\ & \xi_2 & & \\ & & \ddots & \\ & & & \xi_N \end{bmatrix} = \mathbf{Q}\mathbf{T}\mathbf{R}_{x_0}\mathbf{T}^H\mathbf{Q}^H = \mathbf{E}^H\mathbf{R}_{x_0}\mathbf{E}, \quad (48)$$

TABLE 5. NONUNITARY SYNTHESIS RECURSION

$\xi N = \sigma^2_{d_N}$
for $p = N - 1{:}-1{:}1$
$\qquad \xi_p = \sigma^2_{d_p} - \xi^{-1}_{p+1}|\delta_{p+1}|^2$
end

where ξ is diagonal and the coefficients of the upper-triangular Gram matrix $\mathbf{Q}$ are found directly from the recursion in Table 5. Matrix $\mathbf{T}$ is always unitary when formed in this manner while matrix $\mathbf{Q}$ is not necessarily unitary. Therefore matrix $\mathbf{E} = \mathbf{T}^H\mathbf{Q}^H$ in (48) is generally nonunitary and invertible. The diagonalization of the covariance matrix in (48), using the algorithm described in Tables 4 and 5, is obtained in a numerically stable manner. Furthermore, the diagonalization of the tridiagonal covariance matrix in Table 5 is guaranteed to be complete in $N - 1$ recursions, as opposed to the unknown and varying convergence property of the QR algorithm (Table 2) used in the KLT.

The rationale for developing this nonunitary diagonalization is that a correlation-based ranking of the subspaces can be imposed, where the unitary diagonalization would destroy the induced directional preference. Thus, reduced-rank signal representation or compression can be obtained by pruning the decomposition. This implies that the matrix $\mathbf{R}_t$ is reduced from $N \times N$ to $k \times k$ by discarding the last $N - k$ rows and columns (or never computing them). The recursion in Table 5 is then implemented with the value for N replaced by the value for k. Finally, the MMSE ξ_0 is found by applying the recursion in Table 5 one last time:

$$\xi_0 = \sigma^2_{d_0} - \xi^{-1}_1|\delta_1|^2. \quad (49)$$

This solution has an interesting interpretation. The KLT results in a unitary diagonalizing transformation for one process which most compactly represents the modal or component signal energy for that process using the fewest spectral coefficients. The generalized joint-process KLT produces a nonunitary diagonalization, where the spectral coefficients are the *modal mean-square error values* between the observed composite process and the selected reference process. Recall that the motivation for the joint-process KLT is to redirect the original emphasis on the compaction of signal energy in $\mathbf{x}_0$ to the compaction of the estimation energy which minimizes the mean-square error. Therefore, this result seems intuitively satisfying, and it is especially worth noting the fact that an *a priori* knowledge of $\mathbf{R}^{-1}_{x_0}$ is never used.

Now return to the previous numerical example in $\mathscr{R}^2$, with the 2×2 matrix $\mathbf{R}_{x_0}$ in (19), the 2×1 vector $\mathbf{r}_{x_0 d_0}$ in (20), and $\sigma^2_{d_0} = 10$. The vectors $\mathbf{h}_1$ and $\mathbf{B}_1$ are then defined by (20), (42), and (43) to yield

$$\mathbf{T}_1 = \begin{bmatrix} \mathbf{h}^H_1 \\ \mathbf{B}_1 \end{bmatrix} = \frac{1}{\sqrt{82}} \begin{bmatrix} 9 & 1 \\ -1 & 9 \end{bmatrix}. \quad (50)$$

The correct tridiagonal form is then found after one application of the recursion:

$$\mathbf{R}_t = \mathbf{T}_1\mathbf{R}_{x_0}\mathbf{T}^H_1 = \begin{bmatrix} \sigma^2_{d_1} & \delta^*_2 \\ \delta_2 & \sigma^2_{d_2} \end{bmatrix} = \frac{1}{82}\begin{bmatrix} 892 & 320 \\ 320 & 748 \end{bmatrix}. \quad (51)$$

TABLE 6. THE MMSE (dB) FOR EACH WIENER FILTER

$\xi_0 \in \mathscr{R}^2$	$\xi_{JKLT} \in \mathscr{R}^1$	$\xi_{CSM} \in \mathscr{R}^1$	$\xi_{PC} \in \mathscr{R}^1$
0.3951	3.9127	6.6901	8.0811

The diagonal matrix $\boldsymbol{\xi}$ is then given by

$$\boldsymbol{\xi} = \begin{bmatrix} \xi_1 & 0 \\ 0 & \xi_2 \end{bmatrix} = \begin{bmatrix} \sigma_{d_1}^2 - \dfrac{|\delta_2|^2}{\sigma_{d_2}^2} & 0 \\ 0 & \sigma_{d_2}^2 \end{bmatrix} = \frac{1}{82}\begin{bmatrix} 755.1 & 0 \\ 0 & 748 \end{bmatrix}. \quad (52)$$

It is simple in this case to verify (48)

$$\boldsymbol{\xi} = \mathbf{Q}\mathbf{T}_1\mathbf{R}_{x_0}\mathbf{T}_1^H\mathbf{Q}^H, \quad (53)$$

where the upper-triangular matrix $\mathbf{Q}$ is given by

$$\mathbf{Q} = \begin{bmatrix} 1 & -\dfrac{\delta_2^*}{\sigma_{d_2}^2} \\ 0 & 1 \end{bmatrix} = \begin{bmatrix} 1 & -0.4278 \\ 0 & 1 \end{bmatrix}, \quad (54)$$

is easily found from the recursion in (14), (16), (17), and Table 5. The MMSE is now computed by

$$\xi_0 = \sigma_{d_0}^2 - \frac{|\delta_1|^2}{\xi_1}, \quad (55)$$

which yields 0.3951 dB in agreement with the previous full-rank MMSE result. Finally, the rank-k joint-process KLT (JKLT) chooses a rank-k subspace $\mathscr{C}_{jklt}^k$ with an associated MMSE ξ_{JKLT}. Here, $k = 1$, and

$$\xi_{JKLT} = \sigma_{d_0}^2 - \frac{|\delta_1|^2}{\sigma_{d_1}^2}, \quad (56)$$

which yields an MMSE of 3.9127 dB. As shown in Table 6, this represents an improvement of nearly 3 dB over the cross-spectral metric and slightly over 4 dB compared with the principal components.

6.3. Analysis of the JKLT

The basis vectors of the JKLT are determined as a function of both the self-directional preference due to the eigenvectors that are associated with the observed-data covariance matrix $\mathbf{R}_{x_0}$ and the induced directional preference due to the effect of the cross-correlation $\mathbf{r}_{x_0d_0}$. This fact implies the ability to perform low-rank subspace tracking in nonstationary signal environments and optimal reduced-rank Wiener filtering in stationary signal environments. Here optimality is with respect to the minimization of the average MMSE, as a function of rank, over all possible signal environments without knowledge of the full-rank Wiener solution (or of $\mathbf{R}_{x_0}^{-1}$).

To demonstrate these facts, consider again revisiting the numerical 2×2 example. Here, the eigenvectors in (22) are vectors at 45° and −45° in the x-y plane. Now let $\mathbf{r}_{x_0d_0}$ be a unit vector which varies from 0° to 360° in the plane. The resulting Wiener filter $\mathbf{w}$ maps out an ellipse as the cross-correlation vector varies counterclockwise in the plane. These mappings are depicted in Figure 7.

The optimal rank-2 Wiener filter and the different rank-1 solutions are compared as the statistics vary in Figure 8, where the angle associated with $\mathbf{r}_{x_0d_0}$ varies from $\phi = 0°$ to $\phi = 360°$. Note that this experiment tells the entire story

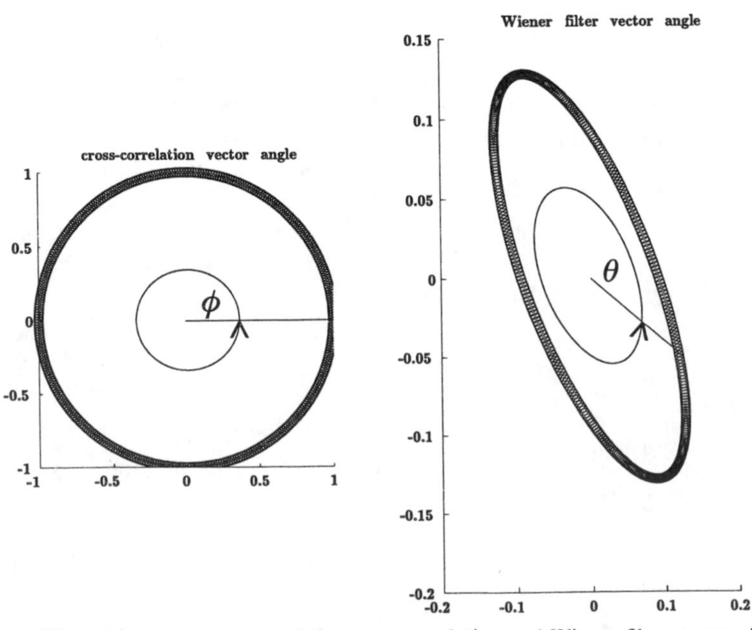

Fig. 7. The angle representation of the cross-correlation and Wiener filter vectors in $\mathscr{R}^2$.

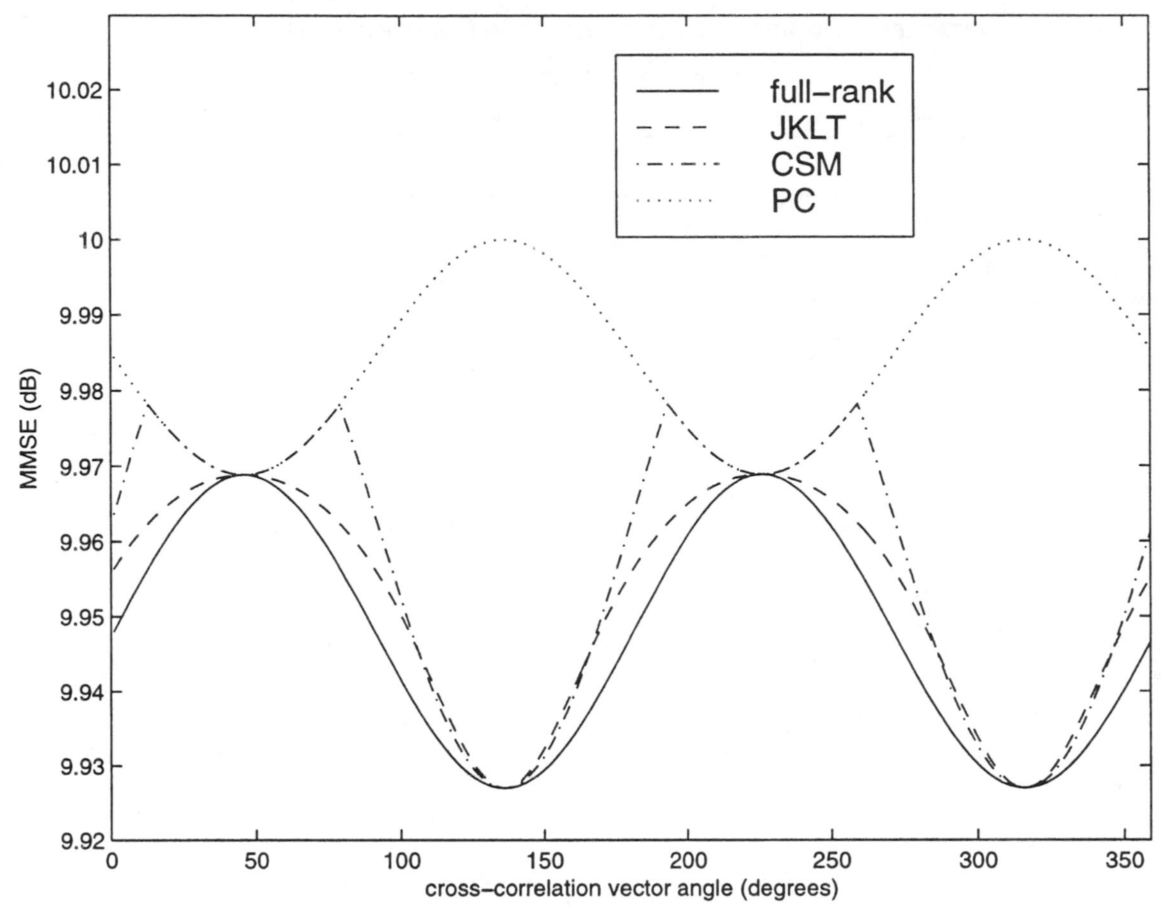

Fig. 8. The performance as a function of statistical variation.

for $\mathscr{R}^2$; that is, the performance comparison is now made over all possible desired signals via the inclusion of all possible induced cross-correlation unit vectors.

The principal-components technique fixes the rank-1 basis vector as that eigenvector which corresponds to the largest eigenvalue. This choice results in poor performance except for the small region where the Wiener filter is approximately colinear with that eigenvector. As seen in Figure 8, the principal-components MMSE obtains the full-rank optimum only twice over 360°, exactly when the Wiener filter is colinear.

The cross-spectral metric introduces the ability to change eigenvectors depending upon the optimal Wiener solution. It is seen in Figure 8 that the cross-spectral metric obtains the optimal solution twice as often, whenever the Wiener filter is colinear with either eigenvector. Thus the cross-spectral metric "switches" between the eigenvectors to optimize the MMSE.

Interestingly, both the principal components and the cross-spectral metric require *a priori* knowledge of the eigenvectors. This is computationally on the same order as calculating $\mathbf{R}_{x_0}^{-1}$ and therefore the Wiener solution. However, neither of these techniques makes very good use of this information. The JKLT has no such requirement for prior knowledge, instead using only the cross-correlation vector or its estimate. With this available information, a Wiener filter and the resulting MMSE for any rank can be computed efficiently. However, the merit of this method is realized in Figure 8. It is seen that the JKLT represents the signal environment using a basis set which tracks the Wiener solution (or converges to it) very efficiently as a function of rank. Note that this is done without ever computing the full-rank Wiener solution. When the slope of the optimal weight vector dynamics goes to zero, corresponding to the dynamics slowing down, the JKLT becomes a unitary decomposition, the basis selection converges to the eigenvectors, and all of the methods obtain the optimal MMSE.

A few notes are now in order with respect to performance trends. First, the cross-spectral metric is demonstrated always to perform as well as or better than the principal-components algorithm in terms of the resulting MMSE as a function of rank. The JKLT always performs as well as or better than the principal-components and nearly always provides a better MMSE performance than the cross-spectral metric. The case where the cross-spectral metric may outperform the JKLT is over small regions where a relatively low-magnitude cross-correlation vector is contained in a subspace spanned by the eigenvectors that correspond to small eigenvalues. Here, the cross-spectral metric can take advantage of the additional information it has due to knowledge of the dominating self-directional

TABLE 7. THE APPROXIMATE AVERAGE MMSE LOSS (dB) FOR EACH WIENER FILTER

$\int_{0^\circ}^{360^\circ} \lvert\xi_0(\omega) - \xi_{JKLT}(\omega)\rvert d\omega$	$\int_{0^\circ}^{360^\circ} \lvert\xi_0(\omega) - \xi_{CSM}(\omega)\rvert d\omega$	$\int_{0^\circ}^{360^\circ} \lvert\xi_0(\omega) - \xi_{PC}(\omega)\rvert d\omega$
3.58	6.68	30.00

preference. When this occurs, the cross-spectral metric can "switch" to a subspace spanned by the low-magnitude eigenvectors faster than the JKLT can "learn" the correct subspace. Notice that this situation does not affect the performance of the algorithms along a principal-components subspace.

Finally, the performance from the numerical example used throughout the chapter may be exploited to demonstrate some general properties relative to the optimality of the JKLT. The JKLT optimizes the MMSE performance on average over all possible signal environments. Here, the area between the full-rank MMSE curve and the other reduced-rank MMSE curves in Figure 8 represents the average MMSE performance loss over all possible signal environments in $\mathscr{R}^2$, and these measurements are presented in Table 7. These results depict that, with a cross-correlation unit vector in $\mathscr{R}^2$, on average the JKLT outperforms the principal components by approximately 26.5 dB and the cross-spectral metric by approximately 3 dB.

7. Radar Example

A radar detection example is now presented to demonstrate the application of intelligent reduced-rank signal processing to a practical problem of interest. Sensor signal processing introduces a unique sample support interpretation due to the dependence of the training data on range. Ergodicity, in the context of sensor signal processing, implies replacing the ensemble statistical average with an average over range rather than time. Therefore the IID and stationarity assumptions on the joint statistics of the training and test data are replaced with clutter and noise homogeneity assumptions throughout the data collected by the radar. Intelligent rank reduction offers the opportunity to reduce the sample support requirements without greatly degrading performance. Intelligent training methods allow the statistical estimation to concentrate on the clutter and noise which are most correlated with that present in the test data set, thereby optimizing the detection problem. The intelligent subspace selection also reduces the sample support requirements when the stationary and IID assumptions are valid, allowing rapid adaptation.

7.1. Radar Signal Processing

Consider now the use of an airborne space-time adaptive processing (STAP) radar for the detection of a target at a particular range. The STAP radar collects angle-Doppler data over multiple range cells, as shown in Figure 9. The use of STAP is required because the two-dimensional extent of ground clutter exhibits both spatial and Doppler dependence, as further explained in Section 7.3. A STAP processor is composed of K antenna elements which provide spatial degrees of freedom and a J-tap Doppler filterbank with time lags that correspond to the radar pulse repetition interval (PRI) in order to provide the spectral degrees of freedom. The total number of adaptive space-time degrees of freedom is then $N = KJ$.

The $N \times 1$ space-time steering vector $\mathbf{s}$ forms a beam at an angle-Doppler location where target presence or absence is going to be tested. The $N \times 1$ space-time snapshot formed from the radar data collected at the range gate of interest is denoted $\mathbf{x}_p$. Assume for the moment that the (unavailable) clairvoyant clutter and noise covariance matrix $\mathbf{R}$ is known *a priori*. Then the optimal STAP weight vector for Gaussian colored noise detection (simultaneous detection and clutter/noise mitigation) is given by [3, 27]

$$\mathbf{w}_a = \frac{\mathbf{R}^{-1}\mathbf{s}}{\mathbf{s}^H\mathbf{R}^{-1}\mathbf{s}} = \mathbf{s} - \mathbf{B}^H\mathbf{w}, \tag{57}$$

where the full row-rank matrix $\mathbf{B}$ is the nullspace of vector $\mathbf{s}$ such that $\mathbf{Bs} = 0$. The weight vector $\mathbf{w}$ is the optimal weight vector for estimating the clutter and noise present in the beamformer output, $d_0 = \mathbf{s}^H\mathbf{x}_p$, from that present in

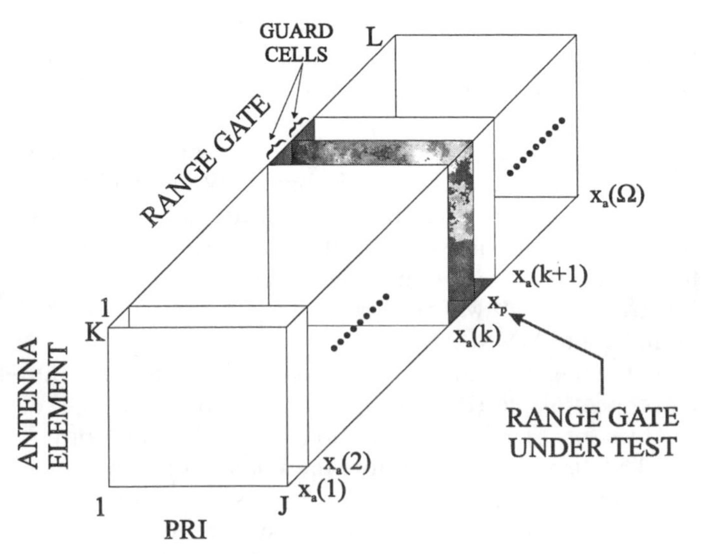

Fig. 9. The three-dimensional STAP data cube.

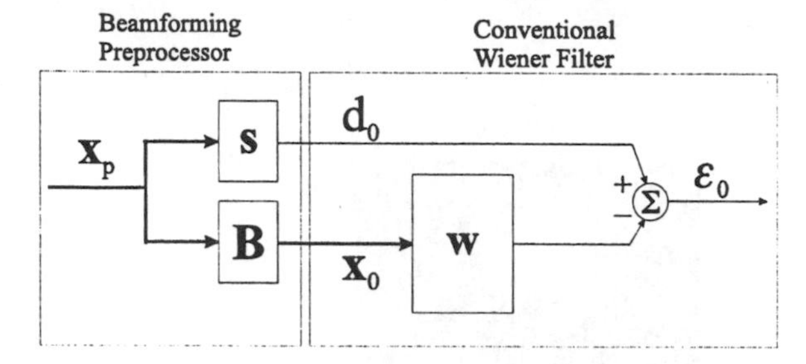

Fig. 10. The filtered-data form of the STAP processor.

the data outside the radar look direction, $\mathbf{x}_0 = \mathbf{B}\mathbf{x}_p$. This weight vector is computed by the Wiener-Hopf equation,

$$\mathbf{w} = \mathbf{R}_{x_0}^{-1}\mathbf{r}_{x_0 d_0}, \tag{58}$$

where $\mathbf{R}_{x_0} = \mathbf{BRB}^H$ and $\mathbf{r}_{x_0 d_0} = \mathbf{BRs}$. The weight vector in the center of (57) is termed the direct form, while the weight vector on the far right is called the filtered-data form.

The STAP filter output is given by

$$\varepsilon_0 = \mathbf{w}_a^H\mathbf{x}_p = \frac{\mathbf{s}^H\mathbf{R}^{-1}\mathbf{x}_p}{\mathbf{s}^H\mathbf{R}^{-1}\mathbf{s}} = (\mathbf{s}^H - \mathbf{w}^H\mathbf{B})\mathbf{x}_p. \tag{59}$$

The direct form output noise power (under the null hypothesis) is computed as follows:

$$P = \mathbf{w}_a^H\mathbf{R}\mathbf{w}_a = \frac{1}{\mathbf{s}^H\mathbf{R}^{-1}\mathbf{s}}. \tag{60}$$

This may also be expressed in the filtered-data form,

$$P = \sigma_{d_0}^2 - \mathbf{w}^H\mathbf{R}_{x_0}\mathbf{w} = \sigma_{d_0}^2 - \mathbf{r}_{x_0 d_0}^H\mathbf{R}_{x_0}^{-1}\mathbf{r}_{x_0 d_0}, \tag{61}$$

where the STAP beamformer output noise power is $\sigma_{d_0}^2 = \mathbf{s}^H\mathbf{Rs}$. It is now seen via Figure 10 that these STAP variables are directly related to the Wiener filter studied extensively earlier in this text. It is also of interest to note that the beamforming preprocessor in Figure 10 represents the first stage of the JKLT which accounts for directional preference, as presented earlier. This structure therefore fits naturally within the theoretical framework of Wiener filtering and the JKLT algorithm.

The detection problem can be evaluated using these Wiener filter variables. One popular adaptive constant false alarm rate (CFAR) detection test, called the adaptive matched filter (AMF) [28, 29], is given by

$$\Lambda = \frac{|\varepsilon_0|^2}{P} = \frac{|\mathbf{s}^H\mathbf{R}^{-1}\mathbf{x}_p|^2}{\mathbf{s}^H\mathbf{R}^{-1}\mathbf{s}} \underset{H_0}{\overset{H_1}{\gtrless}} \eta, \tag{62}$$

where H_1 and H_0 are the target present and target absent hypotheses, respectively. The performance of the AMF CFAR test is a function of the SINR and the false-alarm probability. The SINR is therefore the most frequently evaluated performance measure in assessing STAP performance. The optimization of the weight vector to maximize the SINR results in a locus of solutions with the form $\mathbf{w}_a = \alpha\mathbf{R}^{-1}\mathbf{s}$. The Wiener filters in (57) and (58) represent the operating point which provides a distortionless response of the target test cell out of this locus of solutions. The optimal SINR is calculated as follows:

$$\xi_o = \frac{|\mathbf{w}_a^H\mathbf{s}|^2}{\mathbf{w}_a^H\mathbf{R}\mathbf{w}_a} = \mathbf{s}^H\mathbf{R}^{-1}\mathbf{s} = \frac{1}{\sigma_{d_0}^2 - \mathbf{r}_{x_0 d_0}^H\mathbf{R}_{x_0}^{-1}\mathbf{r}_{x_0 d_0}} = \frac{1}{P}. \tag{63}$$

The traditional method of applying rank reduction for STAP processing is to perform a KLT of the covariance matrix $\mathbf{R}$ and retain the principal components [30–33]. However, an analysis of the SINR in either the direct or the filtered-data form yields the relations

$$\xi_o = \mathbf{s}^H\mathbf{R}^{-1}\mathbf{s} = \sum_{i=1}^{N}\frac{|\mathbf{E}_i^H\mathbf{s}|^2}{\lambda_i}, \tag{64}$$

and

$$\xi_o = \frac{1}{\sigma_{d_0}^2 - \mathbf{r}_{x_0 d_0}^H\mathbf{R}_{x_0}^{-1}\mathbf{r}_{x_0 d_0}} = \frac{1}{\sigma_{d_0}^2 - \sum_{i=1}^{N-1}\frac{|\mathbf{F}_i^H\mathbf{r}_{x_0 d_0}|^2}{d_i}}, \tag{65}$$

respectively. The KLT of the two pertinent covariance matrices in (64) and (65) are $\mathbf{R} = \mathbf{E\Lambda E}^H$, with diagonal matrix $\mathbf{\Lambda}$ composed of elements λ_i, and $\mathbf{R}_{x_0} = \mathbf{FDF}^H$, with diagonal matrix $\mathbf{D}$ composed of elements d_i. It is evident that the cross-spectral metric directly maximizes the reduced-rank SINR expression in (64) relative to the basis vectors $\mathbf{E}_i$.[5] The reduced-rank maximization of the SINR in (65) requires the minimization of a mean-square error expression in the denominator. The optimal KLT-based reduced-rank solution here is also given by a cross-spectral metric with basis vectors $\mathbf{F}_i$.

Finally, the JKLT algorithm can be applied to this problem to introduce a better type of intelligent rank reduction. As previously noted, that the first stage of the JKLT is exactly the transformation from the direct form to the filtered-data form of the Wiener filter using matrix filters $\mathbf{s}$ and $\mathbf{B}$. The implementation of the JKLT algorithm is then identical to that presented earlier for the standard Wiener filter, as evident through the comparison of Figures 2 and 10.

7.2. Estimation of the Statistics and Sample Support

The unknown $N \times N$ noise covariance matrix is estimated in practice by the analysis of $\Omega = L - 2g - 1$ auxiliary range gates, where g is the number of *guard* range gates on each side of the test range gate as shown in Figure 9. These guard cells are excluded to ensure that the target of interest does not extend into the training region. The

[5] This optimization for the direct form may, however, lead to a poor choice of subspaces for estimating statistics with small sample support. This is due to the selection of a nondominant subspace associated with the noise. A modified direct form cross-spectral metric is presented in [34] which properly selects a stable subspace.

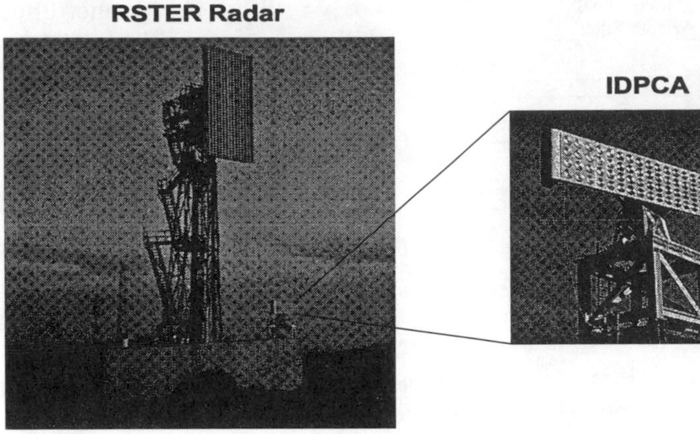

RSTER Radar

IDPCA

Fig. 11. The Mountain Top radar system.

maximum likelihood estimate of the covariance matrix is calculated from the Ω auxiliary range gates by evoking a Gaussian assumption and the aforementioned range ergodicity condition:

$$\hat{\mathbf{R}} = \frac{1}{\Omega} \sum_{k=1}^{\Omega} \mathbf{x}_a(k)\mathbf{x}_a^H(k), \tag{66}$$

where $\mathbf{x}_a(k)$ is the $N \times 1$ space-time auxiliary data snapshot from range cell $k \in \{1, 2, \ldots, \Omega\}$. The STAP filter can then be calculated with (57) and (58), where the clairvoyant covariance $\mathbf{R}$ is replaced with the maximum likelihood sample covariance $\hat{\mathbf{R}}$. Similarly, the AMF CFAR test can then be calculated using either the filtered-data variables or the direct substitution of the sample covariance matrix in (62).

It is assumed that $\Omega \gg 2N$ to satisfy the sample support requirements for the estimation of the covariance matrix in (66). This further assumes that the range gate snapshots $\{\mathbf{x}_a(k)\}_{k=1}^{\Omega}$ contain stationary and IID samples of the clutter and noise present in the test snapshot $\mathbf{x}_p$. The earth is not, in general, homogeneous over the large range extent required by STAP radars. This may preclude the existence of $2N$ IID stationary samples in the auxiliary data set. Therefore rank reduction and an intelligent signal processing measure, such as the JKLT, are needed to provide both a lower sample support and a reasonable amount of training data from the STAP data cube. It is also noted that analogous estimation issues are present even in the stationary Gaussian case with finite training data. Intelligent signal processing techniques are required under these conditions to determine the smallest stable subspace which maximizes the SINR. The reduced sample support requirements then map directly to rapid convergence and the ability to track nonstationarities.

An SINR loss is defined by

$$\xi = \frac{|\hat{\mathbf{w}}_a^H \mathbf{s}|^2}{(\hat{\mathbf{w}}_a^H \mathbf{R} \hat{\mathbf{w}}_a)(\mathbf{s}^H \mathbf{R}^{-1} \mathbf{s})}, \tag{67}$$

where $\hat{\mathbf{w}}_a$ represents the weight vector formed by substituting (66) into (57) and (58). The SINR loss is the normalized SINR first presented in [4], and the interest here is in evaluating the loss due to reduced-rank versions of the estimated Wiener filter $\hat{\mathbf{w}}_a$. The SINR loss is called the *region of convergence for adaptivity* (ROC) when it is evaluated as a function of both the effective rank (as determined by the signal representation) and the amount of training data supplied for sample support [35]. This framework provides an informative way to analyze the potential benefit of intelligent signal processing methods for radar detection.

7.3. Simulation

The DARPA Mountain Top radar is simulated to demonstrate the ROC performance of the intelligent signal processing algorithms as a function of rank and sample support. The Mountain Top radar employs the Radar Surveillance Technology Experimental Radar (RSTER) and the Inverse Displaced-Phase Center Array (IDPCA), both colocated at the same site [36], as shown in Figure 11. The radar consists of $K = 14$ half-wavelength spaced elements and $J = 16$ pulses in the PRI. The elevation angle is fixed (pre-beamformed) and the azimuth angle represents the only free parameter. The dimension of the adaptive processor is $KJ = 224$.

The scenario of interest consists of one target and returns from ground clutter, as depicted in Figure 12. The ground

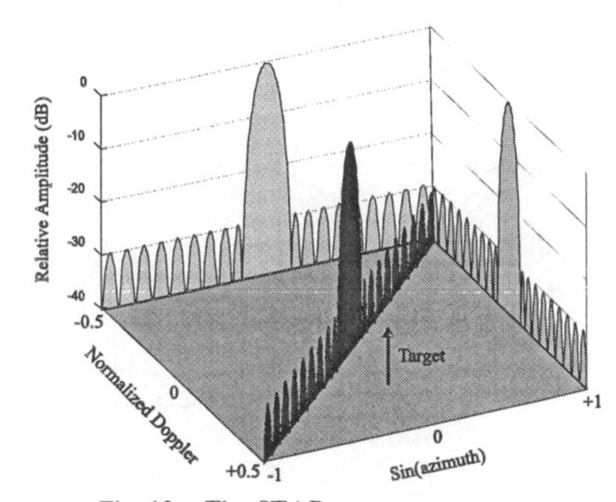

Fig. 12. The STAP power spectrum.

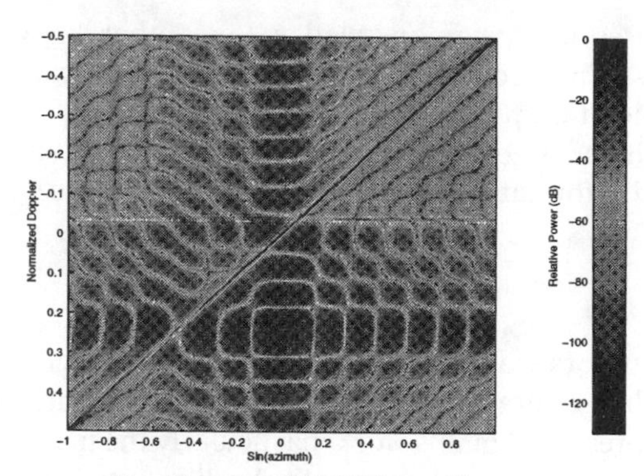

Fig. 13. The STAP Wiener filter.

clutter has a normalized Doppler frequency which is a linear function of the spatial frequency. The locus in angle-Doppler space where the clutter is present is termed the clutter ridge. The slope of the clutter ridge, whose value is denoted β [37], dictates the number of times that the clutter Doppler spectrum aliases into the unambiguous Doppler space. The parameters selected for this simulation correspond to the clutter exactly filling the Doppler space once, or $\beta = 1$. The portion of Doppler space that the clutter ridge spans depends upon the platform velocity, the radar pulse repetition frequency, and the radar operating wavelength. The two-dimensional extent of the clutter means that the mainbeam target competes with mainbeam clutter in the angle domain and the sidelobe clutter in the Doppler domain.

The optimal two-dimensional spectrum of the STAP weight vector, calculated using (57) for this example, is depicted in Figure 13. It is readily seen that the optimal filter applies a unity gain at the angle-Doppler look direction while simultaneously placing a null on the clutter ridge.

A Monte Carlo analysis is now considered to analyze the region of convergence for adaptivity which is obtainable using the principal-components, the cross-spectral metric, and the JKLT algorithms. The effective rank r_c of the clutter can be estimated by Brennan's rule [37],

$$r_c = K + (J - 1)\beta. \tag{68}$$

The Mountain Top radar system, with $\beta = 1$, results in $r_c = 29$. It is therefore expected that the principal-components algorithm would require a rank on the order of 29 and that approximately $2r_c = 58$ samples would be required for convergence. The Monte Carlo simulations consist of 50 independent realizations.

The principal-components ROC is shown in Figure 14. The area of this plot where the sample support is less than the rank (the lower left triangular region) represents the region where the reduced-rank covariance matrix is numerically unstable. The area where the SINR loss is negligible

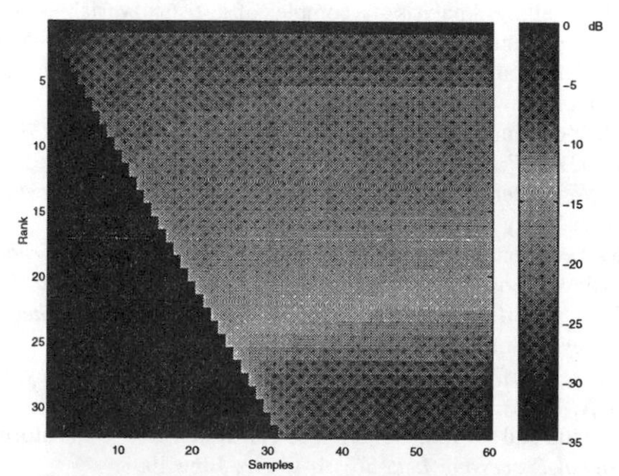

Fig. 14. The ROC for the principal components.

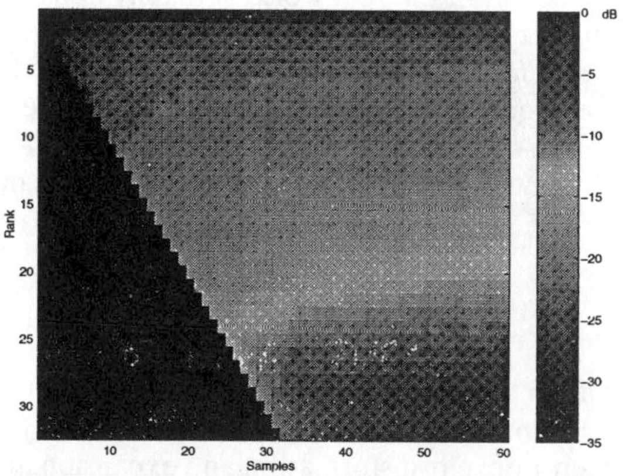

Fig. 15. The ROC for the cross-spectral metric.

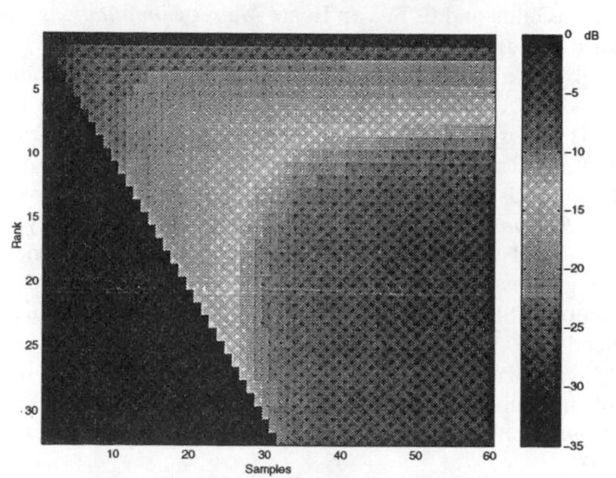

Fig. 16. The ROC for the JKL2.

See insert for color figures.

(the lower right region defined by high sample support and high rank) represents the region of convergence for adaptivity with respect to the principal-components method. The slope or roll-off of the performance surface depicts both the robustness and the sensitivity of the processor as a function of rank and sample support. It is evident in Figure 14 that the principal-components ROC does indeed require a rank of approximately $r_c = 29$, although the necessary sample support is only a little greater than r_c, or about half of the upper bound calculated in [4]. The reduction of rank is seen to result in a quick loss of performance. This is an indication of a strong sensitivity to variations in rank and sample support.

The ROC for the cross-spectral metric is presented in Figure 15. The cross-spectral metric ROC covers a slightly greater area where the sample support is high. This demonstrates that a greater rank reduction may be possible as long as sufficient sample support is available for adaptively estimating the statistics. The robustness and sensitivity to reductions in rank and sample support an nearly equivalent to those of the principal components.

Finally, the ROC for the JKLT is depicted in Figure 16. Here it is seen that the area covered by the ROC is significantly greater than that possible with the principal-components or the cross-spectral metric. The JKLT ROC includes the region of both lower sample support and significantly lower rank. The importance of this result for radar detection in the real world is readily apparent when one considers the greatly reduced range homogeneity requirements for the low-rank JKLT. The dominating area of convergence for adaptivity in Figure 16 represents much better robustness and sensitivity properties in comparison with the principal-components and cross-spectral metric algorithms.

8. Conclusions

This chapter addresses intelligent signal representation and basis selection for detection and estimation problems. The Karhunen-Lóeve transformation and expansion are examined with emphasis on the relevant conditions of optimality. It is demonstrated that the principal components do not provide the best enumeration of the eigenvectors corresponding to the pertinent covariance matrix for Wiener filtering. The use of intelligent subspace selection with the KLT results in the cross-spectral metric. The cross-spectral metric corrects the signal representation optimization relative to the eigenvector basis while simultaneously raising information-theoretic questions about the use of this basis itself. It is then demonstrated that the principal components' lack of optimality results in the KLT not being optimal for problems concerned with signal representation for detection and estimation. Finally, a new intelligent signal processing approach to signal representation, termed the joint Karhunen-Lóeve transform, is derived that uses less prior knowledge than other known techniques. The reduced-rank JKLT is demonstrated to be capable of outperforming the eigenbased algorithms even though the algorithm complexity and the amount of required information are both significantly reduced.

Acknowledgments

The authors would like thank Dr. Dan Dudgeon of MIT Lincoln Laboratory for his significant influence on the ideas presented in this work, particularly the matched filter perspective. The authors would also like to thank Dr. Peter Zulch and Lt. Col. Jan North of the Department of the Air Force, Dr. John Tague of the Office of Naval Research, Mr. Steven Huang of SAIC, and Prof. Michael Zoltowski of Purdue University for many interesting discussions and comments relating to this work.

References

[1] S. P. Applebaum. Adaptive arrays. Technical Report SPL TR 66-1, Syracuse University Research Corporation, August 1966.

[2] B. Widrow, P. E. Mantey, L. J. Griffiths, and B. B. Goode. Adaptive antenna systems. *Proc. IEEE* 55:2143–2159, December 1967.

[3] L. E. Brennan and I. S. Reed. Theory of adaptive radar. *IEEE Trans. Aerosp. Electron. Syst.* 9:237–251, March 1973.

[4] I. S. Reed, J. D. Mallett, and L. E. Brennan. Rapid convergence rate in adaptive arrays. *IEEE Trans. Aerosp. Electron. Syst.* AES-10(6):853–863, November 1974.

[5] H. Hotelling. Analysis of a complex of statistical variables into principal components. *J. Educ. Psychol.* 24:417–441 and 498–520, 1933.

[6] C. Eckart and G. Young. The approximation of one matrix by another of lower rank. *Psychometrica* 1:211–218, 1936.

[7] H. Hotelling. Relations between two sets of variates. *Biometrika* 28:321–377, 1936.

[8] T. W. Anderson. *An Introduction to Multivariate Statistical Analysis.* John Wiley & Sons, New York, 1958.

[9] M. L. Eaton. *Multivariate Statistics: A Vector Space Approach.* John Wiley & Sons, New York, 1983.

[10] R. J. Muirhead. *Aspects of Multivariate Statistical Theory.* John Wiley & Sons, New York, 1982.

[11] L. L. Scharf. *Statistical Signal Processing.* Addison-Wesley, Reading, MA, 1991.

[12] Y. Hua and W. Liu. Generalized Karhunen-Loeve transform. *IEEE Signal Processing Lett.* 5(6):141–142, June 1998.

[13] Y. Yamashita and H. Ogawa. Relative Karhunen-Loeve transform. *IEEE Trans. Signal Processing* 44(2):371–378, February 1996.

[14] G. H. Golub and C. F. Van Loan. *Matrix Computations.* John Hopkins University Press, Baltimore, MD, 1990.

[15] D. S. Watkins. *Fundamentals of Matrix Computations.* John Wiley & Sons, New York, 1991.

[16] D. H. Johnson and D. E. Dudgeon. *Array Signal Processing.* Prentice Hall, Englewood Cliffs, NJ, 1993.

[17] J. S. Goldstein and I. S. Reed. Reduced rank adaptive filtering. *IEEE Trans. Signal Processing* 45(2):492–496, February 1997.

[18] J. S. Goldstein and I. S. Reed. Multidimensional Wiener filtering using a nested chain of orthogonal scalar Wiener filters. Technical Report CSI-96-12-04, University of Southern California, December 1996.

[19] J. S. Goldstein, I. S. Reed, and L. L. Scharf. A new method of Wiener filtering. In *Proc. AFOSR/DSTO Workshop on Defence Applications of Signal Processing,* Victor Harbor, Australia, June 1997.

[20] J. S. Goldstein. *Optimal Reduced-Rank Statistical Signal Processing, Detection and Estimation Theory.* PhD thesis, University of Southern California, November 1997.

[21] J. S. Goldstein, I. S. Reed, and L. L. Scharf. A multistage representation of the Wiener filter based on orthogonal projections. *IEEE Trans. Information Theory* 44(7):2943–2959, November 1998.

[22] J. S. Goldstein and I. S. Reed. A new method of Wiener filtering and its application to interference mitigation for communications. In *Proc. IEEE MILCOM,* Monterey, CA, November 1997.

[23] J. S. Goldstein, I. S. Reed, L. L. Scharf, and J. A. Tague. A low-complexity implementation of adaptive Wiener filters. In *Proc. 31th Asilomar Conf. Signals, Syst. Comput.,* volume 1, pages 770–774, Pacific Grove, CA, November 1997.

[24] J. S. Goldstein, I. S. Reed, and P. A. Zulch. Multistage partially adaptive STAP detection algorithm. *IEEE Trans. Aerosp. Electron. Syst.* 35(2):645–661, April 1999.

[25] J. S. Goldstein, J. R. Guerci, and I. S. Reed. An optimal generalized theory of signal representation. In *Proc. IEEE ICASSP,* Pheonix, AZ, March 1999.

[26] T. Kailath. The innovations approach to detection and estimation theory. *Proc. IEEE* 58:680–695, May 1970.

[27] J. S. Goldstein and I. S. Reed. Theory of partially adaptive radar. *IEEE Trans. Aerosp. Electron. Syst.* 33(4):1309–1325, October 1997.

[28] W. S. Chen and I. S. Reed. A new CFAR detection test for radar. *Digital Signal Proc.* 4:198–214, October 1991.

[29] F. C. Robey, D. R. Fuhrmann, E. J. Kelly, and R. Nitzberg. A CFAR adaptive matched filter detector. *IEEE Trans. Aerosp. Electron. Syst.* 28(1):208–216, January 1992.

[30] D. W. Tufts, R. Kumaresan, and I. Kirsteins. Data adaptive signal estimation by singular value decomposition of a data matrix. *Proc. IEEE* 70(6):684–685, June 1982.

[31] A. A. Shah and D. W. Tufts. Determination of the dimension of a signal subspace from short data records. *IEEE Trans. Signal Processing* 42(9):2531–2535, September 1994.

[32] I. P. Kirsteins and D. W. Tufts. Adaptive detection using a low rank approximation to a data matrix. *IEEE Trans. Aerosp. Electron. Syst.* 30(1):55–67, January 1994.

[33] A. M. Haimovich and Y. Bar Ness. An eigenanalysis interference canceler. *IEEE Trans. Signal Processing* 39(1):76–84, January 1991.

[34] J. R. Guerci, J. S. Goldstein, and I. S. Reed. Optimal and adaptive reduced-rank STAP. *IEEE Trans. Aerosp. Electron. Syst.* April 2000.

[35] P. A. Zulch, J. R. Guerci, J. S. Goldstein, and I. S. Reed. Comparison of reduced-rank signal processing techniques. In *Proc. 31th Asilomar Conf. Signals, Syst. Comput.,* Pacific Grove, CA, November 1998.

[36] G. W. Titi. An overview of the ARPA/Navy Mountain Top Program. In *Proc. IEEE Adapt. Ant. Systems Symp.,* Long Island, NY, November 1994.

[37] L. E. Brennan and F. M. Staudaher. Subclutter visibility demonstration. Technical Report RL-TR-92-21, USAF Rome Laboratory, March 1992.

Chapter 13

Signal Detection in a Nonstationary Environment Reformulated as an Adaptive Pattern Classification Problem

Simon Haykin and David J. Thomson

Abstract

The primary purpose of this paper is the improved detection of a nonstationary target signal embedded in a nonstationary background. Accordingly, the first part of the paper is devoted to a detailed exposition of how to deal with the issue of nonstationarity. The material presented here starts with Loève's probabilistic theory of nonstationary processes. From this principled discussion, three important tools emerge: the dynamic spectrum, the Wigner-Ville distribution as an instantaneous estimate of the dynamic spectrum, and the Loève spectrum. Procedures for the estimation of these spectra are described, and their applications are demonstrated using real-life radar data.

Time, an essential dimension of learning, appears explicitly in the dynamic spectrum and Wigner-Ville distribution and implicitly in the Loève spectrum. In each case, the one-dimensional time series is transformed into a two-dimensional image where the presence of nonstationarity is displayed in a more visible manner than it is in the original time series. This transformation sets the stage for reformulating the signal detection problem as an adaptive pattern classification problem, whereby we are able to exploit the learning property of neural networks. Thus in the second part of the paper we describe a novel learning strategy for distinguishing between the different classes of received signal, such as (1) there is no target signal present in the received signal, (2) the target signal is weak, and (3) the target signal is strong.

In the third part of the paper we present a case study based on real-life radar data. The case study demonstrates that the adaptive approach described in the paper is indeed superior to the classical approach for signal detection in a nonstationary background.

1. INTRODUCTION

The detection of a target signal in a background of noise is basic to signal processing. The term "noise" is used here in a generic sense; its exact description depends on the application of interest. The need for signal detection arises in many diverse fields, as summarized in Table 1. The detection problems summarized in this table may be grouped under two headings, depending on how the raw data originate:

1. *Time series*: radar, sonar, digital communications, and scientific data.
2. *Images*: magnetic resonance images, positron emission tomographs, mamograms, X-ray photographs, nondestructive testing of metals, detection of explosives in carry-on bags, and detection of land mines.

In this paper, we focus attention on the first class of signal detection problems. Nevertheless, many of the ideas described herein also apply to the second class.

The classical approach to the design of a signal detection system is to optimize a time-invariant (i.e., fixed) receiver structure for a known class of signals with unknown parameters that are corrupted by noise of known statistics [1-3]. The parameter used in the design is the likelihood ratio of the received signal. When, however, the noise is nonstationary, but of known form, the receiver must take on a time-varying structure, making its design more difficult [4]. The design becomes even more difficult when the statistics of the noise are unknown. Our interest in this paper lies in the class of signal detection problems described as that of *detecting a nonstationary target signal in a nonstationary environment of unknown statistics*. This is the most difficult class of signal detection problems. For example, in radar, sonar, and mobile communications systems nonstationarity of the received signal arises due to variations in environmental conditions. Although, it may be feasible to view the received signal in such systems as quasi-stationary in the very short term, and possibly stationary in long-term averages, at the intermediate time-scales useful for engineering applications the received signal can be significantly nonstationary. In radio

Table 1: Areas of Application where Signal Detection Arises

Area	Target	Main source of interference
1. Radar	Echo from Aircraft, ship, etc.	Clutter due to radar backscatter from unwanted objects, noise
2. Electronic-counter-counter measures (ECCM)	False radar transmission	Radar clutter, noise
3. Impulse (ground-penetrating) radar	Buried objects	Clutter due to reflections from the ground, noise
4. Sonar	Echo from submarine	Reverberation due to reflections from the ocean body, noise
5. Digital Communications	Symbol 1 or Symbol 0	Intersymbol interference or multipath, noise
Multiuser detection in code-division multiple access systems	Sequence of 1s and 0s	Interference from other users, multipath, noise
6. Biomedical signal processing	Epileptic seizure in the EEG signal of an infant	Normal EEG signal
7. Magnetic resonance images (MRI)	Abnormal distribution of tissue	Neighboring tissues
8. Positron emission tomography (PET)	Regional measurement of metabolism	Normal metabolism in the neighboring tissues
9. Mamograms	Low-contrast anomalies in the tissue	Spatial and contrast resolutions
10. Chest X-ray photographs	Abnormal tissue	Very high dynamic range of spatial resolutions
11. Nondestructive testing of metals	Defects	Clutter due to reflection from main body of metallic object
12. Explosives detection: X-ray image	Explosives, firearms, knives	Presence of harmless objects in carry-on luggage
13. Detection of land mines	Echo from a mine buried in the ground	Reflections from the ground and other objects of no interest
14. Detection of a flaw in textured clothing material	Flaw in the texture	Normal texture

systems, for example, it is usually a reasonable assumption that the continuum (as opposed to impulsive) noise may be approximately stationary for milliseconds to seconds, and also if averaged over months. The solar component of such noise, however, varies on both the 5-minute scale of the solar p-modes [5], and also with the 26-day solar rotation and the 11-year solar cycle [6]. In the radar data used in this paper we find unexpected evidence for cyclostationary, or periodically correlated, behavior in the clutter.

To deal with the issue of unknown statistics of the additive noise, we may use a neural network to compute the likelihood ratio of the received signal by training it on different realizations of the received signal. Such an approach is described in [7-9]. The neural networks described in those references belong to the class of *focused time-lagged feedforward networks*, focused in the sense that a short-term memory structure (used for dealing with time) is confined entirely to the input layer of the neural network. In Sandberg and Xu [10], it is shown that these networks are universal approximators of myopic nonlinear dynamic systems. Unfortunately, however, their use is limited to stationary processes. To deal with a nonstationary process using a neural network approach, the implicit effect of time has to be distributed inside the synaptic structure of the neural network as described in Wan [11], or else a recurrent neural network has to do the learning in a dynamic fashion [12].

In this paper, we take a different approach for dealing with signal detection in a nonstationary environment. We proceed by first computing *two-dimensional image* representations of the received signal that account for the nonstationary nature of the signal from one representation to the next. In so doing, we account for time in an explicit sense, thereby transforming the detection problem into an adaptive pattern classification problem that lends itself to *learning* [13]. Such an approach to adaptive signal detection is described in [14–17]. This approach is philosophically distinct from classical detection procedures. *The parsimonious, but frequently overly-simplified models of signal and noise are replaced by a highly redundant and over-complete representation of the received signal.*

The main body of the paper is organized as follows. Section 2 discusses the nonstationary behavior of a signal and the related issue of time-frequency analysis in general terms. Section 3 presents a theoretical background for dealing with nonstationary signals. This discussion leads naturally to an overview of procedures for estimating the spectrum of a nonstationary signal in Section 4. In Section 5 we present some results on the time-frequency analysis of real-life clutter data as an illustrative example of the procedures described. Section 6 describes a modular learning strategy for the detection of a target signal embedded in a nonstationary background. Section 7 presents highlights of a case study that builds on this detection philosophy. Section 8 discusses the issue of cost functions for supervised training of the pattern classifiers in the adaptive receiver. The paper concludes with some final remarks in Section 9.

2. An Overview of Nonstationary Behavior and Time-Frequency Analysis

The statistical analysis of nonstationary signals has had a rather mixed history. Although the general second-order theory was published during 1946 by Loève [18,19], it has not been applied nearly as extensively as the theory of stationary processes published only slightly previously by Wiener and Kolmogorov. There were, at least, four distinct reasons for this neglect:

(i) Loéve's theory was *probabilistic*, not statistical, and there does not appear to have been successful attempts to find a statistical version of the theory until some time later.

(ii) At that time of publications, the mathematical training of most engineers and physicists in signals and random processes was minimal and, recalling that even Wiener's delightful book was referred to as "The Yellow Peril", it is easy to imagine the reception that a general nonstationary theory would have received.

(iii) Even if the theory had been commonly understood at the time and good statistical estimation procedures had been available, the computational burden would probably have been overwhelming. This was the era when Blackman-Tukey estimates of the stationary spectrum were developed, not because they were great estimates but, primarily, because they were computationally more efficient than other forms.

(iv) Finally, it cannot be denied that the general theory was significantly harder to grasp than that for stationary processes.

Nonetheless, it was realized that many, perhaps most, of the signals being worked with were nonstationary and, starting with the available tools, (i.e., the ability to estimate the spectrum of a stationary signal) the spectrogram was developed. The idea was that, if the process is not "too" nonstationary, then for a relatively short time block a "quasi-stationary" approximation can be used, so that for the length of the block the spectrum can be approximated by its average. It was also recognized that a major drawback of the spectrogram is that the block lengths and offset between blocks are arbitrary. Thus, although speech, underwater sound, radar, and similar communities have much empirical experience to guide such choices, little can be done with a new, possibly unique, data series except "cut and try" methods. Consequently, it is common to regard the spectrogram as a heuristic or ad-hoc method.

To account for the nonstationary behavior of a signal, we have to include *time* (implicitly or explicitly) in a description of the received signal. Given the desirability of working in the frequency domain for well-established reasons, we may include the effect of time by adopting a time-frequency description of the signal. During the last twenty years many papers have been published on various estimates of time-frequency distributions; see, for example, Cohen's book [20] and the references therein. In most of this work, the signal is assumed to be *deterministic*. In addition, many of the proposed estimators are constrained to match time and frequency *marginal* density conditions. If $D(t, f)$ is a time-frequency distribution of a signal $x(t)$, it is required that the time marginal satisfy the condition

$$\int_{-\infty}^{\infty} D(t, f) df = |x(t)|^2$$

and, similarly, if $y(f)$ is the Fourier transform of $x(t)$, the frequency marginal density must satisfy the condition

$$\int_{-\infty}^{\infty} D(t, f) dt = |y(f)|^2$$

where t denotes continuous time and f denotes frequency. Given the large differences observed between waveforms collected on sensors spaced short distances apart, see, for example, Vernon [21], the time marginal requirement is a rather strange assumption. Worse, the frequency marginal distribution is, except for a factor of $1/N$, just the periodogram of the signal. Since it has been known since at least the 1930's that the periodogram is badly biased and inconsistent[1], and we have personally experienced engineering data, Thomson [22], where the periodogram was wrong by more than a factor of 10^{10} over most of the frequency range, imposition of such a constraint must be viewed skeptically as well. Thus we do not consider matching marginal distributions, as commonly defined, to be important.

Similarly, several estimates have been proposed that attempt to reduce the cross-terms[2] in the Wigner-Ville distribution by using the analytic signal instead of the original data. However, as

---

1. A biased estimate is one where the expected value of the estimate differs from the value of the quantity being estimated. An inconsistent estimate is one where the variance of the estimate does not decrease with sample size. The periodogram is an unstable, wrong, answer.

2. *Cross-terms* arise when the Wigner-Ville distribution is applied to the sum of two signals. The Wigner-Ville distribution of such a sum is not equal to the sum of the Wigner-Ville distributions of the two signals; the difference is accounted for by the cross-terms.

the analytic signal is commonly derived by Fourier transforming the data, discarding the negative frequency components, and taking the inverse Fourier transform the frequency-domain bias of the analytic signal is dominated by the periodogram bias. The opinion has been expressed that these concerns apply only in a near-pathological data set and that the sidelobe performance of the Slepian sequences is rarely needed. We consider this opinion ill-advised as we rarely know in advance what sidelobe performance is needed. (Slepian sequences are defined in Section 3.2.) As an example, the dynamic range of the spectrum for the radar data used in this paper exceeds 10^4, so an estimate constrained to match periodogram marginals could easily be in error by an order of magnitude over most of the frequency domain.

This being said, the Wigner-Ville distribution used in the following is computed by the standard, basic form. We have several reasons for leaving the Wigner-Ville distribution unaltered:

(i) As we describe it below, the expected value of the Wigner-Ville distribution is just a coordinate rotation of the Loève spectrum. It is not, however, a particularly good *statistical* estimate.

(ii) The basic Wigner-Ville distribution is a sufficient statistic in that it can be inverted to recover the original data to within a phase constant [23]. Thus, although it is not an attractive estimate from a statistical viewpoint, its completeness properties allow it to be effective in our application (i.e., signal detection).

(iii) The data used here are complex-valued, so there is no need to estimate the analytic signal.

(iv) The cross-terms are visually distinctive and so may be a significant help in recognizing that more than one component is present in the received signal.

In the final analysis, whether we adopt the stochastic or deterministic approach to time-frequency analysis for representing the nonstationary behavior of a signal depends on details of the problem of interest. It is easy to imagine problems where one or the other of these two approaches would be preferable, but there are other problems, such as the radar data used in our examples, where a good case can be made for both viewpoints.

3. Theoretical Background

Suppose we are given data consisting of a single finite realization of N contiguous samples of a discrete-time process $x(t)$ for $t = 0, \cdots, N\text{-}1$; henceforth, t denotes discrete time. We assume that the process is harmonizable, Loève [19], so that it has the Cramér, or spectral, representation:

$$x(t) = \int_{-1/2}^{1/2} e^{j2\pi\nu t} dX(\nu) \tag{1}$$

where $dX(\nu)$ is the *increment process*. In this paper, we also assume that the process has zero mean, that is, $\mathbf{E}\{dX(\nu)\} = 0$, and, correspondingly, $\mathbf{E}\{x(t)\} = 0$. (Note that this is *not* the same as assuming that an average has been subtracted from the data.) As parameters of interest, consider then the covariance function

$$\Gamma_L(t_1, t_2) = \mathbf{E}\{x(t_1)x^*(t_2)\} \tag{2}$$

$$= \int_{-\infty}^{\infty}\int_{-\infty}^{\infty} e^{j2\pi(t_1 f_1 - t_2 f_2)} \gamma_L(f_1, f_2) df_1\, df_2$$

and the generalized spectral density

$$\gamma_L(f_1, f_2) df_1\, df_2 = \mathbf{E}\{dX(f_1) dX^*(f_2)\} \tag{3}$$

where * indicates complex conjugate. Equation (3) describes the essential feature of nonstationary processes, namely, that there is correlation between different frequencies.

If the process is stationary, the covariance $\Gamma_L(t_1, t_2)$, depends by definition only on the time difference t_1 - t_2 and the Loève spectrum $\gamma_L(f_1, f_2)$ becomes $\delta(f_1 - f_2)S(f_1)$ where $S(f)$ is the ordinary power spectrum. Similarly, for a white nonstationary process, the covariance function becomes $\delta(t_1 - t_2)\ P(t_1)$ where $P(t)$ is the expected power at time t. Thus, as both the spectrum and covariance functions include delta function discontinuities in simple cases, neither should be expected to be "smooth"; and continuity properties depend on direction in the (f_1,f_2) or (t_1,t_2) plane. These problems are more easily dealt with by rotating both the time and frequency coordinates of the generalized correlations (2) and spectral densities (3), respectively, by 45°. In the time domain, we define the new coordinates to be a "center" t_0 and a delay τ, as shown here:

$$t_1 + t_2 = 2t_0 \tag{4}$$

$$t_1 - t_2 = \tau$$

Equivalently, we may write

$$t_1 = t_0 + \tau/2$$

$$t_2 = t_0 - \tau/2$$

We denote the covariance function in the rotated coordinates by $\Gamma(\tau, \tau_0)$ and so write

$$\Gamma_L(t_1, t_2) = \Gamma\left(t_1 - t_2, \frac{t_1 + t_2}{2}\right) = \Gamma(\tau, t_0) \tag{5}$$

Similarly, we define new frequency coordinates f and g by writing

$$f_1 + f_2 = 2f \tag{6}$$

$$f_1 - f_2 = g$$

Equivalently, we may write

$$f_1 = f + g/2$$

$$f_2 = f - g/2$$

Denote the rotated spectrum by

$$\gamma(g, f) = \gamma_L\left(f + \frac{g}{2}, f - \frac{g}{2}\right) \tag{7}$$

Substituting these definitions in equation (2) shows that the term $t_1 f_1 - t_2 f_2$ in the exponent of the Fourier transform becomes $(t_0 g + \tau f)$, and so

$$\Gamma(t_0, \tau) = \int_{-\infty}^{\infty}\int_{-\infty}^{\infty} e^{j2\pi(\tau f + t_0 g)} \gamma(g, f) df\, dg \tag{8}$$

Because f is associated with the time difference τ, it corresponds to the ordinary frequency of stationary processes and we refer to it as the "ordinary" or "stationary" frequency. Similarly, because g is associated with the average time t_0, it describes the behavior of the spectrum over long time spans and we refer to g as the "nonstationary" frequency. Now consider the continuity of γ as a function of f and g. On the line $g = 0$, the generalized spectral density γ is just the ordinary spectrum with the usual continuity (or lack thereof) conditions normally applying to stationary spectra. As a function of g, however, we expect to find a δ function discontinuity at g=0 if, for no other reason, that almost all data contain some stationary additive noise. Consequently, smoothers in the (f, g) plane (or, equivalently, the (f_1, f_2) plane) should not be isotropic, but require much higher resolution along the nonstationary frequency coordinate than along the ordinary frequency axis f.

A slightly less arbitrary way of handling the g coordinate is to Fourier Transform $\gamma(g, f)$ with respect to the "nonstationary" frequency, g, and define

$$D(t_0, f) = \int_{-\infty}^{\infty} e^{j2\pi t_0 g} \gamma(f, g) dg$$

as the theoretical "dynamic spectrum" of the process. The motivation is to transform the very rapid variation expected around $g = 0$ into a slowly varying function of t_0 while leaving the usual dependence on f. Because δ functions in frequency transform into a constant in time, in a stationary process $D(t_0, f)$ does not depend on t_0 and becomes $S(f)$. Writing D as

$$D(t_0, f) = \int_{-\infty}^{\infty} e^{j2\pi\tau f} \mathbf{E}\left\{ x\left(t_0 + \frac{\tau}{2}\right) x^*\left(t_0 - \frac{\tau}{2}\right) \right\} d\tau \tag{9}$$

we see that *the rotated Loève spectrum is the expected value of the Wigner-Ville distribution*. This relation has been rediscovered several times, see for example [24]. Note carefully, however, that unlike the standard Wigner-Ville distribution, D is defined to be an expected value.

Stated in another way, the Wigner-Ville distribution is the *instantaneous estimate* of the dynamic spectrum $D(t_0, f)$, and therefore simpler to compute than the dynamic spectrum. Taking the complex conjugate of the last equation shows that D is real, and, as the Fourier transform of a covariance, must be non-negative definite; see [25]. In many ways, current discussions about positivity of the Wigner-Ville and similar distributions are reminiscent of those occurring in papers in the 1945-1970 era on whether the normalization $1/N$ or $1/(N-t)$ was "correct" for estimating lag-τ autocorrelations. The correct answer was that direct calculation of sample autocorrelations was a bad idea in any case and, given that the wrong estimate was being computed, the normalization was more-or-less irrelevant!

3.1 Multiple Window Estimates

Multiple window estimates of the spectrum, [26], are a class of estimates based on approximately solving the integral equation that expresses the projection of $dX(f)$ onto the Fourier transform of the data, $y(f)$. Taking the Fourier transform of the observed data, that is,

$$y(f) = \sum_{t=0}^{N-1} x(t) e^{-j2\pi ft}$$

and using the spectral representation (1) for $x(t)$, we have the fundamental equation of spectrum estimation:

$$y(f) = \int_{-1/2}^{1/2} K_N(f - \nu) dX(\nu) \tag{10}$$

where the *Dirichlet kernel* is given by

$$K_N(f) = \frac{\sin N\pi f}{\sin \pi f} e^{-j2\pi f \frac{N-1}{2}} \tag{11}$$

There are several points that must be remembered about this fundamental equation:

(i) Because we may take the inverse Fourier transform of $y(f)$ and recover $x(t)$ for $0 \le t \le N-1$, $y(f)$ is a *sufficient statistic* and completely equivalent to the original data.

(ii) The *finite* Fourier transform $y(f)$ is **not** equivalent to the spectral generator $dX(\nu)$. Remember that $dX(\nu)$ is assumed to generate the entire data sequence for *all* t, not just the portion observed.

(iii) Despite definitions given in many elementary texts, $(1/N)|y(f)|^2$ is *not*, repeat *not*, the spectrum, even in the limit of large N. It is the periodogram, biased and inconsistent.

(iv) While (10) is formally a convolution of dX with a Dirichlet kernel, it is more constructive to think of it as a *Fredholm integral equation of the first kind*. As such, it does not have a unique solution. It does, however, have useful approximate solutions. We mentioned above that "multiple window estimates" does not refer to a particular estimate, but rather to a class of estimates: the class is defined by the method used to form the necessarily approximate solution of the integral equation. Viewed in this way, spectrum estimation is in reality an *inverse problem.*

As multiple window methods have been described in a book, [27], many papers, and are becoming the "standard" in geophysics [28], elaborate description of their properties is unnecessary; the reader is referred to [29-32] for details and we give only the equations necessary to define notation here.

3.2 Spectrum Estimation as an Inverse Problem

Recall the fundamental equation (10) and attempt eigen-solution of the integral equation on the interval $(f - W, f + W)$ by assuming that the *observable* portion of dX has the expansion

$$d\hat{X}(f - \nu) = \sum_{k=0} x_k(f) V_k^*(\nu) d\nu \qquad (12)$$

on the *local* frequency domain $(f - W, f + W)$. Here $V_k(\nu)$ is a Slepian function (discrete prolate spheroidal wave function, see Appendix A of [30] for definitions). Using the integral equation and properties of the Slepian functions we obtain the raw expansion, or eigencoefficients

$$y_k(f) = \sum_{t=0}^{N-1} e^{-j2\pi fn} \nu_n^{(k)}(N, W) x(t) \qquad (13)$$

as the Fourier transform of the data, $x(t)$, windowed by the kth Slepian sequence, $\nu_n^{(k)}(N, W)$. We retain the $K = 2NW$ coefficients corresponding to functions with eigenvalues $\lambda_k \approx 1$ for subsequent inference. *These eigencoefficients represent the information in the signal projected onto the local frequency domain.* This process resembles conventional, windowed spectrum estimation in that a fast Fourier transform may be used for efficient computation, but differs in that standard estimates are best regarded as the first term of the multiple window expansion.

Because the Slepian sequences are time-limited, they cannot be strictly bandlimited, and the kth sequence has a fraction $1 - \lambda_k(N,W)$ outside the interval $(-W,W)$. Uncorrected, this out-of-band energy contributes bias which can be severe for the higher-order, or transition eigencoefficients, that is those of order $k \approx K$. Among the various ways of dealing with this exterior bias, the best method found to date is by *coherent sidelobe subtraction* as outlined in [29]. Here we choose the bandwidth W to be large enough so that the bias on the lower-order terms is negligible and $x_k(f) \approx y_k(f)$ for $k << K$, thereby estimating the higher-order eigencoefficients by $x_k(f) \approx y_k(f) - \hat{b}_k(f)$ for larger k. The bias estimate $\hat{b}_k(f)$ is formed from an exterior convolution of the Slepian sequence with an estimate of (12) and iterated. Denote the estimated eigencoefficients by $x_k(f)$ and collect them in the vector $\mathbf{X}(f)$:

$$\mathbf{X}(f) = [x_0(f), x_1((f),\cdots,)x_{K-1}(f)]^T \tag{14}$$

To see the dependence on the bandwidth W of the estimate, recall that there are $K \approx \lfloor 2NW \rfloor$ windows with eigenvalues near 1. If the spectrum is flat *within the local domain* the coefficients are uncorrelated because the windows are orthogonal, and each contributes two degrees of freedom so estimates of the form $\mathbf{X}^{\dagger}(f)\ \mathbf{X}(f)$ have $2K$ degrees of freedom, where † denotes Hermitian transposition. If W is too small, we have poor statistical stability, but if W is too large, the estimate has poor frequency resolution. Typically W is chosen between $1.5/N$ and $20/N$ with a time-bandwidth product of 4 or 5 being a common starting point. Thus $W = 4/N$ or $5/N$ with corresponding $K = 6$ or 8 gives estimates with 12 or 16 degrees-of-freedom.

We must emphasize, however, that these only apply to the simplest forms of estimates and both quadratic inverse estimates, see [29-32], and free parameter estimates of the type described therein give high-resolution estimates that are, within reason, largely independent of the choice of W. These estimates also give implicit extrapolations of the time series.

4. High-Resolution Multiple-Window Spectrograms

Beginning with the estimate of $d\hat{X}$ defined in (12), define the narrow-band process

$$X(t, f) = \int_{-W}^{W} e^{j2\pi t\xi} d\hat{X}(f \oplus \xi)$$

where $\oplus$ denotes addition with the constraint that $|\xi| < W$. On taking the inverse transform of the Slepian function, $X(t, f)$ becomes

$$X(t, f) = \sum_{k=0}^{K-1} \lambda_k v_t^{(k)} x_k(f)$$

where λ_k is the kth eigenvalue. Clearly the complex function $X(t,f)$ is not a time-frequency distribution but more akin to the output of a filter bank. Note, however, first, that if we write the approximate impulse response of the implied filters, they go from maximum phase at $t = 0$ through zero phase at $t = (N\text{-}1)/2$, to minimum phase at $t = N\text{-}1$. Second, $X(t, f)$, as defined here *extrapolates* the signal to t outside the interval $[0, N\text{-}1]$, and so resembles the Papoulis estimates [33]. The squared amplitude, $| X(t, f) |^2$, gives power as a function of time and frequency

$$F(t, f) = \frac{1}{K}\left|\sum_{k=0}^{K-1} \lambda_k x_k(f) v_t^{(k)}\right|^2 \tag{15}$$

and is an effective high-resolution spectrogram. Integrating this distribution over time gives the basic multiple window spectrum estimate

$$\sum_{t=0}^{N-1} F(t, f) = \frac{1}{K}\sum_{k=0}^{K-1} \lambda_k |x_k(f)|^2 \tag{16}$$

and so gives a much more accurate distribution of power than time-frequency distributions that simply match $| y(f) |^2$. Similarly, integrating $F(t, f)$ over frequency gives

$$\int_{-\infty}^{\infty} F(t, f)df = \frac{1}{K}\sum_{n=0}^{N-1} \left| \sum_{k=0}^{K-1} \lambda_k v_t^{(k)} v_n^{(k)} \right|^2 |x(n)|^2 \tag{17}$$

or, approximately, the convolution of $|x(t)|^2$ with $[\sin(2\pi Wt)/(\pi t)]^2$. Thus within a resolution interval $\Delta t = 1/(2W)$, power is approximately localized in time. Similarly, the narrow-band properties of the Slepian sequences imply that cross-terms are negligible for components separated in frequency by more than $2W$, so the time-frequency resolution area is of order 1. There are, however, two problems with this estimate: first, its distribution is proportional to χ_2^2, so it is statistically unstable; second, in common with time-frequency distributions of the form $x(t) \cdot \bar{y}(f)$ (see Chapter 14 of Cohen [20]), this estimate can be thought of as

$$\int \gamma_L(\xi, f) e^{j2\pi\xi t} d\xi$$

and so has "mixed" continuity properties caused by integrating across the expected δ-function sheets at 45 degrees in only one of the two variables. A more serious criticism is that, although such estimates satisfy enhanced marginal conditions, they appear to overlook the essential feature of correlation between frequencies more than $2W$ apart. Nonetheless, this "high-resolution" spectrogram represents, in applications where the spectrogram is useful, a vast improvement on the standard version. This estimate, like other multiple window estimates, can obviously be extended to include overlapping data sections, so high-resolution spectrograms of long data sets can be formed by averaging the above estimates. Much more, indeed, is possible. Extending the definition (13) to make the base position b explicit

$$y_k(b, f) = \sum_{n=0}^{N-1} e^{-j2\pi fn} v_n^k(N, W) x(b+n) \tag{18}$$

we have, corresponding to (15),

$$F(b \oplus t, f) = \frac{1}{K} \left| \sum_{k=0}^{K-1} x_k(b, f) v_t^{(k)} \right|^2 \tag{19}$$

where $\oplus$ again represents a restricted sum, with $0 \leq t \leq N-1$. (Given the extrapolation properties of these estimates, mentioned above, this restriction is not strictly necessary, only conservative.) The next subsection, 4.1 on nonstationary quadratic-inverse estimates, discusses another way to improve on the standard spectrogram, and, the following subsection, 4.2, shows how the basic expansion $X(t, f)$ may be used to estimate correlations between frequencies.

4.1 Nonstationary Quadratic-Inverse Theory

The problem of stability in the above estimate can be "solved" by quadratic-inverse theory, [30-32]. This is a way to generate minimum-variance unbiased estimates of second moment quantities directly from the eigencoefficients of the linear inverse solution without going through the ad-hoc procedure of generating the linear inverse, squaring, and then estimating the required second moments from these. Here we compute the eigensequences of the squared kernel (rigorously, the squared *truncated* kernel)

$$\alpha_l A_l(n) = \sum_{m=0}^{N-1} \left[\frac{\sin 2\pi W(n-m)}{\pi(n-m)}\right]^2 A_l(m) \tag{20}$$

These sequences rapidly approach those of the continuous time problem, [34], and there are approximately $4NW$ non-zero eigenvalues. Thus we have, approximately $\alpha_l \sim 2NW - l/2$ for $l = 0, 1, \cdots 4NW$ and, as the variances of the quadratic-inverse coefficients are proportional to α_l^{-1}, the first few coefficients are nearly as stable as the standard multiple window spectrum. The associated bases matrices

$$A_{jk}^{(l)} = \sqrt{\lambda_j \lambda_k} \sum_{n=0}^{N-1} v_n^{(j)} v_n^{(k)} A_l(n) \tag{21}$$

are real, symmetric, and trace-orthogonal; that is

$$\mathrm{tr}\{\mathbf{A}^{(l)}\mathbf{A}^{(m)}\} = \alpha_l \delta_{lm} \tag{22}$$

The expansion coefficients corresponding to $F(t, f)$ are

$$\hat{p}_l(f) = \frac{1}{\alpha_l}\mathbf{X}^{\dagger}(f)\ \mathbf{A}^{(l)}\mathbf{X}(f) \tag{23}$$

and so we have

$$P(t, f) = \sum_{l=0} \hat{p}_l(f) A_l(t) \tag{24}$$

The coefficients $\hat{p}_l(f)$ are often informative in their own right, see [31,35]. In particular, the zero-order function $A_0(t)$ is approximately constant so $\mathbf{A}^{(0)} \approx \mathbf{I}$, and $\hat{p}_0(f)$ is approximately standard multiple-window spectrum. The order 1 function $A_1(t)$ is approximately equal to

t - (N-1)/2. Thus $\mathbf{A}^{(1)}$ is zero on the diagonal and approximately constant on the sub- and super-diagonal, and $\hat{p}_1(f)$ is approximately the first *time-derivative* of the spectrum, and so on. One useful ad-hoc quantity is $\hat{p}_1(f)/\hat{p}_0(f)$ approximately the time-derivative of $\ln S(t, f)$. For example, in [35] $\hat{p}_1(f)$, computed from residuals of a global temperature series from 1854 to 1992, was almost uniformly negative across frequencies. While one must consider the series formally as nonstationary, the most reasonable explanation is not metaphysical, but simply that instrumentation and spatial coverage has improved since 1854. In this example the quadratic inverse estimates are preferable to a spectrogram or the Loève spectrum; the decrease in power is relatively small, the data series has only 138 samples, so computing a spectrogram would be difficult, and could be easily misinterpreted. Here, the negative derivative of the noise spectrum probably reflects little more than the improvements in instrumentation and spatial coverage that have occurred since 1854.

Expanding $F(t, f)$ of (16) in terms of the $A_l(t)$'s, it can be seen that the resulting coefficients, $F_l(f)$, are biased by α_l / K. However $F(t, f)$ is biased and positive, whereas truncation and Gibb's ripples can cause $P(t,f)$ to be negative.

Although spectrograms are insensitive to correlations between widely different frequencies, when the temporal evolution of the spectrum is slow spectrograms form a useful intermediate class of time-frequency distributions. Quadratic-inverse estimates improve on the spectrogram by allowing for changing power within the block, and for tests between blocks.

While the basic theory of multiple-window methods is usually written in terms of a finite block size N, we can obviously apply the same methods to overlapping time blocks to form spectrograms [29]. On each block we estimate a dynamic spectrum $D(t, f)$, it's frequency derivative, $D'(t, f)$, from [30], time derivative, $\dot{D}(t, f)$ from [23], and perhaps higher terms. These low-order terms are very stable with variances proportional to $1/\alpha_l$. Because $\hat{p}_0(f)$ is approximately the spectrum, $\hat{p}_1(f)$ is approximately the first *time-derivative* of the spectrum, etc., we can either make a "smoother" that uses these or, better, given $D(t, f)$ and $\dot{D}(t, f)$, test if an estimate $D(t + \Delta, f)$ is "reasonable". Also, the "Nyquist sampling rate" for $F(t, f)$ is simply $\Delta = 1/(2W)$, so K samples spaced N/K are obtained in each time block. Thus, if the blocks are offset by Δ we have K estimates at each point of the time-frequency plane, so that averages and variances can be computed. The covariances between blocks can be computed, tests for homogeneity of correlated variances are known, so the procedure can be used to test whether a choice of N and W is reasonable. Assuming that the estimate is reasonable, note that the average of the $F(t, f)$'s at each resampling time will be reasonably stable. Because of correlations between blocks, the stability of an average will be much less than $2K$ degrees-of-freedom, but the long lower tails characteristic of $\log\chi_2^2$ distributions are considerably suppressed. We use log spectra because: formally, the information content of a signal is measured by it's Wiener entropy- a logarithmic measure; pragmatically, most engineering applications are designed for human use and both the eye and ear have a logarithmic response. With the exception of helioseismology, it is difficult to find plots of power spectra which are not on a logarithmic (or decibel) scale. Thus we have a spectrogram with both good stability and time resolution!

Incidentally, taking either a log spectrogram or $\frac{\partial}{\partial t}\ln D(t, f)$ into a singular value decomposition and then analyzing the time eigenvectors as standard time series is often very useful; see Thomson [29].

4.2 Multiple Window Estimates of the Loève Spectrum

Taking the complex demodulates at two different frequencies, f_1 and f_2, an obvious estimate of their covariance is

$$\hat{\gamma}(f_1, f_2) = \frac{1}{K}\sum_{t=0}^{N-1} X(t, f_1)X^*(t, f_2) \tag{25}$$

where the normalization is proportional to the number of independent samples. The orthogonality of the Slepian sequences gives

$$\hat{\gamma}(f_1, f_2) = \frac{1}{K}\sum_{k=0}^{K-1} x_k(f_1)x_k^*(f_2) \tag{26}$$

This is the estimate given in [26] and generally works well, see [36,37] or [38]. An alternative motivation is that, if we consider the product of two estimates of the form

$$dX(f \oplus \xi) \sim \sum_{k=0}^{K-1} \hat{x}_k(f)V_k(\xi)d\xi \tag{27}$$

for $|\xi| < W$ then, guided by the continuity arguments of Section (3), we use a weight $W(\xi_1, \xi_2) = \delta(\xi_1 - \xi_2)$ so that smoothing over a bandwidth W is done on the stationary frequency, and no smoothing on the nonstationary direction, the same estimate is obtained. A similar smoothing scheme was proposed in [39] and applied effectively in [40].

It is often useful to plot the dual-frequency spectrum as a dual-frequency coherence, that is, defining

$$C(f_1, f_2) = \frac{\hat{\gamma}(f_1, f_2)}{[S(f_1)S(f_2)]^{1/2}} \tag{28}$$

plot a dual-frequency magnitude-squared coherence, $|C(f_1, f_2)|^2$ and the phase. Significance level calculations for this magnitude-squared coherence (MSC) are exactly the same as they are for ordinary MSC calculations, see [41].

There are far too many extensions of this approach to describe in detail here; however, an indication of some directions should be mentioned:

a) The correlation estimate in (26) can be extended to include a time delay, that is $\mathrm{ave}\{X(t, f_1)(X^*(t+\tau, f_2))\}$ which results in a quadratic form

$$\hat{\gamma}(f_1, f_2, \tau) = \sum_{j=0}^{N-1} \sum_{k=0}^{N-1} x_j(f_1) x_k^*(f_2) B_{jk}(\tau) \tag{29}$$

b) We may use a similar quadratic form with, for example, $\mathbf{A}^{(1)}$, to test for energy transfer between frequencies. More generally, for a specific spectral pattern of interest, a weight $W(\xi_1, \xi_2)$ is chosen to emphasize it, and the integration over $-W < \xi_1, \xi_2 < W$ results in an appropriate weight matrix.

c) Treat $X(t, f)$ as a matrix, possibly scaling by $1/S(f)^{1/2}$, compute its singular value decomposition, then treat the dominant time-eigenvectors as new time series.

d) The same procedures can be applied to multivariate time series; in bivariate problems compute $\mathrm{ave}\{X(t, f_1) Y^*(t, f_2)\}$ or similar, or several series can be "stacked" in the SVD process.

e) In communications signals, it is common to encounter the same signal with sidebands reversed. In this case the appropriate smoother would be perpendicular to the standard one and, as in [26], can be obtained by leaving the second coefficient unconjugated.

5. Spectrum Analysis of Radar Signals

We now apply these ideas to three radar data sets: sea clutter (i.e., radar backscatter from an ocean surface) on its own, weak target signal in clutter, and strong target signal in sea clutter. The target signal was due to the echo from a small piece of ice floating in the ocean under the dynamics of the ocean waves. Note that these are actual data, not simulations, and consequently the clutter components in all three series are necessarily different and the energy in the clutter component of the series varies with conditions. Each series consists of 256 complex samples taken at $\Delta T = 1.0$ ms.

Figures 1(a), (b) and (c) and 3 show high-resolution spectrograms of the data computed by the method of Section 4 but averaging the results of 10 sections of 229 samples each, each offset by 3 samples. A time-bandwidth product of 6 was used with $K = 10$ windows on each section. The bandwidth is thus $\pm 6/(229\Delta T) = \pm 26$ Hz. The section offset used here is smaller than recommended above and the section averaging used to suppress the lower tails of the $\log\chi^2$ distribution. (Normally, we would not attempt to compute a spectrogram from a sample of size 256.) The spectrogram of the clutter, Fig. 1(a) shows a band near -110 Hz with more power than elsewhere, but otherwise the spectrogram is reasonably flat over the clutter spectrum. The contribution due to receiver noise is about 20 dB below the clutter spectrum.

By contrast, the spectrogram of the weak growler, Fig. 1(b), shows a strong, frequency-independent vertical stripe (due to clipping of the time series) near $t = 27$ ms as well as a second frequency stripe at about -25 Hz in addition to the features visible in the clutter spectrogram.

With the strong target, Fig. 1(c), the stripe centered near 0 Hz (representing the Doppler shift of the target signal) is much more obvious, the clutter band is still there, and there is a weaker clutter image band, possibly due to a slight imbalance in the in-phase and quadrature

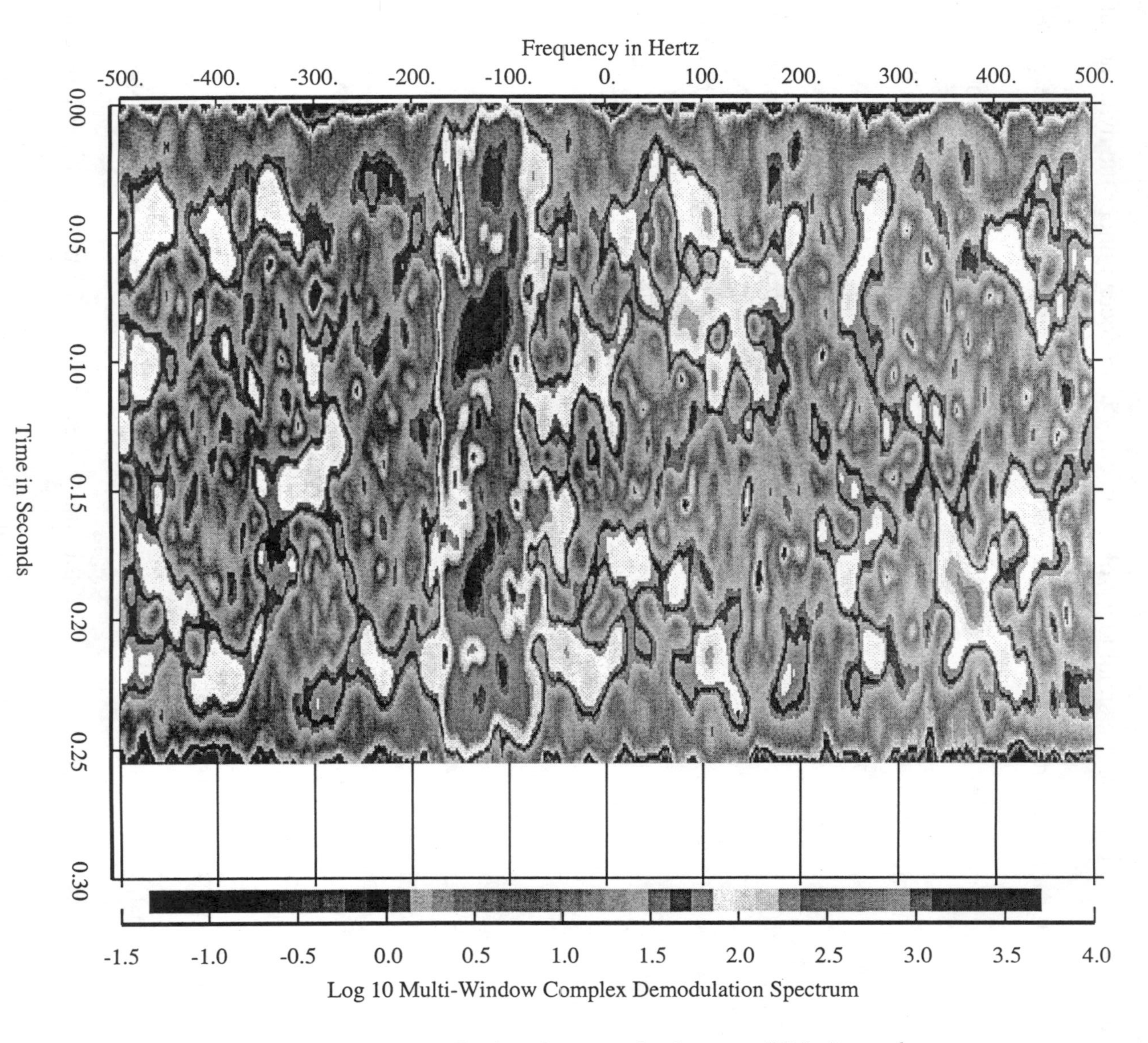

Fig. 1 - a) Radar clutter only data set, HH channel

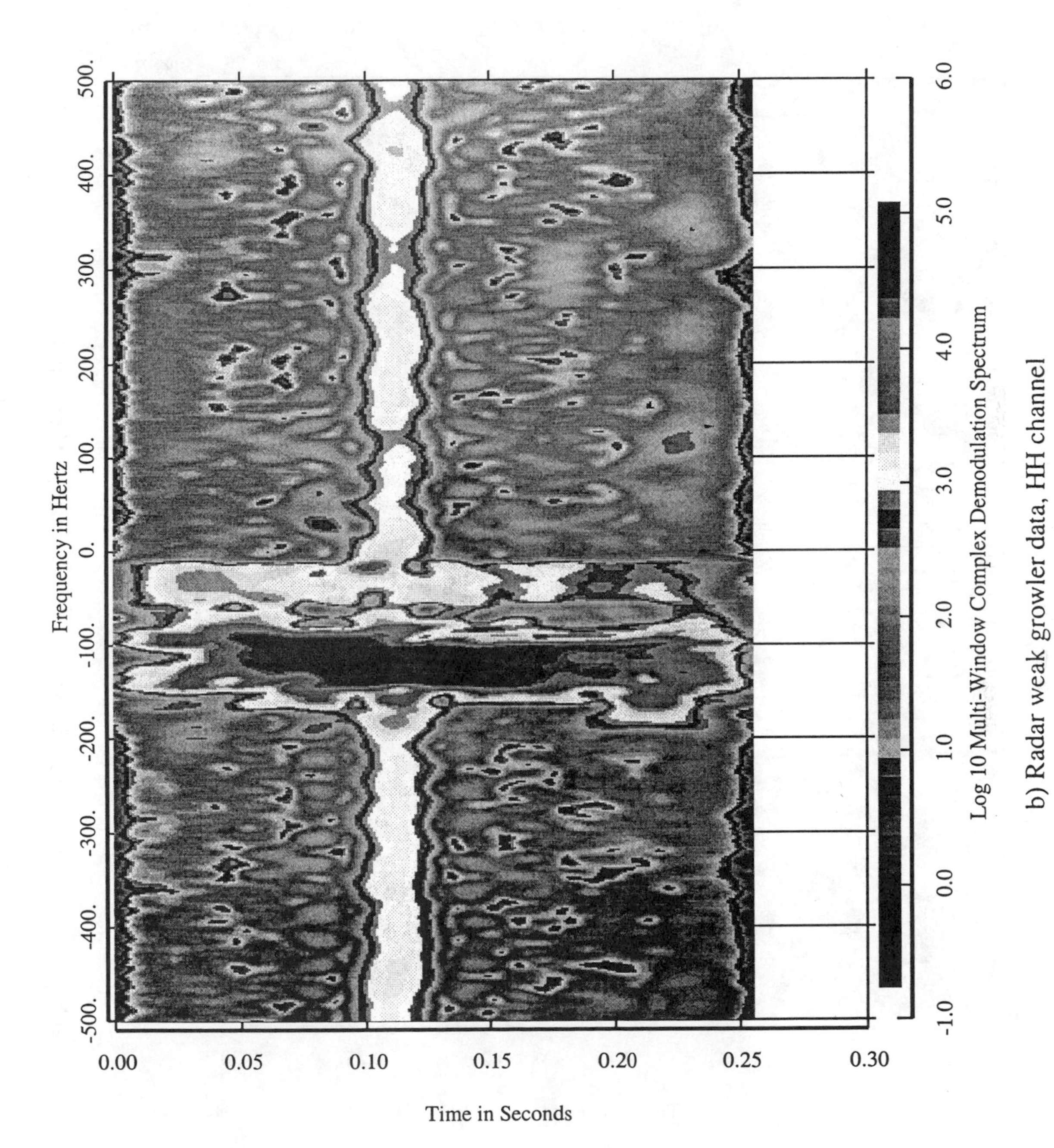

b) Radar weak growler data, HH channel

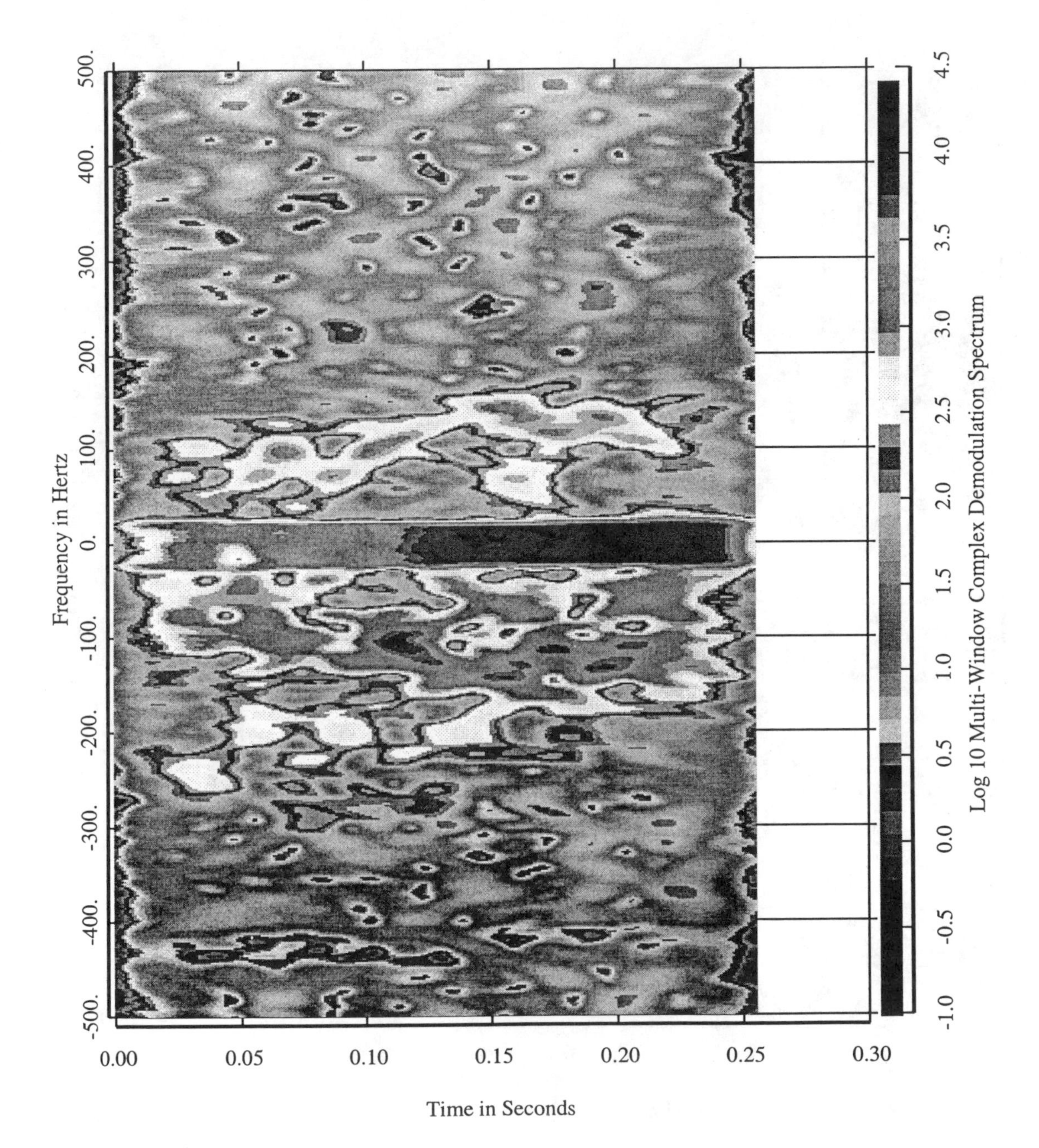

c) Radar strong growler data, HH channel

channels of the coherent receiver.

Figures 2(a), 2(b), and 2(c) show the estimates of the corresponding Loève spectra, and the gain in information beyond that apparent in the spectrogram is striking. First, the diagonal bands evident in all three series show that the data are periodically correlated, or cyclostationary; this was *not* obvious in the spectrograms, nor expected. In Fig. 2(a) it can be seen that the peak of the Loève spectrum for clutter is near -125 Hz, as before. In the weak target signal case, an extra peak centered at about -25 Hz can be seen in Fig. 2(b), while in the strong target signal case shown in Fig. 2(c) the peak near zero is dominant but, in contrast to the spectrogram, the periodic correlation of the clutter is still visible.

Figures 3(a), 3(b) and 3(c) present the corresponding Wigner-Ville distributions of the sea clutter on its own, weak target in sea clutter, and strong target in sea clutter, respectively. The important feature to observe here is the presence of a zebra-like pattern (alternating between dark and bright narrow stripes) in the images due to the presence of a target signal. This pattern occupies an area located between the instantaneous frequency plot of the target near 0 Hz and that of the clutter. This pattern is indeed a manifestation of the cross Wigner-Ville distribution terms due to the combined presence of a target signal and clutter. Most importantly, the presence of this zebra-like pattern is found to be (1) fairly pronounced at relatively low target signal-to-clutter ratios, and (2) relatively robust to variations in the target signal-to-clutter ratio [17].

Although the high-resolution spectra estimates of Figs. 1 and 2 based on the method of multiple windows display the dynamic spectrum of the radar signals in ways that are similar (in some parts) and yet different (in other parts) from the corresponding Wigner-Ville distributions of Fig. 3, the important point to note from these two differently computed sets of images is that both approaches accentuate the differences between the different classes of radar signals in their own

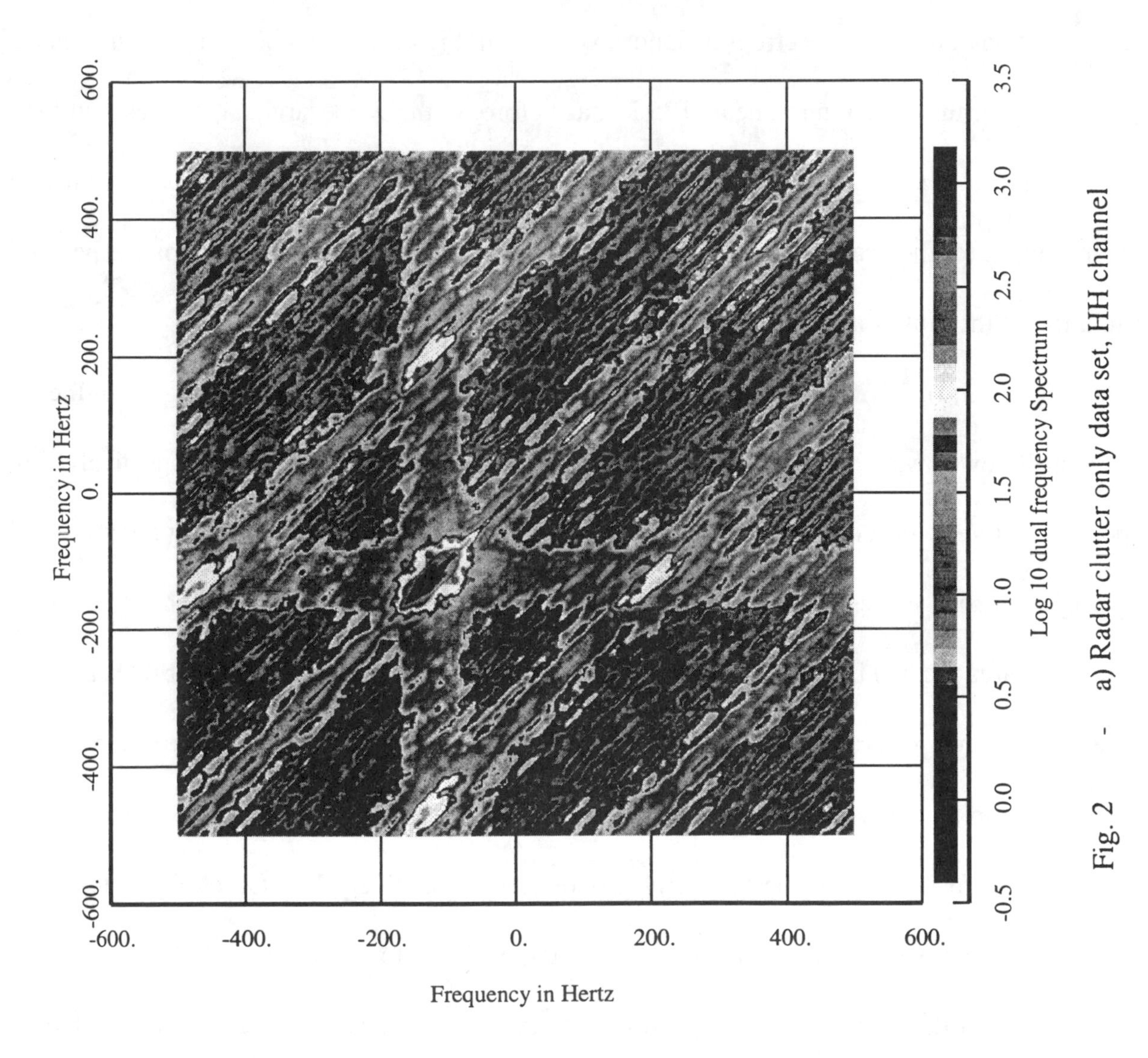

Fig. 2 - a) Radar clutter only data set, HH channel

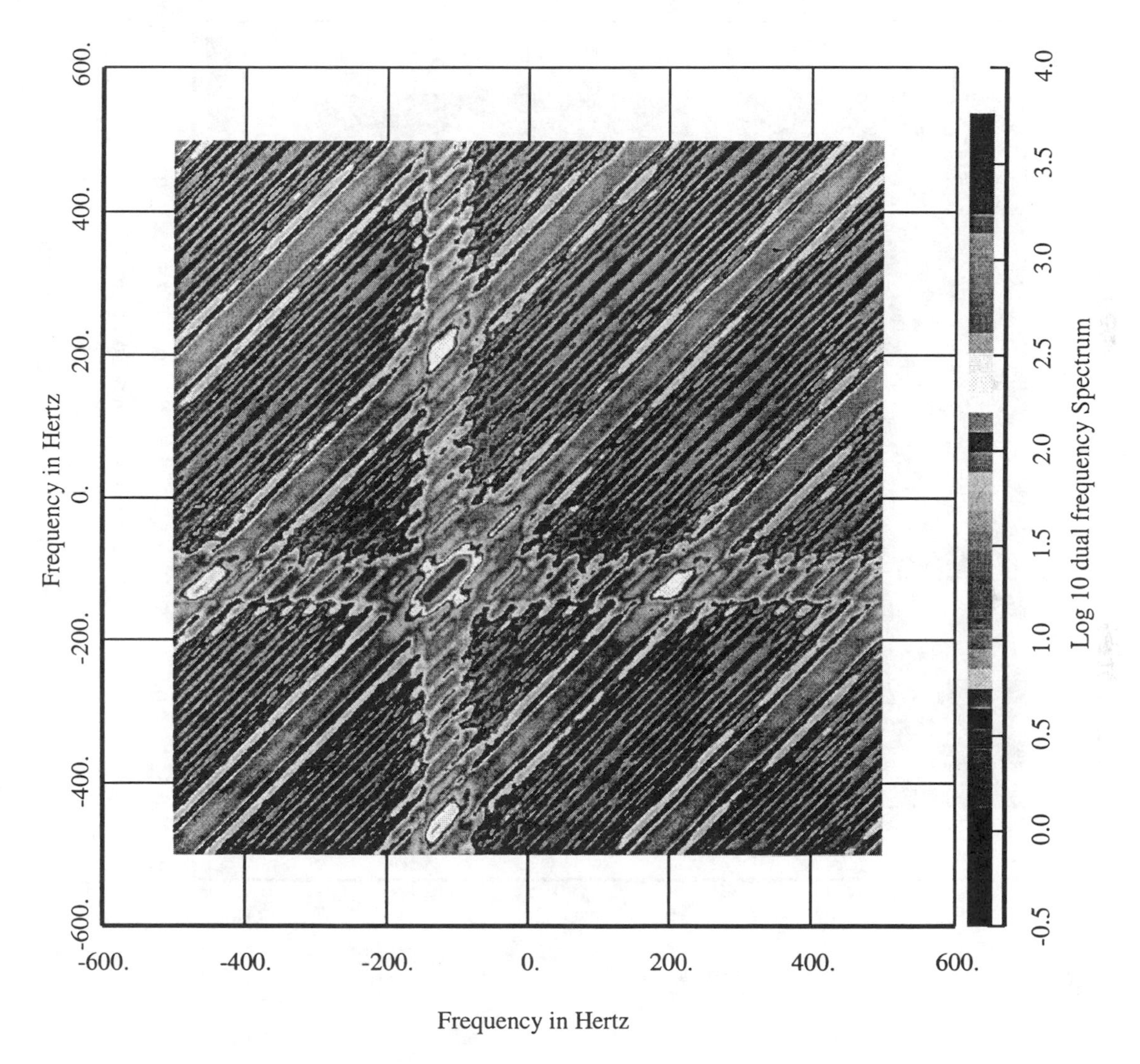

b) Radar weak growler data, HH channel

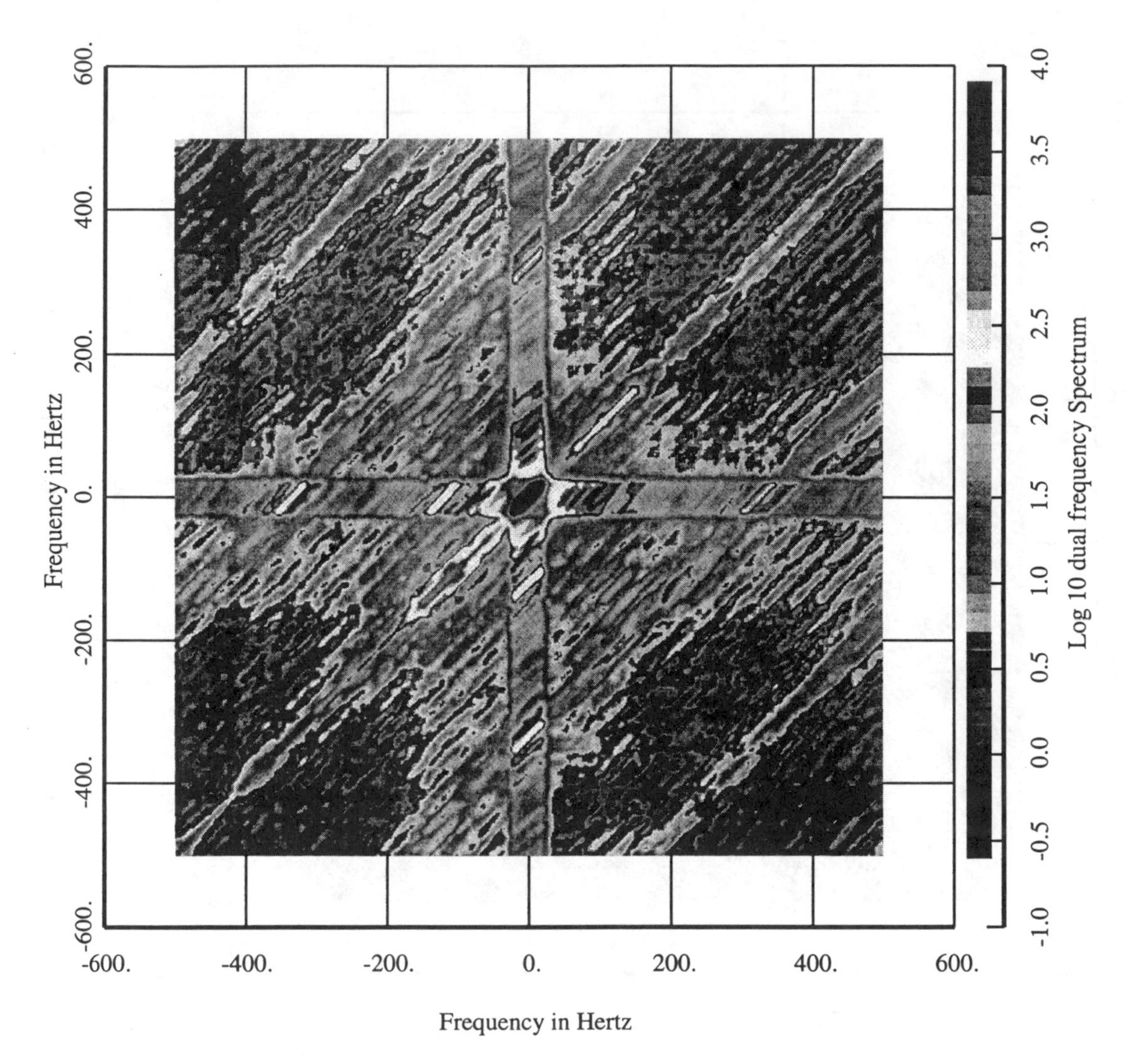

c) Radar strong growler data, HH channel

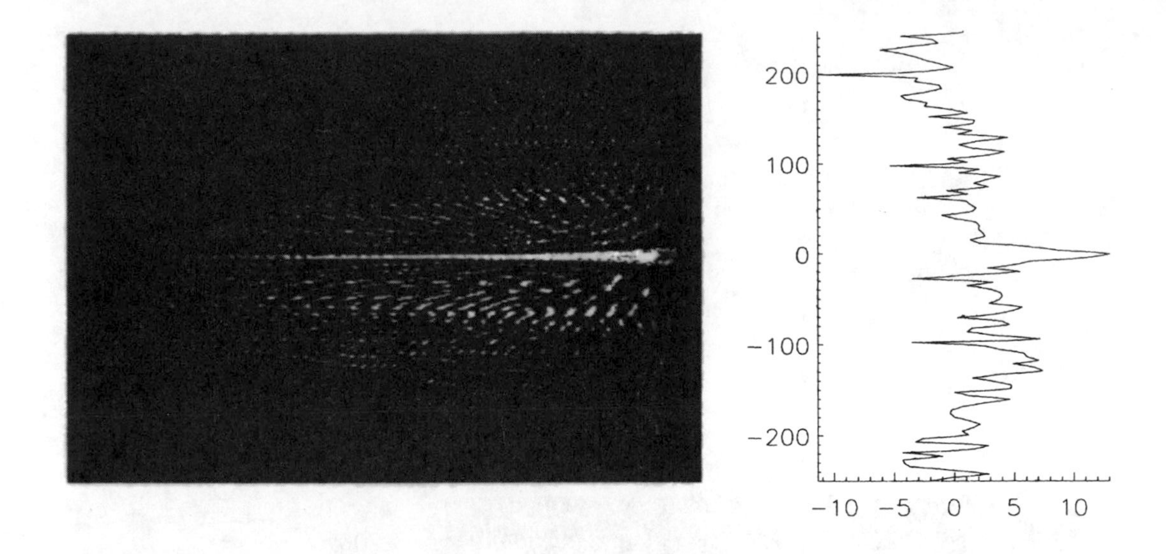

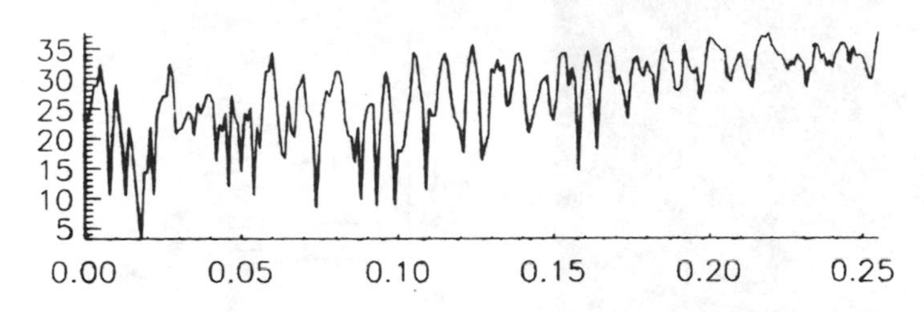

Fig. 3 - a) WVD for a clearly visible growler

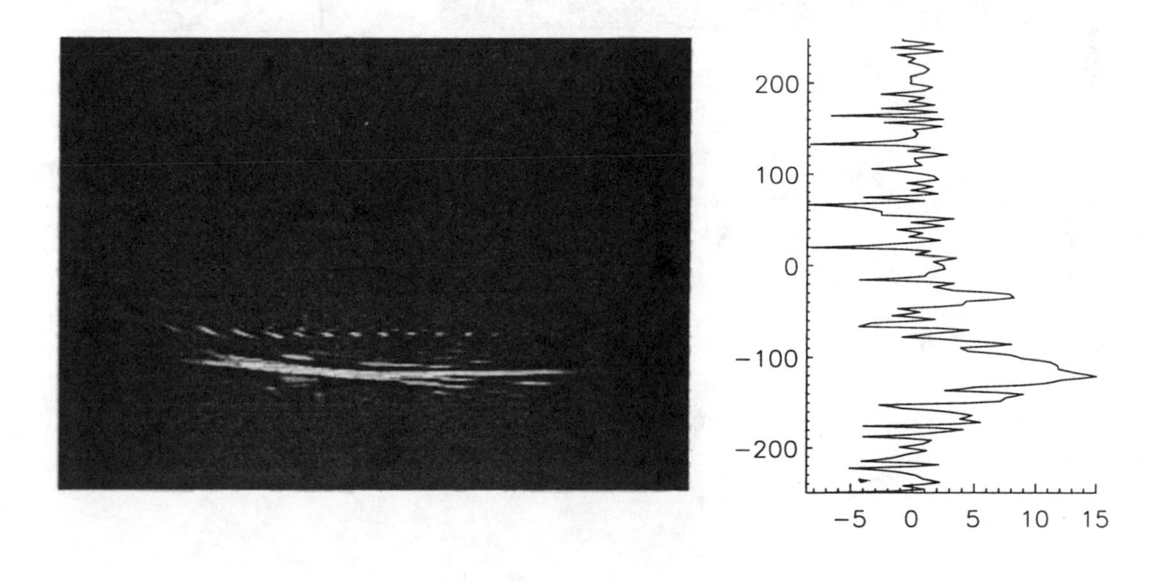

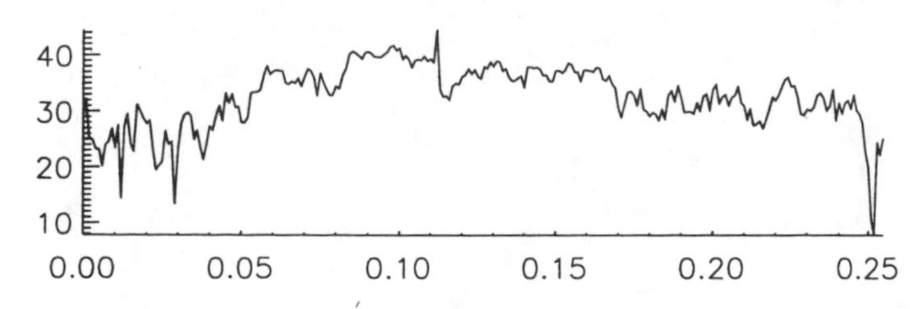

b) WVD for a barely visible growler

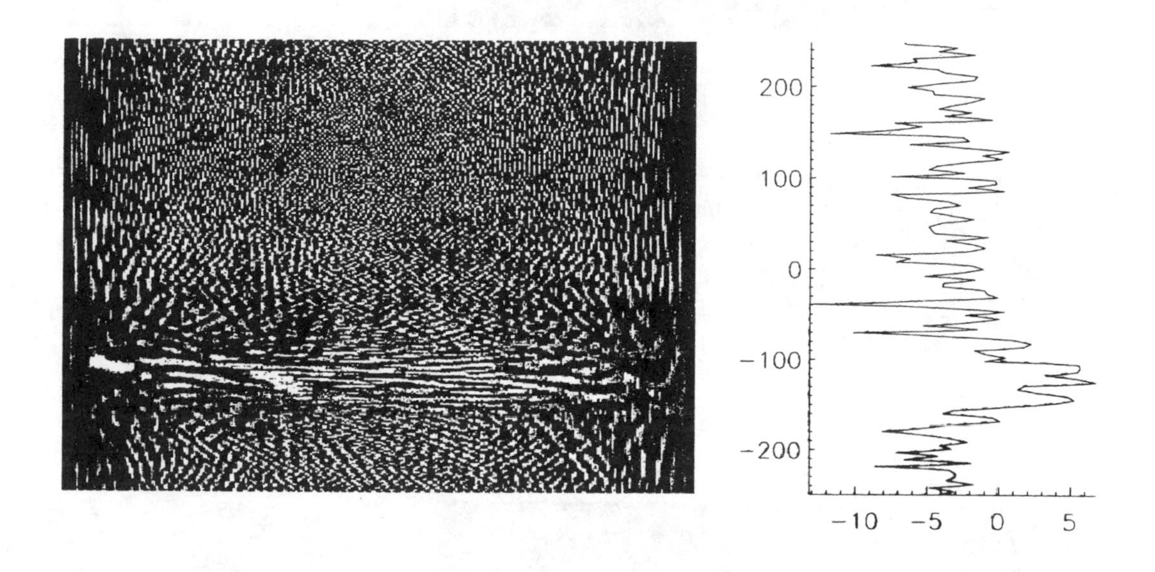

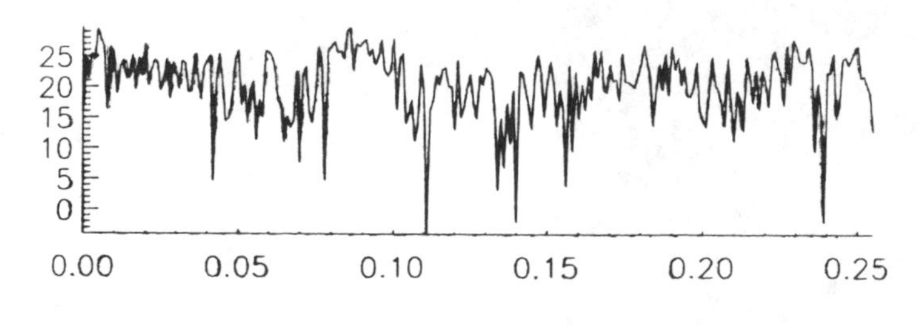

c) WVD for sea clutter

individual ways, making them more visible than the original time series. Simply put, the power of these methods lies in their ability to make a weak target signal buried in a strong clutter background *visible* in signal-processing terms.

6. Modular Learning Machine for Adaptive Signal Detection

Regardless of whether we use the high-resolution images exemplified by Figs. 1 and 2 or the Wigner-Ville distributions of Fig. 3, these two approaches do share a common property: The one-dimensional time series representing the received signal is transformed into a *highly redundant two-dimensional image*. For the detection strategy to be computationally efficient, the redundant information contained in this image would have to be removed by some means. This is not so different from signals such as speech where redundancy is stripped for coding, and then added later for error protection.

In pattern recognition theory, the removal of redundant information is referred to as *feature extraction* [42,43]. Traditionally, feature extraction is followed by *pattern classification.* At the output of the pattern classifier a decision is made as to whether the image applied to the feature extractor, or equivalently the original received signal, belongs to one of two possible (hypotheses) classes:

- Null hypothesis, H_0: The received signal consists of noise alone.
- Other hypothesis, H_1: The received signal consists of a target signal plus noise.

In other words, we have a binary hypothesis testing problem, the solution of which is optimized in some statistical sense.

Figure 4 shows the block diagram of a *modular learning machine* [17,44] for adaptive signal detection, based on the strategy described. At the input end of the machine we have a *time series-to-image transformer* that uses dynamic spectrum analysis or time-frequency analysis for its design. From that point on, the machine splits into two channels, one termed the *noise channel*

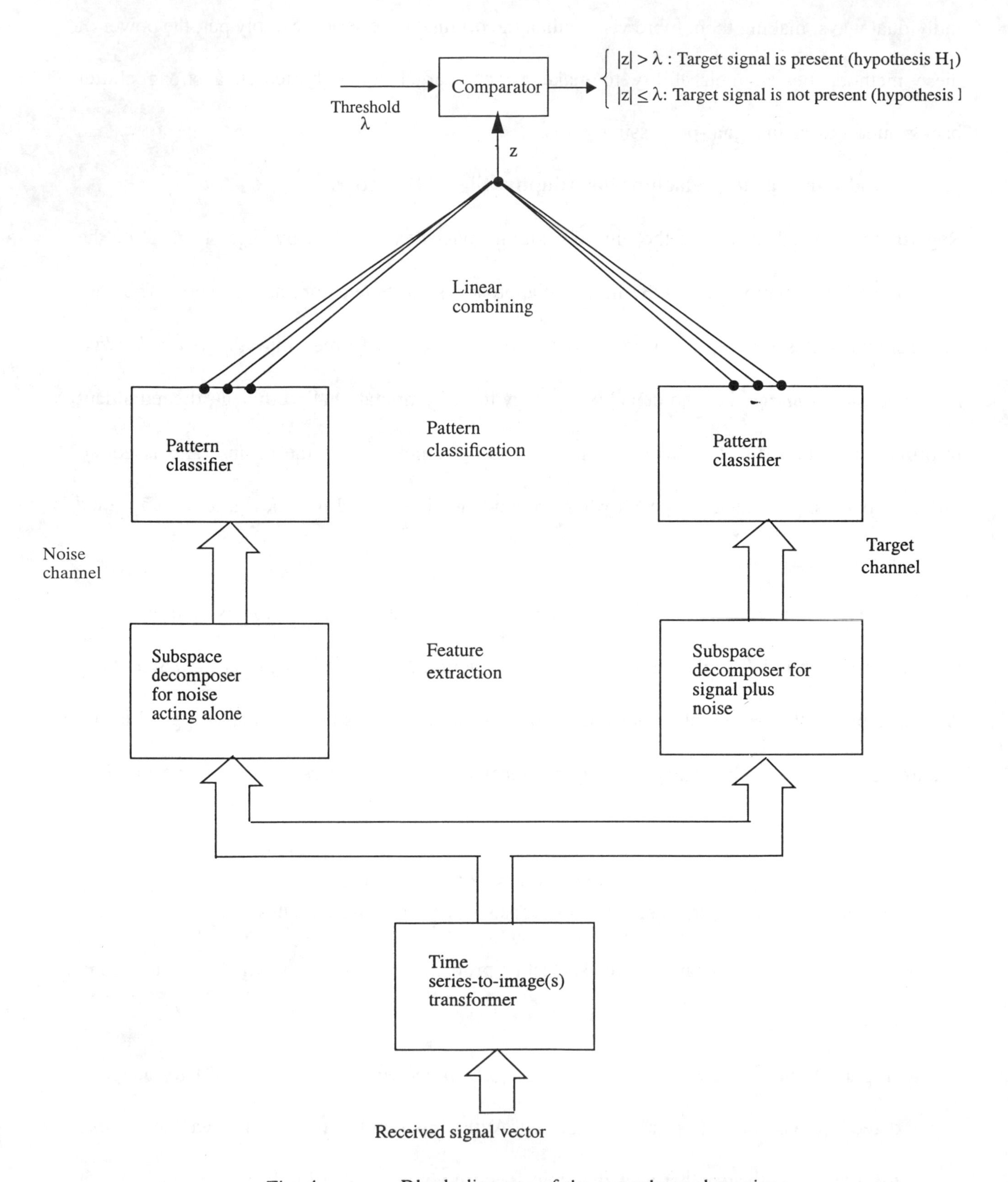

Fig. 4 - Block diagram of the two-channel receiver

and the other termed the *target channel*. The two channels are linearly combined at their outputs, and then a final decision is made as to whether a target signal is present in the received signal or not.

Each channel consists of two functional blocks: feature extractor, and pattern classifier. A popular method for implementing the feature extractor is principal components analysis (PCA), the aim of which is to learn the *dominant eigenvectors that are most representative of the different realizations of the pertinent class of data*. In basic mathematical terms, this operation involves performing an eigendecomposition on the covariance matrix of a data vector obtained by scanning the image on a column by column basis, arranging the eigenvalues in decreasing order, and retaining only those eigenvectors that are associated with the dominant eigenvalues. In effect, the PCA performs a *subspace decomposition* on the images belonging to class H_0 or class H_1, and converges onto a solution that captures the features that are most common to the different physical realizations of the class in question. Accordingly, when an image representing class H_0, say, is projected onto the two subspaces computed by the noise and target channels, the outputs of the feature extractor in the noise channel are relatively small whereas the corresponding outputs of the feature extractor in the target channel are relatively large. The reverse holds for the presentation of an image representing class H_1. We may therefore state that the feature extractor in the noise channel is *adaptively matched to input data known to consist of noise alone*. A similar statement holds for the feature extractor in the target channel. Both feature extractors are designed using a *self-organized learning* procedure that tracks the statistical variations in the received signal.

Turning next to the pattern classifiers and linear combiner, they are designed by using a *supervised learning* procedure. To do this, we may use one of two procedures:

1. The two pattern classifiers are trained separately, with hard decisions being made at their respective outputs. When the machine is presented with a received signal known to consist of noise only, the pattern classifier in the noise channel is trained to classify that signal, for example, as belonging to one of the following three categories:

   - The received signal is definitely noise.
   - The received signal is on the border line of being interpreted as noise or a weak target signal plus noise.
   - The received signal looks like a weak target signal plus noise.

   Correspondingly, when the machine is presented with a received signal known to consist of target signal plus noise, the pattern classifier in the target channel is trained to classify that signal as belonging to one of the following three categories:

   - The received signal contains a strong target signal.
   - The received signal contains a weak target signal.
   - The received signal looks like noise.

   The outputs of the two channels are *linearly combined* to produce the overall output z, where the final decision (in favor of hypothesis H_0 or H_1) is made. To design the linear combiner, the machine goes through another round of supervised training with data not seen before. The categorization of the noise/target channel output in the manner described here may be viewed as a *discrete approximation to soft-decision making*.

2. The two pattern classifiers and linear combiner are all trained simultaneously. The individual outputs of the two pattern classifiers (three each, for example) are now free to assume values set by the activation functions of their output neurons (processing units).

The attractive feature of method (1) is that the decision boundaries between the different sub-classes of the received signal are well defined at the outputs of the two channels. However this advantage, if any, over method (2) is gained at the expense of increased training.

Irrespective of whichever of these two methods is adapted, the adaptive two-channel receiver of Fig. 4 relies on the use of *learning-to-learn procedures* for its design. Most importantly, the target and noise channels tend to *reinforce* each other in their individual decisions, thereby providing for an improved detection performance over that attainable with a single channel. To that end, the use of linearly combining the outputs of the two channels, viewed as a form of ensemble averaging, is usually considered to be superior to the use of majority voting [45].

6.1 How Does the Adaptive Receiver of Fig. 4 Respond to a Nonstationary Environment

An issue that may need further clarification is how the adaptive receiver of Fig. 4 is able to respond to statistical variations in the environment. The answer to this fundamental issue lies in (1) the transformation of a time series into an image or images, (2) the adaptive subspace decompositions of the images so computed, and (3) the training of the pattern classifiers under the tutelage of a "teacher".

The time series-to-image transformers enhance the nonstationary character of the received signal, making it more *visually discernible* and therefore more *readily learnable*. In this context, it is noteworthy to recognize that the sonar-based echolocation system of a bat relies on the computation of time-frequency maps for its own operation [46,47]. Indeed, a bat is able to detect and track its prey (e.g., a flying insect) in a difficult environment with a facility and success rate that would be the envy of a sonar and radar engineer.

The adaptive subspace decompositions performed on the time-frequency images (maps) provide for *compact representations* of a wide variety of different realizations of the two classes of data, that is, under hypotheses H_0 and H_1. This is done by transforming the higher-dimensional spaces of the images to lower-dimensional spaces, subject to the *constraint of information preservation* (i.e., reconstruction of the original data with minimal distortion). Moreover, the features constituting the compact representations are *decorrelated*, thereby paving the way for the efficient training of the pattern classifiers.

Finally, it is in the design of the pattern classifiers where the accounting for nonstationary behavior of the environment comes into focus. Ground-truthed data pertaining to hypotheses H_0 and H_1 are presented alternately to the receiver, and the free parameters of the pattern classifiers are adjusted so as to minimize a statistical criterion of interest. By "ground truthing" we mean that when the data are collected, the prevalent environmental conditions (i.e., the target is present or not) are carefully monitored and recorded by human observers. In the course of this supervised training, information contained in the input data is transferred and stored in the values assigned to the free parameters of the subspace decomposers and pattern classifiers. The net result of the training process is that a nonlinear decision boundary is constructed in the input space between the hypotheses H_0 and H_1, with the data having a direct say in how the decision boundary is adaptively constructed. On the one side of the decision boundary we have data points assigned to hypothesis $H_{0;}$ each such point represents a *different* realization of this hypothesis knowing that no target is present, with the difference arising because of statistical variations in the environment. On the other side of the boundary we have data points assigned to the hypothesis H_1; each such data point represents a *different* realization of this second hypothesis knowing that a target is present, with the difference again being due to statistical variations in the environment. Subject to the proviso *that the training data are representative of the nonstationary environment*, and the pattern classifiers are therefore forced to assign the data points appropriately to class H_0 or class H_1, under the tutelage of a teacher, the receiver should achieve a performance under test that is close to the performance under training. It is assumed here that the test data are different from the training data but drawn from the same environment. Indeed, we can make the following general observation [45]:

The more exhaustive the training data set is in its representation of the nonstationary environment, the more likely it is that the receiver adapts to its environment fully and therefore be able to provide a robust detection performance.

The receiver design described herein differs markedly from the way in which classical receivers are designed. In the *classical approach*, we start with a mathematical model of the received signal and end up testing the receiver performance with real-life data. The success of the classical approach rests largely on how close the mathematical model is to the realities of the data. On the other hand, in the *adaptive approach through learning* described in this paper, the data set is allowed to "speak for itself" in how the receiver is designed. Stated in yet another way, the classical approach relies on mathematical tractability, and the challenge in this approach is to formulate the right mathematical model for the received signal that accounts for the nonstationary behavior of the environment. In the adaptive approach through learning, the need for mathematical modeling is eliminated, and the challenge in this modern approach is twofold: (1) collect a ground-truthed database that is large enough in size and fully representative of the nonstationary environment, and (2) design the receiver using appropriate learning-to-learn procedures that respond to statistical variations of the environment.

7. Case Study: Radar Target Detection of a Small Target in Sea Clutter

The adaptive two-channel receiver of Fig. 4 defies a statistical analysis of detection performance along traditional lines due to its nonlinear nature. Therefore, to evaluate the practical merit of this new receiver, we performed a case study involving the detection of a growler floating in an ocean environment. A *growler* is a small piece of ice that is broken off an iceberg. The above-surface visible portion of it is about the size of a grand piano (i.e., a radar cross-section of about 1 m^2).

However, recognizing that about 90% of the volume of ice lies below the water surface, a growler represents an object large enough to be hazardous to navigation in ice-infested waters, such as those encountered on the East Coast of Canada during the Spring and early Summer. The radar task at hand is that of detecting the radar echo from a growler in the presence of interference represented by sea clutter.

For the collection of radar data representative of this environment, an instrument-quality radar system called the *IPIX radar* was used. The IPIX radar [48] is a fully coherent, polarimetric, X-band radar system, equipped with computer control and digital data acquisition capability. The present study is confined to the use of coherent data collected under the polarimetric condition of horizontal transmit and horizontal receive only. The radar was operating in a staring mode (i.e., pointing onto a patch of the ocean surface). A series of experiments using the IPIX radar were performed at a site located on the East Coast of Canada. The radar was mounted at a height above sea level that would be representative of a ship-mounted radar. Ground-truthing of the data collected was maintained throughout the experiments, thereby providing knowledge of the conditions under which the various datasets were collected. This case study was chosen for the application at hand because both the target of interest (a growler) and the background interference (sea clutter) are known to exhibit nonstationarity, which would require the use of adaptivity. Moreover, the generation of sea clutter is governed by a nonlinear dynamical process, which would therefore require the use of nonlinear processing. Thus, the detection of a growler in sea clutter provides a suitable medium for testing the capabilities of our new detection strategy. Details of this case study were presented in [17]. The material presented here is a summary of the results presented in that paper.

This case study was chosen for two reasons:

1. The ocean environment is highly nonstationary and therefore provides the right setting for validating the adaptive detection procedure through learning described in this paper.
2. A large set of ground-truthed data was available for the study.

7.1 Details of the Receiver

The 2-D Wigner-Ville distribution (WVD) image used in the study had a time dimension $L = 256$ with spacing $T = 1$ ms, and a frequency dimension $M = 256$ with spacing $F = 4$ Hz. Note that the WVD image is real valued even though the received signal is complex valued.

Each of the two PCA networks in Fig. 4 consisted of a *feedforward neural network* with an input layer of $M = 256$ source nodes (fed from the WVD image of the received signal) and a single computation layer of $p = 5$ *linear* neurons. Both networks were *fully connected*, in that each neuron of either network was connected to all the source nodes of its respective input layer. The total number of connections/independent weights for each PCA network was 1280. The training set for $\text{PCA}^{(0)}$ network was made up of $A_0 = 2000$ epochs, representing hypothesis H_0. The training set for $\text{PCA}^{(1)}$ network was made up of $A_1 = 500$ epochs, representing hypothesis H_1. The individual epochs of WVD images were generated using examples of the received signal, each being made up of 256 samples. Both PCA networks were trained using the *generalized Hebbian algorithm* (GHA) due to Sanger [49]. Let $\mathbf{x}(t)$ denote the input vector applied to the algorithm at iteration (time step) t, $\mathbf{y}(t)$ denote the corresponding value of output vector, and $\mathbf{W}(t)$ the weight matrix of the PCA network under training. The change $\Delta\mathbf{W}(t)$ in the weight matrix computed by the algorithm at iteration t is defined by

$$\Delta\mathbf{W}(t) = \eta(\mathbf{y}(t)\mathbf{x}^T(t) - \text{LT}[\mathbf{y}(t)\mathbf{y}^T(t)]\mathbf{W}(t)) \tag{30}$$

where the superscript T denotes matrix transposition, the operator LT makes the matrix enclosed inside the square brackets lower triangular by setting all the elements above its diagonal equal to zero, and the scaling factor η denotes the learning rate parameter of the algorithm. The GHA operates on the WVD directly. Most importantly, it is well suited for the application at hand by virtue of the large size of the training data and the ability of the algorithm to track changes in the input data from one epoch to the next.

Turning next to the pattern classifiers in Fig. 4, we used a type of feedforward neural networks known as *multilayer perceptrons* (MLPs). The input layer of each MLP consisted of an array of p x L source nodes fed from a compressed image with $p = 5$ and $L = 256$. The first hidden layer consisted of an array of 5 x 15 neurons. The network architecture of this layer was *constrained* as shown in Fig. 5. Specifically, it incorporated the following concepts for improved training and perhaps better generalization performance [13,45]:

1. *Receptive field*, which means that a neuron in each row of the first hidden layer is connected only to a certain number of source nodes (denoted by R) that lie in its local neighborhood in the corresponding row of the input layer.
2. *Overlap of receptive fields*, which means that the receptive fields of adjacent neurons in a particular row overlap by a certain number of source nodes (denoted by S).
3. *Weight sharing*, which means that the receptive fields of all the neurons in a particular row share the same set of synaptic weights.

For our present study, we chose $R = 32$ and $S = 16$. In addition, each MLP had a second hidden layer of 25 neurons and output layer of three neurons, both of which were fully connected. The neurons in both MLPs were all *nonlinear*, using a sigmoid activation function defined by the *logistic function*

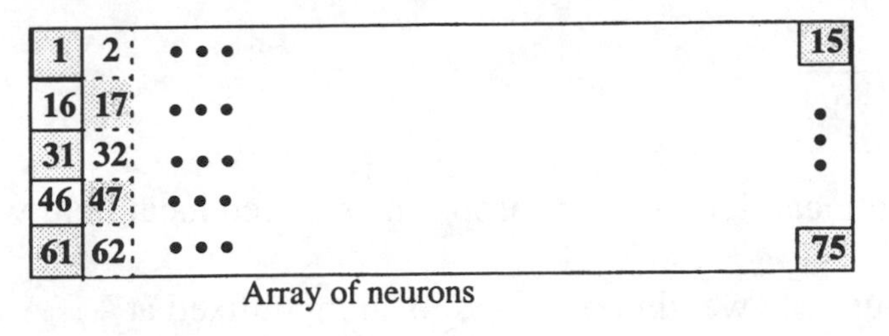

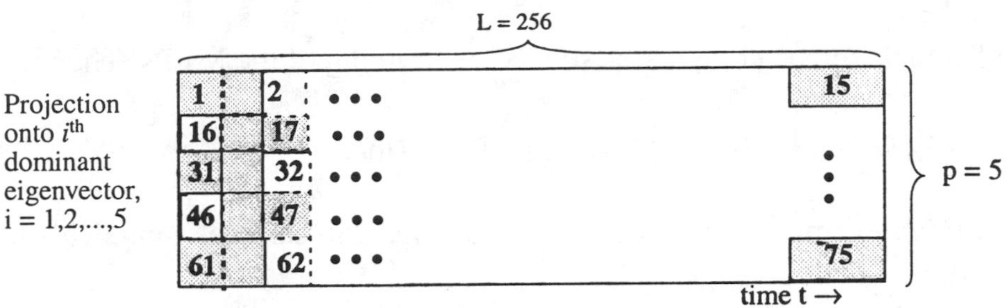

(a) Two-dimensional display of first hidden layer

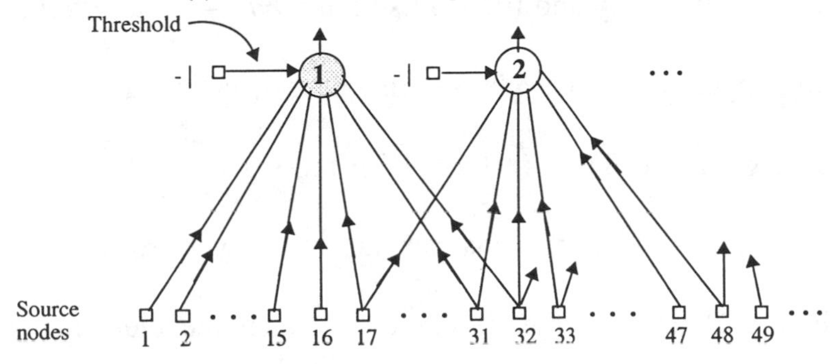

(b) First row of neurons in the first hidden layer

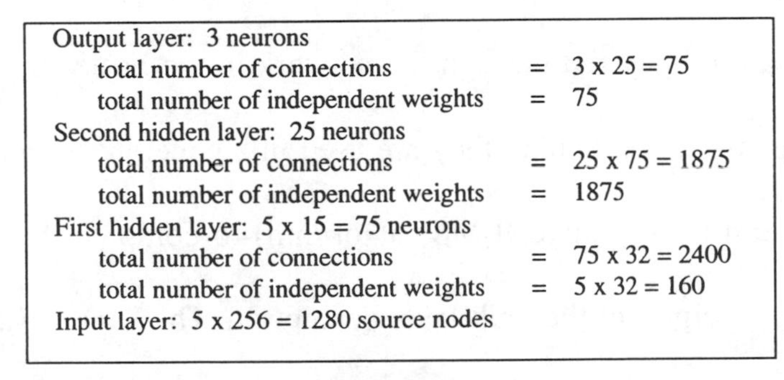

Output layer: 3 neurons
- total number of connections = 3 x 25 = 75
- total number of independent weights = 75

Second hidden layer: 25 neurons
- total number of connections = 25 x 75 = 1875
- total number of independent weights = 1875

First hidden layer: 5 x 15 = 75 neurons
- total number of connections = 75 x 32 = 2400
- total number of independent weights = 5 x 32 = 160

Input layer: 5 x 256 = 1280 source nodes

(c) Connectivity of each MLP

Fig. 5 - Architectural details of each multilayer perceptron (MLP)
a) Two-dimensional display of first hidden layer
b) First row of neurons in the first hidden layer
c) Connectivity of each MLP

$$\varphi(v) = \frac{1}{1 + \exp(-v)} \tag{31}$$

where v is the induced local field of the neuron. The induced local field v includes a bias, which is represented by an adjustable weight connected to an input fixed at +1. A total of 10,156 examples of the received signal were used to do the supervised training of the MLPs. They were made up as follows: B_0 = 7,150 examples representing hypothesis H_0, and B_1 = 3,006 examples representing hypothesis H_1. Each example of the received signal was 256 samples long. This training dataset was completely different from that used to train the PCA networks. The two MLPs and linear combiner were treated as a single entity and trained using the *back-propagation* (BP) algorithm [45,50]. The BP algorithm operates in two phases. In Phase I, called the forward phase, the synaptic weights of the network are fixed. In this phase the signal applied to the input layer propagates through the network in a forward manner, layer by layer. Phase I is completed by calculating the error signals, defined as the difference between the elements of the desired response vector and the corresponding values of the actual output signals of the neurons in the output layer. The error signals are propagated through the network in the backward direction in phase II, called the backward phase. In particular, they are used in a generalized delta rule (i.e., generalization of the popular least-mean-square (LMS) algorithm) to compute the adjustments applied to the individual synaptic weights of the multilayer perceptron. The BP algorithm has two useful properties:

- Simplicity of implementation.
- Stochastic gradients (i.e., derivatives of the error performance surface with respect to the weights in the network).

It is because of these two properties that the BP algorithm has established itself as the workhorse for the training of MLPs intended particularly for pattern classification, hence its use for the design of the pattern classifiers in the two-channel receiver of Fig. 4.

7.2 Detection Results

Figure 6 presents a visual display of the *predetection* performances of two different receivers:

1. Doppler constant false-alarm rate (CFAR) receiver; this system was chosen as a frame of reference because of its widespread use as a conventional radar receiver.
2. Neural network (NN) implementation of the two channel receiver of Fig. 4, which involved the use of three output nodes per channel.

Predetection refers to the receiver output *prior* to thresholding. The results of Fig. 6 were obtained for a long dwell time (approximately 35 s) along a range swath of 200 m and a range gate (resolution) of 5m; the total number of radar samples represented here is 2.68 x 10^6. The test data used here were completely different from the data used to train the PCA networks and those used to train the MLP's. The darkness of the display in Fig. 6 is a measure of the actual power of the receiver output before thresholding. The two parts of the figure have been normalized separately to remove any bias introduced by changes in dynamic ranges of the receivers. The Doppler CFAR and NN receivers paint the two hypotheses H_0 and H_1 in dramatically different colors. In particular, the discrimination between the clutter background (hypothesis H_0) and target (hypothesis H_1) is far more pronounced in the NN receiver than it is in the Doppler CFAR receiver. This is the direct result of the fact that the Doppler CFAR receiver is basically linear, whereas the NN receiver is highly nonlinear thereby capturing the information content of the received signal to a fuller extent.

To further emphasize the performance difference between these two receivers, Fig. 7 shows their *postdetection* performances obtained by comparing the amplitude of each receiver

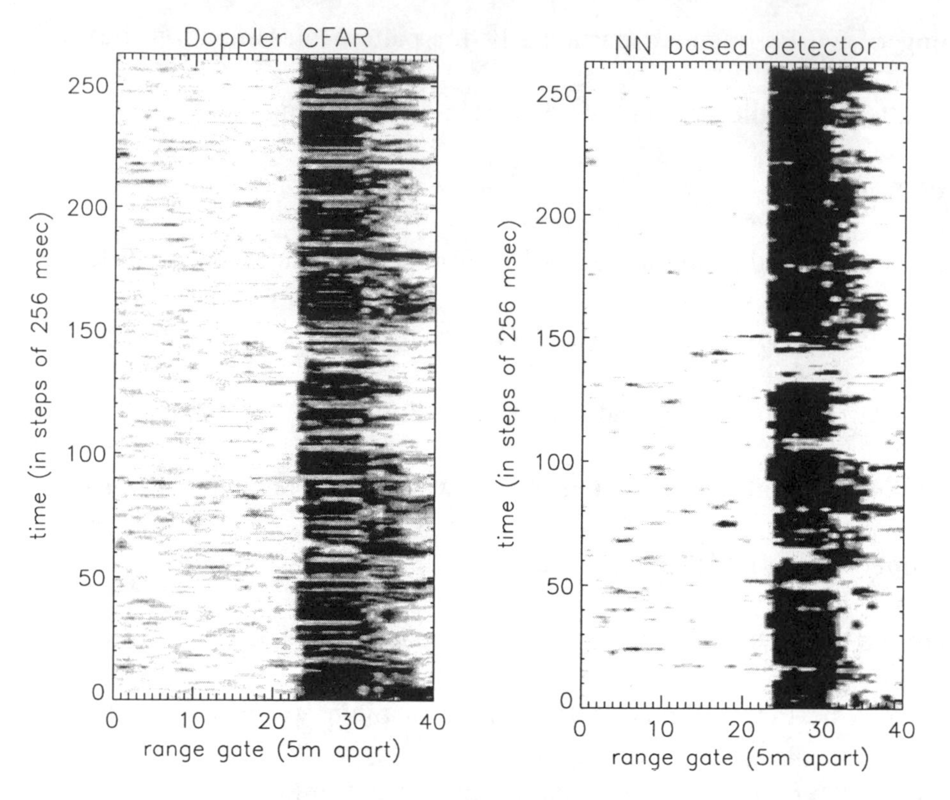

Fig. 6 - Detection statistics for the Doppler CFAR and NN receivers

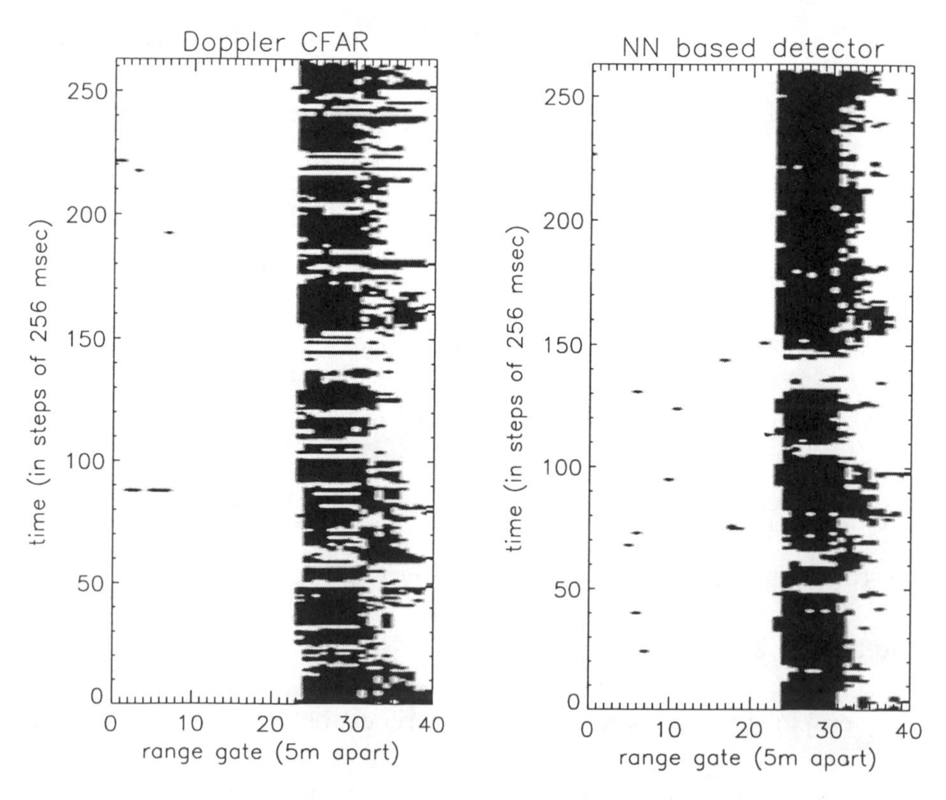

Fig. 7 - Postdetection results for the Doppler CFAR and NN receivers

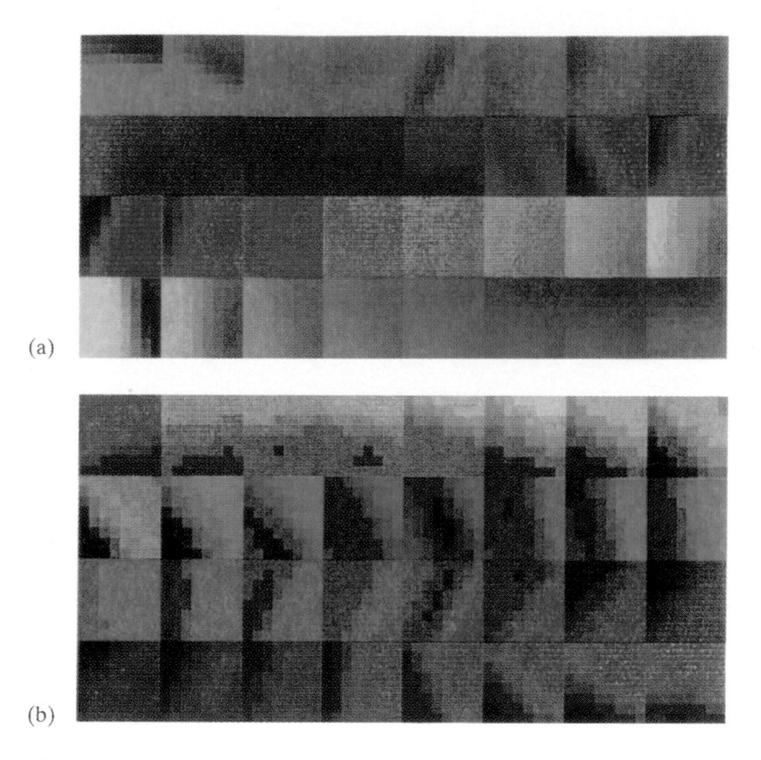

Figure 14.6 Map of component basis blocks for 32-class MPC networks. The class number progresses left to right, top to bottom. (a) 1 Component and (b) 2 Component.

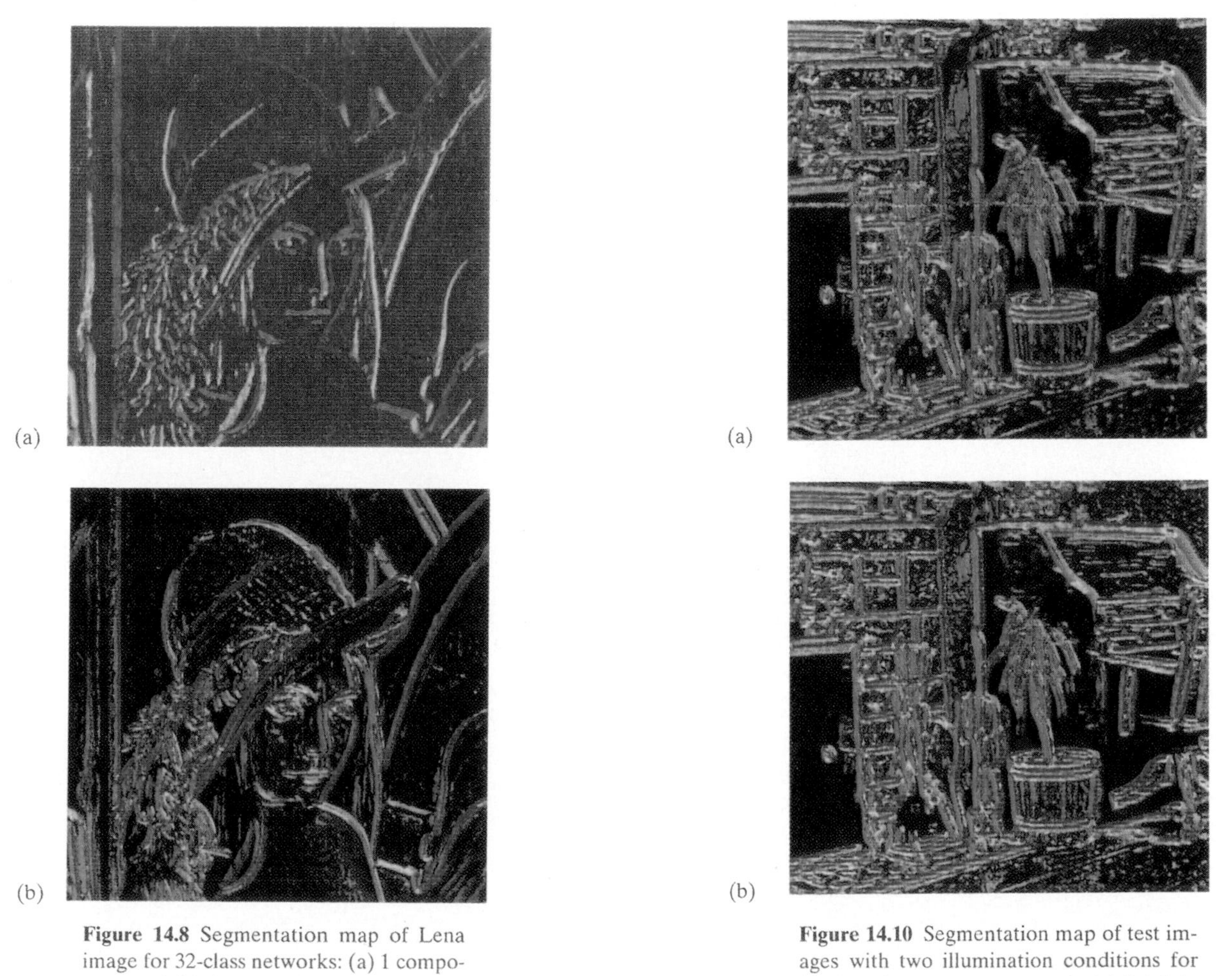

Figure 14.8 Segmentation map of Lena image for 32-class networks: (a) 1 component and (b) 2 component.

Figure 14.10 Segmentation map of test images with two illumination conditions for 32-class, 2-component network: (a) Brightly illuminated and (b) Dimly illuminated.

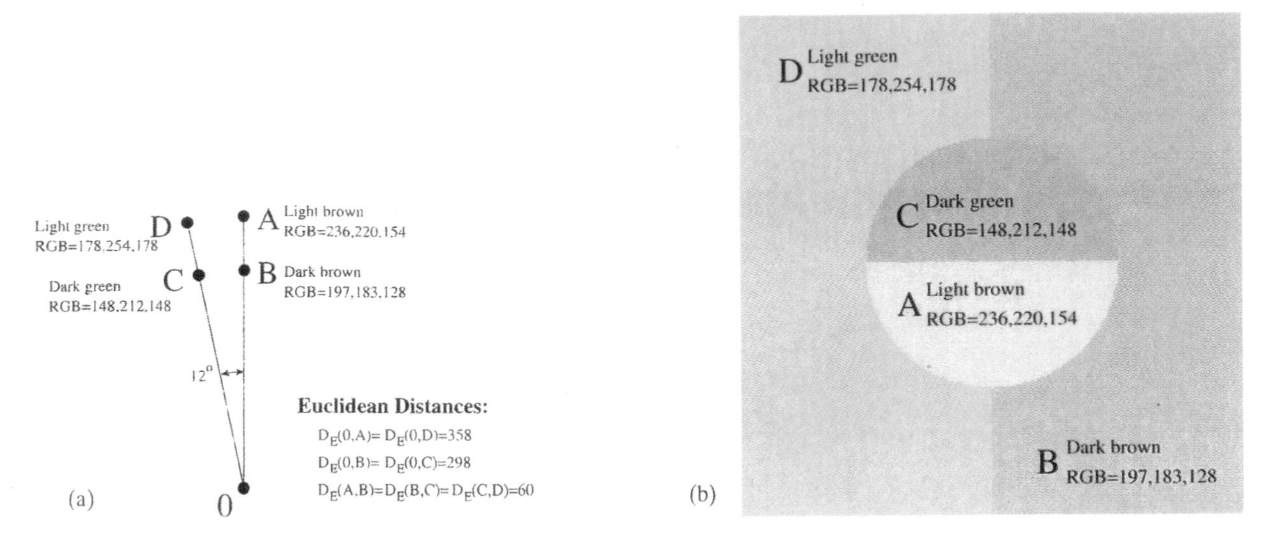

Figure 14.11 Four color examples illustrating differences between Euclidean distance and vector angle similarity measures: (a) Vector representation (b) Colors of the four examples.

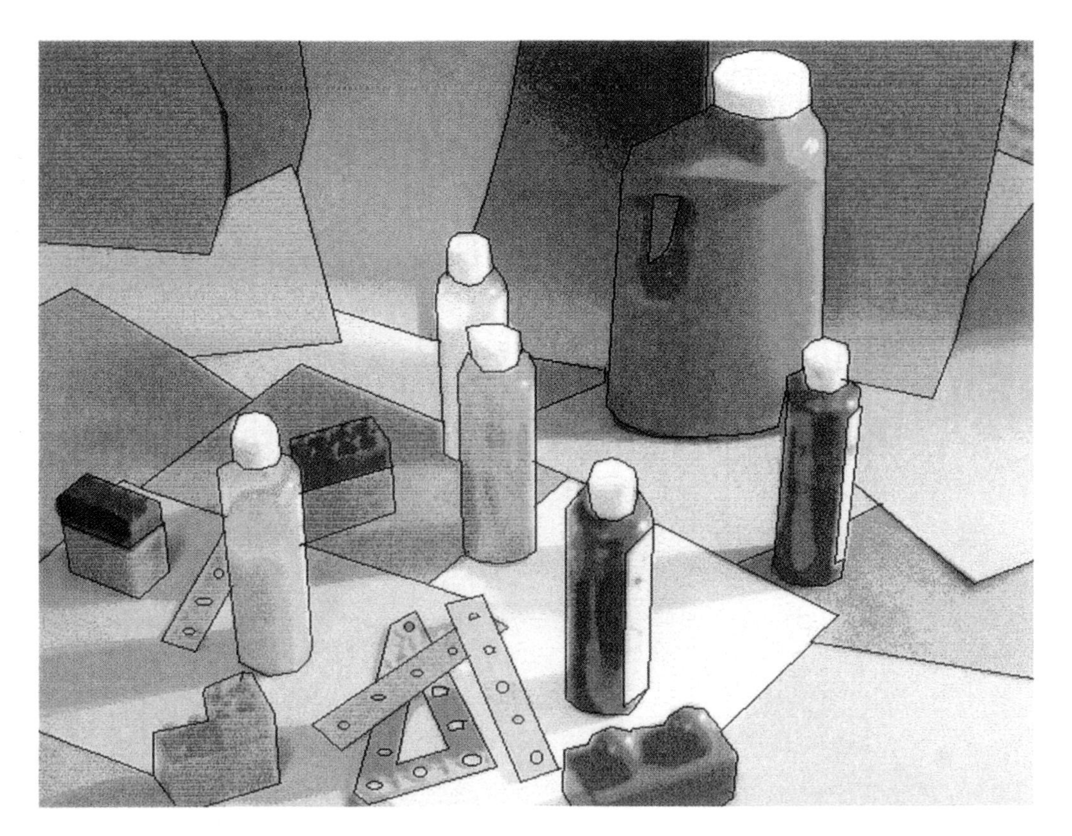

Figure 14.13 Test image for color image segmentation with overlay of manual segmentation.

output against a threshold. The threshold was set for a false alarm rate of 10^{-3}; that is, the probability that a target is present in the received signal when actually it is not was prescribed not to exceed 10^{-3}. This false alarm rate is considered typical for the operation of a surveillance radar. The color black in Fig. 7 signifies the presence of the growler (hypothesis H_1), and white signifies its absence (hypothesis H_0). With the radar operating in a dwelling mode, the growler should ideally be visible to the radar all of the time, that is, we should ideally see a continuous black strip extending all along the time axis. With this ideal picture in mind, we see a remarkable improvement in the behavior of the NN receiver in that it fills in the periods of "silence" frequently seen in the detection performance of the conventional Doppler CFAR receiver. This so-called silence is obviously caused by the partial obstruction of the growler (target) by an ocean wave in front of it or the dipping of the growler in a trough. The detection performance displayed in Fig. 7 is indeed quite remarkable. It shows that the NN receiver is able to perform well, even in a situation when the radar returns from the growler are weak. In other words, a "barely visible target is made visible in signal processing terms." The other important observation is the occasional blanking of a signal from the growler (as seen, for example, in the middle of the plot); in such cases, there is no way any method would be able to detect the target since insofar as the radar is concerned, the target is just not there to be seen.

To describe the detection performance of the receiver of Fig. 4 in traditional quantitative terms, we used a test dataset consisting of a total of 32292 examples with each example consisting of 256 radar samples, which (as mentioned previously) were completely different from the data used to train the PCA networks and those used to train the MLP's. The results of the tests may be summarized as follows:

- For a probability of false alarm $P_{FA} > 0.03$, the NN receiver outperformed the Doppler CFAR receiver.

- For $P_{FA} = 10^{-3}$, the probability of detection, PD, for the two receivers was as follows:

$$P_D = \begin{cases} 0.91 & \text{for the NN receiver} \\ 0.71 & \text{for the Doppler CFAR receiver} \end{cases}$$

7.3 Robustness of the Detector

Table 2 tabulates some relevant radar and environmental parameters in the database that was used for the study. The database was made up of four training datasets and nine test datasets. The test datasets had not been previously seen by the receiver, either for self-organized training of the PCA networks or for supervised training of the MLP classifiers. Although the training datasets corresponded to more-or-less similar environmental conditions, the point to note from this table is that the significant wave height is approximately 1.5 m. However, the test datasets pertained to wave heights varying from 1.5-2.6 m. Since the growler protrudes only about 1 m above the water line, the differences in waveheights are significant. Table 2 thus clearly points to the *robustness* of the neural network-based receiver. The robust behavior of the modular detection strategy is attributed directly to the *adaptive* nature of the receiver, which results from the combined use of self-organized learning and supervised learning for its design.

8. Cost Functions for Supervised Training of the Pattern Classifiers

In the case study on radar detection described in Section 7, we used multilayer perceptrons for the pattern classifiers in the adaptive receiver of Fig. 4. The multilayer perceptrons were trained using the back-propagation algorithm. In its traditional form, this algorithm relies on the minimum mean-square error as the optimization criterion. However, the standard criterion for radar detection is the Neyman-Pearson criterion [51]. According to this latter criterion, the probability of detection is maximized subject to a prescribed upper bound imposed on the probability of false

Table 2

Radar parameters on transmission:

radio frequency:	9.39 GHz
pulse repetition frequency:	2 kHz
pulse duration:	200 ns

Analog-to-digital conversion at the receiver:

sampling rate:	30 MHz
wordlength:	8 bits

Data Sets		Environmental Variables			
Purpose	Number	Significant wave height (m)	Wind speed (kt)	Max. wave height (m)	Peak period, (s)
Training	1	1.59	15	2.6	11.11
	2	1.57	8	2.4	11.11
	3	1.47	18	2.34	11.11
	4	1.46	20	2.27	11.11
Testing	1	1.47	18	2.34	11.11
	2	1.47	21	2.5	11.35
	3	1.46	20	2.27	11.11
	4	2.38	23	3.24	7.69
	5	2.6	20	3.71	9.09
	6	2.23	8	3.47	10.00
	7	2.41	20	3.42	8.00
	8	2.67	18	3.85	7.69
	9	2.42	12	3.61	8.00

alarm [1-3]. Unfortunately, minimization of the mean-square error does not guarantee fulfillment of the Newman-Pearson criterion.

Recognizing that the basic idea of the Newman-Pearson criterion is to treat the two kinds of error (i.e., missed detection and false alarm) differently, Principe et al. [52] have proposed a *mixed-norm* formulation of the cost functions for "batch" learning as follows:

$$\mathcal{E}(t) = \begin{cases} \frac{1}{N_0} \sum_{n=1}^{N_0} |d - y(x(t), \mathbf{w})|^{p_0}, & x(t) \in C_0 \\ \frac{1}{N_1} \sum_{n=1}^{N_0} |d - y(x(t), \mathbf{w})|^{p_1} & x(t) \in C_1 \end{cases} \tag{32}$$

where d is the desired response, N_0 is the number of noise-only samples (i.e., hypothesis H_0 is true), and N_1 is the number of target-plus-noise samples (i.e., hypothesis H_1 is true); p_0 and p_1 are the norms for hypotheses H_0 and H_1, respectively; C_0 and C_1 are the corresponding classes of data; $y(x(t),\mathbf{w})$ is the receiver output in response to the received signal $x(t)$, parameterized by the weight vector $\mathbf{w}$ representing the two pattern classifiers and linear combiner. In order to mimic the Neyman-Pearson criterion with the mixed-norm cost function of Eq. (32), two requirements have to be satisfied:

1. Given that hypothesis H_1 is true, the largest deviation of the output $y(x(t),\mathbf{w})$ from the desired response d should be minimized so as to set the threshold at the receiver output to as high a level as possible.
2. Given that hypothesis H_0 is true, the influence of large errors should be deemphasized so that as few of the corresponding output samples exceed the threshold.

These two objectives can be achieved by using the L_∞ norm (i.e., $p_1 = \infty$) for the training examples for which hypothesis H_1 is true, and the L_1 norm (i.e., $p_0 = 1$), or even fractional norms for the training examples for which hypothesis H_0 is true. In Principe et al. [52], the necessary modifications to the back-propagation algorithm to work with the cost function of Eq. (32) are described.

By training the multilayer perceptrons in the manner described here, we may expect a further improvement in the performance of the adaptive receiver applied to the radar detection problem discussed in Section 7.

8. Summary and Discussion

In this paper we have described an adaptive receiver, based on learning, for the detection of a target signal buried in a nonstationary background of unknown statistics. In a fundamental sense the receiver relies on three functional blocks:

1. The transformation of a one-dimensional received signal into two-dimensional images (maps), whereby the role of *time* (an essential dimension of learning) is highlighted and the time evolution of the frequency content of the received signal therefore made clearly visible. That is, the received signal is preprocessed such that the target signal and noise components are separated to the best advantage in the "extracted feature space".
2. Subspace decomposition for the purpose of dimensionality reduction and therefore efficient learning.
3. Pattern classification to pave the way for reliable decision making.

The learning process is self-organized in performing the subspace decomposition and supervised in performing the pattern classification. The receiver may therefore be succinctly described as an *adaptive detection system based on learning.*

The approach taken here is radically different from the classical approach to receiver design. In particular, the need for mathematical modeling of the received signal is eliminated. Instead, successful design of the receiver rests on *the availability of a sufficient number of real-life examples representative of the nonstationary environment* in which the receiver operates. Part of this database is used to train the receiver, and the remaining part is used to test its performance. The training set is itself split into two parts, one part devoted to design the subspace decomposers and the other part devoted to design the pattern classifiers. Accordingly, the free parameters of the receiver are adjusted in a systematic fashion whereby the information contained in the examples about the environment is extracted and stored in the receiver as parameter values.

In summary, the main virtue of the adaptive two-channel receiver described in this paper is the ability to learn a complex input-output mapping of the environment through a training session. For this learning to be effective, we must have a set of examples representative of the environment: the more representative the examples are, the more robust will the behavior of the receiver be with respect to statistical variations of the environment.

Acknowledgments

The authors wish to thank anonymous reviewers for their many constructive comments, and also wish to thank Dr. Tarun Bhattacharya and Mr. Brian Currie for reading the final version of the paper with comments.

References

[1] Van Trees, H.L. **Detection, Estimation, and Modulation Theory**, Part 1, Wiley, 1968.

[2] McDonough, R.N., and A.D. Whalen, **Detection of Signals in Noise**, Second Edition, Academic Press.

[3] Poor, H.V, **An Introduction to Signal Detection and Estimation**, Springer, 1988.

[4] Glaser, E.M. *Signal detection by adaptive filters*, IRE Trans. Information Theory, Vol IT-7, pp. 87-98, July 1960.

[5] Tersranta, H., S. Urpo, S. Pohjolainen, and E-L. Leskinen, Measurements of Solar Oscillations at 37 HGz, Proc. Symp. Seismology of the Sun and Sun-like Stars, Tenerife, Spain, pp. 235-239, Sept. 1988, ESA SP-286, Dec. 1988.

[6] Lanzerotti, L.J., D.J. Thomson, and C.G. Maclennan, Wireless at high altitudes -- Environmental effects on space-based assets, Bell Labs Technical Journal, Vol. 1, pp. 5-9, 1997.

[7] Lippmann, R.P., and P. Bakman. *Adaptive neural net preprocessing for signal detection in non-Gaussian noise*, Advances in Neural Information Processing Systems, Vol. 1, 1989, pp. 124-132.

[8] Wilson, E., S. Umesh, and D.W. Tufts. *Multistage neural network structure for transient detection and feature extraction*, IEEE International Conference on Acoustics, Speech, and Signal Processing, 1993, pp. 489-492.

[9] Luo, F.-L., and R. Unbehauen. **Applied Neural Networks for Signal Processing**, Cambridge University Press, 1997.

[10] Sandberg, I.W., and L. Xu. *Uniform approximation of multidimensional myopic maps*, IEEE Trans. Circuits and Systems, vol. 44, 1997, pp. 477-485.

[11] Wan, E. *Temporal backpropagation for FIR neural networks*, IEEE International Joint Conference on Neural Networks, Vol. 1, 1990, pp. 575-580.

[12] Feldkamp, L.A., G.V. Puskorius, and P.C. Moore. *Adaptive behavior for fixed weight networks*, Information Sciences, Vol. 98, 1997, pp. 217-235.

[13] LeCun, Y., and Y. Bengio. *Convolution networks for images, speech, and time series*. In M.A. Arbib, editor, The Handbook of Brain Theory and Neural Networks, MIT Press, 1995.

[14] Abeysekera, S.S., and B. Boashash. *Methods of signal classification using the images produced by the Wigner-Ville distribution*, Pattern Recognition Letters, Vol. 12, 1991, pp. 717-729.

[15] Malhoft, D. *A neural network approach to the detection problem using joint-time frequency distribution*, IEEE International conference on Acoustics, Speech, and Signal Processing, 1990, pp. 2739-2742.

[16] Bhattacharya, T.K., and S. Haykin. *Neural network-based radar detection from an ocean environment*, IEEE Trans. Aerospace and Electronic Systems, Vol. 33, pp. 408-420, April 1997.

[17] Haykin, S., and T.K. Bhattacharya. *Modular learning strategy for signal detection in a nonstationary environment*, IEEE Trans. Signal Processing, Vol. 45, 1997, pp. 1619-1637.

[18] Loève, M. *Fonctions aleatoires du second ordre*, Rev. Sc. Paris, t. 84, pp. 195-206, 1946.

[19] Loève, M. **Probability Theory**, 1963, D. Van Nostrand.

[20] Cohen, L. **Time-Frequency Analysis**, 1995, Prentice-Hall.

[21] Vernon, Frank. L, III. **Analysis of Data Recorded on the ANZA Seismic Network**, *PhD Thesis*, **University California, San Diego**, 1989.

[22] Thomson, D.J. *Spectrum Estimation Techniques for Characterization and Development of WT4 Waveguide*, Bell System Tech. J., **56**. *Part I*, pp 1769-1815, *Part II*, pp 1983-2005, 1977.

[23] Hlawatsch, F. *Regularity and unitary of bilinear time-frequency signal representations*, IEEE Trans. Information Theory, Vol. 38, pp. 82-94, January 1992.

[24] Martin, W. T*ime-Frequency Analysis of Random Signals*, Proc. ICASSP pp. 1325-1328, 1982.

[25] Flandrin, P. *On the positivity of the Wigner-Ville Spectrum*, Signal Processing. **11**, pp 187-189, 1986.

[26] Thomson, D.J. *Spectrum Estimation and Harmonic Analysis*, Proc. IEEE **70**, pp 1055-96, 1982.

[27] Percival, D.B., and A. T. Walden. *S***pectral Analysis for Physical Applications; Multitaper and Conventional Univariate Techniques**, 1993, Cambridge Univ. Press.

[28] Tauxe, L.I. *Sedimentary records of relative paleointensity of the geomagnetic field; theory and practice*, Rev. Geophysics. **31**, pp 319-354, 1993.

[29] Thomson, D.J. *Time Series Analysis of Holocene Climate Data*, Phil. Trans. R. Soc. Lond. A. **330**, pp 601-616, 1990.

[30] Thomson, D.J. *Quadratic-Inverse spectrum estimates; applications to paleoclimatology*, Phil. Trans. R. Soc. Lond. A. **332**, pp 539-97.

[31] Thomson, D.J. *Nonstationary Fluctuations in stationary time-series,* Proc. SPIE. **2027**, pp 236-244, 1990.

[32] Thomson, D. J. [1994]. *An Overview of Multiple-Window and Quadratic-Inverse Spectrum Estimation Methods*, Proc. ICASSP. **6**, pp 185-194, 1994.

[33] Papoulis, A. *A New Algorithm in Spectral Analysis and Band-Limited Extrapolation*, IEEE Trans. on Circuits and Systems. **CAS-22**, pp 735-742.

[34] Gori, F. & Palma, C. *On the Eigenvalues of sinc^2 Kernel*, J. Phys. A: Math. Gen. **8**, pp 1709-19, 1975.

[35] Thomson, D. J. *Dependence of global temperatures on atmospheric* CO_2 *and solar irradiance*, Proc. Natl. Acad. Sci. USA. **94**, pp 8370-8377, 1977.

[36] Mellors, R. J., F. Vernon, and D. Thomson. ***Detection of dispersive signals using multitaper dual-frequency coherence***, Proceedings of the 18th Seismic Research Symposium on Monitoring a Comprehensive Test Ban Treaty, Annapolis, Maryland, pp. 745-753, 1996.

[37] Mellors, R.J., F.L. Vernon, and D.J. Thomson. *Detection of dispersive signals using multitaper dual-frequency coherence*, In Press, Geophysical J.Int., 1998.

[38] Schild, R., & D.J. Thomson. T*he Q0957+561 Time Delay, Quasar Structure, and Microlensing*, pp 73-84 of **Astronomical Time Series**, *D. Maoz et al. (Eds.)*, Kluwer Academic Press, 1997.

[39] Hurd, H. L. *Spectral coherence of nonstationary and transient stochastic processes*, Proc. Fourth IEEE ASSP Workshop on Spectrum Estimation and Modeling, Minneapolis, MN, pp 387-390, 1988.

[40] Gerr, N. L., and J. C. Allen. *The Generalized Spectrum and Spectral Coherence of a Harmonizable Time Series*, Digital Signal Processing. **4**, pp 222-238, 1994.

[41] Carter, G. C. (ed.). **Coherence and Time Delay Estimation**, IEEE Press, New York, 1993.

[42] Duda, R.O., and P.E. Hart. **Pattern Classification and Scene Analysis**, 1973. Wiley.

[43] Fukunaga, K. **Statistical Pattern Recognition**, Second Edition, 1990, Academic Press.

[44] Haykin, S. *Neural networks expand SP's horizons*, IEEE Signal Processing Magazine, Vol. 13, No. 2, pp. 24-29.

[45] Haykin, S. **Neural Networks: A Comprehensive Foundation**, 1994, Macmillan.

[46] Suga, N., *Cortical computational maps for auditory imaging*, Neural Networks, Vol. 3, pp. 3-21, 1990.

[47] Simmons, J.A., *Time-frequency transforms and images of targets in the sonary of bats*, NEC Research Institute, Princeton, NJ, 1991.

[48] Haykin, S., C. Krasnor, T. Nohara, B. Currie, and D. Hamburger. *A coherent dual-polarized radar for studying the ocean environment*, IEEE Trans. Geoscience and Remote Sensing, Vol. 29, 1991, pp. 1890191.

[49] Sanger, T.D. *Optimal unsupervised learning in a single-layer linear feedforward network*, Neural Networks, Vol. 1, 1989, pp. 459-473.

[50] Rumelhard, D.E., G.E. Hinton, and R.J. Williams. *Learning representations by back-propagating errors*, Nature, 1986, Vol. 323, pp. 533-536.

[51] Newman, J., and E.S. Pearson, *On the problem of the most efficient tests as statistical hypotheses*, Phil. Trans. Roy. Soc., London, Series A, 1933, Vol. 231, pp. 289-337.

[52] Principe, J.C., M. Kim, and J.W. Fisher III, *Target discrimination in synthetic aperture radar using artificial neural networks*, IEEE Transactions on Image Processing, 1998, Vol. 7, 1136-1149.

Chapter 14

Data Representation Using Mixtures of Principal Components

ROBERT D. DONY
SCHOOL OF ENGINEERING,
UNIVERSITY OF GUELPH,
GUELPH, ONTARIO, CANADA N1G 2W1

SIMON HAYKIN
ELECTRICAL AND COMPUTER ENGINEERING,
MCMASTER UNIVERSITY,
HAMILTON, ONTARIO, CANADA L8S 4K1

Abstract—This chapter presents a new approach to data representation, the mixture of principal components. In general, it can be considered as a portion of a spectrum of representations and has been successfully used in such diverse applications as image compression and nonlinear clustering. At one extreme of the spectrum is vector quantization (VQ). Data are represented by a set of zero-dimensional points, Voronoi centers, within the N-dimensional space. Only data corresponding to the exact values of the centers are represented exactly. As such, it is a nonlinear representation. Euclidean distance is used to measure similarity under this representation. At the other extreme lies principal component analysis (PCA). Here, data are represented by a linear combination of a set of N basis vectors. The representation is complete and continuous because all possible data vectors may be represented exactly. Between these two extremes lies the mixture of principal components (MPC). Data are represented by a set of M-dimensional subspaces where $0 < M < N$. The subspace projection length is used as the similarity measure which reduces to the vector angle for the one-dimensional case. In this approach a data vector within a class (subspace) is represented as a continuous, linear combination of the M basis vectors of the subspace in a manner analogous to the PCA representation. But, because of the partitioning of the data into a discrete number of regions or classes, the MPC effects a nonlinear mapping of the data as does VQ. Applications presented include gray-scale image feature extraction and color segmentation.

1. Introduction

An important feature of intelligent systems is their ability to adapt to the changing chacteristics of data they are processing. One approach is to allow a continuum of adaptation, i.e., allow the parameters of the system to adjust to the statistics of the input data in a time-varying fashion [19]. An alternative approach is to assume a "hard" mixture model of the data. In a general mixture model, an observation may contain a number of sources and is formed, say, as a weighted sum of these sources. Each source has associated with it a set of characteristics or statistics which vary from source to source. For the approach considered in this chapter, data are assumed to originate from only *one* of a finite number of distinct sources. This may be viewed as a special case of the generalized mixture model where only one of the weights is nonzero. When such data are processed, an intelligent system first identifies the source of the present data and then uses a model of processing which is appropriate for that particular source.

To create such a system, *a priori* knowledge about the number of sources and their characteristics can be exploited. With complete knowledge of the statistics of each source, optimal processing methods could be derived for each source. A classifier that uses the difference in the distributions of each source can be constructed to classify the input signals, for example, a Bayesian classifier [12, 43]. However, with limited knowledge of the sources, a learning approach may be employed. For example, if labeled data are available, a multilayer perceptron neural network could be trained using the backpropagation algorithm to design a classifier [18]. Such a system would "learn" the characteristics of each source for classification.

Unfortunately, it may be the case that no knowledge

exists about the sources. As such, labeled data would not be available. Inferring the characteristics of each of the distinct sources without explicitly labeled data reduces to the clustering problem [12, 14]. Here, a measure of similarity between different data samples is used to form clusters of similar samples. K distinct clusters would correspond to K sources in the mixture model. Data are represented in a space appropriate to the problem at hand and a prototype for each class is computed using a clustering algorithm. A data sample to be classified is compared to the class prototypes using a similarity measure and is assigned to the class which best represents the sample.

This chapter develops a new approach to data representation for a mixture model, the mixture of principal components (MPC), and compares it to other representations, e.g., a Euclidean distance–based approach. The chapter is organized as follows. Section 2 defines a spectrum of representations in which the characteristics of a number of representations are compared. Following in Section 3, the method of subspace pattern recognition is discussed and compared to the classical Euclidean distance approach. Section 4 presents the training algorithms for the MPC representation. Two applications of the new representation are presented in Sections 5 and 6: gray-scale feature extraction and color image analysis, respectively. Finally, the chapter concludes with Section 7.

Note: Color versions of Figs. 6, 8, 10, 11, and 13 in this chapter are presented at the end of the book.

2. A Spectrum of Representations

Two of the main representations for data, for example as used in data coding, are principal components analysis (PCA) and vector quantization (VQ) [33]. Both these representations, in effect, are the two limits of a potential spectrum of representations. Vector quantization is a zero-dimensional nonlinear representation of an N-dimensional data set while principal components is a full N-dimensional, linear representation of the same. A new approach has been proposed which combines advantages of these two limiting cases [11].

2.1. Principal Components

The purpose of principal component analysis is to find a transformation that may reduce the dimensionality of the data space while maintaining the maximum variance [22]. The direction of the first basis vector that defines the first, or principal, component is in the direction of maximum variance. The remaining basis vectors are mutually orthogonal and, in order, maximize the remaining variances subject to the orthogonality condition.

The principal components representation can use up to the full N principal components to represent N-dimensional data. This is the basis for the optimal Karhunen-Loève transform (KLT) [6]. The representation is complete, i.e., if all N components are used, the data are represented exactly. Therefore, the representation is an N-dimensional *volume* and so is continuous since all possible input vectors may be represented. An N-dimensional data vector $\mathbf{x}$ is represented by principal components or coefficients, denoted by the vector $\mathbf{y}$, that is defined by

$$\mathbf{y} = [\mathbf{W}]\mathbf{x} \qquad [1]$$

where $[\mathbf{W}]$ is an orthonormal $N \times N$ matrix whose ith row is the ith principal component vector. On reconstruction,

$$\mathbf{x} = [\mathbf{W}]^T\mathbf{y} \qquad [2]$$

These relationships hold true for other orthonormal basis functions such as the discrete cosine transform (DCT) [36]. Because PCA uses all the components, it is a very powerful technique due to its complete and continuous representation. The representation is also a linear mapping of the data. This characteristic affords a high degree of mathematical tractability in the analysis and design of the approach. However, the limitations of linear techniques for image processing and signal processing in general are well known [20]. Further, the human visual system, which can outperform any artificial vision system in all but the most trivial tasks, gains much of its power through the many nonlinear stages of processing and representation.

2.2. Vector Quantization

At the other extreme, VQ is a purely discrete representation of the data. Unlike PCA, which uses up to the full N principal components, VQ uses only one of a number of Voronoi centers (codewords) for each input vector. For a set of K codewords, $\{\mathbf{w}_1, \mathbf{w}_2, \ldots, \mathbf{w}_K\}$, an input vector $\mathbf{x}$ is represented by the kth codeword such that the reconstructed vector, $\hat{\mathbf{x}}$, is

$$\hat{\mathbf{x}} = \mathbf{w}_k, \quad \text{if} \quad \|\mathbf{x} - \mathbf{w}_k\| = \min_{i=1}^{K} \|\mathbf{x} - \mathbf{w}_i\| \qquad [3]$$

Fundamental to this mapping is the use of Euclidean distance to choose the prototypical $\mathbf{w}_k$. Each of the centers or codewords represents a single *point* in the N-dimensional input space. It is therefore a zero-dimensional representation of the data since the representation of any input vector is a single point in the original data space. Contrast this with the basis vector representation of linear transforms. As such, the representation under vector quantization is a highly nonlinear function of the input vector.

2.3. A Mixture of Principal Components

Between these two extremes lies the mixture of principal components (MPC). Like VQ, this approach partitions the data set into a number of nonoverlapping regions. However, each region is represented not by a zero-dimensional point but by an M-dimensional linear subspace. Therefore, the data are represented by one-dimensional *lines,* two-dimensional *planes,* etc., depending on the dimension, M, of the subspace. As with PCA, each subspace is a continuous representation with only M orthogonal components where $0 < M < N$. Each input vector is assigned to the most appropriate partition and then represented by the M basis vectors of that region. This representation can be expressed as

$$\mathbf{y} = [\mathbf{W}]_k \mathbf{x}, \quad \text{if} \quad \mathbf{x} \in C_k \qquad [4]$$

where $[\mathbf{W}]_i$ is an $M \times N$ matrix whose rows are the M principal components of the partition C_i. The reconstructed vector, $\hat{\mathbf{x}}$, is calculated as

$$\hat{\mathbf{x}} = [\mathbf{W}]_k^T \mathbf{y}, \quad \text{if} \quad \mathbf{x} \in C_k \qquad [5]$$

The MPC approach combines the features of both PCA and VQ representations. Within a class, an input vector is represented as a continuous, linear combination of the M basis vectors of the subspace in a manner analogous to the PCA representation. But, because of the partitioning of the data into a discrete number of regions or classes, the MPC effects a nonlinear mapping of the data in a manner analogous to VQ.

Figure 1 illustrates the relation between the three representations for a two-dimensional example. The PCA approach forms a complete, continuous representation using a linear combination of the two basis vectors. With VQ, the input space is partitioned, in this example into 10 regions. Each region is represented by a Voronoi center. Under MPC, the space is also partitioned, in this case into four regions. Within each region, the data are represented by a single basis vector. For higher dimensional input spaces, the number of basis vectors may be two or more, forming planes, hyperplanes, or other higher dimensional subspaces within the input data space.

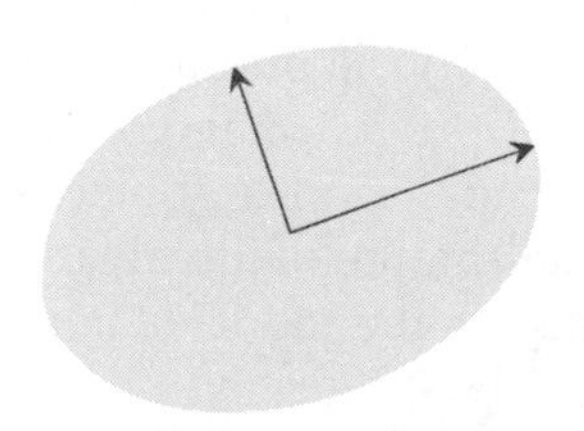

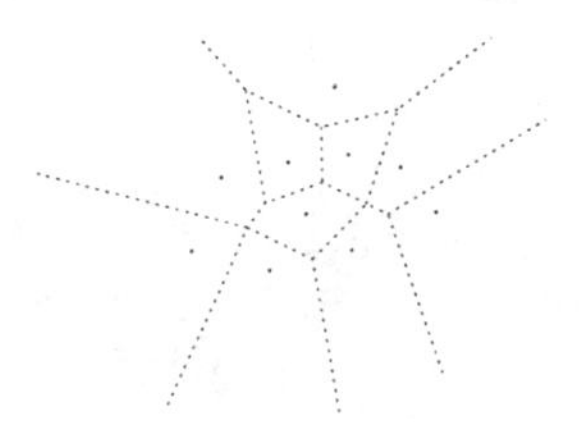

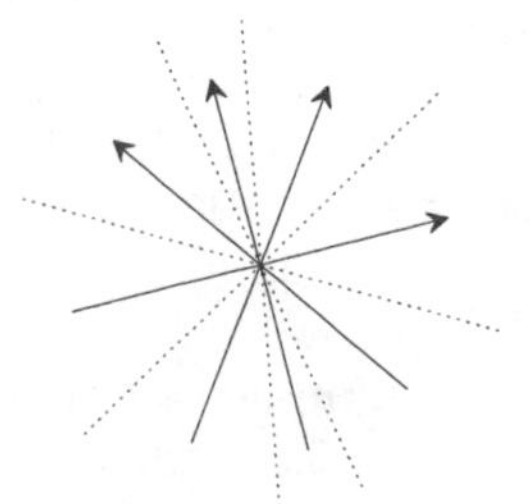

Fig. 1. A spectrum of representations in two dimensions.

3. Subspace Pattern Recognition

3.1. Similarity Measure

There is then the problem of how to represent the classes and assign data vectors appropriately. In many classical pattern recognition techniques, classes are represented by prototypical feature vectors and class membership is determined by some transformed Euclidean distance between an input vector and the prototypes [12, 14]. For example, with the K-means and LBG vector quantization algorithms, the classes are represented by their means and the vector-to-class distance is the Euclidean distance between the class mean and an input vector. The class boundaries form closed regions within the input space. For the VQ representation, this is the case.

However, such a measure of similarity is inappropriate for the MPC representation. Instead, a measure of the similarity between an input vector and a M-dimensional subspace is required. This is the basis of subspace pattern recognition [35]. In subspace pattern recognition, classes are represented as linear subspaces within the original data space and the basis vectors that define the subspace implicitly define the features of the data set. The classification of data is based on the efficiency by which the subspace can represent the data as measured by the norm of the projected data. Each input vector is classified according to the subspace classifier [10]

$$\mathbf{x} \in C_k \quad \text{if} \quad \|[\mathbf{W}]_k^T[\mathbf{W}]_k\mathbf{x}\|^2 = \max_{i=1}^{K} \|[\mathbf{W}]_i^T[\mathbf{W}]_i\mathbf{x}\|^2 \qquad [6]$$

or equivalently

$$\mathbf{x} \in C_k \quad \text{if} \quad \|[\mathbf{W}]_k\mathbf{x}\|^2 = \max_{i=1}^{K} \|[\mathbf{W}]_i\mathbf{x}\|^2 \qquad [7]$$

As pointed out in [10], maximizing the norm of the projection is equivalent to minimizing the mean squared error as illustrated in Figure 2. The projection matrix, $[\mathbf{P}]$ is simply $[\mathbf{P}] = [\mathbf{W}]^T[\mathbf{W}]$. The residual $\tilde{\mathbf{x}}$ is minimized when the angle between the input vector $\mathbf{x}$ and the plane of the subspace defined by $[\mathbf{P}]$ is minimized.

For a one-dimensional subspace, the subspace is defined by a single vector. In this case, the subspace classifier reduces to simply the vector angle between the input and the class basis vector. Notice that although VQ represents a class by a single vector as well, the measure of similarity is fundamentally different. For VQ it is the Euclidean

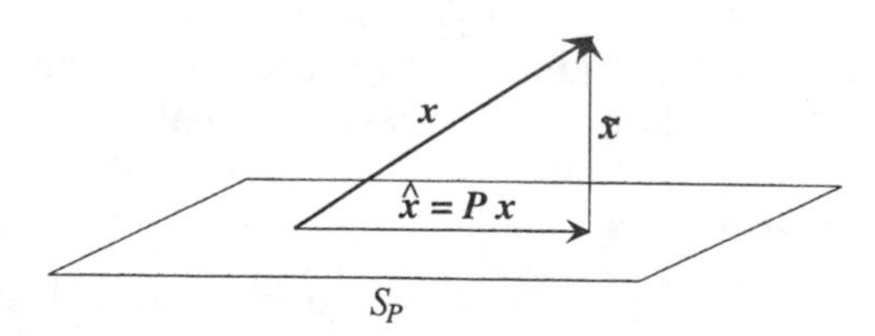

Fig. 2. Projection of x onto subspace S_P defined by $[\mathbf{P}]$.

distance whereas for the one-dimensional MPC, it is the vector angle.

3.2. Class Prototype

The optimal prototypical vector for VQ that minimizes the squared error under Euclidean distance is the class mean as the generalization of the Lloyd-Max quantizer in N dimensions [17]. For the MPC representation, the optimal representation for each cluster or class is the subspace spanned by the M principal components, or equivalently, the M eigenvectors corresponding to the M largest eigenvalues of the cluster's covariance matrix. Given that the matrix $[\mathbf{W}]$ is orthonormal, i.e., $[\mathbf{W}]^T[\mathbf{W}] = [\mathbf{I}]$, the error can be calculated as:

$$\begin{aligned} E[\|\hat{\mathbf{x}} - \mathbf{x}\|^2] &= E[(\hat{\mathbf{x}} - \mathbf{x})^T(\hat{\mathbf{x}} - \mathbf{x})] \\ &= E[([\mathbf{W}]\hat{\mathbf{y}} - [\mathbf{W}]\mathbf{y})^T([\mathbf{W}]\hat{\mathbf{y}} - [\mathbf{W}]\mathbf{y})] \\ &= E[(\hat{\mathbf{y}} - \mathbf{y})^T[\mathbf{W}]^T[\mathbf{W}](\hat{\mathbf{y}} - \mathbf{y})] \\ &= E[(\hat{\mathbf{y}} - \mathbf{y})^T(\hat{\mathbf{y}} - \mathbf{y})] \\ &= E[\|\hat{\mathbf{y}} - \mathbf{y}\|^2] \end{aligned} \quad [8]$$

which is the error of the coefficients. Due to the removal of the $N - M$ lowest order coefficients, i.e., $y_i = 0$ for $i = M + 1, \ldots, N$, the error is

$$\begin{aligned} E[\|\hat{\mathbf{y}} - \mathbf{y}\|^2] &= E\left[\frac{1}{N}\sum_{i=1}^{N}(y_i - \hat{y}_i)^2\right] \\ &= \frac{1}{N}E\left[\sum_{i=1}^{M}(y_i - y_i)^2 + \sum_{i=M+1}^{N}(y_i - 0)^2\right] \\ &= \frac{1}{N}E\left[\sum_{i=M+1}^{N} y_i^2\right] \\ &= \frac{1}{N}\sum_{i=M+1}^{N}\sigma_i^2 \\ &= \frac{1}{N}\sum_{i=M+1}^{N}\lambda_i \end{aligned} \quad [9]$$

Since the matrix $[\mathbf{W}]$ whose rows are the eigenvectors of the cluster's covariance matrix diagonalizes the covariance matrix, the variances of the coefficients, σ_i^2, are the eigenvalues of the covariance matrix, λ_i. To minimize the expected squared error, the M coefficients corresponding to the M largest eigenvalues should be kept. Therefore, the M basis vectors of $[\mathbf{W}]$ are the principal components.

Again note the difference between the MPC and VQ representations. The prototype of the latter is the class mean while the prototype for the one-dimensional subspace case of the latter is the class principal component.

3.3. Norm Invariance

The MPC representation is also independent of the norm of the input vector. Under Euclidean distance, two input vectors that differ only by a scalar magnitude may be represented by different classes, whereas under the subspace measure, the representation is invariant of the vector norm. The issue of invariance is not new, e.g., [16], but it is important that the appropriate measure be employed if invariance is desired. One could argue that simply normalizing the input and taking the centroids under the normalized space as the prototypes results in an invariant representation. In effect the space is reduced to the planar surface of an N-dimensional sphere. Under some specific conditions it can be equivalent to the MPC representation for the special case of the subspace dimension being $M = 1$. However, this approach completely ignores the characteristics of the distribution due to the extra dimension not represented by the $(N - 1)$-dimensional surface of the sphere. Further, it does not generalize to incorporate higher dimensional representations where the dimension of the subspace is $M > 1$.

A simple example can be constructed to illustrate this point as shown in Figure 3. The vectors marked with lines and +'s are constructed such that their angles are taken from a uniform distribution $[0, 2\pi)$ while the distribution of the radii varies with the angle. The points around the unit circle are the vectors normalized to $\|\mathbf{x}\| = 1$. The distribution of the normalized points around the unit circle is relatively uniform with some random variation from the ideal distribution. However, the raw data show a distinct two-dimensional structure with two classes being quite evident: one in the horizontal direction and the other in the vertical direction. This structure is not evident in the normalized data. Clearly, then, under the normalization the two-dimensional structure of the data set is lost.

As for accounting for the mean as an offset from the origin in the specification of the subspace, this may or may not be appropriate depending on the application. For example, images are a record of the luminance values detected by a sensor which are formed by the product of the illumination falling on a scene and the reflectance of the objects in the scene [41]. The luminance of an object in an image will vary linearly with the amount of illumination falling upon it. The subspaces which fit this model of image formation must therefore include the origin.

4. Training

The problem, now, is to calculate the appropriate set of basis vectors or weights. Without knowing *a priori* the required classes, their defining projectors $[\mathbf{P}]_i$, and their corresponding transformation bases, a learning algorithm is required to extract the appropriate parameters from the dataset.

For VQ, a standard approach to the calculation of the codebook is by way of the Linde, Buzo, and Gray (LBG) algorithm or, in effect, the K-means algorithm. Initially, K codebook entries are set to random values. On each iteration, each block in the input space is classified based on its nearest codeword. Each codeword is then replaced

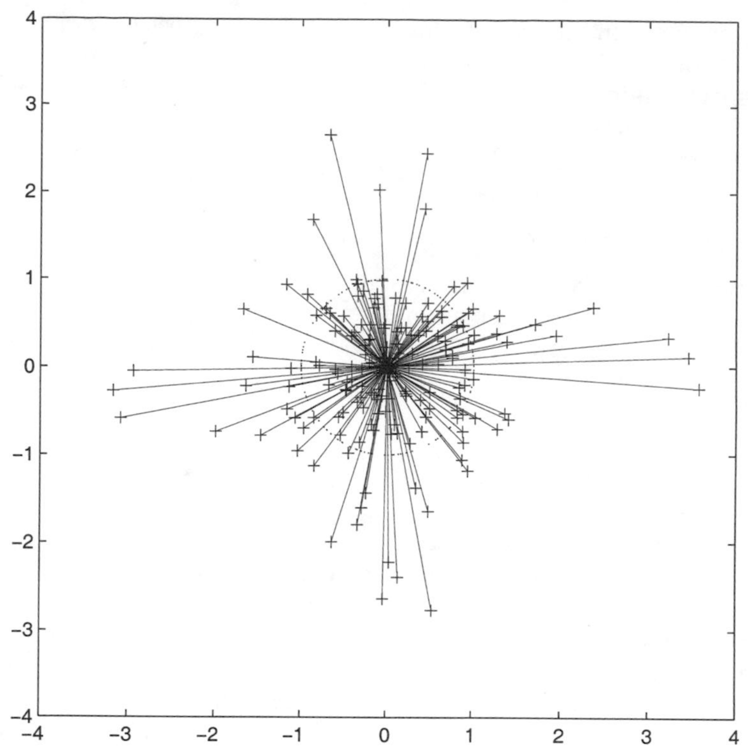

Fig. 3. Effect of normalizing data. Vectors marked + are original data, dots are normalized to $\|\mathbf{x}\| = 1$.

by the mean of its resulting class. The iterations continue until a minimum acceptable error is achieved. This algorithm minimizes the mean square error over the training set. However, the minimum at convergence is not guaranteed to be the global minimum. Variations exist which overcome local minima and convergence speed problems [17].

The LBG algorithm is designed for a Euclidean distance–based representation of the data as it determines the means for each class as the class prototypes. For the MPC representation, the iterative form of the LBG algorithm can be adapted to calculate the K sets of principal components as follows [10]:

1. Initialize K transformation matrices $\{[\mathbf{W}]_1, [\mathbf{W}]_2, \ldots, [\mathbf{W}]_K\}$.
2. For each training input vector $\mathbf{x}$:
   (a) classify the vector based on the subspace classifier

$$\mathbf{x} \in C_i \quad \text{if} \quad \|[\mathbf{P}]_i\mathbf{x}\| = \max_{j=1}^{K} \|[\mathbf{P}]_j\mathbf{x}\| \qquad [10]$$

   where $[\mathbf{P}]_i = [\mathbf{W}]_i^T[\mathbf{W}]_i$, and
   (b) update transform matrix $[\mathbf{W}]_i$ according to:

$$[\mathbf{W}]_i = [\mathbf{W}]_i + \alpha[\mathbf{Z}](\mathbf{x}, [\mathbf{W}]_i) \qquad [11]$$

   where α is a learning parameter, and $[\mathbf{Z}](\mathbf{x}, [\mathbf{W}]_i)$ is a learning rule such that Eq. 11 converges to the M principal components of $\{\mathbf{x}|\mathbf{x} \in C_i\}$.
3. Repeat for each training vector until the transformations converge.

In the first step, some care must be taken in the choice of the initial set of transformation matrices. They should be representative of the distribution space of the training data. If some of the $[\mathbf{W}]_i$'s were to be initialized to values corresponding to regions outside the distribution space, then they would never be used. Hence, the resulting partition would be clearly suboptimal. There are a number of methods to reduce the possibility of this occurring as described here:

Arbitrarily partition the training set into K classes and estimate the corresponding transformations using either iterative learning rules or batch eigendecomposition.

Use a single fixed-basis transformation such as the DCT and add a small amount of random variation to each class to produce a set of unique transformations.

Use an estimate of the global principal components of the data with a small amount of random variation added to each class.

Algorithms based on this outline will produce K transformation matrices $\{[\mathbf{W}]_1, [\mathbf{W}]_2, \ldots, [\mathbf{W}]_K\}$. Given the appropriate learning rule $[\mathbf{Z}](\mathbf{x}, [\mathbf{W}]_i)$ in Eq. 11, each matrix will converge to the principal components for that particular class of data. Since the principal components minimize the mean squared error, each $[\mathbf{W}]_i$ is optimal for its class. The classification rule in Eq. 10 is equivalent to finding the transformation which results in the minimum squared error for the particular vector. The combination of these two rules, therefore, produces the optimal set of linear transformations for the resulting partition. Conversely, for the resulting set of linear transformations, the partitioning of the data is optimal with respect to minimizing the MSE.

There exist a number of learning rules which compute the M principal components of a data set, $[\mathbf{Z}](\mathbf{x}, [\mathbf{W}]_i)$. For example, Oja's linear Hebbian rule [34]

$$\mathbf{w}(t+1) = \mathbf{w}(t) + \alpha[y(t)\mathbf{x}(t) - y^2(t)\mathbf{w}] \qquad [12]$$

which has been shown to converge to the largest principal component [18] could be used with a recursive calculation of the M principal components. Sanger's generalized Hebbian algorithm (GHA) [38] extends Oja's model to compute the leading M principal components as

$$[\mathbf{W}](t+1) = [\mathbf{W}](t) + \alpha(t)(\mathbf{y}(t)\mathbf{x}^T(t) - \mathbf{LT}[\mathbf{y}(t)\mathbf{y}^T(t)][\mathbf{W}](t)) \qquad [13]$$

where $\mathrm{LT}[\cdot]$ is the lower triangular operator, i.e., it sets all elements above the diagonal to zero. Similarly, other algorithms exist with various convergence properties [1–5, 25, 38–40, 47, 48] with the resulting set of transformations being equivalent. In fact, if the algorithm were implemented in a batch mode, the explicit calculation of the eigenvectors of the class covariance matrices would also produce the same transformation bases.

4.1. Topological Organization

In some applications, it may be advantageous to have some similarity between "neighboring" classes. Kohonen [23] introduced the concept of classes ordered in a "topological map" of features. In such a map, each class has associated with it a neighborhood of similar classes. For example, in a one-dimensional (linear) topology class C_i would be adjacent to classes C_{i-1} and C_{i+1}. Numerous examples exist in the neurosciences where sensor features are ordered in the brain such that locations for similar features are located spatially close to one another [13, 23, 37, 45].

For an artificial network to have such a similarity, the training algorithm must account for the neighborhood of the features being learned. In many standard clustering algorithms such as K-means, each input vector $\mathbf{x}$ is classified and only the "winning" class is modified during each iteration. In Kohonen's self-organizing map (SOM), the vector $\mathbf{x}$ is used to update not only the winning class but also its neighboring classes. Initially, the neighborhood may be quite large during training, e.g., half the number of classes or more. As the training progresses, the size of the neighborhood shrinks until, eventually, it includes only the one class.

Kohonen's algorithm has been shown to extract the mean vectors as the prototypes for each class. So, for VQ the LBG algorithm for codebook design and the SOM algorithm are closely related [28]. In fact, the LBG algorithm is the *batch* equivalent of the SOFM algorithm for a neighborhood size of one, $N(C_i) = \{C_i\}$. A number of researchers [29–32] have successfully used the SOM algorithm to generate VQ codebooks and have demonstrated a number of advantages of using the SOM algorithm over the classical LBG algorithm [9].

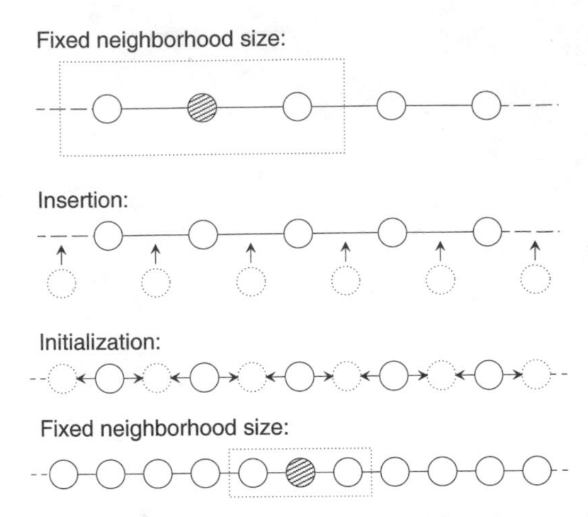

Fig. 4. Method of network growing by insertion of new nodes.

To extract the MPC representation on a topologically ordered map, the SOM approach can be modified from a Euclidean-based algorithm to a subspace-based algorithm by changing the similarity measure to that of Eq. 10 and the update rule to the class of Eq. 11 (e.g., Eq. 13) [8, 24].

Alternatively, instead of starting with a large update neighborhood that shrinks during training, the same topological ordering of features can be achieved by growing the network while fixing the neighborhood size [18, 27, 44]. This results in significant computational savings. Initially the network consists of a small number of classes. Once the network has converged for a given stage, the number of classes is increased by inserting new modules between existing ones as illustrated in Figure 4. The new weights are initialized to the mean of the neighboring weights and the new network is retrained. Of course the new elements can be inserted only in a manner that preserves the network topology. For example, a linear network can have a single insertion at any connection between two weights whereas a two-dimensional network can have only insertions of complete rows or columns. Further, it may be unclear where to insert new weights. A simple approach, then, is simply to double the network to achieve growth.

Again, the network-growth approach to training a topologically ordered network can be simply extended to the MPC representation [11].

5. Gray-Scale Feature Extraction

One application which is particularly well suited to the MPC representation is the extraction of features in a gray-scale image. Different features or regions in an image may have significantly different statistical characteristics. For example the correlation in either the x or y directions is quite different for horizontal and vertical edges. For an image compression scheme which adapts to different regions in an image an MPC network can classify the input

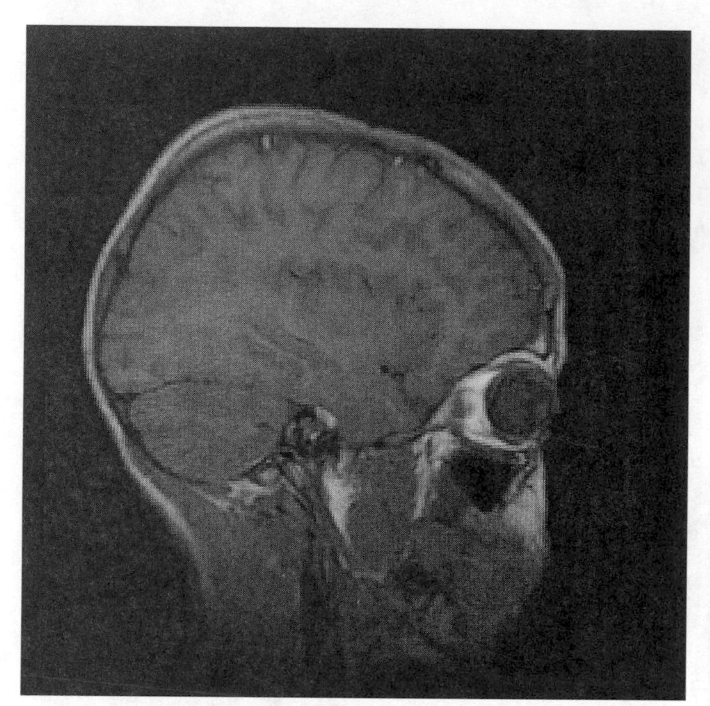

Fig. 5. MR image for training.

based on the features extracted and process the features with a transformation that is optimal with respect to those features [10]. The MPC network can extract such features without *a priori* knowledge of their importance or distribution.

5.1. Training Data

Figure 5 shows the magnetic resonance image (MRI) used for training. The image consists of 256 × 256 pixels with the dynamic range of 8 bits or 256 gray levels. During training, the two networks were presented randomly chosen blocks of 8 × 8 pixels from the training image.

Two sets of networks were trained using the preceding training algorithm. The first set had two components per class ($M = 2$) and the second had a single component ($M = 1$). A simple one-dimensional topology was used in this investigation. At the ends of the topology, wrap-around was used so that classes C_0 and C_{K-1} were considered adjacent. The initial network size for both configurations was chosen to be four classes. The network growing method was used until networks of $K = 32$ classes were trained. The training was terminated at 32 classes because the improvement in the resulting segmentation maps from 16 classes to 32 classes was negligible.

5.2. Network Basis Vectors

Figure 6a shows the basis blocks for the one-component network. Similarly, Figure 6b shows the second component basis blocks for the two-component network. The class number progresses left to right, top to bottom with the top left class being arbitrarily chosen as class 1.

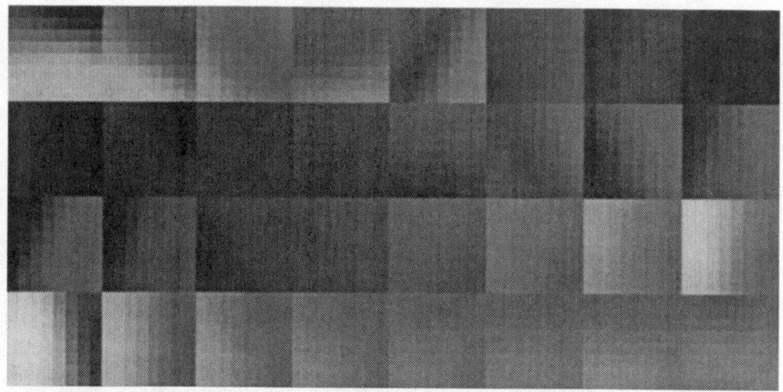

(a) 1 Component

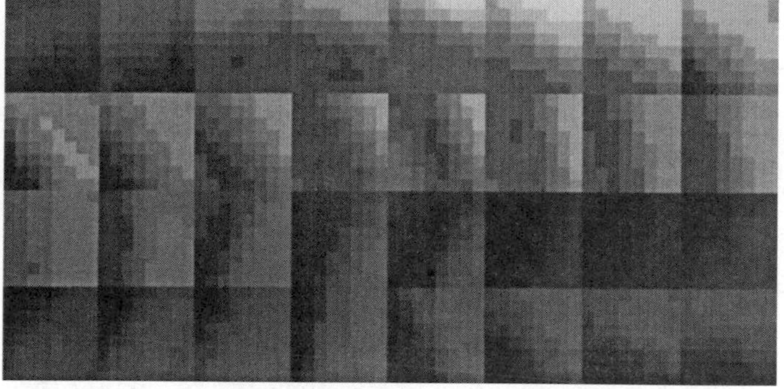

(b) 2 Component

Fig. 6. Map of component basis blocks for 32-class MPC networks. The class number progresses left to right, top to bottom. (See insert for color figures)

For both networks, the progression of features across adjacent classes is quite regular and the orientations of the features of adjacent classes are similar. This shows that the network growing approach used to train the networks produces a topologically ordered map of features. There are parallels between the organization of this feature map and the regular progression in the orientation sensitivity of the columns in the visual cortex [21, 42]. No *a priori* conditions were imposed as to what features were important in the image yet the networks show a preference for edges and lines, features which are of perceptual significance. Also, the distribution of features in the networks corresponds to the relative distribution of the same features in the training image.

5.3. Segmentation Results

The networks were used to segment the Lena test image of Figure 7. The segmentation was performed by taking the surrounding 8 × 8 block for each pixel in the image, classifying the block, and replacing the central pixel by the resulting class value. Since the class topology was circular, the class value of each pixel in the resulting segmentation map was represented by a color value. For the two-component example, the intensities of the segmentation color were weighted by the magnitude of the second coefficient for each block. Figures 8a shows the basis blocks for the

Fig. 7. Lena image for testing segmentation.

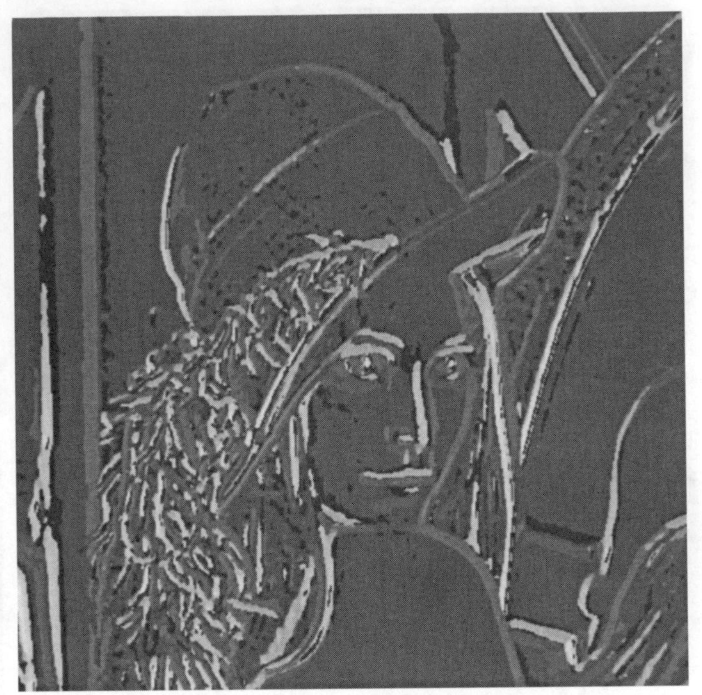

(a) 1 Component

(b) 2 Component

Fig. 8. Segmentation map of Lena image for 32-class networks. (See insert for color figures)

one-component network and 8b shows the results for the two-component network.

Despite the fact that the networks were trained on a very different image, they have identified perceptually important features in the image. The figures clearly show that the networks have captured the essential information required to represent an image since the details in the original image are readily identifiable in the segmentation maps. They also show the preference of the segmentor for edge and line features as well as flat regions in the one-component case. The edges around the hat, face, and mirror are dramatically shown. The continuity of the segmentation transitions shows the high degree of similarity between neighboring classes.

5.4. Illumination Variations

To test the performance of the segmentation under different illumination conditions, the two-component network was used to segment the images shown in Figure 9a and b. One image was captured with the scene being brightly illuminated; the other was captured after the lights were turned off. The respective segmentation maps are shown in Figure 10a and 10b. As these two segmentation maps show, a substantial change in illumination has very little effect on the assignment of the features in an image to the various classes.

6. Color Image Analysis

6.1. Color Representation

Fundamental to the human perception of color is the trichromatic model of color representation, which requires three variables to describe a point in color space completely. There exist a number of equivalent three-dimensional color spaces with varying characteristics [15]. One group of spaces uses Cartesian coordinates to represent points in the space. Examples include the three primary

(a) 1 Component

(b) 2 Component

Fig. 9. Test image with two illumination conditions.

(a) Brightly illuminated

(b) Dimly illuminated

Fig. 10. Segmentation map of test images with two illumination conditions for 32-class, 2-component network. (See insert for color figures)

illumination colors RGB, the complementary colors CMY, and the opponent color representation YIQ. An alternative set of spaces employ polar coordinates and include the HSI and HSV spaces [15, 26].

While the RGB model corresponds most closely with the physical sensors for colored light such as the cones in the human eye or the red, green, and blue filters in most color CCD sensors, the perception of color qualities more closely follows the HSI model. The intensity, I, represents the overall brightness, which is independent of color and is a linear value. Hue, H, represents the fundamental or dominant color and is measured as an angle on a color circle with the three primary colors spaced 120° apart. Finally, saturation, S, represents the purity of a color, where pastel shades have low saturation values and pure spectral colors are completely saturated. The latter two values specify the chromaticity of a color point.

6.2. Vector Angle Versus Euclidean Distance

Color spaces which use polar coordinate systems such as HSI incorporate the perceptually important hue component directly in the space as one of the coordinates. As such, such spaces may be more appropriate for color processing. However, a nonlinear transformation involving sinusoids is required to convert the raw RGB data to HSI [15]. For high-speed color imaging applications this extra computational overhead and delay may not be feasible. More important, these spaces do not generalize well to multispectral data with more than three components. Therefore, processing such as edge detection and segmentation based on the raw RGB or multispectral coordinates would have an advantage.

A simple approach would be to use the Euclidean distance as a measure of similarity between the representation of color pixels as three-dimensional data points. However, as Figure 11a illustrates, the Euclidean distance measure of similarity may not correspond to a perceptual notion of similarity. Colors A and B differ only in intensity, having the same chromatic values. The same is true of colors C and D. In this example, the Euclidean distance between B and C, differing in both intensity and chromaticity, is the same as that between A and B and between C and D. But, it is obvious that, at least perceptually, the color difference between B and C is more pronounced than the difference for the other two pairs.

The figure also illustrates the effect of illumination variations on the resulting color pixel value. The image formation process can be modeled as the product of the illumination falling on a scene, $\mathbf{E}$, and the reflectance, ρ, [41]

$$\mathbf{L}(x, y) = \mathbf{E}(x, y)\rho(x, y) \qquad [14]$$

where $\mathbf{L}$ is the luminance of the formed image. In RGB coordinates, two pixel values from areas of the same reflectivity characteristics but under different illumination intensities (assuming equal color balance in the illumination) are related as

$$\mathbf{v}_1 = \alpha\mathbf{v}_2 \qquad [15]$$

with $\alpha > 0$. Even though the two pixels come from areas with the same intrinsic color, the Euclidean distance in RGB space would not be zero due to the variation in illumination. Typically, the goal in image analysis is to determine characteristics about the underlying physical properties of the scene being imaged. These are inferred from the reflectivity of the scene. Therefore, it is the reflectivity which conveys the information about the scene, and variations in the illumination can be considered as noise.

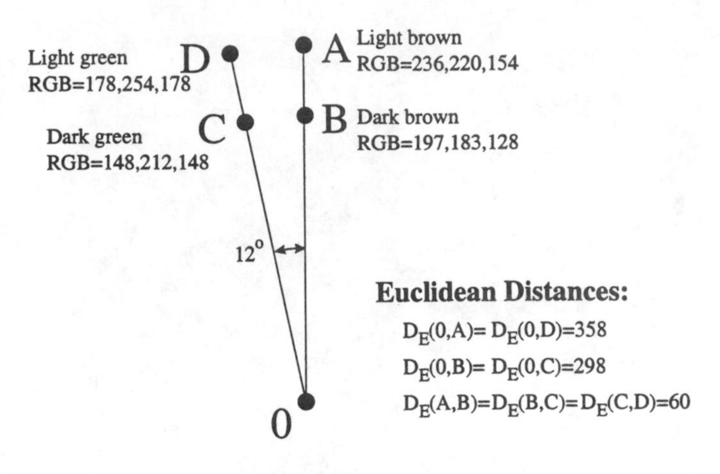

(a) Vector representation

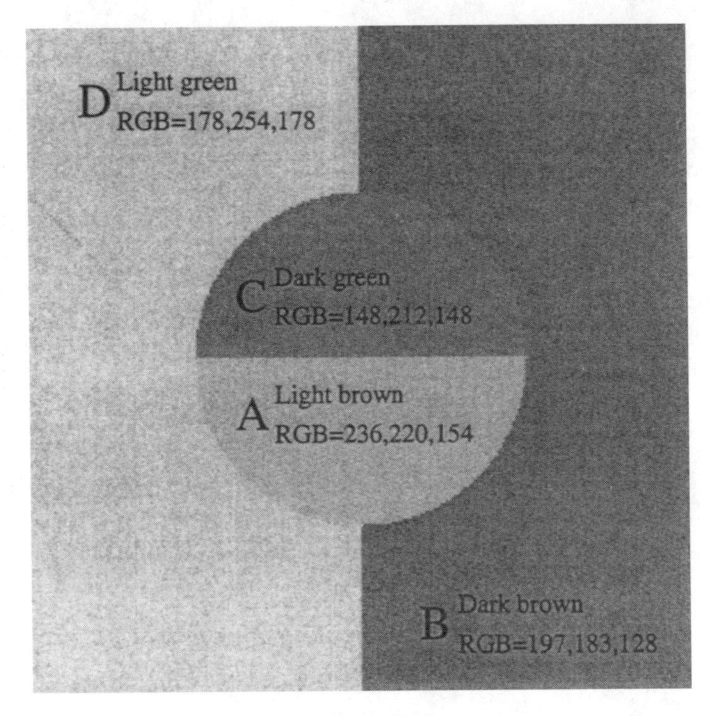

(b) Colors of the four examples

Fig. 11. Four color examples illustrating differences between Euclidean distance and vector angle similarity measures. (See insert for color figures)

As discussed previously, the subspace classifier reduces to a vector angle measure for the one-component case of the MPC representation. From Figure 11a, it is clear that the angle between colors A and B is zero whereas the angle between colors A and D is nonzero (12°). Examining Figure 11b, it appears that our perception of similarity more closely corresponds to the vector angle measure of similarity than to the Euclidean distance measure. Colors

A and B differ simply by their shades whereas colors A and D are fundamentally different.

6.3. Edge Detection

One application for such a similarity measure is edge detection on color images [7].

A very simple edge operator is the Roberts operator, which calculates the maximum absolute difference between diagonally adjacent pixels in a 2×2 block. This operator can be generalized to multidimensional pixel values as:

$$E_R = \max(\|\mathbf{v}(x, y) - \mathbf{v}(x + 1, y + 1)\|, \|\mathbf{v}(x, y + 1) - \mathbf{v}(x + 1, y)\|) \quad [16]$$

where $\mathbf{v}(x, y)$ is the vector containing the multiple values of the pixel at coordinate (x, y), and $\|\cdot\|$ is the L_2 vector norm. The pixels are represented by RGB values.

The Roberts operator can be extended to the proposed vector angle difference. The vector angle Roberts can be calculated using the maximum sine of the angles between diagonally adjacent pixels in a 2×2 block as

$$S_R^2 = 1 - \max\left(\frac{\mathbf{v}^T(x, y)\mathbf{v}(x + 1, y + 1)}{\|\mathbf{v}(x, y)\| \, \|\mathbf{v}(x + 1, y + 1)\|}, \frac{\mathbf{v}^T(x, y + 1)\mathbf{v}(x + 1, y)}{\|\mathbf{v}(x, y + 1)\| \, \|\mathbf{v}(x + 1, y)\|}\right)^2 \quad [17]$$

Figure 12a shows the results of operating on the test image with the Euclidean distance Roberts operator. Around the circle, all the edge values are 60. The vertical edge inside the circle between the light brown and dark green areas is 89. The horizontal edges between the dark brown and light green areas is 89. As expected, the Euclidean distance weighted the differences between the light and dark versions of the same color as equally important as the difference between the dark green and dark brown colors. Contrast these results with those for the vector angle operator as shown in Figure 12b. This operator produces a value of 0.20 at the edges of the regions that differ in hue and saturation. At the boundary of the regions that differ only in intensity, the operator produces a zero edge value.

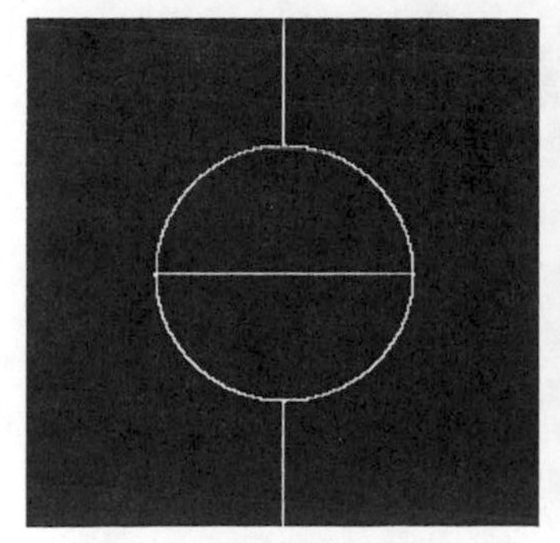

(a) Euclidean distance

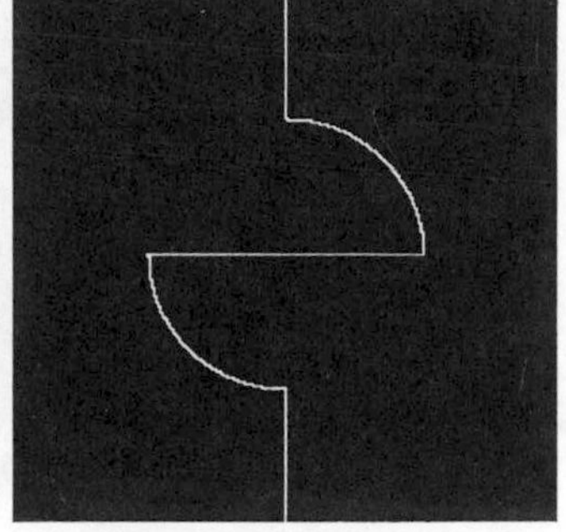

(b) Vector angle

Fig. 12. Results for Roberts edge detectors on test color image.

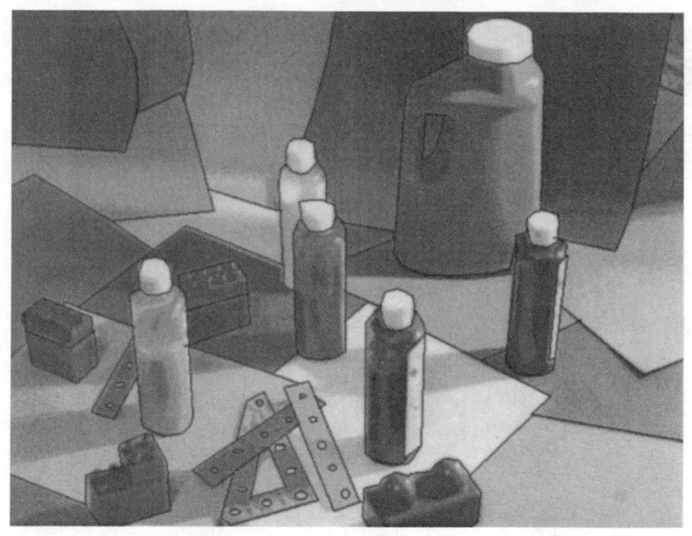

Fig. 13. Test image for color image segmentation with overlay of manual segmentation. (See insert for color figures)

6.4. Segmentation

The alternative to edge detection, which looks at differences between pixels, is segmentation, which clusters similar pixels. To compare the effects of using the two different similarity measures for segmentation, the test image shown in Figure 13 was used for evaluation [46]. It has several important features: shadows, areas with the same hue but varying saturation or intensity, areas of differing hue but similar intensity. The image has been overlaid with manually determined segmentation boundaries to show clearly the different regions.

For the MPC representation using vector angle, an eight-class, one-component network was trained using 3×3 pixel blocks of RGB values for a total input dimension of 27. The blocks were randomly chosen from the image. Similarly, an eight-class K-means algorithm was used on the same data to create a Euclidean or "VQ" representation of the image data.

The difference between the two approaches is shown in Figure 14. It is clear that the MPC network representation is significantly less sensitive to variations in illumination and shadows. For the K-means result of Figure 14b, it can be seen that many shadows are segmented as separate regions as is the case of other areas in the image where the illumination changes. However, in Figure 14a, for the most part, shadows do not affect the segmentation as regions of constant color (hue) remain contiguous under a variety of illumination variations.

7. Conclusion

A new approach to data representation has been discussed, namely the mixture of principal components (MPC). Together with vector quantization and principal components analysis at the two extremes, it forms a spectrum of repre-

(a) MPC with vector angle

(b) K-means with Euclidean distance

Fig. 14. Segmentation maps for eight classes with overlay of manual segmentation.

sentations. The measure of similarity for MPC is the subspace distance which results in the minimum squared error representation. Unlike Euclidean distance, the subspace distance is invariant of the vector norm. The class prototypes are the M principal components of the class data and can be computed using a clustering algorithm based on Hebbian learning. When used as a gray-scale feature extractor, it extracts perceptually important features in a completely self-organizing fashion. Further, the segmentation maps are the same for differing illumination conditions. For color image edge detection, the use of one-dimensional subspace projection, which reduces to the vector angle, results in edge maps which correspond more closely to our perception of color differences. When MPC is used to segment color images, the effects of shadows are significantly reduced compared to the Euclidean distance method. Further, all the color processing was performed on the raw RGB data and the method can generalize to higher order multispectral data.

In conclusion, then, the MPC representation using subspace distances represents an alternative representation to the classical Euclidean distance–based frameworks. One of the important differences between the two is the invariance of the MPC to the norm of the input vector data. For the image processing applications presented, this is of advantage due to the presence of illumination variations as an unwanted noise source.

References

[1] C. C. Chatterjee, V. P. Roychowdhury, and E. K. P. Chong. On relative convergence properties of principal component analysis algorithms. *IEEE Trans. Neural Networks* 9(2):319–329, March 1998.

[2] H. Chen and R. Liu. Adaptive distibuted orthogonalization process for principal components analysis. In *Proc. IEEE Int. Conf. on Acoustics, Speech, and Signal Processing '92,* pages II 283–296, San Francisco, March 23–26, 1992.

[3] H. Y. Chen Tianping and Y. Wei-Yong. Global convergence of Oja's subspace algorithm for principal component extraction. *IEEE Trans. Neural Networks* 9(1):58–67, January 1998.

[4] P. Comon and G. H. Golub. Tracking of a few extreme singular values and vectors in signal processing. *Proceedings of the IEEE* 78(8):1327–1343, August 1990.

[5] K. I. Diamantaras. *Principal Component Learning Networks and Applications.* Ph.D. thesis, Princeton University, October 1992.

[6] R. Dony. *CRC Handbook of Transforms and Data Compression,* chapter 1: Karhunen-Loève transform. K. R. Rao and P. C. Yip, editors. Boca Raton, FL: CRC Press, 2000.

[7] R. Dony and S. Wesolkowski. Edge detection on color images using RGB vector angles. In *1999 Canadian Conference on Electrical and Computer Engineering,* Edmonton, AB, Canada, May 1999.

[8] R. D. Dony and S. Haykin. Self-organizing segmentor and feature extractor. In *Proc. IEEE Int. Conf. on Image Processing,* pages III 898–902, Austin, TX, November 13–16, 1994.

[9] R. D. Dony and S. Haykin. Neural network approaches to image compression. *Proc. IEEE* 83(2):288–303, February 1995.

[10] R. D. Dony and S. Haykin. Optimally adaptive transform coding. *IEEE Trans. Image Processing* 4(10):1358–1370, October 1995.

[11] R. D. Dony and S. Haykin. Image segmentation using a mixture of principal components representation. *IEE Proceedings—Vision, Image and Signal Processing* 144(2):73–80, April 1997.

[12] R. O. Duda and P. E. Hart. *Pattern Classification and Scene Analysis.* New York: John Wiley & Sons, 1973.

[13] B. Fritzke. Growing cell structures—a self-organizing network for unsupervised and supervised learning. *Neural Networks* 7(9):1441–1460, 1994.

[14] K. Fukanaga. *Introduction to Statistical Pattern Recognition,* 2nd ed. San Diego, CA: Academic Press, 1990.

[15] R. C. Gonzalez and R. E. Woods. *Digital Image Processing.* Reading, MA: Addison-Wesley, 1993.

[16] G. Granlund and H. Knutsson. *Signal Processing for Computer Vision.* Boston: Kluwer Academic Publishers, 1995.

[17] R. M. Gray. *Source Coding Theory.* Norwell, MA: Kluwer Academic Publishers, 1990.

[18] S. Haykin. *Neural Networks: A Comprehensive Foundation.* New York: Macmillan, 1994.

[19] S. Haykin. *Adaptive Filter Theory.* 3rd ed. Prentice Hall, 1996.

[20] S. Haykin. Neural networks expand SP's horizons. *IEEE Signal Processing Magazine* 13(2):24–49, March 1996.
[21] D. H. Hubel and T. N. Wiesel. Brain mechanisms of vision. *Scientific American,* pages 130–146, September 1979.
[22] I. Jolliffe. *Principal Component Analysis.* New York: Springer-Verlag, 1986.
[23] T. Kohonen. The self-organizing map. *Proc. IEEE* 78(9):1464–1480, September 1990.
[24] T. Kohonen. Emergence of invariant-feature detectors in the adaptive-subspace SOM. *Biological Cybernetics* 75:281–291, 1996.
[25] S. Y. Kung and K. I. Diamantaras. A neural network learning algorithm for adaptive principal component extraction (APEX). In *Proc. IEEE Int. Conf. Acoustics, Speech, and Signal Processing 90,* pages 861–864, Albuquerque, NM, April 3–6, 1990.
[26] H. Levkowitz. *Color Theory and Modeling for Computer Graphics, Visualization, and Multimedia Applications.* Kluwer Academic Publishers, 1997.
[27] S. P. Luttrell. Image compression using a neural network. In *Proc. IGARSS '88,* pages 1231–1238, Edinburgh, September 13–18, 1988.
[28] S. P. Luttrell. Self-organization: A derivation from first principle of a class of learning algorithms. In *IEEE Conference on Neural Networks,* pages 495–498, Washington, DC, 1989.
[29] M. Manohar and J. C. Tilton. Compression of remotely sensed images using self organized feature maps. In H. Wechsler, editor, *Neural Networks for Perception,* volume 1: Human and Machine Perception, pages 345–367. San Diego, CA: Academic Press, 1992.
[30] J. D. McAuliffe, L. E. Atlas, and C. Rivera. A comparison of the LBG algorithm and Kohonen neural network paradigm for image vector quantization. In *Proc. IEEE Int. Conf. Acoustics, Speech, and Signal Processing '90,* pages 2293–2296, Alburquerque, NM, April 3–6, 1990.
[31] N. M. Nasrabadi and Y. Feng. Vector quantization of image based upon a neural-network clustering algorithm. In *SPIE Vol. 1001 Visual Communications and Image Processing '88,* pages 207–213, 1988.
[32] N. M. Nasrabadi and Y. Feng. Vector quantization of image based upon the Kohonen self-organizing feature map. In *Proc. IEEE Int. Conf. on Neural Networks,* pages I 101–105, San Diego, CA, July 24–27, 1988.
[33] A. N. Netravali and B. G. Haskell. *Digital Pictures: Representation and Compression.* New York: Plenum, 1988.
[34] E. Oja. A simplified neuron model as a principal component analyzer. *J. Math. Biol.* 15:267–273, 1982.
[35] E. Oja. *Subspace Methods of Pattern Recognition.* Letchworth, UK: Research Studies Press, 1983.
[36] K. R. Rao and P. Yip. *Discrete Cosine Transform: Algorithms, Advantages, Applications.* New York: Academic Press, 1990.
[37] H. Ritter, T. Martinetz, and K. Schulten. *Neural Computation and Self-organizing Maps: An Introduction.* Reading, MA: Addison-Wesley, 1992.
[38] T. D. Sanger. Optimal unsupervised learning in a single-layer linear feedforward neural network. *Neural Networks* 2:459–473, 1989.
[39] T. D. Sanger. An optimality principle for unsupervised learning. In D. S. Touretzky, editor, *Advances in Neural Information Processing Systems 1,* pages 11–19, 1989.
[40] V. Solo and X. Kong. Performance analysis of adaptive eigenanalysis algorithms. *IEEE Trans. Signal Processing* 46(3):363–645, March 1998.
[41] T. G. Stockham Jr. Image processing in the context of a visual model. *Proc. IEEE* 60(7):828–842, July 1972.
[42] N. V. Swindale. The development of topography in the visual cortex: A review of models. *Network Computation in Neural Systems* 7(2):161–248, 1996.
[43] S. Theodoridis and K. Koutroumbas. *Pattern Recognition.* San Diego, CA: Academic Press, 1999.
[44] K. K. Truong. Multilayer Kohonen image codebooks with a logarithmic search complexity. In *Proc. IEEE Int. Conf. Acoustics, Speech, and Signal Processing '91,* pages 2789–2792, Toronto, Canada, May 14–17, 1991.
[45] T. Villmann, R. Der, T. M. Herrmann, and M. Martinetz. Topology preservation in self-organizing feature maps: Exact definition and measurement. *IEEE Transactions on Neural Networks* 8(2):256–266, March 1997.
[46] S. Wesolkowski, R. Dony, and M. Jernigan. Global color image segmentation strategies: Euclidean distance vs. vector angle. In *Neural Networks for Signal Processing IX: Proceedings of the 1999 IEEE Signal Processing Society Workshop,* pages 419–428, Madison, WI, August 23–25, 1999.
[47] L. Xu and A. Yuille. Robust principal component analysis by self-organizing rules based on statistical physics approach. Technical Report 92-3, Harvard Robotics Laboratory, February 1992.
[48] L. Xu and A. L. Yuille. Robust principal component analysis by self-organizing rules based on statistical physics approach. *IEEE Trans. Neural Networks* 6(6):131–143, January 1995.

Chapter 15

Image Denoising by Sparse Code Shrinkage

AAPO HYVÄRINEN, PATRIK HOYER, AND ERKKI OJA
NEURAL NETWORKS RESEARCH CENTRE
HELSINKI UNIVERSITY OF TECHNOLOGY
P.O. BOX 5400, FIN-02015 HUT, FINLAND
http://www.cis.hut.fi/projects/ica/

Abstract—Sparse coding is a method for finding a representation of data in which each of the components of the representation is only rarely significantly active. Such a representation is closely related to independent component analysis (ICA) and has some neurophysiological plausibility. In this chapter, we show how sparse coding can be used for image denoising. We model the noise-free image data by independent component analysis and denoise a noisy image by maximum likelihood estimation of the noisy version of the ICA model. This leads to the application of a soft-thresholding (shrinkage) operator on the components of sparse coding. Our method is closely related to the method of wavelet shrinkage and coring methods, but it has the important benefit that the representation is determined solely by the statistical properties of the data. In fact, our method can be seen as a simple rederivation of the wavelet shrinkage method for image data, using just the basic principle of maximum likelihood estimation. On the other hand, it allows the method to adapt to different kinds of data sets.

1. Introduction

Sparse coding [3, 14, 32, 33] is a method for finding a neural network representation of multidimensional data in which only a small number of neurons is significantly activated at the same time. Equivalently, this means that a given neuron is activated only rarely. In this chapter, we assume that the representation is linear. Denote by $\mathbf{x} = (x_1, x_2, \ldots, x_n)^T$ the observed n-dimensional random vector that is input to a neural network and by $\mathbf{s} = (s_1, s_2, \ldots, s_n)^T$ the vector of the transformed component variables, which are the n linear outputs of the network. Denoting further the weight vectors of the neurons by $\mathbf{w}_i$, $i = 1, \ldots, n$, and by $\mathbf{W} = (\mathbf{w}_1, \ldots, \mathbf{w}_n)^T$ the weight matrix whose rows are the weight vectors, the linear relationship is given by

$$\mathbf{s} = \mathbf{W}\mathbf{x} \tag{1}$$

We assume here that that the number of sparse components, i.e., the number of neurons, equals the number of observed variables, but this need not be the case in general [33, 21]. Sparse coding can now be formulated as a search for a weight matrix $\mathbf{W}$ such that the components s_i are as "sparse" as possible. A zero-mean random variable s_i is called sparse when it has a probability density function with a peak at zero and heavy tails (i.e., the density is relatively large far from zero). For all practical purposes, sparsity is equivalent to supergaussianity [22] or leptokurtosis (positive kurtosis) [25].

Sparse coding is closely related to independent component analysis (ICA) [1, 4, 7, 18, 22, 23, 24, 31]. In the data model used in ICA, one postulates that $\mathbf{x}$ is a linear transform of independent components: $\mathbf{x} = \mathbf{A}\mathbf{s}$. Inverting the relation, one obtains (1), with $\mathbf{W}$ being the (pseudo)inverse of $\mathbf{A}$. Moreover, it has been proved that the estimation of the ICA data model can be reduced to the search for uncorrelated directions in which the components are as nongaussian as possible [7, 18]. If the independent components are sparse (more precisely, supergaussian), this amounts to the search for uncorrelated projections which have as sparse distributions as possible. Thus, estimation of the ICA model for sparse data is roughly equivalent to sparse coding if the components are constrained to be uncorrelated. This connection to ICA also shows clearly that sparse coding may be considered as a method for redundancy reduction, which was indeed one of the primary objectives of sparse coding in the first place [3, 14].

Sparse coding of sensory data has been shown to have advantages from both physiological and information processing viewpoints [3, 14]. In this chapter, we present and analyze a denoising method [20] based on sparse coding, thus increasing the evidence in favor of such a coding strategy. Given a signal corrupted by additive gaussian noise, we attempt to *reduce gaussian noise* by soft thresholding ("shrinkage") of the sparse components. Intuitively, because only a few of the neurons are active (i.e., significantly nonzero) simultaneously in a sparse code, one may assume that the activities of neurons with small absolute values are purely noise and set them to zero, retaining just a few components with large activities. This method is then shown to be very closely connected to the wavelet shrinkage method [11], as well as bayesian wavelet coring [36]. In fact, sparse coding may be viewed as a principled, adaptive way for determining an orthogonal wavelet-like basis based on data alone. Another advantage of our method is that the shrinkage nonlinearities can be adapted to the data as well.

This chapter is organized as follows. In Section 2, the basic problem is formulated as maximum likelihood estimation of a nongaussian variable corrupted by gaussian noise. In Section 3, the optimal sparse coding transformation is derived using maximum likelihood estimation of a linear generative model (ICA). Section 4 discusses the alternative approach of minimum mean-square estimation. The resulting algorithm of sparse code shrinkage is summarized in Section 5, and connections to other methods are discussed in Section 6. Extensions of the basic theory are discussed in Section 7. Section 8 contains experimental results, and some conclusions are drawn in Section 9.

2. Maximum Likelihood Denoising of Nongaussian Random Variables

2.1. Maximum Likelihood Denoising

The starting point of a rigorous derivation of our denoising method is the fact that the distributions of the sparse components are nongaussian. Therefore, we shall begin by developing a general theory that shows how to remove gaussian noise from (scalar) nongaussian variables, making minimal assumptions on the data. Our method is based on maximum likelihood (ML) estimation of nongaussian variables which are corrupted by gaussian noise.

Denote by s the original (scalar) nongaussian random variable and by ν gaussian noise of zero mean and variance σ^2. Assume that we observe only the random variable y:

$$y = s + \nu \tag{2}$$

and we want to estimate the original s. Denoting by p the probability density of s and by $f = -\log p$ its negative log-density, the maximum likelihood method gives the following estimator[1] for s:

$$\hat{s} = \arg\min_u \left[\frac{1}{2\sigma^2}(y-u)^2 + f(u) \right]. \tag{3}$$

[1] This might also be called a maximum *a posteriori* estimator.

Assuming f to be strictly convex and differentiable, this minimization is equivalent to solving the following equation:

$$\frac{1}{\sigma^2}(\hat{s} - y) + f'(\hat{s}) = 0 \tag{4}$$

which gives

$$\hat{s} = g(y) \tag{5}$$

where the inverse of the function g is given by

$$g^{-1}(u) = u + \sigma^2 f'(u). \tag{6}$$

Thus, the ML estimator is obtained by inverting a certain function involving f' or the score function [35] of the density of s. For nongaussian variables, the score function is nonlinear, and so is g.

In the general case, even if (6) cannot be inverted, the following first-order approximation of the ML estimator (with respect to noise level) is always possible:

$$\hat{s}^* = y - \sigma^2 f'(y), \tag{7}$$

still assuming f to be convex and differentiable. This estimator is derived from (4) simply by replacing $f'(\hat{s})$, which cannot be observed, by the observed quantity $f'(y)$; these two quantities are equal to first order. The problem with the estimator in (7) is that the sign of $\hat{s}^*$ is often different from the sign of y even for symmetrical zero-mean densities. Such counterintuitive estimates are possible because f' is often discontinuous or even singular at 0, which implies that the first-order approximation is quite inaccurate near 0. To alleviate this problem of "overshrinkage" [13], one may use the following modification:

$$\hat{s}^o = \text{sign}(y) \max(0, |y| - \sigma^2 |f'(y)|). \tag{8}$$

Thus we have obtained the exact maximum likelihood estimator (5) of a nongaussian random variable corrupted by gaussian noise and its two approximations in (7) and (8).

2.2. Modeling Sparse Densities

To use the estimator defined by (5) in practice, the densities of the s_i need to be modeled with a parameterization that is rich enough. We have developed two parameterizations that seem to describe very well most of the densities encountered in image denoising. Moreover, the parameters are easy to estimate, and the inversion in (6) can be performed analytically. Both models use two parameters and are thus able to model different degrees of supergaussianity, in addition to different scales, i.e., variances. The densities are here assumed to be symmetric and of zero mean.

2.2.1 Laplace Density. First we review the classical Laplace (or double exponential) density, which is the classical sparse density. The density of a Laplace distribution of unit variance [14, 29] is given by

$$p(s) = \frac{1}{\sqrt{2}} \exp(-\sqrt{2}|s|). \tag{9}$$

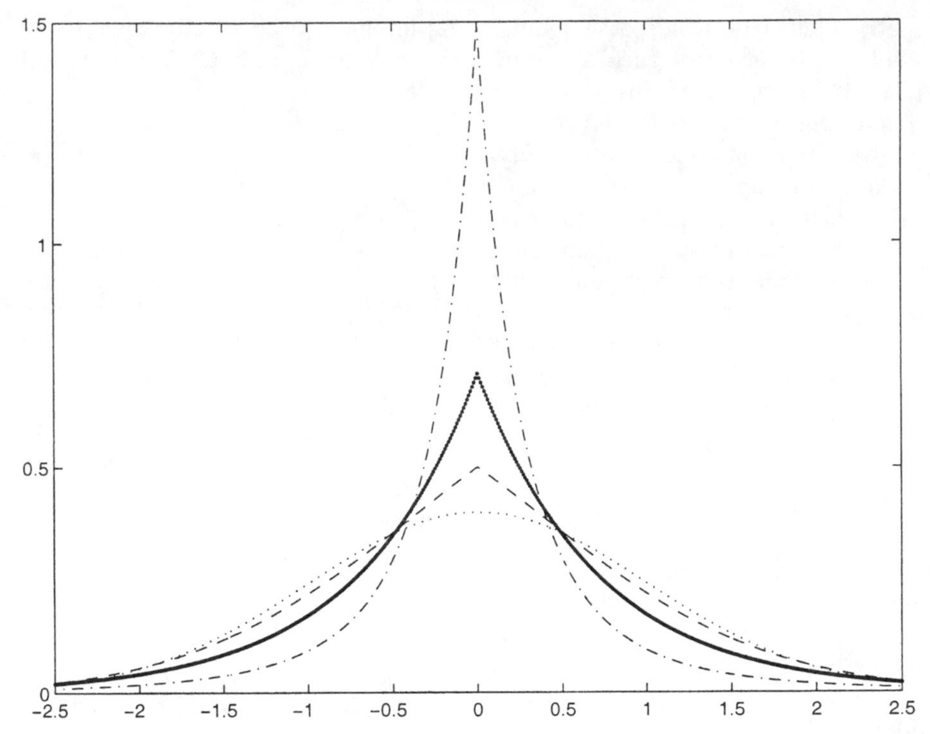

Fig. 1. Plots of densities corresponding to different models of the sparse components. Solid line: Laplace density in (9). Dashed line: a typical moderately supergaussian density given by (11). Dash-dotted line: a typical strongly supergaussian density given by (14). For comparison, gaussian density is given by the dotted line.

The Laplace density is plotted in Figure 1. For this density, the ML denoising nonlinearity g takes the form[2]

$$g(y) = \operatorname{sign}(y)\max(0, |y| - \sqrt{2}\sigma^2). \qquad (10)$$

The function in (10) is a *shrinkage* function that reduces the absolute value of its argument by a fixed amount, as depicted in Figure 2. Intuitively, the utility of such a function can be seen as follows. Since the density of a supergaussian random variable (e.g., a Laplace random variable) has a sharp peak at zero, it can be assumed that small values of y correspond to pure noise, i.e., to $s = 0$. Thresholding such values to zero should thus reduce noise, and the shrinkage function can indeed be considered a soft thresholding operator.

2.2.2 Mildly Sparse Densities. Our first density model is suitable for supergaussian densities that are not sparser than the Laplace distribution [20] and is given by the family of densities

$$p(s) = C\exp(-as^2/2 - b|s|), \qquad (11)$$

where $a, b > 0$ are parameters to be estimated, and C is an irrelevant scaling constant. The classical Laplace density is obtained when $a = 0$, and gaussian densities correspond to $b = 0$. Indeed, since the score function (i.e., f') of the

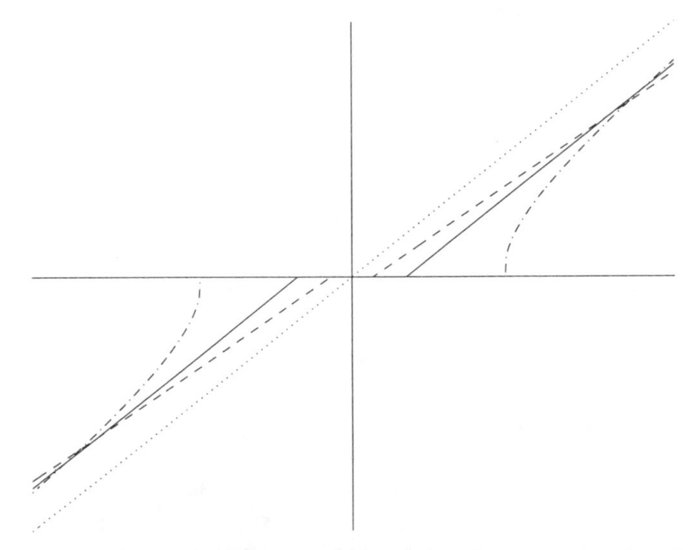

Fig. 2. Plots of the shrinkage functions. The effect of the functions is to reduce the absolute value of its argument by a certain amount which depends on the noise level. Small arguments are set to zero. This reduces gaussian noise for sparse random variables. Solid line: shrinkage corresponding to Laplace density as in (10). Dashed line: typical shrinkage function obtained from (13). Dash-dotted line: typical shrinkage function obtained from (17). For comparison, the line $x = y$ is given by the dotted line. All the densities were normalized to unit variance. For illustration purposes, the densities are normalized to unit variance, but these parameterizations allow the variance to be choosen freely. Noise variance was fixed to .3.

[2] Rigorously speaking, the function in (6) is not invertible in this case, but approximating it by a sequence of invertible functions, (10) is obtained as the limit.

gaussian distribution is a linear function and the score function of the typical supergaussian distribution, the Laplace density, is the sign function, it seems reasonable to approximate the score function of a symmetric, mildly supergaussian density of zero mean as a linear combination of these two functions. Figure 1 shows a typical density in the family.

A simple method for estimating a and b was derived in [20]. This was based on projecting (using a suitable metric) the score function of the observed data on the subspace spanned by the two functions (linear and sign). Thus we obtain [20]

$$b = \frac{2p_s(0)E\{s^2\} - E\{|s|\}}{E\{s^2\} - [E\{|s|\}]^2} \tag{12}$$

$$a = \frac{1}{E\{s^2\}}[1 - E\{|s|\}b]$$

where $p_s(0)$ is the value of the density function of s at zero. Corresponding estimators of a and b can then be obtained by replacing the expectations in (12) by sample averages; $p_s(0)$ can be estimated, e.g., using a single kernel at 0.

For densities in the family (11), the nonlinearity g takes the form:

$$g(u) = \frac{1}{1+\sigma^2 a}\operatorname{sign}(u)\max(0, |u| - b\sigma^2) \tag{13}$$

where σ^2 is the noise variance. This function is a shrinkage with additional scaling, as depicted in Figure 2.

2.2.3 Strongly Sparse Densities. In fact, most densities encountered in image denoising are sparser than the Laplace density. Therefore, we have developed a second model that describes such very sparse densities:

$$p(s) = \frac{1}{2d}\frac{(\alpha+2)[\alpha(\alpha+1)/2]^{(\alpha/2+1)}}{[\sqrt{\alpha(\alpha+1)/2} + |s/d|]^{(\alpha+3)}}. \tag{14}$$

Here, d is a scale parameter, and α is a sparsity parameter. When $\alpha \to \infty$, the Laplace density is obtained as the limit. The strong sparsity of the densities given by this model can be seen, e.g., from the fact that the kurtosis [14, 22] of these densities is always is always larger than the kurtosis of the Laplace density and reaches infinity for $\alpha \leq 2$. Similarly, $p(0)$ reaches infinity as α goes to zero. Figure 1 shows a typical density in the family.

A simple consistent method for estimating the parameters d, $\alpha > 0$ in (14) can be obtained from the relations

$$d = \sqrt{E\{s^2\}}, \tag{15}$$

$$\alpha = \frac{2 - k + \sqrt{k(k+4)}}{2k-1} \tag{16}$$

with $k = d^2 p_s(0)^2$.

The resulting shrinkage function can be obtained as[3]

$$g(u) = \operatorname{sign}(u)\max\left(0, \frac{|u| - ad}{2} + \frac{1}{2}\sqrt{(|u| + ad)^2 - 4\sigma^2(\alpha+3)}\right) \tag{17}$$

where $a = \sqrt{\alpha(\alpha+1)/2}$, and $g(u)$ is set to zero in case the square root in (17) is imaginary. This is a shrinkage function that has a certain hard-thresholding flavor, as depicted in Figure 2.

2.2.4 Choice of Model. Given a sparse density, we thus model it using one of the preceding models. Choosing whether model (11) or (14) should be used can be based on moments of the distributions. We suggest that if

$$\sqrt{E\{s^2\}}p_s(0) < \frac{1}{\sqrt{2}}, \tag{18}$$

the first model in (11) be used; otherwise use the second model in (14). The limit case $\sqrt{E\{s^2\}}p_s(0) = \sqrt{1/2}$ corresponds to the Laplace density, which is contained in both models.

2.2.5 Some Other Models. For the sake of completeness, we also give two classical models for sparse densities. These models are not, however, well suited for our method. The first alternative is the *generalized Laplacian* density [36]

$$p(s) = C\exp(-|s/d|^{\alpha}) \tag{19}$$

where d adjusts the scale, $\alpha \in (0,2]$ is a sparsity parameter, and C is a normalizing constant. As $\alpha \to 0$ the density becomes sparser and sparser, while at the other end $p = 2$ gives the Gaussian density. Estimation of the parameters d and α can be based on the second and fourth moments of the data as in [36]. It should be noted that the fourth moment is quite sensitive to outliers and thus the estimate is not very robust. No closed-form solution of the shrinkage function is available, so if one wants to use this parametrization then these estimators have to be calculated numerically.

A second alternative model for a sparse density can be obtained by using a Gaussian mixture model. A Gaussian mixture model is a parametrization

$$p(s) = \sum_{i=1}^{N} P_i G_i(s) \tag{20}$$

[3] Strictly speaking, the negative log-density of (14) is not convex, and thus the minimum in (5) might be obtained in a point not given by (17): in this case, the point 0 might be the true minimum. To find the true minimum, the value of likelihood at $g(y)$ should be compared with its value at 0, which would lead to an additional thresholding operation. However, such a thresholding changes the estimate only very little for reasonable values of the parameters d and α, and therefore we omit it, using (17) as a simpler and very accurate approximation of the minimization in (3).

where $\sum_{i=1}^{N} P_i = 1$, and each G_i is a Gaussian kernel

$$G_i(s) = \frac{1}{\sqrt{2\pi}d_i} \exp(-(s - \mu_i)^2/2d_i^2). \tag{21}$$

To represent sparse densities, we can choose a mixture of two gaussians ($N = 2$), both zero-mean ($\mu_i = 0$) but of different variance. Choosing the right parameters, such a mixture can be made sparse, so that a significant amount of data is found at or near zero, while much of the remaining data is far from zero. The estimation of the parameters (P_i and d_i) could be performed using the expectation maximization (EM) algorithm [10]. Such a parameterization does not, however, seem to describe image components very well for our purposes.

3. Finding the Sparse Coding Transformation

In the previous section, it was shown how to reduce additive gaussian noise in scalar nongaussian random variables by means of ML estimation. To denoise random vectors, we could apply such a shrinkage operation separately on each component. But before shrinkage, we would like to linearly transform the data so that the component-wise denoising is as efficient as possible. We shall restrict ourselves here to the class of linear, orthogonal transformations, for reasons that will be explained later. This restriction will be partly relaxed in Section 7.

Let us consider the estimation of the generative data model of independent component analysis (ICA) in the presence of noise. The noisy version of the conventional ICA model is given by

$$\mathbf{x} = \mathbf{A}\mathbf{s} + \boldsymbol{\nu} \tag{22}$$

where the latent variables s_i are assumed to be independent and nongaussian (usually supergaussian), $\mathbf{A}$ is a constant square matrix, and $\boldsymbol{\nu}$ is a gaussian noise vector. The noise-free ICA model has been shown to describe some important aspects of the basic higher order structure of image data [32, 33, 5].

Thus, modeling image data with the ICA model, an intuitively simple method for denoising the whole vector $\mathbf{x}$ could be obtained as follows: First, find estimates $\hat{s}_i$ of the (noise-free) independent components, and then reconstruct $\mathbf{x}$ as $\hat{\mathbf{x}} = \hat{\mathbf{A}}\hat{\mathbf{s}}$, as proposed in [17, 27]. Unfortunately, estimating $\mathbf{s}$ in this way is, in general, computationally very demanding. However, in [17] it was proved that if the covariance matrix of the noise and the mixing matrix $\mathbf{A}$ fulfill a certain relation, the estimate $\hat{\mathbf{s}}$ can be obtained simply by applying a shrinkage nonlinearity on the components of $\hat{\mathbf{A}}^{-1}\mathbf{x}$. This relation is fulfilled, e.g., if $\mathbf{A}$ is orthogonal, and noise covariance is proportional to identity.

Thus, restricting $\mathbf{W}$ to be orthogonal, we can find the optimal sparsifying transformation by *estimating the matrix* $\mathbf{A}$ *in the model* (22), under the constraint of orthogonality of $\mathbf{A}$. The sparsifying transformation is then given by $\mathbf{W} = \mathbf{A}^{-1}$. We adopt this method in the rest of this chapter.

To estimate $\mathbf{A}$ under the constraint of orthogonality, we could use the following approximative procedure. First we find an estimate $\tilde{\mathbf{A}}$ of $\mathbf{A}$ using any conventional ICA method and then transform its inverse $\mathbf{W}_0 = \tilde{\mathbf{A}}^{-1}$ by

$$\mathbf{W} = \mathbf{W}_0(\mathbf{W}_0^T\mathbf{W}_0)^{-1/2} \tag{23}$$

to obtain an orthogonal transformation matrix. The utility of this approximative method resides in the fact that there exist algorithms for ICA that are computationally highly efficient [18, 22]. Therefore, the preceding procedure enables one to estimate the basis even for data sets of high dimensions. Empirically, we have found that this approximation does not cause significant deterioration of the statistical properties of the obtained sparse coding transformation.

In practice, we can further simplify the estimation of $\mathbf{W}$ by assuming that we have access to a random variable $\mathbf{z}$ that has the same statistical properties as $\mathbf{x}$ and can be observed without noise. Thus we can estimate $\mathbf{W}$ using ordinary noise-free ICA algorithms. This assumption is not unrealistic on many applications: for example, in image denoising it simply means that we can observe noise-free images that are somewhat similar to the noisy image to be treated, i.e., they belong to the same environment or context. We make this assumption because estimating the sparsifying matrix from noisy data too problematic: noisy ICA [12, 19, 17] has proved to be an inherently difficult problem. However, in principle we could estimate $\mathbf{W}$ from noisy data directly by any method of noisy ICA estimation, as discussed in Section 7.

4. Mean-Square Error Approach

4.1. Minimum Mean-Square Estimator in Scalar Case

In the preceding text, we developed a denoising method based on maximum likelihood estimation of a generative model. It could also be interesting to derive a corresponding result using the criterion of minimum mean-square error. Let us consider the basic scalar denoising problem of Section 2. The mean-square error of the estimator would be minimized if we define $\hat{s}$ as the conditional expectation:

$$\hat{s} = E\{s|y\}. \tag{24}$$

Unfortunately, such a minimum mean-square error (MMSE) estimator cannot be obtained in closed form for any interesting nongaussian distributions. It could be obtained in closed form if the distribution of s was modeled by a mixture of gaussians, but such models do not seem to be suitable here.

A first-order approximation of $\hat{s}$ can be obtained, however. This is in fact equal to the estimator in (7), as proved in the Appendix. As a tractable first-order approximation, we thus obtain the same estimator as with ML estimation.

4.2. Analysis of Mean-Square Error

In this subsection, we analyze the denoising capability of the scalar MMSE estimator given in (24). We show that, roughly, the more nongaussian the variable s is, the better gaussian noise can be reduced. Nongaussianity is here measured by Fisher information. Due to the intractability of the general problem, we consider here the limit of infinitesimal noise, i.e., all the results are first-order approximations with respect to noise level. Due to the first-order equality between the ML and MMSE estimators, the analysis is also valid for the ML estimator.

To begin with, recall the definition of Fisher information [8] of a random variable s with density p:

$$I_F(s) = E\left\{\left[\frac{p'(s)}{p(s)}\right]^2\right\}. \tag{25}$$

The Fisher information of a random variable (or, strictly speaking, of its density) equals the conventional, "parametric" Fisher information [35] with respect to a hypothetical location parameter [8].

Fisher information can be considered as a measure of nongaussianity. It is well known [15] that in the set of probability densities of unit variance, Fisher information is minimized by the gaussian density, and the minimum equals 1. Fisher information is not, however, invariant to scaling; for a constant a, we have

$$I_F(as) = \frac{1}{a^2} I_F(s). \tag{26}$$

The main result on the performance of the ML estimator is the following theorem [20], whose proof is reproduced in the Appendix:

Theorem 1 *Define by (5) and (6), or alternatively by (24), the estimator $\hat{s} = g(y)$ of s in (2). For small σ, the mean-square error of the estimator $\hat{s}$ is given by*

$$E\{(s - \hat{s})^2\} = \sigma^2[1 - \sigma^2 I_F(s)] + o(\sigma^4), \tag{27}$$

where σ^2 is the variance of the gaussian noise ν.

To get more insight into the theorem, it is useful to compare the noise reduction of the ML estimator with the best *linear* estimator in the minimum mean-square error sense. If s has unit variance, the best linear estimator is given by

$$\hat{s}_{lin} = \frac{y}{1 + \sigma^2}. \tag{28}$$

This estimator has the following mean-square error:

$$E\{(s - \hat{s}_{lin})^2\} = \frac{\sigma^2}{1 + \sigma^2}. \tag{29}$$

We can now consider the ratio of these two errors, thus obtaining an index that gives the percentage of additional noise reduction due to using the nonlinear estimator $\hat{s}$:

$$R_s = 1 - \frac{E\{(\hat{s} - s)^2\}}{E\{(\hat{s}_{lin} - s)^2\}}. \tag{30}$$

The following corollary follows immediately:

Corollary 1 *The relative improvement in noise reduction obtained by using the nonlinear ML estimator instead of the best linear estimator, as measured by R_s in (30), is given by*

$$R_s = (I_F(s) - 1)\sigma^2 + o(\sigma^2), \tag{31}$$

for small noise variance σ^2, and for s of unit variance.

Considering the above-mentioned properties of Fisher information, Theorem 1 thus means that the more nongaussian s is, the better we can reduce noise. In particular, for sparse variables, the sparser s is, the better the denoising works. If s is gaussian, $R = 0$, which follows from the fact that the ML estimator is then equal to the linear estimator $\hat{s}_{lin}$. This shows again that for gaussian variables, allowing nonlinearity in the estimation does not improve the performance, whereas for nongaussian (e.g., sparse) variables, it can lead to significant improvement.[4]

4.3. Minimum Mean Squares Approach to Basis Estimation

We could also use the mean square as a criterion for choosing the sparsifying basis. Assume that we observe a multivariate random vector $\tilde{\mathbf{x}}$ which is a noisy version of the nongaussian random vector $\mathbf{x}$:

$$\tilde{\mathbf{x}} = \mathbf{x} + \boldsymbol{\nu}. \tag{32}$$

where the noise $\boldsymbol{\nu}$ is gaussian and of covariance $\sigma^2\mathbf{I}$. Again, we would like to find an orthogonal transformation of the data so that the shrinkage method reduces noise as much as possible. Given an orthogonal (weight) matrix $\mathbf{W}$, the transformed vector equals

$$\mathbf{W}\tilde{\mathbf{x}} = \mathbf{W}\mathbf{x} + \mathbf{W}\boldsymbol{\nu} = \mathbf{s} + \tilde{\boldsymbol{\nu}}. \tag{33}$$

The covariance matrix of $\tilde{\boldsymbol{\nu}}$ equals the covariance matrix of $\boldsymbol{\nu}$, which means that the noise remains essentially uncharged.

The noise reduction obtained by the scalar MMSE method is, according to Theorem 1, proportional to the sum of the Fisher informations of the components $s_i = \mathbf{w}_i^T\mathbf{x}$. Thus, the optimal orthogonal transformation $\mathbf{W}_{opt}$ can be obtained as

$$\mathbf{W}_{opt} = \arg\max_{\mathbf{W}} \sum_{i=1}^{n} I_F(\mathbf{w}_i^T\mathbf{x}) \tag{34}$$

where $\mathbf{W}$ is constrained to be orthogonal, and the $\mathbf{w}_i$ are the rows of $\mathbf{W}$. To estimate $\mathbf{W}$ using this method, one can use the approximation of Fisher information derived in [20].

To estimate the optimal orthogonal transform $\mathbf{W}_{opt}$, we need to assume that we have access to a random variable $\mathbf{z}$ that has the same statistical properties as $\mathbf{x}$ and can be observed without noise.

[4] For multivariate gaussian variables, however, improvement can be obtained by Stein estimators [13].

In image denoising, however, the preceding result needs to be slightly modified. These modifications are necessary because of the well-known fact that ordinary mean-square error is a rather inadequate measure of errors in images. Perceptually more adequate measures can be obtained, e.g., by weighting the mean-square error so that components corresponding to lower frequencies have more weight. Since the variance of the sparse and principal components is larger for lower frequencies, such a perceptually motivated weighting can be approximated simply by the following objective function:

$$J = \sum_{i=1}^{n} E\{(\mathbf{w}^T\mathbf{z})^2\} I_F(\mathbf{w}_i^T\mathbf{z}). \tag{35}$$

Using (26), this can be expressed as

$$J = \sum_{i=1}^{n} I_F\left(\frac{\mathbf{w}_i^T\mathbf{z}}{\sqrt{E\{(\mathbf{w}_i^T\mathbf{z})^2\}}}\right). \tag{36}$$

This is the normalized Fisher information, which is a scale-invariant measure of nongaussianity.

In fact, maximizing (36) is very closely related to the estimation of the ICA model, as in (3). Maximizing the sum of nongaussianities of $\mathbf{w}^T\mathbf{z}$ is one intuitive method of estimating the ICA data model [18]. Therefore, it can be seen that using the perceptually weighted mean-square error, we rediscover the basis estimation method of Section 3.

5. Sparse Code Shrinkage

Now we summarize the algorithm of sparse code shrinkage as developed in the preceding sections. In this method, the ML noise reduction is applied on sparse components, first choosing the orthogonal transformation so as to maximize the sparseness of the components. This restriction to sparse variables is justified by the fact that in many applications, such as image processing, the distributions encountered are sparse. The algorithm is as follows:

1. Using a representative noise-free set of data $\mathbf{z}$ that has the same statistical properties as the n-dimensional data $\mathbf{x}$ that we want to denoise, estimate the sparse coding transformation $\mathbf{W} = \mathbf{W}_{opt}$ by first estimating the ICA transform matrix and then orthogonalizing it. (See Section 3.)
2. For every $i = 1, \ldots, n$, estimate a density model for $s_i = \mathbf{w}_i^T\mathbf{z}$, using the models described in Section 2.2. (Choose by (18) whether model (11) or (14) is to be used for s_i. Estimate the relevant parameters, e.g., by (12) or (15), respectively.) Denote by g_i the corresponding shrinkage function, given by (13) or by (17), respectively.
3. Observing $\tilde{\mathbf{x}}(t)$, $t = 1, \ldots, T$, which are samples of a noisy version of $\mathbf{x}$ as in (32), compute the projections on the sparsifying basis:
$$\mathbf{y}(t) = \mathbf{W}\tilde{\mathbf{x}}(t). \tag{37}$$
4. Apply the shrinkage operator g_i corresponding to the density model of s_i on every component $y_i(t)$ of $y(t)$, for every t, obtaining
$$\hat{s}_i(t) = g_i(y_i(t)), \tag{38}$$
where σ^2 is the noise variance (see later on estimating σ^2).
5. Transform back to original variables to obtain estimates of the noise-free data $\mathbf{x}(t)$:
$$\hat{\mathbf{x}}(t) = \mathbf{W}^T\hat{\mathbf{s}}(t). \tag{39}$$

If the noise variance σ^2 is not known, one might estimate it, following [11], by multiplying by 0.6475 the mean absolute deviation of the y_i corresponding to the very sparsest s_i.

6. Comparison with Wavelet and Coring Methods

The resulting algorithm of sparse code shrinkage is closely related to wavelet methods [2], in particular wavelet shrinkage [11], with the following differences:

1. Our method assumes that one first estimates the orthogonal basis using noise-free training data that has similar statistical properties. Thus our method could be considered as a principled method of choosing the wavelet basis for a given class of data: instead of being limited to bases that have certain abstract mathematical properties (like self-similarity), we let the basis be determined by the data alone, under the sole constraint of orthogonality.
2. In sparse code shrinkage, the shrinkage nonlinearities are estimated separately for each component, using the same training data as for the basis. In wavelet shrinkage, the form of shrinkage nonlinearity is fixed, and the shrinkage coefficients are either constant for most of the components (and perhaps set to zero for certain components) or constant for each resolution level [11]. (More complex methods such as cross-validation [30] are possible, though.) This difference stems from the fact that wavelet shrinkage uses minimax estimation theory, whereas our method uses ordinary ML estimation. Note that point 2 is conceptually independent of point 1 and further shows the adaptive nature of sparse code shrinkage.
3. Our method, although primarily intended for sparse data, could be directly modified to work for other kinds of nongaussian data.
4. An advantage of wavelet methods is that very fast algorithms have been developed to perform the transformation [29], avoiding multiplication of the data by the matrix $\mathbf{W}$ (or its transpose).

5. Of course, wavelet methods avoid the computational overhead, and especially the need for additional, noise-free data required for estimating the matrix **W** in the first place. The requirement for noise-free training data is, however, not an essential part of our method, as shown in Section 7.

The connection is especially clear if one assumes that both steps 1 and 2 of sparse code shrinkage in Section 5 are omitted, using a wavelet basis and the shrinkage function (13) with $a_i = 0$ and a b_i that is equal for all i (except perhaps some i for which it is zero). Such a method would be essentially equivalent to wavelet shrinkage.

A related method is Bayesian wavelet coring, introduced in [36]. In Bayesian wavelet coring, the shrinkage nonlinearity is estimated from the data to minimize mean-square error. Thus the method is more adaptive than wavelet shrinkage but still uses a predetermined sparsifying transformation.

7. Extensions of the Basic Theory

In this section, we present some extensions to the basic theory already given.

7.1. Nongaussian Noise

In fact, in the preceding derivation noise was always considered as noise in the sparse components s_i, i.e., after the transformation. Therefore, if the noise in the x_i is weakly nongaussian (e.g., Laplace noise), then the noise in s_i is a sum of independent noise components, and thus, due to the central limit theorem, more gaussian than the original noise. This means that the noise in the components may still be considered approximately gaussian, and the basic method presented before can still be expected to work satisfactorily even for weakly nongaussian noise. Modifications of the method are thus necessary only in the case of strongly nongaussian (e.g., impulsive) noise. With such noise, however, it may be more useful to use methods based on reconstructing missing pixels [34]. Related work using wavelets can be found in [26].

7.2. Estimation of Parameters from Noisy Data

We have already seen that using the ML principle of basis estimation as in Section 3, it is possible to estimate the sparse code transformation directly from noisy data. We emphasize this here because this important point was missing in our earlier work. Basis estimation from noisy data can thus be accomplished by any method that can estimate the noisy ICA model. In practice, however, this may be very problematic, because the estimation of the noisy ICA model is still very much an open research problem, and it seems probable that the performance that can be obtained will necessarily be considerably lower than what can be obtained from noise-free data. Therefore, in many practical applications, it may not be useful to estimate the basis from noisy data. For the same reasons, estimation of the optimal nonlinearities from noisy data may be very problematic; this is also a future research problem. Nonlinearities that are fixed by prior knowledge may often give satisfactory results, however.

7.3. Nonorthogonal Bases

No restriction to orthogonal bases is necessary, either, if the basis is estimated by the ML principle of Section 3. In fact, the transformation can be taken to be simply the inverse of the mixing matrix **A**, i.e., $\mathbf{W} = \mathbf{A}^{-1}$, where **A** is estimated by any ordinary ICA estimation method. On the other hand, this has the unpleasant side effect that the noise in the transformed sparse components will be correlated. To be able to estimate the noise-free components without computationally complex operations, we must then approximate the noise structure by a diagonal matrix. Thus the preceding shrinkage functions can be simply adapted to the case of nonorthogonal transformations by replacing the constant noise covariance by an estimate of the noise variance in the direction of each of the sparse components. Thus we obtain a component-wise shrinkage as before, ignoring all correlations between noise components. However, the advantage gained by relaxing the restriction of orthogonality is diminished by the need to approximate noise covariance.

8. Experiments

8.1. Generation of Image Data

It would be interesting to test the sparse code shrinkage method on real data (i.e., data which is not available free of noise). However, evaluating such results are difficult, thus we decided to test the performance on images which were artificially corrupted with noise. We chose two separate datasets so as to be able to compare the performance and see the differences of the results for different datasets.

The first set of images consisted of *natural scenes* previously used in [16] in which ICA was applied to image data. These images were hoped to reflect truly natural images, i.e., images void of human-imposed structure. An example of the images used is shown in Figure 3. These images may be obtained from our Web pages.[5] The second set consisted of demo images from Kodak's PhotoCD system. These are royalty free and may be downloaded from the Internet by FTP.[6] These are intended to represent images of the human-built world, which have quite different statistics from the natural scenes of the first set. Figure 4 displays an example of these images which we will call *man-made* scenes.

[5] http://www.cis.hut.fi/projects/ica/data/images

[6] ftp://ipl.rpi.edu/pub/image/still/KodakImages/

Fig. 3. A representative image from the first set of image data (natural scenes).

From both sets, 10 images were picked at random for estimation of the transforms and the densities, and a separate image from both sets was picked for the actual denoising experiments.

8.2. Remarks on Image Data

8.2.1 Windowing. Thus far, we have considered the problem of denoising a random vector of arbitrary dimension. When applying this framework to images, certain problems arise.

The simplest way to apply the method to images would be simply to divide the image into $N \times N$ windows and denoise each such window separately. This approach, however, has a couple of drawbacks: statistical dependences across the synthetic edges are ignored, resulting in a blocking artifact, and the resulting procedure is not translation invariant: The algorithm is sensitive to the precise position of the image with respect to the windowing grid.

We have solved this problem by taking a *sliding window* approach. This means that we do not divide the image into distinct windows, rather we denoise every possible $N \times N$ window of the image. We then effectively have N^2 different suggested values for each pixel and select the final result as the mean of these values. Although originally chosen

Fig. 4. A representative image from the second set of image data (man-made scenes).

on rather heuristic grounds, the sliding window method can be justified by two interpretations.

The first interpretation is *spin-cycling.* The basic version of the recently introduced wavelet shrinkage method is not translation invariant, because this is not a property of the wavelet decomposition in general. Thus, Coifman and Donoho [6] suggested performing wavelet shrinkage on all translated wavelet decompositions of the data and taking the mean of these results as the final denoised signal, calling the obtained method spin-cycling. It is easily seen that our sliding window method is then precisely spin-cycling of the distinct window algorithm.

The second interpretation of sliding windows is due to the *method of frames.* Consider the case of decomposing a data vector into a linear combination of a set of given vectors, where the number of given vectors is larger than the dimensionality of the data; i.e., given $\mathbf{x}$ and $\mathbf{A}$, where $\mathbf{A}$ is an m by n matrix ($m < n$), find the vector $\mathbf{s}$, in $\mathbf{x} = \mathbf{As}$. This has an infinite number of solutions. The classical way is to select the solution $\mathbf{s}$ with minimum norm, given by $\hat{\mathbf{s}} = \mathbf{A}^{\dagger}\mathbf{x}$, where $\mathbf{A}^{\dagger}$ is the pseudoinverse of $\mathbf{A}$. This solution is often referred to as the method of frames solution [9].

Now consider each basis window in each possible window position of the image as an overcomplete basis for the whole image. Then, if the transform we use is orthogonal, the sliding window algorithm is equivalent to calculating the method of frames decomposition, shrinking each component, and reconstructing the image.

8.2.2 The Local Mean. ICA applied to image data usually gives one component representing the local mean image intensity, or the DC component. This component normally has a distribution that is not sparse, often even subgaussian. Thus, it must be treated separately from the other, supergaussian components.

One could estimate a suitable density model for this component and denoise it just as the others. However, since the component generally has a large variance it is relatively unaffected by the noise, and a simplification is simply to leave it alone. This is the approach we have chosen to take. This means that in all experiments we first subtract the local mean (and drop the dimension using PCA) and then estimate a suitable sparse coding basis for the rest of the components. In restructuring the image after denoising, we again add the local means.

8.2.3 Normalizing the Local Variance. Some image processing methods normalize the local variance in an image. We can consider such a variant of our method as well. We can incorporate normalization into our method by dividing each image window by its norm. This can be done in both the estimation of parameters and the denoising procedure.[7]

[7] When considering noisy windows, one must estimate the norm before the noise was added and divide by this quantity. This estimation can be done reasonably well when the size of the image window is large enough, e.g., 8×8.

8.3. Transform Estimation

8.3.1 Methods. Estimating an ICA transform from patches of natural image data has previously been shown to give a transform mainly consisting of local filters, resembling somewhat the so-called Gabor filters or wavelets [16, 32, 33, 5], and this is what we find here as well.

Each image was first linearly normalized so that pixels had zero mean and unit variance. A set of 10,000 image patches (windows) were taken at random locations from the images. From each patch the local mean was subtracted as explained before. This resulted in a linear dependence between the pixels of each patch, and thus we reduced the dimension by one using standard PCA. The preprocessed dataset was used as the input to the FastICA algorithm [22, 18], using the hyperbolic tangent nonlinearity [18].

8.3.2 Results. Figure 5 shows the results for the first data set, which was patches from natural scenes with 8×8 window size. The PCA transform consists of global features and resembles strongly the 2D Fourier transform. The transform given by ICA, however, finds features which resemble local edges and bars. They can thus be said to be more representative of the image data.

There is an interesting difference between the ICA filters (comprising the separating matrix **W**) and the basis vectors (making up the mixing matrix **A**). The basis vectors constitute the "building blocks" from which the data is thought to be generated. Thus they have the same type of features that we are used to seeing in images. Looking closely, it is easily seen that the filters are similar to the basis vectors, in that each filter has the same position and orientation as its corresponding basis vector. However, the separating filters are clearly more "spiky." This is in essence a consequence of the whitening, since whitening must amplify high frequencies (because these have the smallest variance).

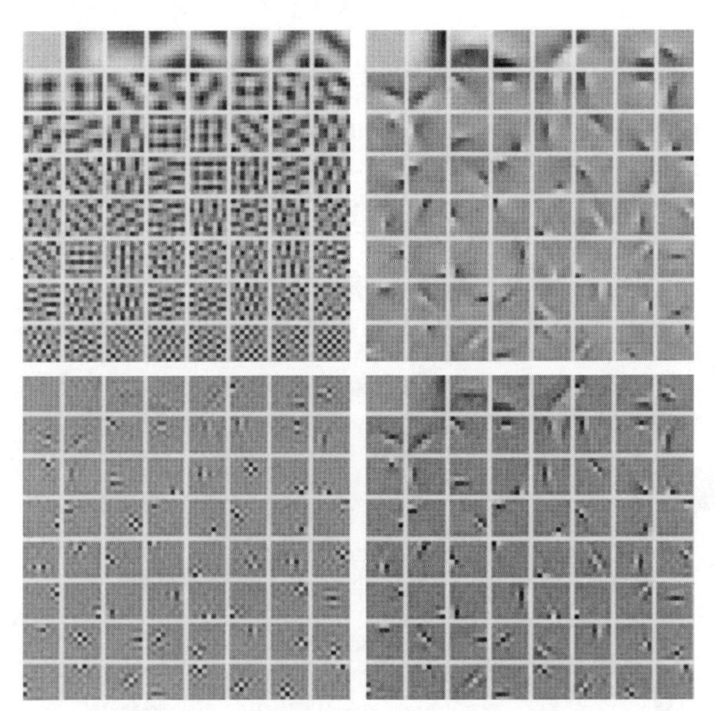

Fig. 5. Transforms estimated for natural scene data (8-by-8 patches). The windows have been ordered according to mean frequency for visualization purposes. (This ordering is irrelevant in the denoising algorithm.) Top left: PCA transform (orthogonal, so basis and filters are the same). Top right: ICA basis. Bottom left: ICA separating filters. Bottom right: Orthogonalized ICA transform.

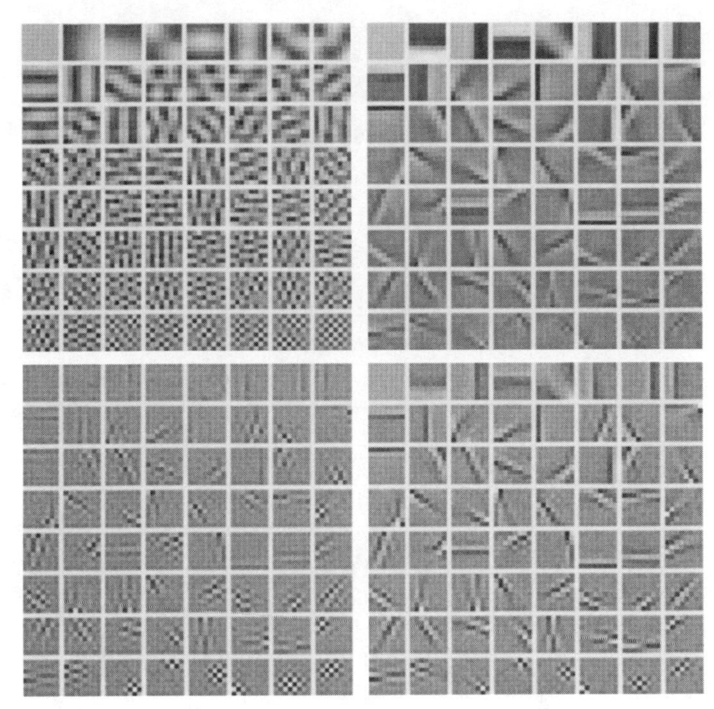

Fig. 6. Transforms estimated for man-made scene data (8 by 8). The windows have been ordered according to mean frequency. Top left: PCA transform. Top right: ICA basis. Bottom left: ICA separating filters. Bottom right: Orthogonalized ICA transform.

In our basic method the ICA transform is orthogonalized, and the transform thus obtained is simply the ICA transform for zero-phase whitened data. This transform is also depicted in Figure 5 and can be seen as being "in between" the separating matrix **W** and the mixing matrix **A**. The features of the transform are not essentially changed by this orthogonalization.

In the second dataset, the image patches were gathered from the man-made scenes. The results are shown in Figure 6. The difference from the ICA transforms from natural scenes is quite clear. In man-made scenes, there are much stronger continuous lines and edges, which shows up in the ICA decomposition as basis vectors which are more edgelike than the "Gabor-like" basis vectors from the natural scenes.

We also experimented with a larger window size of 16 by 16 pixels. The ICA basis found (not shown) was qualitatively similar to that estimated from the smaller windows, consisting of lines and edges at various locations and at various orientations.

For comparison, we show a Daubechies wavelet basis (filter length = 4) in Figure 7.

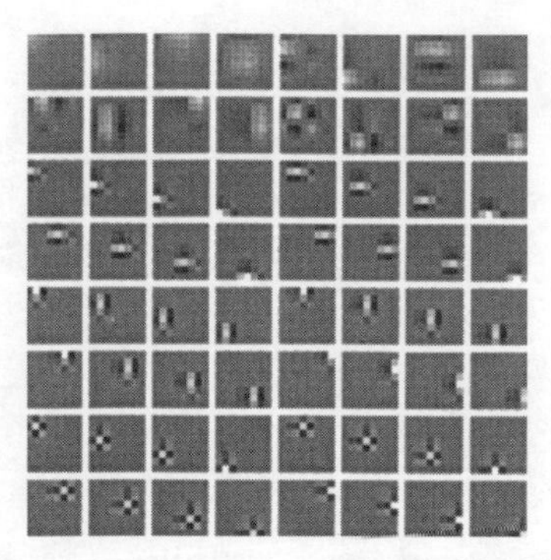

Fig. 7. A wavelet basis, to be compared with the transforms in the preceding figures.

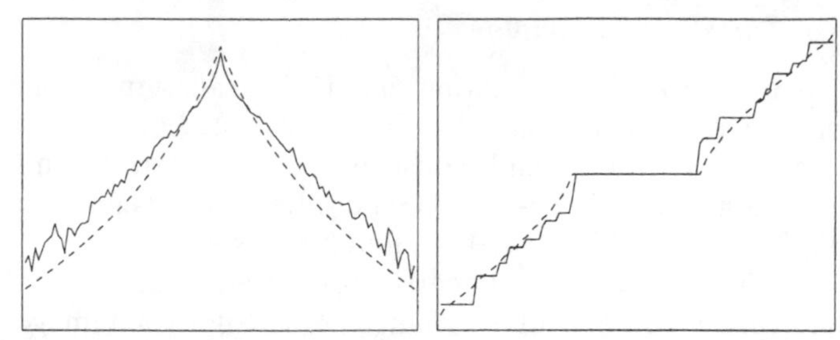

Fig. 9. Analysis a randomly selected component from the orthogonalized ICA transforms of natural scenes, with window size 8 × 8. Top: Nonparametrically estimated log-densities (solid curve) vs. the best parametrization (dashed curve). Bottom: Nonparametric shrinkage nonlinearity (solid curve) vs. that given by our parametrization (dashed curve).

8.4. Component Statistics

Because the denoising procedure is based on the property that individual components in the transform domain have sparse distributions, it is necessary to test how well this requirement holds. Also, the method requires selecting suitable parametrizations and estimating the parameters; thus in these experiments we also evaluate how well the proposed parametrizations of Section 2.2 approximate the underlying densities.

Measuring the sparseness of the distributions can be done by almost any nongaussianity measure. We have chosen the most widely used measure, the normalized kurtosis. Recall that normalized kurtosis is defined as

$$\kappa(s) = \frac{E\{s^4\}}{(E\{s^2\})^2} - 3. \tag{40}$$

The average sparseness of each of the three transform sets (estimated in the previous section) was calculated in the following way: First, we sampled 30,000 image patches from the same dataset which was used for estimation of the transform. Then, we transformed these samples using the estimated PCA, ICA, and orthogonalized ICA transforms and calculated the normalized kurtosis for each component of each basis separately. In Figure 8 the mean of these component kurtoses is displayed for each transform of each dataset. Because of the sparse structure in the images, all three transforms show supergaussian distributions; indeed even the individual pixel values show a mildly supergaussian distribution when the local mean has been subtracted. From the graph, it is seen that the ICA transform clearly finds a sparser representation than PCA. Also, note that the orthogonalized ICA representation is not quite as sparse as that given by standard ICA but it is still far more supergaussian than PCA on average.

Next, we attempted to compare various parametrizations in the task of fitting the observed densitics. We picked one component at random from the orthogonal 8 × 8 sparse coding transform for both natural scenes. First, using a nonparameteric histogram technique, we estimated the density of the component, and from this representation derived the log density and the shrinkage nonlinearity shown in Figure 9. Next, we fitted the parametrized densities discussed in Section 2.2 to the observed density. Note that in each case, the densities were sparser than the Laplace density, and thus the very sparse parametrization in (14) was used. In can be seen that the density and the shrinkage nonlinearity derived from the density model match quite well those given by nonparametric estimation. We did the corresponding experiments for a window size 16 × 16, shown in Figure 10; the results are not different

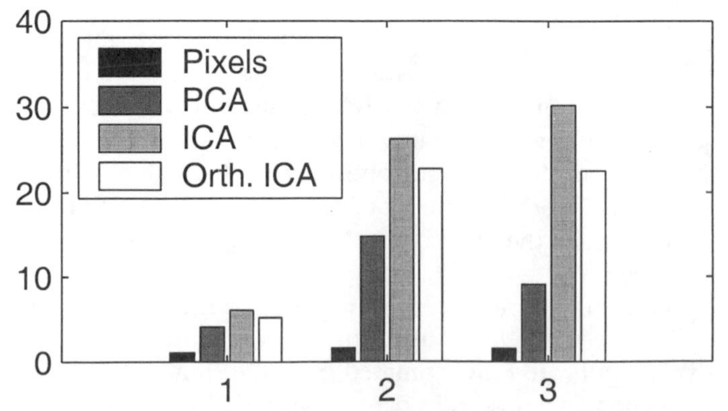

Fig. 8. Mean of normalized kurtosis of components for the different datasets and different transforms within each dataset. (1) Natural scenes, 8-by-8 patch bases. (2) Man-made scenes, 8-by-8 bases. (3) Man-made scenes, 16-by-16 bases.

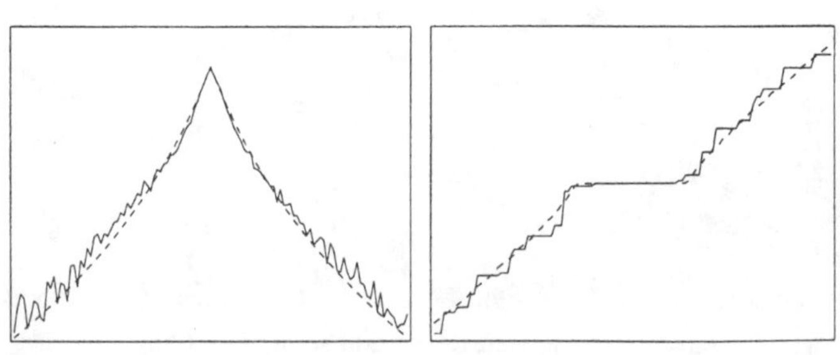

Fig. 10. Analysis a randomly selected component from the orthogonalized ICA transforms of natural scenes, this time with window size 16 × 16. Top: Nonparametrically estimated log-densities (solid curve) vs. the best parametrization (dashed curve). Bottom: Nonparametric shrinkage nonlinearity (solid curve) vs. that given by our parametrization (dashed curve).

in this case. The results for man-made scenes (not shown) were similar as well.

The gaussian mixture model in (20) (not shown) gave very poor matches to the observed densities. The generalized Laplace density did give good matches, as observed by others [29, 36], but the drawback is that it does not allow an analytical form for the shrinkage nonlinearity, which is why we prefer our parameterization.

As already mentioned, a possible variant of the method is to normalize the local image variance by dividing each input window by its norm. To experiment with such a denoising method, we had to estimate density parameters for such data. Thus, we proceeded as in the preceding experiments but normalized each observed image patch to unit length before transforming it by the orthogonalized ICA transform. Now, the densities were not nearly as sparse, and the mildly sparse density model (11) was the appropriate parametrization.

In conclusion, we have seen that the components of the sparse coding bases found are highly supergaussian for natural image data in the basic case. This is true whether the images depict nature or man-made scenes.

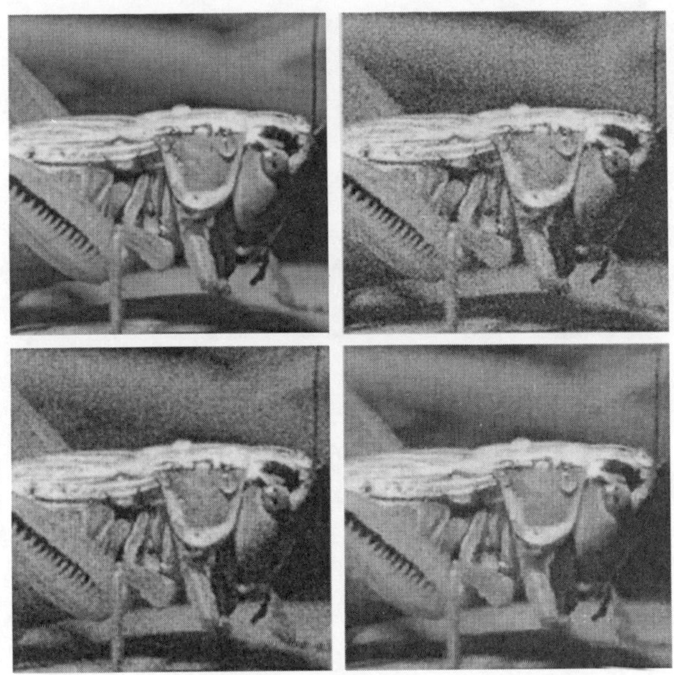

Fig. 11. Denoising a natural scene (noise level 0.3). Top left: The original image. Top right: Noise added. Bottom left: After Wiener filtering. Bottom right: Results after sparse code shrinkage.

8.5. Denoising Results

To begin the actual denoising experiments, a random image from the natural scene collection was chosen for denoising, and gaussian noise of standard deviation 0.3 was added (compared to a standard deviation of 1.0 for the original image). This noisy version was subsequently denoised using the Wiener filter to give a baseline comparison. Then, the sparse code shrinkage method was applied using the estimated orthogonalized ICA transform (8×8 windows), with the component nonlinearities as given by the appropriate estimated parametrization. Figure 11 shows the results of the first experiments. Visually, it seems that sparse code shrinkage gives the best noise reduction while retaining the features in the image. The Wiener filter does not really eliminate the noise. It seems as though our method is performing like a feature detector, in that it retains those features which are clearly visible in the noisy data but cuts out anything which is probably a result of the noise. Thus, it reduces noise effectively due to the nonlinear nature of the shrinkage operation.

In Figure 12, exactly the same parameters have been used with the exception that the noise level has been raised to 0.5. The results are qualitatively very similar to those for the lower noise level.

The same experiments were then performed for data consisting of man-made scenes. We show the results for the noise level of 0.5 in Figure 13. This data seems to suit the method even better than the data from natural scenes. The image consists mainly of even areas with sharp edges. These are the type of features that the basis and density parameters have been adapted to, and the results are quite good.

To see the effect of window size, we performed the same experiments with a larger window size. The results (not shown) did not really change at all.

Since the theory was derived under the strict assumption that the noise was gaussian, it is interesting to see how the method performs when the noise is slightly nongaussian, e.g., Laplacian. The results of such an experiment are

Fig. 12. Denoising a natural scene (noise level 0.5). Top left: The original image. Top right: Noise added. Bottom left: After Wiener filtering. Bottom right: Results after sparse code shrinkage.

Fig. 13. Denoising a man-made scene (noise level 0.5). Top left: The original image. Top right: Noise added. Bottom left: After Wiener filtering. Bottom right: Results after sparse code shrinkage.

shown in Figure 14. Perhaps a bit surprisingly, the method performed quite well. For a possible explanation, see Section 7.1. Note, however, that when the noise becomes increasingly supergaussian (i.e., impulsive) the assumptions of the method begin to break down, and it would be better

Fig. 14. Denoising a man-made scene with *Laplacian noise* (noise level 0.5). Top left: The original image. Top right: Noise added. Bottom left: After Wiener filtering. Bottom right: Results after sparse code shrinkage.

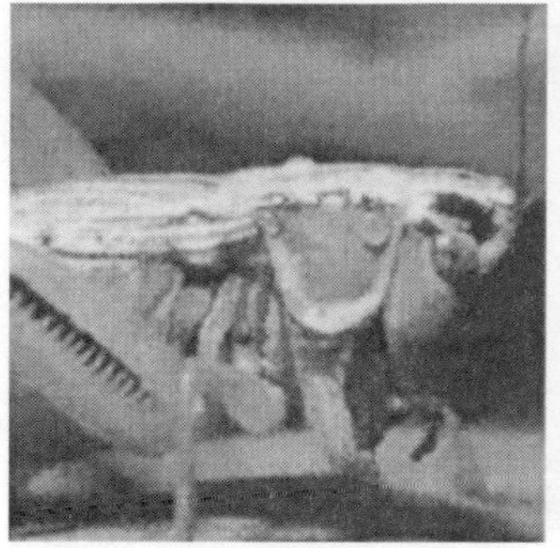

Fig. 15. Left: Denoising using *local variance normalization,* with Gaussian noise at level 0.3 (compare with Figure 11). Right: Denoising using local variance normalization, when the noise level has been raised to 0.5 (compare with Figure 12).

to use some regression-based methods as discussed in Section 7.1.

We also experimented with the variant of the method where normalization of the local variance is performed; see Section 8.2.3. The results are displayed in Figure 15. These results should be compared with those of Figures 11 and 12. This variant cuts noise in even areas very well but leaves some noise in the area around edges.

We performed the experiments with a median filter technique as well (not shown), which gave results that were qualitatively similar to the Wiener filter results. The results obtained by using the standard ICA transform (not shown) instead of the orthogonalized were clearly inferior to those obtained by the orthogonalized method.

To conclude, visual inspection of the results shows that the method performs very well. The denoising result is qualitatively quite different from those given by traditional filtering methods and more along the lines of wavelet shrinkage and coring results [36, 37].[8]

9. Conclusion

We presented the method of sparse code shrinkage [20] using maximum likelihood estimation of nongaussian random variables corrupted by gaussian noise. In the method, we first determine an orthogonal basis in which the components of given multivariate data have the sparsest distributions possible. The sparseness of the components is utilized in ML estimation of the noise-free components; these estimates are then used to reconstruct the original noise-free data by inverting the transformation. This is an approximation of the estimation of the noisy ICA model. The resulting method of sparse code shrinkage is closely connected to wavelet shrinkage and coring methods; in fact, it can be considered as a principled way of choosing the

[8] There is a large number of different variants of the wavelet shrinkage method, differing in choice of wavelet basis as well as the choice of shrinkage function. No one choice would have made a fair comparison, and thus we chose not to compare our method explicitly to wavelet shrinkage here.

orthogonal wavelet-like basis based on data alone as well as an alternative way of choosing the shrinkage nonlinearities.

References

[1] S.-I. Amari, A. Cichocki, and H. H. Yang. A new learning algorithm for blind source separation. In *Advances in Neural Information Processing Systems 8*, pages 757–763. MIT Press, Cambridge, MA, 1996.

[2] A. Antoniadis and G. Oppenheim, editors. *Wavelets in Statistics.* Springer, 1995.

[3] H. B. Barlow. What is the computational goal of the neocortex? In C. Koch and J. L. Davis, editors, *Large-scale neuronal theories of the brain.* Cambridge, MA: MIT Press, 1994.

[4] A. J. Bell and T. J. Sejnowski. An information-maximization approach to blind separation and blind deconvolution. *Neural Computation* 7:1129–1159, 1995.

[5] A. J. Bell and T. J. Sejnowski. The 'independent components' of natural scenes are edge filters. *Vision Research* 37:3327–3338, 1997.

[6] R. R. Coifman and D. L. Donoho. Translation-invariant de-noising. Technical report, Department of Statistics, Stanford University, 1995.

[7] P. Comon. Independent component analysis— a new concept? *Signal Processing* 36:287–314, 1994.

[8] T. M. Cover and J. A. Thomas. *Elements of Information Theory.* New York: John Wiley & Sons, 1991.

[9] Ingrid Daubechies. *Ten Lectures on Wavelets.* Philadelphia: Society for Industrial and Applied Mathematics., 1992.

[10] A. P. Dempster, N. Laird, and D. Rubin. Maximum likelihood for incomplete data via the EM algorithm. *J. Royal Statistical Society Ser. B,* 39:1–38, 1977.

[11] D. L. Donoho, I. M. Johnstone, G. Kerkyacharian, and D. Picard. Wavelet shrinkage: asymptopia? *J. Royal Statistical Society Ser. B,* 57:301–337, 1995.

[12] S. C. Douglas, A. Cichocki, and S. Amari. A bias removal technique for blind source separation with noisy measurements. *Electronics Letter.* 34:1379–1380, 1998.

[13] B. Efron and C. Morris. Data analysis using Stein's estimator and its generalizations. *J. American Statistical Association* 70:311–319, 1975.

[14] D. J. Field. What is the goal of sensory coding? *Neural Computation* 6:559–601, 1994.

[15] P. J. Huber. Projection pursuit. *Annals of Statistics* 13(2):435–475, 1985.

[16] J. Hurri, A. Hyvärinen, and E. Oja. Wavelets and natural image statistics. In *Proc. Scandinavian Conf. on Image Analysis '97,* Lappenranta, Finland, 1997.

[17] A. Hyvärinen. Independent component analysis in the presence of gaussian noise by maximizing joint likelihood. *Neurocomputing* 22:49–67, 1998.

[18] A. Hyvärinen. Fast and robust fixed-point algorithms for independent component analysis. *IEEE Trans. Neural Networks* 10(3):626–634, 1999.

[19] A. Hyvärinen. Gaussian moments for noisy independent component analysis. *IEEE Signal Processing Letters* 6(6):145–147, 1999.

[20] A. Hyvärinen. Sparse code shrinkage: Denoising of nongaussian data by maximum likelihood estimation. *Neural Computation* 11(7):1739–1768, 1999.

[21] A. Hyvärinen, R. Cristescu, and E. Oja. A fast algorithm for estimating overcomplete ICA bases for image windows. In *Proc. Int. Joint Conf. on Neural Networks,* Washington, DC, 1999.

[22] A. Hyvärinen and E. Oja. A fast fixed-point algorithm for independent component analysis. *Neural Computation* 9(7):1483–1492, 1997.

[23] C. Jutten and J. Herault. Blind separation of sources, part I: An adaptive algorithm based on neuromimetic architecture. *Signal Processing* 24:1–10, 1991.

[24] J. Karhunen, E. Oja, L. Wang, R. Vigário, and J. Joutsensalo. A class of neural networks for independent component analysis. *IEEE Trans. Neural Networks* 8(3):486–504, 1997.

[25] M. Kendall and A. Stuart. *The Advanced Theory of Statistics.* Charles Griffin & Company, 1958.

[26] H. Krim and I. C. Schick. Minimax description length for signal denoising and optimized representation. *IEEE Trans. Info. Theory,* 1999.

[27] M. Lewicki and B. Olshausen. A probabilistic framework for the adaptation and comparison of image codes. *J. Opt. Soc. Am. A Optics, Image Science, and Vision* 16(7):1587–1601, 1998.

[28] D. G. Luenberger. *Optimization by Vector Space Methods.* New York: John Wiley & Sons, 1969.

[29] S. G. Mallat. A theory for multiresolution signal decomposition: The wavelet representation. *IEEE Trans. PAMI* 11:674–693, 1989.

[30] G. P. Nason. Wavelet shrinkage using cross-validation. *J. Royal Statistical Society Ser. B* 58:463–479, 1996.

[31] E. Oja. The nonlinear PCA learning rule in independent component analysis. *Neurocomputing* 17(1):25–46, 1997.

[32] B. A. Olshausen and D. J. Field. Natural image statistics and efficient coding. *Network* 7(2):333–340, 1996.

[33] B. A. Olshausen and D. J. Field. Sparse coding with an overcomplete basis set: A strategy employed by V1? *Vision Research* 37:3311–3325, 1997.

[34] J. J. K. O'Ruanaidh and W. J. Fitzgerald. *Numerical Bayesian Methods Applied to Signal Processing.* New York: Springer, 1996.

[35] M. Schervish. *Theory of Statistics.* New York: Springer, 1995.

[36] E. P. Simoncelli and E. H. Adelson. Noise removal via bayesian wavelet coring. In *Proc. Third IEEE Int. Conf. on Image Processing,* pages 379–382, Lausanne, Switzerland, 1996.

[37] T. Yu, A. Stoschek, and D. Donoho. Translation- and direction-invariant denoising of 2-d and 3-d images: Experience and algorithms. In *Proceedings of the SPIE, Wavelet Applications in Signal and Image Processing IV,* pages 608–619, 1996.

Appendix

A. Proof of Theorem 1

From (5) we have

$$\hat{s} = y - \sigma^2 f'(y) + O(\sigma^4) \tag{41}$$

where f' is the derivative of the negative log-density f. Thus we obtain

$$\begin{aligned} \hat{s} - s &= \nu - \sigma^2 f'(y) + O(\sigma^4) \\ &= \nu - \sigma^2[f'(s) + f''(s)\nu] + O(\sigma^4) \end{aligned} \tag{42}$$

and

$$\begin{aligned} E\{(\hat{s} - s)^2\} &= E\{\nu^2\} + \sigma^4 E\{f'(s)^2\} - 2\sigma^2 E\{\nu^2\}E\{f''(s)\} \\ &\quad + o(\sigma^4) = \sigma^2 - \sigma^4 I_F(s) + o(\sigma^4) \end{aligned} \tag{43}$$

where we have used the property [35]

$$E\{\nabla^2 f(s)\} = I_F(s) \tag{44}$$

B. The 1D MMSE Estimator

Here we derive the first-order approximation of the MMSE estimator in (24), showing that it is identical to the one in (7). Let us denote $\tilde{g}(s) = g(s) - s$. Clearly, $\tilde{g}$ is infinitesimal, so let us denote by $o(\tilde{g})$ terms that are of lower order than $E\{\tilde{g}(s)\}$. We have:

$$E\{(s - g(y))^2\} = E\{[s - g(s) - g'(s)\nu - \frac{1}{2}g''(s)\nu^2]^2\} \quad (45)$$

$$+ o(\sigma^2) = E\{[-\tilde{g}(s) - (1 + \tilde{g}'(s))\nu - \frac{1}{2}\tilde{g}''(s)\nu^2]^2\} + o(\sigma^2)$$

$$= E\{\tilde{g}(s)^2\} + E\{(1 + \tilde{g}'(s))^2\}\sigma^2 + E\{\tilde{g}(s)g''(s)\}\sigma^2 + o(\sigma^2).$$

From the second line it can be seen that the approximation error incurred by this Taylor expansion is of order $o(\tilde{g})o(\sigma^2)$. Next we havc

$$(1 + \tilde{g}(s))^2 = 1 + 2\tilde{g}(s) + o(\tilde{g}) \quad (46)$$

and

$$\tilde{g}''(s) = g''(s) \quad (47)$$

which means that (45) can be written as

$$E\{(s - g(y))^2\} = \sigma^2 + E\{\tilde{g}(s)^2\} + 2E\{\tilde{g}'(s)\}\sigma^2 - \sigma^2 o(\tilde{g}) + o(\sigma^2). \quad (48)$$

We can now use variational calculus [28] to find the $\tilde{g}$ that minimizes the mean-square error. Neglecting terms of smaller order and equating the variational derivative of the right-hand side of (48) with respect to $\tilde{g}$ to zero, we obtain

$$p(s)g(s) - p(s)s + \sigma^2 p'(s) = 0 \quad (49)$$

which gives

$$g(s) = s - \sigma^2 f'(s) \quad (50)$$

with $f = p'/p$, and we have indeed (7).

Index

T

U

V

W

About the Editors

Simon Haykin is University Professor in the department of electrical and computer engineering at McMaster University, Hamilton, Ontario, Canada. This new prestigious faculty position was established by the Senate of McMaster University in 1995, and Dr. Haykin is the first appointee from the faculty of engineering. He is the founding technical editor of a series of books on "Adaptive and Learning Systems for Signal Processing, Communications and Control" published by John Wiley & Sons, Inc. He is a Fellow of the IEEE, a Fellow of the Royal Society of Canada, and recipient of the Honorary Doctor of Technical Sciences, ETH, Zurich, Switzerland. Dr. Haykin's research interests include adaptive space-time processing for wireless communications, chaos for radar target detection, and intelligent hearing aid systems.

Bart Kosko is a faculty member of electrical engineering at the University of Southern California. He holds degrees in philosophy, economics, mathematics, and electrical engineering and is an award-winning composer. Dr. Kosko is a past director of USC's Signal and Image Processing Institute and an elected governor of the International Neural Network Society. He is author of the trade books *Fuzzy Thinking* (Hyperion, 1993) and *Heaven in a Chip* (Random House, 2000), the novel *Nanotime* (Avon Books, 1997), and the textbooks *Neural Networks and Fuzzy Systems, Neural Networks for Signal Processing* (Prentice Hall, 1992), and *Fuzzy Engineering* (Prentice Hall, 1997).